Differential Geometry, Calculus of Variations, and Their Applications

LECTURE NOTES

IN PURE AND APPLIED MATHEMATICS

1. *N. Jacobson,* Exceptional Lie Algebras
2. *L. -Å. Lindahl and F. Poulsen,* Thin Sets in Harmonic Analysis
3. *I. Satake,* Classification Theory of Semi-Simple Algebraic Groups
4. *F. Hirzebruch, W. D. Newmann, and S. S. Koh,* Differentiable Manifolds and Quadratic Forms (out of print)
5. *I. Chavel,* Riemannian Symmetric Spaces of Rank One (out of print)
6. *R. B. Burckel,* Characterization of C(X) Among Its Subalgebras
7. *B. R. McDonald, A. R. Magid, and K. C. Smith,* Ring Theory: Proceedings of the Oklahoma Conference
8. *Y.-T. Siu,* Techniques of Extension on Analytic Objects
9. *S. R. Caradus, W. E. Pfaffenberger, and B. Yood,* Calkin Algebras and Algebras of Operators on Banach Spaces
10. *E. O. Roxin, P.-T. Liu, and R. L. Sternberg,* Differential Games and Control Theory
11. *M. Orzech and C. Small,* The Brauer Group of Commutative Rings
12. *S. Thomeier,* Topology and Its Applications
13. *J. M. Lopez and K. A. Ross,* Sidon Sets
14. *W. W. Comfort and S. Negrepontis,* Continuous Pseudometrics
15. *K. McKennon and J. M. Robertson,* Locally Convex Spaces
16. *M. Carmeli and S. Malin,* Representations of the Rotation and Lorentz Groups: An Introduction
17. *G. B. Seligman,* Rational Methods in Lie Algebras
18. *D. G. de Figueiredo,* Functional Analysis: Proceedings of the Brazilian Mathematical Society Symposium
19. *L. Cesari, R. Kannan, and J. D. Schuur,* Nonlinear Functional Analysis and Differential Equations: Proceedings of the Michigan State University Conference
20. *J. J. Schäffer,* Geometry of Spheres in Normed Spaces
21. *K. Yano and M. Kon,* Anti-Invariant Submanifolds
22. *W. V. Vasconcelos,* The Rings of Dimension Two
23. *R. E. Chandler,* Hausdorff Compactifications
24. *S. P. Franklin and B. V. S. Thomas,* Topology: Proceedings of the Memphis State University Conference
25. *S. K. Jain,* Ring Theory: Proceedings of the Ohio University Conference
26. *B. R. McDonald and R. A. Morris,* Ring Theory II: Proceedings of the Second Oklahoma Conference
27. *R. B. Mura and A. Rhemtulla,* Orderable Groups
28. *J. R. Graef,* Stability of Dynamical Systems: Theory and Applications
29. *H.-C. Wang,* Homogeneous Branch Algebras
30. *E. O. Roxin, P.-T. Liu, and R. L. Sternberg,* Differential Games and Control Theory II
31. *R. D. Porter,* Introduction to Fibre Bundles
32. *M. Altman,* Contractors and Contractor Directions Theory and Applications
33. *J. S. Golan,* Decomposition and Dimension in Module Categories
34. *G. Fairweather,* Finite Element Galerkin Methods for Differential Equations
35. *J. D. Sally,* Numbers of Generators of Ideals in Local Rings
36. *S S. Miller,* Complex Analysis: Proceedings of the S.U.N.Y. Brockport Conference
37. *R. Gordon,* Representation Theory of Algebras: Proceedings of the Philadelphia Conference
38. *M. Goto and F. D. Grosshans,* Semisimple Lie Algebras
39. *A. I. Arruda, N. C. A. da Costa, and R. Chuaqui,* Mathematical Logic: Proceedings of the First Brazilian Conference

40. *F. Van Oystaeyen,* Ring Theory: Proceedings of the 1977 Antwerp Conference
41. *F. Van Oystaeyen and A. Verschoren,* Reflectors and Localization: Application to Sheaf Theory
42. *M. Satyanarayana,* Positively Ordered Semigroups
43. *D. L. Russell,* Mathematics of Finite-Dimensional Control Systems
44. *P.-T. Liu and E. Roxin,* Differential Games and Control Theory III: Proceedings of the Third Kingston Conference, Part A
45. *A. Geramita and J. Seberry,* Orthogonal Designs: Quadratic Forms and Hadamard Matrices
46. *J. Cigler, V. Losert, and P. Michor,* Banach Modules and Functors on Categories of Banach Spaces
47. *P.-T. Liu and J. G. Sutinen,* Control Theory in Mathematical Economics: Proceedings of the Third Kingston Conference, Part B
48. *C. Byrnes,* Partial Differential Equations and Geometry
49. *G. Klambauer,* Problems and Propositions in Analysis
50. *J. Knopfmacher,* Analytic Arithmetic of Algebraic Function Fields
51. *F. Van Oystaeyen,* Ring Theory: Proceedings of the 1978 Antwerp Conference
52. *B. Kedem,* Binary Time Series
53. *J. Barros-Neto and R. A. Artino,* Hypoelliptic Boundary-Value Problems
54. *R. L. Sternberg, A. J. Kalinowski, and J. S. Papadakis,* Nonlinear Partial Differential Equations in Engineering and Applied Science
55. *B. R. McDonald,* Ring Theory and Algebra III: Proceedings of the Third Oklahoma Conference
56. *J. S. Golan,* Structure Sheaves over a Noncommutative Ring
57. *T. V. Narayana, J. G. Williams, and R. M. Mathsen,* Combinatorics, Representation Theory and Statistical Methods in Groups: YOUNG DAY Proceedings
58. *T. A. Burton,* Modeling and Differential Equations in Biology
59. *K. H. Kim and F. W. Roush,* Introduction to Mathematical Consensus Theory
60. *J. Banas and K. Goebel,* Measures of Noncompactness in Banach Spaces
61. *O. A. Nielson,* Direct Integral Theory
62. *J. E. Smith, G. O. Kenny, and R. N. Ball,* Ordered Groups: Proceedings of the Boise State Conference
63. *J. Cronin,* Mathematics of Cell Electrophysiology
64. *J. W. Brewer,* Power Series Over Commutative Rings
65. *P. K. Kamthan and M. Gupta,* Sequence Spaces and Series
66. *T. G. McLaughlin,* Regressive Sets and the Theory of Isols
67. *T. L. Herdman, S. M. Rankin, III, and H. W. Stech,* Integral and Functional Differential Equations
68. *R. Draper,* Commutative Algebra: Analytic Methods
69. *W. G. McKay and J. Patera,* Tables of Dimensions, Indices, and Branching Rules for Representations of Simple Lie Algebras
70. *R. L. Devaney and Z. H. Nitecki,* Classical Mechanics and Dynamical Systems
71. *J. Van Geel,* Places and Valuations in Noncommutative Ring Theory
72. *C. Faith,* Injective Modules and Injective Quotient Rings
73. *A. Fiacco,* Mathematical Programming with Data Perturbations I
74. *P. Schultz, C. Praeger, and R. Sullivan,* Algebraic Structures and Applications Proceedings of the First Western Australian Conference on Algebra
75. *L. Bican, T. Kepka, and P. Nemec,* Rings, Modules, and Preradicals
76. *D. C. Kay and M. Breen,* Convexity and Related Combinatorial Geometry: Proceedings of the Second University of Oklahoma Conference
77. *P. Fletcher and W. F. Lindgren,* Quasi-Uniform Spaces
78. *C.-C. Yang,* Factorization Theory of Meromorphic Functions
79. *O. Taussky,* Ternary Quadratic Forms and Norms
80. *S. P. Singh and J. H. Burry,* Nonlinear Analysis and Applications
81. *K. B. Hannsgen, T. L. Herdman, H. W. Stech, and R. L. Wheeler,* Volterra and Functional Differential Equations

82. *N. L. Johnson, M. J. Kallaher, and C. T. Long,* Finite Geometries: Proceedings of a Conference in Honor of T. G. Ostrom
83. *G. I. Zapata,* Functional Analysis, Holomorphy, and Approximation Theory
84. *S. Greco and G. Valla,* Commutative Algebra: Proceedings of the Trento Conference
85. *A. V. Fiacco,* Mathematical Programming with Data Perturbations II
86. *J.-B. Hiriart-Urruty, W. Oettli, and J. Stoer,* Optimization: Theory and Algorithms
87. *A. Figa Talamanca and M. A. Picardello,* Harmonic Analysis on Free Groups
88. *M. Harada,* Factor Categories with Applications to Direct Decomposition of Modules
89. *V. I. Istrătescu,* Strict Convexity and Complex Strict Convexity: Theory and Applications
90. *V. Lakshmikantham,* Trends in Theory and Practice of Nonlinear Differential Equations
91. *H. L. Manocha and J. B. Srivastava,* Algebra and Its Applications
92. *D. V. Chudnovsky and G. V. Chudnovsky,* Classical and Quantum Models and Arithmetic Problems
93. *J. W. Longley,* Least Squares Computations Using Orthogonalization Methods
94. *L. P. de Alcantara,* Mathematical Logic and Formal Systems
95. *C. E. Aull,* Rings of Continuous Functions
96. *R. Chuaqui,* Analysis, Geometry, and Probability
97. *L. Fuchs and L. Salce,* Modules Over Valuation Domains
98. *P. Fischer and W. R. Smith,* Chaos, Fractals, and Dynamics
99. *W. B. Powell and C. Tsinakis,* Ordered Algebraic Structures
100. *G. M. Rassias and T. M. Rassias,* Differential Geometry, Calculus of Variations, and Their Applications
101. *R.-E. Hoffmann and K. H. Hofmann,* Continuous Lattices and Their Applications
102. *J. H. Lightbourne, III and S. M. Rankin, III,* Physical Mathematics and Nonlinear Partial Differential Equations

Other Volumes in Preparation

Differential Geometry, Calculus of Variations, and Their Applications

Edited by
GEORGE M. RASSIAS
THEMISTOCLES M. RASSIAS
Athens, Greece

MARCEL DEKKER, INC. New York and Basel

Library of Congress Cataloging in Publication Data

Main entry under title:

Differential geometry, calculus of variations, and
their applications.

(Lecture notes in pure and applied mathematics ; 100)
(Pure and applied mathematics)
Includes index.
1. Geometry, Differential. 2. Calculus of
variations. I. Rassias, George M., [date].
II. Rassias, Themistocles M., [date]. III. Series:
Lecture notes in pure and applied mathematics ; v. 100.
IV. Series: Monographs and textbooks in pure and applied
mathematics.
QA641.D4144 1985 516.3'6 85-13017
ISBN 0-8247-7267-9

MARCEL DEKKER, INC.
270 Madison Avenue, New York, New York 10016

Current printing (last digit):
10 9 8 7 6 5 4 3 2 1

PRINTED IN THE UNITED STATES OF AMERICA

Léonhard Euler (1707-1783)

(Courtesy of the Archives of the USSR Academy of Sciences, Leningrad.)

Tom. XII

Praes. d. 20. Aug. 1739.
Prelect. d. 21. Sept.
24.
28.
5. Octb.
8.
12 Octb. — fin.

6

N. I

Tab. I. II. III

Investigatio Curvarum quae evolutae sui similes producunt.

A. L. E.

§. 1. In hac dissertatione nomini evolutarum aliquanto latiorem significationem tribuo, quam vulgo fieri solet, ac non solum eam curvam ex cujus evolutione data curva nascitur, hujus evolutam appello, sed insuper evolutam hujus evolutae, similiterque universam curvarum seriem, quarum quaelibet praecedentis est evoluta. Interim tamen hoc discrimen in denominatione observabo, ut ipsam datae curvae evolutam, quae hoc nomine insigniri consuevit, ejus evolutam primam appellem; hujus vero evolutam secundam, eamque ex cujus evolutione ista nascitur, tertiam atque ita porro. Sic si datae curvae A evoluta sit curva B, curvae autem B evoluta C, atque hujus curvae C evoluta D, hujusque E et ita deinceps; erit mihi respectu curvae A curva B evoluta prima, curva C evoluta secunda, curva D evoluta tertia, E quarta atque ita porro.

§. 2. Hac vocis evolutae significatione praemissa, in hac dissertatione in eas curvas inquirere constitui, quarum evolutae vel primae vel secundae vel tertiae etc. ipsis sint similes. Quod quidem ad evolutas primas attinet a Viro Clarissimo Prof. Krafft jam est ostensum, praeter spiralem logarithmicam et cycloidem alias non dari curvas, quae cum suis evolutis primis conveniant; atque idem alia methodo hic sum demonstraturus, quae simul viam praeparet ad eas curvas investigandas, quae similes sint suis evolutis vel secundis vel tertiis etc. Neque vero in hoc negotio viam simplicissimam sum secuturus, quae facillime ad cognitionem curvarum quaesitarum manuducat, sed praecipue mihi propositum est relationem inter arcum curvae et respondentem radium curvedinis investigare, quae in differentialibus altioris gradus est involuta, quae alia methodo

(Courtesy of the Archives of the USSR Academy of Sciences, Leningrad.)

XII

Praes. d. 9. Septb. 1738.
Prael. 18 Octb.
16 dito

Solutio Problematis cuiusdam a Celeb. Dan. Bernoullio propositi.

A. L. Euler.

§. 1. In postremis litteris, quas Cel. Daniel Bernoulli ad me dedit Basileâ d. 24 Maji huius anni, mentionem fecit alicuius problematis, in quod occasione alicuius Problematis mechanici incidisset. Quaerebat autem inter omnes curvas isoperimetricas eisdem terminis contentas eam in qua $\int r^m ds$ haberet maximum minimumve valorem, denotantibus s arcum curvae, r vero eius radium osculi. Annunciat vero Vir Celeb. simul, se duplicem solutionem esse nactum, ad quarum alteram ponendo elementum arcus ds, ad alteram autem ponendo elementum abscissae dx constans pervenisset: at ambas hasce solutiones ita esse comparatas, ut inter se non conspirare videantur. Casu quidem quo $m = 1$ hanc mihi perscripsit aequationem a se inventam, $ds = \frac{2rdr}{\sqrt{(g + 4nr - 4rr)}}$ qua curvae natura exprimatur. Sin autem inter omnes omnino curvas quaereretur ea, in qua esset $\int r\,ds^2$ minimum, se invenisse scribit fore $r = 0$, cum tamen ego jam ante Ipsi scripsissem cycloidem huic quaestioni satisfacere tanquam solam. Hanc ob rem me rogavit ut problema hoc pariter aggrederer, solutionemque quam fuero consecutus, secum communicarem.

(Courtesy of the Archives of the USSR Academy of Sciences, Leningrad.)

PREFACE

Léonhard Euler, one of the most productive mathematicians of all times, died on September 18, 1783 at the age of 76.

His work was devoted to several different areas of pure and applied mathematics and he made highly significant contributions in every field of mathematics that existed in his day. In 1744 there appeared his Methodus inveniendi lineas curvas maximi minimive proprietate gaudentes. This was the first exposition of the calculus of variations; it contained "Euler's equations" with many applications, including the discovery that the catenoid and the right helicoid are minimal surfaces.

The origins of topology lie in Euler's solution of the Königsberg bridge problem, and his formula $V - E + F = 2$ connecting the number of vertices (V), edges (E), and faces (F) of a simple polyhedron. This formula—Euler's characteristic—is extremely useful in a great variety of situations. It yields almost everything known about the famous four-color problem: given a map, can it be colored with four colors so that no two faces touching along an edge have the same color?

In his treatise of 1736, Euler was the first to introduce explicitly the concept of mass point or particle, and was also the first to introduce the concept of acceleration of a particle moving along any curve and to use the notion of vector in connection with velocity and acceleration. A large portion of his work was devoted to celestial mechanics, the three-body problem, and the perturbations of the planets, leading to the momentous problem of the stability of the solar system. In 1739 there appeared his new theory of music and in 1760-61, in his Letters to a German Princess, his philosophical exposition of the most important problems of natural science. Lagrange, Laplace, and Gauss were influenced by Euler's work. Gauss stated that "the study of Euler's works will remain the best school for the different fields of mathematics and nothing else can replace it."

This volume contains a series of papers, written by internationally recognized authorities in their fields, dedicated to the memory of Léonhard Euler on the 200th anniversary of his death (1783-1983). These papers deepen our understanding of some of the longstanding research problems of geometry, calculus of variations, and their applications at a level suitable for advanced graduate students and research mathematicians in a variety of fields. We have attempted to provide a source of the latest developments in this important and growing subject. The book's presentation of concepts and methods makes it an invaluable reference for teachers and other professionals in mathematics who are interested in applied research and technology, philosophy of mathematics, and mathematics education.

It is a pleasure to express our warmest thanks to all of the scientists who contributed to this volume. We would also like to acknowledge the superb assistance in editing and composition that the staff of Marcel Dekker, Inc. has provided in the preparation of this publication.

George M. Rassias
Themistocles M. Rassias

CONTENTS

CONTRIBUTORS

RALPH H. ABRAHAM Mathematics Board, University of California, Santa Cruz, California

FADHEL AL-MUSALLAM[a] Department of Mathematics, West Virginia University, Morgantown, West Virginia

JOHN ARGYRIS Institut für Statik und Dynamik der Luft- und Raumsfahrtkonstruktionen, Universität Stuttgart, Stuttgart, Federal Republic of Germany

MICHEL BAUDERON Department of Information Sciences, I.U.T.A., Université de Bordeaux I, Bordeaux, France

JOHN K. BEEM Department of Mathematics, University of Missouri—Columbia, Columbia, Missouri

MELVYN S. BERGER Center for Applied Mathematics and Mathematical Science, University of Massachusetts at Amherst, Amherst, Massachusetts

SAMUEL BOURNE Department of Mathematics, University of California at Berkeley, Berkeley, California

CLAUDE BREZINSKI Laboratoire d'Analyse Numérique et d'Optimisation, Université de Lille 1, UER IEEA, Villeneuve d'Ascq Cedex, France

GOTTFRIED BRUCKNER Institut für Mathematik, Akademie der Wissenschaften der DDR, Berlin, German Democratic Republic

MURRAY R. CANTOR[b] Department of Mathematics, University of Texas at Austin, Austin, Texas

LOKENATH DEBNATH Department of Mathematics, University of Central Florida, Orlando, Florida

J. ST. DOLTSINIS Institut für Statik und Dynamik der Luft- und Raumfahrtkonstruktionen, Universität Stuttgart, Stuttgart, Federal Republic of Germany

HAMID DRLJEVIĆ Ekonomski Fakultet, Mostar, Yugoslavia

HALLDÓR I. ELÍASSON Faculty of Engineering and Science, University of Iceland, Reykjavík, Iceland

ARTHUR E. FISCHER Department of Mathematics, University of California, Santa Cruz, California

HANS R. FISCHER Department of Mathematics and Statistics, University of Massachusetts at Amherst, Amherst, Massachusetts

Current Affiliations:

[a]Department of Mathematics, Arizona State University, Tempe, Arizona
[b]Department of Geophysical Research, Shell Development Company, Houston, Texas

S. H. GOULD Department of Mathematics, Brown University, Providence, Rhode Island

DIEGO BRICIO HERNÁNDEZ[a] Department of Process Engineering and Hydraulics, Universidad Autónoma Metropolitana—Iztapalapa, México D.F., Mexico

DONALD H. HYERS[b] Department of Mathematics, University of Southern California, Los Angeles, California

RICHARD D. JÄRVINEN[c] Department of Mathematics, College of St. Benedict's/St. John's University, St. Joseph, Minnesota

BORIS A. KUPERSHMIDT[d] Department of Mathematics, The University of Michigan, Ann Arbor, Michigan

ERNESTO A. LACOMBA Department of Mathematics, Universidad Autónoma Metropolitana—Iztapalapa, México D. F., Mexico

ERNST WOLFGANG LAEDKE Department of Physics, Universität Essen—GHS, Essen, Federal Republic of Germany

JOSHUA LESLIE Department of Mathematics, Northwestern University, Evanston, Illinois

L. LOSCO Department of Mechanics, Ecole Nationale Superieure de Mècanique et Microtechniques, Besançon, France

JOHN C. MATOVSKY[e] Department of Mathematics, University of Texas at Austin, Austin, Texas

SAM B. NADLER, JR. Department of Mathematics, West Virginia University, Morgantown, West Virginia

MITSURU NAKAI Department of Mathematics, Nagoya Institute of Technology, Nagoya, Japan

PHILIP E. PARKER[f] Department of Mathematics, University of Missouri—Columbia, Columbia, Missouri

HAROLD R. PARKS Department of Mathematics, Oregon State University, Corvallis, Oregon

JEAN-PAUL PENOT Department of Mathematics, Faculté des Sciences, Université de Pau, Pau, France

ROY PLASTOCK Department of Mathematics, New Jersey Institute of Technology, Newark, New Jersey

GEORGE M. RASSIAS International Scientific Center, Ltd., Athens, Greece

THEMISTOCLES M. RASSIAS Department of Mathematics, University of La Verne, Kifissia, Athens, Greece

HANNO RUND Program in Applied Mathematics, University of Arizona, Tucson, Arizona

Current Affiliations:

[a]Department of Mathematics

[b]Retired

[c]Department of Mathematics and Statistics, St. Mary's College, Winona, Minnesota

[d]The University of Tennessee Space Institute, Tullahoma, Tennessee

[e]Department of Mathematics and Statistics, Louisiana Tech University, Ruston, Louisiana

[f]Department of Mathematics, Wichita State University, Wichita, Kansas

LEO SARIO Department of Mathematics, University of California, Los Angeles, California

MAU-HSIANG SHIH Department of Mathematics, Chung Yuan University, Chung-Li, Taiwan

KARL H. SPATSCHEK Department of Physics, Universität Essen—GHS, Essen, Federal Republic of Germany

GERHARD STRÖHMER Institut für Mathematik, Rheinisch-Westfallischen Technische Hochschule, Aachen, Federal Republic of Germany

KOK-KEONG TAN Department of Mathematics, Statistics and Computing Science, Dalhousie University, Halifax, Nova Scotia, Canada

ENZO TONTI Istituto di Scienza della Costruzione, Università di Trieste, Trieste, Italy

FLOYD L. WILLIAMS Department of Mathematics, University of Massachusetts at Amherst, Amherst, Massachusetts

Differential Geometry, Calculus of Variations, and Their Applications

1 CATEGORIES OF DYNAMICAL MODELS

Ralph H. Abraham / Mathematics Board, University of California, Santa Cruz, California

HISTORICAL INTRODUCTION

In 1966, my manuscript [10] was nearing completion. Its Conclusion [1], inspired by Duhem [14] and Velikovsky [22], was an essay on the stability of the solar system.[1] During the writing, a package arrived from Rene Thom. It was the first few chapters of his own manuscript.[2] Subtitled "An Outline of a General Theory of Models," it diverted me to this subject, and my Conclusion expanded accordingly. Despite these ambitious early efforts, little has passed to clarify the strategies of dynamical modeling, save the appearance of an increasing number of examples of the art.[3] In this paper, we will try again to move towards a general theory of dynamical models.

Since Newton created dynamical models, there have been two catastrophes in the basic paradigm. From Newton to Poincaré, a simple strategy guided all modeling. This strategy, which we shall call the classical quantitative scheme, evolved to suit the needs of the physical sciences. This classical scheme, clearly defined by Duhem [14], has been the subject of most of the literature on models in the philosophy of science.[4] It is described in detail in Section B1.

The crisis of celestial mechanics at the end of the last century (see [10] for the whole story) prompted Poincaré to create a new paradigm, which we shall call the modern qualitative scheme, in 1882. This was the subject of the Conclusion [1] and is also described in Sec. A5 below.

The needs of theoretical biology led Thom to create a new paradigm, which we shall call the dynamical bifurcation scheme, in 1966. His book [21], although now known primarily as the source of Elementary Catastrophe Theory, contains the more general General Catastrophe Theory (that is, the dynamical bifurcation scheme) and more general schemes as well. He related these schemes to morphogenesis in the abstract. Although there are exciting applications of the dynamical bifurcation scheme throughout the sciences, convincing models for biological morphogenesis have not yet appeared.[5] We review the dynamical bifurcation scheme in Section C3.

The complex bifurcation schemes introduced recently [8] are ramifications of Thom's simple bifurcation scheme, designed particularly for physiological modeling. Hopefully, they will be applicable to biological morphogenesis as well.

This paper is a tutorial on these schemes. Their applications in physiology will be discussed in a later paper. In Part A, we explain exactly what is meant by a scheme. In Parts B and C, we review the historical schemes. These are extended, in Part D, to the complex schemes which have been developed for physiological modeling. Our primary goal is to make these schemes, or dynamical modeling strategies, sufficiently

---

1. This is not included in the second edition.
2. Now available in English [21].
3. The only mention of these events in the literature of the philosophy of science which we have come across is in Garfinkel ([16], p. 31).
4. See Hesse [18] and Garfinkel ([16], p. 170) for example.
5. For a summary, see Rosen [19].

clear that the reader may follow them in making new models. Secondarily, we hope to advance toward a general theory of models. Eventually, this would provide a synthesis of general systems theory, control theory, homeokinetics, dissipative structures, industrial and urban dynamics, and related concepts.

A. DYNAMICAL MODELS, SCHEMES, AND TYPES

A1. What Is a Model?

The sparse literature of the theory of models is troubled by this question. We wish to dispose of it at the outset by means of a naive map of the noosphere introduced earlier in this series [7].

What is the difference between a train and a model train? Between a solar system and an orrery? Between a Toyota and a toy auto? Besides having differing mechanisms, the two elements in any of these pairs of analogues differ in their degree of susceptibility to our influence. If we somehow made a mathematical model of the entire phenomenal universe, in which a particular phenomenon was represented by a point, we might imagine a real-valued function defined upon it, which is proportional to this susceptibility. This function would have a higher value at "model train" than at "train." We will call this a reality function. The level sets (contours) of this function would consist of equisusceptible phenomena. We will call them levels of reality.

And we further may imagine our model to include the noosphere, or noumenal universe, with its own reality function. We imagine this as one model, and one reality function, with all noumena higher than any phenomena, and a gap in between. Thus, there is room in this vision for "train," "model train," and "mathematical model of a train," in an ascending sequence.

To simplify discussion, we may further imagine that the levels of reality have been discretized into a finite set. At the lowest level are the most intractable phenomena, such as "cosmos" or "solar system." These we call hardware. Up a level we find "orrery," "physics lab," "stirred chemical reacter," "meristem culture dish," and so on. These we call labware. All of these lower levels comprise phenomena. So a bit higher, we find software, or programs for labware (whether real, analog, or digital). Higher yet, there are "Schwartzschild universe," "Lorenz attractor," and other mathematical objects, or etherware. And at the top level, we may find "armchair experiment," "cognitive model," "scientific theory," and other flights of pure fantasy and science fiction, or knoware. See Figure 1. As we have noumena and phenomena in one bag here, we call the things in the bag objects, as in "mathematical objects" or "objects of thought." We are not trying to outrage Kant, but simply to create a practical framework for the working dynamicist.

Now should we observe objects on two different levels which are analogous, according to some pattern matching process among the cognitive skills of our collective nous, we say the higher one is a model of the lower one, the domain. This may occur by design or by chance. Accidents might occur because of some higher design, or concordance, as described earlier in the series [7]. In any case, we shall use the word concord for this perceived quality of analogy instead of analogy, metaphor, model, isomorph, etc. By modeling we shall mean the assumption of concord, especially between a well-known object and a poorly understood one, for the sake of study. In particular, dynamical modeling will denote the architecture of a mathematical model concordant with given dynamical phenomena.

The concord of two objects, as far as science knows at present, is observable solely through the medium of the human brain. The function of modeling (as defined above) is a cognitive one. It may be that we must be able to observe a concordance across several levels of reality in order to understand our surroundings in the phenomenal universe. This is the essence of theory. In any case, science is the pursuit of concordance across several levels, and mathematical modeling is part of this essence. Our goal here is to provide explicit schemes, or strategies, for dynamical modeling.

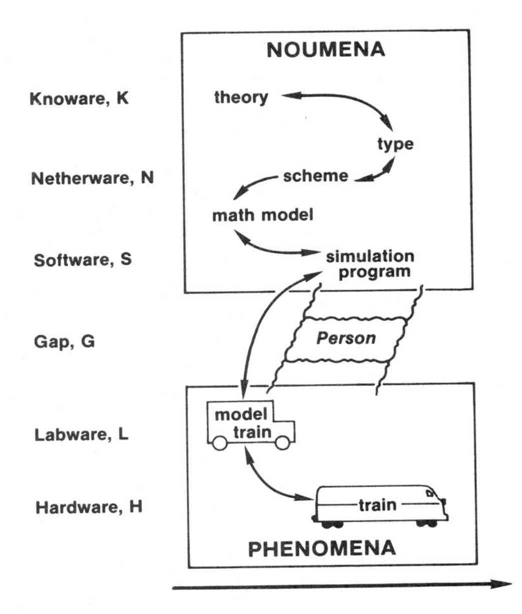

Fig. 1

So far, we have tried to give our naive answer to the question: what is a model? It is hard to be more precise without resorting to the concept of a scheme. We will outline it in the next Section.

A2. What Is a Scheme?

To Duhem,[6] a physical theory consisted of an experimental domain, D, together with a mathematical model, M, connected by a conventional interpretation, A. This connection is to be regarded as a contract, agreement, or accord, between discussers.[7] This accord identifies terms of D (especially observable physical parameters like voltage) with functions or variables of M. This picture of a theory is symbolized in Fig. 2.

In this context, Duhem identified two useful properties of a theory: adequacy, and stability. Adequacy refers to the quality of concordance between the predictions of M and the observations of D, as identified by A. Stability means the insensitivity of the predictions to changes in the model. In 1966, we tried to make precise these ideas of Duhem, and particularly these two useful properties, in a mathematical context appropriate to the physical sciences.[8] In this formalization, a problem develops.

First of all, as we have already described at length what we mean by a model, we shall use the word scheme for a mathematical object intended for modeling an unspecified phenomenon, or a whole family of virtual phenomena, of which an actual phenomenon may correspond to a single member. Schemes will occur in different types, which

---

6. This idea must go back centuries, but we find a particularly clear discussion in Duhem [14]. As far as we know, this is the first use of a scheme in this context. This is summarized in the Conclusion [1] in mathematical terms, with explicit examples from celestial mechanics.
7. It is tempting to describe this in the language of categories and functors. In this account, we will be satisfied with a naive description.
8. See the Conclusion [1] for precise statements, and the illustrative models for the solar system.

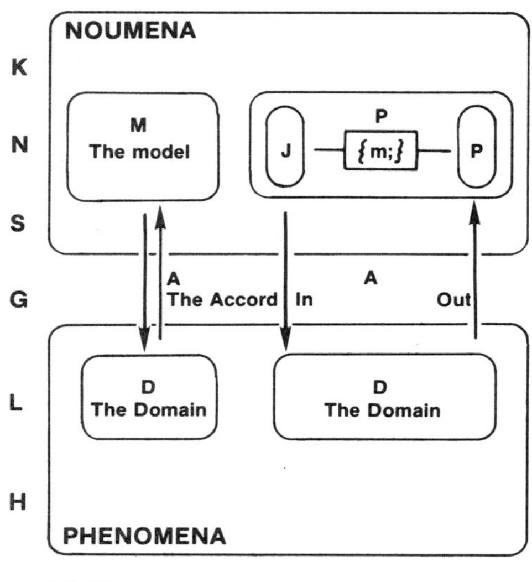

Fig. 2

we will not try to enumerate at present.[9] For example, Duhem's picture of a scientific theory, shown in Fig. 2(a), is a scheme of the singleton (S) type. It consists of two objects (a noumenon modeling a phenomenon) and a conventional interpretation.

But stability is not a property of a model, but of a scheme. It has nothing to do with the experimental domain. Further, it is not a property of an isolated scheme of singleton type, but only of a scheme belonging to a family of schemes. We wish to regard such a family as another type of scheme. We propose to call such a family a scheme of the family-of-singleton (F/S) type.[10]

Thus, Duhem's idea of a scientific theory belongs in the context of a particular scheme-type, which we shall now describe in more detail.

In a scheme of the family-of-singleton (F/S) type, there must be (among other things) two topological spaces. One of these, J, is a geometric model of the observed states of the experimental domain or device,. The other, P, corresponds to the predictions of the models. For each point j of J there is a model M(j), and a prediction of M(j), which is a point p(j) of P. Thus the predictions of the family of models $\{M(j)\}$ comprise a function $p: J \to P$. A model M(j) of this family is stable in the sense of Duhem if p is continuous at j. This scheme is shown in Fig. 2. Note that the accord, A, is represented here in two parts: input,[11] In(A), and output or observation, Out(A). The adequacy in the sense of Duhem of a model in this scheme becomes a mathematical question. For the observations and the predictions belong to the same topological space.[12]

This scheme-type of Duhem is sufficient for celestial mechanics, Thom's theory of abstract morphogenesis, and many important applications. But it is too limited for

9. The types we know so far may all be described as categories.
10. Readers familiar with category theory will recognize here a functor: any type may be extended into a larger type, through the functor family-of, which is a universal construction.
11. This corresponds to the preparation procedure of Thom ([21], p. 15).
12. The two functions p and $\mathrm{Out} \circ D \circ \mathrm{In}$ end up in the space P (D denotes the experimental process). Thus adequacy is defined by a neighborhood of the diagonal (uniformity), in the product space $P \times P$.

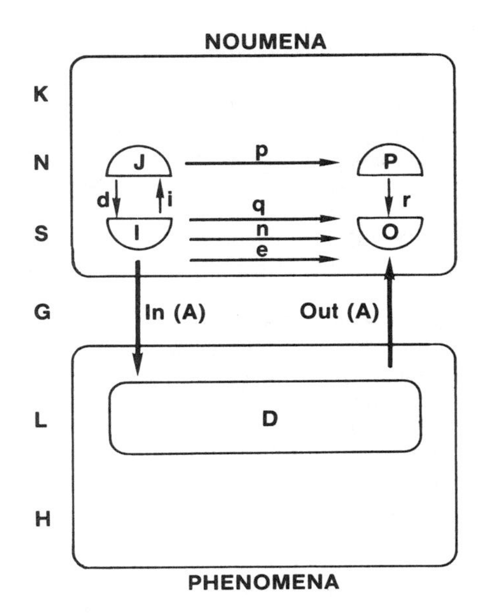

Fig. 3

others (especially self-organization[13]). Hence we proposed the complex scheme-types [8], which are developed further in Section D of this paper.

A3. The Essentials of a Dynamical Scheme

A dynamical model is a very special type of mathematical object. It may be in concord with dynamical phenomena in a physical, biological, or social realm. For the sake of definiteness, we will consider the target (domain) of the model to be phenomena on the labware level of reality. Thus, we will speak of the lab system, or the experimental domain, and use metaphors such as initial preparation, control parameters, observation of the device, etc. This should not be construed as the entire context of this theory, however, So when we say device, we may mean equally the social system, biological organism, or whatever is the target of the modeling, even if it belongs to the lowest (least susceptible) of the hardware levels of reality.

The device and its mathematical model are to be discussed in the context of a given theory: vocabulary, cognitive strategy, rules of inference, and so on. This idea of a theory is more general than Duhem's, and is conventional nowadays. What Duhem called a scientific theory may be closer to what we call here a scheme.

Among all schemes, there is a class of particular recipes which are most suitable for modeling dynamical phenomena on the lab level (or any other more-or-less susceptible level) or reality, which we call dynamical schemes. These might occur in any scheme-type. We now enumerate the essential ingredients of a dynamical scheme.

In a dynamical scheme, we must always have four geometric models: I, J, O, P, for the virtual states of the lab system or device. These will usually be topological spaces or differentiable manifolds of finite or infinite dimension. Beyond these, there must be four functions: p, q, r, s, relating the models in the commutative diagram shown in Fig. 3, in the box labeled "noumena." Finally, there must be the device, D, or lab system, shown in the box labeled "phenomena" in Fig. 3, and two interfaces: In(A), Out(A), across the noumenal/phenomenal gap.

There may be other ingredients, as we shall see in the examples in Part B. In this section, we discuss these essentials, one at a time.

---

13. For a thorough discussion of the problems, see [9].

The space J is a geometric model for the virtual states of the system, in mathematical idealization. This means, primarily, as a continuum. In practice, the idealized states are subdivided in classes, such as S, phase space, a manifold of internal parameters (which evolve during a process according to the dynamics of the scheme), and C, control space, a space of control parameters which stay constant during the processes of the device. In this case, the topological space J is the Cartesian product of the differentiable manifold S and the topological space C.

The space P is a geometric model for the predictions of the scheme, and the function p from J to P is the prediction algorithm. In a dynamical scheme, this algorithm will typically involve the integration of a vectorfield.

These complete the essentials on the etherware level. We will see subsequently that the components of a scheme on this level comprise a subscheme called the N-scheme.

In 1966, this satisfied us as a precise description of the family of models. But in the meanwhile, the vagueness of the conventional interpretation has emerged as a practical problem in the workshops of applied dynamics. So we now add to this picture a parallel scheme on the software level, called the S-scheme.

Thus we have the space I of practical inputs to the device. This may be, for example, a finite set. The function d from J to I is the discretization algorithm, a decision procedure for sorting idealized, continuous, virtual states into bins. These bins, the elements of the space I, correspond more directly to actual states of the device, which can be prepared by the experimentalist as initial states for the dynamical process of the device.

The function i is the (partial) inverse of d. It is an injection algorithm, placing the bin labels within the cellular structure of J determined by d. While the elements of I are still noumenal (that is, belonging to mathematical reality, or to the world of thought) they are closer to the reality of the lab system than are the points of J.

Likewise we have the space O of practical outputs from the device, defined with due respect to the limitations of the observational system. The function r is the reduction algorithm relating idealized predictions and virtual observations.

The function q from I to O is simply the composite of i, p, and r. It is the numerical version of the prediction function. The function n from I to O is an (optional) numerical procedure, emulating the mathematical model, and executable by a machine (or a room full of people). In any case, q and n should be very similar.

We note here that the parallel subschemes (N and S) comprise a simple hierarchical scheme, and further levels could be added when necessary. For example, Thom's idea of a reduction from a metabolic model to a static one ([21], p. 52) could be expressed in this way.

The function s from I to O is not defined intrinsically to the mathematical scheme, but simply represents on the software level the result of three other processes: In(A), D, and Out(A). The first of these, In(A), is the actual input to the device of the initial state specified by a point of I. This process, which carries information across the gap, must be accomplished (barring psychic phenomena) by a person. Thus we shall call it a personal process. The second, D, is the dynamical process of the device (organism, etc.), in lab reality. The third, Out(A), is the transit across the gap of the actual output of the process D, observed and interpreted as a point of O, by another person of the experimental team. The two personal processes, In(A) and Out(A), comprise the conventional interpretation in this scheme. The composition of the three processes defines the mathematical function s, which we call the experimental system, or sometimes, the experimental team. Its precise specification requires the accord of the experimentalists, and their perfect performance according to the agreement. Thus, it may be a little optimistic to describe the experimental system as a mathematical function. So while we informally think of these as mathematical functions, we show them on Fig. 4 as half-arrows, to distinguish them from functions.

Finally, we may remark the similarity of this picture, as shown in Fig. 3, with the schemes of Duhem, as shown in Fig. 2.

We have emphasized the bilevel structure of the mathematical scheme, etherware and software, out of a practical necessity: an object is not a model, without a functional

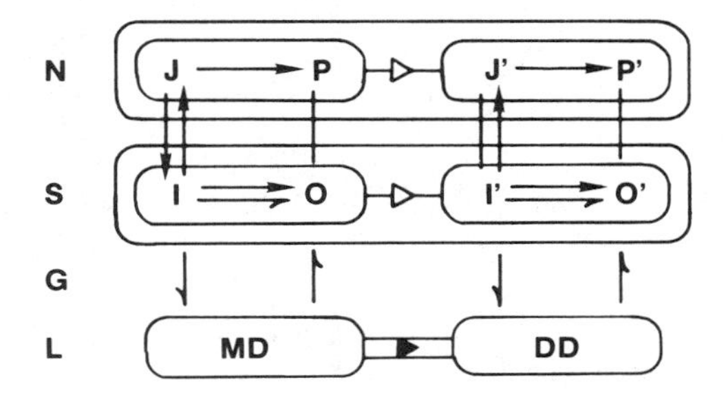

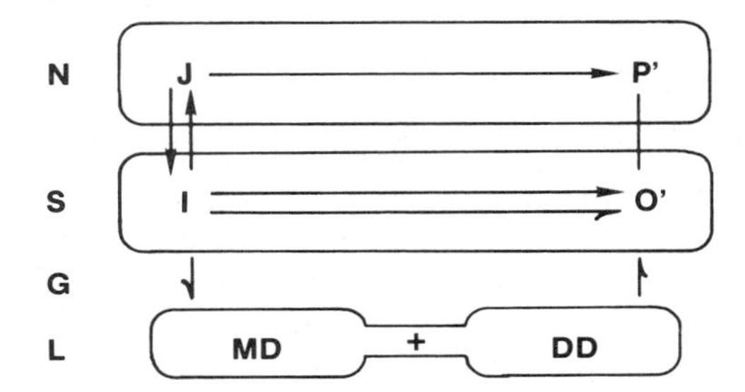

Fig. 4

prediction algorithm. At the present state of the art, this means the simulation of the etherware on an analog, digital, or hybrid computing machine.[14] For if the mathematical prediction, p, is not constructively defined, or effectively determinable, then the functional prediction algorithm will be the numerical procedure, n, which emulates it. Thus, the prediction algorithm on the software level, n, may involve a process in lab reality, and two personal transits across the gap, like the experimental system, s. And therefore n, like s, may be somewhat fuzzier than a mathematical function. So far, the theory (ambient cognitive scheme on the knoware level) has not taken this too seriously, although we have made some proposals along this line [12].

A4. A Simple Static Scheme

Besides dynamical schemes, there are many other strategies for modeling phenomenal systems. Thom [21] has described some very elaborate ones, called static models, based on singularities of mappings. Here we describe the simplest of these, which will be useful in constructing networks.

Suppose several systems, having total state spaces,

$$J_1, J_2, \ldots, J_k$$

jointly influence another system, through its control space, C. It may miraculously happen that this control space is segmented as a Cartesian product,

$$C = C_1 \times C_2 \times \cdots \times C_k$$

with one factor space for each of the controlling systems. Then the joint influence may be expressed by a set of functions,

$$f_1: J_1 \to C_1, \quad f_2: J_2 \to C_2, \quad \ldots, \quad f_k: J_k \to C_k$$

These functions may be combined in a single function,

14. As recently as 1945, this functional prediction was carried out "by hand" by the experimental team, running around on a giant piece of paper in their socks. Thus, although personal processes were involved, no transits across the gap were necessary.

$$g: J_1 \times \cdots \times J_k \rightarrow C = C_1 \times \cdots \times C_k$$

defined by

$$g(j_1, \ldots, j_k) = (f_1(j_1), \ldots, f_k(j_k))$$

But more generally, the inputs may be combined by a function

$$g: J_1 \times J_2 \times \cdots \times J_k \rightarrow C$$

without any segmentation of C as a Cartesian product. This is what we call a static scheme. The inputs are instantaneously translated into an output by a (nonlinear) function of several variables. No dynamical evolution is required within the algorithm of the scheme.

A static scheme may also depend on control parameters, which change the algorithm of the scheme (that is, the function). This is simply a function of more variables,

$$g: J_1 \times \cdots \times J_k \times K \rightarrow C$$

This is a static scheme of family-of-singleton type, which will find application in Part D.

B. EXEMPLARY SCHEMES, SINGLETON TYPE

B1. The Classical Quantitative Scheme

This is historically the first dynamical scheme, created by Newton for the apple. It belongs to the singleton type. Its essential components are the following.

First, we have a manifold representing the virtual states of the device, S. The basic datum of the scheme is a smooth vectorfield, v, on S. To simplify the description, we will assume it is complete, so every integral curve may be prolonged indefinitely into the past and the future.[15] Then $J = S \times R$, and the prediction space, P, is just S. The prediction map, $p: S \times R \rightarrow S$, is defined by the motion of an initial state along the trajectory of the vectorfield for the prescribed time.[16] Thus, $p(s,t) = c(t)$, where c is the unique integral curve of v with $c(0) = s$. This function is continuous everywhere.

This completes the components on the etherware level, E. We may regard this as a subscheme, the N-scheme, and record its data as the object (J, v, P, $p: J \rightarrow P$).[17]

The balance of the scheme, on the software level, S, involves the integration of the vectorfield by numerical methods, whether by power series or polygonal approximation, by hand or by machine. Thus, as described at the end of Section A3, I is a finite set, corresponding to cells of a partition of J. The discretization map, $d: J \rightarrow I$, assigns to a virtual state the label of its cell. The injection map, $i: I \rightarrow J$, is a right-inverse of d, defined by the specification of a preferred point, or nucleus, in each cell. Likewise, the space O is a finite set associated to a partition of P, and the reduction map, $r: P \rightarrow O$, is a discretization like d. As used in modern times, this classical scheme

---

15. Otherwise we would take J to be the domain of the complete solution of v ([11], p. 68).
16. Alternatively, we could take, for the prediction space, P, a function space of curves in S, fibered over J by the map $b: P \rightarrow J$ which evaluates a curve at zero. Thus, the fiber P(j) is a function space of curves beginning at j. If H is the open half-plane of points (x,y) such that $x < y$, then P is also fibered over $H \times J$ by the map $a: P \rightarrow H \times J$ which assigns to a curve the endpoints of its domain, and its beginning point. We need not specify which function space, as we describe a scheme, not a model, but there are standard choices. Finally, the prediction function, $p: J \rightarrow P$ is defined by the maximal integral curves of the vectorfield v. This function will be continuous except at exceptional points.
17. These describe the objects of a category, for which the morphisms may be prescribed variously.

will have the optional numerical algorithm, $n: I \to O$, emulating the prediction function, $p: J \to P$, and approximating its parallel numerical version, $q = r \circ p \circ i: I \to O$.

The data of the full scheme described here comprise the object (S, V, $J = S \times R$, $P = S$, $p: J \to P$, I, $d: J \to I$, $i: I \to J$, O, $r: P \to O$, $n: I \to O$). The first five components comprise the E-scheme. The components (I, O, $n: I \to O$) comprise the optional S-scheme, a subscheme which could serve as a model without any E-scheme, as is frequently done in the computer simulation field.

B2. The Modern Qualitative Scheme

This is the revolutionary paradigm introduced by Poincaré in 1882. As in the singleton classical scheme, ($J = S \times R$, v, $P = S$, $p: J \to P$), the basic datum is a single vectorfield, v, on the virtual state space, S. This is called the dynamic of the scheme. We drop the time factor R in J, as the prediction will be not for a particular time in the future, but forever. Thus, the prediction is not a point of S, but a closed subset, the omega-limit set.[18] We imagine the phase portrait of v, behind the scenes, as a guide to the prediction algorithm. This contains an oriented, nonparameterized curve connecting each point s of S to its omega-limit set, $p(s)$. We choose for P an appropriate space of subsets of S.[19] This completes the E-scheme, ($J = S$, v, P, $p: S \to P$). Note that p is constant on the Insets of the various limit sets of the dynamic, and discontinuous between them. The extension of this subscheme to the S-level is routine and we omit the description of the discretization processes. The numerical procedure, n, is particularly problematic; see [12].

B3. The Attractor/Basin Scheme, Singleton Type

This is a slightly simplified variant of the Poincaré scheme of the preceding section, introduced recently to represent the minimum scheme in some sense [12]. Having fixed the basic datum, v in $V(S)$, let A be the union of the attractors (suitably defined by one of the various definitions in vogue), B be the union of their basins, and X be the complement of B in S. We call A the locus of attraction, and X the locus of separation of v. The idea of this scheme is to ignore points in X, and then to proceed as in the preceding scheme. Thus we let $J = B$, P and $p: J \to P$ as before. Behind the scenes, we envision the attractor/basin portrait of v, instead of the phase portrait. This is particularly reasonable if X is rare, that is, of probability measure zero. For then, after discretization, S cannot be distinguished from B, as X is experimentally invisible.

C. EXEMPLARY SCHEMES, FAMILY-OF-SINGLETON TYPE

C1. The Modern Quantitative Scheme

Here we have a simple example of the functor family-of. We begin with a classical singleton scheme as in Section B1, ($J = S \times R$, $v \in V(S)$, $P = S$, $p: J \to P$), and a new topological space or manifold, C. This is to model the control (static) states of the target device or system. This is the classical system of first order differential equations, with coefficients depending on parameters.

---

18. Actually, this is the coarse prediction. In many applications, one would want a fine prediction. This would be the omega-limit set (coarse prediction) together with some information about the dynamic on it. In the case of a limit point, there is none. For a limit cycle, one might want to predict some qualitative information, such as the period and amplitude. And for a chaotic limit set, one could ask for topological invariants (equivalence class of the limit dynamic, entropy, fractal dimension) and perhaps some qualitative information, such as the characteristic exponents, amplitude, etc. But we will not formalize this variation here.
19. For example, the set of all subsets of S, with the Hausdorff pseudometric topology induced by a metric on S, see ([11], p. 515).

Let V(S) denote the space of smooth vectorfields on S. The basic datum, replacing the vectorfield v of V(S) in the singleton scheme, is a function, $F: C \to V(S)$. Like the metabolic field of Thom ([21], pp. 40, 46, 52; [23], pp. 293, 616, 633), this assigns a dynamical system (vectorfield) on S to each point of C. We will prefer to call this the dynamical field, or just the dynamic, of the scheme.

The prediction space is again S, and the prediction p(c, s, t) is the position at time t along the trajectory of F(c) beginning at s. The complete E-scheme in this case is $(J = C \times S \times R,\ F: C \to V(S),\ P = S,\ p: J \to P)$. This is completed to the full scheme exactly as before, except that C must also be discretized. This is a scheme of family-of-singleton type because for each point c of C, we have a singleton classical scheme, $(S \times R,\ v = F(c),\ P = S,\ p: S \times R \to P)$. Each of these will be stable in the sense of Duhem, as the function F (and thus also p) is continuous everywhere.

C2. The Structural Stability Scheme

This idea, introduced by Andronov and co-workers in 1937, is a maximal family construction. Let S be the dynamical state space, or phase space, and V(S) a space of smooth vectorfields on S. Once again, the basic datum of the scheme is a chosen vectorfield, v, of V(S). But as one is never sure to choose the right one, we choose an open neighborhood, U, of v in V(S). Further, it is traditional to regard the phase portrait as the prediction, even though this is much richer in information than any experimental process. Thus we take $J = U$, P the space of phase portraits, and $p: U \to P$ assigns to each vectorfield its phase portrait. But what is P exactly?

Although we think of the points of P as phase portraits (a system of oriented, nonparameterized paths) in S, it is easiest to define P as the underlying set of V(S), with a new topology. This topology makes any two vectorfields close, which have close phase portraits in the following sense: there is a homeomorphism of S carrying one to the other, which is close to the identity (in an appropriate topology on the group of homeomorphisms of S).[20]

Thus the components of P are equivalence classes of phase portraits, and the interior of these components are the structurally stable vectorfields. The function p is continuous on these interiors. The scheme may be considered a reasonable one if p is continuous on $J = U$, so this space should be in the interior of one of the equivalence classes. Such an E-scheme, $(J,\ P,\ p: J \to P)$, is a structural stability scheme. This is a scheme of the family-of-singleton type, in that for each point of u of $J = U$, we have a singleton qualitative scheme (as in Section B2), $(S \times R,\ u,\ S,\ p(u): S \times R \to S)$. And each of these is stable in the sense of Duhem, as p is continuous by construction: $J = U$ is contained in one component of P. If J is not restricted to one interior, it is a bifurcation scheme, to which we now turn.

C3. The Dynamical Bifurcation Scheme

This is essentially the scheme proposed by Thom in 1966, under the name metabolic model ([21], pp. 40, 46). We prefer to call this a dynamical model. In any case, this is in contrast to his static models, little used outside of catastrophe theory, but potentially very useful in a more general setting.

In a dynamical bifurcation scheme, we have, as in the family-type quantitative scheme (Section C1), two spaces of virtual state variables: the control space C of control (static) parameters, and the phase space S, a manifold of internal (dynamic) variables. The virtual state space, then, is $J = C \times S$. Originally, Thom chose the control space, C, belonging to space-time ([21], pp. 15, 40). Later he generalized this to an open set of Euclidean space ([21], p. 52). At some point, this was further generalized to a finite-dimensional manifold. The first explicit description of this case we have

---

20. The uniform topology induced by a metric on S may be a good choice here. Thus for a positive real, r, the homeomorphisms within uniformity r of the identity define an r-disk in the phase portrait topology on V(S).

found is Zeeman ([23], p. 289). And finally, infinite-dimensional manifolds have been considered as virtual state spaces, for both static and dynamic states ([4], p. 142).

Next, we have for basic datum, the dynamical field (metabolic field, in the language of Thom), $F: C \to V(S)$, which replaces the inclusion $U \to V(S)$ of the open set U, in the preceding scheme. (Thus, the scheme of Andronov is a special case of the scheme of Thom, if we allow infinite-dimensional spaces for C.)

Finally, for the prediction, we could take the phase-portrait as in the preceding scheme (C2), or the omega-limit set, as in the singleton qualitative scheme (B2). The later is more practical, if the scheme is to be useful in the sense of Zeeman, that is, constructively predictive ([23], p. 293). Thus, let P be the appropriate space of closed subsets of S, as before (B2). The complete E-scheme is ($J = C \times S$, $F: C \to V(S)$, P, $p: C \times S \to P$). This scheme is of the family-of-singleton type, in that for each virtual control setting, $c \in C$, we have a singleton qualitative scheme (S, F(c), P, $p: S \to P$). One of these will be stable in the sense of Duhem if c is a regular point of the field: F(c) is structurally stable. Otherwise, c is a bifurcation point of F.

A useful conceptual model in this scheme is the geometric picture composed of all the phase portraits (or simpler yet, the attractor-basin portraits) stacked side-by-side over a picture of C. This is called the bifurcation diagram. The literature of dynamics (e.g., [11]) and catastrophe theory (e.g., [23]) abound in these diagrams, which contain all the predictions of the bifurcation scheme.

C4. The Stable Family Scheme

Just as the first family-type scheme was introduced to deal with the question of stability of a singleton model, the first family-of-families scheme was introduced, by Sotomayor [20], to deal with the question of stability of a family-type scheme, the dynamical bifurcation scheme.

Ignoring the technical details (which in any case are not yet firmly fixed) we have an additional control space, B, a function space of dynamical fields, $M = M(C, V(S))$, and the basic datum is a map, $G: B \to M$. Alternatively, this may be considered as a dynamical field, $F: B \times C \to V(S)$. Thus the scheme is the object ($J = B \times C \times S$, $F: B \times C \to V(S)$, P, $p: B \times C \times S \to P$). This is exactly the same as a dynamical bifurcation scheme, except for the interpretation: a derived family-of-singleton scheme, ($C \times S$, $G(b): c \to V(S)$, P, $p: C \times S \to P$) is a stable family if G is continuous at b. The tricky part, the definition of the topology of M in generalization of the phase-portrait topology of V(S) described previously (Section C2), is omitted. The general idea is due to Thom ([21], pp. 44, 320), and the details to Sotomayor [20].

D. EXEMPLARY SCHEMES, COMPLEX TYPE

D1. The Serial Coupling Scheme

Suppose we have two lab devices, and a satisfactory model for each. Then we couple them together in such a way that the first (master) is almost totally unaffected by the work it is doing in changing the control (static) state variables of the second (slave, or driven) device or system.[21] We call this a serial coupling. Now we want to construct a model for the system composed of the coupled devices. We may simply model the combined system, using a dynamical bifurcation scheme. But this ignores the important information concerning the coupling. Thus, we want a scheme with more structure for modeling the combined system, which allows the serial coupling to be modeled as well. Such is a serial coupling scheme, the first example of the complex scheme-type. In

---

21. The outstanding example is the forced oscillator, discussed in the next section. More extensive discussions may be found in the literature of nonlinear dynamics [13].

this Section, we describe the serial coupling of two dynamical bifurcation schemes. Figure 4 shows both schemes for the combined device: with and without explicit coupling.

Now let $MD = [C \times S,\ F: C \to V(S),\ P,\ p: C \times S \to P]$ be the E-scheme of the master device, and $DD = [D \times E \times T,\ G: D \times E \to V(T),\ Q,\ q: D \times E \times T \to Q]$ be the E-scheme of the driven device.

Here we have deliberately expressed the control space of the driven system as a Cartesian product of two spaces, D and E. This is to allow the enslavement of some controls of the driven system by the master device (those of D) while others (those of E) remain free. The coupling is now expressed by a function $d: C \times S \to D$ from the total, virtual state space of the driving system to the slave variables of the driven system.[22] The effect of this coupling is to remove the control space, D, from the scheme. Thus our serial coupling scheme is the object,

$$[(C \times S,\ F,\ P,\ p: C \times S \to P);\ (D \times E \times T,\ G,\ Q,\ q: D \times E \times T \to Q);$$
$$d: C \times S \to D,\ H: C \times E \to V(S \times T),\ r: (C \times E) \times (S \times T) \to R]$$

or equally,

$$[MD;\ DD;\ d: C \times S \to D,\ H: C \times E \to V(S \times T),\ r: (C \times E) \times (S \times T) \to R]$$

where $H(c,e)(s,t) = (F(c)(s);\ G(d(c,s),\ e))(t)$, and r is the prediction algorithm of the coupled scheme: for given values of the free controls $(C \times E)$ and preparable initial states $(S \times T)$ it assigns the omega-limit set of the combined dynamic. Thus R is a space of closed subsets of $S \times T$.[23]

Essentially, the serially coupled scheme is completely specified by the data [MD; DD; RC], where RC denotes the rigid coupling, d, because H and r are derivable from these.

On the other hand if the combined system is regarded as a simple device, the dynamical bifurcation scheme for it would be the object:

$$[C \times E \times S \times T,\ H: C \times E \to V(S \times T),\ Q,\ r: C \times E \times S \times T \to R]$$

We regard $C \times E$ as the control space in this case, and $S \times T$ as the dynamical (phase) space. Thus the prediction, $r: C \times E \times S \times T \to R$, is the same as in the coupled scheme above. Both are shown in Fig. 4.

This scheme forgets the full coupled structure of the combined system. It is obtained from the serial coupling scheme by a _forgetful functor_.[24]

So far, we have described the serial coupling scheme previously proposed for complex systems [8]. Finally, we must add to this scheme a new complication: the coupling function, $d: C \times S \to D$, may be changed by a parameter in an additional control space, B, belonging to the coupling itself. This is something like the _static model_ of Thom ([21], p. 40), and is analogous to _flexible coupling_ in the context of the conventional (parallel) coupling of dynamical systems [5]. We will call this one an _adjustable coupling_, as opposed to the rigid coupling of the preceding paragraph. Thus we replace the rigid coupling function, $RC = d: C \times S \to D$, in the scheme above, by an _adjustable coupling function_, $d: B \times C \times S \to D$. This produces the object:

$$[(C \times S,\ F,\ P,\ p: C \times S \to P);\ (D \times E \times T,\ G,\ Q,\ q: D \times E \times T \to Q);$$
$$B,\ d: B \times C \times S \to D,\ H: B \times C \times E \to V(S \times T),\ r: (B \times C \times E) \times (S \times T) \to R]$$

or equally,

$$[MD; DD; B,\ d: B \times C \times S \to D; H: B \times C \times E \to V(S \times T),\ r: (B \times C \times E) \times (S \times T) \to R]$$

---

22. We have adopted this terminology from Haken [17].
23. Warning: this is larger than $P \times Q$.
24. See [15] for an introduction to local category theory, which is the background for our treatment of schemes.

where $H(b,c,e)(s,t) = (F(c)(s); G(d(b,c,s), e)(t))$ and $r((b,c,e), (s,t))$ is the omega-limit set of (s,t) in the dynamical system $H(b,c,e)$. As in the rigid case, these are derivable from the essential data [MD; DD; AC], where AC = [B, d: $B \times C \times S \to D$], the adjustable coupling.

Some symbols for these schemes are introduced in Figure 5 (see also [8]).

D2. The Canonical Example: Forced Oscillation

Here at last is an actual model. The lab devices are a robust oscillator (master device, MD) and a weighted spring or pendulum with damping (driven device, DD). These are illustrated in the figure on p. 133 of [13]. Nonlinear dynamics texts explain all one would want to know about these devices, and the result of coupling them [13]. Our formulation is a slight variation on the classical theme.

The model for the master device (MD) is:

control space: $C = R$, the real numbers, representing the angular velocity,
phase space: $S = T^1$, a circle, representing the phases of the oscillator,
dynamical field: $F: C \to V(C)$, where $F(c)$ is the vectorfield (expressed as a first order ordinary differential equation for the sake of familiarity)

$$s' = c$$

The model for the damped oscillator (DD; we choose the spring, for the sake of definiteness) is:

control space: $D = R$, the real numbers, representing the acceleration of the support point of the spring,
phase space: $T = R^2$, the Euclidean plane, representing the extension and velocity of the spring, (x,y)

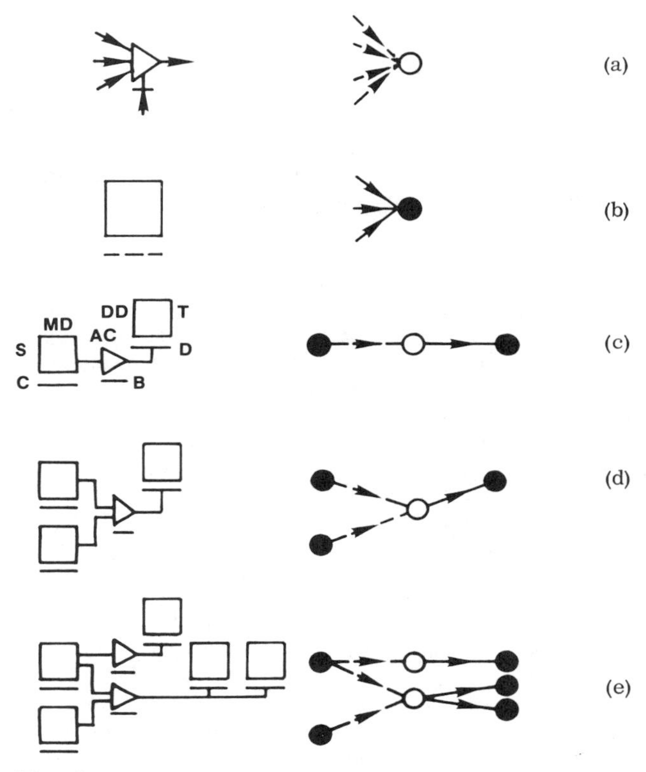

Fig. 5

dynamical field: $G: D \to V(T)$, where G(d) is the vectorfield (expressed as a system of two first order, ordinary differential equations, more or less as in the books [13], p. 212)

$$x' = y$$

$$y' = -(1/m)f(x) + ky + d$$

where f is the restoring force function of the spring, k the coefficient of friction or damping, and d the assumed uniform acceleration of the support point.

The adjustable coupling (AC) is given by:

control space: $B = R$, the real numbers, for amplitude,
coupling function: $d: B \times C \times S \to D$, defined by

$$d(b, c, s) = b \sin (s)$$

which does not depend directly on c, the angular velocity of the oscillator. But integrating the oscillator equation at the initial condition, $s = 0$, we find $s = ct$ (phase = angular velocity times time) so c controls the frequency of the forcing term,

$$d(b, c, s) = b \sin (ct)$$

so the flexibly coupled device (CD)—found by replacing d in (DD) by the function (AC)—is:

control space: $B \times C = R \times R$, representing amplitude and angular velocity of the driving oscillator (MD),
phase space: $S \times T = T^1 \times R^2$, for phase of the driver, extension of the spring, and velocity of the weight,
dynamical field: $H: B \times C \to V(S \times T)$, expressed as a system of equations,

$$s' = c$$

$$x' = y$$

$$y' = -(1/m)f(x) + ky + b \sin (s)$$

which are the standard equations of this classical model ([13], p. 213). Other examples of serial coupling schemes, intermittency for example, have been proposed [8].

D3. Serial Network Schemes

It is now our goal to extend the considerations of the preceding section to complex systems. We might consider a fixed quiver (directed graph), with a dynamical bifurcation scheme at each vertex and a serial coupling on each directed edge. When several edges depart (spread) from one vertex, we have a plausible situation: the instantaneous state of the master device is coupled simultaneously to different driven systems, through different serial couplings, rigid or adjustable. But when several directed edges arrive (fan) to a single vertex, how can we represent the cooperation of several masters in driving a single scheme? We must have a multicoupling scheme to represent this cooperation. This is an example of a static scheme (see Sec. A4).

For example, if two masters, $MD1 = (C \times S, F, P, p)$ and $MD2 = (D \times T, G, Q, q)$, share control of a single slave, $DD = (E \times U, H, R, r)$, we must have a serial multicoupling function, $e: C \times S \times D \times T \to E$, to represent the determination of the driven control, e, by the instantaneous states (c, s) of MD1 and (d, t) of MD2, through the function, e(c, s, d, t). Or in the case of an adjustable multicoupling, we must have a scheme, $AC = (B, e: B \times C \times S \times D \times T \to E)$, where B is an additional control space for the multicoupling.

Then for a fan of directed edges to arrive at a common vertex (dynamical bifurcation scheme), they must converge first as inputs to a common multicoupler, from which a single output directs the target dynamical scheme. Thus, as a diagram of the network, we must consider an odd mathematical object: a directed graph with two kinds of vertices—say green (for dynamical bifurcation schemes) and red (for static multicoupler schemes).

This diagram will be easiest to understand if we associate all directed edges arriving at a vertex with that vertex. Although not essential, it simplifies our discussion. Thus, edges arriving at a red vertex are red. This red fan represents a static multicoupling scheme, as shown in Fig. 5(a). The red edges represent the arguments of the multicoupling function, whether to be used as inputs from dynamical systems, or adjustable coupling controls. Similarly, edges arriving at a green vertex are green. A green fan represents a dynamical bifurcation scheme, as shown in Fig. 5(b). The green edges represent factors of the segmented control space (see Section A4). As segmentation of a control space is a special structure adapted to a particular multicoupling situation, it is more desirable to represent all multicouplings by a serial multicoupling function (red fan). Thus, we prefer to consider green fans having only one or two input segments. One of these is for the controlling input, arriving from the output of a multicoupling function (red vertex). The other (optional) represents free controls.

The fans are to be connected in a schematic diagram for a network. The rules are:

1. Red vertices may be joined only to the input ends of green edges.
2. Green vertices may be joined only to the input ends of red edges.

Some examples are shown in Fig. 5, both in pictographic and in schematic diagrams. Let us choose a specific schematic diagram, such as the one shown in Fig. 5(e).

Corresponding to each such schematic, there is a category of complex schemes. Each of these schemes is determined by essential data: a dynamical bifurcation scheme for each dynamic vertex (green fan), and an adjustable static multicoupling scheme for each static vertex (red fan). Thus, each schematic diagram determines a category of schemes. And all these together comprise the complex type of schemes.

Surely this seems too arduous, yet we shall see in later papers that complex systems in nature, especially in ecology and physiology, present such schematic diagrams in a very natural way.

D4. Serial Cycles and Parallel Coupling

While the coupling concept in the context of dynamical systems theory is classical, it is commonly applied only to the parallel coupling of two systems. To further clarify the serial coupling concept we have emphasized here and elsewhere [8, 9] we contrast the two in this section.

Recall that for two vectorfields, F in V(M) and G in V(N), their direct product is a vectorfield (F, G) in V(M × N), defined by

$$(F, G)(m, n) = (F(m), G(n))$$

Let P be any very small vectorfield on M × N. Then the sum (F, G) + P is a weak parallel coupling of F and G. If further P depends on a parameter, $P: C \to V(M \times N)$, the dynamical field $(F, G) + P: C \to V(M \times N)$ is a flexible coupling of F and G [5]. Typically, C is a neighborhood of the origin in Euclidean space, and P(0) = 0. A strong parallel coupling refers to an arbitrary (possibly enormous) perturbation of the direct product, (F, G). In other words, this could mean any vectorfield in V(M × N).

Now let us express both of these coupling concepts in the specific case of the forced oscillation model of Sec. D2. In the notation of first order systems of ordinary differential equations, we have:

The uncoupled system:

$$a' = w$$

$$x' = y$$

$$y' = -(1/m)f(x) + ky + d$$

With serial coupling, $d(a,b) = b \sin(a)$:

$$a' = w$$

$$x' = y$$

$$y' = -(1/m)f(x) + ky + b \sin(a)$$

With parallel coupling, $P = (p, q, r)$:

$$a' = w + p(a,b,x,y)$$

$$x' = y + q(a,b,x,y)$$

$$y' = -(1/m)f(x) + ky + d + r(a,b,x,y)$$

So we see that adjustable serial coupling is a special case of flexible parallel coupling. But serial coupling provides a natural way to specify the very restricted class of perturbations which occur in the serial coupling situation so prevalent in the phenomenal universe, while weak parallel coupling provides a natural way to model the nonspecific perturbations of unknown coupling mechanisms.

We may end here with a serial coupling situation which comes closer to parallel coupling in its specific form. Consider a typical network of serially coupled schemes. Very frequently, this will contain a serial cycle. That is, the schematic diagram contains a directed subgraph which is a closed cycle, as shown in Fig. 6(a). These occur naturally, for example, in the ring of cells used by Rashevsky and Turing in modeling biological morphogenesis ([13], p. 109). The serial cycle implies a very intimate coupling. The shortest case is a serial coupling from the output of a dynamical bifurcation scheme to its own control input, as shown in Fig. 6(b). Expressed in equations, this is:

$$\text{MD} = \text{DD: } x' = F(c, x)$$

$$\text{AC: } c = c(x)$$

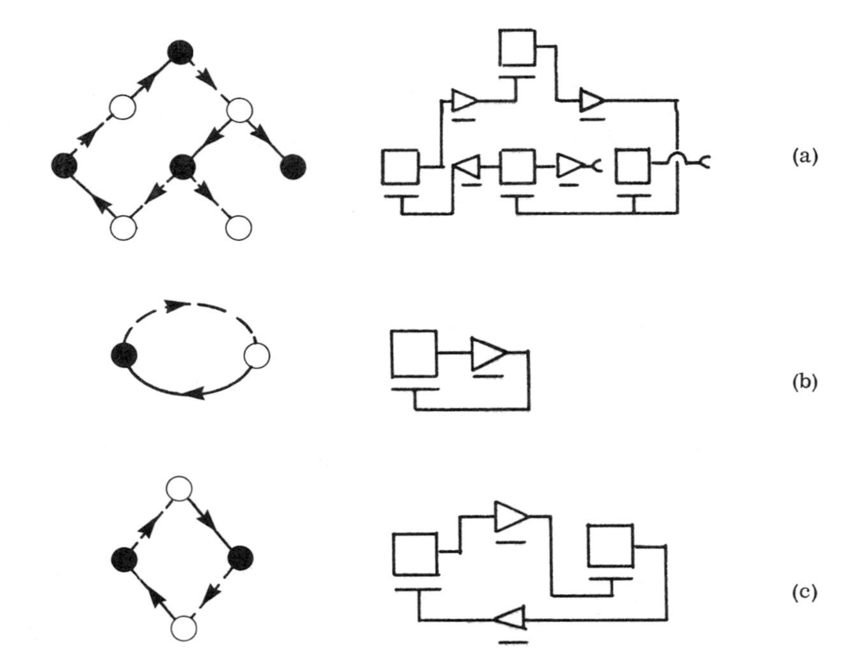

Fig. 6

Bicoupled system:

$$x' = F(c(x), x) = f(x)$$

The effect of the short feedback is to eliminate the control parameters.

Now suppose we consider the next shortest case, as shown in Fig. 6(c). In a simple version, d does not specifically depend on c nor vice versa, and both are rigid. Expressed in equations, this means:

$$D1: x' = F(c, x)$$

$$D2: y' = G(d, y)$$

$$C1: d = d(x)$$

$$C2: c = c(y)$$

Bicoupled system:

$$x' = F(c(y), x) = f(x, y)$$

$$y' = G(d(x), y) = g(x, y)$$

This is essentially a very strong parallel coupling, and the controls have been eliminated. In conclusion, serial coupling is a very useful concept, which may replace parallel coupling in many applications, giving additional structure and better modeling.

Previously, we have described a kind of multiplication table of attractors, based upon flexible parallel coupling ([5], Sec. 2; [8], Fig. O1; [9], Sec. 12). In the context of serial coupling networks, there is an analogue, the composition of bifurcations. For example, if two dynamical bifurcation schemes are serially coupled, and each exhibits a single canonical bifurcation, what are the possible bifurcation diagrams of the coupled system? As we have suggested earlier, this may be found to depend on generic properties of the adjustable coupling function, which maps the total state space of the master scheme (containing the locus of attraction) onto the control space of the driven scheme (containing the bifurcation set). The desired generic property is the transversal intersection of the locus of attraction of the master with the bifurcation set of the slave [9], Sec. 13). Here is a rich source of problems for dynamicists, fundamental for understanding the behavior of complex systems.

ACKNOWLEDGMENTS

This entire series, On Morphodynamics, has pushed against the grain of traditional specialized scholarship. Without the support of the handful of scientists who found it not wholly unreasonable, I could not have persisted. Thus, it is a special pleasure to acknowledge them here: Fred Abraham, Lawrence Domash, Walter Freeman, Alan Garfinkel, Werner Guttinger, Herman Haken, Nick Herbert, Morris Hirsch, Erich Jantsch, Hans Jenny, Jean-Michel Kantor, Paul Kramerson, Terrence McKenna, Charles Musès, Timothy Poston, Jacques Ravatin, Otto Rossler, Dan Sunday, Stanley White, Arthur Winfree, and F. E. Yates. Finally, we owe special thanks to René Thom for inspiration, and to Fred Abraham, Alan Garfinkel, and Tim Poston for critical comments on the manuscript.

REFERENCES

1. Abraham, R. H.: Introduction and Conclusion, in [10].
2. Abraham, R. H.: Introduction to Morphology, Univ. Claude-Gernard, Lyons (1972).
3. Abraham, R. H.: Psychotronic vibrations, in Proceedings, First International Congress of Psychotronics and Parapsychology, Prague (1973).
4. Abraham, R. H.: Vibrations and the realization of form, in Evolution and Consciousness (E. Jantsch and C. H. Waddington, eds.). Addison-Wesley (1976).

5. Abraham, R. H.: The macroscopy of resonance, in Structural Stability, the Theory of Catastrophes, and Applications (P. Hilton, ed.). Springer (1976).
6. Abraham, R. H.: Simulation of cascades by videofeedback, in Structural Stability, the Theory of Catastrophes, and Applications (P. Hilton, ed.). Springer (1976).
7. Abraham, R. H.: The function of mathematics in the evolution of the noosphere, in The Evolutionary Vision (E. Jantsch, ed.). AAAS Selected Symposium, vol. 61 (1981).
8. Abraham, R. H.; Shaw, C. D.: Dynamics, a visual introduction, in Self-Organizing Systems (F. E. Yates, ed.). Plenum (to appear).
9. Abraham, R. H.: Dynamics and self-organization, in Self-Organizing Systems (F. E. Yates, ed.). Plenum (to appear).
10. Abraham, R. H.: Foundations of Mechanics, First Edition. Benjamin (1967).
11. Abraham, R. H.; Marsden, J. E.: Foundations of Mechanics, Second Edition. Benjamin/Cummings (1978).
12. Abraham, R. H.: DYNASIM: exploratory research in bifurcations using interactive computer graphics, in Bifurcation Theory and Applications in the Scientific Disciplines (O. Gurel and O. Rossler, eds.); Ann. New York Acad. Sci. 316, 676-684 (1979).
13. Abraham, R. H.; Shaw, C. D.: Dynamics, the Geometry of Behavior, Part 1: Periodic Behavior. Aerial, Santa Cruz, CA (1982).
14. Duhem, P.: The Aim and Structure of Physical Theory. Paris (1914).
15. Eilenberg, S.; Cartan, H.: Foundations of fiber bundles, in Intern. Sym. on Algebraic Topology. UNAM, Mexico City (1958).
16. Garfinkel, A.: Forms of Explanation, Rethinking the Questions in Social Theory. Yale Univ. (1981).
17. Haken, H.: Synergetics, An Introduction. Springer (1977).
18. Hesse, M. B.: Models and Analogies in Science, Univ. Notre Dame, In. (1966).
19. Rosen, R.: Dynamical System Theory in Biology, vol. 1. Wiley (1970).
20. Sotomayor, J.: Bull. Am. Math. Soc. 74, 722-726 (1968).
21. Thom, R.: Structural Stability and Morphogenesis, an Outline of a General Theory of Models, tr. D. Fowler. Benjamin-Cummings (1972).
22. Velikovsky, I.: World in Collision.
23. Zeeman, E. C.: Catastrophe Theory and Its Applications. Addison-Wesley (1977).

2 ALMOST UNIVERSAL MAPS AND THE ALMOST FIXED POINT PROPERTY

Fadhel Al-Musallam* and Sam B. Nadler, Jr. / Department of Mathematics, West Virginia University, Morgantown, West Virginia

1. INTRODUCTION

A metric space (Y, d) is said to have the almost fixed point property (written AFPP) provided that if f: Y → Y is a continuous function, then, for each $\epsilon > 0$, there exists a point $y_\epsilon \in Y$ such that $d(f(y_\epsilon), y_\epsilon) < \epsilon$. Clearly, the almost fixed point property and the (usual) fixed point property are equivalent for compact metric spaces. Practically the only noncompact metric spaces which are known to have AFPP are the open unit ball in Euclidean n-space R^n and any bounded open topological disk in R^2 whose boundary is locally connected ([5] and [13], p. 99).

This paper has two related purposes. First, in Sec. 3, we introduce the concept of almost universal mappings and show that it is related in an intrinsic way to AFPP. Thus, we obtain a technique for studying AFPP which has not been used before. We mention that the notion of almost universal mappings is a generalization of the notion of universal mappings, and that universal mappings have been used to study the fixed point property [8]. In Sec. 3 we not only establish some basic facts about almost universal mappings, but we also obtain some useful sufficient conditions in order that a mapping be almost universal [see (3.5), (3.9), (3.13), and (3.14)]. For the most part we confine our attention to those results which will be used later. However, we mention that some results in Sec. 3, as well as results in Sec. 4, would seem to be of interest independent of their applications in this paper. The second purpose of the paper is to use almost universal mappings to determine classes of metric spaces which have AFPP. Most of these results are in Sec. 5, though some very general results concerning spaces which have AFPP are in Secs. 3 and 4 [e.g., (3.15) and (4.2)]. For our discussion, we state one of the results from Sec. 5 (the definitions of geometric cone and geometric suspension are in Sec. 2): If Y is a nonempty connected subset of a chainable continuum, then Y, the geometric cone $\kappa(Y)$, and the geometric suspension $\Sigma(Y)$ have AFPP [(5.4)]. The fact that such spaces Y have AFPP is a generalization of Hamilton's theorem [6] and provides, for the first time, a class of one-dimensional noncompact metric spaces having AFPP [note: a stronger result for Y is in (5.3)]. The fact that the spaces $\kappa(Y)$ and $\Sigma(Y)$ have AFPP is related, in the following sense, to results in [5]: For each $\epsilon > 0$, there are ϵ-mappings from $\kappa(Y)$ and $\Sigma(Y)$ onto an open disk or onto an open disk together with a connected piece of its boundary [this follows from the proof of (5.2)]—thus, $\kappa(Y)$ and $\Sigma(Y)$ are generalizations of the spaces in [5] and they have AFPP. We give examples and discussions in Sec. 5 which show that certain natural generalizations of our results are not possible. For instance, in relation to the result above, we give an example of a topological line L in a tree-like continuum in the plane such that L, the geometric cone over . L, and the geometric suspension over L do not have AFPP [see (5.8)]. Other results in Sec. 5 include generalizations of the cone part of (5.4), results about punctured suspensions [(5.11) and (5.12)], and a result about geometric cones over sphere-like spaces which are not necessarily bounded [see (5.13)].

---

*Current affiliation: Department of Mathematics, Arizona State University, Tempe, Arizona

We mention that some results about almost universal mappings are stronger than some results about AFPP [e.g., cf. (5.2) and the proof of (5.4); also, see Sec. 4]. We also note that several of our results about AFPP are new even with the additional assumption that the space is compact (in which case, we have the fixed point property).

2. PRELIMINARIES

If $f: X \to Y$ is a function and if $A \subset X$, then $f \mid A$ denotes the restriction of f to A. By a <u>mapping</u> we mean a continuous function.

Let X and Y be topological spaces. Then, X is <u>contractible with respect to</u> Y ([10], p. 370) provided that every happing from X into Y is homotopic to a constant mapping (i.e., is <u>nullhomotopic</u>). A space which is contractible with respect to itself is said to be <u>contractible</u>. For basic facts about homotopy, see [4, 10].

A mapping f from a metric space Z_1 into a space Z_2 is said to be an <u>ϵ-mapping</u> (for some $\epsilon > 0$) provided that for each point $z_2 \in f[Z_2]$, $\operatorname{diam}[f^{-1}(z_1)] < \epsilon$ where <u>diam</u> denotes diameter. The following well-known fact is easy to prove using [14], 7.1, p. 11.

<u>2.1. Lemma</u>. Let Z_1 and Z_2 be compact metric spaces and let $\epsilon > 0$. If $f: Z_1 \to Z_2$ is an ϵ-mapping, then there exists a $\delta > 0$ satisfying the following: If $K \subset f[Z_1]$, $K \neq \emptyset$, and $\operatorname{diam}[K] < \delta$, then $\operatorname{diam}[f^{-1}(K)] < \epsilon$.

A <u>Hausdorff continuum</u> is a compact connected Hausdorff space consisting of more than one point. A <u>continuum</u> is a Hausdorff continuum which is metrizable.

Let P be a collection of spaces. Then, a metric space X is said to be <u>P-like</u> provided that for each $\epsilon > 0$ there exists an ϵ-mapping from X onto some space $P_\epsilon \in P$. If $P = \{\text{arc}\}$ and if X is a P-like continuum, then X is called <u>chainable</u> or <u>snakelike</u> (it is easy to see that this definition is equivalent to the one in [10], p. 224). We can obtain nonmetric analogues of these notions by replacing ϵ-mappings with U-mappings where U is an open cover.

For each $n = 1, 2, \ldots$, let R^n denote Euclidean n-space with its usual norm $\|\cdot\|$ given by

$$\|x\| = (x_1^2 + \cdots + x_n^2)^{\frac{1}{2}} \quad \text{for each } x = (x_1, \ldots, x_n) \in R^n$$

let $B^n = \{x \in R^n: \|x\| \leq 1\}$, and let $S^{n-1} = \{x \in B^n: \|x\| = 1\}$. By an <u>n-cell</u> we mean any space which is homeomorphic to B^n. If Q is an n-cell, then $\partial Q = h^{-1}(S^{n-1})$ where h is any homeomorphism from Q onto B^n. The following lemma has been used many times. It is proved by first assuming (without loss of generality) that $Q = B^n$, and then letting $\gamma(z)$ be the point in S^{n-1} which lies on the directed ray $\overrightarrow{\beta(z)\alpha(z)}$ from $\beta(z)$ through $\alpha(z)$—see, for example, [8], p. 435.

<u>2.2. Lemma</u>. Let Z be a topological space and let Q be an n-cell. Let $\alpha, \beta: Z \to Q$ be mappings such that $\alpha(z) \neq \beta(z)$ for all $z \in Z$. Let $Y \subset \alpha^{-1}(\partial Q)$. Then, $\alpha \mid Y: Y \to \partial Q$ can be extended to a mapping $\gamma: Z \to \partial Q$.

A mapping r from a space X into a space Y is called an <u>r-map</u> provided that there is a mapping $k: Y \to X$ such that $r \circ k$ is the identity map on Y ([14], p. 7). If $Y \subset X$ and if $r: X \to Y$ satisfies $r(y) = y$ for all $y \in Y$, then r is called a <u>retraction</u> and Y is said to be a <u>retract</u> of X. Clearly, homeomorphisms and retractions are r-maps.

If $(E, \|\cdot\|)$ is a normed linear space, then we let $(E \times R^1, |||\cdot|||)$ denote the normed linear space which is the cartesian product of E with R^1 and whose norm is given by

$$|||(z,t)||| = (\|z\|^2 + t^2)^{\frac{1}{2}} \quad \text{for all } (z,t) \in E \times R^1$$

<u>We assume for the rest of this section that any metric space under consideration lies in the subspace $E \times \{0\}$ of a normed linear space of the form</u> $(E \times R^1, |||\cdot|||)$. We remark that any metric space can be embedded in this manner and, by the proof of [4], 8.1,

p. 79, such an embedding is an isometry if the metric on the original space is bounded (note that isometries preserve AFPP).

Now let X be a metric space (as above). Let $v_X^+ = (e, t)$ and $v_X^- = (e', t')$ be vectors in $E \times R^1$ such that $t > 0$ and $t' < 0$. By the geometric suspension over X with vertices v_X^+ and v_X^- we mean the union of all convex arcs from points of X to v_X^+ and v_X^- (such convex arcs will be denoted by $\widehat{xv_X^+}$ and $\widehat{xv_X^-}$, respectively, for $x \in X$). The geometric cone over X with vertex v_X^+ is the union of all convex arcs $\widehat{xv_X^+}$ for $x \in X$. The vertex for a geometric cone over X will sometimes be denoted by v_X instead of v_X^+. We let $\Sigma(X)$ and $\kappa(X)$ denote any given geometric suspension and geometric cone over X, respectively. This last notation does not take into account the vertices—indeed, all of our results are valid for any choice of vertices. We mention that we will frequently write x when we mean the point $(x, 0)$ in $\Sigma(X)$ or $\kappa(X)$.

We note that if X is compact, then the geometric suspension and the geometric cone over X are homeomorphic to the topological suspension and the topological cone over X, respectively. However, this is false when X is not compact since the topological cone need not even be metrizable (see [15], 16B, p. 113).

If X and Y are metric spaces and if $f: X \to Y$ is a mapping, then the induced mapping $\tilde{f}: \Sigma(X) \to \Sigma(Y)$ is defined by letting, for any $x \in X$ and $\lambda \in [0, 1]$,

$$\tilde{f}(\lambda \cdot x + [1 - \lambda] \cdot v_X^+) = \lambda \cdot f(x) + [1 - \lambda] \cdot v_Y^+$$

$$\tilde{f}(\lambda \cdot x + [1 - \lambda] \cdot v_X^-) = \lambda \cdot f(x) + [1 - \lambda] \cdot v_Y^-$$

The induced mapping $f^*: \kappa(X) \to \kappa(X)$ is the restriction of $\tilde{f}$ to $\kappa(X)$. The continuity of $\tilde{f}$ and f^* follows easily from the continuity of f.

For use later on, let us note the following two elementary lemmas.

2.3. Lemma. For any nonempty metric space X, $\kappa(X)$ is contractible and, if X is connected, $\Sigma(X)$ is contractible with respect to S^1.

Proof. The contractibility of $\kappa(X)$ is clear by using the homotopy $h: \kappa(X) \times [0, 1] \to \kappa(X)$ defined by letting $h(z, s) = [1 - s] \cdot z + s \cdot v_X$ for each point $(z, s) \in \kappa(X) \times [0, 1]$. Now assume that X is connected. Let $Y_1 = \kappa(X)$ and let $Y_2 = [\Sigma(X) - \kappa(X)] \cup X$. Since Y_1 and Y_2 are homeomorphic and since Y_1 is contractible, Y_1 and Y_2 are each contractible with respect to S^1. Also: Y_1 and Y_2 are closed subsets of $\Sigma(X)$, $Y_1 \cup Y_2 = \Sigma(X)$, and, since $Y_1 \cap Y_2 = X$, $Y_1 \cap Y_2$ is connected. Therefore, by [14], (5.2), p. 221, it follows that $\Sigma(X)$ is contractible with respect to S^1.

2.4. Lemma. If X and Y are metric spaces and if $f: X \to Y$ is an ϵ-mapping for some $\epsilon > 0$, then $f^*: \kappa(X) \to \kappa(Y)$ and $\tilde{f}: \Sigma(X) \to \Sigma(Y)$ are ϵ-mappings (for the same ϵ).

Proof. The lemma is easy to prove using the formulas for f^* and $\tilde{f}$. The details are omitted.

3. ALMOST UNIVERSAL MAPPINGS

A discussion of the objectives of this section is in Sec. 1.

Let X and Y be topological spaces. A mapping $f: X \to Y$ is universal [8] provided that given any mapping $g: X \to Y$, there exists $x_0 \in X$ such that $f(x_0) = g(x_0)$. Now assume that Y is a metric space with metric d. Then we say that a mapping $f: X \to Y$ is almost universal provided that given any mapping $g: X \to Y$ and any $\epsilon > 0$, there exists $x_\epsilon \in X$ such that $d(f(x_\epsilon), g(x_\epsilon)) < \epsilon$. Let us note the following:

3.0. Proposition. Universal mappings into metric spaces are almost universal. Furthermore: If X is a compact topological space, (Y, d) is a metric space, and $f: X \to Y$ is almost universal, then f is universal.

Proof. The first part of the proposition is obvious. Let X and (Y, d) be as in the second part of the proposition, and assume that $f: X \to Y$ is a mapping which is not universal. Then, there is a mapping $g: X \to Y$ such that $f(x) \neq g(x)$ for any $x \in X$. Define a mapping $\psi: X \to R^1$ by $\psi(x) = d(f(x), g(x))$ for each $x \in X$. Since X is compact, ψ attains its minimum value at some point $p \in X$. Let $\epsilon = \psi(p)$. Since $f(p) \neq g(p)$, $\epsilon > 0$. Thus, since $d(f(x), g(x)) \geq \psi(p) = \epsilon$ for all $x \in X$, f is not almost universal. This completes the proof of (3.0).

The following two elementary propositions are important for this paper since they relate almost universal mappings to the almost fixed point property (defined in Sec. 1).

3.1. Proposition. A metric space (X, d) has AFPP if and only if the identity map $i_X: X \to X$ is almost universal.

The result in (3.1) follows immediately from definitions. It is analogous to Proposition 3 of [8] which says that a topological space X has the fixed point property if and only if the identity map $i_X: X \to X$ is universal.

If $f: X \to Y$ is a universal mapping, then Y has the fixed point property [8, Prop. 2] and $f[X] = Y$ [8], Prop. 1. These results are false for almost universal mappings since, as is easy to see, the inclusion map $j: (0,1) \to [0,1)$ is almost universal (cf. (3.5)). However, let us note the following two results, the first of which strengthens half of (3.1).

3.2. Proposition. Let X be a topological space and let (Y, d) be a metric space. If $f: X \to Y$ is almost universal, then Y has AFPP.

Proof. Let $g: Y \to Y$ be a mapping and let $\epsilon > 0$. Since $f: X \to Y$ is almost universal, there exists $x_\epsilon \in X$ such that $d(f(x_\epsilon), g \circ f(x_\epsilon)) < \epsilon$. Thus, the point $f(x_\epsilon)$ moves less than ϵ under g and, therefore, we have proved (3.2).

3.3. Proposition. Let X be a topological space and let (Y, d) be a metric space. If $f: X \to Y$ is almost universal, then $f[X]$ is dense in Y.

Proof. Let $y_0 \in Y$ and define $g: X \to Y$ by letting $g(x) = y_0$ for all $x \in X$. Then, for each $\epsilon > 0$, there exists $x_\epsilon \in X$ such that $d(f(x_\epsilon), g(x_\epsilon)) < \epsilon$. Hence, $d(f(x_\epsilon), y_0) < \epsilon$. It now follows that $f[X]$ is dense in Y.

Let us note the following corollary to (3.2):

3.4. Corollary. Let X be a topological space and let (Y, d) be a metric space. If $f: X \to Y$ is almost universal, then Y is connected.

Proof. It is easy to see that any metric space having AFPP must be connected. Therefore, (3.4) follows from (3.2).

The next result concerns mappings from connected topological spaces into bounded intervals. It shows that the converse of (3.3) holds for such mappings, and it will be used in the proof of (5.3). It is easy to see using (3.2) and (3.4) that if X is any topological space and if $f: X \to J \subset R^1$ is almost universal, then J must be a bounded interval. Thus, the assumption in (3.5) that J be a bounded interval is essential. With respect to the assumption in (3.5) that X is connected, see (3.7). Let us also note that if the interval J in (3.5) is compact and if f maps the connected space X onto J, then f is universal ([8], Prop. 8).

3.5. Proposition. Let X be a connected topological space, let J be a nonempty bounded interval in R^1, and let $f: X \to J$ be a mapping. Then, f is almost universal if and only if $f[X]$ is dense in J.

Proof. If $f: X \to J$ is almost universal, then, by (3.3), $f[X]$ is dense in J. Conversely, assume that $f[X]$ is a dense subset of J. Suppose that $f: X \to J$ is not almost universal. Then there is a mapping $g: X \to J$ and an $\epsilon > 0$ such that $|f(x) - g(x)| \geq \epsilon$ for all $x \in X$. Let

$$A = \{x \in X: f(x) - g(x) \geq \epsilon\}$$

and let

$$B = \{x \in X: g(x) - f(x) \geq \epsilon\}$$

Note that $A \cup B = X$ and that A and B are mutually disjoint closed subsets of X. We now show that $A \neq \emptyset$ and $B \neq \emptyset$. Let $s_0 = \text{glb}(J)$ and let $t_0 = \text{lub}(J)$. Since $f[X]$ is a dense subset of J, there exist points $p, q \in X$ such that $t_0 - f(p) < \epsilon$ and $f(q) - s_0 < \epsilon$. Since $g(p) \leq t_0$, $g(p) - f(p) < \epsilon$. Thus, $p \notin B$ and therefore, since $A \cup B = X$, $p \in A$. Similarly, since $g(q) \geq s_0$, it follows that $q \in B$. Hence, $A \neq \emptyset$ and $B \neq \emptyset$. Therefore, from the properties of A and B, X is not connected. This is a contradiction and, therefore, $f: X \to J$ is almost universal. This completes the proof of (3.5).

3.6. Corollary. Let J be a nonempty bounded interval in R^1. If f is any mapping from a connected topological space X onto J, then f is almost universal.

The following example shows that if X is not connected, then a mapping from X onto a bounded interval need not be almost universal.

3.7. Example. Let X be a topological space which is the union of two disjoint nonempty closed subsets X_1 and X_2 for which there are mappings f_i from X_i onto the closed interval $[i - 1, i]$. Define $f: X \to [0, 2]$ by letting $f(x) = f_i(x)$ if $x \in X_i$. Define $g: X \to [0, 2]$ by letting $g(x) = 2$ if $x \in X_1$ and $g(x) = 0$ if $x \in X_2$. Then f is a mapping from X onto $[0, 2]$, but f is not almost universal since $|f(x) - g(x)| \geq 1$ for all $x \in X$.

In general, retractions and homeomorphisms are not almost universal (e.g., the identity map on R^1). Note the following four results about r-maps (see Sec. 2 for definition).

3.8. Proposition. Let X be a topological space and let (Y, d) be a metric space. If $r: X \to Y$ is an r-map and if (Y, d) has AFPP, then r is almost universal.

Proof. There exists $k: Y \to X$ such that $r \circ k$ is the identity map i_Y on Y. Let $g: X \to Y$ be a mapping and let $\epsilon > 0$. Since Y has AFPP, i_Y is almost universal by (3.1). Thus, there exists $y_\epsilon \in Y$ such that $d(y_\epsilon, g \circ k(y_\epsilon)) < \epsilon$. Hence, $d(r(k(y_\epsilon)), g(k(y_\epsilon))) < \epsilon$ and, therefore, we have proved (3.8).

3.9. Proposition. Let X be a topological space and let (Y, d_1) and (Z, d_2) be metric spaces. If $f: X \to Y$ is an almost universal mapping and if $r: Y \to Z$ is a uniformly continuous r-map, then $r \circ f: X \to Z$ is almost universal.

Proof. There is a mapping $k: Z \to Y$ such that $r \circ k$ is the identity map i_Z on Z. Let $g: X \to Z$ be a mapping and let $\epsilon > 0$. There exists $\delta > 0$ such that if $y_1, y_2 \in Y$ and $d_1(y_1, y_2) < \delta$, then $d_2(r(y_1), r(y_2)) < \epsilon$. Since $f: X \to Y$ is almost universal, there exists $x_\delta \in X$ such that $d_1(f(x_\delta), k \circ g(x_\delta)) < \delta$. Hence, $r \circ k = i_Z$,

$$d_2(r \circ f(x_\delta), g(x_\delta)) = d_2(r \circ f(x_\delta), r(k \circ g(x_\delta))) < \epsilon$$

and, therefore, we have proved (3.9).

3.10. Corollary. Any uniformly continuous r-map defined on a metric space having AFPP is almost universal.

Proof. The result follows from (3.9) by letting $X = Y$, letting $f: Y \to Y$ be the identity map, and assuming that Y has AFPP (thus, f is almost universal by (3.1)).

3.11. Corollary. Uniformly continuous r-maps preserve AFPP.

Proof. Use (3.10) and (3.2).

3.12. Remark. Without the assumption of uniform continuity, (3.9)-(3.11) would be false (e.g., any homeomorphism from $(0, 1)$ onto R^1).

The following two results, as well as some previous results, give useful sufficient conditions in order that a mapping be almost universal.

3.13. Proposition. Let Z be a topological space such that Z is contractible with respect to S^{n-1} for some n. Let (M, d) be a metric space such that for each $\eta > 0$ there is a mapping m_η from M into an n-cell $M_\eta \subset M$ such that $d(m_\eta(y), y) < \eta$ for all $y \in M$. Let f be a mapping from Z into M satisfying the following: For each $\eta > 0$, there is a (nonempty) subset Z_η of $(m_\eta \circ f)^{-1}(\partial M_\eta)$ such that

$$m_\eta \circ f \mid Z_\eta \to \partial M_\eta$$

is not nullhomotopic. Then $f: Z \to M$ is almost universal.

Proof. Suppose that $f: Z \to M$ is not almost universal. Then there exist a mapping $g: Z \to M$ and an $\epsilon > 0$ such that $d(f(z), g(z)) \geq \epsilon$ for all $z \in Z$. Let $\eta = \epsilon/3$ and let m_η, M_η, and Z_η be as in the hypotheses. If there exists $z_0 \in Z$ such that

$$d(m_\eta \circ f(z_0), m_\eta \circ g(z_0)) < \eta$$

then

$$d(f(z_0), g(z_0)) \leq d(f(z_0), m_\eta(f(z_0))) + d(m_\eta(f(z_0), m_\eta(g(z_0))) + d(m_\eta(g(z_0)), g(z_0))$$
$$< \eta + \eta + \eta = \epsilon$$

a contradiction. Thus, $d(m_\eta \circ f(z), m_\eta \circ g(z)) \geq \eta$ for all $z \in Z$. Hence, by (2.2), the mapping $m_\eta \circ f \mid Z_\eta: Z_\eta \to \partial M_\eta$ can be extended to a mapping $h: Z \to \partial M_\eta$. Since Z is contractible with respect to S^{n-1}, h is nullhomotopic. Thus, since

$$h \mid Z_\eta = m_\eta \circ f \mid Z_\eta$$

the mapping $m_\eta \circ f \mid Z_\eta: Z_\eta \to \partial M_\eta$ is nullhomotopic, a contradiction. Therefore, $f: Z \to M$ is almost universal and we have proved (3.13).

By using (3.0) and (2.1) we see that the following proposition and Lemma 1 of [8] are equivalent when the spaces are compact and metric.

3.14. Proposition. Let X be a topological space, let (Y, d) be a metric space, and let $f: X \to Y$ be a mapping. Assume that for each $\epsilon > 0$ there exists a mapping g_ϵ from (Y, d) into a metric space $(Z_\epsilon, \rho_\epsilon)$ such that (1) and (2) below hold:

(1) for some $\delta = \delta(g_\epsilon) > 0$, $\operatorname{diam}[g_\epsilon^{-1}(K)] < \epsilon$ whenever $K \subset g_\epsilon[Y]$, $K \neq \emptyset$, and $\operatorname{diam}[K] < \delta$;

(2) $g_\epsilon \circ f: X \to Z_\epsilon$ is almost universal.

Then, $f: X \to Y$ is almost universal.

Proof. Let $g: X \to Y$ be a mapping and let $\epsilon > 0$. Let δ be as in (1). Then, by (2), there exists a point $x_\delta \in X$ such that

$$\rho_\epsilon(g_\epsilon \circ f(x_\delta), g_\epsilon \circ g(x_\delta)) < \delta$$

Hence, letting $K = \{g_\epsilon \circ f(x_\delta), g_\epsilon \circ g(x_\delta)\}$ and using (1), we have that $d(f(x_\delta), g(x_\delta)) < \epsilon$. Therefore, we have proved (3.14).

Let us note that when the spaces are not compact, (1) in (3.14) can not (in general) be replaced by the weaker condition that g_ϵ be an ϵ-mapping. For example, let $f: R^1 \to R^1$ be the identity map and, for each $\epsilon > 0$, let g_ϵ be a homeomorphism from R^1 onto $(0, 1)$. The fact that (2) of (3.14) holds follows from (3.6). Clearly, f is not almost universal since R^1 does not have AFPP.

The following consequence of (3.14) gives conditions in terms of almost universal mappings which imply that their domain has AFPP. Thus, the result is a dual theorem to (3.2).

3.15. Corollary. Let (Y, d) be a metric space. Assume that for each $\epsilon > 0$ there is an almost universal mapping g_ϵ from (Y, d) into a metric space $(Z_\epsilon, \rho_\epsilon)$ such that g_ϵ satisfies (1) of (3.14). Then, (Y, d) has AFPP.

Proof. First use (3.14) with $X = Y$ and f being the identity map, and then use (3.1).

4. GENERAL THEOREMS ABOUT ALMOST UNIVERSAL INDUCED MAPS

Recall from Sec. 2 the definitions of geometric cones, geometric suspensions, and the induced maps f^* and $\tilde{f}$. In particular, we assume throughout this section that our metric spaces lie in a normed linear space in the manner specified in Sec. 2.

In this section we prove three general theorems about maps and induced maps being almost universal, and we derive three corollaries about AFPP. These results will be used extensively in proofs, examples, and discussions in Sec. 5. We have included them in a separate section since they are different types of results from those in Sec. 5.

4.1. Theorem. Let X and Y be metric spaces and let $f: X \to Y$ be a mapping. If $\tilde{f}: \Sigma(X) \to \Sigma(Y)$ is almost universal, then f is almost universal.

Proof. Let $|||\cdot|||$ denote the norm on the linear space containing $\Sigma(Y)$. Assume that $f: X \to Y$ is not almost universal. Then there is a mapping $g: X \to Y$ and an $\epsilon > 0$ such that $|||f(x) - g(x)||| \geq \epsilon$ for all $x \in X$. Define a mapping $G: \Sigma(X) \to \Sigma(Y)$ as follows: For each $x \in X$ and each $\lambda \in [0, 1]$,

$$G(\lambda \cdot x + [1 - \lambda] \cdot v_X^+) = \lambda \cdot g(x) + [1 - \lambda] \cdot v_Y^-$$

$$G(\lambda \cdot x + [1 - \lambda] \cdot v_X^-) = \lambda \cdot g(x) + [1 - \lambda] \cdot v_Y^+$$

We now show that there is a $\delta > 0$ such that $\tilde{f}$ and G are more than δ apart at any point of $\Sigma(X)$. Recall from Sec. 2 that $v_Y^+ = (e, t)$ for some $t > 0$ and $v_Y^- = (e', t')$ for some $t' < 0$. Let

$$\mu = \max\{1 - [\epsilon/(3|||e - e'|||)], 2/3\}$$

Let $\delta = \min\{\epsilon/3, (1 - \mu) \cdot (t - t')\}$. Now fix $x \in X$ and $\lambda \in [0, 1]$. Since $f(x), g(x) \in Y$ and $Y \subset E \times \{0\}$, the second coordinate of f(x) and of g(x) is zero. Using this fact and the formula for $|||\cdot|||$ in Sec. 2, routine calculations show that

$$
\begin{aligned}
&|||G(\lambda\cdot x+[1-\lambda]\cdot v_X^+)-\tilde{f}(\lambda\cdot x+[1-\lambda]\cdot v_X^+)|||\\
&\quad=|||\lambda\cdot g(x)+[1-\lambda]\cdot v_Y^- -\lambda\cdot f(x)-[1-\lambda]\cdot v_Y^+|||\\
&\quad\geq \max\{\,|\lambda|||g(x)-f(x)|||-(1-\lambda)|||e-e'|||\,|,\ (1-\lambda)\cdot(t-t')\}
\end{aligned}
$$

We denote this last maximum by M. Assume first that $\lambda \geq \mu$. Then, $\lambda|||g(x) - f(x)||| \geq \mu\cdot\epsilon \geq (2/3)\cdot\epsilon$ and $(1-\lambda)|||e - e'||| \leq (1-\mu)|||e - e'||| \leq [\epsilon/(3|||e - e'|||)]\,|||e - e'||| = \epsilon/3$. Thus,

$$|\lambda|||g(x) - f(x)||| - (1-\lambda)|||e - e'|||\,| \geq \epsilon/3$$

and, hence, $M \geq \epsilon/3 \geq \delta$. Next assume that $\lambda \leq \mu$. Then,

$$(1-\lambda)\cdot(t-t') \geq (1-\mu)\cdot(t-t') \geq \delta$$

and, hence, $M \geq \delta$. Therefore,

$$|||G(\lambda\cdot x+[1-\lambda]\cdot v_X^+)-\tilde{f}(\lambda\cdot x+[1-\lambda]\cdot v_X^+)||| \geq \delta$$

Similar computations show that

$$|||G(\lambda\cdot x+[1-\lambda]\cdot v_X^-)-\tilde{f}(\lambda\cdot x+[1-\lambda]\cdot v_X^-)||| \geq \delta$$

Therefore, $\tilde{f}$ is not almost universal. This completes the proof of (4.1).

4.2. Corollary. If Y is a metric space such that $\Sigma(Y)$ has AFPP, then Y has AFPP.

Proof. Let $f: Y \to Y$ denote the identity map. Note that $\tilde{f}: \Sigma(Y) \to \Sigma(Y)$ is the identity map. Thus, since $\Sigma(Y)$ has AFPP, we have by (3.1) that $\tilde{f}$ is almost universal. Hence, by (4.1), f is almost universal. Therefore, by (3.1), Y has AFPP.

4.3. Theorem. Let X be a bounded subset of a normed linear space $(E', \|\cdot\|')$, let Y be a contractible bounded subset of a normed linear space $(E, \|\cdot\|)$, and let $f: X \to Y$ be a mapping. If $f^*: \kappa(X) \to \kappa(Y)$ is almost universal, then f is almost universal.

Proof. Assume that $f: X \to Y$ is not almost universal. Then there is a mapping $g: X \to Y$ and an $\epsilon > 0$ such that $\|f(x) - g(x)\| \geq \epsilon$ for all $x \in X$. Since Y is contractible, there is a homotopy $h: X \times [0,1] \to Y$ such that, for all $x \in X$, $h(x,0) = g(x)$ and $h(x,1) = p$ for some point $p \in Y$. Since Y is bounded in $(E \times R^1, |||\cdot|||)$, $\mathrm{lub}\{|||v_Y - y|||: y \in Y\} = b$ exists (and $b \neq 0$ since $v_Y \notin E \times \{0\}$). Choose and fix μ_1 and μ_2 such that $0 < \mu_1 < \mu_2 < 1$ and $\mu_1 < \epsilon/(2b)$. For each $x \in X$ and each $\lambda \in [0,1]$, let

$$k([1-\lambda]\cdot x+\lambda\cdot v_X)=\begin{cases} g(x), & \text{if } 0\leq\lambda\leq\mu_1\\ h(x,\ [\lambda-\mu_1]/[\mu_2-\mu_1]), & \text{if } \mu_1\leq\lambda\leq\mu_2\\ p, & \text{if } \mu_2\leq\lambda\leq 1\end{cases}$$

Since X is a bounded subset of $(E', \|\cdot\|')$ and since $\mu_2 < 1$, it follows using the formula for the norm on $E' \times R^1$ that the middle part of the formula for k is continuous. Thus, k is continuous. Now we show that there is a $\delta > 0$ such that f^* and k are more than δ apart at any point of $\kappa(X)$. Let t denote the second coordinate of v_Y, and recall from Sec. 2 that $t > 0$. Let $\delta = \min\{\epsilon/2, \mu_1\cdot t\}$. Let $x \in X$ and let $\lambda \in [0,1]$. First assume that $0 \leq \lambda \leq \mu_1$. Then,

$$
\begin{aligned}
|||f^*([1-\lambda]\cdot x+\lambda\cdot v_X)-k([1-\lambda]\cdot x+\lambda\cdot v_X)||| &= |||[1-\lambda]\cdot f(x)+\lambda\cdot v_Y - g(x)|||\\
&\geq |\,|||f(x)-g(x)|||-\lambda|||f(x)-v_Y|||\,|
\end{aligned}
$$

Since $f(x) \in Y$, $|||f(x) - v_Y||| \leq b$. Thus, since $0 \leq \lambda \leq \mu_1$, we have that $\lambda|||f(x) - v_Y||| \leq \mu_1\cdot b < \epsilon/2$. Hence, since $|||f(x) - g(x)||| = \|f(x) - g(x)\| \geq \epsilon$,

$$|\,|||f(x) - g(x)||| - \lambda|||f(x) - v_Y|||\,| > \epsilon/2$$

Thus, $||| f^*([1 - \lambda] \cdot x + \lambda \cdot v_X) - k([1 - \lambda] \cdot x + \lambda \cdot v_X) ||| > \epsilon/2 \geq \delta$. Next assume that $\mu_1 < \lambda \leq 1$. Then, since $Y \subset E \times \{0\}$ and k maps $\kappa(X)$ into Y, the second coordinate of $k([1 - \lambda] \cdot x + \lambda \cdot v_X)$ is zero. Hence, using the formula for $||| \cdot |||$, we see that

$$||| f^*([1 - \lambda] \cdot x + \lambda \cdot v_X) - k([1 - \lambda] \cdot x + \lambda \cdot v_X) ||| \geq \lambda \cdot t > \mu_1 \cdot t \geq \delta$$

Thus, we have shown that f^* and k are more than δ apart at any point of $\kappa(X)$. Therefore, $f^*: \kappa(X) \to \kappa(Y)$ is not almost universal. This completes the proof of (4.3).

Since AFPP and the fixed point property are equivalent for compact metric spaces, the following corollary is a generalization of a result in [9], p. 96.

4.4. Corollary. Let Y be a bounded subset of a normed linear space. If Y is contractible and if $\kappa(Y)$ has AFPP, then Y has AFPP.

Proof. Use (4.3) and (3.1) in a manner analogous to the way (4.1) and (3.1) were used in the proof of (4.2).

4.5. Theorem. Let X and Y be metric spaces and let $f: X \to Y$ be a mapping. If $\tilde{f}: \Sigma(X) \to \Sigma(Y)$ is almost universal, then $f^*: \kappa(X) \to \kappa(Y)$ is almost universal.

Proof. Let $||| \cdot |||$ denote the norm on the linear space containing $\Sigma(Y)$. Assume that $f^*: \kappa(X) \to \kappa(Y)$ is not almost universal. Then there is a mapping $g: \kappa(X) \to \kappa(Y)$ and an $\epsilon > 0$ such that for all $z \in \kappa(X)$, $||| f^*(z) - g(z) ||| \geq \epsilon$. Define a mapping $G: \Sigma(X) \to \Sigma(Y)$ as follows: For each $x \in X$ and each $\lambda \in [0, 1]$,

$$G(\lambda \cdot x + [1 - \lambda] \cdot v_X^+) = g(\lambda \cdot x + [1 - \lambda] \cdot v_X^+)$$

$$G(\lambda \cdot x + [1 - \lambda] \cdot v_X^-) = g(\lambda \cdot x + [1 - \lambda] \cdot v_X^+)$$

Let $t > 0$ and $t' < 0$ denote the second coordinates of v_Y^+ and v_Y^-, respectively. Let $\delta = (\epsilon/2) \cdot (|t'| / ||| v_Y^+ - v_Y^- |||)$ and note that $0 < \delta \leq \epsilon/2$. We now show that $\tilde{f}$ and G are at least δ apart at any point of $\Sigma(X)$. Let $x \in X$ and let $\lambda \in [0, 1]$. Since $G | \kappa(X) = g$ and $\tilde{f} | \kappa(X) = f^*$ and since

$$||| f^*(z) - g(z) ||| \geq \epsilon > \delta \quad \text{for all } z \in \kappa(X)$$

we only need to verify that $||| \tilde{f}(z) - G(z) ||| \geq \delta$ at the point $z = \lambda \cdot x + [1 - \lambda] \cdot v_X^-$. Let

$$\mu = 1 - [\epsilon/(2 ||| v_Y^+ - v_Y^- |||)]$$

First assume that $0 \leq \lambda \leq \mu$. The second coordinate of $\tilde{f}(z)$ is $[1 - \lambda] \cdot t'$ and the second coordinate of G(z) is of the form $\alpha \cdot t$ for some $\alpha \in [0, 1]$. Hence, using the formula for $||| \cdot |||$, we have that

$$||| \tilde{f}(z) - G(z) ||| \geq |[1 - \lambda] \cdot t' - \alpha \cdot t|$$

and thus, since $[1 - \lambda] \cdot t' < 0$ and $\alpha \cdot t \geq 0$,

$$||| f(z) - G(z) ||| \geq |[1 - \lambda] \cdot t'| = [1 - \lambda] \cdot |t'|$$

Since $0 \leq \lambda \leq \mu$, $[1 - \lambda] \cdot |t'| \geq [1 - \mu] \cdot |t'| = \delta$ and, hence,

$$||| f(z) - G(z) ||| \geq \delta$$

Next assume that $\mu \leq \lambda \leq 1$. Since $\tilde{f}(z) = \lambda \cdot f(x) + [1 - \lambda] \cdot v_Y^-$ and $G(z) = g(\lambda \cdot x + [1 - \lambda] \cdot v_X^+)$, we see that

$$\begin{aligned}|||\tilde{f}(z) - G(z)||| &= |||\tilde{f}(z) + [1-\lambda]\cdot v_Y^+ - [1-\lambda]\cdot v_Y^+ - G(z)||| \\ &= |||\tilde{f}(\lambda\cdot x + [1-\lambda]\cdot v_X^+) - g(\lambda\cdot x + [1-\lambda]\cdot v_X^+) - [1-\lambda]\cdot(v_Y^+ - v_Y^-)||| \\ &\geq \big| |||\tilde{f}(\lambda\cdot x + [1-\lambda]\cdot v_X^+) - g(\lambda\cdot x + [1-\lambda]\cdot v_X^+)||| - [1-\lambda]\cdot |||v_Y^+ - v_Y^-||| \big|\end{aligned}$$

Thus, since $|||\tilde{f}(\lambda\cdot x + [1-\lambda]\cdot v_X^+) - g(\lambda\cdot x + [1-\lambda]\cdot v_X^+)||| \geq \epsilon$ and $[1-\lambda]\cdot|||v_Y^+ - v_Y^-||| \leq [1-\mu]\cdot|||v_Y^+ - v_Y^-||| = \epsilon/2$,

$$|||\tilde{f}(z) - G(z)||| \geq \epsilon/2 \geq \delta$$

Thus, we have shown that $\tilde{f}$ and G are at least δ apart at any point of $\Sigma(X)$. Therefore, $\tilde{f}: \Sigma(X) \to \Sigma(Y)$ is not almost universal. This completes the proof of (4.5).

4.6. Corollary. If Y is a metric space such that $\Sigma(Y)$ has AFPP, then $\kappa(Y)$ has AFPP.

Proof. Let f: Y   Y denote the identity map. Note that $\tilde{f}$ is the identity map on $\Sigma(Y)$ and f^* is the identity map on $\kappa(Y)$. Since $\Sigma(Y)$ has AFPP, we have by (3.1) that $\tilde{f}$ is almost universal. Hence, by (4.5), f^* is almost universal. Therefore, by (3.1), $\kappa(Y)$ has AFPP.

5. APPLICATIONS

In this section we use previous results to determine some specific classes of metric spaces which have AFPP. For a general discussion of the material in this section, see Sec. 1. Other comments will be made at appropriate places.

Our first main result about AFPP is (5.4). The key ingredient in the proof of this result is the recognition that certain maps and induced maps are almost universal. Thus, as in Sec. 4, we obtain results about almost universal mappings [(5.2) and (5.3)] which are stronger than results about AFPP. The proof of (5.2) uses the following lemma.

5.1. Lemma. Let X be a nonempty connected metric space and let k be a mapping from X onto a dense subset of a bounded interval J in R^1. Then the mapping $\tilde{k}: \Sigma(X) \to \Sigma(J)$ is almost universal.

Proof. We will use (3.13) with $Z = \Sigma(X)$ and $M = \Sigma(J)$. Let d denote the metric that M inherits as a subset of R^2. Assume that J consists of more than one point since, otherwise, M is an arc, $\tilde{k}$ maps Z onto M, and, thus, $\tilde{k}$ is universal ([8], Prop. 8). To use (3.13), note that, since X is connected, Z is contractible with respect to S^1 by (2.3). Now let $\eta > 0$. Since k[X] is a dense connected subset of J and since J is a bounded interval containing more than one point, there is a closed subinterval $I_\eta = [t_1, t_2]$, $t_1 < t_2$, of k[X] for which there is a retraction r_η from J onto I_η such that $|r_\eta(s) - s| < \eta$ for all $s \in J$. Since $I_\eta \subset k X$, there are points $x_1, x_2 \in X$ such that $k(x_1) = t_1$ and $k(x_2) = t_2$. Let $M_\eta = \Sigma(I_\eta)$ where we assume that the vertices v_η^+ and v_η^- of M_η are the same as the vertices of M. Then we see that $\tilde{r}_\eta$ is a retraction from M onto the 2-cell M_η such that $d(\tilde{r}_\eta(y), y) < \eta$ for all $y \in M$. Note that

$$\partial M_\eta = \widehat{t_1 v_\eta^-} \cup \widehat{t_1 v_\eta^+} \cup \widehat{t_2 v_\eta^-} \cup \widehat{t_2 v_\eta^+}$$

Now we let

$$Z_\eta = \widehat{x_1 v_X^-} \cup \widehat{x_1 v_X^+} \cup \widehat{x_2 v_X^-} \cup \widehat{x_2 v_X^+}$$

Recalling our choices for x_1, x_2, v_η^+ and v_η^-, and using the formula in Sec. 2 for $\tilde{k}$, we see easily that $\tilde{k}|Z_\eta$ is a homeomorphism from Z_η onto ∂M_η. Since $\tilde{r}_\eta|\partial M_\eta$ is the identity map,

$$\tilde{r}_\eta \circ \tilde{k}|Z_\eta = \tilde{k}|Z_\eta$$

Thus, $\tilde{r}_\eta \circ \tilde{k}|Z_\eta: Z_\eta \to \partial M_\eta$ is a homeomorphism and, hence, is not nullhomotopic. We have now shown that all the hypotheses of (3.13) are satisfied. Therefore, by (3.13), $\tilde{k}: Z \to M$ is almost universal and we have proved (5.1).

5.2. Theorem. Let f be a mapping from a nonempty connected metric space X into a chainable continuum C. Let Y be a subset of C such that f X is a dense subset of Y. Then the mappings $\tilde{f}: \Sigma(X) \to \Sigma(Y)$, $f^*: \kappa(X) \to \kappa(Y)$, and $f: X \to Y$ are almost universal.

Proof. We will use (3.14), so let $\epsilon > 0$. Since C is a chainable continuum, there is an ϵ-mapping h_ϵ from C onto $[0, 1]$. By (2.4), $\tilde{h}_\epsilon: \Sigma(C) \to \Sigma([0, 1])$ is also an ϵ-mapping (same ϵ). Thus, since $\Sigma(C)$ and $\Sigma([0, 1])$ are compact, there exists a $\delta > 0$ satisfying (2.1) for the mapping $\tilde{h}_\epsilon$. Let $g_\epsilon = h_\epsilon | Y$ and let $J_\epsilon = g_\epsilon[Y]$. Assume, without loss of generality, that the vertices of $\Sigma(Y)$ and $\Sigma(C)$ are the same and that the vertices of $\Sigma(J_\epsilon)$ and $\Sigma([0, 1])$ are the same. Then, $\tilde{g}_\epsilon = \tilde{h}_\epsilon | \Sigma(Y)$. Hence, since δ satisfies (2.1) for $\tilde{h}_\epsilon$, clearly δ satisfies (1) of (3.14) for $\tilde{g}_\epsilon$. We now show that $\tilde{g}_\epsilon \circ \tilde{f}$ satisfies (2) of (3.14). Since f[X] is a dense subset of Y, $g_\epsilon(f[X])$ is a dense subset of J_ϵ. Thus, since X is connected, J_ϵ is connected. Hence, letting $k = g_\epsilon \circ f$, we have that k maps X onto a dense subset of the bounded interval J_ϵ. Therefore, by (5.1), $\tilde{k}: \Sigma(X) \to \Sigma(J_\epsilon)$ is almost universal. Since $k = g_\epsilon \circ f$, it is easy to see that $\tilde{k} = \tilde{g}_\epsilon \circ \tilde{f}$. Hence, $\tilde{g}_\epsilon \circ \tilde{f}$ satisfies (2) of (3.14). Thus, by (3.14), $\tilde{f}: \Sigma(X) \to \Sigma(Y)$ is almost universal. Therefore, by (4.5) and (4.1), $f^*: \kappa(X) \to \kappa(Y)$ and $f: X \to Y$ are almost universal. This completes the proof of (5.2).

Let us note the following more general version of part of (5.2). The other parts can not be so generalized since, by definition, $\Sigma(X)$ and $\kappa(X)$ are subsets of normed linear spaces and, thus, are metric spaces.

5.3. Theorem. Let f be a mapping from a nonempty connected topological space X into a chainable continuum C. Let Y be a subset of C such that f X is a dense subset of Y. Then, $f: X \to Y$ is almost universal.

Proof. The proof uses ideas in the proof of (5.2). Let $\epsilon > 0$. Let h_ϵ, g_ϵ, and J_ϵ be as in the proof of (5.2). A $\delta > 0$ satisfying (1) of (3.14) for g_ϵ is the same δ which satisfies (2.1) for h_ϵ. As in the proof of (5.2), $g_\epsilon(f[X])$ is a dense subset of J_ϵ and J_ϵ is a nonempty bounded interval. Thus, by (3.5), $g_\epsilon \circ f: X \to J_\epsilon$ satisfies (2) of (3.14). Therefore, by (3.14), $f: X \to Y$ is almost universal.

In connection with (5.3), we state the following known result ([8], Thm. 3, p. 437): Any mapping from a compact connected Hausdorff space onto a chainable Hausdorff continuum is universal (note: compactness and Hausdorffness are assumed and used in [8] beginning on p. 436). Using (3.0), we see that (5.3) is a generalization of [8], Theorem 3 for the case when the chainable Hausdorff continuum is metric.

The following easy consequence of (5.2) is one of our main results about AFPP. It can be used to obtain many interesting examples of spaces which have the almost fixed point property. A discussion of this result is in Sec. 1.

5.4. Theorem. If Y is a nonempty connected subset of a chainable continuum, then Y, $\kappa(Y)$, and $\Sigma(Y)$ have AFPP.

Proof. The theorem follows easily from (5.2) and (3.1) by letting $f: Y \to Y$ be the identity map and by noting that $f^*: \kappa(Y) \to \kappa(Y)$ and $\tilde{f}: \Sigma(Y) \to \Sigma(Y)$ are also identity maps.

Recall that a compact metric space has AFPP if and only if it has the fixed point property. Thus, for example, (5.4) implies Hamilton's theorem that any chainable continuum has the fixed point property [6]. Furthermore [by (5.4)], the suspension over any chainable continuum has the fixed point property, and this implies Hamilton's theorem by (4.2).

In (5.6) and (5.7) we generalize the cone part of (5.4) to certain types of nonconnected subsets of chainable continua. The other parts of (5.4) can not be generalized in this fashion, as can be seen using (4.2) and the fact that a metric space must be connected if it has AFPP. The proof of (5.6) uses (5.4) and the following variation of Theorem 6 in [2].

5.5. Lemma. Let (X, d) be a metric space. Assume that X_1 and X_2 are closed subsets of X such that $X = X_1 \cup X_2$ and $X_1 \cap X_2 = \{w\}$ for some point w. If X_1 and X_2 each has AFPP (with the subspace metric), then X has AFPP (and conversely).

Proof. Let $g\colon X \to X$ be a mapping and let $\epsilon > 0$. We will show that there exists $p \in X$ such that $d(g(p), p) < \epsilon$. Assume that $g(w) \neq w$ (otherwise, we are done). Then, $g(w) \in X_1 - X_2$ or $g(w) \in X_2 - X_1$. Without loss of generality, assume that $g(w) \in X_1 - X_2$. Then, since $X_1 - X_2$ is an open subset of X, there exists $\delta > 0$ such that $\delta \leq \epsilon$ and such that if $x \in X$ and $d(w, x) < \delta$, then $g(x) \in X_1 - X_2$. Define a retraction r from X onto X_1 by

$$r(x) = \begin{cases} x, & \text{if } x \in X_1 \\ w, & \text{if } x \in X_2 \end{cases}$$

Since $r \circ (g \mid X_1)$ is a mapping from X_1 into x_1 and since X_1 has AFPP, there exists $p \in X_1$ such that $d(r(g(p)), p) < \delta$. Suppose that $g(p) \in X_2$. Then, since $r(g(p)) = w$, $d(w, p) < \delta$. Hence, by the choice of δ, $g(p) \in X_1 - X_2$, a contradiction. Thus, $g(p) \in X_1$. Hence, $r(g(p)) = g(p)$. Therefore, since $d(r(g(p)), p) < \delta \leq \epsilon$, $d(g(p)), p) < \epsilon$. This completes the proof of half of (5.5). The converse is proved by applying (3.11) to the mapping r defined above.

5.6. Theorem. Let Y be a nonempty subset of a chainable continuum. If Y has only finitely many components, then $\kappa(Y)$ has AFPP.

Proof. By (5.4) the theorem is true if Y has only one component. Assume inductively that the theorem is true if Y has only n components. Now let Y be as in the theorem such that Y has exactly $n + 1$ components $Y_1, \ldots, Y_{n+1}$. Let $X = \kappa(Y)$ with vertex v_Y, and let $X_1 = \kappa(\bigcup_{i=1}^{n} Y_i)$ and $X_2 = \kappa(Y_{n+1})$ with the same vertex v_Y. Then we see that X_1 and X_2 are closed subsets of X, $X = X_1 \cup X_2$, and $X_1 \cap X_2 = \{v_Y\}$. By the induction assumption, X_1 has AFPP and, by (5.4), X_2 has AFPP. Therefore, by (5.5) X has AFPP. This completes the proof of (5.6).

We do not know if (5.6) can be extended to include all nonempty subsets of chainable continua. In connection with this problem, let us note the following simple generalization of (5.6).

5.7. Theorem. Let Y be a nonempty subset of a chainable continuum (X, d). Assume that for each $\epsilon > 0$, there is a mapping m_ϵ from Y into a subset Y_ϵ of Y, where Y_ϵ has only finitely many components, such that $d(m_\epsilon(y), y) < \epsilon$ for all $y \in Y$. Then, $\kappa(Y)$ has AFPP.

Proof. Let $f\colon \kappa(Y) \to \kappa(Y)$ be a mapping and let $\epsilon > 0$. Let $g = f \mid \kappa(Y_\epsilon)\colon \kappa(Y_\epsilon) \to \kappa(Y)$ and let $m_\epsilon^*\colon \kappa(Y) \to \kappa(Y_\epsilon)$ be the induced map. By (5.6), $\kappa(Y_\epsilon)$ has AFPP. Thus, there is a point $p_\epsilon \in \kappa(Y_\epsilon)$ such that $|||\, m_\epsilon^* \circ g(p_\epsilon) - p_\epsilon \,||| < \epsilon$. It is easy to see that $|||\, m_\epsilon(g(p_\epsilon)) - g(p_\epsilon) \,||| < \epsilon$. Hence, by the triangle inequality, $|||\, g(p_\epsilon) - p_\epsilon \,||| < 2\epsilon$, i.e., $|||\, f(p_\epsilon) - p_\epsilon \,||| < 2\epsilon$. Therefore, we have proved (5.7).

In (5.8) and (5.9) we give examples which show that certain natural generalizations of (5.4) are not possible.

Since chainable continua are tree-like and are embeddable in the plane [3], it is natural to wonder if (5.4) can be generalized to nonempty connected subsets of planar tree-like continua. In (5.8) we give an example of a topological line L in an hereditarily decomposable tree-like continuum in the plane such that L does not have AFPP even for homeomorphisms from L onto L. We will also see that $\kappa(L)$ and $\Sigma(L)$ do not have AFPP. Thus, the example shows that (5.4) can not be generalized to subsets of simple tree-like continua in the plane. In relation to the example, let us also note the following facts. Hereditarily decomposable tree-like continua have the fixed point property ([12]; also, see [11])—it is an old theorem that they have the fixed point property for homeomorphisms ([7], Thm. II, p. 21). It is not known if tree-like continua in the plane have the fixed point property (though tree-like continua in general do not [1]).

5.8. Example. The tree-like continuum X is drawn in Fig. 1. It is composed of a (topological) line L and two simple triods T_1 and T_2. We consider L as being the union of two half-lines L_1 and L_2 which intersect at a_1. For each i = 1 and 2, L_i compactifies on T_i in a counterclockwise direction. We describe a homeomorphism h from L onto L such that for some $\epsilon > 0$, $\|h(y) - y\| \geq \epsilon$ for all $y \in L$. For this purpose we consider nine sequences—the first three terms of these sequences are pictured in Fig. 1. The midpoint of the segment from O_1 to a (b, c respectively) is denoted by a/2 (b/2, c/2 respectively). We assume that as $i \to \infty$, $a_i \to a$, $b_i \to b$, $c_i \to c$, $r_i \to a/2$, $s_i \to a/2$, $t_i \to b/2$, $u_i \to b/2$, $v_i \to c/2$, and $w_i \to c/2$. If $p, q \in L$, $p \neq q$, then the juxtaposition pq denotes the arc in L with endpoints p and q. The symbol h(pq) = yz denotes a homeomorphism from an arc pq in L onto an arc yz in L such that h(p) = y and h(q) = z. With this notation in mind, the following equalities describe our desired homeomorphism through one revolution of L_1 around

$$h(a_1 s_1) = b_1 v_1, \quad h(s_1 b_1) = v_1 c_1, \quad h(b_1 u_1) = c_1 r_1$$

$$h(u_1 c_1) = r_1 a_2, \quad h(c_1 w_1) = a_2 t_2, \quad h(w_1 a_2) = t_2 b_2$$

The pattern is then repeated, i.e., $h(a_2 s_2) = b_2 v_2$, $h(s_2 b_2) = v_2 c_2$, etc. The desired homeomorphism h is thus defined on L_1. Let d and e be the points in L_2 pictured in Fig. 1. On the arc $a_1 e$ we let our desired homeomorphism be fixed point free such that $h(a_1 e) = b_1 d$. Then, the procedure that was used to define h on L_1 is used to define h on the rest of L_2. Having thus defined our homeomorphism h, it is easily seen that h has the required properties. Therefore, L does not have AFPP even for homeomorphisms from L onto L. Since L does not have AFPP, $\Sigma(L)$ does not have AFPP by (4.2) and, since L is contractible, $\kappa(L)$ does not have AFPP by (4.4). This completes (5.8).

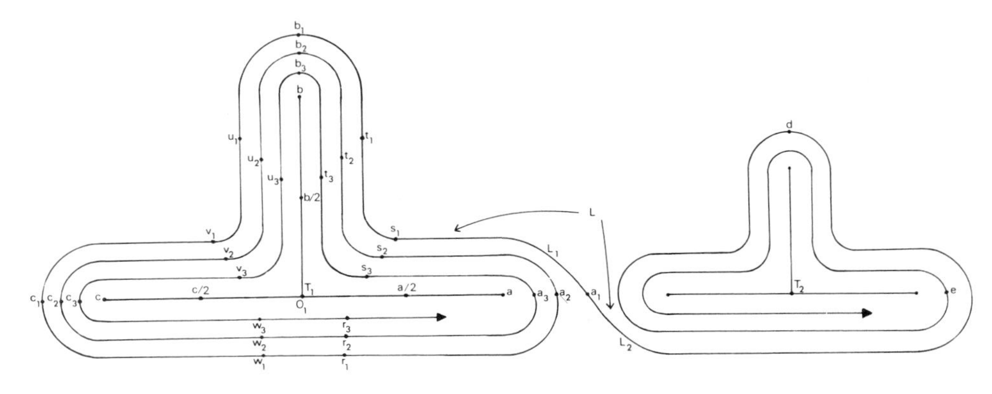

Fig. 1

Our next example involves subsets of circle-like continua. Note the following facts: Every continuum which is properly contained in a circle-like continuum is chainable, every nonempty connected subset of a chainable continuum has AFPP (by (5.4)), and, as is easy to see, every nonempty connected proper subset of the circle itself has AFPP. However, the following example shows that nonempty connected proper subsets of circle-like continua need not have AFPP.

5.9. Example. Let X be the circle-like continuum drawn in Fig. 2—we assume that X is symmetric about the origin in R^2. Let p and -p be the points of X pictured in Fig. 2. Let $Y = X - \{p, -p\}$ and note that Y is a nonempty connected proper subset of X. Define $f: Y \to Y$ by letting $f(y) = -y$ for all $y \in Y$. As the mapping f clearly shows, Y does not have AFPP (in addition, f is a homeomorphism of Y onto Y). We remark that, by (4.2), $\Sigma(Y)$ does not have AFPP but that, by (5.13), $\kappa(Y)$ does have AFPP.

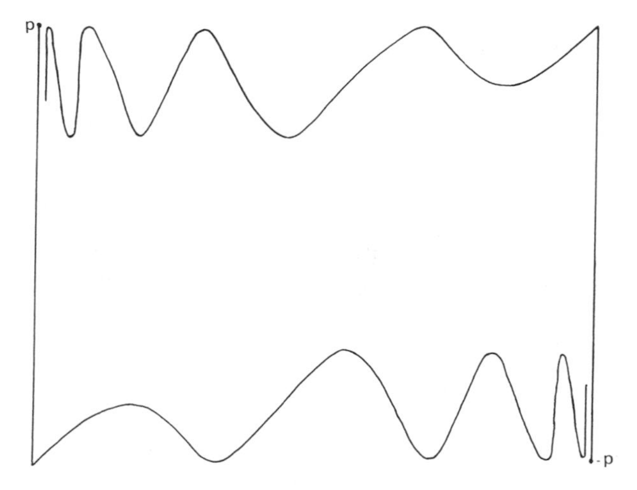

Fig. 2

5.10. Remark. Let H be a topological half-line (i.e., a homeomorph of $[0, +\infty)$) lying in a circle-like continuum X. It follows easily that the closure $\overline{H}$ of H is not all of X and, thus, $\overline{H}$ is chainable. Hence, by (5.4), H has AFPP. Using this fact, an easy argument shows that any topological line L lying in a circle-like continuum X has AFPP (even though the closure of L may be all of X). Therefore, unlike (5.8), an example having the properties in (5.9) is not possible for topological lines.

Our next two results concern "punctured suspensions." Note that $\Sigma(S^n)$ does not have AFPP for any $n = 0, 1, \ldots$. However, if we remove one of the vertices (or any given point) from $\Sigma(S^n)$, the resulting punctured suspension has AFPP. This observation generalizes as follows:

5.11. Theorem. If Y is a subset of a compact metric space X and if $\kappa(Y)$ has AFPP, then $\Sigma(Y) - \{v_Y^+\}$ and $\Sigma(Y) - \{v_Y^-\}$ have AFPP.

Proof. We first prove that $Z = \kappa(Y) - Y$ has AFPP. Let $g: Z \to Z$ be a mapping and let $\epsilon > 0$. Recall our general assumption that $X \subset E \times \{0\}$, a subspace of the normed linear space $(E \times R^1, |||\cdot|||)$. Thus, since X is compact, Y is a bounded subset of the normed linear space $(E \times R^1, |||\cdot|||)$. Hence, $\text{lub}\{|||y - v_Y^+||| : y \in Y\} = b$ exists (and $b \neq 0$ since

$v_Y^+ \notin E \times \{0\}$). Choose and fix $\mu_0 < \epsilon/(2b)$ such that $0 < \mu_0 \leq 1$. For each $y \in Y$ and each $\lambda \in [0,1]$, let

$$r([1-\lambda]\cdot y + \lambda \cdot v_Y^+) = \begin{cases} [1-\mu_0]\cdot y + \mu_0 \cdot v_Y^+, & \text{if } 0 \leq \lambda \leq \mu_0 \\ [1-\lambda]\cdot y + \lambda \cdot v_Y^+, & \text{if } \mu_0 \leq \lambda \leq 1 \end{cases}$$

We see that r is continuous and that, since $0 < \mu_0 \leq 1$, r maps $\kappa(Y)$ into Z. Hence, $g \circ r$ is a mapping defined on all of $\kappa(Y)$. Thus, since $\kappa(Y)$ has AFPP, there exists a point $p \in \kappa(Y)$ such that $|||g \circ r(p) - p||| < \epsilon/2$. Using the fact that $\mu_0 < \epsilon/(2b)$, easy calculations show that $|||p - r(p)||| < \epsilon/2$. Therefore, by the triangle inequality, $|||g(r(p)) - r(p)||| < \epsilon$. This completes the proof that Z has AFPP. Now we can prove the theorem for the case of $\Sigma(Y) - \{v_Y^-\}$. Let k: $\kappa(X) \to \Sigma(X)$ be the mapping which takes a given convex arc $\widehat{xv_X^+}$ in $\kappa(X)$ onto $\widehat{xv_X^+} \cup \widehat{xv_X^-}$ as follows: Let m_X denote the midpoint of $\widehat{xv_X^+}$; then $k(v_X^+) = v_X^+$, $k(m_X) = x$, $k(x) = v_X^-$, k is linear between v_X^+ and m_X, and k is linear between m_X and x. By the compactness of X, k is uniformly continuous. Thus, from the way k is defined, $k \mid Z$ is a uniformly continuous homeomorphism from Z onto $\Sigma(Y) - \{v_Y^-\}$. Therefore, since Z has AFPP, we have by (3.11) that $\Sigma(Y) - \{v_Y^-\}$ has AFPP. This completes the proof of (5.11).

We have already observed that (5.6) and (5.7) are false for geometric suspensions (see the paragraph preceding (5.5)). However, (5.6) and (5.7) are true for punctured suspensions:

5.12. Corollary. Let Y be a nonempty subset of a chainable continuum. If Y has only finitely many components or, more generally, if Y is as in (5.7), then $\Sigma(Y) - \{v_Y^+\}$ and $\Sigma(Y) - \{v_Y^-\}$ have AFPP.

Proof. Use (5.6), (5.7), and (5.11).

We do not require the space Y in the following theorem to be contained in a compact set or even to be bounded. Thus, the theorem is of a different type from many of the results about AFPP in this section and in Sec. 4. An example is given in (5.14).

5.13. Theorem. If (Y, d) is a metric space such that, for each $\epsilon > 0$, there is a non-nullhomotopic ϵ-mapping f_ϵ: $Y \to S^{n-1}$ (where n is allowed to vary with ϵ), then $\kappa(Y)$ has AFPP.

Proof. Let g: $\kappa(Y) \to \kappa(Y)$ be a mapping and let $\epsilon > 0$. Let f_ϵ: $Y \to S^{n-1}$ be as in the hypothesis of the theorem and let f_ϵ^*: $\kappa(Y) \to \kappa(S^{n-1})$ be the induced mapping. Note that $\kappa(S^{n-1})$ is an n-cell. Also note that $f_\epsilon^* \mid Y$: $Y \to S^{n-1}$ is not nullhomotopic since $f_\epsilon^* \mid Y = f_\epsilon$. Hence, since $\kappa(Y)$ is contractible (by (2.3)), f_ϵ^*: $\kappa(Y) \to \kappa(S^{n-1})$ is universal by [8, Prop. 10]. Thus, there exists a point $z_\epsilon \in \kappa(Y)$ such that $f_\epsilon^*(z_\epsilon) = f_\epsilon^* \circ g(z_\epsilon)$. Therefore, since f_ϵ^* is an ϵ-mapping (by (2.4)), the point z_ϵ moves less than ϵ under g. This completes the proof of (5.13).

Let us note that (5.13) implies that the cone over any circle-like continuum X has the fixed point property [if X is also chainable, then use (5.4)]. A special case is in Theorem 4 of [8].

5.14. Example. The theorem in (5.13) is applicable to spaces Y which are not bounded and which do not themselves have AFPP. Let Y be the space drawn in Fig. 3—we assume that Y is symmetric about the origin in R^2. As the waves approach one or the other of the dotted lines, they become unbounded (in one direction or the other). It is easy to see that there is a one-to-one mapping from Y onto S^1 and that any such mapping is

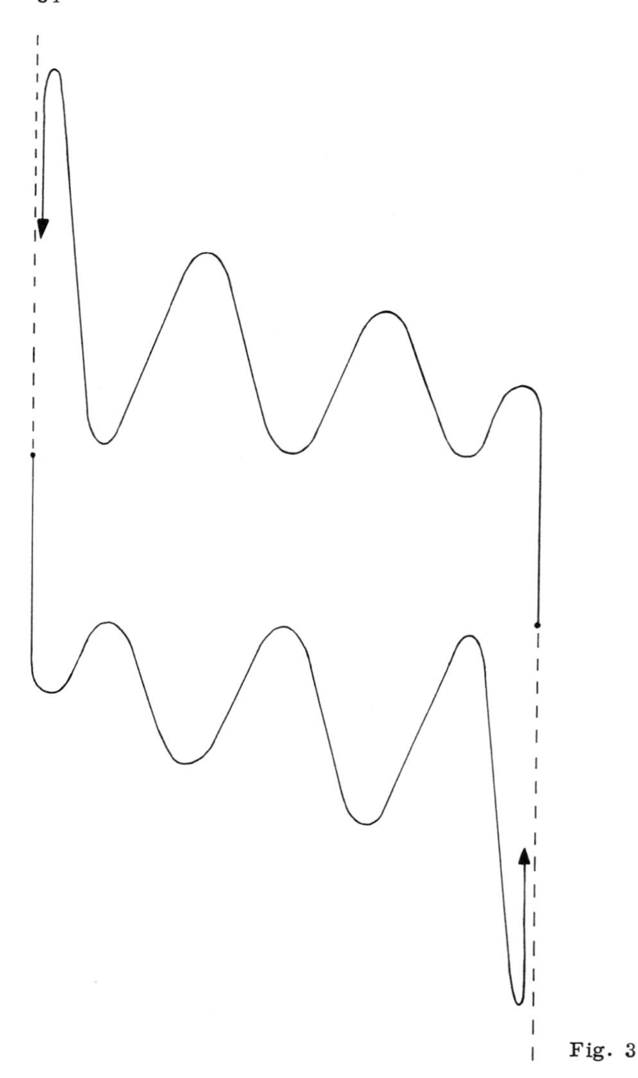

Fig. 3

non-nullhomotopic. Thus, by (5.13), $\kappa(Y)$ has AFPP. Note that Y does not have AFPP (by letting $f(y) = -y$ for all $y \in Y$).

REFERENCES

1. Bellamy, D. P.: A tree-like continuum without the fixed-point property. Preprint.
2. Bing, R. H.: Am. Math. Monthly 76, 119-132 (1969).
3. Bing, R. H.: Duke Math. J. 18, 653-663 (1951).
4. Borsuk, K.: Theory of Retracts, Polish Scientific Publishers (1967).
5. Fort, M. K., Jr.: J. Indian Math. Soc. 18, 23-26 (1954).
6. Hamilton, O. H.: Proc. Amer. Math. Soc. 2, 173-174 (1951).
7. Hamilton, O. H.: Trans. Amer. Math. Soc. 44, 18-24 (1938).
8. Holsztyński, W.: Bull. Pol. Acad. Sci. 15, 433-438 (1967).
9. Kinoshita, S.: Fund. Math. 40, 96-98 (1953).

10. Kuratowski, K.: Topology, vol. II. Academic Press (1968).
11. Maćkowiak, T.: Bull. Pol. Acad. Sci. 26, 61-64 (1978).
12. Mańka, R.: Fund. Math. 91, 105-121 (1976).
13. Van der Walt, T.: Fixed and Almost Fixed Points, Mathematical Centre Tracts, 2nd ed. (1967).
14. Whyburn, G. T.: Analytic Topology. Amer. Math. Soc. (1942).
15. Willard, S.: General Topology. Addison-Wesley (1970).

3 ON THE NATURAL APPROACH TO FLOW PROBLEMS

John Argyris and J. St. Doltsinis / Institut für Statik und Dynamik der Luft- und Raumfahrtkonstruktionen, Universität Stuttgart, Stuttgart, Federal Republic of Germany

1. INTRODUCTION

The present paper surveys recent work on fluid dynamics performed at the ISD, University of Stuttgart. The paper serves in the main as a survey on modern developments in finite element methods for fluid motion, and is particularly devoted to a natural description of the relevant phenomena. Its main attention is focused on incompressible media. First draft of the theory has been presented at a lecture given at the Conference on Finite Elements in Water Resources in Hanover in 1982.

In Sec. 2, the natural terminology [1, 2] is introduced and methodically applied to the formulation of field quantities characteristic of fluid motion, such as the scalar pressure field and the vectorial velocity field. The condition of conservation of mass is derived in natural terms and natural measures for the stress and the rate of deformation are connected by the appropriate constitutive relations. Aiming at the analysis of fluid motion coupled with thermal phenomena, the natural approach is subsequently extended to the consideration of the temperature field and the heat flow [3].

Section 3 indicates the transition to finite domains as a foundation for the development of the finite element theory of the flow problem. The streamline upwind/Petrov-Galerkin formulation of [4] may be used for the discretization technique in connection with either the strict fulfillment of the incompressible statement or with the penalty approach to the condition of incompressibility. In a subsequent step the finite element discretization of the thermally coupled fluid flow problem is considered and the governing equations are established.

For typographical brevity, we omit in this paper a discussion of numerical integration schemes in the time domain. Also the important task of an effective solution of the equations governing the flow problem is not handled in the present contribution. For this purpose the reader is referred to the presentation in [28].

The theory presented in the paper is applied in Sec. 4 to the numerical analysis of some typical examples of viscous fluid motion. Thus, the convection dominated flow over a step is considered for the two- and the three-dimensional case, and the solution of thermally coupled flows is demonstrated on the Bénard instability phenomenon in a fluid between two planes of different temperatures. The interested reader may consult [28] for an analysis of cavity flows involving free and forced heat convection with a change from liquid to solid phase of the material.

2. ON THE NATURAL APPROACH TO FLUID MOTION

2.1. Natural Approach

In the natural methodology of continuum mechanics, all considerations are established on or derived from an infinitesimal tetrahedron element which replaces the classical parallelepiped applied in the traditional cartesian point of view. For comparison purposes both elements are shown in Fig. 1 together with the associated coordinate systems. An elegant application of the tetrahedron element demands the use of supernumerary or homogeneous reference systems. One of these may be defined by the directions of the six edges of the tetrahedron. The natural formulation of the mechanics and thermomechanics of solids [5, 3, 2] may be based on the Lagrangian approach in the sense that the tetrahedron constitutes then a moving and deforming material element. In our present considerations of fluid motion, however, we prefer to adopt the Eulerian description [6, 7] in which the tetrahedron represents a fixed geometrical element in space.

Before developing the natural concept we first review alternative representations of a vector in three-dimensional space [3, 2] and illustrate then the argument on the two-dimensional case depicted in Fig. 2. Consider the vector $\underline{r}$ defined by cartesian entries

$$\underline{r} = \{r_x \; r_y \; r_z\} = \{r^1 \; r^2 \; r^3\} = \{r^i\} \quad i = 1, 2, 3 \tag{2.1}$$

In the natural terminology the vector $\underline{r}$ may, on the one hand, be composed from non-unique independent vectorial contributions taken along the tetrahedal edges

$$\underline{r}_c = \{r_c^\alpha \; r_c^\beta \; r_c^\gamma \; r_c^\delta \; r_c^\epsilon \; r_c^\zeta\} = \{r_c^\vartheta\} \quad \vartheta = \alpha, \ldots, \zeta \tag{2.2}$$

This forms the so-called component description of a vector. On the other hand, we may introduce as measures the unique orthogonal projections of the vector $\underline{r}$ onto the natural coordinate axes,

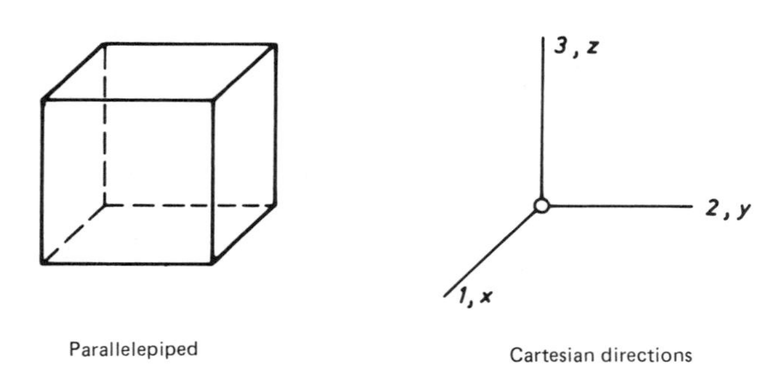

(a) Cartesian approach

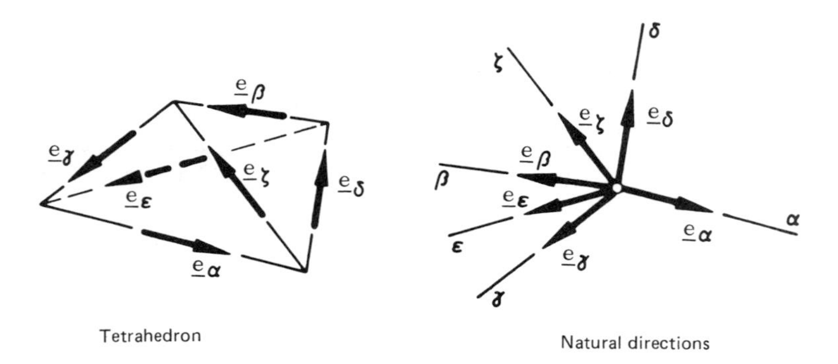

(b) Natural approach

Fig. 1 Cartesian and natural system of reference.

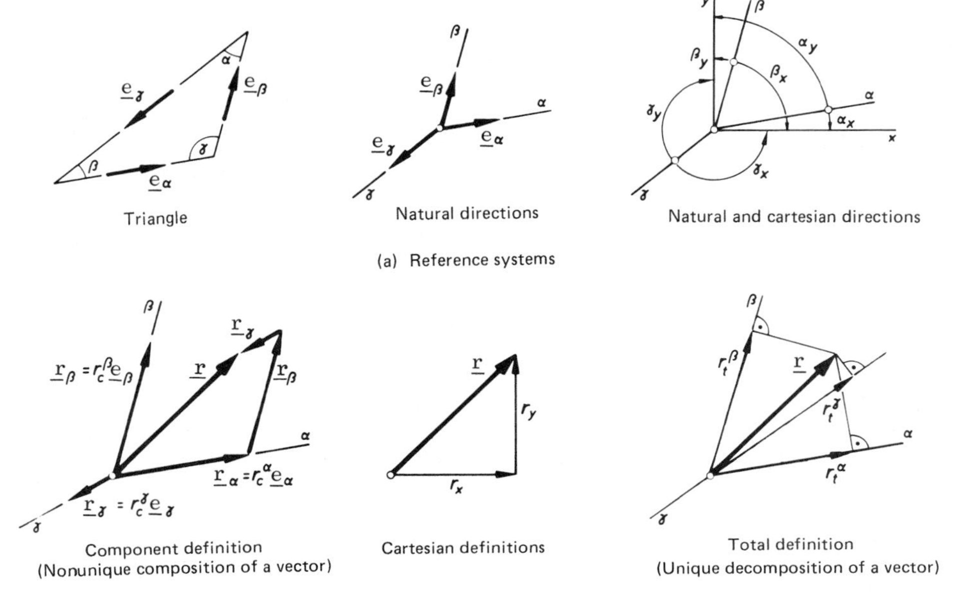

Fig. 2 Natural and cartesian specifications of a vector for the two-dimensional case.

$$\underline{r}_t = \{r_t^\alpha \; r_t^\beta \; r_t^\gamma \; r_t^\delta \; r_t^\epsilon \; r_t^\zeta\} = \{r_t^\vartheta\} \qquad \vartheta = \alpha, \ldots, \zeta \tag{2.3}$$

This forms the total description of a vector.

Considering Fig. 2, we observe that for a given component representation $\underline{r}_c$ of the vector its cartesian form $\underline{r}$ is deduced by the transformation

$$\underline{r} = \underline{B}_{ND}^t \underline{r}_c \tag{2.4}$$

with the matrix

$$\underline{B}_{ND} = [\cos(\vartheta, i)] \qquad \vartheta = \alpha, \ldots, \zeta; \; i = 1, 2, 3 \tag{2.5}$$

The total natural entries $\underline{r}_t$ of (2.3) are then obtained through the relation

$$\underline{r}_t = \underline{B}_{ND}\underline{r} = \underline{B}_{ND}\underline{B}_{ND}^t\underline{r}_c = \underline{B}_{tc}\underline{r}_c \tag{2.6}$$

where (2.4) has been used. The symmetric matrix

$$\underline{B}_{tc} = \underline{B}_{ND}\underline{B}_{ND}^t = [\cos(\vartheta, \varphi)] \qquad \vartheta = \alpha, \ldots, \zeta; \; \varphi = \alpha, \ldots, \zeta \tag{2.7}$$

establishes the direct connection between total and component definitions. It is evident that due to the redundancy of the natural quantities, the above transformations are not invertible.

Finally, we note that the scalar product of two arbitrary vectors $\underline{a}$ and $\underline{b}$ may be given in one of the equivalent forms

$$\underline{a}^t\underline{b} = \underline{a}_c^t\underline{b}_t = \underline{a}_t^t\underline{b}_c \tag{2.8}$$

as is easily confirmed with the aid of (2.4) and (2.6).

2.2. Pressure Field

We proceed next to the description of a scalar field, e.g., the pressure p in the fluid which for arbitrary non-steady conditions is a function of the time t and the position vector $\underline{x}$. We express this by

$$p = p(t, \underline{x}) \tag{2.9}$$

and examine the consequences of the different representations of the vector $\underline{x}$ on the description of the scalar field.

Positions may be defined by component coordinates $\underline{x}_c$, cf. (2.2). From the associated dependence in (2.9) the pressure gradient then reads

$$\underline{g}_t(p) = \left\{ \frac{\partial p}{\partial x_c^{\vartheta}} \right\} = \frac{\partial p}{\partial \underline{x}_c^t} = \left[\frac{\partial p}{\partial \underline{x}_c} \right]^t = \left[\frac{\partial p}{\partial \underline{x}} \frac{d\underline{x}}{d\underline{x}_c} \right]^t = \underline{B}_{ND}\, \underline{g}(p) \tag{2.10}$$

where the chain rule confirms the transformation (2.6) between total natural and cartesian specification of a gradient vector and justifies the total notation of (2.10).

Actually, $\underline{g}_t(p)$ comprises the rates of change of p in the natural directions (and hence the corresponding orthogonal projections of the gradient vector).

Consider next the transformation rule (2.4) leading from the component to the cartesian definition of the gradient vector,

$$\underline{g}(p) = \left\{ \frac{\partial p}{\partial x^i} \right\} = \frac{\partial p}{\partial \underline{x}^t} = \left[\frac{\partial p}{\partial \underline{x}} \right]^t = \left[\frac{\partial p}{\partial \underline{x}_t} \frac{d\underline{x}_t}{d\underline{x}} \right]^t = \underline{B}_{ND}^t\, \underline{g}_c(p) \tag{2.11}$$

Here the component vector $\underline{g}_c(p)$ is clearly

$$\underline{g}_c(p) = \left\{ \frac{\partial p}{\partial x_t^{\vartheta}} \right\} = \frac{\partial p}{\partial \underline{x}_t^t} = \left[\frac{\partial p}{\partial \underline{x}_t} \right]^t \tag{2.12}$$

In fact, $\underline{g}_c(p)$ merely comprises the formal derivatives of p with respect to a non-unique dependence on $\underline{x}_t$ and represents component contributions to the gradient vector. Applying next the chain rule to the gradient of (2.10) we obtain the relation

$$\underline{g}_t(p) = \left[\frac{\partial p}{\partial \underline{x}_c} \right]^t = \left[\frac{\partial p}{\partial \underline{x}_t} \frac{d\underline{x}_t}{d\underline{x}_c} \right]^t = \underline{B}_{tc}\, \underline{g}_c(p) \tag{2.13}$$

which agrees with the transformation of (2.6) between component natural and total natural quantities. In conclusion we list the invariance of the expression

$$\underline{g}_t^t(p)\, d\underline{x}_c = \underline{g}^t(p)\, d\underline{x} = \underline{g}_c^t(p)\, d\underline{x}_t \tag{2.14}$$

which furnishes the increment of the scalar p associated with a change of spatial location, and may be verified by the chain rule, or via an appropriate interpretation of (2.8).

2.3. Velocity Field

The extension of the above terminology to the description of a vector field is straightforward. Consider for instance the velocity field,

$$\underline{v} = \underline{v}(t, \underline{x}) \tag{2.15}$$

which is in general unsteady. The acceleration of a certain particle may be obtained by the so-called material differentiation of the velocity vector with respect to time as

$$\dot{\underline{v}} = \frac{d\underline{v}}{dt} = \frac{\partial \underline{v}}{\partial t} + \frac{\partial \underline{v}}{\partial \underline{x}} \underline{v} \tag{2.16}$$

The first term of the expression in (2.16) represents the local derivative of the velocity with respect to time and is to be evaluated at a fixed location. The second term represents the contribution of convection and is dependent on the gradient of the velocity field.

Disregarding for the time being a particular representation of the velocity vector $\underline{v}$, its gradient may be measured with respect to one of the different specifications of the location vector $\underline{x}$. Taking component natural coordinates $\underline{x}_c$ we obtain in analogy to (2.10) the total natural gradient

$$\mathscr{G}_t(\underline{v}) = \left\{ \frac{\partial \underline{v}}{\partial x_c^\vartheta} \right\} = \frac{\partial \underline{v}}{\partial \underline{x}_c^t} = \left[\frac{\partial \underline{v}}{\partial \underline{x}_c} \right]^t = \left[\frac{\partial \underline{v}}{\partial \underline{x}} \frac{d\underline{x}}{d\underline{x}_c} \right]^t = \underline{B}_{ND} \mathscr{G}(\underline{v}) \tag{2.17}$$

In (2.17) the cartesian gradient

$$\mathscr{G}(\underline{v}) = \left\{ \frac{\partial \underline{v}}{\partial x^i} \right\} = \frac{\partial \underline{v}}{\partial \underline{x}^t} = \left[\frac{\partial \underline{v}}{\partial \underline{x}} \right]^t = \left[\frac{\partial \underline{v}}{\partial \underline{x}_t} \frac{d\underline{x}_t}{d\underline{x}} \right]^t = \underline{B}_{ND}^t \mathscr{G}_c(\underline{v}) \tag{2.18}$$

may be related in analogy to (2.11) to the component gradient of $\underline{v}$,

$$\mathscr{G}_c(\underline{v}) = \left\{ \frac{\partial \underline{v}}{\partial x_t^\vartheta} \right\} = \frac{\partial \underline{v}}{\partial \underline{x}_t^t} = \left[\frac{\partial \underline{v}}{\partial \underline{x}_t} \right]^t \tag{2.19}$$

which represents an extension of (2.12) and is derived from a functional dependence of $\underline{v}$ on $\underline{x}_t$. Applying once more the chain rule to (2.17) we deduce

$$\mathscr{G}_t(\underline{v}) = \left[\frac{\partial \underline{v}}{\partial \underline{x}_c} \right]^t = \left[\frac{\partial \underline{v}}{\partial \underline{x}_t} \frac{d\underline{x}_t}{d\underline{x}_c} \right]^t = \underline{B}_{tc} \mathscr{G}_c(\underline{v}) \tag{2.20}$$

which relates directly the total to the component natural gradient and represents an extension of (2.13) to a vector field $\underline{v}$. We also note the invariance of the expression

$$\mathscr{G}_c^t(\underline{v})\underline{v}_t = \mathscr{G}^t(\underline{v})\underline{v} = \mathscr{G}_t^t(\underline{v})\underline{v}_c \tag{2.21}$$

which is analogous to (2.14) and represents the convective acceleration term of (2.16). Here $\underline{v}$ symbolizes one of the three differently defined representations of the velocity vector. The invariance of (2.21) may be confirmed by the chain rule.

We next turn our attention to the particle acceleration of (2.16) and observe that it may be represented by one of the alternative forms adopted for the description of the velocity vector $\underline{v}$. Thus, in component natural terms we have

$$\dot{\underline{v}}_c = \frac{\partial \underline{v}_c}{\partial t} + \mathscr{G}_t^t(\underline{v}_c)\underline{v}_c \tag{2.22}$$

with the total gradient matrix of $\underline{v}_c$

$$\mathscr{G}_t(\underline{v}_c) = \frac{\partial \underline{v}_c}{\partial \underline{x}_c^t} = \left[\frac{\partial \underline{v}_c}{\partial \underline{x}_c} \right]^t \tag{2.23}$$

The cartesian form of the acceleration on the other hand may be expressed as

$$\dot{\underline{v}} = \frac{\partial \underline{v}}{\partial t} + \underline{\mathscr{G}}^t(\underline{v})\underline{v} = \underline{B}_{ND}^t \dot{\underline{v}}_c \tag{2.24}$$

Here $\underline{v}$ comprises the cartesian components of the velocity, cf. (2.1). Here the cartesian gradient matrix

$$\underline{\mathscr{G}}(\underline{v}) = \frac{\partial \underline{v}}{\partial \underline{x}^t} = \left[\frac{\partial \underline{v}}{\partial \underline{x}}\right]^t \tag{2.25}$$

should not be confused with the expression of (2.18), which is not limited to a particular representation of the vector $\underline{v}$. The total natural formulation of the acceleration is given by

$$\dot{\underline{v}}_t = \frac{\partial \underline{v}_t}{\partial t} + \underline{\mathscr{G}}_c^t(\underline{v}_t)\underline{v}_t = \underline{B}_{ND}\dot{\underline{v}} = \underline{B}_{tc}\dot{\underline{v}}_c \tag{2.26}$$

with the associated component gradient matrix of $\underline{v}_t$

$$\underline{\mathscr{G}}_c(\underline{v}_t) = \frac{\partial \underline{v}_t}{\partial \underline{x}_t^t} = \left[\frac{\partial \underline{v}_t}{\partial \underline{x}_t}\right]^t \tag{2.27}$$

We observe that relations (2.4) and (2.6) between the alternative vector specifications also apply to the acceleration, as confirmed by (2.24) and (2.26).

2.4. Continuity Condition

We proceed to the natural formulation of the continuity condition. To this end, consider in Fig. 3 the infinitesimal tetrahedron element defined by the lengths of the six edges

$$\underline{l} = \left[l^\alpha \; l^\beta \; l^\gamma \; l^\delta \; l^\epsilon \; l^\zeta \right] = \left[l^\vartheta \right] \qquad \vartheta = \alpha, \ldots, \zeta \tag{2.28}$$

with a volume V. When determining the flow of mass through the element as induced by component natural velocities we stipulate the column matrix

$$\underline{\omega}_c = \{\omega_c^\vartheta\} = \left\{ \frac{\partial v_c^\vartheta}{\partial x_c^\vartheta} \right\} \qquad \vartheta = \alpha, \ldots, \zeta \tag{2.29}$$

containing the rate of change of all component velocities along the edges ϑ.

Consider next a component natural velocity characterized by the intensity v_c^α and assume for the time being an incompressible fluid with density ρ. Under these conditions mass permeates at a rate $\rho v_c^\alpha A^\alpha$ through the center of the face A^α not containing l^α into the element and is discharged at a rate $\rho(v + dv)_c^\alpha A^\alpha$ through the opposite face at a distance $l^\alpha/3$ from the point of input. The balance of input-output of fluid mass due to the component natural velocity v_c^α is seen to be simply

$$-\rho\, dv_c^\alpha A^\alpha = -\rho\, \omega_c^\alpha \frac{l^\alpha}{3} A^\alpha = -\rho\, \omega_c^\alpha V \tag{2.30}$$

where

$$\omega_c^\alpha = \frac{\partial v_c^\alpha}{\partial x_c^\alpha} \tag{2.31}$$

denotes the rate of change of v_c^α along α. Generalizing to all natural directions ϑ and applying the column matrix $\underline{\omega}_c$ of (2.29) we obtain the rate of mass supply by summation

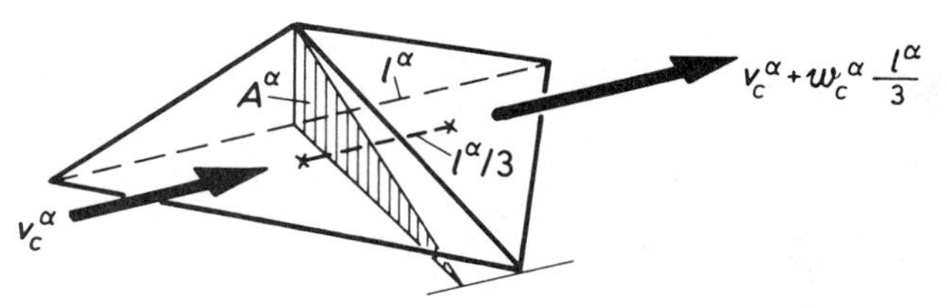

Fig. 3 Mass supply to a natural element due to a component velocity vector v_c^α.

of the individual contributions defined by (2.30) as arising for all component natural velocities of $\underline{v}_c$. Hence the condition of conservation of mass for an incompressible fluid is given by

$$\underline{e}_6^t \underline{w}_c = \underline{e}_3^t \underline{w} = \text{div } \underline{v} = 0 \tag{2.32}$$

where

$$\underline{w} = \{w^i\} = \left\{\frac{\partial v^i}{\partial x^i}\right\} \quad i = 1, 2, 3 \tag{2.33}$$

is the cartesian counterpart of $\underline{w}_c$. Note also the summation vectors

$$\underline{e}_6 = \{1\ 1\ 1\ 1\ 1\ 1\}, \quad \underline{e}_3 = \{1\ 1\ 1\} \tag{2.34}$$

In the case of a compressible fluid, expression (2.30) for a typical rate of the component mass supply must account for a change of the density ρ along α. Consequently, the condition of conservation of mass (2.32) is modified into

$$\rho \underline{e}_6^t \underline{w}_c + \underline{g}_t^t(\rho) \underline{v}_c = \rho \underline{e}_3^t \underline{w} + \underline{g}^t(\rho) \underline{v} = \text{div } (\rho \underline{v}) = -\frac{\partial \rho}{\partial t} \tag{2.35}$$

Here the density gradients

$$\underline{g}_t(\rho) = \frac{\partial \rho}{\partial \underline{x}_c^t} = \left[\frac{\partial \rho}{\partial \underline{x}_c}\right]^t = \underline{B}_{ND} \underline{g}(\rho) \tag{2.36}$$

and

$$\underline{g}(\rho) = \frac{\partial \rho}{\partial \underline{x}^t} = \left[\frac{\partial \rho}{\partial \underline{x}}\right]^t \tag{2.37}$$

correspond to the definitions of (2.10) and (2.11).

2.5. Rate of Deformation and Stress

A specification of the behavior of fluid flow demands the introduction of suitable stress and deformation measures. In classical continuum mechanics (see, e.g., [8]), the rate of deformation is defined by the symmetric part of the cartesian gradient of the instantaneous velocity field. With reference to (2.18) we may thus write

$$\frac{\partial \underline{v}}{\partial \underline{x}} = \left[\frac{\partial \underline{v}}{\partial \underline{x}}\right]_S + \left[\frac{\partial \underline{v}}{\partial \underline{x}}\right]_A = \underline{\mathcal{G}}^t(\underline{v}) \tag{2.38}$$

The associated instantaneous material spin is then

$$\underline{W} = \frac{1}{2}\left\{\frac{\partial \underline{v}}{\partial \underline{x}} - \left[\frac{\partial \underline{v}}{\partial \underline{x}}\right]^t\right\} = \left[\frac{\partial \underline{v}}{\partial \underline{x}}\right]_A = -\underline{\mathscr{G}}_A(\underline{v}) \tag{2.39}$$

and is defined by the antisymmetric part. The rate of deformation of the material

$$\underline{D} = \begin{bmatrix} D^{11} & D^{12} & D^{13} \\ D^{21} & D^{22} & D^{23} \\ D^{31} & D^{32} & D^{33} \end{bmatrix} = \frac{1}{2}\left\{\frac{\partial \underline{v}}{\partial \underline{x}} + \left[\frac{\partial \underline{v}}{\partial \underline{x}}\right]^t\right\} = \left[\frac{\partial \underline{v}}{\partial \underline{x}}\right]_S = \underline{\mathscr{G}}_S(\underline{v}) \tag{2.40}$$

is correspondingly specified by the symmetric part of the cartesian velocity gradient matrix. In what follows we refer to the column matrix

$$\underline{\delta} = \{D^{11}\ D^{22}\ D^{33}\ \sqrt{2}\,D^{12}\ \sqrt{2}\,D^{23}\ \sqrt{2}\,D^{13}\} \tag{2.41}$$

as the cartesian rate of deformation.

Natural measures of the rate of deformation were originally defined by reference to the deformation of the fluid material instantaneously occupying the tetrahedron element [9, 7]. They may be expressed in terms of the natural definitions of the velocity gradient [10]. Thus, the total natural rate of deformation is given by

$$\underline{\delta}_t = \left\{\frac{\partial v_t^{\alpha}}{\partial x_c^{\alpha}}\ \frac{\partial v_t^{\beta}}{\partial x_c^{\beta}}\ \frac{\partial v_t^{\gamma}}{\partial x_c^{\gamma}}\ \frac{\partial v_t^{\delta}}{\partial x_c^{\delta}}\ \frac{\partial v_t^{\epsilon}}{\partial x_c^{\epsilon}}\ \frac{\partial v_t^{\zeta}}{\partial x_c^{\zeta}}\right\} \tag{2.42}$$

The column matrix $\underline{\delta}_t$ comprises the rates of extension of the material along the six natural directions [5]. It may be related to the cartesian definition of the rate of deformation via

$$\underline{\delta}_t = \underline{C}^t\underline{\delta} \tag{2.43}$$

where the detailed structure of the transformation matrix $\underline{C}$ may be found in [9]. The component natural rate of deformation is defined in analogy to the total one in (2.42) as

$$\underline{\delta}_c = \left\{\frac{\partial v_c^{\alpha}}{\partial x_t^{\alpha}}\ \frac{\partial v_c^{\beta}}{\partial x_t^{\beta}}\ \frac{\partial v_c^{\gamma}}{\partial x_t^{\gamma}}\ \frac{\partial v_c^{\delta}}{\partial x_t^{\delta}}\ \frac{\partial v_c^{\epsilon}}{\partial x_t^{\epsilon}}\ \frac{\partial v_c^{\zeta}}{\partial x_t^{\zeta}}\right\} \tag{2.44}$$

The column matrix $\underline{\delta}_c$ may be related uniquely to the total natural rate of deformation by

$$\underline{\delta}_t = \underline{\mathscr{A}}\,\underline{\delta}_c \tag{2.45}$$

where the transformation matrix

$$\mathscr{A} = \underline{C}^t\underline{C} = [\cos^2(\vartheta, \varphi)] \qquad \vartheta = \alpha, \ldots, \zeta;\ \varphi = \alpha, \ldots, \zeta \tag{2.46}$$

is also given in [9] and presumes that component velocities vary only along the direction of their action [10]. In this case, $\underline{\delta}_c$ of (2.44) and $\underline{\omega}_c$ of (2.29) are identical.

For stresses we must adopt a corresponding definition to the rate of deformation so that their scalar product satisfies the condition of invariance for the virtual rate of work. Thus the column matrix

$$\underline{\sigma} = \{\sigma^{11}\ \sigma^{22}\ \sigma^{33}\ \sqrt{2}\,\sigma^{12}\ \sqrt{2}\,\sigma^{23}\ \sqrt{2}\,\sigma^{13}\} \tag{2.47}$$

comprises the Cauchy stresses in their cartesian form and corresponds to the rate of deformation $\underline{\delta}$ of (2.41). The natural component stresses

$$\underline{\sigma}_c = \{\sigma_c^{\alpha}\ \sigma_c^{\beta}\ \sigma_c^{\gamma}\ \sigma_c^{\delta}\ \sigma_c^{\epsilon}\ \sigma_c^{\zeta}\} \tag{2.48}$$

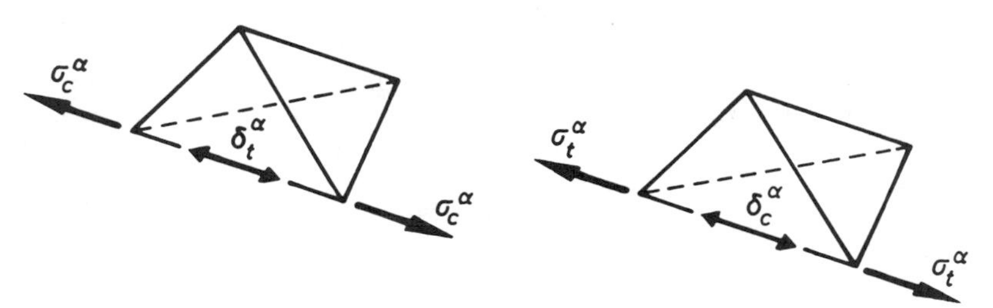

Fig. 4 Corresponding definitions of natural stresses and rates of deformation.

correspond following [9, 5] to the total natural rate of deformation $\underline{\delta}_t$ of (2.42), while the total natural stresses

$$\underline{\sigma}_t = \{\sigma_t^\alpha \; \sigma_t^\beta \; \sigma_t^\gamma \; \sigma_t^\delta \; \sigma_t^\epsilon \; \sigma_t^\zeta\} \tag{2.49}$$

correspond to the component natural rate of deformation $\underline{\delta}_c$ of (2.44) (cf. Fig. 4). The invariance of the virtual rate of work may now be expressed as

$$\underline{\delta}_t^t \underline{\sigma}_c = \underline{\delta}^t \underline{\sigma} = \underline{\delta}_c^t \underline{\sigma}_t \tag{2.50}$$

Bearing in mind (2.43) and (2.45) we easily confirm the relation

$$\underline{\sigma}_c = \underline{C}^{-1} \underline{\sigma} = \mathscr{A}^{-1} \underline{\sigma}_t \tag{2.51}$$

connecting the different representations of the stress state.

2.6. Constitutive Relations for Incompressible Viscous Fluids

In formulating the stress-strain relations appertaining to the fluid motion, it is convenient to split the stress state into hydrostatic and deviatoric contributions. We may ignore here an account of the standard cartesian approach (see, e.g., [11]) and apply instead the natural approach to this subject as developed in [7, 9]. Considering first total stresses we write

$$\underline{\sigma}_t = \underline{\sigma}_{tH} + \underline{\sigma}_{tD} \tag{2.52}$$

and obtain the hydrostatic part of the total stress in the form

$$\underline{\sigma}_{tH} = \frac{1}{3} \underline{E}_6 \underline{\sigma}_c \tag{2.53}$$

where the matrix

$$\underline{E}_6 = \underline{e}_6 \underline{e}_6^t \tag{2.54}$$

performs the summation of the component stresses in each row and yields the total hydrostatic stress in each of the natural directions. The deviatoric part of the total stress follows then from (2.52) as

$$\underline{\sigma}_{tD} = [\underline{\mathscr{A}} - \frac{1}{3} \underline{E}_6] \underline{\sigma}_c \tag{2.55}$$

in which relation (2.51) between total and component definitions is used. Partitioning next the component stress as

$$\underline{\sigma}_c = \underline{\sigma}_{cH} + \underline{\sigma}_{cD} \qquad (2.56)$$

we may derive the hydrostatic and the deviatoric part by application of (2.51) to the total quantities of (2.53) and (2.55), respectively. A decomposition of the total natural rate of deformation

$$\underline{\delta}_t = \underline{\delta}_{tV} + \underline{\delta}_{tD} \qquad (2.57)$$

into volumetric parts

$$\underline{\delta}_{tV} = \frac{1}{3}\underline{E}_6\underline{\delta}_c \qquad (2.58)$$

and deviatoric parts

$$\underline{\delta}_{tD} = [\underline{\mathcal{A}} - \frac{1}{3}\underline{E}_6]\underline{\delta}_c \qquad (2.59)$$

proceeds along the same argument. Also the component natural rate of deformation

$$\underline{\delta}_c = \underline{\delta}_{cV} + \underline{\delta}_{cD} \qquad (2.60)$$

may be partitioned analogously.

Consider next an incompressible fluid, i.e., a fluid undergoing only isochoric deformations. In this case the volumetric rate of deformation must vanish. This yields

$$\delta_V = \frac{1}{3}\underline{e}_6^t\underline{\delta}_c = \frac{1}{3}\underline{e}_6^t\underline{\mathcal{A}}^{-1}\underline{\delta}_t = 0 \qquad (2.61)$$

which is equivalent to (2.32). In the absence of viscous effects the incompressible fluid is described as an ideal one for which deviatoric stresses are absent. Then the stress field derives simply from a static pressure p. Consequently, the total stresses reduce to

$$\underline{\sigma}_t \equiv \underline{\sigma}_{tH} = -p\underline{e}_6 \qquad (2.62)$$

and by (2.51) the component stresses become

$$\underline{\sigma}_c \equiv \underline{\sigma}_{cH} = -p\underline{\mathcal{A}}^{-1}\underline{e}_6 \qquad (2.63)$$

In a viscous incompressible fluid, on the other hand, a rate of deformation—which is exclusively deviatoric because of (2.61)—leads to deviatoric total stresses of the form

$$\underline{\sigma}_{tD} = 2\mu\underline{\delta}_{tD} = 2\underline{\mathcal{A}}\underline{\delta}_{cD} \qquad (2.64)$$

or to deviatoric component stresses,

$$\underline{\sigma}_{cD} = \underline{\mathcal{A}}^{-1}\underline{\sigma}_{tD} = 2\mu\underline{\delta}_{cD} = 2\mu\underline{\mathcal{A}}^{-1}\underline{\delta}_{tD} \qquad (2.65)$$

where μ denotes the viscosity coefficient of the fluid.

For the viscous case the stress is ultimately obtained by a superposition of a hydrostatic contribution arising from (2.62), (2.63), and the deviatoric contribution of (2.64), (2.65). Thus, the total natural stress reads

$$\underline{\sigma}_t = 2\mu\underline{\mathcal{A}}\underline{\delta}_{cD} - p\underline{e}_6 \qquad (2.66)$$

and the component one

$$\underline{\sigma}_c = \underline{\mathcal{A}}^{-1}[2\mu\underline{\delta}_{tD} - p\underline{e}_6] \qquad (2.67)$$

We observe that the constitutive relations in (2.66) and (2.67) are expressed in terms of corresponding stress and rate of deformation measures.

If standard computational procedures are to be applied to the analysis of the isochoric condition motion of an incompressible fluid, one may use the so-called penalty approach. The isochoric (2. 61) can then be relaxed and the pressure p is related to the volumetric rate of deformation as follows

$$p = -3\bar{\kappa}\delta_V \tag{2.68}$$

where

$$\bar{\kappa} = \frac{2\mu}{3}\,\frac{1+\bar{\nu}}{1-2\bar{\nu}} \rightarrow \infty \tag{2.69}$$

represents the penalty parameter. In (2.68), $\bar{\kappa}$ may be interpreted as a modulus of viscous compressibility and is expressed in (2.69) in an analogous manner to the well-known elastic bulk modulus. The strictly incompressible constitutive relations (2.66), (2.67) may now be modified accordingly. For instance, (2.67) assumes in the penalty approach the form

$$\underline{\sigma}_c = 2\mu\left[\underline{\mathscr{A}}^{-1} + \frac{\bar{\nu}}{1-2\bar{\nu}}\,\underline{\mathscr{A}}^{-1}\underline{E}_6\underline{\mathscr{A}}^{-1}\right]\underline{\delta}_t \tag{2.70}$$

in which

$$\bar{\nu} = \frac{1}{2} \tag{2.71}$$

may be used as an alternative penalty parameter.

2.7. Fluid Motion Coupled with Thermal Phenomena

In this subsection we consider fluid motion coupled to thermal phenomena. To this purpose we assume the following unsteady temperature field

$$T = T(t, \underline{x}) \tag{2.72}$$

where $\underline{x}$ denotes the position vector. In extension of the argument in Sec. 2.2 the time rate of the temperature of a particle may be expressed in the alternative forms

$$\begin{aligned} \dot{T} &= \frac{\partial T}{\partial t} + \underline{g}_t^t(T)\,\underline{v}_c \\ &= \frac{\partial T}{\partial t} + \underline{g}^t(T)\,\underline{v} \\ &= \frac{\partial T}{\partial t} + \underline{g}_c^t(T)\,\underline{v}_t \end{aligned} \tag{2.73}$$

The different formulations of the temperature gradient in (2.73) may be compared to the definitions in (2.10), (2.11), and (2.12), respectively. In the present case the invariance condition of (2.14) becomes

$$\underline{g}_t^t(T)\,\underline{v}_c = \underline{g}^t(T)\,\underline{v} = \underline{g}_c^t(T)\,\underline{v}_t \tag{2.74}$$

The time rate of the temperature is associated with a rate of heat stored in the fluid material. The latter may be expressed per unit material volume as

$$\dot{q}_m = \rho c\dot{T} = \dot{q}_s + \dot{q}_k \tag{2.75}$$

where c denotes the specific heat capacity of the fluid. In accordance with (2.73) the rate of heat stored in the material may be composed in the Eulerian approach of two parts. Thus, the rate of heat stored in a unit volume when fixed in space reads

$$\dot{q}_s = \rho c \frac{\partial T}{\partial t} \qquad (2.76)$$

and is associated with the temperature rate obtained at a fixed location. The contribution

$$\dot{q}_k = \rho c \underline{g}_t^t(T) \underline{v}_c = \rho c \underline{g}^t(T) \underline{v} = \rho c \underline{g}_c^t(T) \underline{v}_t \qquad (2.77)$$

is the heat convection term due to the motion of the fluid and may be presented in one of the alternative formulations, natural or cartesian, as shown in (2.77).

We proceed next to the specification of the heat supply to a unit volume of space due to a directed heat flow, i.e., conduction. Following Sec. 2.1, the heat flow vector $\dot{\underline{q}}$ with cartesian entries

$$\dot{\underline{q}} = \{\dot{q}_x\ \dot{q}_y\ \dot{q}_z\} = \{\dot{q}^1\ \dot{q}^2\ \dot{q}^3\} = \{\dot{q}^i\} \quad i = 1, 2, 3 \qquad (2.78)$$

may alternatively be represented by the component natural contributions

$$\dot{\underline{q}}_c = \{\dot{q}_c^\alpha\ \dot{q}_c^\beta\ \dot{q}_c^\gamma\ \dot{q}_c^\delta\ \dot{q}_c^\epsilon\ \dot{q}_c^\zeta\} = \{\dot{q}_c^\vartheta\} \quad \vartheta = \alpha \ \ldots, \zeta \qquad (2.79)$$

or by the total natural quantities

$$\dot{\underline{q}}_t = \{\dot{q}_t^\alpha\ \dot{q}_t^\beta\ \dot{q}_t^\gamma\ \dot{q}_t^\delta\ \dot{q}_t^\epsilon\ \dot{q}_t^\zeta\} = \{\dot{q}_c^\vartheta\} \quad \vartheta = \alpha, \ldots, \zeta \qquad (2.80)$$

The reader is reminded of the interrelations between the alternative representations of the heat flow vector in accordance with (2.4) and (2.6).

When determining the heat flow through an infinitesimal tetrahedron element shown in Fig. 5, as arising from component natural heat fluxes [3], we have to introduce the column matrix

$$\dot{\underline{c}}_c \quad \{c_c^\vartheta\} = \left\{ \frac{\partial \dot{q}_c^\vartheta}{\partial x_c^\vartheta} \right\} \quad \vartheta = \alpha, \ldots, \zeta \qquad (2.81)$$

Consider now in Fig. 5 a component natural heat flux characterized by the intensity $\dot{q}_c^\alpha$ progressing through the tetrahedron. Noting the input $\dot{q}_c^\alpha A^\alpha$ and the output $(\dot{q} + d\dot{q})_c^\alpha A^\alpha$ of the heat rate emerging at a distance $l^\alpha/3$ from the point of input we deduce for the rate of heat supply to the element as contributed by the component natural heat flux $\dot{q}_c^\alpha$

$$-d\dot{q}_c^\alpha A^\alpha = -\dot{c}_c^\alpha \frac{l^\alpha}{3} A^\alpha = -\dot{c}_c^\alpha V \qquad (2.82)$$

where

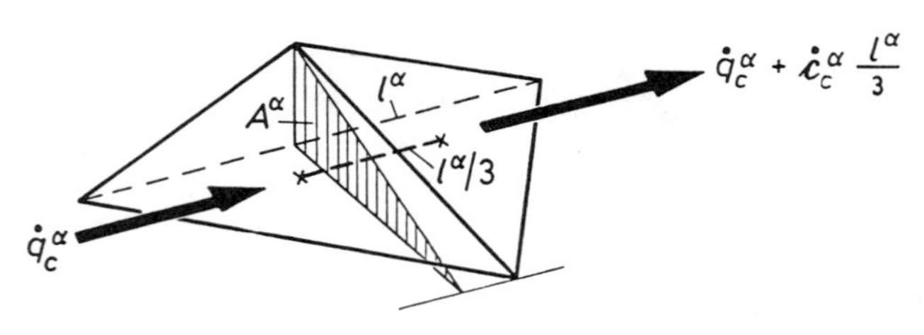

Fig. 5 Heat supply to a natural element due to a component heat flux vector $\dot{q}_c^\alpha$.

$$\dot{c}_c^{\alpha} = \frac{\partial \dot{q}_c^{\alpha}}{\partial x_c^{\alpha}} \qquad (2.83)$$

denotes the rate of change of $\dot{q}_c^{\alpha}$ along α. Generalizing for all natural directions we apply the column matrix $\dot{\underline{c}}_c$ of (2.81) and obtain the rate of heat supply by summation of all individual contributions as expressed by (2.82). Hence, we find

$$\dot{q}_l = -\underline{e}_6^t \dot{\underline{c}}_c = -\underline{e}_3^t \dot{\underline{c}} = -\operatorname{div} \dot{\underline{q}} \qquad (2.84)$$

where $\dot{c}$ is the cartesian counterpart of the column matrix $\dot{\underline{c}}_c$ and reads

$$\dot{\underline{c}} = \{\dot{c}^i\} = \left\{ \frac{\partial \dot{q}^i}{\partial x^i} \right\} \quad i = 1, 2, 3 \qquad (2.85)$$

We conclude this subsection by presenting a natural counterpart to the Fourier's law relating the heat flow to the temperature gradient [3]. Starting with the cartesian form

$$^l\dot{\underline{q}} = -\lambda \left[\frac{\partial T}{\partial \underline{x}}\right]^t = -\lambda \underline{g}(T) \qquad (2.86)$$

where λ denotes the thermal conductivity of the fluid, we deduce the natural relation

$$\dot{\underline{q}}_t = -\lambda \underline{B}_{ND} \underline{B}_{ND}^t \underline{g}_c(T) = -\lambda \underline{B}_{tc} \underline{g}_c(T) = -\underline{\lambda}_{tc} \underline{g}_c(T) \qquad (2.87)$$

by an appropriate application of (2.6) and (2.4) or (2.11). We note that

$$\underline{\lambda}_{tc} = \lambda \underline{B}_{tc} \qquad (2.88)$$

symbolizes the natural thermal conductivity matrix connecting via (2.87) the total natural heat flow vector $\dot{\underline{q}}_t$ with the component temperature gradient $\underline{g}_c(T)$. We also observe that the connection between $\dot{\underline{q}}_t$ and the total temperature gradient $\underline{g}_t(T)$ is simply given by the thermal conductivity λ of the material as in (2.86).

3. DISCRETIZATION BY FINITE ELEMENTS

3.1. Weak Form of the Equations Governing Fluid and Thermal Flow

Bearing in mind our prospective application of the finite element technique to the flow problem we write in the following the basic equations in their weak form assuming a finite volume V bounded by the surface S. Thus, a weak form of the momentum balance may be expressed in natural terms as

$$\int_V (\rho \underline{\tilde{v}}_t^t \dot{\underline{v}}_c + \underline{\tilde{\delta}}_t^t \underline{\sigma}_c)\, dV = \int_V \underline{\bar{v}}_t^t \underline{f}_c \, dV + \int_S \underline{\bar{v}}_t^t \underline{n} \underline{\sigma}_c \, dS \qquad (3.1)$$

where $\underline{\tilde{v}}$ symbolizes a virtual velocity field and $\underline{\tilde{\delta}}$ the associated rate of deformation. Also, $\underline{f}$ denotes the body force vector acting per unit volume and $\underline{n}$ a normal operator yielding the surface tractions. Alternative formulations of (3.1) in natural or in cartesian terms are possible as outlined in Sec. 2. The virtual rate of kinetic energy, for instance, on the left-hand side in (3.1) is given in terms of one of the expressions

$$\underline{\tilde{v}}_t^t \dot{\underline{v}}_c = \underline{\tilde{v}}_c^t \dot{\underline{v}}_t = \underline{\tilde{v}}^t \dot{\underline{v}} \qquad (3.2)$$

offered in (2.8) for the scalar product of two vectors. Clearly, the acceleration $\dot{\underline{v}}$ consists, in the Eulerian approach adopted here of a local part and a convective part and may be specified in the component natural form $\dot{\underline{v}}_c$ of (2.22), the cartesian form $\dot{\underline{v}}$ of

(2.24), or the total natural form of (2.26). We also observe that the component natural stress $\underline{\sigma}_c$ in (3.1) obeys the constitutive laws of Sec. 2.6.

For a weak formulation of the isochoric condition we rely on expression (2.61) and write in natural terms

$$\int_V 3\tilde{p}\delta_V \, dV = \int_V \tilde{p}\underline{e}_6^t\underline{\delta}_c \, dV = \int_V \tilde{p}\underline{e}_6^t\underline{\mathscr{A}}^{-1}\underline{\delta}_t \, dV = 0 \tag{3.3}$$

where $\tilde{p}$ represents the virtual pressure field.

We next turn our attention to the heat flow as occurring concurrently with the fluid motion. The heat balance of the volume in question may be expressed in natural terms as

$$\int_V \tilde{T}\rho c \left[\frac{\partial T}{\partial t} + \underline{g}_c^t(T)\underline{v}_t\right] dV + \int_V \tilde{T}\underline{e}_6^t\underline{\dot{c}}_c \, dV = \int_V \tilde{T}\underline{\delta}_t^t\underline{\sigma}_c \, dV \tag{3.4}$$

where $\tilde{T}$ denotes a virtual temperature field. In (3.4), the first integral on the left-hand side is due to the rate of heat stored in the material, in accordance with (2.75). It is specified through the local term of (2.76) and the convective term of (2.77). The second integral reproduces the rate of heat supply (2.84) by heat conduction. It balances the stored heat expression with due consideration of the rate of dissipation in the material as given by the right-hand side of (3.4). The second integral in (3.4) associated with the heat flux may be transformed as follows (c.f. [3])

$$\int_V \tilde{T}\underline{e}_6^t\underline{\dot{c}}_c \, dV = -\int_V \underline{g}_t^t(\tilde{T})\underline{\dot{q}}_c \, dV + \int_S \tilde{T}\dot{q}_n \, dS \tag{3.5}$$

where due to (2.8) and (2.87)

$$\underline{g}_t^t(\tilde{T})\underline{\dot{q}}_c = \underline{g}_c^t(\tilde{T})\underline{\dot{q}}_t = -\underline{g}_c^t(\tilde{T})\underline{\lambda}_{tc}\underline{g}_c(T) \tag{3.6}$$

Furthermore, the boundary condition

$$\dot{q}_n = \alpha(T - T_\infty) \tag{3.7}$$

expresses the local heat exchange between the surface S under the temperature T and the surrounding medium under a temperature T_∞; the associated heat transfer coefficient is denoted by α. Thus (3.5) may be brought into the final form

$$\int_V \tilde{T}\underline{e}_6^t\underline{\dot{c}}_c \, dV = \int_V \lambda\,\underline{g}_t^t(\tilde{T})\,\underline{g}_c(T) \, dV + \int_S \tilde{T}\alpha(T - T_\infty) \, dS \tag{3.8}$$

By substitution of (3.8) in (3.4) one obtains the fundamental expression for the derivation of the relevant finite element relations.

3.2. Natural Finite Element Equations for Fluids

To set up a finite element formulation of the flow problem consider first the weak momentum equation (3.1) in conjunction with an approximate representation of the velocity field within each finite element expressed by

$$\underline{v}_c = \underline{\omega}_N(\underline{x}_t)\,\underline{V}_c(t) \tag{3.9}$$

The column matrix

$$\underline{V}_c = \{\underline{V}_{c1}\,\underline{V}_{c2} \cdots \underline{V}_{cn}\} = \{\underline{V}_{cj}\} \quad j = 1, \ldots, n \tag{3.10}$$

comprises the component natural contributions to the velocity vector at any one of the n nodes of an element

$$\underline{V}_{cj} = \{V_c^\alpha \ V_c^\beta \ V_c^\gamma \ V_c^\delta \ V_c^\epsilon \ V_c^\zeta\}_j = \{V_c^\vartheta\}_j \qquad \vartheta = \alpha, \ldots, \zeta \tag{3.11}$$

Correspondingly, the matrix

$$\underline{\omega}_N = [\underline{\omega}_{N1} \ \underline{\omega}_{N2} \ \cdots \ \underline{\omega}_{Nn}] = [\underline{\omega}_{Nj}] \qquad j = 1, \ldots, n \tag{3.12}$$

contains the diagonal matrices

$$\underline{\omega}_{Nj} = \lceil \omega_j \rfloor \tag{3.13}$$

of dimensions 6×6 which interpolate the velocities $\underline{V}_{cj}$. Note also that the ω_j's depend only on the total natural coordinates $\underline{x}_t$.

The local part of the acceleration within the element may now be established immediately via (3.9) as

$$\frac{\partial \underline{v}_c}{\partial t} = \underline{\omega}_N \frac{d\underline{V}_c}{dt} = \underline{\omega}_N \dot{\underline{V}}_c \tag{3.14}$$

Before entering into the derivation of the convective part of the acceleration we observe that the velocity field (3.9) may alternatively be described by

$$\underline{v}_c = \underline{\underline{V}}_c(t)\,\underline{\underline{\omega}}(\underline{x}_t) \tag{3.15}$$

where $\underline{\underline{V}}_c$ is here the super row matrix of the component nodal velocities,

$$\underline{\underline{V}}_c = [\underline{V}_{c1} \ \underline{V}_{c2} \ \cdots \ \underline{V}_{cn}] = [\underline{V}_{cj}] \qquad j = 1, \ldots, n \tag{3.16}$$

and $\underline{\underline{\omega}}$ the column matrix,

$$\underline{\underline{\omega}} = \{\omega_1 \ \omega_2 \ \cdots \ \omega_n\} = \{\omega_j\} \qquad j = 1, \ldots, n \tag{3.17}$$

Hence, the velocity gradient may be written as

$$\underline{\mathscr{G}}_c(\underline{v}_c) = \left[\frac{\partial \underline{v}_c}{\partial \underline{x}_t}\right]^t = \left[\underline{\underline{V}}_c \frac{d\underline{\underline{\omega}}}{d\underline{x}_t}\right]^t = \underline{\mathscr{G}}_c(\underline{\underline{\omega}})\,\underline{\underline{V}}_c^t \tag{3.18}$$

with

$$\underline{\mathscr{G}}_c(\underline{\underline{\omega}}) = \left[\frac{d\underline{\underline{\omega}}}{d\underline{x}_t}\right]^t \tag{3.19}$$

The convective term of the acceleration (cf. (2.21)) may now be expressed as

$$\underline{\mathscr{G}}_c^t(\underline{v}_c)\,\underline{v}_t = \underline{\underline{V}}_c\,\underline{\mathscr{G}}_c^t(\underline{\underline{\omega}})\,\underline{v}_t \tag{3.20}$$

in which the velocity gradient is represented by (3.18). The total natural velocity appearing in (3.20) obeys the interpolation rule of (3.9) in the form

$$\underline{v}_t = \underline{\omega}_N \underline{V}_t \tag{3.21}$$

with the column matrix (cf. (3.10))

$$\underline{V}_t = \{\underline{V}_{t1} \ \underline{V}_{t2} \ \cdots \ \underline{V}_{tn}\} = \{\underline{V}_{tj}\} \qquad j = 1, \ldots, n \tag{3.22}$$

Here $\underline{V}_t$ comprises the field of total natural velocities at each of the n element nodes

$$\underline{V}_{tj} = \{V_t^\alpha \ V_t^\beta \ V_t^\gamma \ V_t^\delta \ V_t^\epsilon \ V_t^\zeta\}_j = \{V_t^\vartheta\}_j \qquad \vartheta = \alpha, \ldots, \zeta \tag{3.23}$$

Applying next expression (2.22), we obtain the acceleration by a summation of the local part (3.14) and the convective part (3.20) in the form

$$\dot{\underline{v}}_c = \frac{\partial \underline{v}_c}{\partial t} + \underline{\mathscr{G}}_c^t(\underline{v}_c)\underline{v}_t = \underline{\omega}_N \dot{\underline{V}}_c + \underline{\underline{V}}_c \underline{\mathscr{G}}_c^t(\underline{\underline{\omega}})\underline{\omega}_N \underline{V}_t \tag{3.24}$$

We now proceed to the rate of deformation within the finite element. To this purpose we consider the total natural rate of deformation $\underline{\delta}_t$ of (2.42) and rewrite it in the form

$$\underline{\delta}_t = \left[\frac{\partial}{\partial x_c^{\vartheta}}\right]\underline{v}_t = \underline{\mathscr{D}}_c \underline{v}_t \qquad \vartheta = \alpha, \ldots, \zeta \tag{3.25}$$

where the operator $\mathscr{D}_c$ is the (6 × 6) diagonal matrix

$$\mathscr{D}_c = \left[\frac{\partial}{\partial x_c^{\alpha}} \; \frac{\partial}{\partial x_c^{\beta}} \; \frac{\partial}{\partial x_c^{\gamma}} \; \frac{\partial}{\partial x_c^{\delta}} \; \frac{\partial}{\partial x_c^{\epsilon}} \; \frac{\partial}{\partial x_c^{\zeta}}\right] = \left[\frac{\partial}{\partial x_c^{\vartheta}}\right] \qquad \vartheta = \alpha, \ldots, \zeta \tag{3.26}$$

and

$$\frac{\partial}{\partial x_c^{\vartheta}} = \frac{\partial}{\partial \underline{x}_t}\frac{\partial \underline{x}_t}{\partial x_c^{\vartheta}} = \frac{\partial}{\partial \underline{x}_t}\underline{b}_{t\vartheta} \tag{3.27}$$

Here $\underline{b}_{t\vartheta}$ symbolizes the ϑ-th column of the matrix $\underline{B}_{tc}$ in (2.6), respectively in (2.7). Application of the interpolation rule (3.21) furnishes the total natural rate of deformation within the element as,

$$\underline{\delta}_t = \underline{\mathscr{D}}_c \underline{\omega}_N \underline{V}_t = \underline{a}_t \underline{V}_t \tag{3.28}$$

Turning our attention to the virtual velocity field $\tilde{\underline{v}}_t$ introduced in (3.1), we set

$$\tilde{\underline{v}}_t = \tilde{\underline{\omega}}_N \tilde{\underline{V}}_t \tag{3.29}$$

The definition of the column matrix $\tilde{\underline{V}}_t$ is in line with that of $\underline{V}_t$ in (3.22). As to $\tilde{\underline{\omega}}_N$ its formation is that of $\underline{\omega}_N$ of (3.12) but may be based on different interpolation functions $\tilde{\omega}_j \neq \omega_j$. The associated virtual rate of deformation reads then in analogy to (3.28)

$$\tilde{\underline{\delta}}_t = \underline{\mathscr{D}}_c \tilde{\underline{\omega}}_N \tilde{\underline{V}}_t = \tilde{\underline{a}}_t \tilde{\underline{V}}_t \tag{3.30}$$

In finite element theory, forces are assumed to be transmitted exclusively through the element nodes. Let the column matrix

$$\underline{P}_c = \{\underline{P}_{c1} \; \underline{P}_{c2} \cdots \underline{P}_{cn}\} = \{\underline{P}_{cj}\} \qquad j = 1, \ldots, n \tag{3.31}$$

comprise the component natural element contribution to the force vector at each of the n element nodes,

$$\underline{P}_{cj} = \{P_c^{\alpha} \; P_c^{\beta} \; P_c^{\gamma} \; P_c^{\delta} \; P_c^{\epsilon} \; P_c^{\zeta}\}_j = \{P_c^{\vartheta}\} \qquad \vartheta = \alpha, \ldots, \zeta \tag{3.32}$$

In accordance with the invariance rule (2.8), the component natural representation of the nodal force vector $\underline{P}_{cj}$ of (3.32) corresponds to the total natural definition of the nodal velocity vector $\underline{V}_{tj}$ of (3.23). Disregarding for simplicity the volume forces on the right-hand side of (3.1) and expressing the surface integral through the nodal quantities the virtual work expression (3.1) assumes for a finite element of volume V the form

$$\int_V (\rho \tilde{\underline{v}}_t^t \dot{\underline{v}}_c + \tilde{\underline{\delta}}_t^t \underline{\sigma}_c)\, dV = \tilde{\underline{V}}_t^t \underline{P}_c \tag{3.33}$$

Introducing the kinematic relations (3.29), (3.30) in (3.33) we obtain the component natural forces at the element nodes as

$$\underline{P}_c = \int_V (\rho \,\underline{\tilde{\omega}}_N^t \dot{\underline{v}}_c + \underline{\tilde{a}}_t^t \underline{\sigma}_c)\, dV \tag{3.34}$$

The acceleration term on the right-hand side of (3.34) may be transformed with the aid of (3.24) into

$$\int_V \rho \underline{\tilde{\omega}}_N^t \dot{\underline{v}}_c \, dV = [\int_V \rho \underline{\tilde{\omega}}_N^t \underline{\omega}_N \, dV] \dot{\underline{V}}_c + [\int_V \rho \underline{\tilde{\omega}}_N^t \underline{\underline{V}}_c \mathscr{G}_c^t (\underline{\underline{\omega}}) \, \underline{\omega}_N \, dV] \, \underline{V}_t$$

$$= \underline{m}_N \dot{\underline{V}}_c + \underline{n}_N \underline{V}_t \tag{3.35}$$

where

$$\underline{m}_N = \int_V \rho \underline{\tilde{\omega}}_N^t \underline{\omega}_N \, dV \tag{3.36}$$

corresponds to the Lagrangian mass matrix while

$$\underline{n}_N = \int_V \rho \underline{\tilde{\omega}}_N^t \underline{\underline{V}}_c \mathscr{G}_c^t (\underline{\underline{\omega}}) \, \underline{\omega}_N \, dV \tag{3.37}$$

accounts for the nonlinear convective contribution inherent to the present Eulerian approach.

To specify the stress term on the right-hand side of (3.34) we call upon expression (2.67) for the component natural stresses and obtain

$$\int_V \underline{\tilde{a}}_t^t \underline{\sigma}_c \, dV = \int_V 2\mu \underline{\tilde{a}}_t^t \underline{\mathscr{A}}^{-1} \underline{\delta}_{tD} \, dV - \int_V \underline{\tilde{a}}_t^t \underline{\mathscr{A}}^{-1} \underline{e}_6 p \, dV \tag{3.38}$$

Carrying out the deviatoric operation (2.59) in conjunction with (2.45) and using the kinematics as prescribed in (3.28), the first integral on the right-hand side of (3.38) is transformed into

$$\int_V 2\mu \underline{\tilde{a}}_t^t \underline{\mathscr{A}}^{-1} \underline{\delta}_{tD} \, dV = \{\int_V 2\mu \underline{\tilde{a}}_t^t \underline{\mathscr{A}}^{-1} [\underline{\mathscr{A}} - \frac{1}{3} \underline{E}_6] \underline{\mathscr{A}}^{-1} \underline{a}_t \, dV\} \underline{V}_t = \underline{d}_N \underline{V}_t \tag{3.39}$$

where

$$\underline{d}_N = \int_V 2\mu \underline{\tilde{a}}_t^t \underline{\mathscr{A}}^{-1} [\underline{\mathscr{A}} - \frac{1}{3} \underline{E}_6] \underline{\mathscr{A}}^{-1} \underline{a}_t \, dV \tag{3.40}$$

represents the viscosity matrix of the element and reflects the deviatoric response of the isochoric fluid. With respect to the second integral on the right-hand side of (3.38) we introduce the approximation

$$p = \underline{\pi} \underline{p} \tag{3.41}$$

to describe the pressure field within the element. In (3.41) $\underline{p}$ is the column matrix

$$\underline{p} = \{p_1 \; p_2 \cdots p_m\} = \{p_k\} \qquad k = 1, \ldots, m \tag{3.42}$$

and contains the pivotal values of the pressure and $\underline{\pi}$ the interpolation functions within the row matrix

$$\underline{\pi} = [\pi_1 \; \pi_2 \cdots \pi_m] = [\pi_k] \qquad k = 1, \ldots, m \tag{3.43}$$

Introducing (3.41) into (3.38) one obtains

$$\int_V \underline{\tilde{a}}_t^t \underline{\mathscr{A}}^{-1} \underline{e}_6 p \, dV = [\int_V \underline{\tilde{a}}_t^t \underline{\mathscr{A}}^{-1} \underline{e}_6 \underline{\pi} \, dV] \underline{p} = \underline{h}_N \underline{p} \tag{3.44}$$

where

$$\underline{h}_N = \int_V \underline{\tilde{a}}_t^t \underline{\mathscr{A}}^{-1} \underline{e}_6 \underline{\pi} \, dV \tag{3.45}$$

is the hydrostatic element matrix. Using (3.44), (3.39), and (3.35), the component natural forces (3.34) of the element may ultimately be expressed as

$$\underline{P}_c = \underline{m}_N \dot{\underline{V}}_c + \underline{n}_N \underline{V}_t + \underline{d}_N \underline{V}_t - \underline{h}_N \underline{p} \tag{3.46}$$

3.3. Transition to Cartesian Definitions; Discretized Navier-Stokes Equations

Before proceeding to the assembly of the element contributions (3.46) within the region considered, we transform (3.46) into a global cartesian system of reference. Denoting the respective cartesian element nodal forces by

$$\underline{P} = \{\underline{P}_1 \; \underline{P}_2 \cdots \underline{P}_n\} = \{\underline{\ }_j\} \qquad j = 1, \ldots, n \tag{3.47}$$

and the corresponding velocities by

$$\underline{V} = \{\underline{V}_1 \; \underline{V}_2 \cdots \underline{V}_n\} = \{\underline{V}_j\} \qquad j = 1, \ldots, n \tag{3.48}$$

we may apply relation (2.4) connecting natural and cartesian definitions of vectors to obtain on the element level

$$\underline{P} = \lceil \underline{B}_{ND}^t \rfloor \underline{P}_c \tag{3.49}$$

and

$$\underline{V} = \lceil \underline{B}_{ND}^t \rfloor \underline{V}_c \tag{3.50}$$

We note also that in

$$\underline{v} = \underline{\omega} \underline{V} \tag{3.51}$$

the interpolation matrix $\underline{\omega}$ corresponds to the definition of $\underline{\omega}_N$ in (3.12) but with entries $\underline{\omega}_j$ (cf. (3.13) of dimension 3×3, in order to maintain consistency with the cartesian definition. One may now substitute (3.50) in (3.51) to express $\underline{v}$ in terms of $\underline{v}_c$. Relating on the other hand $\underline{v}$ to $\underline{v}_c$ via (2.4) and expressing the latter through the interpolation (3.9) we deduce a second expression for $\underline{v}$. Thus,

$$\underline{v} = \underline{\omega} \lceil \underline{B}_{ND}^t \rfloor \underline{V}_c = \underline{B}_{ND}^t \underline{\omega}_N \underline{V}_c \tag{3.52}$$

and hence

$$\underline{\omega} \lceil \underline{B}_{ND}^t \rfloor = \underline{B}_{ND}^t \underline{\omega}_N \tag{3.53}$$

Applying next the transformation (2.6) to the velocities we obtain for the total elemental velocities

$$\underline{V}_t = \lceil \underline{B}_{ND} \rfloor \underline{V} \tag{3.54}$$

An analogous argument to that used in (3.52) yields in the present case

$$\underline{v}_t = \underline{\omega}_N \lceil \underline{B}_{ND} \rfloor \underline{V} = \underline{B}_{ND} \underline{\omega} \underline{V} \tag{3.55}$$

and hence

$$\underline{\omega}_N\lfloor\underline{B}_{ND}\rfloor = \underline{B}_{ND}\underline{\omega} \tag{3.56}$$

Substituting in (3.49) $\underline{P}_c$, as given in (3.46), and $\underline{V}_t$ as defined in (3.54) furnishes the cartesian forces

$$\begin{aligned}\underline{P} &= \lfloor\underline{B}^t_{ND}\rfloor\underline{P}_c\\ &= \lfloor\underline{B}^t_{ND}\rfloor\underline{m}_N\dot{\underline{V}}_c + \lfloor\underline{B}^t_{ND}\rfloor\underline{n}_N\lfloor\underline{B}_{ND}\rfloor\underline{V} + \lfloor\underline{B}^t_{ND}\rfloor\underline{d}_N\lfloor\underline{B}_{ND}\rfloor\underline{V} - \lfloor\underline{B}^t_{ND}\rfloor\underline{h}_N\underline{p}\\ &= \underline{m}\dot{\underline{V}} + \underline{n}\,\underline{V} + \underline{d}\,\underline{V} - \underline{h}\,\underline{p}\end{aligned} \tag{3.57}$$

Using (3.56) and (3.53) as well as (3.36) and (3.50) we may verify that the first term in the second expression of (3.57) reduces to

$$\begin{aligned}\lfloor\underline{B}^t_{ND}\rfloor\underline{m}_N\dot{\underline{V}}_c &= \lfloor\underline{B}^t_{ND}\rfloor[\int_V \rho\tilde{\underline{\omega}}^t_N\underline{\omega}_N\, dV]\,\dot{\underline{V}}_c\\ &= [\int_V \rho\tilde{\underline{\omega}}^t\underline{\omega}\, dV]\lfloor\underline{B}^t_{ND}\rfloor\dot{\underline{V}}_c\\ &= [\int_V \rho\tilde{\underline{\omega}}^t\underline{\omega}\, dV]\,\dot{\underline{V}} = \underline{m}\,\dot{\underline{V}}\end{aligned} \tag{3.58}$$

Note the expression for the elemental mass matrix

$$\underline{m} = \int_V \rho\,\tilde{\underline{\omega}}^t\underline{\omega}\, dV \tag{3.59}$$

Consider next the second term on the right-hand side of (3.58). Application of (3.37) for the natural convectivity matrix yields the cartesian counterpart

$$\begin{aligned}\underline{n} &= \lfloor\underline{B}^t_{ND}\rfloor\underline{n}_N\lfloor\underline{B}_{ND}\rfloor\\ &= \lfloor\underline{B}^t_{ND}\rfloor[\int_V \rho\tilde{\underline{\omega}}^t_N\underline{\underline{V}}_c\underline{\mathscr{G}}^t_c(\underline{\underline{\omega}})\underline{\omega}_N\, dV]\lfloor\underline{B}_{ND}\rfloor\\ &= \int_V \rho\tilde{\underline{\omega}}^t\underline{\underline{V}}\,\underline{\mathscr{G}}^t(\underline{\underline{\omega}})\underline{\omega}\, dV\end{aligned} \tag{3.60}$$

in which use is made of the relation (2.18) connecting $\underline{\mathscr{G}}$ and $\underline{\mathscr{G}}_c$. Also, $\underline{\underline{V}}$ denotes the super row matrix of the cartesian nodal velocities.

Finally, the cartesian viscosity and hydrostatic elemental matrix

$$\underline{d} = \lfloor\underline{B}^t_{ND}\rfloor\underline{d}_N\lfloor\underline{B}_{ND}\rfloor \tag{3.61}$$

and

$$\underline{h} = \lfloor\underline{B}^t_{ND}\rfloor\underline{h}_N \tag{3.62}$$

represent standard transformations and do not require further elaboration.

We observe in the above finite element idealization that identity of $\tilde{\omega}_j$, the weighting functions, with ω_j, the interpolation functions, reduces the discretization procedure to that of Galerkin. In most structural applications, this method leads to symmetric matrices and the associated solutions are known to possess the property of best approximation. In convection dominated flow problems, however, we adopt a suggestion of [4] and prefer to apply the streamline upwind/Petrov-Galerkin technique. In this case $\tilde{\omega}_j$ and ω_j are taken to be different. Bearing in mind the aforementioned publication in which a detailed description of the method is given we restrict our present account to an elaboration of the alternative natural formulation. Following [4], the weighting functions $\tilde{\omega}_j$ are formed as

$$\tilde{\omega}_j = \omega_j + s_j \tag{3.63}$$

where ω_j is the standard interpolation function at the j-th element node and s_j a perturbation defined by

$$s_j = k \frac{\underline{v}^t \underline{\mathcal{g}}(\omega_j)}{\underline{v}^t \underline{v}} = k \frac{\underline{v}_t^t \underline{\mathcal{g}}_c(\omega_j)}{\underline{v}_t^t \underline{v}_c} \tag{3.64}$$

which induces an upwinding in the streamline direction. The scalar coefficient k is specified in [4] as a function of the velocity and the element dimensions. The natural expression for s_j in (3.64) may be seen to simply rely on the invariance of alternative expressions of scalar products as shown in (2.8). In (3.64) ω_j is assumed to be a function of the total natural coordinates $\underline{x}_t$. The associated gradient $\underline{\mathcal{g}}_c$ follows then the definition of (2.12) with ω_j in place of the pressure. In conclusion we note that the upwind technique introduces an additional dependence on the velocity into the finite element characteristics. As outlined in [4] under certain conditions the upwind scheme affects merely the weighting of the acceleration term in (3.34) but not that of the stress term. In this case the element viscosity matrix is symmetric.

Turning next our attention to the entire flow domain, the element contributions to the nodal forces as given by (3.57) may be summed up and yield the global relation

$$\underline{R} = \underline{M}\,\dot{\underline{V}} + \underline{N}\,\underline{V} + \underline{D}\,\underline{V} - \underline{H}\,\underline{p} \tag{3.65}$$

which represents the discretized form of the Navier-Stokes equations. In (3.65) $\underline{R}$ denotes the column matrix of the nodal forces applied to the flow domain, $\underline{V}$ and $\dot{\underline{V}}$ are the corresponding velocities and accelerations, and the column matrix $\underline{p}$ defines the pressure field in the entire flow domain. The matrices $\underline{M}$, $\underline{N}$, $\underline{D}$, and $\underline{H}$ may be deduced by a straightforward assembly procedure from the matrices $\underline{m}$, $\underline{n}$, $\underline{d}$, and $\underline{h}$ of the individual elements.

3.4. Isochoric Condition. Exact Analysis and Approximate Penalty Formulation

We now proceed to the discretization of the isochoric condition using the natural methodology and consider to this end the last expression in (3.3). Introducing a relation analogous to that of (3.41) for the variation of $\tilde{p}$ and expressing $\underline{\delta}_t$ as in (3.28) we deduce for a finite element the condition

$$\int_V \tilde{p}\,\underline{e}_6^t \underline{\mathscr{A}}^{-1} \underline{\delta}_t \, dV = \tilde{\underline{p}}^t [\int_V \tilde{\underline{\pi}}\,\underline{e}_6^t \underline{\mathscr{A}}^{-1} \underline{a}_t \, dV]\, \underline{V}_t = \tilde{\underline{p}}^t \underline{g}_N^t \underline{V}_t = o \tag{3.66}$$

where the matrix

$$\underline{g}_N = [\int_V \tilde{\underline{\pi}}\,\underline{e}_6^t \underline{\mathscr{A}}^{-1} \underline{a}_t \, dV]^t = \int_V \underline{a}_t^t \underline{\mathscr{A}}^{-1} \underline{e}_6 \tilde{\underline{\pi}}^t \, dV \tag{3.67}$$

coincides with the matrix $\underline{h}_N$ in (3.45) for the case when $\tilde{\omega}_j = \omega_j$ and $\tilde{\pi}_j = \pi_j$. To obtain the cartesian form of (3.66) we refer to (3.54) and deduce

$$\underline{g}_N^t \underline{V}_t = \underline{g}_N^t \lceil \underline{B}_{ND} \rfloor \underline{V} = \underline{g}^t \underline{V} = \underline{o} \tag{3.68}$$

Hence the cartesian counterpart of the natural matrix $\underline{g}$ is

$$\underline{g} = \lceil \underline{B}^t_{ND} \rfloor \underline{g}_N \qquad (3.69)$$

The isochoric condition for the entire flow domain may now be symbolized by (cf. (3.68))

$$\underline{G}^t \underline{V} = \underline{o} \qquad (3.70)$$

where the column matrix $\underline{V}$ comprises the velocities at the nodal points of the finite element mesh, and $\underline{G}$ is composed by the individual element matrices $\underline{g}$ in (3.69).

In the penalty approach the isochoric condition is relaxed in accordance with (2.68). As a consequence the weak formulation in (3.3) is correspondingly affected. Adopting the finite element approximation in (3.66) one obtains in the penalty approach

$$\underline{g}^t_N \underline{V}_t = -\underline{\bar{\kappa}}^{-1} \underline{p} \qquad (3.71)$$

The matrix $\underline{\bar{\kappa}}^{-1}$ may be seen to represent the integral expression

$$\underline{\bar{\kappa}}^{-1} = \int_V \frac{1}{\kappa} \underline{\tilde{\pi}}^t \underline{\pi}\, dV \qquad (3.72)$$

Solution of (3.71) for the pressure yields

$$\underline{p} = -\underline{\bar{\kappa}}\, \underline{g}^t_N \underline{V}_t = -\underline{\bar{\kappa}}\, \underline{g}^t_N \lceil \underline{B}_{ND} \rfloor \underline{V} = -\underline{\bar{\kappa}}\, \underline{g}^t \underline{V} \qquad (3.73)$$

where use is made of (3.54), (3.69) when forming the alternative cartesian expression on the right-hand side of (3.73). Substitution of (3.73) in (3.57) determines a pure velocity formulation. Isolating the two last terms in the final expression in (3.57) we consequently have

$$\underline{d}\,\underline{V} - \underline{h}\,\underline{p} = \underline{d}\,\underline{V} + \underline{h}\,\underline{\bar{\kappa}}\,\underline{g}^t \underline{V} = [\underline{d} + \underline{h}\,\underline{\bar{\kappa}}\,\underline{g}^t] = \underline{\bar{d}}\,\underline{V} \qquad (3.74)$$

The matrix

$$\underline{\bar{d}} = \underline{d} + \underline{h}\,\underline{\bar{\kappa}}\,\underline{g}^t \qquad (3.75)$$

represents the elemental viscosity in the penalty approach and is a symmetric matrix in an ordinary Galerkin approximation. The above procedure corresponds to the mixed finite element technique of [15] in which velocity and pressure field are approximated independently. An alternative penalty formulation of the viscous incompressible problem may be obtained by substitution of (2.68) in (3.38). This leads to a pure velocity formulation in (3.46) or (3.57) without the need of a separate approximation for the pressure. On the other hand, this advantage involves necessarily a reduced integration scheme for the volumetric part of the associated viscosity matrix $\underline{\bar{d}}$ [13, 14]. Summarizing, the discretized Navier-Stokes equations for the entire flow domain may be written in the penalty approach as

$$\underline{R} = \underline{M}\dot{\underline{V}} + \underline{N}\underline{V} + \underline{\bar{D}}\underline{V} \qquad (3.76)$$

where the relaxed isochoric constraint in the viscosity matrix is included in $\underline{\bar{D}}$ in accordance with one or the other approximation technique.

3.5. Finite Element Equations for Heat Flow

As a final item we consider the finite element approximation of the heat balance in the fluid as governed by (3.4) and (3.8). To this end we write the temperature field within the element as

$$T = \underline{\tau}(\underline{x}_t)\underline{T}(t) \qquad (3.77)$$

where the column matrix

$$\underline{T} = \{T_1 \ T_2 \cdots T_n\} = \{T_j\} \qquad j = 1, \ldots, n \tag{3.78}$$

comprises the temperatures at the element nodes and the row matrix

$$\underline{\tau} = [\tau_1 \ \tau_2 \cdots \tau_n] = [\tau_j] \qquad j = 1, \ldots, n \tag{3.79}$$

the interpolation functions. Analogously, we express the virtual temperature field as

$$\tilde{T} = \underline{\tilde{\tau}}(\underline{x}_t)\underline{\tilde{T}}(t) \tag{3.80}$$

where the weighting functions $\tilde{\tau}_j$ in $\underline{\tilde{\tau}}$ may be constructed in accordance with the streamline upwind/Petrov-Galerkin concept, as detailed in (3.63) for $\tilde{\omega}_j$. Applying (3.77) we may obtain the local part of the temperature rate as

$$\frac{\partial T}{\partial t} = \underline{\tau}\,\underline{\mathring{T}} \tag{3.81}$$

Correspondingly the convective part becomes

$$\underline{g}_c^t(T)\underline{v}_t = \underline{v}_t^t \underline{g}_c(T) = \underline{V}_t^t \underline{\omega}_N^t \underline{\mathscr{g}}_c(\underline{\tau}^t)\underline{T} \tag{3.82}$$

where use is made of (3.21) for $\underline{v}_t$.

With the aid of (3.80), (3.81), and (3.82) the first integral in the heat balance of (3.4) may be transformed into

$$\int_V \tilde{T}\rho c\left[\frac{\partial T}{\partial t} + \underline{g}_c^t(T)\underline{v}_t\right] dV = \underline{\tilde{T}}^t[\int_V \rho c\underline{\tilde{\tau}}^t\underline{\tau}\, dV]\,\underline{\dot{T}} + \underline{\tilde{T}}^t[\int_V \rho c\underline{\tilde{\tau}}^t \underline{V}_t^t \underline{\omega}_N^t \underline{\mathscr{g}}_c(\underline{\tau}^t)\, dV]\,\underline{T}$$

$$= \underline{\tilde{T}}^t[\underline{c}\,\underline{\dot{T}} + \underline{k}\,\underline{T}] \tag{3.83}$$

The matrix

$$\underline{c} = [\int_V \rho c\underline{\tilde{\tau}}^t\underline{\tau}\, dV] \tag{3.84}$$

represents the heat capacity matrix of the element in a Lagrangian approach and must be supplemented in the present Eulerian presentation by the convective contribution associated in (3.83) with the coefficient matrix

$$\underline{k} = \int_V \rho c\underline{\tilde{\tau}}^t \underline{V}_t^t \underline{\omega}_N^t \underline{\mathscr{g}}_c(\underline{\tilde{\tau}}^t)\, dV = \int_V \rho c\underline{\tilde{\tau}}^t \underline{V}^t \underline{\omega}^t \underline{\mathscr{g}}(\underline{\tau}^t)\, dV \tag{3.85}$$

The second integral expression $\underline{k}$ in (3.85) refers to a cartesian specification, the transition from the first natural expression being a consequence of (2.74).

The second integral on the left-hand side of (3.4) may be put as a consequence of (3.8) into the finite element form

$$\int_V \underline{g}_c^t(\tilde{T})\underline{\lambda}_{tc}\underline{g}_c(T)\, dV + \int_S \tilde{T}\alpha(T - T_\infty)\, dS = \underline{\tilde{T}}^t[\underline{\ell}\,\underline{T} - \underline{\dot{q}}_S] \tag{3.86}$$

S being the element surface. Application of (3.80) and (3.77) yields the equivalent expression

$$\int_V \underline{g}_c^t(\tilde{T})\underline{\lambda}_{tc}\underline{g}_c(T)\, dV + \int_S \tilde{T}\alpha T\, dS$$

$$= \underline{\tilde{T}}^t[\int_V \underline{\mathscr{g}}_c^t(\underline{\tilde{\tau}}^t)\underline{\lambda}_{tc}\underline{\mathscr{g}}_c(\underline{\tau}^t)\, dV]\,\underline{T} + \underline{\tilde{T}}^t[\int_S \alpha\underline{\tilde{\tau}}^t\underline{\tau}\, dS]\,\underline{T} = \underline{\tilde{T}}^t\,\underline{\ell}\,\underline{T} \tag{3.87}$$

The element conductivity matrix is thus given by

$$\underline{\ell} = \int_V \underline{\mathscr{g}}_c^t(\underline{\tilde{\tau}}^t)\underline{\lambda}_{tc}\underline{\mathscr{g}}_c(\underline{\tau}^t)\, dV + \int_S \alpha\underline{\tilde{\tau}}^t\underline{\tau}\, dS \tag{3.88}$$

Its transcription into the cartesian form may be established by substitution of (2.88) for $\underline{\lambda}_{tc}$ and application of the gradient relations (2.20), (2.17), and (2.18). We find

$$\int_V \underline{\mathscr{g}}_c^t(\tilde{\underline{\tau}}^t)\underline{\lambda}_{tc}\underline{\mathscr{g}}_c(\underline{\tau}^t)\, dV = \int_V \lambda\underline{\mathscr{g}}^t(\tilde{\underline{\tau}}^t)\underline{\mathscr{g}}(\underline{\tau}^t)\, dV \qquad (3.89)$$

Furthermore, we observe in (3.86) that

$$\dot{\underline{q}}_s = \int_S \alpha\tilde{\underline{\tau}}^t T_\infty ds = [\int_S \alpha\tilde{\underline{\tau}}^t \underline{\tau}\, dS]\, \underline{T}_\infty \qquad (3.90)$$

represents a prescribed heat rate through the element surface.

Concerning the rate of dissipation defined by the integral on the right-hand side of (3.4), one may write

$$\int_V \tilde{T}\underline{\delta}_t^t\underline{\sigma}_c\, dV = \tilde{\underline{T}}^t[\int_V \tilde{\underline{\tau}}^t\underline{\delta}_t^t\underline{\sigma}_c\, dV] = \tilde{\underline{T}}^t\dot{\underline{q}}_d \qquad (3.91)$$

and

$$\dot{\underline{q}}_s = \int_V \tilde{\underline{\tau}}^t\underline{\delta}_t^t\underline{\sigma}_c\, dV = \int_V \tilde{\underline{\tau}}^t\underline{\delta}^t\underline{\sigma}\, dV \qquad (3.92)$$

Here $\underline{\delta}_t$, $\underline{\sigma}_c$ and $\underline{\delta}$, $\underline{\sigma}$ may be deduced from the mechanical account of the flow problem in subsection 3.2.

Collecting the contributions (3.83), (3.86), and (3.91) into the overall heat balance of the element as expressed by (3.4) we obtain

$$\underline{c}\,\dot{\underline{T}} + \underline{k}\,\underline{T} + \underline{\ell}\,\underline{T} = \dot{\underline{q}} \qquad (3.93)$$

where

$$\dot{\underline{q}} = \dot{\underline{q}}_s + \dot{\underline{q}}_d \qquad (3.94)$$

is a generalized heat rate. The finite element equations for the entire flow domain assume then the form

$$\underline{C}\,\dot{\underline{T}} + \underline{K}\,\underline{T} + \underline{L}\,\underline{T} = \dot{\underline{Q}} \qquad (3.95)$$

in which $\underline{T}$, $\dot{\underline{T}}$, $\dot{\underline{Q}}$ are column matrices comprising quantities at the nodes of the finite element mesh and $\underline{C}$, $\underline{K}$, $\underline{L}$ are the relevant global matrices deduced by assembly of the respective element matrices.

4. NUMERICAL EXAMPLES

In this section we present some examples illustrating the application of the preceding theory on the solution of pure and thermally coupled flow problems. Details of the numerical solution methods, omitted in this paper, may be studied in [28]. There, the numerical aspects are discussed taking account of the pertinent literature on the subject [16-25], which include recent developments. We should stress here that the streamline upwind/Petrov-Galerkin scheme is applied to all our examples. The capability of this method is demonstrated in what follows for convection dominated flow in two and three dimensions. The solution of thermally coupled flows is illustrated on the Bénard type instability. Cavity flows with free and forced convection including a change of phase are treated in [28].

4.1. Flow Over a Step

The transient incompressible flow over a step demonstrates the applicability of the independent p-$\underline{v}$ formulation and a two stage solution strategy as described in [4, 28]. Due to the high Reynolds number a turbulent flow field develops necessitating the use of upwind techniques.

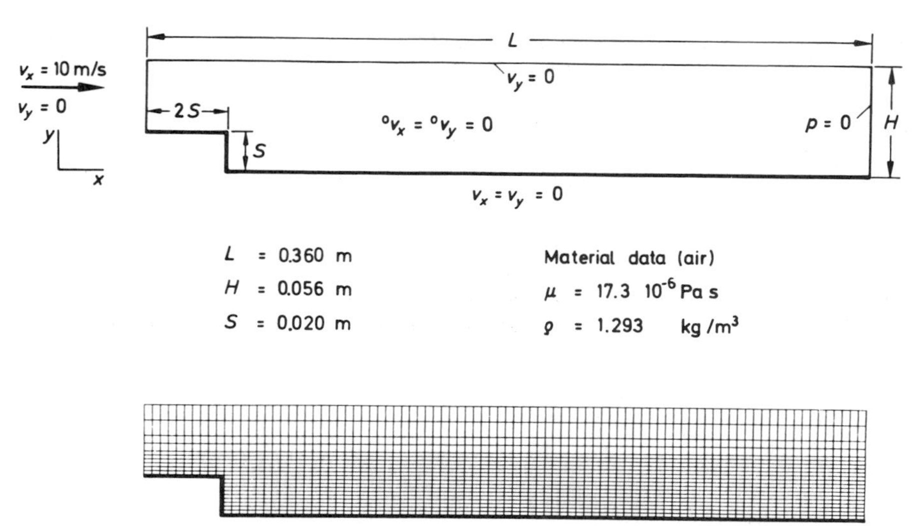

Fig. 6 Flow over a step. Description and finite element discretization.

The geometry of the flow domain and the boundary conditions used in the calculation are sketched in Fig. 6 together with the material data of the medium (air). At the inlet a constant velocity profile is prescribed which yields a Reynolds number of 14950 based on on the step height. A zero velocity component in cross flow direction is assumed at the upper side and zero pressure at the outlet of the channel. The flow region is discretized by a mesh of 1700 bilinear plane elements QUAP4 as indicated in Fig. 6. Starting from a

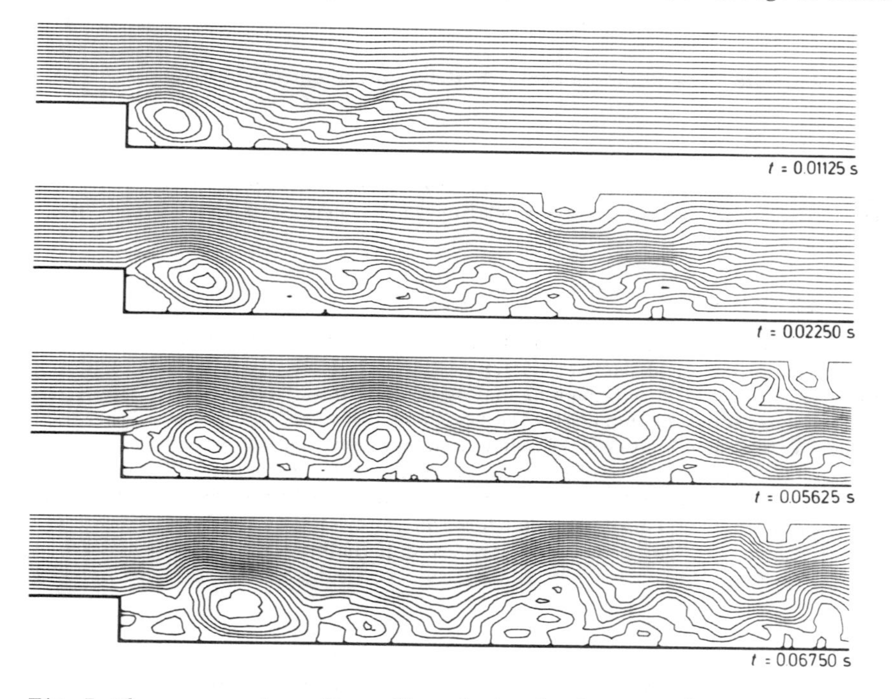

Fig. 7 Flow over a step. Streamlines during development of turbulent flow.

Fig. 8 Visualization of flow over a step by aluminum powder in water [26].

quiescent initial condition the development of the turbulent flow is investigated up to a total duration of t = 67.5 ms using 900 time increments. In Fig. 7 the onset of turbulent flow is shown in the upper two streamline plots while the other plots depict the fully turbulent flow field. The disturbances in the flow field near the outlet may be caused by the somewhat unrealistic pressure boundary condition. Also, the zero cross flow condition at the upper side of the flow region seems to be not well adjusted to the process. Despite all these shortcomings the long-time exposure of the flow over a step (water, visualized by aluminum powder) shown in Fig. 8 and extracted from [26] compares quite well with the streamlines at the instant t = 56.25 ms of the numerical investigation.

4.2. Flow in a Quadratic Duct with a Step

The efficiency of the upwind scheme and its three-dimensional generalization involving the two stage solution algorithm is demonstrated in this example. The flow region, the boundary conditions and the data of the fictitious material are depicted in Fig. 9. At the inlet cross-section a constant flow velocity is assumed, the Reynolds number of 200 being based on the duct dimension H. A zero pressure condition is adopted at the outlet. The flow domain is discretized by 1368 linear volumetric elements HEXE8 as shown in Fig. 10. Calculations were performed in 60 time steps from the initial conditions to a steady state at dimensionless times t = 6. At the final stage, projections of the nodal point velocity vectors onto the xy- and yz-planes are shown in Fig. 11. The following example is concerned with the solution of a coupled fluid/thermal problem.

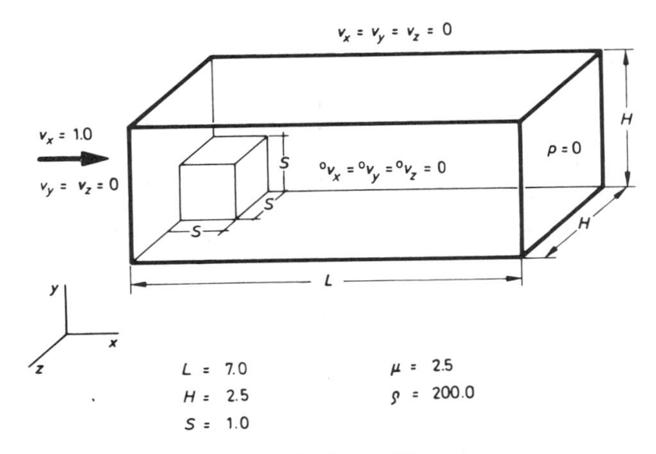

Fig. 9 Flow is a quadratic duct with a step.

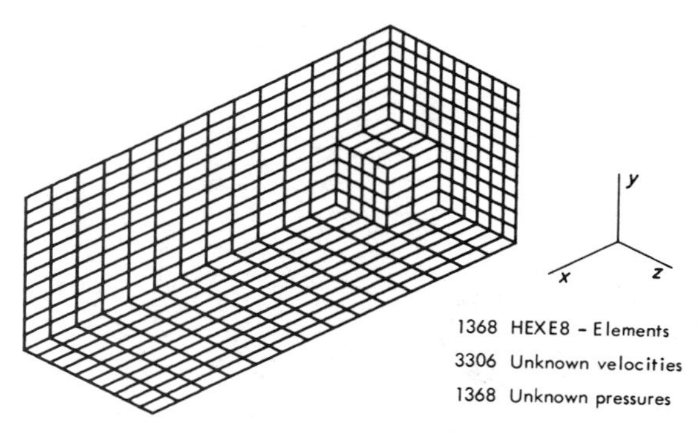

Fig. 10 Flow in a duct. Discretization.

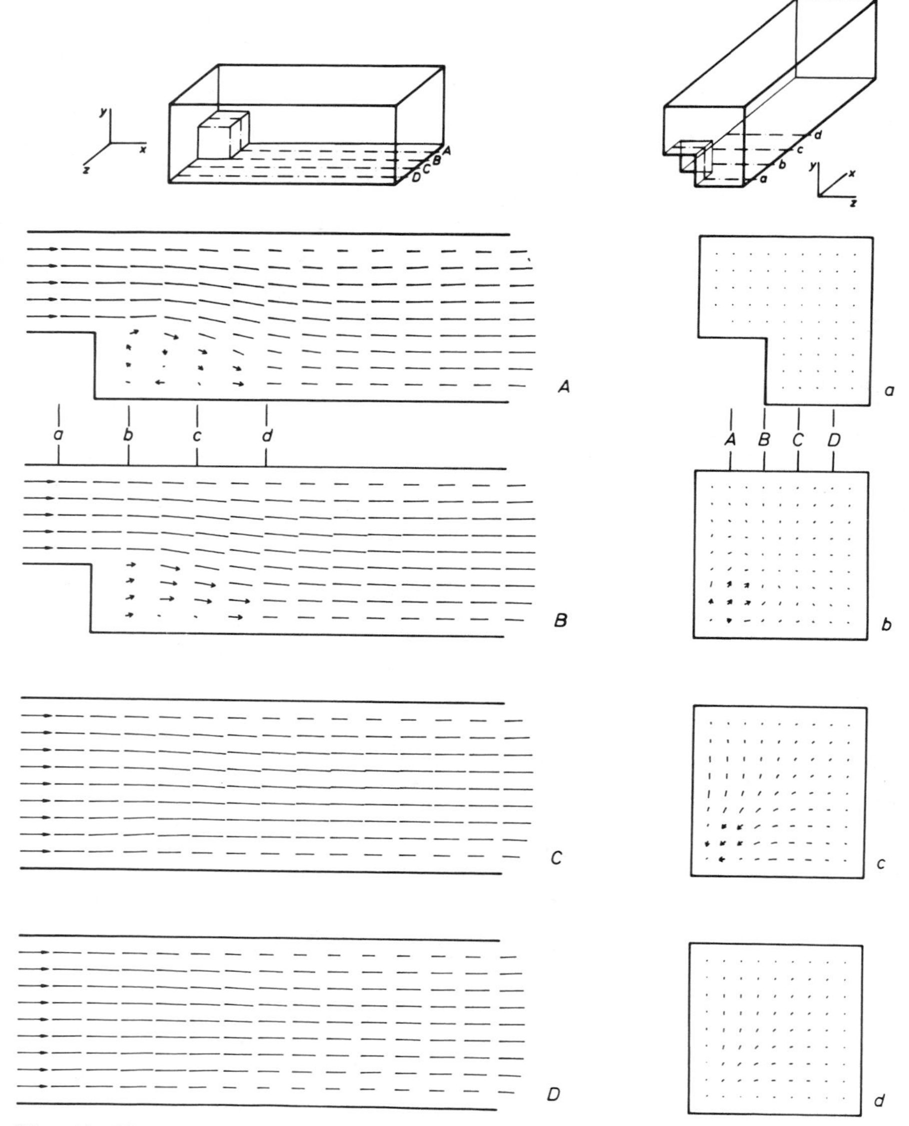

Fig. 11 Flow in a duct. Projections of nodal point velocity vectors at a stationary state.

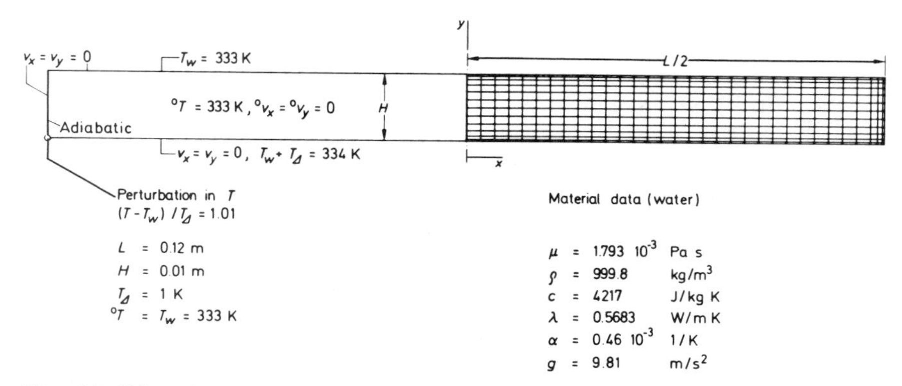

Fig. 12 Bēnard convection in a rectangular box.

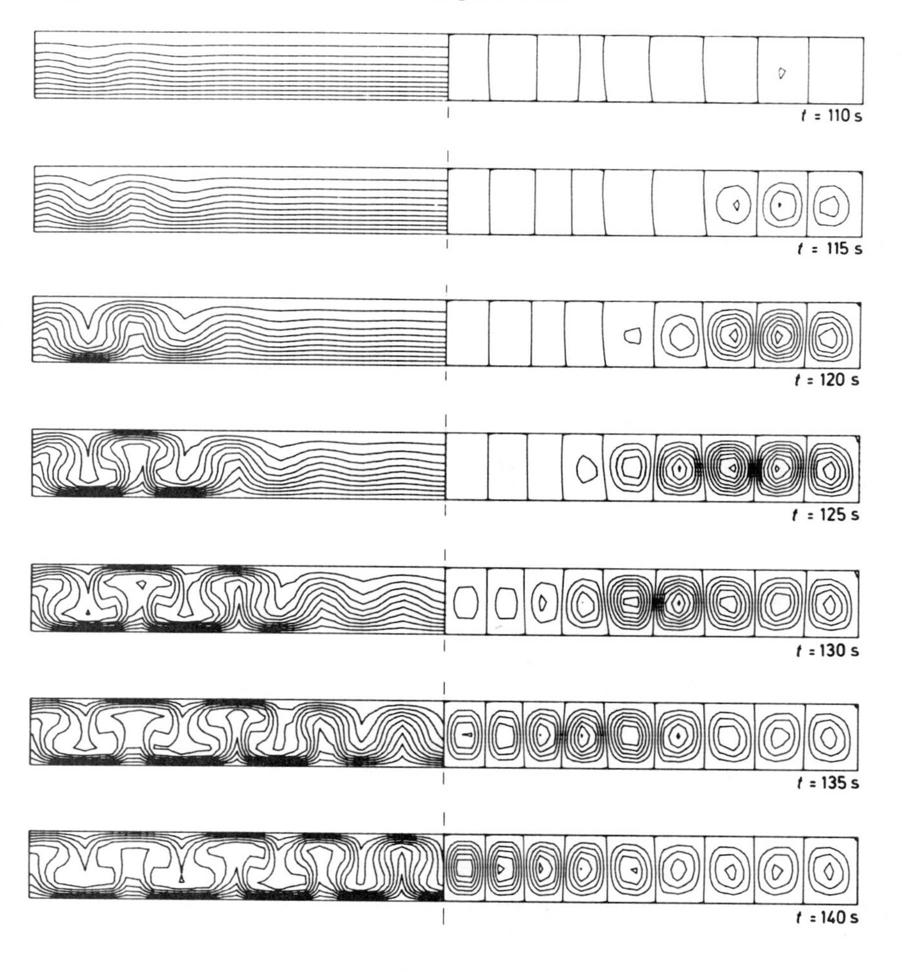

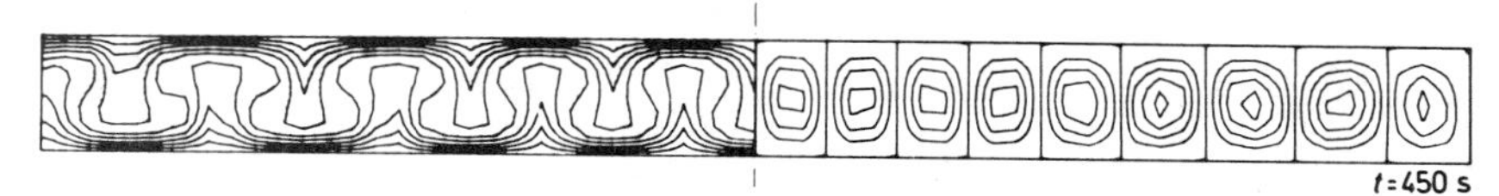

Fig. 13 Bēnard convection. Distribution of temperature (left) and stream function (right) at transient and quasi-stationary conditions.

4.3. Bénard Convection in a Rectangular Box

In a fluid heated from below buoyancy driven convection rolls will develop above a critical value of the Rayleigh number (cf. Fig. 12)

$$\mathrm{Ra} = \frac{g\alpha T_{\Delta} H^3}{(\mu/\rho)(\lambda/\rho c)} \tag{4.1}$$

This process is analyzed for water enclosed in a rectangular box, disregarding any three-dimensional effects. The lower and upper plate of the box are held at a constant temperature, but the vertical side walls are assumed to be subject to an adiabatic state. The fluid is initially set at the same temperature as the upper plate and is then heated from below. As soon as the critical Rayleigh number Ra = 1708 is exceeded convection rolls begin to develop. To avoid the difficulties associated with the bifurcation phenomenon at the critical Rayleigh number a perturbation in temperature is applied which determines the rotational sense of the first vortex. The analysis is continued until stationary conditions are attained.

The mechanical and thermal data of the fluid (water) are quoted in Fig. 12 together with the discretization by QUAP4 plane elements. The Rayleigh number is evaluated to be 20250 which exceeds by far the critical value. This fact facilitates the generation of an unstable process. The calculation extends over t = 450 s and involves 60 time steps, varying between 2.5 s and 20 s. The small increments prove necessary in the initial process of the formation of the convection rolls within the time interval between 100 s and 150 s. The temperature perturbation applied for the initiation of the convective flow is removed after 150 s when all vortices are formed. Figure 13 exhibits isotherms and streamlines at different stages of the process. The development of the convection rolls and the steady state condition is in good agreement with experimental and analytical investigations [27]. The time between the initiation of convection up to the fully developed flow corresponds to the predictions. The series of differential interferograms repro-

Fig. 14 Differential interferogram of transient convection of silicone oil [27].

duced in Fig. 14 shows the formation of convection rolls in silicone oil under similar conditions.

The calculation of the coupled fluid and thermal problems was performed by an iterative sequential solution of the two individual problems (cf. [3]). The thermal equation, dominant in the Bénard convection phenomenon, was solved first followed by the solution of the flow problem. All coupling quantities were taken into account, i.e., the convective terms in the thermal problem and the buoyancy forces in the flow problem, the latter being calculated using the Boussinesq approximation. The iterative solution of the discretized equations leads to linear equation systems with non-symmetric coefficient matrices due to the convection terms. The equation system of the thermal problem was solved using the QR-factorization for the non-symmetric coefficient matrix.

For the flow problem the penalty approach was applied with the convection terms on the right-hand side so that standard solution methods were eligible. Upwinding was used in both problems with an upwind parameter of 0.258 [4]. Convergence below the limit $\epsilon = 10^{-2}$ in the heat rates and velocities respectively was required to terminate the iteration of the individual problems. The sequential solutions were continued until both the velocity and the temperature increments were reduced below the convergence limit of $\epsilon = 10^{-2}$ between consecutive iterations.

REFERENCES

1. Argyris, J. H., et al.: Comput. Meths. Appl. Mech. Engrg. 17/18, 1-106 (1979).
2. Argyris, J. H.; Doltsinis, J. St.; Pimenta, P. M.; Wüstenberg, H.: Comput. Meths. Appl. Mech. Engrg. 32, 3-57 (1982).
3. Argyris, J. H.; Doltsinis, J. St.: Comput. Meths. Appl. Mech. Engrg. 25, 195-253 (1981).
4. Brooks, A. N.; Hughes, T. J. R.: Comput. Meths. Appl. Mech. Engrg. 32, 199-259 (1982).
5. Argyris, J. H.; Doltsinis, J. St.: Comput. Meths. Appl. Mech. Engrg. 20, 213-252 (1979); 21, 91-128 (1980).
6. Argyris, J. H. et al.: Eulerian and Lagrangian techniques for elastic and inelastic deformation processes, in Computational Methods in Nonlinear Mechanics (J. T. Oden, ed.), North-Holland (1980), pp. 13-66.
7. Argyris, J. H.; Doltsinis, J. St.; Wüstenberg, H.: Analysis of thermo-plastic forming processes—natural approach. Computers and Structures 19 (1984).
8. Prager, W.: Introduction to Mechanics of Continua, Ginn and Co. (1961).
9. Argyris, J. H.: Ing. Archiv 34, 33-55 (1965).
10. Argyris, J. H.; Doltsinis, J. St.: A primer on superplasticity in natural formulation, Comput. Meths. Appl. Mech. Engrg. 46, 83-131 (1984).
11. Hohenemser, K.; Prager, W.: ZAMM 12, 216-226 (1932).
12. Argyris, J. H.; Mareczek, G.: Ing. Archiv 43, 92-109 (1974).
13. Malkus, D. S.; Hughes, T. J. R.: Comput. Meths. Appl. Mech. Engrg. 15, 63-81 (1975).
14. Oden, J. T.: RIP-methods for Stokesian flows, in Finite Elements in Fluids, vol. 4 (R. H. Gallagher et al., eds.). Wiley (1982).
15. Taylor, R. L.; Zienkiewicz, O. C.: Mixed finite element solution of fluid flow problems, in Finite Elements in Fluids, vol. 4 (R. H. Gallagher et al., eds.). Wiley (1982).
16. Felippa, C. A.; Park, K. C.: Comput. Meths. Appl. Mech. Engrg. 17/18, 277-313 (1979).
17. Glowinski, R.; Dinh, Q. V.; Periaux, J.: Domain decomposition methods for nonlinear problems in fluid dynamics, Fenomech '81, Comput. Meths. Appl. Mech. Engrg. 41, 27-109 (1983).
18. Hestenes, M. R.; Stiefel, E.: J. Res. Nat. Bur. Stand. 49, 409-436 (1952).
19. Jennings, A.; Malik, G. M.: Int. J. Num. Meths. Engrg. 12, 141-158 (1978).
20. Dennis, J. E.; Moré, J. J.: SIAM Review 19, 46-89 (1977).

21. Matthies, H.; Strang, G.: Int. J. Num. Meths. Engrg. 14, 1613-1626 (1974).
22. Thomasset, F.: Implementation of Finite Element Methods for Navier-Stokes Equations. Springer (1981).
23. Hughes, T. J. R.; Winget, J.; Levit, I.; Tesduyer, T. E.: New alternating direction procedures in finite element analysis based upon EBE approximate factorization, in Recent Developments in Computer Methods for Nonlinear Solid and Structural Mechanics (S. N. Atluri and N. Perrone, eds.). ASME Applied Mechanics Symposium Series (1983).
24. Marchuk, G. J.: Methods of Numerical Mathematics. Springer (1975).
25. Hughes, T. J. R.; Levit, I.; Winget, J.: Comput. Meths. Appl. Mech. Engrg. 36, 241-254 (1983).
26. Tani, I.: Experimental investigation of flow separation over a step, in IUTAM Symposium Freiburg 1957, Grenzschichtforschung/Boundary Layer Research (H. Görtler, ed.). Springer (1958), pp. 377-386.
27. Kirchartz, K. R.; Oertel, H.: ZAMM 62, T211-T213 (1982).
28. Argyris, J.; Doltsinis, J. St.; Pimenta, P. M.; Wüstenberg, H.: Natural finite element techniques for fluid motion, Comput. Meths. Appl. Mech. Engrg. 45, 3-55(1984).

4 DIFFERENTIAL GEOMETRY AND LAGRANGIAN FORMALISM IN THE CALCULUS OF VARIATIONS

Michel Bauderon / Department of Information Sciences, I.U.T.A., Université de Bordeaux I, Bordeaux, France

I. INTRODUCTION

Since the first mathematical discussion of variational problems by Bernouilli (1697) and the first derivation of a necessary condition of extremality by Euler (1744), the calculus of variations has known a very rich and fruitful history. Even in the limited range of one-independent variable calculus, it has been for more than two centuries one of the basic tools of Mechanics, were it classical, relativistic, or quantum mechanics.

Its evolution roughly followed two directions. First, the basic search for solutions of variational problems, led through the works of Lagrange, Legendre, Jacobi, Weierstrass, and many others, to a development along the lines of (partial) differential equations and functional analysis. This way was greatly renewed by the works of Morse, Palais, and Smale. But it was also broadened by the consideration of more than one independent variable through the so-called Plateau problem and the search for minimal submanifolds. A thorough account of this still very vivid research field may be found in the book by Morrey [20]. See also Rassias [25] for more recent results and further references.

On the other hand, through its study of the Hamiltonian formalism in Mechanics, Elie Cartan introduced differential geometry and his exterior calculus in the field of the calculus of variations. That was the beginning of the great renewal of classical mechanics which took place in the last decades. Good references on this point are the books of Herman [14] or Abraham and Marsden [1]. Multiple integrals appeared much later in this differential geometrical setting (see Dedecker [7]), through the growing interest of both mathematicians and physicists in field theories. A differential geometrical treatment appeared to be quite necessary, as most of these problems were of a global nature. Moreover, the very efficient "canonical formalism" of Hamilton was known (since the works of Cartan) to have its most natural expression in this language. Some authors tried to generalize directly the symplectic geometric framework through an axiomatic definition of a new "multisymplectic" structure (Kijowski and Szczyrba [16]). But most tried to give a complete generalization—including Lagrangian theory—of the classical formalism. This gave rise to numerous works, more or less general, more or less involved. One could cite works by Garcia [10], Horndecki [15], Kuperschmidt [17], Ragionari and Ricci [24], Szapiro [27], Takens [28], Trostel [34], Tsujishita [35], Vinogradov [39, 40], and Goldschmidt and Sternberg [12].

A detailed study of symmetries and conservations laws was initiated by Trauman [33] (following Noether) and was pursued and sometimes generalized by Aldaya and de Azcarraga [2], Olver [22], and Takens [29].

A last problem of great interest in many of the recent works is the so-called inverse problem of the calculus of variations. A very detailed exposition of the single integral problem may be found in the book by Santilli [26] who shows the great importance of the concept of variational self adjointness (which appeared in the works of Vainberg [38] followed by Tonti [32]). This problem is to be found in most of the papers referred to earlier. See also Bauderon [3], Dedecker and Tulczyjew [8], Olver [21], Takens [29], and Tulczyjew [36]. It must be noted that this problem is still of a great physical interest (see Santilli [26] and Telega [30, 31] for many examples) and has not been really solved yet, even in the simplest cases.

In what follows, we discuss a very general setting for the Lagrangian formalism in the calculus of variations. We shall deal with problems involving a finite number of independent variables, arbitrarily many dependent functions and partial derivatives of these functions up to any order. This will be done in the formalism of jet bundles introduced by Ehresmann and which has proved since to be very valuable (see, for instance, books by Pommaret [23] or Kumpera and Spencer [18]). Main results of this theory are recalled in Sec. II, where notations are settled. Part III is devoted to the construction of a natural exact sequence (the Euler-Lagrange sequence) which generalizes and unifies most of the classical approaches to variational self-adjointness. As some of the results are straightforward generalizations of the finite-dimensional case, their proofs will be omitted, and the reader will be referred to our previous paper [3]. In part IV, we define a notion of the variational problem and give a discussion of most classical topics (symmetries and the inverse problem).

It must be noted that the list of bibliographical references given at the end of this paper is by no means exhaustive.

II. JET BUNDLES

In this section, we describe a very general geometrical setting which we believe to be highly relevant for the calculus of variations. Most of the results are well known and will be recalled without proofs.

All manifolds, fibered manifolds, bundles, and differentiable maps will be assumed to be smooth (of class C^∞), although many proofs may obviously be carried on with weaker hypotheses.

We shall try—as far as possible—to keep our notations consistent with those widely used (see, for instance, Abraham and Marsden [1], Lang [19], or Bourbaki [5]).

1. Fibered Manifolds

For a manifold M modelled on a real Banach space $\underline{M}$ (also called an $\underline{M}$-manifold), $\mathcal{F}(M)$ and $\mathcal{T}_M$: TM → M will denote respectively the algebra of smooth real valued functions on M and the tangent bundle of M.

Let $M = (M, B, \pi)$ or π: M → B be a <u>fibered manifold</u>, i.e., M and B are manifolds (respectively the <u>total</u> and <u>base space</u>) and π is a surjective submersion. For any open set U of B, $\Gamma_U(\pi)$ is the set of smooth sections of the fibered manifold over U; $\Gamma(\pi)$ is the sheaf of local sections of π. In case $M = (M, B, \pi)$ is a vector bundle, global sections are defined and the set $\Gamma_B(\pi)$ is a vector space.

The vertical tangent bundle of M is the subbundle $T^vM = (T^vM, M, \mathcal{T}_M)$ of the tangent bundle of M defined as the kernel of the bundle map $T\pi$: TM → TB (the tangent map of π). If necessary we shall call it more precisely the π-vertical tangent bundle.

We shall make extensive use of the following vector bundles: the bundle of tangent q-vectors

$$T_qM = (\Lambda^q(TM), M, \mathcal{T}^q_M)$$

and the subbundle of tangent vertical q-vectors

$$T^v_qM = (\Lambda^q(T^vM), M, \mathcal{T}^{qv}_M)$$

the bundle of p-covectors (exterior p-forms)

$$\omega^p(M) = (\Lambda^p(T^*M), M, \mathcal{T}^{*p})$$

and the subbundle of ω^{i+j}

$$\omega^i_j(M) = ((\Lambda^i(T^vM)^*)\wedge(\pi^*\Lambda^j(T^*B)),\ M,\ \mathcal{T}^{*i}_j)$$

consisting of those (i + j)-forms whose inner product with (at least) i + 1 vertical vectors gives zero.

To each of these vector bundles is attached a vector space of smooth sections over M: thus

$$\mathcal{T}_q(M) = \Gamma(T_qM) \quad \text{with } \mathcal{T}_1(M) = \mathscr{X}(M)$$

$$\mathcal{T}^v_q(M) = \Gamma(T^v_qM)$$

$$\Omega^p(M) = \Gamma(\omega^p(M)) \quad \text{and}$$

$$\Omega^i_j(M) = \Gamma(\omega^i_j(M))$$

are the corresponding vector spaces of q-vector fields and differential p-forms. Note that $\Omega^i_j(M)$ is obviously a linear subspace of $\Omega^{i+j}(M)$.

Last, let $\Omega(M) = \bigoplus_p \Omega^p(M)$ be the exterior algebra of M and $(\Omega(M), d)$ the associated cochain complex (graded vector space with a differential of degree 1). Let H be the usual functor sending a cochain complex (C, d) into its vector space of cohomology H(C) = ker d/Im d. Then, H sends $(\Omega(M), d)$ into the de Rham cohomology vector space of M which will be denoted $H_{dR}(M)$.

From now on, we shall assume that the base space B is a finite dimensional smooth manifold with dimension n. We need this in order to have a good definition of integration on B.

2. Symmetries

Let us recall that an automorphism φ of the fibered manifold $\pi: M \to B$ is a collection (φ^M, φ^B) where φ^M and φ^B are respectively diffeomorphisms of M and B, such that $\pi \circ \varphi^M = \varphi^B \circ \pi$.

Such an automorphism will be called a symmetry of the fibered manifold. Let $S(\pi)$ denote the group of symmetries of the fibered manifold. For any section s of π over an open set U of B, set $s' = \varphi^M \circ s \circ (\varphi^B)^{-1}$. This defines an action of $S(\pi)$ on the sheaf of local sections of π.

An infinitesimal symmetry of the fibered manifold is a projectable vectorfield on M, i.e., a vectorfield **X** of M is an infinitesimal symmetry of π if there exist a vectorfield $\mathbf{X}^B$ on B such that for any point m in M, $\mathbf{X}^B(\pi(m)) = T_m\pi(\mathbf{X}(m))$.

Let PM denote the Lie algebra of infinitesimal symmetries of π. Then, the bundle map $T\pi$ induces a linear map $\mathcal{T}\pi$: PM $\to \mathcal{T}B$ defined by $\mathcal{T}\pi(\mathbf{X}) = \mathbf{X}^B$. The kernel of this linear map will be called the space of vertical (projectable) vectorfields on M, vertical infinitesimal symmetries of π.

In a neighborhood of any point m of M, an infinitesimal symmetry **X** defines a one-parameter (semi-)group of symmetries of the fibered manifold $\varphi_t = (\varphi^M_t, \varphi^B_t)$ (with t in a neighborhood of 0 in $\mathbb{R}$). The action of φ_t defines a one parameter family of section of π:

$$s_t = \varphi^M_t \circ s \circ \varphi^B_{-t}$$

which is called a smooth deformation of the section s along the infinitesimal symmetry **X**. It satisfies clearly: $s_0 = s$.

Moreover, we have:

$$\frac{d}{dt} s_t\Big|_{t=0} = \mathbf{X}_0 s - Ts_0 \mathbf{X}^B$$

This relation defines a vectorfield on s(U), which will be called the vertical part of the infinitesimal symmetry X along the section s, and will be denoted $\bar{X}^v$. Note that if X is vertical, we have $\bar{X}^v = X$.

It is easily shown that:

Proposition 1. The linear space P^vM of vertical vectorfields on M is an ideal of the Lie algebra PM, that is, for any vertical vectorfield Y and any infinitesimal symmetry X, [X, Y] and [Y, X] are vertical.

Proof. Let $X \in PM$ and $Y \in P^vM$. Let $(\varphi_t^M, \varphi_t^B)$ be the local flow of X in a neighborhood of some point $m \in M$.

We have:

$$[X, Y](m) = L_X Y(m) = \frac{d}{dt}(T_{\varphi_t^M(m)}\varphi_{-t}^M(Y(\varphi_t^M(m))\Big|_{t=0}$$

But for any t in some neighborhood of 0, we have:

$$\begin{aligned} T_m \pi(T_{\varphi_t^M(m)}\varphi_{-t}^M(Y(\varphi_t^M(m))) &= T_{\varphi_t^M(m)}(\pi \circ \varphi_{-t}^M)(Y(\varphi_t^M(m))) \\ &= T_{\varphi_t^M(m)}(\varphi_{-t}^B \circ \pi)(Y(\varphi_t^M(m))) \\ &= T_{\pi(\varphi_t^M(m))}\varphi_{-t}^B(T_{\varphi_t^M(m)}\pi(Y(\varphi_t^M(m))) \\ &= 0 \text{ as Y is vertical} \end{aligned}$$

Q.E.D.

3. Jets of Sections

In this paragraph, we recall some definitions and a few results about spaces of jets of sections of a fibered manifold $\pi: M \to B$. Proofs of these results and more details about jet bundles (as well as different definitions) may be found—in the finite dimensional case—in Bourbaki [5], Goldschmidt [11], Pommaret [23], and Kumpera and Spencer [18]. These proofs may be extended in a straightforward way to our slightly more general setting (see for instance Gutkin [13]).

Let x be a point in B and s, s' be two germs of local sections of π at x. We say that s and s' define the same k-jet at x if s(x) = s'(x) and the derivatives of s and s' coincide up to the k-th order (the first k terms of their Taylor expansions are the same). We thus have an equivalence relation whose classes are the k-jets of sections of π with source x, denoted by $J_x^k(\pi)$. Let $J^k(\pi) = \bigcup_x J_x^k(\pi)$ be the space of k-jets of sections of π.

Main results are summarized in the following theorem:

Theorem I. (i) For any integer k, the space $J^k(\pi)$ may be endowed with a canonical structure of Banach manifold. Moreover, the map sending a k-jet into the ℓ-jet it defines and the map sending a k-jet into its source give rise to fibered manifolds denoted by π_k^ℓ: $J^k(\pi) \to J^\ell(\pi)$ and π_k: $J^k(\pi) \to B$ for $k \geq \ell \geq 0$; note that $J^0(\pi)$ will be identified with M. If π is a vector bundle, π_k are vector bundles.

(ii) A section s of π over an open set U of B defines a section of $J^k(\pi)$ over U, $j^k s$: $x \to j^k s(x) = j_x^k s(x) = j_x^k s$. The map $s \to j^k s$ induces a sheaf morphism j^k: $\Gamma(\pi) \to \Gamma(J^k(\pi))$, which reduces to a linear map in case π is a vector bundle.

(iii) For any integer r, the bundle π^r_{r+1} is an affine bundle (cf. Pommaret [23] or [3]) whose associated vector bundle is the vector bundle $\pi^*_r S^{r+1}(T * B) \times \pi^0_r * T^v M$ over $J^r(\pi)$. S^r denotes here the r-th symmetric power. Q.E.D.

An immediate consequence of Theorem 1 (iii) is:

Proposition 2. The de Rham cohomology of the jet manifolds is constant: for any integer k, one has $H_{dR}(J^k(\pi)) = H_{dR}(M)$.

Proof. It follows from the fact that the total space of an affine bundle is clearly diffeomorphic with the total space of its associated vector bundle. A vector bundle having the same cohomology than its base manifold, we have for any integer k: $H_{dR}(J^{k+1}(\pi)) = H_{dR}(J^k(\pi))$. The proposition then follows. Q.E.D.

It is clear from the definition of the jet-jet mappings π^ℓ_k that for any $k \geq \ell \geq m \geq 0$ we have: $\pi^m_k = \pi^\ell_k \circ \pi^m_\ell$ and hence that the family $(J^k(\pi), \pi^\ell_k)$ is an inverse system whose inverse limit is called the space of ∞-jets of sections and will be denoted by $(J(\pi), \pi^\ell)$, where π^ℓ is the limit projection π^ℓ: $J(\pi) \to J^\ell(\pi)$. The space $J(\pi)$ may be endowed with the inverse limit topology (cf. Takens [28] and Gutkin [13]) but is not a well-behaved smooth manifold. In the finite dimensional case (for M), Takens [29] has proved that $J(\pi)$ is paracompact and admits smooth partitions of unity. In our more general case this should be true only with restrictive hypotheses on M. We shall not however need this to define a differential calculus on $J(\pi)$.

4. Calculus on $J(\pi)$

The chain rule obviously makes the system $(TJ^k(\pi), T\pi^\ell_k)$ into an inverse system whose projective limit may be called the tangent bundle of $J(\pi)$. Unfortunately, this definition does not provide us with the expected notion of vectorfields on $J(\pi)$ as some kind of inverse limit of vectorfield on the $J^k(\pi)$'s.

The best way to define a differential calculus on $J(\pi)$ is to define first its structure sheaf, i.e., the sheaf of smooth functions on $J(\pi)$. Suitable notions of vectorfields, derivations and differential forms now follow just like in the finite dimensional case. The interested reader may refer to the papers by Boardman [4], Takens [28], and Gutkin [13].

We shall limit ourselves to the definition of the kind of vectorfields we shall need later.

Definition 1. A vectorfield X_k on $J^k(\pi)$ is projectable if for any integer $\ell \geq k$, there exists a vectorfield X_ℓ on $J^\ell(\pi)$ such that:

$$\forall \bar{s}_k \in J^k(\pi) \qquad X_\ell(\pi^\ell_k(\bar{s}_k)) = T\pi^\ell_k(X_k(\bar{s}_k))$$

We let $PJ^k(\pi)$ be the linear subspace of $J^k(\pi)$ consisting of projectable vectorfields on $J^k(\pi)$. These vectorfields are straightforward generalizations of the infinitesimal symmetries of the fibered manifold.

The tangent mapping $T\pi^\ell_k$ now induces a linear mapping $\mathcal{T}\pi^\ell_k$ from $PJ^k(\pi)$ to $PJ^\ell(\pi)$ such that $(PJ^k(\pi), \mathcal{T}\pi^\ell_k)$ is an inverse system whose limit will be, by definition, the linear space of projectable vectorfields on $J(\pi)$ and will be denoted by $PJ(\pi)$. It is clear that such a vectorfield may be considered as a suitable smooth section of $TJ(\pi)$, whatever the definition of such a section might be.

On the other hand, defining differential forms is much easier, as the pull-back mappings π^ℓ_{k*}: $\Omega(J^\ell(\pi)) \to \Omega(J^k(\pi))$ define a direct system of differential graded algebras. Its limit will be the differential graded algebra of differential forms of $J(\pi)$ (one may easily check that the wedge product and exterior derivative commute with direct limits and may so be defined on $\Omega(J(\pi))$; see for instance Tulczyjew [36, 37]). For any integers

$k \geq \ell \geq 0$ we shall systematically identify $\Omega(J^{\ell}(\pi))$ with its direct images in $\Omega(J^k(\pi))$ and $\Omega(J(\pi))$. Note that $(\Omega(J(\pi)), d)$ is a cochain complex of which $(\bigoplus_{p \in N} \Omega^p_n(J(\pi)), d)$ is a subcomplex, which will be denoted by A.

By the same limiting process, we are now able to define the notions of inner product of a differential p-form and a projectable q-vectorfield (belonging to $P_qJ(\pi) = \Lambda^q PJ(\pi) = \varprojlim \Lambda^q P^k J(\pi)$), of Lie bracket of two projectable vectorfields and Lie derivative of a and Lie derivative of a p-form by a projectable vectorfield as the limits of the usual notions on ordinary Banach manifolds. They will be denoted respectively by $i(X)\omega$, $[X, Y]$, and $L(X)\omega$. All the usual identities are satisfied (see Takens [28], Tulczyjew [37], and Gutkin [13]).

Proposition 3. Let ω be a differential p-form on $J(\pi)$. If, for any simple projectable q-vectorfield $X_1 \wedge \cdots \wedge X_q$ one has: $i(X_1 \wedge \cdots \wedge X_q)\omega = 0$ the form ω identically vanishes on $J(\pi)$. Q.E.D.

The proof is straightforward and shows the importance of projectable vectorfields. From now on the adjective "projectable" will be omitted.

5. Cohomology of $J(\pi)$

Definition 2. The de Rham cohomology vector space of $J(\pi)$ is the linear space $H(\Omega(J(\pi)))$. We write $H_{dR}(J(\pi)) = H(\Omega(J(\pi)))$.

The consistency of our definition follows from the commutation of the cohomology functor with direct limits.

Indeed, we have $\Omega(J(\pi)) = \varinjlim \Omega(J^k(\pi))$ and so:

$$H_{dR}(J(\pi)) = H(\varinjlim \Omega(J^k(\pi)) = \varinjlim H(\Omega(J^k(\pi))$$

The right-hand term is exactly the de Rham cohomology space of $J^k(\pi)$, hence:

$$H_{dR}(J(\pi)) = \lim H_{dR}(J^k(\pi))$$

Proposition 4. The spaces $J(\pi)$ and M have the same de Rham cohomology, that is to say: $H_{dR}(J(\pi)) = H_{dR}(M)$.

The proof follows from Proposition 2 and Definition 2. Q.E.D.

6. Prolongations of Symmetries

The action of (infinitesimal) symmetries of the fibered manifold is extended to the bundle of jets through the action of (infinitesimal) symmetries on the sheaf of local sections defined in Sec. II.2.

Let $\bar{s}_k$ be a point in $J^k(\pi)$ and s be a section whose k-jet at $x = \pi^k(x)$ is precisely $\bar{s}_k$. Let now φ be a symmetry of the fibered manifold. The k-jet prolongation of φ is the automorphism φ_k of the k-jet bundle defined by: $\varphi_k(\bar{s}_k) = j^k s'(\varphi^B(x))$.

The k-jet prolongation of the infinitesimal symmetry X is defined in the following way: let s_t be the smooth deformation of the section s along the infinitesimal symmetry X and $x_t = \varphi_t^B(x)$. We then set:

$$X_k(\bar{s}_k) = \frac{d}{dt} j^k s_t(x_t)\Big|_{t=0}$$

Let Δ_k denote the map sending the vectorfield X to the vectorfield X_k. It is easily seen that:

Theorem II. The map Δ_k is a Lie algebras homomorphism from PM to $PJ^k(\pi)$ which sends the ideal P^vM into $P^vJ^k(\pi)$. Q.E.D.

A detailed study may be found in Takens [28] for the finite dimensional case or in Gutkin [13].

The map Δ_k is the k-jet prolongation map. A vectorfield on $J^k(\pi)$ is said to be integrable if it belongs to the image I_k of Δ_k and vertical integrable if it belongs to $I_k \cap P^vJ^k(\pi)$.

It follows immediately from the definition of Δ_k that it satisfies the following relation:

$$\text{for } k \geq \ell \geq 0 \qquad \Delta_\ell(X) = \mathscr{T}\pi_k^\ell(\Delta_k(X))$$

which allows us to define a Lie algebras homomorphism Δ from PM to $PJ(\pi)$ as the inverse limit of the Δ_k's. Again this homomorphism sends P^vM into $P^vJ(\pi)$. The inverse limit of the family (I_k) will be simply denoted I.

III. THE EULER-LAGRANGE EXACT SEQUENCE

1. Some Endomorphisms of $PJ(\pi)$

We define the map D^1 to be the composed map: $D^1 = \Delta \circ \mathscr{T}\pi^0$. It follows from Theorem II that D^1 is a Lie algebra endomorphism of $PJ(\pi)$ which leaves (globally) invariant the ideal $P^vJ(\pi)$ of vertical projectable vectorfield on $J(\pi)$. Clearly, Im D^1 = Im Δ = I.

This map may also be seen as the inverse limit of the family of Lie algebras homomorphism $D_k^1 = \Delta_k \circ \mathscr{T}\pi_{2k}^0$ (see [3]). Moreover, D^1 is a projector of $PJ(\pi)$ (as $D^1 \circ D^1 = D^1$) and sends any vectorfield of $PJ(\pi)$ onto its integrable component.

This projector is naturally extended to projectable p-vectorfields by setting $D^p = \Lambda^p D^1$ (p-th exterior power of the map D^1) and defines a linear endomorphism of $P_pJ(\pi)$ leaving invariant $P_p^vJ(\pi)$. It is also a projector whose image I_p is the vector space of integrable p-vectorfields on $J(\pi)$.

Proposition 5. The linear map D^p commutes with the action of any integrable vectorfield ("commutation with the Lie derivative or bracket").

Proof. Let Y be a projectable p-vectorfield. Through linearity of the map D^p, we can suppose that Y is a decomposable p-vectorfield, i.e., $Y = Y^1 \wedge \cdots \wedge Y^p$. For any k, set $\bar{Y}^k = D^1Y^k$.

Let now $X \in I$ be any integrable vectorfield. The action of X on Y is defined by:

$$L(X)Y = \sum_{i=1}^{p} Y \wedge \cdots \wedge [X, Y^i] \wedge \cdots \wedge Y^p$$

Now,

$$dD^p(L(X)Y) = \sum_{1\leq i\leq p} \bar{Y}^1 \wedge \cdots \wedge D^1([X, Y^i]) \wedge \cdots \wedge \bar{Y}^p$$

and the proof follows from the definition of D^p and from $D^1X = X$. Q.E.D.

2. Adjoint Operators

We shall now make use of the fact that the base space B is a finite dimensional manifold in order to define the adjoints of the operators D^p, through the canonical pairing between $P_pJ(\pi)$ and $\Omega_n^p(J(\pi))$ given by integration of the inner product over any suitable domain of integration on B (for instance, any compact oriented submanifold with boundary). Such a domain N will be called an n-chain in B without more precision.

Theorem III. Let $\pi: M \to B$ be a fibered manifold, with B a finite dimensional manifold with dimension n. Then, for every integer $p \geq 1$, there exists a unique linear mapping (the adjoint of D^p)

$$D^{p*}: \Omega^p_n(J(\pi)) \to \Omega^p_n(J(\pi))$$

such that for any (n + p)-form ω, for any n-chain N of B, for any section s defined on an open neighborhood of N, for any vertical p-vectorfield X on $J(\pi)$ vanishing on a neighborhood of $js(\partial N)$, one has:

$$\int_N js^*(i(D^p(X))\,\omega) = \int_N js^*(i(X)D^{p*}\omega)$$

Proof. A detailed proof of the finite dimensional case was given in a previous paper [3]. It is done on $J^k(\pi)$ for every $k \geq 0$ and needs three lemmata. The reader will easily check that the proof of the third does not really use the finite dimensionality of the fiber and find the Banach analogs of the first two.

Let us just recall that it is in fact proven that for any $p \geq 1$ there exists a bilinear mapping G^p from $P_pJ(\pi) \times \Omega^p_n(J(\pi))$ into $\Omega^0_{n-1}(J(\pi))$ such that, with the notations of Theorem III:

$$js^*(i(D^p(X)\,\omega) - js^*(i(X)D^{p*}\omega) = js^*\,dG^p(X, \omega)$$

The mapping G^p will be called the Green function of operator D^p. We have thus defined a graded linear endomorphism D^* of $\bigoplus_{p \in \mathbb{N}^*} \Omega^p_n(J(\pi))$, and a family of Green functions G^p.

3. Euler-Lagrange Sequence

To each of the operators D^{p*} $(p \geq 1)$ is naturally attached an image-kernel exact sequence of vector spaces. Let $G^p_nJ(\pi)$ and Λ^p denote respectively the kernel and the image of D^{p*}. The kernels are clearly characterized by the following proposition:

Proposition 6. With the assumptions and notations of Theorem III, a (n + p)-form ω of $\Omega^p_n(J(\pi))$ belongs to the linear space $G^pJ(\pi)$ if an only if, for any simple vertical integrable p-vectorfield $\bar{X}^1 \wedge \cdots \wedge \bar{X}^p$, one has:

$$\int_N js^*(i(\bar{X}^1 \wedge \cdots \wedge \bar{X}^p)\,\omega) = 0$$

Proof. Evident from the definitions. Q.E.D.

Moreover, we can see easily that the linear map D^{1*} is in fact a projector of $\Omega_n(J(\pi))$ so that we may write:

$$\Omega_n(J(\pi)) = G_nJ(\pi) \oplus \Lambda^1$$

The image Λ^1 of D^{1*} is exactly the vector space of π^0-semi-basic forms on $J(\pi)$. Note that this is not true on $J^k(\pi)$ at every order k, as the map D^{p*}_k is defined from $\Omega^p_n(J^k(\pi))$

to $\Omega^p(J^{2k}(\pi))$ (see [3] for details).

We shall now extend our construction for $p = 0$.

Definition 4. Let $G_n^0J(\pi)$ be the linear subspace of $\Omega_n^0(J(\pi))$ consisting of those forms ω such that, for any n-chain N of B and for any pair (s_1, s_2) of sections of π equals on a neighborhood of N and smoothly homotopic, one has:

$$\int_N js_1^*\omega = \int_N js_2^*\omega$$

Let now Λ^0 denote the quotient of $\Omega_n^0(J(\pi))$ by $G_n^0J(\pi)$ and D^0* the canonical projection. We have thus extended D* to a graded linear mapping (which will still be denoted D*) mapping $A(J(\pi))$ onto $\Lambda = \bigoplus_{p \in N} \Lambda^p$ with kernel $G = \bigoplus_{p \in N} G_n^pJ(\pi)$ (with $A(J(\pi)) = \bigoplus_{p \in N} \Omega_n^p(J(\pi))$).

The subspaces $G_n^pJ(\pi)$ were previously introduced by Takens [28] whose definitions we followed, but did not appear in an exact sequence:

Definition 5. The exact sequence of graded linear space

$$0 \to G \to A(J(\pi)) \xrightarrow{D^*} \Lambda \to 0$$

is the Euler-Lagrange sequence of the fibered manifold π: M $\to$ B.

Let us recall that the cochain complex $(A(J(\pi)), d)$ was simply denoted by A.

Theorem IV. The Euler-Lagrange sequence is a split exact sequence of graded differential vector spaces (cochain complexes).

Proof. We need the following:

Lemma. For any integer $p \geq 0$, the exterior derivative d maps $G_n^pJ(\pi)$ into $G_n^{p+1}J(\pi)$, i.e., $G = (G_n^pJ(\pi), d)$ is a subcomplex of A.

Proof of this lemma was given in the finite dimensional case by Takens (Theorem 3.6.d in [28]) and extended to the Banach case by Gutkin [13].

This allows us to define for any integer $p \geq 0$ a unique linear mapping $\delta^p: \Lambda^p \to \Lambda^{p+1}$ such that:

(i) $\delta^p \circ D^{p*} = D^{p+1*} \circ d$ and

(ii) $\delta^{p+1} \circ \delta^p = 0$

by letting $\delta^p(\omega) = D^{p+1*} \circ d(\omega_1)$ where ω belongs to Λ^p and ω_1 is any antecedent of ω by D^{p*} (in $G_n^pJ(\pi)$). The lemma shows that δ^p is well defined and makes (Λ^p, δ^p) into a graded differential vector space still denoted by Λ. Relation (ii) now shows that the Euler-Lagrange sequence is an exact sequence of cochain complexes which is split as a sequence of vector spaces. Q.E.D.

This theorem may be viewed as showing the existence of a commutative diagram with exact rows and "differential columns":

$$\begin{array}{ccccccccc}
 & & & & & & 0 & & \\
 & & & & & & \downarrow & & \\
0 & \to & G_n^0 J(\pi) & \to & \Omega_n^0 J(\pi) & \xrightarrow{D^{0*}} & \Lambda^0 & \to & 0 \\
 & & \downarrow & & \downarrow & & \downarrow & & \\
0 & \to & G_n^1 J(\pi) & \to & \Omega_n^1 J(\pi) & \xrightarrow{D^{1*}} & \Lambda^1 & \to & 0 \\
 & & \downarrow & & \downarrow & & \downarrow & & \\
 & & \vdots & & \vdots & & \vdots & & \\
 & & \downarrow & & \downarrow & & \downarrow & & \\
0 & \to & G_n^p J(\pi) & \to & \Omega_n^p J(\pi) & \xrightarrow{D^{p*}} & \Lambda^p & \to & 0 \\
 & & \downarrow & & \downarrow & & \downarrow & & \\
 & & \vdots & & \vdots & & \vdots & &
\end{array}$$

in which the mapping $\delta^0 : \Lambda^0 \to \Lambda^1$ is injective as follows from a remark by Takens (Remark 4.3 in [28]).

Theorem V. The Euler-Lagrange sequence is an exact sequence of graded differential PM-modules (or I-modules).

Proof. It follows from the fact that the exterior derivative commutes with the Lie de v-ative that G is a graded differential submodule of the PM-module A (as the action of PM on G and A is defined by the Lie derivative). We must only prove that the map D* commutes with the Lie derivative. This is obvious for D^{0*}. Now, for $p \geq 1$, the map D^{p*} is defined by: $\int_{js(N)} i(D^p(X))\omega = \int_{js(N)} i(X)D^{p*}\omega$, where we may limit X to be of the form $X = X^1 \wedge \cdots \wedge X^p$.

The action of $Y \in PJ(\pi)$ on the first integrand is defined by:

$$\begin{aligned}
L(\bar{Y})(i(D^p(X))\omega) &= i(L(\bar{Y})(D^p(X))).L(\bar{Y})\omega \qquad \text{with } \bar{Y} = D^1(Y) \\
&= i(D^p(L(\bar{Y})X)).L(\bar{Y})\omega \qquad \text{(Proposition 5)} \\
&= i(L(\bar{Y})X).D^{p*}L(\bar{Y})\omega
\end{aligned}$$

But, for the second integrand:

$$L(\bar{Y})(i(X)D^{p*}\omega) = i(L(\bar{Y})X).L(\bar{Y})D^{p*}\omega$$

and the proof follows from the unicity of the adjoint operator. Q.E.D.

4. Cohomology

As an immediate consequence of Theorem IV, the Euler-Lagrange sequence being split, we have the following:

Theorem VI. The Euler-Lagrange exact sequence gives rise in cohomology to an exact sequence of graded vector spaces:

$$0 \to H(G) \to H(A) \to H(\Lambda) \to 0$$

Q.E.D.

More precisely, we have for every integer $p \geq 0$, an exact sequence

$$0 \to H^p(G) \to H^p(A) \to H^p(\Lambda) \to 0$$

We have noticed in Sec. 4 that Λ is a subcomplex of the cochain complex $(\Omega(J(\pi)), d)$ defining the de Rham cohomology of $J(\pi)$. This shows that $H(\Lambda)$ is a graded linear subspace of $H_{dR}(J(\pi)) = H_{dR}(M)$, i.e., for any integer $p \geq 0$, $H^p(\Lambda)$ is a linear subspace of $H_{dR}^{n+p}(M)$.

Corollary 1 (Poincaré lemma). If the total space M is contractible, the complex (Λ, δ) is acyclic in all dimensions $(H(\Lambda) = 0)$.

Corollary 2. If $H_{dR}^{n+1}(M) = 0$ then $H^1(\Lambda) = 0$.

In fact these two corollaries give us only sufficient conditions and not the best ones. Indeed, the cohomology of Λ may be imbedded in the "vertical cohomology" of the jet bundles which is itself a subspace of the de Rham cohomology of $J(\pi)$. This is done in most recent papers dealing with the inverse problem (see Olver [22], Takens [29], Tsujishita [35], Dedecker and Tulczyjew [8], Vinogradov [40], Kuperschmidt [17]), all these authors using more or less a special case of the general theory on p.d.e. (see Pommaret [23]). However, it is to be noticed that they do not obtain any significantly stronger result, although they have to use much more sophisticated tools. Moreover, Takens is the only one to prove a global result (our Corollary 2).

IV. CALCULUS OF VARIATIONS

1. A General Framework

As is shown in many textbooks, classical mechanics may be easily described (from the Lagrangian viewpoint) using the theory of bundles (see Abraham and Marsden [1] or Hermann [14]). The base space is then the real line $\mathbb{R}$, the total space is $TQ \times \mathbb{R}$, where TQ is the velocity phase space of the system under study, the bundle map being the second projection. A Lagrangian is now a real-valued map depending on time, coordinates in Q and TQ. Therefore, it may be viewed as a functional on $J^1(Q \times \mathbb{R})$ which is precisely $TQ \times \mathbb{R}$.

Mechanics of continua may now be described by allowing the fiber Q to be an infinite dimensional (Banach) manifold (see Chernoff and Marsden [6]).

Most field theories used in presentday physics are Lagrangian field theories. The real reason—intrinsic physical necessity, or well known efficiency of classical mechanics and optics—is a point of deep interest (see Santilli [26]), and active research (see Hadronic Journal). Anyway this is enough to make important a thorough analysis of Lagrangian field theories in their higher generality, to develop corresponding Hamiltonian or multisymplectic theories and to study symmetries of such theories. Then, it must be found when a given physical system may be described by a Lagrangian field theory—this is the inverse problem.

2. Lagrangian Theory

A reasonable generality will be attained in the following setting.

Let $\pi: M \to B$ be a smooth fibered manifold, with finite dimensional base space.

Definition 6. A Lagrangian on M is an horizontal n-form $\mathcal{L}$ in $\Omega_n^0(J(\pi))$ (with n = dim B). The action integral defined for any open set U in B on $\Gamma_U(\pi)$ is given by $I(N, s) = \int_N js^* \mathcal{L}$, for any n-chain N such that $\bar{N} \subset U$.

One must note that generally, the set $\Gamma_U(\pi)$ has no algebraic structure and does not have any well-behaved differential structure. Even in the case of a finite dimensional vector bundle, $\Gamma_B(\pi)$ is just a Fréchet space (see Dieudonné [9]). So, in order to really solve such a problem, one will have to restrict the study to smaller spaces of sections. However, this is absolutely not necessary for all that follows.

Definition 7. A smooth deformation with fixed boundary of the section s defined on a neighborhood U of the n-chain N is a one-parameter family $t \to s_t$ of sections such that $s_0 = s$ and $s_t = s$ on some neighborhood of ∂N.

Any such deformation of s defines a vertical vectorfield X on M by $X(s(x)) = (d/dt)\, s_t(x)|_{t=0}$ for x in U, such that X = 0 on a neighborhood on ∂N. Reciprocally, any vertical vectorfield X vanishing on a neighborhood of ∂N, defines clearly a smooth deformation of s with fixed boundary.

It is now clear that a necessary condition for a section s to be an extremum of the action integral, is that for any smooth deformation with fixed boundary s_t of the section s, one has:

$$\frac{d}{dt} I(N, s_t)\Big|_{t=0} = 0$$

Proposition 7. A necessary condition for a section s defined on a neighborhood of $\bar{N}$ to be an extremum of the variational problem with Lagrangian $\mathcal{L}$ is that, for any integrable vertical vectorfield $\bar{X}$ vanishing on a neighborhood of $js(\partial N)$,

$$\int_N js^*(i(\bar{X})\, d\mathcal{L}) = 0$$

Proof. This follows immediately from:

$$\frac{d}{dt} I(N, s_t)\Big|_{t=0} = \frac{d}{dt}\int_N js_t^* \mathcal{L}\Big|_{t=0} = \int_N js^*(i(\bar{X}(d\mathcal{L}) \quad \text{with } \bar{X} = \Delta X$$

X being the vectorfield associated with the deformation of s.

Definition 8. The Euler-Lagrange form of the problem with Lagrangian $\mathcal{L}$ is the (n + 1)-form $\mathcal{E} = D^{1*} \circ d(\mathcal{L}) = \delta^0 \circ D^{0*}(\mathcal{L})$.

It now follows from the Banach analog of the classical Du Bois-Reymond lemma:

Proposition 8. A necessary condition for a section s defined on a neighborhood of $\bar{N}$ to be a relative extremum is that $js^*\mathcal{E}$ vanishes on N.

From the full expression of $\mathcal{E}$ in local coordinates it is easy to see that Proposition 8 is the straightforward generalization of the classical Euler-Lagrange equations (see [3]).

All we have said before may be extended to the case of weaker hypotheses upon the differentiability of the sections. In fact, if s is only of class C^{2r} (and this might be true also for the fibered manifold) one may define a variational problem of class C^r with a Lagrangian belonging to $\Omega_n^0(J^r(\pi))$. Then the Euler-Lagrange form $\mathcal{E}$ will belong to Λ^1 in $\Omega_n^1(J^{2r}(\pi))$.

3. Symmetry Transformations

We shall follow the definitions given by Trautman in [33]. Let $\pi: M \to B$ be a fibered manifold. We shall say that a symmetry φ of π is a symmetry transformation of the variational problem with Lagrangian $\mathcal{L}$ if it carries solutions of the Euler-Lagrange equations into solutions of this same equation. That is to say:

Definition 9. A symmetry of π is a symmetry transformation if, for any section s defined on a neighborhood of the n-chain N and such that $js^*\mathcal{E}$ vanishes on N, $js'^*\mathcal{E}$ vanishes on $\varphi^B(N)$ (with $s' = \varphi^M \circ s \circ (\varphi^B)^{-1}$).

It follows from Definition 9 that if φ_t is a one-parameter family of symmetry transformations of the variational problem given by $\mathcal{L}$, leaving invariant the Euler-Lagrange form of the problem, its infinitesimal generator X satisfies $L(\bar{X})\mathcal{E} = 0$.

From Theorem V, we can write:

$$L(\bar{X})\mathscr{L} = L(\bar{X})D^{1*}(d\mathscr{L}) = D^{1*}(dL(\bar{X})\mathscr{L}) = \delta^0 \circ D^{*0}(L(\bar{X})\mathscr{L})$$

that is, $L(\bar{X})\mathscr{L}$ belongs to the space $G_n^0 J(\pi)$.

Definition 9'. An infinitesimal symmetry transformation (i.s.t.) of the variational problem is an infinitesimal symmetry X such that: $L(\bar{X})\mathscr{L}$ belongs to $G_n^0 J(\pi)$ (cf. Takens [28]).

It is easily seen that the set of i.s.t. of a given Lagrangian is a Lie subalgebra $T(\mathscr{L})$ of PM.

A special case of symmetry transformations is given by:

Definition 10. A symmetry φ of π is an invariant transformation if it leaves invariant the action integral, i.e., for any n-chain and any section defined on a neighborhood of N,

$$\int_N js^*\mathscr{L} = \int_{\varphi^B(N)} js'^*\mathscr{L}$$

This definition is clearly equivalent to: for any section s, one has $\varphi^{B*}(js'^*\mathscr{L}) = js^*\mathscr{L}$.

It may be extended to:

Definition 11. A symmetry φ of π is a generalized invariant transformation if there exist a (n - 1) form α on the base space B, such that: for any section s, one has: $\varphi^{B*}(js'^*\mathscr{L}) + d\alpha = js^*\mathscr{L}$.

We consider now a one-parameter family φ_t of generalized invariant transformations of a given variational problem. Let α_t denote the one-parameter family of (n - 1) forms given with φ_t. For any section s of π we have the relation:

$$\varphi_t^{B*}(js_t^*\mathscr{L}) + d\alpha_t = js^*\mathscr{L} \qquad \text{where} \quad s_t = \varphi_t^M \circ s \circ \varphi_{-t}^B$$

Differentiating this relation with respect to t and setting t = 0, we obtain:

$$d(i(X^B)js^*\mathscr{L}) + js^*(i(\bar{X}^V)d\mathscr{L}) + d\mathring{\alpha} = 0$$

where $\dot{\alpha} = \frac{d}{dt}\alpha_t\Big|_{t=0}$; the "vertical" part X^V of the infinitesimal generator (X, X^B) of φ_t being given by $X^V = \frac{d}{dt}s_t\Big|_{t=0} = X \circ s - Ts \circ X^B$, we set $X^V = \Delta X^V$.

Definition 12. An infinitesimal symmetry (X, X^B) is an infinitesimal generalized invariant transformation (i.g.i.t.) of the variational problem if, for any section s, one has, for some (n - 1) form α on B:

$$d(i(X^B)js^*\mathscr{L}) + js^*(i(\bar{X}^V)d\mathscr{L}) + d\alpha = 0$$

Proposition 9. The set $I(\mathscr{L})$ of i.g.i.t. of the variational problem defined by the Lagrangian $\mathscr{L}$ is a Lie subalgebra of the Lie algebra PM of symmetries of the fibered manifold.

Proof. Let X and Y be two i.g.i.t. We have:

$$i([\bar{X}^V, \bar{Y}^V])d\mathscr{L} = L(\bar{X}^V)i(\bar{Y}^V)d\mathscr{L} - i(\bar{Y}^V)L(\bar{X}^V)d\mathscr{L}$$

Now, as $js^*i(\bar{X}^V)d\mathscr{L} = d(-\alpha - i(X^B)js^*\mathscr{L})$, we have $js^*(L(\bar{X}^V)d\mathscr{L}) = 0$. Hence,

$$js^*(i([\bar{X}^V, \bar{Y}^V]) d\mathcal{L}) = js^*(L(\bar{X}^V)i(\bar{Y}^V) d\mathcal{L})$$

$$= js^*(di(\bar{X}^V)i(\bar{Y}^V) d\mathcal{L}) \quad \text{(same reason)}$$

which shows that the right-hand side form is closed. The proof follows immediately.

Q.E.D.

Let now (X, X^B) be an i.g.i.t. From Theorem III, the formula in Definition 12 may be rephrased as:

$$d(i(X^B)js^*\mathcal{L}) + js^*(i(\bar{X}^V)\mathcal{E}) + js^*\, dG^1(\bar{X}^V, d\mathcal{L}) + d\alpha = 0$$

or:

$$d(i(X^B)js^*\mathcal{L} + js^*G^1(\bar{X}^V, d\mathcal{L}) + \alpha) + js^*(i(\bar{X}^V)\mathcal{E}) = 0$$

which proves the following:

Theorem VII (Noether's theorem). Let φ_t be a one parameter group of generalized invariant transformations of the problem, with infinitesimal generator (X, X^B). Let N be an n-chain of B. For any extremal s of the variational problem defined on a neighborhood of N, the (n - 1) form on B

$$C(N, s, X) = js^*(G^1(\bar{X}^V, d\mathcal{L})) + i(X^B)js^*\mathcal{L} + \dot{\alpha}$$

is a closed form on N.

Still following Trauman, we can say that the theorem of Noether provides us with a conservation law associated with any infinitesimal generalized invariant transformation (or with any one-parameter group of generalized invariant transformation).

4. The Inverse Problem

In Sec. IV.2 we have shown how to associate with any Lagrangian defined on a fibered manifold $\pi: M \to B$ its so-called Euler-Lagrange form which provides us with a necessary condition of extremality. From the general study (Sec. III.3), it follows that this form belongs to the linear subspace Λ^1 and depends only on the equivalence class of the Lagrangian in the space Λ^0.

Now, most actual physical problems do not provide the physicist with a Lagrangian as a basic datum of the problem. It is the so-called "evolution equation" which appears as the most natural, and physicists are quite happy when this equation appears to be the Euler-Lagrange equation associated with some Lagrangian.

Definition 13. An element of Λ^1 is called an evolution equation or a source equation (Takens [28]), or an equation of motion (Santilli [26]). Such a form is (locally) variational or potential if it belongs (locally) to the image of δ^0.

Definition 14. An element of Λ^1 is self-adjoint if it belongs to the kernel of δ^1.

Looking for conditions of potentialness (conditions for an evolution equation to derive from a Lagrangian) is the inverse problem of the calculus of variations. We have the following results:

Theorem VIII. (i) An evolution equation is locally potential if and only if it is self-adjoint (locally: on any simply connected open set of M).

(ii) If $H^1(\Lambda)$ vanishes then any locally potential evolution equation is globally potential. A sufficient condition for this is that $H^{n+1}_{dR}(M)$ vanishes.

This theorem is a mere traduction of the corollaries to Theorem VI. As we already pointed out, many results of the type i) were proved earlier but under stronger hypotheses or much more involved technicalities (see, for instance, Horndecki [15], Takens [28], or Tulczyjew [37]). In [3] we have shown that our condition of self-adjunction is a generalization of the one introduced by Vainberg in a slightly different setting [38]. The global result was proved independently by Takens [29]. It shows, for instance, that if M is a fibered manifold with contractible fiber (this is the case of a vector or affine bundle), any locally variational problem is globally defined, the potentialness of a source equation being then a pure local matter. But this is true for more complicated bundles, provided that $H^{n+1}_{dR}(M)$ vanishes (Hopf fibrations with fiber S^3 or S^7 are well-known examples).

5. Conclusion

Further developments of the differential geometrical theory of the calculus of variations may be obtained along a few lines, two of which we believe to be of special importance.

First, a natural question is the following: How can this formalism be extended to the Hamiltonian theory and to a generalized symplectic geometry? A few tentative answers have been given in many recent works (mainly motivated by study in field theories) among which one could cite: Dedecker [7], Garcia [10], Goldschmidt and Sternberg [12], Kuperschmidt [17], Ragionieri and Ricci [24], or Szapiro [27]. However, we believe that none of these appears as yet to be entirely satisfactory.

A second question is: Under which conditions an evolution equation possessing symmetries or conservation laws is the Euler-Lagrange equation of some variational problem? Some results have been given by Takens [28], but no general answer has been given to this special form of the inverse problem. We think that some interesting results may be expressed in terms of the cohomology groups of a Lie algebra of infinitesimal symmetries with coefficients in the Euler-Lagrange exact sequence.

REFERENCES

1. Abraham, R.; Marsden, J. E.: Foundations of Mechanics, 2nd edition. Benjamin (1980).
2. Aldaya, V.; De Azcarraga, J. A.: J. Math. Phys. 19, 1876-1880, 1869-1875 (1978).
3. Bauderon, M.: Ann. Inst. Henri Poincaré XXXVI, No. 2, 159-179 (1982).
4. Boardman, J.: Pub. Math. IHES 33, 21-57 (1967).
5. Bourbaki, N.: Variétés différentiables et analytiques. Fascicules de résultats. Hermann, Paris.
6. Chernoff, P.; Marsden, J. E.: Properties of Infinite Dimensional Hamiltonian Systems. Springer (1974).
7. Dedecker, P.: On the Generalization of Symplectic Geometry to Multiple Integrals in the Calculus of Variations. Springer (1977).
8. Dedecker, P.; Tulczyjew, W.: Spectral Sequences and the Inverse Problem of the Calculus of Variations. Springer (1980).
9. Dieudonne, J.: Elements d'analyse, tomes 1-4. Gauthier-Villars.
10. Garcia, P. L.: The Poincaré-Cartan invariant in the calculus of variation, in Symp. Math., vol. 14. Istituto Nazion. di Alta Mate, Roma, pp. 219-246.
11. Goldschmidt, H.: J. Diff. Geom. I, 269-307 (1967).
12. Goldschmidt, H.; Sternberg, S.: Ann. Inst. Fourier 23, No. 1, 203-267 (1973).
13. Gutkin, D.: Espace des jets et calcul des variations en dimension infinie, Publ. UER Math. Pures Appl. IRMA 3, No. 2, Exp. No. 4.
14. Hermann, R.: Differential Geometry and the Calculus of Variations. Math. Sci. Press, vol. XVII.
15. Horndecki, G. W.: Differential operators associated with the Euler-Lagrange operator. Tensor N.S. 28 (1974).

16. Kijowski, J.; Szczyrba, W.: Multisymplectic manifolds and the geometrical construction of the Poisson brackets in the classical field theory. Colloque d'Aix, C.N.R.S. (1974).
17. Kuperschmidt, B.: Geometry of Jet Bundles and the Structure of Lagrangian and Hamiltonian Formalism. Springer (1980).
18. Kumpera, A. K.; Spencer, D. C.: Lie Equations. Ann. Math. Studies No. 73 (1972).
19. Lang, S.: Differentiable Manifolds. Addison-Wesley (1972).
20. Morrey, C. B.: Multiple Integrals in the Calculus of Variations. Springer (1966).
21. Olver, P. J.: Math. Proc. Camb. Phil. Soc. 88, 71-88 (1980).
22. Olver, P. J.: Math. Proc. Camb. Phil. Soc. 85, 143-160 (1979).
23. Pommaret, J. F.: Systems of Partial Differential Equations and Lie Pseudogroups. Gordon and Breach (1978).
24. Ragionieri, R.; Ricci, R.: Hamiltonian formalism in the calculus of variations, Bollettino U.M.I. 18-B (5), 119-130 (1981).
25. Rassias, T. M.: Morse theory and Plateau's problem, in Selected Studies: Physics-Astrophysics, Mathematics, History of Science, North-Holland (1982), pp. 261-292.
26. Santilli, R. M.: Foundations of Theoretical Mechanics, I, The Inverse Problem in Newtonian Mechanics. Springer (1978).
27. Szapiro, T.: Geodesic Fields in the Calculus of Variations of Multiple Integrals Depending on Derivatives of Higher Order. Springer (1980).
28. Takens, F.: Symmetries, Conservation Laws and Variational Principles. Springer (1977).
29. Takens, F.: J. Diff. Geom. 14, 543-562 (1979).
30. Telega, J. J.: J. Inst. Math. Appl. 24, 175-195 (1979).
31. Telega, J. J.: Int. J. Eng. Sci. 20, No. 8, 913-933, 935-946.
32. Tonti, E.: Bull. Sci. Acad. Roy. Belg. 55, 137-165 (1969).
33. Trautman, A.: Commun. Math. Phys. 6, 248-261 (1967).
34. Trostel, R.: Hadronic Journal 5, 1023-1119 (1982).
35. Tsujishita, T.: Osaka J. Math. 19, 311-363 (1982).
36. Tulczyjew, W.: Bull. Soc. Math. France 105, 419-431 (1977).
37. Tulczyjew, W.: The Euler-Lagrange Resolution. Springer (1980).
38. Vainberg, M.: Variational Methods for the Study of Nonlinear Operators. Holden-Day (1964).
39. Vinogradov, A. M.: Sov. Math. Dokl. 18, 1200-1204 (1977).
40. Vinogradov, A. M.: Sov. Math. Dokl. 19, 146-148 (1977).

5 THE GEOMETRY OF BICHARACTERISTICS AND STABILITY OF SOLVABILITY

John K. Beem and Philip E. Parker* / Department of Mathematics, University of Missouri—Columbia, Columbia, Missouri

1. INTRODUCTION

In a previous paper [2] we studied the geometric consequences of two conditions on Lorentzian manifolds which together imply the solvability of the Klein-Gordon equation in the sense of distributions. In this paper and a successor [3] we shall consider the stability of the solvability of pseudodifferential equations of real principal type which satisfy the pseudoconvexity condition. Although our original motivation was the stability of solvability for the Klein-Gordon equation, most of our results have a much wider range of validity.

From the point of view of PDE theory, elliptic operators are the simplest. After them, operators of real principal type are the next most tractable. Indeed, Fourier integral operators do for them what pseudodifferential operators do for elliptic differential operators (e.g., provide parametrices). As is by now well known, pseudoconvexity is a sufficient condition for solvability for operators of real principal type. Stability of solvability for these operators supports the validity of (at least the results of) certain perturbation techniques used in such applications as quantum field theory.

We shall prove in the sequel [3] that the two conditions of real principal type and pseudoconvexity are jointly stable in the Whitney (or C^∞ fine) topology. We shall also study the Whitney stability of certain sectional curvature conditions as part of the applications. In this paper we prove that pseudoconvexity separately and the two conditions jointly are stable in the $F\mathscr{D}$ topology of Michor [13], and in special cases are jointly stable in an asymptotically defined topology which we call the $F\mathscr{E}$ topology by analogy; see Section 3 for details. The $F\mathscr{E}$ topology is defined in terms of asymptotic classes of background Riemannian metrics, the study of which comprises Section 2. These classes provide a rigorous foundation for the study of general asymptotic properties of structures on noncompact manifolds. They are probably related to those introduced by Hörmander [9], but we have not pursued this here. We note only that both allow one to make precise estimates concerning the asymptotics of symbols.

In Sec. 3 we show that if β_0 is the principal symbol of the d'Alembertian $\square$ on Minkowski spacetime, then there is an $F\mathscr{E}$ neighborhood U of β_0 in the space of principal symbols of order 2 such that each $\beta \in U$ is both of real principal type and pseudoconvex. This shows, among other things, that the Klein-Gordon equation $(\square - m^2)u = 0$ is solvable for any spacetime $(\mathbb{R}^n, \beta)$ sufficiently close to Minkowski spacetime $(\mathbb{R}, \beta_0)$. Alternatively, it shows that solvability of the Klein-Gordon equation is not destroyed by small perturbations in the $F\mathscr{E}$ topology. This stability result is actually valid for Lorentzian manifolds having ends with certain types of metrics (Theorem 3.6).

Our notation is mostly standard. We use <u>smooth</u> to mean C^∞, $\mathscr{E}$ denotes the smooth functions, $\mathscr{D}$ denotes those with compact support, Γ denotes smooth sections, and $\mathring{U}$ denotes the interior of U. Finite differentiability classes are indicated by superscripts; e.g., C^r, $r \geq 0$; $C^0 = C$; $F\mathscr{D}^r$. Manifolds are smooth, paracompact, connected, and

*Current affiliation: Department of Mathematics, Wichita State University, Wichita, Kansas

usually noncompact. We always consider pseudoriemannian structures as (2, 0)-tensors, rather than (0, 2)-tensors, so they are naturally principal symbols.

2. ENDS AND ORDER CLASSES

In applications, one is frequently interested in studying properties "at infinity"; e.g., asymptotically flat spaces [5]. When the underlying space is $\mathbb{R}^n$, there is the space $\mathcal{S}$ of rapidly decreasing functions ([17], p. 92) which has been most useful; cf. [10]. It is well known (and easy to see) that as a set of smooth functions, $\mathcal{S}$ is not invariant under changes of coordinates. In this section we generalize $\mathcal{S}$ to noncompact manifolds in preparation for our subsequent work.

Let X be a manifold and consider the set of ends

$$E(X) := \lim\ \mathrm{inv}\ \pi_0\ (X \backslash K)$$

where K runs over the compact sets in X. Since $E(X) = \emptyset$ when X is compact, we shall henceforth assume that X is noncompact unless otherwise specified. For appropriate K we can say that components of $X\backslash K$ represent one or more ends, or that components of the boundary of K represent ends. Consider now an end $e \in E(X)$. We say that e is a collar iff there exists a compact $K \subseteq X$ such that: (1) the component of the boundary of K that represents e is an embedded compact submanifold Y, called the type of e; (2) the component of $X\backslash K$ that represents e is diffeomorphic to $Y \times (0, \infty)$. The name comes from collaring theorems, which imply that one can attach a smooth boundary to e iff e is a collar. When Y above is a sphere, torus, etc., we speak of spherical, toral, etc., ends. If $e \in E(X)$ we write $e(X\backslash K)$ to denote that component (implicitly required to exist) of $X\backslash K$ which represents e.

Definition 2.1. Two Riemannian metrics g and h on X are order related at e, denoted by $h_X = O(g_X)$ as $x \to e$, iff for some compact $K \subseteq X$ and for all $x \in e(X\backslash K)$

$$h_x(\xi, \xi) \leq Cg_x(\xi, \xi)$$

for some constant $C > 0$ which depends only on K, g, h, and e, uniformly for $\xi \in T^*_xX$.

In order to define our analogue of rapidly decreasing functions, we require a brief digression on jets. Details may be found, for example, in [14]. Given $x \in X$, let $\mathcal{I}_x \subseteq \mathcal{E}$ be the ideal of smooth functions vanishing at x and set $\mathcal{Z}^r_x = \mathcal{I}^{r+1}_x \mathcal{E}$. Define

$$J^r_x(X) := \mathcal{E}(X)/\mathcal{Z}^r_x(X)$$

$$J^r(X) := \bigcup_{x \in X} J^r_x(X)$$

and denote the natural linear map $\mathcal{E} \to J^r_x$ by j^r_x. Define the r-jet extension map j^r: $\mathcal{E} \to \Gamma(J^r)$: $j^r\phi(x) := j^r_x(\phi)$. There is a unique smooth vector bundle structure on J^r which is compatible with the jet extension maps. Recall that $J^r(X) \subseteq (T^*)^rX \oplus \underline{1}$, where $(T^*)^r$ denotes the r-fold iteration of T^* and $\underline{1}$ denotes the trivial line bundle over X. By dualizing and iterating the usual Sasaki construction [16], we can lift a Riemannian metric g on X to a Riemannian metric $g^{(r)}$ on $(T^*)^rX$. The usual numerical absolute value induces a fiber metric $|\cdot|$ on $\underline{1}$, and thus $g^{(r)} \oplus |\cdot|$ provides a fiber metric on $(T^*)^rX \oplus \underline{1}$. The restriction of this fiber metric to $J^r(X)$ will henceforth be denoted simply by $|\cdot|_g$. Note that if g is complete then so is $|\cdot|_g$. From now on we shall only be concerned with complete Riemannian metrics.

Definition 2.2. Let g be a (complete) Riemannian metrix on X with associated distance function d. The set of g-rapidly decreasing functions on X is

$$\mathcal{S}(X, g) := \{\phi \in \mathcal{E}(X);\ j^r\phi(x) = O(d_0(x)^{-k}) \text{ as } x \to e, \text{ for each end } e,$$
$$\text{for all } r, k \in \mathbb{N}, \text{ and for any } x_0 \in X\}$$

where $d_0(x) := d(x_0, x)$, and "$j^r\phi(x) = O(d_0(x)^{-k})$ as $x \to e$" means that for some compact $K \subseteq X$ and for all $x \in e(X\setminus K)$, $|j^r\phi(x)|_g \leq C\, d(x_0, x)^{-k}$ for some constant $C > 0$ which depends only on K, g, x_0, ϕ, and e.

It will be convenient to use the phrase "as $x \to \infty$" to mean "as $x \to e$, for each end e." If X has finitely many ends then we can choose C above independent of e; in this case, "as $x \to \infty$" will be interpreted as including this uniformity.

Lemma 2.3. If $g_x = O(h_x)$ as $x \to \infty$ then $\mathcal{S}(X, g) \subseteq \mathcal{S}(X, h)$.

Proof. Letting d and δ be the respective distance functions, it follows from Definition 2.1 that $d_0(x) = O(\delta_0(x))$ as $x \to \infty$ (with the obvious meaning for this notation) for any $x_0 \in X$. Substituting into Definition 2.2, the conclusion follows immediately. Q.E.D.

Lemma 2.4. $\mathcal{S}(X, g)$ is invariant under quasi-isometries.

Proof. By definition, g and h are quasi-isometric iff $g_x = O(h_x)$ and $h_x = O(g_x)$ uniformly as $x \to \infty$. Now apply Lemma 2.3. Q.E.D.

This motivates

Definition 2.5. We say that g and h are of the same order at ∞ (or at e) iff $g_x = O(h_x)$ and $h_x = O(g_x)$ as $x \to \infty$ (or as $x \to e$). The set of all (complete) Riemannian metrics on X of the same order at ∞ as g is denoted by $\mathcal{O}(g)$ and called the order class at ∞ of g.

The following lemma summarizes the basic properties of order classes. The proofs are easy and are omitted.

Lemma 2.6. In the space of Riemannian metrics on X,

(1) being of the same order at ∞ is an equivalence relation;
(2) $\mathcal{O}(g)$ is a convex positive cone;
(3) $\phi g \in \mathcal{O}(g)$ iff ϕ is bounded.

Also,

(4) when X has only finitely many ends, $\mathcal{O}(g)$ is the set of all metrics quasi-isometric to g.

The precise invariance of the space $\mathcal{S}(X, g)$ has not been determined and is under investigation for a future report. The following, however, is suggestive.

Lemma 2.7. Let g and h be two (complete) Riemannian metrics on X with distance functions d and δ. If $x_0 \in X$ is fixed and $d_0(x) = O(\delta_0(x)^k)$ as $x \to \infty$ for some $k \in \mathbb{N}$ and $\delta_0(x) = O(d_0(x)^m)$ as $x \to \infty$ for some $m \in \mathbb{N}$, then $\mathcal{S}(X, g) = \mathcal{S}(X, h)$.

Proof. This follows immediately from the transitivity of the O relation and Definition 2.2. Q.E.D.

Of course, as is easy to check, $\mathcal{S}(X, g)$ is a vector space. We now topologize it via a set of seminorms. For each pair k, $m \in \mathbb{N}$ define for each $\phi \in \mathcal{S}(X, g)$

$$\mathcal{P}_{m,k}(\phi) := \sup_{r \leq m} \sup_{X} [(1 + d_0(x))^k |j^r \phi(x)|_g]$$

This topology may be intuitively described as that of global convergence of the function and all its derivatives, rapidly at ∞. Since this topology is strictly finer than that induced from the Schwartz topology on $\mathcal{E}(X)$, q.v. Sec. 3, one may show that $\mathcal{S}(X, g)$ becomes a nuclear Fréchet space [17].

Now if Φ is any diffeomorphism of X then g and $\Phi_* g$ are isometric, so $\mathcal{S}(X,g)$ and $\mathcal{S}(X,\Phi_* G)$ are isomorphic topological vector spaces. But they are not necessarily equal as subsets of $\mathcal{E}(x)$.

Example 2.8. Let g_0 be the usual Euclidean metric on $\mathbb{R}^n$. Then

$$\mathcal{S} = \mathcal{S}(\mathbb{R}^n, g_0) \cong \mathcal{S}(\mathbb{R}^n, e^{|x|^2} g_0)$$

but they are not equal.

Thus there is a certain amount of "coordinate dependence" in $\mathcal{S}(X,g)$. Also, classification by order classes is not finer than isometry classes.

Example 2.9. For $n \geq 2$, $\mathbb{R}^n$ carries metrics g_c of constant curvature $c \leq 0$. We have $g_c \in \mathcal{O}(g_0)$ for every $c < 0$.

One may think of an order class $\mathcal{O}(g)$ as a concept of "bounded at ∞" for tensors. Let E be any tensor bundle over X. Then we can extend the fiber metric $|\cdot|_g$ from $J^r(X)$ to $J^r(E)$, and define g-rapidly decreasing sections of E as

$$\{\sigma \in \Gamma(E);\ j^r\sigma(x) = O(d_0(x)^{-k}) \text{ as } x \to \infty \text{ for all } r,k \in \mathbb{N}\}$$

One can show that this space is a finitely generated projective $\mathcal{S}(X,g)$-module, topologizible as a nuclear Fréchet space; we omit the details as we shall not need these results.

Finally, we present some examples which are not necessary for the remainder but which may help in understanding our spaces $\mathcal{S}(X,g)$.

Remark 2.10. Let g_0 denote the usual Euclidean metric on $\mathbb{R}^n$. If g is any constant metric on $\mathbb{R}^n$ then $g \in \mathcal{O}(g_0)$. Thus $\mathcal{S}(\mathbb{R}^n, g) = \mathcal{S}$.

Recall that a pole on a Riemannian manifold X is $@ \in X$ such that $\exp_@ : T_@ X \to X$ is a diffeomorphism. Normal coordinates at @ provide a linear isomorphism $\mathbb{R}^n \to T_@ X$, and the composition with $\exp_@$ yields a diffeomorphism $\Phi: \mathbb{R}^n \to X$. Rays from $0 \in T_@ X$ are mapped by $\exp_@$ into radial geodesics. The outward (unit) tangent vector to a radial geodesic, or (for convenience) any (unit) vector in $T_@ X$, is called a (unit) radial vector. A radial curvature is the sectional curvature of any plane containing a radial vector.

Theorem 2.11. Let (X, g) be Riemannian with pole @ and $\Phi: \mathbb{R}^n \to X$ the diffeomorphism as above. Define $K, k: [0,\infty) \to [0,\infty)$ by

$$-k(r) := \max\{0, \text{ radial curvatures on } S_r(@)\}$$

$$K(r) := \max\{0, \text{ radial curvatures on } S_r(@)\}$$

where $S_r(@) = \{x; d(@,x) = r\}$. If $\int_0^\infty rk(r)dr < \infty$ and $\int_0^\infty rK(r)\,dr \leq 1$ then $\mathcal{S}(\mathbb{R}^n, \Phi^* g) = \mathcal{S}$.

Proof. The hypotheses are those of Theorem C of [7] which concludes (among other things) that $\exp_@$ is a quasiisometry. Applying Lemma 2.4 and Remark 2.10 we obtain the desired result. Q.E.D.

This theorem applies, for example, to Cartan-Hadamard spaces (complete, simply connected, nonpositive curvature) whose curvature is bounded below by $-C d_0^{-2} (\ln d_0)^{-1-\epsilon}$ where d_0 denotes distance from some fixed point (every point is a pole). Note, however, that the curvature need not keep the same sign everywhere, nor be bounded, and that only one pole is necessary.

If we think of diffeomorphisms of X as analogous to global coordinate changes, then the choice of an order class $\mathcal{O}(g)$ determines some admissible "coordinates": a set of diffeomorphisms which preserve $\mathcal{S}(X, g)$ as a subset of $\mathcal{E}(X)$. The preceding theorem can be interpreted as giving examples of some such diffeomorphisms.

As one would expect, there is an associated space $\mathcal{S}'(X, g)$ of g-tempered distributions. It can be topologized as a nuclear space and we have the inclusions $\mathcal{E}'(X) \subseteq \mathcal{S}'(X, g) \subseteq \mathcal{D}'(X)$. Also, one can easily show that all the functions in $\mathcal{S}(X, g)$ are absolutely integrable with respect to the Riemannian density (volume, if X is orientable) $đg := (\det(g))^{-\frac{1}{2}}$. Now on the usual space $\mathcal{S} = \mathcal{S}(\mathbb{R}^n, g_0)$ there is defined the Fourier transform

$$\mathcal{F}\phi(y) = \int_{\mathbb{R}^n} \exp(-i\, g_0(x, y))\, \phi(x)\, đg_0$$

Observe that we can rewrite the phase in terms of the g_0-distance function d via the polarization identity:

$$g_0(x, y) = \frac{1}{2}(d(0, x)^2 + d(0, y)^2 - d(x, y)^2)$$

This latter expression makes sense in any Riemannian manifold. In particular, if (X, g) is Riemannian with pole @, then we can use Φ as before Theorem 2.11 to transfer the Fourier transform to X. The result is an isomorphism of $\mathcal{S}(X, g)$ which depends on g and @:

$$\mathcal{F}_{@}\phi(y) = \int_X \exp\left[-\frac{i}{2}(d(@, x)^2 + d(@, y)^2 - d(x, y)^2)\right] \phi(x)\, đg$$

where d now denotes the distance function of g.

We can use this formula to define a transform on $\mathcal{S}(X, g)$ which depends on the choice of $@ \in X$ for <u>any</u> Riemannian (X, g). One expects that this transform will share many of the most useful properties of the Fourier transform on $\mathbb{R}^n$; e.g., one should obtain an isomorphism of $\mathcal{S}'(X, g)$ in general, as it follows from above we have for (X, g) with pole @. We shall pursue this elsewhere, and for now offer only the following.

<u>Example 2.12</u>. On $\mathbb{R}^n$ the inverse Fourier transform is given by

$$\mathcal{F}^{-1}\phi(x) = C^{-2} \int_{\mathbb{R}^n} \exp(i\, g_0(x, y))\, \phi(y)\, đg_0$$

where the constant comes from Gaussian normalization,

$$C = \int_{\mathbb{R}^n} \exp\left(-\frac{1}{2} g_0(x, x)\right) đg_0 = (2\pi)^{n/2}$$

On (X, g) with pole @ this becomes

$$\mathcal{F}^{-1}\phi(x) = C^{-2} \int_X \exp\left[\frac{i}{2}(d(@, x)^2 + d(@, y)^2 - d(x, y)^2)\right] \phi(y)\, đg$$

$$C = \int_X \exp\left(-\frac{1}{2} d(@, x)^2\right) đg$$

(This is one traditional placement of the constant; one can be symmetric and replace the Riemannian density $đg$ with the Gaussian density $C^{-1}đg$.)

Reverting to the usual covariant notation $|x|^2 = g_0(x, x)$ in $\mathbb{R}^n$, recall ([6], p. 40f, misprinted) the space form X of constant curvature $K < 0$:

$$X = \{|x|^2 < -\frac{4}{K}\} \subseteq \mathbb{R}^n$$

$$g = (1 + \frac{K}{4}|x|^2)^{-2} g_0$$

Then $\det(g) = (1 + \frac{K}{4}|x|^2)^{-2n}$ and, using the arc length formula

$$d(0,x) = \frac{1}{\sqrt{-K}} \ln \frac{2 + \sqrt{-K}|x|}{2 - \sqrt{-K}|x|}$$

the Gaussian normalization constant is

$$C = \int_X \exp(-\frac{1}{2} d(0,x)^2)\, dg$$

$$= \int_X \exp\left[\frac{1}{2K}\left(\ln \frac{2 + \sqrt{-K}|x|}{2 - \sqrt{-K}|x|}\right)^2\right] (1 + \frac{K}{4}|x|^2)^{-n} dx$$

$$= \frac{2\pi^{n/2}}{\Gamma(\frac{n}{2})} \int_0^{2/\sqrt{-K}} \exp\left[\frac{1}{2K}\left(\ln \frac{2 + \sqrt{-K}|x|}{2 - \sqrt{-K}|x|}\right)^2\right] (1 + \frac{K}{4}r^2)^{-n} r^{n-1} dr$$

Making the change of variable $u = \ln \frac{2 + \sqrt{-K}\, r}{2 - \sqrt{-K}\, r}$ and simplifying, we obtain

$$C = \frac{2^{2-n}}{\Gamma(\frac{n}{2})} \left(\frac{\pi}{-K}\right)^{n/2} \int_0^\infty e^{\frac{u^2}{2K} + (1-n)u} (e^{2u} - 1)^{n-1} du$$

We now expand the binomial in the integrand, complete the square in the exponent, and change the variable again to obtain

$$C = \frac{2^{2-n}\pi^{n/2}}{\Gamma(\frac{n}{2})(-K)^{(n-1)/2}} \sum_{i=0}^{n-1} (-1)^i \binom{n-1}{i} e^{(-K/4)(2i+1-n)^2} \int_{\sqrt{-K}(2i+1-n)}^{\infty} e^{-t^2} dt$$

The integral in this last formula is the complementary error function Erfc $(\sqrt{-K}(2i + 1 - n))$, a confluent hypergeometric function ([12], p. 223, formula (30)). This allows us to simplify the formula for C slightly:

$$\text{n odd:}\quad C = \frac{\pi^{n/2}\left[\sum_{i=1}^{(n-3)/2} (-1)^i \binom{n-1}{i} e^{-K(2i+1-n)^2} + (-1)^{(n-1)/2}\binom{n-2}{(n-1)/2}\right]}{2^{(n-3)/2}(-K)^{(n-1)/2} \prod_{j=1}^{(n-1)/2} (n - 2j)}$$

$$\text{n even:}\quad C = \frac{\pi^{n/2} \sum_{i=0}^{(n-2)/2} (-1)^i \binom{n-1}{i} e^{-(K/4)(n-2i-1)^2} [\sqrt{\pi} - 2\, \text{Erfc}\,(\sqrt{-K}(n - 2i - 1))]}{2^{n-2}(-K)^{(n-1)/2}\left(\frac{n}{2} - 1\right)!}$$

In particular, for n = 3 we have

$$C = \frac{\pi^{3/2}}{-K}(e^{-K} - 1)$$

It is now straightforward to check that $\mathcal{F}$ provides an isomorphism $\mathcal{S}(X, g) \to \mathcal{S}(X, g)$ and thus extends to an isomorphism $\mathcal{S}'(X, g) \to \mathcal{S}'(X, g)$ for (X, g) of constant curvature $K < 0$. It is not much more difficult to prove that $\mathcal{F}$ provides isomorphisms of $\mathcal{S}(X, g)$ and $\mathcal{S}'(X, g)$ for manifolds with pole. This supports our expectations for the general case.

3. TOPOLOGIES AND STABILITY

In this section we discuss the C^∞-coarse, $F\mathcal{S}$, C^∞-fine, and $F\mathcal{D}$ topologies, listed in order from the coarsest to the finest. Stability theorems are established for the $F\mathcal{D}$ and $F\mathcal{S}$ topologies.

A very general result is obtained for the $F\mathcal{D}$ topology. Real principal type and pseudoconvexity are shown to be jointly $F\mathcal{D}$-stable in the space $\mathrm{Smbl}_r(X)$ of principal symbols of order r. In general, these conditions fail to be stable on $\mathrm{Smbl}_r(X)$ using the $F\mathcal{S}$-topology. Nevertheless, we are able to establish $F\mathcal{S}$-stability for a certain class of symbols. If $\beta \in \mathrm{Smbl}_2(X)$ is a Lorentzian metric on X such that (X, β) has what we term uniform ends, then both real principal type and pseudoconvexity are $F\mathcal{S}$-stable at β. This demonstrates the $F\mathcal{S}$-stability of the solvability of the Klein-Gordon equation on these manifolds.

One of the standard topologies is the <u>Schwartz</u> or <u>C^∞-coarse</u> topology. Intuitively, this is the topology of uniform convergence of the function and all of its derivatives on compact sets. Let $\{K_i\}$ be a countable family of compact sets in X such that $K_i \subseteq \mathring{K}_{i+1}$ and $X = \bigcup K_i$. Choose any compatible fiber metric ρ_r on $J^r(X)$ for each $r \geq 0$ and define seminorms

$$\mathcal{P}_{i,k}(\phi) := \sup_{r \leq k} \sup_{x \in K_i} \rho_r(j^r_x(\phi), O_x)$$

The countable set of seminorms makes the space $\mathcal{E}$ of smooth functions on X into a nuclear Fréchet space.

Another standard topology is the <u>Whitney</u> or <u>C^∞-fine</u> topology. Whereas the Schwartz topology gives no control of the convergence at ∞, the Whitney topology gives arbitrary amounts of such control. In doing so, however, it is no longer a vector topology on $\mathcal{E}$; thus it cannot be described in terms of seminorms. To compare these topologies, recall that $j^r: \mathcal{E} \to C(X, J^r(X))$, and taking the inverse limit over r yields $j^\infty: \mathcal{E} \to C(X, J^\infty(X))$. Now on the latter space we can place either the compact-open topology or the graph topology. Via j^∞, these induce the Schwartz, respectively Whitney, topologies on $\mathcal{E}$: see [8] or [13] for details.

On the space $\mathcal{D}$ of smooth functions with compact support we consider the usual Schwartz topology. First, let $\mathcal{E}(K) := \{\phi \in \mathcal{E}(X);\ \mathrm{supp}\ \phi \subset K\}$ for K compact. Using the sequence $\{K_i\}$ above, we topologize $\mathcal{E}(K_i)$ with the seminorms $\mathcal{P}_{i,k}$ for $i, k \in \mathbb{N}$, and then $\mathcal{D} = \varinjlim \mathcal{E}(K_i)$ gets the direct limit topology. This makes $\mathcal{D}$ a nuclear LF-space [17]. Intuitively this describes uniform convergence of the function and all its derivatives and Hausdorff convergence of the support, all within some compact set.

Now regard $\mathcal{E}$ as an affine space over $\mathcal{D}$; we decompose $\mathcal{E}$ into cosets $\phi + \mathcal{D}$. On each coset put the Schwartz topology of $\mathcal{D}$, and on $\mathcal{E}$ put the weak topology generated by these cosets. Thus $U \subseteq \mathcal{E}$ is open iff $U \cap (\phi + \mathcal{D})$ is open for each coset $\phi + \mathcal{D}$. This topology is due to Michor and is called the $F\mathcal{D}$-topology [13, p. 40] on $\mathcal{E}$. Although it is finer than the Whitney topology, it has the advantage that $\mathcal{E}$ becomes a smooth manifold modelled on $\mathcal{D}$. See [13] and [15] for applications.

A <u>real principal symbol of order r</u> on the manifold X is a real-valued function $p \in \mathcal{E}(T^*X \backslash 0)$ which is positive homogeneous of degree r in the fiber variable. The space of all real principal symbols of order r will be denoted by $\mathrm{Smbl}_r(X)$.

Recall that $T^*X \backslash 0$ is a natural symplectic manifold with respect to the canonical 2-form ω [11, 18]. The <u>Hamiltonian vector field</u> H^p is defined by

$$dp = \omega(H^p, \cdot)$$

In local induced coordinates (x, ξ) on $T^*X\backslash 0$ the vector field H^p is given by

$$H^p = \sum_{i=1}^{n} \frac{\partial p}{\partial \xi_i}\frac{\partial}{\partial x_i} - \frac{\partial p}{\partial x_i}\frac{\partial}{\partial \xi_i}$$

The value of p is constant along the integral curves of H^p, and those along which $p = 0$ are called the bicharacteristic strips of p. The projections of these strips to X yield the bicharacteristic curves of p. We say p is of real principal type iff p is real-valued and no complete (i.e., inextendible) bicharacteristic strip projects to a bicharacteristic curve which stays in a compact set. The principal symbol p is pseudoconvex iff for every compact set $K \subseteq X$ there exists a compact set $K' \subseteq X$ such that every bicharacteristic segment with both endpoints in K lies in K'. These terms are also applied to any pseudodifferential operator with principal symbol p.

Since principal symbols are homogeneous in the fiber variables, the $F\mathscr{D}$ topology on them is discrete. Thus we shall modify it to take homogeneity in account. Letting h be an auxiliary Riemannian metric on X, p is completely determined by its order of homogeneity r and its restriction to the h-unit cosphere bundle S^*X. Thus given r, there is a bijection I between $\mathscr{E}(S^*X)$ and the space $\mathrm{Smbl}_r(X)$ of principal symbols of order r. We shall say that a set $U \subseteq \mathrm{Smbl}_r(X)$ is open in the $F\mathscr{D}$ topology iff the corresponding set $I^{-1}(U) \subseteq \mathscr{E}(S^*X)$ is open in the $F\mathscr{D}$ topology on $\mathscr{E}(S^*X)$. Intuitively, two symbols $p, p' \in \mathrm{Smbl}_r(X)$ are close in the $F\mathscr{D}$ topology if there is some compact subset $K \subseteq X$ such that $p = p'$ on $T^*(X\backslash K)$ and the values of p, p', and their corresponding derivatives are close on S^*K.

We now show pseudoconvexity is $F\mathscr{D}$-stable for real valued principal symbols of order r.

Theorem 3.1. If $p \in \mathrm{Smbl}_r(X)$ is pseudoconvex, then there is an $F\mathscr{D}$ open set U(p) about p such that each $p' \in U(p)$ is pseudoconvex.

Proof. Using the above bijection $I: \mathscr{E}(S^*X) \to \mathrm{Smbl}_r(X)$, we regard all symbols as elements of $\mathscr{E}(S^*X)$. We will show every $p' \in \mathscr{E}(S^*X)$ with supp $(p - p')$ compact in S^*X is pseudoconvex.

Let $K_0 = \pi(\mathrm{supp}\,(p - p'))$ be a compact subset of X and choose any compact subset K of X. Then there exists a compact set K_1 with $K_0 \cup K$ contained in the interior $\mathring{K}_1$ of K_1. Since p is pseudoconvex, there is a compact set K_1' which contains any bicharacteristic segment of p with endpoints in K_1. We claim that K_1' will serve as the required compact set for p' and K. If not, there is a bicharacteristic curve $\gamma: [0, a] \to X$ of p' with $\gamma(0)$, $\gamma(a) \in K$ and $\gamma(t_0) \notin K_1'$ for some $0 < t_0 < a$. Choose $t_1 = \inf \{t; \gamma(t) \notin K_1$ for $t_1 < t < t_0\}$ and $t_2 = \sup \{t; \gamma(t) \notin K_1$ for $t_0 < t < t_2\}$. Then $0 \leq t_1 < t_0 < t_2 \leq a$ and $\gamma(t_1), \gamma(t_2) \in K_1$. Since $K_0 \subseteq K_1$, $\gamma\,[t_1, t_2]$ is a bicharacteristic segment for p with both endpoints in K_1, whence it must lie in K_1', which contradicts $\gamma(t_0) \notin K_1'$. Q.E.D.

The homogeneity of the principal symbol p implies that if γ_1 and γ_2 are complete bicharacteristic strips of p with $\gamma_1(0) = (x_0, \xi_0)$ and $\gamma_2 = (x_0, \lambda\xi_0)$ for some positive λ, then γ_1 and γ_2 only differ by a reparametrization. Thus for our purposes it suffices to consider bicharacteristic strips with $\gamma(0) \in S^*X$.

We now state the $F\mathscr{D}$-stability of real principal type at symbols which are both of real principal type and pseudoconvex.

Proposition 3.2. If $p \in \mathrm{Smbl}_r(X)$ is both of real principal type and pseudoconvex, there is an $F\mathscr{D}$ open neighborhood U(p) of p such that each $p' \in U(p)$ is of real principal type.

In [3] we will prove that the conditions of real principal type and pseudoconvexity are jointly stable in the Whitney C^1 topology on $\mathrm{Smbl}_r(X)$. Since the $F\mathscr{D}$ topology is finer than the Whitney topology, Proposition 3.2 will follow from [3].

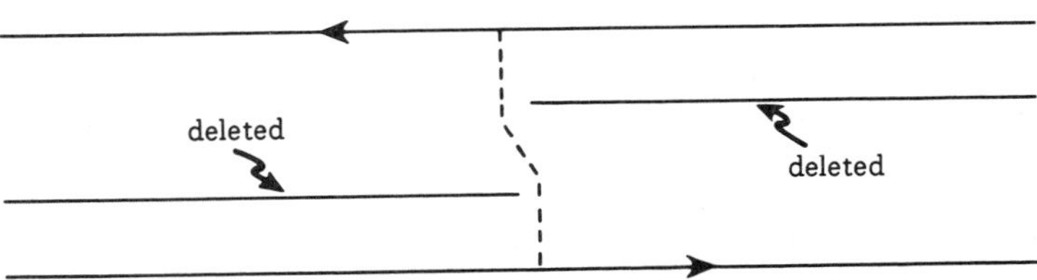

Fig. 1. A Lorentzian manifold (X, g) with g of real principal type but not pseudoconvex. Each F$\mathscr{D}$ neighborhood U(g) of g in $\mathrm{Smbl}_2(X)$ contains some $g' \in U(g)$ with a periodic null geodesic. A typical periodic null geodesic of some g' is shown as a dashed curve.

We will now give a counterexample to show that pseudoconvexity cannot be dropped from the hypothesis of Proposition 3.2. This counterexample shows that the set of all symbols in $\mathrm{Smbl}_r(X)$ of real principal type is not, in general, open in the F$\mathscr{D}$ topology. Let X_0 be the open Möbius strip $\{(x,y); -\infty < x < \infty,\ 0 \leq y \leq 4\}$ with the identification $(x,0) \sim (-x,4)$ and the metric tensor $g_0 = \xi_1 \xi_2$ (i.e., $ds^2 = dxdy$). The bicharacteristic curves of g_0 lie on the null geodesics of g_0 which are the Euclidean lines y = const. and the circles $|x|$ = const. The manifold X will be X_0 less the two closed half lines $L_1 = \{(x,1); x \leq 0\}$ and $L_2 = \{(x,3); x \geq 0\}$. The metric tensor on X will be $g = g_0 | X$. The Lorentzian manifold (X, g) has no imprisoned null geodesics and hence g is a symbol of real principal type. On the other hand, any F$\mathscr{D}$ neighborhood U(g) of g in $\mathrm{Smbl}_2(X)$ will contain some $g' \in U(g)$ such that g' has a periodic null geodesic and hence is not of real principal type. The g' may be chosen WLOG to equal g off some small compact set containing the point (0, 2) in its interior. This counterexample is illustrated in Fig. 1.

Using Theorem 3.1 and Proposition 3.2 we obtain the following result.

<u>Theorem 3.3</u>. The collection of all $p \in \mathrm{Smbl}_r(x)$ which are both of real principal type and pseudoconvex forms an open set in the F$\mathscr{D}$ topology.

There are finite versions of all of the topologies reviewed above: C^r-coarse, C^r-fine, and F$\mathscr{D}^r$. Here r refers to the highest order of differentiability considered in defining the topology. Intuitively, our proofs show that in the class of pseudoconvex symbols the property of real principal type is F$\mathscr{D}^1$-stable, but that pseudoconvexity is F$\mathscr{D}^{-1}$-stable; i.e., it only depends on supports and <u>not</u> at all on values. In [3] we shall show that neither property is either C^r-coarse, $0 \leq r$, or C^0-fine stable, but that both together are C^r-fine stable for $r \geq 1$.

In analogy with the F$\mathscr{D}$ topology on $\mathscr{E}$ we define the <u>F$\mathscr{S}$ topology</u> on $\mathscr{E}$. For a manifold X and an order class $\mathcal{O}(g)$ we have the space $\mathscr{S}(X,g) \subseteq \mathscr{E}(X)$. Regard $\mathscr{E}$ as an affine space over $\mathscr{S}(X,g)$ via the coset decomposition. On each coset $\phi + \mathscr{S}(X,g)$ put the topology of Section 2 and on $\mathscr{E}$ the weak topology generated by these cosets. Thus a set $U \subseteq \mathscr{E}$ is open iff $U \cap (\phi + \mathscr{S}(X,g))$ is open in the coset $\phi + \mathscr{S}(X,g)$. This F$\mathscr{S}$ topology is finer than the Schwartz topology on $\mathscr{E}$ but coarser than the Whitney topology. Thus for a given property of interest, F$\mathscr{S}$-stable ⇒ Whitney stable ⇒ F$\mathscr{D}$ stable. F$\mathscr{S}$-stability only holds on a special class of manifolds; but it is a considerable sharpening of Whitney stability there.

Let $(\mathbb{R}^n, \beta_0)$ be the standard n-dimensional Minkowski space with $\beta_0 = \xi_1^2 - (\xi_2^2 + \dots + \xi_n^2)$. The Klein-Gordon equation $(\Box - m^2)u = 0$ then has principal symbol β_0. Since the bicharacteristic curves of a principal symbol which is a pseudoriemannian structure are its null geodesics, it follows that the bicharacteristics of β_0 lie on Euclidean straight lines. Consequently, β_0 is both of real principal type and pseudoconvex. We will prove that any principal symbol on $\mathbb{R}^n$ which is sufficiently close to β_0 in the F$\mathscr{S}$-topology is both of real principal type and pseudoconvex. This will demonstrate the F$\mathscr{S}$-stability of both real principal type and pseudoconvexity at β_0. It will also show that the solvability of the Klein-Gordon equation is stable with respect to the F$\mathscr{S}$-topology.

Actually, there is $F\mathcal{S}$-stability of real principal type and pseudoconvexity for a class of principal symbols on manifolds which includes β_0 on $\mathbb{R}^n$ as a special case.

Let (H, h) be a positive definite Riemannian manifold. The manifold (H, h) is said to be homogeneous iff for each pair of points in H there is an isometry of H taking the first point to the second. In the rest of this section we will always assume (H, h) is either compact or homogeneous. Furthermore, $(\mathbb{R}, \sigma)$ will always represent $\mathbb{R}^1$ with the standard Euclidean metric $\sigma = \xi_1^2$. The product manifold $\mathbb{R} \times H$ then has the Lorentzian product metric $G = \sigma \oplus (-h)$ which is globally hyperbolic (cf. [1], p. 64 and p. 127). The geodesics of $\mathbb{R} \times H$ are of the form $c(t) = (c_1(t), \gamma(t))$ where c_1 is a geodesic of $\mathbb{R}$ and γ a geodesic of H. Thus c_1 is either a strictly increasing or strictly decreasing (affine) function of t.

Let $g = \sigma \oplus h$ be an auxiliary complete (positive definite) Riemannian metric on $\mathbb{R} \times H$. We will always assume that g has been used to define the $F\mathcal{S}$ topology on $\mathrm{Smbl}_2(\mathbb{R} \times H)$. Any $g' \in \mathcal{O}(g)$ determines the same $F\mathcal{S}$ topology.

Notice that if $F: H \to H$ is an isometry of (H, h), then there is an induced isometry $\bar{F}: \mathbb{R} \times H \to \mathbb{R} \times H$ of $(\mathbb{R} \times H, G)$ defined by $\bar{F}(t, y) = (t, F(y))$. Thus if (H, h) is homogeneous the isometry group of $(\mathbb{R} \times H, G)$ is transitive on the hypersurfaces $\{t_0\} \times H$.

When (H, h) is homogeneous we let B_0 be a locally finite countable collection of charts (U_1, ϕ_1), (U_2, ϕ_2), ..., such that for each i there is an isometry $F_i: U_1 \to U_i$ with $\phi_i = \phi_1 \circ F_i^{-1}$. We will assume without loss of generality that $\bar{U}$ is a compact set contained in a chart (V, ϕ). If (H, h) is compact we simply let B_0 be any finite collection (U_1, ϕ_1), (U_2, ϕ_2), ... of charts which covers H. After fixing B_0 let B be all charts on $\mathbb{R} \times H$ of the form $(\mathbb{R} \times U_i, \mathrm{id} \times \phi_i)$ for $(U_i, \phi_i) \in B_0$. For the rest of this section all local coordinate representations will be with respect to charts of B.

If G has components G^{ij}, let W(G, A) be the symbols $\beta \in \mathrm{Smbl}_2(\mathbb{R} \times H)$ such that if β has components β^{ij}, then

$$|\beta^{ij} - G^{ij}| < A \tag{3.1}$$

for all charts of B. The set W(G, A) is open in the $F\mathcal{S}$ topology.

Lemma 3.4. Let $\mathbb{R} \times H$ have the Lorentzian product metric $G = \sigma \oplus (-h)$. There is a neighborhood U(G) of G in $\mathrm{Smbl}_2(\mathbb{R} \times H)$ such that if $\beta \in U(G) \cap C^\infty(T^*(\mathbb{R} \times H))$ then:

(1) β is a Lorentzian metric on $\mathbb{R} \times H$
(2) β is globally hyperbolic
(3) all sets of the form $\{t_0\} \times H$ are Cauchy surfaces for β
(4) β is both of real principal type and pseudoconvex

Proof: For each $x_0 \in \mathbb{R} \times H$ there is a neighborhood $U(x_0)$ such that if $A > 0$ is chosen sufficient small and

$$|\beta^{ij}(x) - G^{ij}(x)| < A$$

for all $x \in U(x_0)$, then $\beta^{ij}(x)$ represent the components of a Lorentzian metric at x and the hypersurfaces $(\{t\} \times H) \cap U(x_0)$ are all spacelike for the metric $\beta^{ij}(x)$. It follows that for fixed $t_1 \in \mathbb{R}$ there is a continuous function $A: H \to (0, \infty)$ such that for each $(t_1, y) \in \mathbb{R} \times H$, all $\beta^{ij}(t_1, y)$ which satisfy

$$|\beta^{ij}(t_1, y) - G^{ij}(t_1, y)| < A(y)$$

form the components of a Lorentzian metric at (t_1, y) such that the hypersurface $\{t_1\} \times H$ is spacelike. If (H, h) is compact, set $A_1 = \min\{A(y) \mid y \in H\}$, and if (H, h) is homogeneous, set $A_1 = \min\{A(y) \mid y \in \bar{U}_1\}$. We will show that $U(G) = W(G, A_1)$ satisfies the conclusion of the lemma.

The fact that $G^{ij}(t + c, y) = G^{ij}(t, y)$ for all $c \in \mathbb{R}$ implies that if $|\beta^{ij}(x) - G^{ij}(x)| < A_1$, then the components $\beta^{ij}(x)$ form the components of a Lorentzian metric with the surfaces

$\{t\} \times H$ spacelike for all $t \in \mathbb{R}$. If $\beta \in U(G)$ let $c: (a,b) \to \mathbb{R} \times H$ be a complete causal geodesic of β with components $c(s) = (c_1(s), \gamma(s))$ for $c_1: (a,b) \to \mathbb{R}$ and $\gamma: (a,b) \to H$. Since the surfaces $\{t\} \times H$ are spacelike, the component c_1 must be either strictly increasing or strictly decreasing. Thus we may reparametrize the curve c to obtain a curve $\tilde{c}: (a_1, b_1) \to \mathbb{R} \times H$ of the form $\tilde{c}(t) = (t, \tilde{\gamma}(t))$. The form of $G^{ij}(\underset{\approx}{x})$ and the definition of A_1 implies that there is some positive constant δ such that $h(\dot{\tilde{\gamma}}(t), \dot{\tilde{\gamma}}(t)) < \delta$ for all $t \in (a_1, b_1)$. In order to show that each hypersurface $\{t\} \times H$ is a Cauchy surface we need only show that $a_1 = -\infty$ and $b_1 = \infty$. Assume for example that $b_1 < \infty$. Then for $t_0 \in \mathbb{R}$ we have $\tilde{\gamma} | [t_0, b_1]$ of h-length at most $(b_1 - t_0) \cdot \delta^{\frac{1}{2}}$. Since (H, h) is either compact or homogeneous it must be complete and therefore $\tilde{\gamma}(t) \to y_0 \in H$ as $t \to b_1^-$ by Lemma 2.52 of [1, p. 64]. But then $\tilde{c}(t) \to (b_1, y_0)$ as $t \to b_1^-$ and $\tilde{c}$ is extendible which implies $\tilde{c}$ is not complete. Hence b_1 must be equal to ∞. A similar argument shows $a_1 = -\infty$ and this establishes the fact that all $\{t\} \times H$ are Cauchy surfaces. It follows that $(\mathbb{R} \times H, \beta)$ is globally hyperbolic and this global hyperbolicity implies that each $\beta \in U(G)$ is both of real principal type and pseudoconvex. Q.E.D.

Minkowski spacetime is of the form $\mathbb{R} \times H$ with $\beta_0 = \sigma \oplus (-h)$ where $H = \mathbb{R}^{n-1}$ with h the usual Euclidean metric. Consequently, using similar techniques we obtain

Theorem 3.5. Let $(\mathbb{R}^n, \beta_0)$ be Minkowski spacetime. There is an $F\mathscr{S}$-neighborhood $U(\beta_0)$ of β_0 in $\mathrm{Smbl}_2(\mathbb{R}^n)$ such that each $\beta \in U(\beta_0)$ is both of real principal type and pseudoconvex.

Let (X, β) have a finite number of ends $e_1, \ldots, e_k$. There is a decomposition $X = X_0 \cup E_1 \cup \cdots \cup E_k$ of X into disjoint sets where X_0 is compact (possibly empty) and each E_i represents the respective end e_i. We will say that (X, β) has uniform ends if each E_i is a collar of the form $M_i \times H_i$, where each $M_i = \{x | a_i < x < \infty\}$ with $\sigma_i = \xi^2$ and each (H_i, h_i) is a Riemannian manifold which is either compact or homogeneous. The metric on E_i is $\sigma_i \oplus (-h_i)$.

Using a standard result from differential equations ([4], p. 155) as in Section 2 together with Lemma 3.4, one may establish a generalization of Theorem 3.5 to manifolds with uniform ends. This generalization shows the $F\mathscr{S}$-stability of real principal type and pseudoconvexity for these manifolds. The $F\mathscr{S}$-topology is defined using a Riemannian metric g on X such that $g | E_i$ is of the form $\sigma_i + h_i$.

Theorem 3.6. Let (X, β) be a Lorentzian manifold with uniform ends.

(1) If (X, β) is of real principal type, then there is an $F\mathscr{S}$-neighborhood $U_1(\beta)$ of β in $\mathrm{Smbl}_2(X)$ such that each symbol in $U_1(\beta)$ is of real principal type.
(2) If (X, β) is pseudoconvex then there is an $F\mathscr{S}$-neighborhood $U_2(\beta)$ of β in $\mathrm{Smbl}_2(X)$ such that each symbol in $U_2(\beta)$ is pseudoconvex.

Remark 3.7. The conclusions of Theorem 3.6 remain valid if we allow the uniform ends E_i to be of the form $(M_i \times H_i) \backslash K_i$ for K_i a compact subset of $M_i \times H_i$.

Lemma 3.4, Theorem 3.5, and Theorem 3.6 all make essential use of the special nature of pseudo-Riemannian manifolds of Lorentzian signature. Theorem 3.5 and Theorem 3.6 may be generalized to spaces that do not have the Lorentzian signature but the proofs involved use quite different techniques. Here we shall merely state without proof the generalization of Theorem 3.5 to pseudoeuclidean spaces. The $F\mathscr{S}$ topology will be induced by the usual Euclidean metric on $\mathbb{R}^n$.

Proposition 3.8. Let $\mathbb{R}^n_s$ denote $\mathbb{R}^n$ with the standard pseudoeuclidean metric $\beta_0 = \Sigma_{i=1}^{s} \xi_i^2 - \Sigma_{i=s+1}^{n} \xi_i^2$ where $0 < s < n$. Using the $F\mathscr{S}$-topology on $\mathrm{Smbl}_2(\mathbb{R}^n)$ induced by the usual Euclidean metric on $\mathbb{R}^n$, there is an $\mathcal{F}\mathcal{S}$-neighborhood $U(\beta_0)$ of β_0 in $\mathrm{Smbl}_2(\mathbb{R}^n)$ such that each $\beta \in U(\beta_0)$ is both of real principal type and pseudoconvex.

Remark 3.9. In terms of the $\mathcal{F}\mathcal{S}^r$-topologies (which involve only derivatives up to and including order r) Lemma 3.4, Corollary 3.5, and Theorem 3.6 are results valid in the $\mathcal{F}\mathcal{S}^0$ topology. On the other hand, Proposition 3.8 is valid in the finer $\mathcal{F}\mathcal{S}^1$ topology.

In [5] Brill and Cantor considered manifolds with spherical ends of the form $e \cong \{(x_1, \ldots, x_n); |x| > r\} \subseteq \mathbb{R}^n$ having fixed coordinate representation and implicit metrizations. As a final comment we note that many of their results only depend on the order class $\mathcal{O}(g)$ and not the particular metric g chosen.

REFERENCES

1. Beem, J. K.; Ehrlich, P. E.: Global Lorentzian Geometry. Marcel Dekker (1981).
2. Beem, J. K.; Parker, P. E.: Klein-Gordon solvability and the geometry of geodesics. Pac. J. Math. 107, 1-14 (1983).
3. Beem, J. K.; Parker, P. E.: Whitney stability of solvability (to appear).
4. Birkhoff, G.; Rota, G. C.: Ordinary Differential Equations, 2nd ed. Blaisdell (1969).
5. Brill, D.; Cantor, M.: Compos. Math. 43, 317-330 (1981).
6. Cheeger, J.; Ebin, D. G.: Comparison Theorems in Riemannian Geometry. North-Holland (1975).
7. Greene, R. G.; Wu, H. H.: Function Theory on Manifolds Which Possess a Pole. Springer (1979).
8. Hirsch, M.: Differential Topology. Springer (1976).
9. Hörmander, L.: Comm. Pure Appl. Math. 32, 359-443.
10. Kohn, J. J.; Nirenberg, L.: Comm. Pure Appl. Math. 18, 269-305 (1965).
11. Lang, S.: Differential Manifolds. Addison-Wesley (1972).
12. Luke, Y.: The Special Functions and Their Approximations, vol. I. Academic Press (1969).
13. Michor, P. W.: Manifolds of Differentiable Mappings. Shiva (1980).
14. Palais, R. S.: Seminar on the Atiyah-Singer Index Theorem, Ann. Math. Study 57, Princeton (1965).
15. Parker, P. E.: New directions in relativity and quantization of manifolds, in Quan-Theory and Gravitation, A. R. Marlow, ed. Academic Press (1980), pp. 137-146.
16. Sasaki, S.: Tôhoku Math. J. 10, 338-353 (1958).
17. Trèves, F.: Topological Vector Spaces, Distributions, and Kernels. Academic Press (1967).
18. Trèves, F. Introduction to Pseudodifferential and Fourier Integral Operators, vols. 1 and 2. Plenum (1980).

6 THE SIMPLEST NONLINEAR YANG-MILLS THEORY THAT WORKS*

Melvyn S. Berger / Center for Applied Mathematics and Mathematical Science, University of Massachusetts at Amherst, Amherst, Massachusetts

Yang-Mills theory presents an attempt to understand subtle quantum phenomena via certain nonlinear partial differential equations arising as Euler-Lagrange equations of certain real-valued functionals that are invariant under a given gauge group G. Thus it seems natural to inquire whether there are known physical problems that can be successfully and quantitatively treated by such an approach and to explore (as fully as possible) the associated mathematical structures involved. Perhaps the simplest and most successful theory of this type to date is that of Ginzberg and Landau [1] first proposed in 1950 to describe certain nonlinear macroscopic effects in superconductivity. The equilibrium states of this theory are determined by solving parameter dependent boundary value problems for a system of nonlinear elliptic equations combining a nonlinear Schrödinger equation and nonlinear version of Helmholtz's equation. The purely nonlinear mathematic aspects of the theory however, yielded unexpected physical predictions for sufficiently large parameter values. In particular, Abrikosov [2] found that when the relevant parameter, λ say, exceeds a certain critical number $\lambda_0 = 1$ (say), new stable secondary solutions (termed "vortices") appear. Mathematically, these solutions turned out to be classified by an integer n related to $\pi_1(S^1)$ implying that discrete quantum states could be classified by homotopy theory. In contemporary science, high values of the parameter λ have proved extremely important (Type II superconductivity), and Abrikosov's pioneering predictions based on an interesting combination of physical and mathematical reasoning have proved of fundamental importance.

Recently Taubes and Jaffe [3, 4] have carefully explored the case with $\lambda = 1$ using self-duality arguments as in the studies of instantons by Atiyah, Hitchin, and Singer [5]. However, when $\lambda > 1$ (the physically relevant case), this self-duality property of the Ginzberg-Landau equation breaks down and new results are called for. One hint is Abrikosov's observation (based on physical reasoning) that for n = 1 in the limit as $\lambda \to \infty$, the Ginzberg-Landau equations become linear and the associated vortex solution is given by the Green's function of an associated linear Helmholtz equation. In fact, we indicate below that for n = 1 when $1 < \lambda < \infty$ there is a one parameter family of smooth solutions of the Ginzberg-Landau equations interpolating between the vortex solution with n = 1 and $\lambda = 1$ and the n = 1, $\lambda = \infty$ vortex solution given by the Green's function mentioned above. Vortex solutions with n > 1 can be shown to be unstable in the sense that the second variation of the associated energy functional F_λ can be made negative definite, so that for $\lambda > 1$ the vortex solution with n = 1 is more stable, and in fact these n = 1 vortices are exactly the ones physically observed.

*Research partially supported by an NSF grant and an AFOSR grant. Presented at the Eastern AMS Meeting, Providence, R.I., October 1980.

1. MATHEMATICAL FORMULATION AND STATEMENT OF RESULTS

The equilibrium states of the Ginzberg-Landau equations are the finite energy smooth critical points of the following energy functional defined over $\mathbb{R}^2$.

$$F_\lambda(\emptyset, A) = \int_{\mathbb{R}^2} \left\{\frac{1}{2}|\mathrm{curl}\ A|^2 + \sum_j |(\partial_j - iA_j)\phi|^2 + \frac{\lambda}{4}(|\phi|^2 - 1)^2\right\} \tag{1}$$

Here ϕ is a complex scalar field and A is the vector potential of the magnetic field h so that h = curl A. The associated critical points can be classified by the "vortex number" N, an integer-valued topological invariant given by the integral formula

$$N = \frac{1}{2\pi}\int_{\mathbb{R}^2} [\mathrm{curl}\ A]\ dV \tag{2}$$

Finite energy smooth critical points of (1) with $N \neq 0$ are called "vortex solutions" (or vortices) in analogy with the theory of ideal fluids. These solutions appear to have the most physical relevance since when N = 1 they minimize the associated Gibbs free energy functional. In fact, quantitative results on these N = 1 vortex solutions are generally obtained by assuming $\lambda = \infty$ and studying solutions (tending to zero at infinity) of the Helmholtz equation

$$\Delta h - h = -2\pi\delta(x) \tag{3}$$

where $\delta(x)$ is the Dirac delta function.

We now state two results on this situation.

Theorem 1 (on vortex families). For each finite $\lambda \geq 1$, the functional $F_\lambda(\emptyset, A)$ has a smooth rotationally symmetric vortex solution $(A_\lambda, \emptyset_\lambda)$, with N = 1, that is an absolute minimum of (1) among rotationally symmetric smooth finite energy fields $(\emptyset, A)$ with fixed vortex number N. Moreover, as $\lambda \to 1$, these vortex solutions tend to the N = 1 vortex solution of Taubes [4]. (Here rotationally symmetric means $\underline{A} = \alpha(r)\underline{\theta}$ and $\emptyset = \Phi(r)e^{i\theta}$).

Theorem 2 (on nonlinear desingularization). There is a sequence of numbers $\lambda \to \infty$ such that for the N = 1 vortex solutions (described in Theorem 1) $h_\lambda = \mathrm{curl}\ A_\lambda$ tends in the Sobolev space $W_{1,p}(\mathbb{R}^2)$ $(1 < p < 2)$ to the Green's function solution of (3) above.

2. IDEA OF PROOFS

First we rewrite the functional F_λ defined by (1) in terms of the vortex number N (utilizing integration by parts) as follows, via [8],

$$F_\lambda(\emptyset, A) = N + \int_{\mathbb{R}^2} [E_1^2(\emptyset, A) + E_2^2(A, \emptyset) + E_3^2(A, \emptyset)] + \left(\frac{\lambda - 1}{2}\right)\int_{\mathbb{R}^2} (|\phi|^2 - 1)^2 \tag{4}$$

where E_1, E_2, and E_3 are first order differential operators. The vortex number N defined by (2) is a "natural constraint" as in ([6], p. 331) for variational problems associated with (1). Now, assuming $\lambda > 1$, N = 1 and rotational symmetry, the direct method of the calculus of variations shows that the minimum of $F_\lambda(\emptyset, A)$ is attained. Moreover, as $\lambda \to 1$, min $F_\lambda(\emptyset, A) \to 1$ and the associated minimizing solutions tend to a N = 1 critical point of (1) with $\lambda = 1$. By the results of [3, 4], this limit must be the N = 1 solution of Taubes. Theorem 2 follows by considering the isoperimetric problem of minimizing $F_1(\emptyset, A)$, assuming radial symmetry, with fixed N = 1 and fixed $\int_{\mathbb{R}^2} (|\phi|^2 - 1)^2 = R$ (say), as $R \to 0$. We show that the associated critical points coincide with those of Theorem 1 for certain values of $\lambda(R)$ and that $\lambda(R) \to \infty$ as $R \to 0$. Then for such $\lambda(R) \to \infty$ we rewrite the associated Euler-Lagrange equations of (1) in terms Φ_{λ_R} and $n_{\lambda_R} = \text{curl } A_{\lambda_R}$ taking care to study their dependence on λ_R. In particular, $\Phi_{\lambda_R}(r) \to 1$ except at $r = 0$ and $h_{\lambda_R} \to$ Green's function solution of (3), as desired, in accord with our ideas of nonlinear desingularization [7].

REFERENCES

1. Ginzberg, V. L.; Landau, L. D.: J.E.T.P. 20, 1064-1073 (1950).
2. Abrikosov, A. A.: Sov. Phys. J.E.T.P. 5, 1174-1182 (1957).
3. Jaffe, A.; Taubes, G.: Vortices and Monopolies, Progress in Physics 2. Birkhauser, Boston (1980).
4. Taubes, C.: Comm. Math. Phys. 72, 274-291 (1980).
5. Atiyah, M.; Hitchin, N.; Singer, I.: Proc. Roy. Soc. London Ser. A 362, 425-461 (1978).
6. Berger, M. S.: Nonlinearity and Functional Analysis. Academic Press (1977).
7. Berger, M. S.; Fraenkel, L. E.: Comm. Math. Phys. 77, 149-172 (1980).
8. Bogomolny, E.: Sov. J. Nucl. Phys. 24, 449-454 (1976).

7 EULER, MORSE, AND THE CALCULUS OF VARIATIONS

Samuel Bourne / Department of Mathematics, University of California at Berkeley, Berkeley, California

Léonhard Euler (1707-1783) created a new branch of mathematics known in his day as the Calculus of Variations and known today as Variational Analysis. Motivated by geometric considerations he deduced its first general principle which is now referred to as Euler's differential equation for the determination of the maximizing or minimizing arcs. As part of preparing this article, I went back to the works [1] of Joseph-Louis Lagrange (1736-1813) to his commentary on Euler's work on the Calculus of Variations.

Let me now turn to a more recent giant in this field. The New York Times of June 9, 1956 stated that "Professor Morse is known as the originator of a branch of mathematics. His variational theory in the large has been applied extensively in other branches of mathematical physics."

1. LAGRANGE ON EULER

Sur la Méthode des Variations. (Miscellanea Taurinensia, t. IV, 1766-1769.)

J'ai donné, dans le second volume des Miscellanea Taurinensia (*) une nouvelle méthode pour la solution des Problèmes où il s'agit de trouver les courbes qui jouissent de quelque propriété du maximum ou du minimum. Cette méthode, qu'on peut très-bien appeler, d'après M. Euler, méthode des variations, avait déjà été communiquée dès 1755 à ce grand Géomètre, qui l'avait jugée digne de son attention et de son suffrage, comme il parait par les différentes lettres qu'il m'a écrites sur ce sujet, et que je conserve encore. Dans une de ces lettres, datée du 2 octobre 1759, il s'exprime en ces termes:

> Analitica tua solutio Problematis isoperimetrici continet, ut video, quidquid in hae materia desiderari potest, et ego maxime gaudeo, hoc argumentum, quod fere solus, post primos conatus, tractaveram, a te potissimum ad summum perfectionis fastigium esse erectum. Rei dignitas me excitavit, ut tuis luminibus adjutus ipse solutionem analiticam conscripserim, quam autem celare statui, donee ipse tuas meditationes publici juris feceris, ne ullam partem gloriæ tibi debitæ præripiam.

En effet, M. Euler a donné depuis, dans le tome X des Nouveaux Commentaires de Pétersbourg, imprimé en 1766, deux Mémoires assez étendus sur cette matière, dans lesquels, après m'avoir fait honneur de la méthode dont il s'agit, il en explique les principes et les usages avec beaucoup de détail et de précision (**). Après des témoignages aussi formels de la part d'un Géomètre tel que M. Euler, j'ai dû être surpris du peu de justice que m'ont rendue d'autres Géomètres, qui se sont depuis peu occupés du même sujet. M. Fontaine vient de donner, dans le volume de l'Académie des Sciences de Paris pour l'année 1767, un Mémoire intitulé: Addition à la méthode pour la

(*) Oeuvres de Lagrange, t. 1, p. 335.
(**) Voyez les pages 12 et 97 du tome cité.

solution des Problèmes de maximis et minimis. L'Auteur débute par avancer sans aucun fondement que "je me suis égaré dans la route nouvelle que j'ai prise, pour ' n'en avoir pas connu la vraie théorie." Ensuite, pour suppléer au défaut prétendu de ma méthode, il en donne deux autres qu'il regarde comme nouvelles et fort supérieures à toutes les méthodes connues pour le même objet. Je ne crois pouvoir rien faire de mieux pour ma justification que d'inviter les connaisseurs à lire l'Ouvrage même de M. Fontaine et à le comparer avec le mien et avec celui de M. Euler. On verra, si je ne me trompe, que des deux méthodes de M. Fontaine, l'une n'est autre chose que celle que M. Euler avoit donée dans son excellent Ouvrage intitulé Methodus inveniendi lineas curvas, etc., et qu'il a ensuite abandonnée pour adopter la mienne, et que l'autre est la même, quant au fond, que ma méthode, dont elle diffère seulement par la manière vague et imparfaite dont elle est présentée.

Les autres Géomètres dont j'aurais aussi en quelque façon sujet de me plaindre, quoique par une raison bien différente de la précédente, sont les Pères minimes Le Seur et Jacquier, qui viennent de publier à Parme un très-bon Traité de Calcul intégral.

Ces savants, ayant eu pour objet de rassembler les principales méthodes relatives au Calcul intégral, n'ont pas oublié la nouvelle méthode des variations, à laquelle ils ont même destiné un Chapitre entier du second volume de leur Ouvrage. Il aurait été naturel et même équitable qu'ils eussent fait quelque mention de mon Mémoire de 1762, surtout après en avoir transcrit, comme ils ont fait, plusieurs pages entières (*); cependant je serais bien éloigné de leur reprocher cette omission, s'ils s'étaient contentés d'exposer la méthode dont il s'agit, sans citer personne, comme ils en ont usé dans d'autres endroits du même volume (**); mais comme, par la citation des Mémoires de M. Euler dont nous avons parlé plus haut, ils paraissent vouloir lui attribuer cette méthode, je crois pouvoir faire remarquer que j'en suis le premier auteur, et que je n'en partage la possession avec personne.

Je dois encore observer que MM. Le Seur et Jacquier ne s'expriment pas exactement quand ils disent (page 531 du tome II) que M. Euler a démontré que dans les trajectoires décrites par un nombre de corps quelconque, l'intégrale de la vitesse multipliée par l'élément de la courbe est toujours un maximum ou un minimum. M. Euler n'a donné sur ce sujet que ce que l'on trouve dans un Appendice ajouté à son excellent Traité sur les isopérimètres, où il fait voir que la trajectoire qu'un corps doit décrire par des forces centrales quelconques est la même que la courbe qu'on trouverait en supposant que l'intégrale de la vitesse multipliée par l'élément de la courbe fût un maximum ou un minimum.

L'application de ce beau théorème à un système quelconque de corps, et surtout la manière de s'en servir pour résoudre avec la plus grande simplicité et généralité tous les problèmes de Dynamique, m'est entièrement due, et ce qui doit le prouver invinciblement, c'est que cette théorie dépend des mêmes principes que celle des variations, et que l'une et l'autre ont paru dans le même volume des Miscellanea Taurinensia pour les années 1760 et 1761. Je pourrais ajouter que j'avais aussi communiqué cette découverte à M. Euler dès 1756, et comme ce grand Géomètre a bien voulu l'honorer alors de son approbation, je ne doute pas qu'il ne fût très-porté, si l'occasion s'en présentait, à me rendre sur ce sujet la même justice qu'il a bien voulu me rendre à l'égard de la méthode de maximis et minimis.

(*) Voyez les pages 521 et suiv. du volume cité, et les pages 174 et suiv. du tome II des Miscellanea Taurinensia (OEuvres de Lagrange, t. I, p. 336 et suiv.).

(**) Voyez les pages 448 et suiv. de ce volume, et les pages 179 et suiv. du tome III des Miscellanea Taurinensia (OEuvres de Lagrange, t. I, p. 471 et suiv.).

On the Method of Variations (modern: On Variational Calculus)

I gave, in the second volume of Miscellanea Taurinensia, a new method for the solution of problems where it is desired to find the curves that possess certain proper ties of maximum or minimum. That method, which, following Euler, one can very well call the method of variations, had already been communicated early as 1755 to this famous geometer who adjudged it worthy of his attention and his approbation, as is apparent in the several letters he wrote to me on the subject, letters that are still in my possession. In one of these letters, dated October 2, 1759, he expresses himself thus:

> Your solution of the isoperimetric problem contains, as I see it, all that could be desired on this topic. I am very pleased that the proposition which I was almost alone in developing beyond the very earliest attempts, has been extended by you to the highest peak of perfection. The quality of your contribution has inspired me to write up, aided by your insight, a full analytic solution. However, I have to keep it private until you have made public your thoughts, less I snatch away any part of the glory due to you.

Indeed, M. Euler has since given, in volume X of the Nouveaux commentaires de Petersbourg, printed in 1766, two rather lengthy memoirs on this matter. In these, after honoring me with credit for the method in question, he explains its principles and its uses with great detail and precision. After such formal testimony from a Geometer of M. Euler stature, I had to register surprise at the little justice rendered to me by other Geometers who, since then, have worked on the same subject. M. Fontaine has just given, in the 1767 volume of the Academy of Science of Paris, a Memoir entitled: Addition to the method for the solution of problems of maxima and minima. The author starts by proposing without any foundation that "I went astray in the new path I took because I did not know the true theory." Then, to overcome the pretended defects of my method, he supplies two other ones regarded by him as new and vastly superior to all known methods devised for the same end.

I believe I cannot do any better to justify myself than to invite knowledgeable persons to read the (very) work of M. Fontaine and compare it with mine and with Euler's. One will see, if I am not mistaken, that of the two methods proposed by M. Fontaine, one is nothing but that already given by M. Euler in his excellent paper entitled Methodus inveniendi lineas curves, etc. a method he later on abandoned to adopt mine. The other method of M. Fontaine is the same, as its principle, as my method from which it differs only by the vague and imperfect way of its presentation.

The other Geometers who gave me in some way a reason to complain, albeit a very different reason from the preceding one are the Lesser Fathers, Le Soeur, and Jacquier, who just published in Parma a very good Treatise on Integral Calculus.

These scholars, having in mind collecting together the principal methods relating to Integral Calculus, did not forget the new method of variations, to which they even described an entire chapter of the second volume of their Treatise. It would have be natural and even just that they would mention in some way my Memoir of 1762, especially after copying, as they did, several entire pages. Nevertheless, I would remain far from giving them reproach for this omission, if they had contented themselves with exposing the method in question without any reference, as they did in other places in the same volume. However, since by the citations of the memoirs of Euler discussed above, they seem to want to attribute that method to him, I believe I am entitled to remark that I am the first author of it and that I do not share ownership of it with anybody.

I must also observe that Le Soeur and Jacquier do not express themselves with exactitude when they say (page 531 of volume II) that M. Euler has proved that for trajectories described by an arbitrary number of material bodies, the integral of the product of the velocity by the line element is always a maximum or a minimum. M. Euler has given on this subject only what can be found in an Appendix added to his excellent Treatise on isoperimetries, where he shows that the trajectory described by a material body subjected to arbitrary forces is the same as the curve one would find by assuming that the integral of the velocity multiplied by the line element is a maximum or a minimum.

The application of this beautiful theorem to an arbitrary system of material bodies, and above all the way to use it to solve with the utmost simplicity and generality all the problems of Dynamics, is entirely due to us and what can prove it beyond any possible attack is that this theory depends on the same principles as that of variations and that both appeared in the same volume of Miscellanea Taurinensia for the year 1760 and 1761.

I could add that I had also communicated this discovery to M. Euler as early as 1756, and, since this famous Geometer has kindly honored it with his approval, I do not doubt that he would be very inclined, if the occasion presented itself, to render to me the same justice that he consented to render with respect to the method of maxima and minima.

2. (HAROLD) MARSTON MORSE (1892-1977)

After taking my Doctoral examination at the Johns Hopkins University in the Fall of 1949 I was appointed a member of the Institute of Advanced Study in Princeton, New Jersey, to commence the following January. Having studied General Relativity, under F. D. Murnaghan, Hopkins' applied mathematician, and my chairman, my hope was to study under and to work with its originator, Albert Einstein, then in residence at the Institute as Professor-Emeritus in its School of Physics. This hope was realized and my association and friendship with Professor Einstein lasted all of five years.

The School of Mathematics to which I was attached was then run by Professor Marston Morse with the assistance of Professor Deane Montgomery. Although Morse's immense reputation in the field of mathematics preceded my coming to the Institute, my contacts with him were of a limited nature, for my interests were at that time far afield from his. Our acquaintance did not ripen to a first name friendship until the last few years of this life. It was during the period that Themistocles Rassias shared my office in Evans Hall of the University of California at Berkeley, that a correspondence developed between Morse and myself concerning the studies by George and Themistocles M. Rassias and myself on Morse's Global Variational Analysis: (Weierstrass) Integrals on a Riemannian Manifold [2]. One in which Morse conjectures that the Fréchet numbers for a compact connected smooth manifold are finite (cf. T. Rassias [5]) and the other in 1977 in which he bemoans the fact that he may not complete his work in his lifetime and unfortunately he died two months later; a premonition of his death on June 22, 1977. I quote from T. Rassias [4]: "M. Morse while working on the problem of Plateau realized the need for a new framework of the variational problems by making use of the theory of functions spaces. One of the most important questions concerning the problem of Plateau and especially the existence and non-uniqueness theory of the problem is the one studied by M. Morse and C. Tompkins [3]."

I end this tribute to Marston Morse by reprinting in its entirety from the New York Times Book Review of November 12, 1950, "Speaking of Books" by J. Donald Adams.

"Speaking of Books" by J. Donald Adams

With something not far distant from fear and trembling, this department pulls itself together to approach a topic about which, probably, it has no business to be writing at all. Three Sundays ago, in reporting on the Kenyon College conference held in honor of Robert Frost, I ended by saying that in my opinion the most interesting contribution to the conference was a paper read by the eminent mathematician, Marston Morse, entitled, "Reflections on Evaluations in Mathematics and the Arts," and that I proposed to make it the subject of another article.

Even if the mathematics with which Professor Morse is concerned were on the high-school level, I should be starting with a heavy handicap, remembering, as I vividly do, my almost complete incapacity to deal understandingly with the elementary concepts of algebra and geometry; but worse still, Professor Morse, a colleague of Einstein at the Institute for Advanced Studies in Princeton, is one of those men who sail serenely through the mathematical stratosphere, in that lofty region where poetry and mathematics become sisters under the skin. In any case, the reflections set forth in this column will be, for much the most part, his, not mine. I undertake to summarize them because I think they are deeply suggestive concerning the underlying unity with which the human mind operates creatively, whatever kind of creation it is, whether in science, philosophy and religion, or the arts.

The widest-ranging minds have always been aware of that unity; Leonardo's for one, Goethe's for another. It was Goethe who gathered strength as a poet from his scientific consideration of Nature, and who contended that if art undertakes to create works which can compete in truth and beauty with those of Nature, the artist must "learn at least to some extent from Nature the way in which she proceeds in the forming of her works." As Prof. Karl Vietor has pointed out, Goethe's work reveals how "his insights in one field clarify and confirm in another."

Was it not his recognition of this unity which caused Robert Frost once to remark that "a poet must lean hard on facts, so hard, sometimes, that they hurt?" And is there not, in Frost's much-quoted and profound remarks about the figure a poem makes more than a slight resemblance to a mathematical theorem when he says that a poem "begins in delight and ends in wisdom. *** It begins in delight, it inclines to the impulse, it assumes direction with the first laid down, it runs a course of lucky events, and ends in a clarification of life—not necessarily a great clarification, such as sects and cults are founded on, but in a momentary stay against confusion."

Professor Morse, as a matter of fact, uses Frost's conception of the figure a poem makes as evidence of the first essential bond he finds between mathematics and the arts: the fact that "discovery in mathematics is not a matter of logic." Rather, he contends, it is "the result of mysterious powers which no one understands, and in which the unconscious recognition of beauty must play an important part. Out of an infinity of designs a mathematician chooses one pattern for beauty's sake, and pulls it down to earth, no one knows how. Afterward the logic of words and forms sets the pattern right. Only then can one tell someone else." This is much what Frost was saying of writing poetry: "I tell how there may be a better wildness of logic, than of inconsequence. But the logic is backward, in retrospect after the act." And, "For me the initial delight is in the surprise of remembering something I didn't know I knew. I am in a place, in a situation, as if I had materialized from cloud, or risen out of the ground."

These statements were compared by Professor Morse with the accounts given by the great mathematicians Poincaré and Gauss, of how they came to make some of their discoveries. I wish there were space for me to quote from them. Previously he had drawn similar evidence for his contention from the work and statements of Albrecht Dürer, the geometer-artist, who, as Professor Morse observed, more than any other artist "formulated the rules of symmetry, perspective and proportion, and used them in his art."

The underlying unity of creation between mathematics and music, which is more generally accepted (being more obvious), was also touched upon by Professor Morse, who framed his thesis in these words: "The basic affinity between mathematics and the arts is psychological and spiritual and not metrical or geometrical."

Also, he found further evidence of this affinity in the comparative history of the arts, which he described as "the history of recurring cycles and sharp antitheses. These antitheses set pure art against mixed art, restraint against lack of restraint, the transient against the permanent, the abstract against the non-abstract." These antitheses, he pointed out, are found also in the history of mathematics.

He ended by regretting that science is so often attacked without perception of its creative unity with the arts; that it is so often viewed "without its penumbra or its radiance." The creative scientist, he argued, "lives in 'the wildness of logic' where reason is the handmaiden and not the master. *** It is the hour before the break of day when science turns in the womb."

REFERENCES

1. Lagrange, J. L.: Oeuvres, t. II. Gauthier-Villars (1868), pp. 37-39.
2. Morse, M.: Global Variational Analysis. Princeton University Press (1976).
3. Morse, M.; Tompkins, C.: Ann. Math. 40, 443 (1939).
4. Rassias, T. M.: Morse theory and Plateau's problem, in Selected Studies. North-Holland (1982), pp. 261-292.
5. Rassias, T. M.: Morse Theory in Global Variational Analysis, Global Analysis—Analysis on Manifolds. Teubner (1983), pp. 7-18.

8 THE BIRTH AND EARLY DEVELOPMENTS OF PADÉ APPROXIMANTS*

Claude Brezinski / Laboratoire d'Analyse Numérique et d'Optimisation, Université de Lille 1, UER IEEA, Villeneuve d'Ascq Cedex, France

1. THE BIRTH OF PADÉ APPROXIMANTS

1.1. Daniel Bernoulli

Daniel Bernoulli (Groningen, 2/2/1700—Basel, 3/17/1782) is well known among numerical analysts for his method for computing the largest zero of a polynomial (1728). Bernoulli was led to his method by a study of linear difference equations. In two memoirs published in 1730 he tries to extend the method to the smallest zero of an infinite power series [3].

To the equation

$$1 = a_1 x + a_2 x^2 + \cdots + a_n x^n + \cdots$$

he associates the difference equation of infinite order

$$X_n = a_1 X_{n-1} + a_2 X_{n-2} + \cdots$$

Thus the computation of the sequence (X_n) needs an infinite number of initial conditions. Bernoulli uses an infinite number of zeros followed by one. The following terms are then obtained from the recurrence relation and he gets

$$\ldots\ 0,\ 0,\ 1,\ a_1,\ a_1^2 + a_2,\ a_1(a_1^2 + a_2) + a_1 a_2 + a_3,\ \ldots$$

The ratio of two consecutive terms of the sequence gives an approximate value of the smallest zero of the equation.

Bernoulli uses the same method for inverting a power series. Let

$$y = x + ax^2 + bx^3 + \cdots$$

He wants to express x in terms of y. He first writes x as a power series in y. The method of indeterminate coefficients gives

$$x = y - ay^2 + (2a^2 - b)y^3 + \cdots$$

On the other hand one has

$$1 = \frac{1}{y}x + \frac{a}{y}x^2 + \frac{b}{y}x^3 + \cdots$$

Using the preceding technique for difference equations of infinite order he obtains the sequence

$$\ldots,\ 0,\ 0,\ 1,\ \frac{1}{y},\ \frac{1}{y^2} + \frac{a}{y},\ \frac{1}{y^3} + \frac{2a}{y^2} + \frac{b}{y},\ \ldots$$

---

*Presented at the 86th summer meeting of the American Mathematical Society, Toronto, August 23-27, 1982.

The ratio of two consecutive terms of this sequence gives an approximate value of x. For example he has

$$x = \frac{y + 3ay^2 + (a^2 + 2b)y^3 + cy^4}{1 + 4ay + 3(a^2 + b)y^2 + (2ab + c)y^3 + cy^4}$$

Thus x is approximated by a rational fraction in y. If this rational fraction is developed into an ascending power series in y (by effecting the division) it matches the series obtained from the indeterminate coefficients method up to the term whose degree equals the degree of the numerator.

This kind of approximation is weaker than Padé approximation whose degree of approximation is equal to the sum of the degrees of the numerator and the denominator of the rational fraction.

Such approximations are now called Padé-type approximations. They were introduced in [5] and developed in [6].

To illustrate his method, Bernoulli considers the computation of π. 0 is the smallest zero of sin x. Thus π is the smallest zero of sin x/x.

$$\frac{\sin x}{x} = 1 - \frac{x^2}{3!} + \frac{x^4}{5!} - \frac{x^6}{7!} + \cdots$$

which leads to the equation

$$1 = \frac{\pi^2}{3!} - \frac{\pi^4}{5!} + \frac{\pi^6}{7!} - \cdots$$

Setting $q = \pi^2/6$ he finds q = 4191/2555 and $\pi = 3.1376$ while the series gives q = 1048067/673750 and $\pi = 3.0550$.

1.2. Euler

On several occasions Leonhard Euler (Basel, 4/15/1707—St. Petersburg, 9/18/1783) was led to use rational approximations to series which are, in fact, Padé approximants.

An important source about Euler's work is his correspondence with Christian Goldbach (Königsberg, 3/18/1690—Moscow, 11/20/1764) which runs over for many years [17]. In a letter dated October 17, 1730, Euler considers the series

$$S = x + \frac{1}{2.3}\frac{x^3}{b^2} + \frac{1.3}{2.4.5}\frac{x^5}{b^4} + \frac{1.3.5}{2.4.6.7}\frac{x^7}{b^6} + \cdots$$

where b is the diameter of a circle, x is the chord and S the corresponding arch. He gives, without any explanations, the following approximations of S

$$\frac{60b^2x - 17x^3}{60b^2 - 27x^2}$$

$$x + \frac{840b^2x^3 - 122x^5}{120b^2(42b^2 - 25x^2)}$$

It is easy to check that the first approximation satisfies $S + 0(x^7)$ and thus is identical with the Padé approximant [3/2]. The second approximation satisfies $S + 0(x^9)$ and thus is identical to [5/2].

However, Euler does not seem to be conscious of the characteristic property of his approximations to match the initial series up to a certain degree.

Another quite important letter for our purpose was written by Euler to the German astronomer Tobias Mayer (Marbach, 2/17/1723—Göttingen, 2/20/1762) [14]. On July 4, 1751 Mayer wrote to Euler

> I have come across an equation which, though it would appear to be simple, I am unable to integrate. It is namely the following: dy = dx/log x, or this: $dz = e^{e^x} dx$, where log e = 1. In the treatise in which you, sir, have explained

> Mr. Robin's rules of artillery [Neue Grundsätze der Artillerie, Berlin, 1745 was a translation with commentaries of Robin's New Principles of Gunnery, London, 1742] there is a similar equation on page 131 which, however, is only integrated by approximation, an advantage not permitted to me with the above equation. I could not let pass the opportunity of seeking your advice regarding this equation, sir, and I beg you not to interpret such [a desire] unkindly.

On July 27, Euler answered:

> This equation $dy = dx/\ell x$ is in no way integrable, and if one does not want to use the approximation I know of no other method than that the value of y be determined through the Quadrature Mechanics [that is a numerical quadrature method] which can in fact be regarded simply as an approximation, and a very incomplete one at that. I wanted to determine the value of y from the following set-up through a series:
>
> $$y = \frac{x}{\ell x} - x\int x\, d\frac{1}{\ell x} = \frac{x}{\ell x} + \int \frac{dx}{(\ell x)^2};$$
>
> further
>
> $$\int \frac{dx}{(\ell x)^2} = \frac{x}{(\ell x)^2} - \int x\, d\frac{1}{(\ell x)^2} = \frac{x}{(\ell x)^2} + 2\int \frac{dx}{(\ell x)^3}.$$
>
> Further
>
> $$\int \frac{dx}{(\ell x)^3} = \frac{x}{(\ell x)^3} - \int x\, d\frac{1}{(\ell x)^3} = \frac{x}{(\ell x)^3} + 3\int \frac{dx}{(\ell x)^4}$$
>
> and so on, from which one finally obtains
>
> $$y = \frac{x}{\ell x} + \frac{x}{(\ell x)^2} + \frac{1.2.x}{(\ell x)^3} + \frac{1.2.3.x}{(\ell x)^4} + \frac{1.2.3.4.x}{(\ell x)^5} + \text{etc.}$$
>
> Then, if $x = 0$, $y = 0$: but if $x = 1$, $y = \infty$, from which it follows that since x cannot be negative (otherwise ℓx would be imaginary), there cannot be other values to seek for y except those for which x is less than unity; and then ℓx becomes a negative quantity.
>
> Thus $\frac{1}{\ell x} = -U$; and so
>
> $$y = -Ux(1 - 1.U + 1.2.U^2 - 1.2.3.U^3 + 1.2.3.4.U^4 - \text{etc.})$$
>
> and then the whole problem depends upon the sum of the bracketed series which, since it is strongly divergent, is subject to new difficulties.
>
> I have, however, discovered that the series
>
> $$1 - 1U + 2U\ \ - 6U^3 + 24U^4 - 120U^5 + 720U^6 - \text{etc.} = S$$
>
> is equal to the following continuous fraction
>
> $$S = \frac{1|}{|1} + \frac{U|}{|1} + \frac{U|}{|1} + \frac{2U|}{|1} + \frac{2U|}{|1} + \frac{3U|}{|1} + \frac{3U|}{|1} + \frac{4U|}{|1} + \frac{4U|}{|1} + \frac{5U|}{|1} + \frac{5U|}{|1} + \text{etc.}$$
>
> which always closely determines the limits of the value of S, and thus one can approximate to the value of S as closely as one will. Then the values approximating to S are:
>
> $$1;\ \frac{1}{1+U};\ \frac{1+U}{1+2U};\ \frac{1+3U}{1+4U+2UU};\ \frac{1+5U+2UU}{1+6U+6UU};\ \frac{1+8U+11UU}{1+9U+18UU+6U^3};$$
>
> $$\frac{1+11U+26UU+6U^3}{1+12U+36UU+24U^3};\ \frac{1+15U+58UU+50U^3}{1+16U+72UU+96U^3+24U^4};\ \text{etc.},$$
>
> of which every alternate one is greater than S.

It is easy to check that the rational fractions given by Euler are the Padé approximants [0/0], [0/1], [1/1], [1/2], ... of S.

Euler was quite familiar with the process of transforming a formal power series into a continued fraction. His method for performing this transformation is simply the division process which is quite similar to Euclid's algorithm for obtaining the g.c.d. of two positive integers. He used this technique in at least two papers: one in 1755 and another in 1776. Since this method had a great historical influence, let me explain it in detail and prove some results.

Let f be a formal power series

$$f(t) = c_0 + c_1 t + c_2 t^2 + \cdots$$

f can always be written as

$$f(t) = P_0(t) + t^{m_1} g_1(t)$$

where P_0 is a polynomial with the exact degree n_0 (P_0 nonidentically zero), where $m_1 > n_0$ and where g_1 is a power series such that $g_1(0) \neq 0$. If g_1 is identically zero then f is a rational fraction (a polynomial in that case) and the process has to be stopped.

Let f_1 be the reciprocal series of g_1, that is, the series formally defined by $f_1(t) g_1(t) = 1$. Since $g_1(0) \neq 0$, f_1 always exists and $f_1(0) \neq 0$. Then

$$f(t) = P_0(t) + \frac{t^{m_1}}{f_1(t)}$$

f_1 can always be written as

$$f_1(t) = P_1(t) + t^{m_2} g_2(t)$$

where P_1 is a polynomial with the exact degree n_1 ($P_1(0) \neq 0$), where $m_2 > n_1$ and where g_2 is a power series such that $g_2(0) \neq 0$. If g_2 is identically zero then f is a rational fraction and the process has to be stopped.

Let f_2 be the reciprocal series of g_2 which exists since $g_2(0) \neq 0$ and is such that $f_2(0) \neq 0$. Then

$$f(t) = P_0(t) + \frac{t^{m_1}|}{|P_1(t)} + \frac{t^{m_2}|}{|f_2(t)}$$

The same process can be repeated as far as we don't get a series g_i which is identically zero, in which case f is a rational fraction and the process has to be stopped. Finally we obtain a continued fraction expansion of f of the form

$$f(t) = P_0(t) + \frac{t^{m_1}|}{|P_1(t)} + \frac{t^{m_2}|}{|P_2(t)} + \cdots$$

It must be noticed that at each stage of the process n_k can be arbitrarily chosen (P_k is formed by the series f_k up to the degree n_k inclusively) but m_{k+1} is imposed by the form of f_k since m_{k+1} is the degree of the first nonzero term after n_k.

We shall now study the conditions under which the successive convergents of the preceding continued fraction are the Padé approximants of f.

Let us first look at the degrees of the numerators and the denominators of these convergents. Let C_k be the k-th convergent of the continued fraction. We set

$$C_k(t) = \frac{A_k(t)}{B_k(t)}$$

and let N_k and M_k be the degrees of A_k and B_k, respectively. The recurrence relationships of the continued fraction are

$$A_k(t) = P_k(t)A_{k-1}(t) + t^{m_k}A_{k-2}(t)$$

$$B_k(t) = P_k(t)B_{k-1}(t) + t^{m_k}B_{k-2}(t)$$

with

$$A_0(t) = P_0(t) \qquad A_1(t) = P_1(t)P_0(t) + t^{m_1}$$

$$B_0(t) = 1 \qquad B_1(t) = P_1(t)$$

From these relations we immediately get the

Theorem 1.

$$N_k \leq \max(n_k + N_{k-1}, m_k + N_{k-2})$$

$$M_k \leq \max(n_k + M_{k-1}, m_k + M_{k-2})$$

with

$$N_0 = n_0 \qquad N_1 \leq \max(n_0 + n_1, m_1)$$

$$M_0 = 0 \qquad M_1 = n_1$$

Remark. The inequalities can only hold when the quantities in parentheses are equal because, in that case, the degree can be reduced by cancellation of the terms of the highest degree.

Let us now study the approximating properties of the convergents of our continued fraction.

We have

$$C_{k+1}(t) - C_k(t) = (-1)^k \frac{t^{m_1} t^{m_2} \cdots t^{m_{k+1}}}{B_k(t)\, B_{k+1}(t)}$$

But $B_k(0) = P_0(0) \cdots P_k(0)$ and thus we have the

Theorem 2.

$$C_k(t) = f(t) + 0(t^{m_1 + \cdots + m_{k+1}})$$

We immediately obtain, from theorems 1 and 2:

Theorem 3. If

$$N_k + M_k + 1 \leq m_1 + \cdots + m_{k+1}$$

then

$$C_k(t) = [N_k/M_k]_f(t)$$

Let us now look at a particular case in which this condition is satisfied.

Theorem 4. If $\forall k$, $m_k = m$ and $n_k = n \leq m/2$ then

$$C_{2k}(t) = [n + km/km]_f(t) \quad \text{and} \quad C_{2k+1}(t) = [(k+1)m/n + km]_f(t)$$

Proof. $N_0 = m$, $M_0 = 0$, $N_0 + M_0 + 1 = n + 1 \leq m$ and thus the condition of theorem 3 is satisfied.

$$N_1 \leq \max(2n, m) \qquad M_1 = n$$

$$N_1 + M_1 + 1 \leq \max(3n + 1, n + m + 1)$$

But $n + 1 \leq m$ and thus $3n + 1 \leq m + 2n \leq 2m$ since $2n \leq m$.

Moreover, $n + m + 1 \leq 2m$, which shows that the condition of theorem 3 is satisfied. It is easy to prove by induction that

$$N_{2k} \leq n + km \qquad M_{2k} \leq km$$

$$N_{2k+1} \leq (k+1)m \qquad M_{2k+1} \leq n + km$$

Q.E.D.

In many cases if one chooses $n_k = 0$ then $m_{k+1} = 1$ and thus:

$$C_{2k}(t) = [k/k]_f(t) \quad \text{and} \quad C_{2k+1}(t) = [k + 1/k]_f(t)$$

Then C is the continued fraction corresponding to the series f, which explains why Euler and others have encountered Padé approximants.

1.3. George Anderson

George Anderson (fl. 1740) was an English mathematician, about whom nothing is known beyond what is contained in eight letters written to William Jones (L'lanfihangel Tw'r Beird, Wales, 1675—London, 7/3/1749). It seems that he was somebody of an humble origin since, although his handwriting was tolerably good, his spelling and the grammar have occasionally to be corrected. On the other hand he gave proof of "singular ability in treating the most advanced mathematical problems of the time, and by many indications show to have occupied a respectable position in life" [45]. He also studied in Leyden around 1740 and expressed a strong desire (not realized) to become a member of the Royal Society.

On September 16/27, 1740 he wrote to Jones from Leyden [42]:

> But to come to the theorems I am speaking of; let a - x, a, a + x be any three numbers in arithmetic proportion, whose logarithms I denote by L̦, L, L', respectively, then putting n = 0,4342944819 etc., to find L', from a being given, we have the three following theorems, and putting L' - L = ℓ, the value of x is had by the last three theorems.

$$\left.\begin{array}{ll} 1. & L' = L + \dfrac{2nx}{2a + x} \\ 2. & L' = L + \dfrac{3n(2a + x)}{6a + x + 6a^2/x} \\ 3. & L' = L + \dfrac{n}{3}\,\dfrac{15(2a + x)^2 - 4x^2}{(2a + x)10a + x + 10a^2/x} \end{array}\right\} \text{too little by}$$

$$1.\quad \frac{2nx^3}{3(2a+x)^3}\ \text{etc.} \qquad 2.\quad \frac{4nx^5}{45(2a+x)^5}\ \text{etc.} \qquad 3.\quad \frac{4nx^7}{175(2a+x)^7}\ \text{etc.}$$

and on the contrary

$$x = \frac{2\ell a}{2n-\ell} \qquad x = \frac{6\ell a}{3(2n-\ell) + 2\ell^2/n} \qquad x = \frac{(30n^2 + 8\ell^2)\ell a}{15n^2(2n-\ell) + 4\ell^2(12n-\ell)}$$

Where I must observe the lower theorem both for L' and x is always nearer the truth than that immediately above, and consequently the lowest in each is the nearest of all. I have also reduced each of them to that form which will give the least trouble in practice; for in the second and third for finding L', the divisor of $6a^2$ and $10a^2$ (viz. x) is commonly unity, because we may commonly suppose the logarithm of the number, immediately preceding that required, to be given.

The second in each is sufficient for practice, since thereby we can commonly have the logarithm or number true to ten or fifteen places. For example, let $a = 100$, $x = 1$, then $L = 2,000000$ etc., whence by the second theorem L' = $2 + \frac{3n \times 201}{600 + 1 + 60,000} = 2.0043213737824$, erring but 2 in the last place; the error being somewhat more than $\frac{4n}{45 \times (201)^5}$, and by the first term of the errors in each, we may always know how near we are to [the] truth. I have not set down those for the values of x, but they may easily be had by reducing their values to a series, and comparing it with

$$a \times \left(-1 + \frac{\ell}{n} - \frac{\ell^3}{3n^3} + \frac{2\ell^5}{45n^5} - \frac{17\ell^7}{315n^7} + \frac{62\ell^9}{2835n^9} - \text{etc.}\right)$$

which is equal to x.

It is easy to check, by taking for example $a = 1$, that the preceding expressions of L' are the Padé approximants of Log $(1 + x)$.

Thus Anderson went one step farther than Euler since he gave the first term of the error. However, he did not realize either that the degree of this first term is equal to the sum of the degrees of the numerator and the denominator of the approximant.

The method used by Anderson to derive these approximations is unknown.

1.4. Johann Heinrich Lambert

Johann Heinrich Lambert (Mulhouse, 8/26/1728—Berlin, 9/25/1777) contributed to continued fractions on several occasions. The most well known is the proof, which makes use of the continued fraction for tan x obtained by a division process, that π is an irrational number.

As we shall see, Lambert also gave the direct approach to Padé approximants in a paper published in 1758 [36].

His fundamental remark is that if the geometrical series

$$y = x + mx^2 + m^2x^3 + \cdots$$

is multiplied by $(1 - ax)$ with $a = m$ then all the terms vanish except the first one, which Lambert wrote as

$$(1 - mx)y = x + * + * + \text{etc.}$$

And thus one obtains

$$y = \frac{x}{1 - mx}$$

Then he considers the series

$$x = \frac{v^2}{2!} - \frac{v^4}{4!} + \frac{v^6}{6!} - \frac{v^8}{8!} + \cdots$$

He multiplies this series by $(1 + mv \ \ + nv^4)$ and he gets

$$(1 + mv^2 + nv^4)x = \frac{v^2}{2!} - \frac{v^4}{4!} + \frac{v^6}{6!} - \frac{v^8}{8!} + \text{etc} \cdots$$

$$+ \frac{m}{2!}v^4 - \frac{m}{4!}v^6 + \frac{m}{6!}v^8 - \text{etc} \cdots$$

$$+ \frac{n}{2!}v^6 - \frac{n}{4!}v^8 + \text{etc} \cdots$$

Then he chooses m and n such that the third and the fourth terms vanish, that is, the terms of degrees six and eight. Thus he obtains the system

$$\frac{1}{6!} - \frac{m}{4!} + \frac{n}{2!} = 0$$

$$\frac{1}{8!} - \frac{m}{6!} + \frac{n}{4!} = 0$$

whose solution is

$$m = \frac{11}{4\cdot7\cdot9} \qquad n = \frac{13}{2\cdot3\cdot5\cdot7\cdot8\cdot9}$$

and he has

$$x\left(1 + \frac{11}{4\cdot7\cdot9}v^2 + \frac{13}{2\cdot3\cdot5\cdot7\cdot8\cdot9}v^4\right) = \frac{1}{2}v^2 - \frac{5}{4\cdot7\cdot9}v^4 \ ** $$

$$+ \frac{59}{2\cdot3\cdot7\cdot10!}v^{10} - \text{etc} \cdots$$

Then Lambert adds, "adeoque termino quinto et sequentibus omissis, erit proxime

$$x = \frac{7560v^2 - 300v^4}{15120 + 660v^2 + 13v^4}."$$

This is exactly the direct approach to Padé approximants. Lambert also treats the case of the series

$$y = \sqrt{1 - x^2} = 1 - \frac{1}{2}x^2 - \frac{1}{8}x^4 - \frac{1}{16}x^6 - \frac{5}{128}x^8 - \frac{7}{256}x^{10} - \cdots$$

for which he obtains the approximant

$$y = \frac{16 - 20x^2 + 5x^4}{16 - 12x^2 + x^4}$$

and he concludes

> Methodus hactenus exposita eo nititur fundamento universaliori, ut a serie data alia series aut plures subtrahantur, quarum termini terminis homologis seriei datae proxime sint aequales. Hoc pacto enim residuum erit series, cuius singuli termini terminis seriei datae sunt minores. Hac conditione servata, vel me tacente patet seriem assumi posse qualemcunque ipsi satisfacientem, nec adeo opus esse, ut multiplicationis ope eruatur.
>
> Hinc deducere licebit methodum sequentem. Sit series data quaecunque. Sumatur alia, cuius summa sit nota, termini vero a terminis analogis seriei datae quam minime differant. Differentia utriusque seriei erit series data magis convergens et summarum differentiae aequalis . . .

Ex omnibus exemplis hactenus allatis (§ 48-63) patet in singulis casibus, dispari quidem successu, obtineri series propositis magis convergentes, saepissime tamen convergentiam initio tantum serierum maxime esse notabilem, cum in plurimis seriebus hoc modo erutis termini citissime ad rationem aequalitatis accedant, quod caveri non poterit, nisi series mutari possit in aliam serie geometrica magis convergentem.

1.5. Joseph Louis Lagrange

Many contributions to continued fractions are due to Joseph Louis Lagrange (Turin, 1/25/1736—Paris, 4/10/1813) but one of them is of a particular interest for our purpose.

In 1776, Lagrange published a paper on the use of continued fractions in integral calculus where he developed a general method for obtaining the continued fraction expansion of the solution of a differential equation in one variable. Let there be given a differential equation relating y and t. We assume that y behaves as a_0 when t tends to zero. We write $y = a_0/(1 + y_1)$ and substitute into the differential equation. We get a new differential equation relating y_1 and t. We assume that y_1 behaves as a_1 when t tends to zero. We write $y_1 = a_1/(1 + y_2)$ and repeat the same process again. Thus he obtains y as a continued fraction

$$y = \frac{a_0|}{|1} + \frac{a_1|}{|1} + \cdots \qquad \text{where } a_i = b_i t^{r_i}$$

Lagrange gives some examples and he says:

Comme la forme de ces fractions continues est peu commode pour les opérations algébriques, nous allons réduire ces fractions en fractions ordinaires, ce qui donnera lieu à des conséquences importantes sur la nature même de ces fractions.

Thus he reduces these continued fractions to ordinary fractions by calculating their convergents. Let us quote his last example:

. . . on aura les approximations suivantes vers la valeur de $(1 + x)^m$

$$1 + mx$$

$$\frac{1 + \frac{m+1}{2}x}{1 - \frac{m-1}{2}x}$$

$$\frac{1 + \frac{2(m+1)}{3}x + \frac{(m+1)m}{3\cdot 2}x^2}{1 - \frac{m-2}{3}x}$$

$$\frac{1 + \frac{2(m+2)}{4}x + \frac{(m+2)(m+1)}{4\cdot 3}x^2}{1 - \frac{2(m-2)}{4}x + \frac{(m-2)(m-1)}{4\cdot 3}x^2}$$

$$\frac{1 + \frac{3(m+2)}{5}x + \frac{3(m+2)(m+1)}{5\cdot 4}x^2 + \frac{(m+2)(m+1)m}{5\cdot 4\cdot 3}x^3}{1 - \frac{2(m-3)}{5}x + \frac{(m-3)(m-2)}{5\cdot 4}x^2}$$

$$\frac{1 + \frac{3(m+3)}{6}x + \frac{3(m+3)(m+2)}{6\cdot 5}x^2 + \frac{(m+3)(m+2)(m+1)}{6\cdot 5\cdot 4}x^3}{1 - \frac{3(m-3)}{6}x + \frac{3(m-3)(m-2)}{6\cdot 5}x^2 - \frac{(m-3)(m-2)(m-1)}{6\cdot 5\cdot 4}x^3}$$

----

et ces expressions sont exactes, aux quantités près des ordres x^2, x^3, x^4, x^5, x^6, x^7, ... c'est à dire qu'elles sont exactes jusqu'à la puissance de x inclusivement qui est le produit des deux plus hautes puissances de x dans le numérateur et dans le dénominateur; c'est de quoi on pourra si l'on veut se convaincre a posteriori en résolvant les fractions précédentes en séries et en les comparant avec la série

$$1 + mx + \frac{m(m-1)}{2} x^2 + \frac{m(m-1)(m-2)}{2 \cdot 3} x^3 + \cdots$$

On peut traiter de même les autres fractions continues que nous avons trouvées dans le cours de ce Mémoire et en tirer les conclusions semblables; c'est sur quoi nous ne nous arrêterons pas, puisque ce n'est qu'une affaire de pur calcul.

This is really the birth certificate of Padé approximants since Lagrange mentions their fundamental algebraic property.

Every year Lagrange was sending the complete volume of the Nouveaux Mémoires de l'Académie Royale des Sciences et Belles-Lettres de Berlin to Jean Le Rond d'Alembert (Paris, 11/16/1717—Paris, 10/29/1783). On December 12, 1778 he sent him the volume for 1776, which had just appeared, with the following remark [32]:

Il y a comme de raison quelquechose de moi, mais rien qui puisse mériter votre attention . . .

This was not a prophetic view!

2. THE EARLY DEVELOPMENTS

During the nineteenth century continued fractions became more and more popular among the mathematical community. This is, of course, due to the increasing number of exchanges, by letters or visits, between the mathematicians, but it is mainly due to the literature.

To my knowledge, Euler's celebrated book Introductio in analysis infinitorum, published in Lausanne in 1748, contains the first extensive and systematic exposition of the theory of continued fractions (chapter 18).

After Euler's work the subject has been explained in most of the encyclopedias that begin to appear at that time, the principal being L'Encyclopédie or Dictionnaire raisonné des sciences, des arts et des métiers published from 1751 to 1772 under the direction of Jean Le Rond d'Alembert and Denis Diderot (Langres 10/5/1713—Paris, 7/31/1784). The Encyclopédie contains an article on continued fractions written by Marie Jean Antoine Nicolas Caritat, Marquis de Concorcet (Ribemont, 9/17/1743—Bourg-La-Reine, 3/8/1794) which is almost word by word Euler's book.

An exposition of the theory of continued fractions can be also found in the Additions written by Lagrange to Euler's Algebra in 1774.

Following these predecessors many mathematicians became interested in continued fractions. All those who worked on the transformation of a formal power series into a continued fraction, by using, for example, the division process, have in fact obtained Padé approximants since, in most of the cases, the division process leads to the continued fraction corresponding to the power series whose successive convergents are [0/0], [0/1], [1/1], [1/2], [2/2], ...

I shall only quote here the most important contributions. The interested reader is referred to [7] for an extensive study.

In 1803, Basilius Viscovatoff (1778—St. Petersburg, 10/8/1812), who was lieutenant in the Russian army and professor of mathematics at the Institut Wegebaumeister in St. Petersburg, proposed a method for transforming the quotient of two power series into a continued fraction.

A quite similar method was also given by Charles Hutton (Newcastle-upon-Tyne, 8/14/1737—London, 1/27/1823) in the tract IX of the first volume of his Tracts on Mathematical and Philosophical Subjects published in London in 1812.

The division process was also used by Scubert in 1820, Jean-Jacques Bret (Mercurial, 9/25/1781—Grenoble, 1/29/1819) and Joseph Diaz Gergonne (Nancy, 6/19/1771—Montpellier, 5/4/1859).

As I mentioned above, such a process indirectly leads to Padé approximants since it is possible to transform the series $c_0 + c_1x + c_2x^2 + \cdots$ into the continued fraction $b_0 + \frac{x|}{|b_1} + \frac{x|}{|b_2} + \cdots$. This problem was studied in 1833 by Maritz Abraham Stern (Frankfurt-am-Main, 6/29/1807—Zürich, 1/30/1894) who found a recurrence relationship for the b_i's and by Johann Bernhard Hermann Heilermann (Waltrop, 1/13/1820—Godesborg bei Bonn, 9/28/1899) in 1846 who gave them explicitly. He also studied the continued fraction expansion $b_0 - \frac{a_1|}{|x + b_1} - \frac{a_2|}{|x + b_2} - \cdots$ for the ratio of two series in x^{-1}. The case of one single series in x^{-1} was treated by Hermann Hankel (Halle, 2/14/1839—Schramberg near Tübingen, 8/29/1873) in 1862 and by Thomas Jan Stieltjes (Zwolle, 12/29/1865—Toulouse, 12/31/1894) in 1889.

The contraction of a continued fraction was introduced in 1855 by Philipp Ludwig von Seidel (Zweibrücken, 10/24/1821—Munich, 8/13/1896). The contraction of the continued fraction corresponding to a power series leads to the associated continued fraction whose convergents are [0/0], [1/1], [2/2], ... The relationship between the corresponding and associated continued fractions was considered by J. B. H. Heilermann in 1860.

As the history of Padé approximants is very much interlaced with that of continued fractions, we shall not follow that way and we shall only look now at the direct approaches to Padé approximants that do not make use of continued fractions. However, I would like to mention one more contribution of that type since it opened a very important new chapter of mathematics.

In his very celebrated paper on Gaussian quadrature methods presented to the Göttingen Society on September 16, 1814, Carl Friedrich Gauss (Brunswick, 4/23/1777—Göttingen, 2/22/1855) proved that

$$\mathrm{Log}\,\frac{1+x}{1-x} = x + \frac{x^3}{3} + \frac{x^5}{5} + \cdots$$

$$= \frac{1|}{|x^{-1}} - \frac{1/3|}{|x^{-1}} - \frac{2.2/3.5|}{|x^{-1}} - \frac{3.3/5.7|}{|x^{-1}} - \cdots$$

The convergents of this continued fraction are the Padé approximants [1/0], [1/2], ... of the series. The denominators of the convergents are the Legendre orthogonal polynomials as proved by Pafnuti Lvovich Chebyshev (Okatovo, 5/14/1821—St. Petersburg, 1894).

It must be recalled that the general theory of orthogonal polynomials originated from the continued fractions associated with a power series, an approach which has almost been forgotten nowadays.

In a paper published in 1837 but dating from November 1834, Ernst Eduard Kummer (Sorau, 1/29/1810—Berlin, 5/14/1893) made use of Padé approximants for summing slowly convergent series. He wrote:

Weil nun oben gesetzt worden ist:

$$\frac{a_1k^{n-1} + a_2k^{n-2} + \cdots + a_n}{k^n + b_1k^{n-1} + b_2k^{n-2} + \cdots + b_n} = \frac{c_2}{k} + \frac{c_3}{k^2} + \frac{c_4}{k^3} + \cdots,$$

so erhält man, indem man mit dem Nenner multipliert und die gleichen Potenzen von k mit einander vergleicht, folgende Gleichnungen:

$$(10)\quad \begin{cases} a_1 = c_2, \\ a_2 = c_3 + b_2 c_1, \\ a_3 = c_4 + b_3 c_1 + c_2 b_2, \\ \text{-------------------} \\ a_n = c_{n+1} + c_n b_1 + c_{n-1} b_2 + \cdots + c_2 b_{n-1}, \end{cases}$$

$$0 = c_{n+2} + c_{n+1} b_1 + c_n b_2 + \cdots + c_2 b_n,$$

$$0 = c_{n+3} + c_{n+2} b_1 + c_{n+1} b_2 + \cdots + c_3 b_n,$$

$$0 = c_{n+4} + c_{n+3} b_1 + c_{n+2} b_2 + \cdots + c_4 b_n,$$

$$\text{------------------------------------}$$

$$0 = c_{2n+1} + c_{2n} b_1 + c_{2n-1} b_2 + \cdots + c_{n+1} b_n.$$

These are exactly the equations defining the $[n - 1/n]$ Padé approximant. Kummer gives several examples but he does not prove any theoretical result.

In 1845, Carl Gustav Jacob Jacobi (Potsdam, 1804—Berlin, 2/18/1851) proved his celebrated determinantal formula for Padé approximants. In the same paper he gives several representations for the numerators and the denominators of Padé approximants all derived as special cases of interpolating rational fractions studied by Augustin Louis Cauchy (Paris, 8/21/1789—Sceaux, 5/22/1857). Jacobi's representations are based on the systems of linear equations defining the Padé approximants.

Georg Friedrich Bernhard Riemann (Breselenz, 9/17/1826—Selasca, 7/20/1866) proved in October 1863 the convergence of the corresponding continued fraction given by Gauss for the ratio of two hypergeometric series. The proof was found in Riemann's papers after his death. It uses integration in the complex domain. The proof was completed and edited by Hermann Amandus Schwarz (Hermdorf untern Kynast, 1/25/1843—Berlin, 12/1/1921). According to Henri Eugène Padé (Abbeville, 12/17/1863—Aix-en-Provence, 7/9/1953) this is the first proof of convergence for Padé approximants.

Let us quote Riemann when he defines Padé approximants:

Lorsque l'on a une fraction continue infinie de la forme

$$a + \frac{b_1 x|}{|1} + \frac{b_2 x|}{|1} + \cdots$$

qui, pour des valeurs de x suffisamment petites, est convergente et représente la fonction f(x), on voit facilement que la $m^{\text{ième}}$ réduite est égale au quotient p_m/q_m de deux fonctions entières p_m et q_m, toutes deux de degré n lorsque $m = 2n + 1$ et de degré n et $n - 1$ lorsque $m = 2n$. La différence entre la réduite et la fonction f(x), lorsque x est infiniment petit, est infiniment petite d'ordre m. Mais, pour que cela est lieu, autant de conditions doivent être satisfaites qu'il est renfermé de quantités arbitraires dans la fonction fractionnaire égale à la réduite. Par consequent, la $m^{\text{ième}}$ réduite peut être déterminée au moyen de la condition qu'elle coincide avec la fonction en les m premiers termes du développement suivant les puissances de x en tenant compte des degrés du numérateur et du dénominateur, qui sont tous deux égaux à n pour $m = 2n + 1$, et égaux à n et $n - 1$ pour $m = 2n$.

In his thesis dated 1870, Georg Ferdinand Frobenius (Berlin, 10/20/1849—Berlin, 8/3/1917) showed that the numerators, the denominators, and the errors of the convergents of the continued fraction

$$C(x) = \frac{1|}{|a_0 x} - \frac{1|}{|a_1 x} - \frac{1|}{|a_2 x} - \cdots$$

are related by three terms recurrence relationships. These results were extended in a paper published in 1881 where he gave the relations linking the numerators and the denominators of three adjacent approximants in the Padé table. Some of these identities, now known as the Frobenius identities, are connected with Jacobi's determinantal formulas for the coefficients of the continued fraction $a_0 + \frac{x|}{|a_1} + \frac{x|}{|a_2} + \cdots$ whose successive convergents form the main diagonal of the Padé table. A recursive method for computing a_0, a_1, a_2, ... is given by Frobenius who, in fact, gave the first systematic study of Padé approximants and placed their theory on a rigorous basis.

Numerous contributions to Padé approximants are due to Edmond Nicolas Laguerre (Bar Le Duc, 3/9/1834—Bar Le Duc, 8/14/1886). In his first paper of 1876 he says:

> Le principe que j'ai mis en usage est le suivant: F(x) étant une fonction donnée et $\phi(x)/f(x)$ la fraction rationnelle dont le dénominateur est d'un degré donné et qui se rapproche le plus de la valeur de la fonction, je cherche à établir entre $\phi(x)$, f(x) et leur dérivées du premier ordre une relation algébrique qui soit linéaire par rapport à l'une de ces fonctions, $\phi(x)$ par exemple, et par rapport à sa dérivée.

He treats the cases $(x^2 - 1)^{-\frac{1}{2}}$, $(x + a)^m/(x + b)^m$ and $e^{F(x)}$ where F is a polynomial. In his second note of 1876 he studies $\exp(\arctan x^{-1})$ and in 1879, he works out the case of the series

$$\frac{1}{x} - \frac{1!}{x^2} + \frac{2!}{x^3} - \frac{3!}{x^4} + \cdots$$

He shows the convergence of the sequence $([k/k])_{k\in\mathbb{N}}$ to $e^x \int_x^\infty e^{-t} t^{-1} dt$. He also treats the case of $\int_x^\infty e^{-t^2} dt$.

In 1881, Leopold Kronecker (Liegnitz, 12/7/1823—Berlin, 12/29/1891) considered the problem of finding a rational fraction p/q having the same derivatives at a given point that a given function g. He used two techniques for solving this problem. The first one is the Euclidean division algorithm for finding the continued fraction expansion of f/g. The second method is to solve the system of linear equations obtained by imposing that the first coefficients of the power series expansion of gq - fp vanish.

I already mentioned P. L. Chebyshev. In 1865 he wrote:

> En effet, à l'ordinaire elles [les fractions continues] servent à trouver les systèmes des polynômes X, Y, qui rendent la différence UX - Y la plus proche possible de zéro, en supposant bien entendu que la fonction U soit développable en série ordonnée suivant les puissances entières et décroissantes de la variable, et que le degré d'approchement se détermine par sa plus haute puissance dans le reste.

This is the connection between the direct approach to Padé approximants and the approach which uses continued fractions.

In 1885, Chebyshev wrote:

> Développons l'integrale $\int_a^b \frac{f(x)}{z - x} dx$ en série procédant selon les puissances décroissantes de z:

$$\int_a^b \frac{f(x)}{z-x}\,dx = \frac{1}{z}\int_a^b f(x)dx + \frac{1}{z^2}\int_a^b x\,f(x)dx + \cdots$$

Posant

$$\int_a^b f(x)\,dx = A_0,\ \int_a^b x\,f(x)\,dx = A_1,\ \ldots,\ \int_a^b x^{2m-1} f(x)\,dx = A_{2m-1}$$

nous obtenons donc pour l'intégrale $\int_a^b \frac{f(x)}{z-x}\,dx$ l'expression approximative suivante, rigoureuse jusqu'au terme $1/z^{2m}$ inclusivement:

$$\frac{A_0}{z} + \frac{A_1}{z^2} + \cdots + \frac{A_{2m-1}}{z^{2m}}.$$

En désignant donc par

$$\frac{1}{|\alpha_1 z + \beta_1} - \frac{1}{|\alpha_2 z + \beta_2} - \cdots - \frac{1}{|\alpha_m z + \beta_m}$$

la fraction continue que l'on obtient en développant l'expression $\frac{A_0}{z} + \frac{A_1}{z^2} + \cdots + \frac{A_{2m-1}}{z^{2m}}$ en fraction continue et en s'arrêtant à la $m^{\text{ième}}$ réduite nous aurons par conséquent, aussi avec un degré d'approximation jusqu'au terme $1/z^{2m}$ inclusivement

$$\int_a^b \frac{f(x)}{z-x}\,dx = \frac{1}{|\alpha_1 z + \beta_1} - \frac{1}{|\alpha_2 z + \beta_2} - \cdots - \frac{1}{|\alpha_m z + \beta_m}.$$

Let $\{P_n\}$ be a family of orthogonal polynomials on $[a,b]$ with respect to $d\alpha$, that is,

$$\int_b^a P_n(x)\,P_k(x)\,d\alpha(x) = 0, \qquad n \neq k$$

These polynomials satisfy a three terms recurrence relationship

$$P_n(x) = (A_n x + B_n)\,P_{n-1}(x) - C_n P_{n-2}(x)$$

Let us consider the continued fraction

$$\frac{1}{|A_1 x + B_1} - \frac{C_2}{|A_2 x + B_2} - \frac{C_3}{|A_3 x + B_3} - \cdots$$

whose convergents $R_n(x)/S_n(x)$ are the Padé approximants [0/1], [1/2], ... Then $S_n(x) = \sqrt{c_0}\,P_n(x)$ where $c_k = \int_a^b x^k\,d\alpha(x)$. It has been proved in 1896 by Andrei Andrevitch Markov (Ryazan, 6/14/1856—Petrograd, 6/22/1922), a student of Chebyshev, that

$$\lim_{n\to\infty} \frac{R_n(x)}{S_n(x)} = c_0^{-2}(c_0 c_2 - c_1^2)^{\frac{1}{2}} \int_a^b \frac{d\alpha(t)}{x-t}$$

if x is an arbitrary point in the complex plane cut along [a, b] and that the convergence is uniform on every close set of the complex plane having no point in common with [a, b].

Markov's result is a consequence of Stieltjes' theorem on the convergence of Gaussian quadrature methods.

The connection between orthogonal polynomials, continued fractions, and Padé approximants was also explicated by Karl Heun (Wiesbaden, 4/3/1859—Karlsruhe, 1/10/1929) in his Inaugural Dissertation in 1881.

Let us mention that in his book on elliptic functions published in 1888, Georges Henri Halphen (Rouen, 10/30/1844—Versailles, 5/21/1889) was leaded to Padé approximants. He said:

> Soit $\sqrt{X}$ [X is a polynomial in x] développé suivant les puissances ascendantes de $x - \xi$, et déterminons deux polynômes G_m, H_m, le premier du degré $m - 1$, le second du degré m, de manière que le développement de leur quotient $G_m : H_m$ coïncide, le plus loin possible, avec celui de $\sqrt{X}$. La coïncidence pourra aller jusqu'au terme de rang $2m + 1$, en sorte que l'on aura
>
> $$G_m - H_m \sqrt{X} = c(x - \xi)^{2m+1} + c'(x - \xi)^{2m+2} + \cdots$$

Thus when Padé undertook his study the idea of Padé approximants was very much in the air, and several important contributions to their theory had already been done.

The last contribution I would like to present is that of Charles Hermite (Dieuze, 12/24/1822—Paris, 1/14/1901). The first reason for that choice is that he was Padé's advisor, the second reason is that he defined the approximants which are now called the Padé-Hermite approximants. The third reason is that he proved the fundamental result that e is a transcendental number and that the proof used Padé approximants.

Hermite's proof is as follows. e is assumed to be an algebraic number that satisfies

$$a_0 + a_1 e + \cdots + a_n e^n = 0$$

where $a_0, \ldots, a_n$ are integers. Hermite looks for the polynomials $Q, P_0, \ldots, P_n$ of degree k such that

$$e^{jx}Q(x) - P_j(x) = 0(x^{(n+1)k+1}), \quad j = 0, \ldots, n$$

Then

$$T(x) = \sum_{j=0}^{n} a_j P_j(x) - Q(x) \sum_{j=0}^{n} a_j e^{jx} = 0(x^{(n+1)k+1})$$

Since $|T(1)| < 1$ and is an integer for k sufficiently large it follows that

$$T(1) = \sum_{j=0}^{n} a_j P_j(1) = 0$$

Giving to k the values $k, k + 1, \ldots, k + n$, Hermite proves that

$$a_0 + a_1 e + \cdots + a_n e^n \neq 0$$

which contradicts the assumption. This last part of the proof was quite long and difficult and, in a letter to C. A. Borchardt, Hermite declines to enter on a similar search for π. He says

> Je ne me hasarderai point à la recherche d'une démonstration de la transcendance du nombre π. Que d'autres tentent l'entreprise; mais croyez m'en, mon cher ami, il ne laissera pas que de leur en coûter quelques efforts.

This last step was to be passed by Carl Louis Ferdinand Lindemann (Hannover, 4/12/1852—Munich, 3/6/1939) who in 1882 proved that π is a transcendental number, thus ending by a negative answer a question opened for more than 2000 years! The idea of the proof, which uses Padé approximants, is as follows.

If r, s, t, ..., z are distinct real or complex algebraic numbers and a, b, c, ..., n are real or complex algebraic numbers, at least one of which differs from zero, then

$$ae^r + be^s + ce^t + \cdots + ne^z \neq 0$$

But $e^{i\pi} + 1 = 0$ and in the preceding result $a = b = 1$ and $c = \cdots = n = 0$; $s = 0$ is algebraic; $r = i\pi$ is the only cause why $e^{i\pi} + 1 = 0$. Since i is algebraic, $i\pi$ is transcendental and it follows that π is transcendental.

The last fact I would like to mention about the history of Padé approximants is that the name "Padé table" has been used for the first time, as far as I know, by Edward Burr Van Vleck (Middletown, 6/7/1863—Madison, 6/2/1943) in 1903.

On Padé and his work the reader is referred to [8].

ACKNOWLEDGMENTS

I would like to thank J. Dutka for pointing out Lambert's paper, J. Gammel who found Anderson's letter and P. R. Graves-Morris who sent me a copy of it, and G. Mühlbach who looked for Heun's dissertation.

REFERENCES

1. Bernoulli, D.: Comm. Acad. Sci. Imp. Petropolitanae IV, 1728 (1732), pp. 85-100.
2. Bernoulli, D.: Comm. Acad. Sci. Imp. Petropolitanae V, 1730-1731 (1738), pp. 63-69, 70-82.
3. Bouckaert, L. P.: Introduction aux travaux d'analyse de D. Bernoulli, in The Collected Scientific Papers of the Mathematicians and Physicists of the Bernoulli Family, L. P. Bouckaert and B. L. van der Waerden, ed., Birkhäuser Verlag, Basel (to appear).
4. Bret, J. J.: Ann. Math. 9, 45-49 (1818).
5. Brezinski, C.: J. Approx. Theory 25, 295-317 (1979).
6. Brezinski, C.: Padé-type approximation and general orthogonal polynomials, in ISNM Vol. 50, Birkhäuser Verlag, Basel (1980).
7. Brezinski, C.: The long history of continued fractions and Padé approximants, in Padé Approximation and Its Applications, Amsterdam 1980, M. G. de Bruin and H. Van Rossum, eds., Springer (1981).
8. Brezinski, C.: Documents sur la vie et l'oeuvre de Henri Padé, in H. Padé, Oeuvres, C. Brezinski, ed., Librairie Scientifique et Technique A. Blanchard (1984).
9. Cauchy, M. A. L.: Cours d'Analyse, vol. 1. L'imprimerie Royale, Paris (1821).
10. Chebyshev, P. L.: J. Math. Pures Appl (2) 10, 353-358 (1865).
11. Chebyshev, P. L.: Acta Math. 9, 35-56 (1886).
12. Diderot, D.; Le Rond d'Alembert, J.: Dictionnaire raisonné des sciences, des arts et des métiers (1751-1772).
13. Euler, L.: Introductio in Analysis Infinitorum, 2 vols. Lausanne (1748).
14. Forbes, E. G.: The Euler-Mayer Correspondence (1751-1755). Macmillan (1971).
15. Frobenius, F. G.: De Functionum Analyticarum Unius Variabilis per Series Infinitas Repraesentatione. Diss., Berlin (1870).
16. Frobenius, F. G.: J. Reine Angew. Math. 90, 1-17 (1881).
17. Fuss, P. H.: Correspondance mathématique et physique de quelques célèbres géomètres du XVIIIème siècle précédée d'une notice sur les travaux de Léonard Euler, tant imprimés qu'inédits, et publiée sous les auspices de l'Académie Impériale des Sciences de Saint-Petersbourg, 2 vols. St. Petersburg (1843).
18. Gauss, C. F.: Comm. Societatis Regiae Scientiarium Gottingensis Recentiores 15, 39-76 (1814).

19. Gergonne, J. D.: Ann. Math. 9, 261 (1818-1819).
20. Halphen, G. H.: Traité des fonctions elliptiques et de leurs applications. Gauthier-Villars (1888).
21. Hankel, H.: Z. Math. Phys. 7, 338-343 (1862).
22. Heilermann, J. B. H.: J. Reine Angew. Math. 33, 174-188 (1846).
23. Heilermann, J. B. H.: Z. Math. Phys. 5, 362-363 (1860).
24. Hermite, C.: C. R. Acad. Sci. Paris 77, 18-24, 74-79, 226-233, 285-293 (1873).
25. Heun, K.: Die Kugelfunctionen und Lamé' schen Functionen als Determinanten. Inaugural Dissertation, Göttingen (1881).
26. Hutton, C.: Tracts on Mathematical and Philosophical Subjects. London (1812).
27. Jacobi, C. G. J.: J. Reine Angew. Math. 30, 127-156 (1845).
28. Kronecker, L.: Berlin Berichte, 535-600 (1881).
29. Kummer, E. E.: J. Reine Angew. Math. 16, 206-214 (1837).
30. Lagrange, J. L.: Additions aux Eléments d'Algèbre d'Euler. Lyon (1774).
31. Lagrange, J. L.: Nouveaux Mém. Acad. Roy. Sci. Belles-Lett. Berlin 7, 236-264 (1776).
32. Lagrange, J. L.: Oeuvres, J. A. Serret and G. Darboux, eds., Paris (1867-1892).
33. Laguerre, E.: Bull. Soc. Math. France 5, 78-82 (1877).
34. Laguerre, E.: Bull. Soc. Math. France 5, 95-99 (1877).
35. Laguerre, E.: Bull. Soc. Math. France 7, 72-81 (1879).
36. Lambert, J. H.: Acta Helv. 3, 128-168 (1758).
37. Lambert, J. H.: Histoire de l'Académie de Berlin 1761 (1768), pp. 265-322.
38. Lindemann, F.: Math. Ann. 20, 213-225 (1882).
39. Markov, A. A.: Acta Math. 19, 93-104 (1895).
40. Padé, H.: Ann. Ec. Norm. Sup. 9, Supp. 1-93 (1892).
41. Riemann, B.: Sur le développement du quotient de deux séries hypergéométriques en fraction continue infinie, in Oeuvres. Gauthier-Villars, Paris (1898), pp. 369-377.
42. Rigaud, S. J.: Correspondence of Scientific Men of the Seventeenth Century. Georg Olms Verlagsbuchhandlung, Hildesheim (1965).
43. Seidel, L.: Abh. Akad. Wiss. München, Zweite Kl. 7 (1855).
44. Smith, H. J. S.: Proc. London Math. Soc. 8, 6-29 (1876-1877).
45. Stephen, L.; Lee, S., eds.: The Dictionary of National Biography. Smith, Elder and Co. (1908).
46. Stern, M. A.: J. Reine Angew. Math. 10, 1-22, 154-166, 241-274, 364-376 (1833); 11, 33-66, 142-168, 277-306, 311-350 (1834).
47. Stieltjes, T. J.: Ann. Fac. Sci. Toulouse 3, 1-17 (1889).
48. Van Vleck, E. B.: Trans. Am. Math. Soc. 4, 297-332 (1903).
49. Viscovatoff, B.: Nova Acta Acad. Scient. Imp. Petropolitanae 15, 181-191 (1802).

9 ON THE APPROXIMATION OF SOLUTIONS OF QUASIVARIATIONAL INEQUALITIES WITH APPLICATION TO AN ABSTRACT OBSTACLE PROBLEM

Gottfried Bruckner / Institut für Mathematik, Akademie der Wissenschaften der DDR, Berlin, German Democratic Republic

In this paper approximation procedures for a certain type of abstract monotoneous, stationary quasivariational inequalities are found. However, also instationary quasi-variational inequalities, quasiminimum problems, and other implicit problems could be treated in a similar way.

Our method consists in considering the optimization problem

$$\|x - u\|^2 = \inf_{[y,v]\in G(S)} \|y - v\|^2 \tag{1}$$

where G(S) is the graph of the solution operator S of the parametric variational inequality

$$\begin{cases} y \in C(v) \\ (A(y,v) - y^*,\ z - y) \geq h(y,v) - h(z,v) \quad \text{for every } z \in C(v) \end{cases} \tag{2}$$

Then, the quasivariational inequality

$$\begin{cases} u \in C(u) \\ (A(u,u) - y^*,\ z - u) \geq h(u,u) - h(z,u) \quad \text{for every } z \in C(u) \end{cases} \tag{3}$$

is obviously solvable if and only if (1) is solvable and $\inf_{G(S)} \|y - v\|^2 = 0$. If (3) is solvable then every converging approximation procedure for (1) converges to a solution of (3) (cf. [1]).

This method gives us the possibility to apply optimization theory to the approximation of solutions of a quasivariational inequality. Especially, we use here the theory of identification as it has been developed by R. Kluge [7] and further papers. Our approach is only an example how one can proceed. Certainly the theory could be extended and generalized, the conditions weakened.

Some of the results are partially contained in [1-4], but to the reader's convenience they are presented here once more without or with sketched proofs.

The Contents. In Sec. 1 an existence theorem for (1) (Theorem 1.1) and a convergence theorem for a Tikhonov regularization (Theorem 1.2) are given. Further, it is shown that under rather general conditions (3) is equivalent to a fixed point problem with a single-valued operator (Theorem 1.3).

In the first part of Sec. 2 (1) is approximated by a sequence of finite-dimensional unconstrained minimum problems (Theorem 2.1). Here the method of penalty operators in connection with the projection method are applied. In the second part of Sec. 2 an iteration procedure is founded for the further simplification of the unconstrained minimum problems of the first part, and the differentiability and local Lipschitz continuity have to be assumed.

In Sec. 3 as a general example the obstacle-problem in an abstract setting is treated. The (solution depending) obstacle can be situated in the domain or on the boundary. To this special case the theorems of Sec. 2 are applied (Theorems 3.1 and 3.3). At last, for typical sets C(u) the corresponding penalty operators are constructed and their differentiability and Lipschitz continuity proved.

Let us refer to some changes in the notation that, however, can rarely cause any confusion. Throughout the paper the symbols C, C_n, $C(v)$, $C_n(v)$ (in the last part of Sec. 3 also K, K_n) are reserved for closed and convex subsets of a Banach or Hilbert space; only in the last part of Sec. 3 by $C = C(0,1)$, as usual, the space of continuous functions is denoted. The letter S symbolizes in Sec. 1 the solution operator of a parametric problem while in the last part of Sec. 3, S, S_n are symbols for penalty operators.

1. EXISTENCE AND REGULARIZATION PROCEDURES

Let Y be a reflexive Banach space, $\|\cdot\|$ its norm, $(\cdot,\cdot)$ the dual pairing, $U \subset Y$ a weakly closed set,

$$C: U \to 2^Y$$

$C(v) \neq \phi$, convex, closed for every $v \in U$. Let further A be a mapping from $Y \times U$ into Y^*, h an admissible functional on $Y \times U$, and $y^* \in Y^*$. We consider the parametric variational inequality

$$\begin{cases} y \in C(v) \\ (A(y,v) - y^*, z - y) \geq h(y,v) - h(z, v) \quad v \in U, \ \forall z \in C(v) \end{cases} \tag{1.1}$$

and the optimization problem

$$\|x - u\|^2 = \inf_{[y,v] \in G(S)} \|y - v\|^2 \tag{1.2}$$

where $[y,v] \in G(S)$ means that y solves (1.1) for the parameter v ($G(S)$ is the graph of the solution operator of (1.1)). C_n being subsets of Y we define as usual

$$w(s) - \overline{\text{Lim}}\, C_n = \{y, \exists \{n_i\} \subset \{n\} \text{ and } y_{n_i} \in C_{n_i} \text{ s.t. } y_{n_i} \xrightarrow{w(s)} y\}$$

$$w(s) - \underline{\text{Lim}}\, C_n = \{y, \exists y_n \in C \text{ s.t. } y_n \xrightarrow{w(s)} y, \ n \geq n_0(y)\}$$

The question of the existence for (1.2) is answered by

Theorem 1.1. Let the following assumptions be fulfilled

$A(\cdot,v)$ is single-valued, monotoneous, coercive, hemicontinuous, defined on the whole of Y (1.3)

$h(\cdot,v)$ is convex for every $v \in U$ (1.4)

If $v_n \in U$, $v_n \rightharpoonup v$, then $w\text{-}\overline{\text{Lim}}\, C(v_n) \subset C(v) \subset s\text{-}\underline{\text{Lim}}\, C(v_n)$ (1.5)

$A(z_n, v_n) \to A(z,v)$ if $v_n \rightharpoonup v$, $z \in C(v)$ and $z_n \to z$ in the sense of (1.5) (1.6)

$h(\cdot,\cdot)$ is (w,w)-l.s.c. and (s,w)-u.s.c. on $Y \times U$ (1.7)

$$\begin{cases} \text{There is an element } Nu \in C(u) \text{ such that} \\ \|Nu\| \leq b\|u\| + c, \ b \geq 0, \\ |h(Nu,u)| \leq d\|u\| + c, \ d \geq 0, \\ [y,u] \in G(S), \ \|y - u\| \leq c_1, \ \|y\| \to \infty \text{ imply} \\ ((A(y,u), y - Nu) + h(y,u))/\|y\| \to \infty \end{cases} \tag{1.8}$$

Then the problem (1.2) has a solution.

Proof. From (1.3), (1.4), and (1.7) we conclude that (1.1) has a solution for every $v \in U$, and (1.3)-(1.7) imply the weak closedness of G(S) (cf. [7]). Finally, (1.8) implies the boundedness of $\{[y,v],\ [y,v] \in G(S),\ \|y - v\| \leq c\}$ (cf. [4]). The theorem then follows from the fact that a w-l.s.c. functional attains its inf on a bounded and w-closed set (Weierstrass).

Remark. For example, condition (1.8) is fulfilled if $h \equiv 0$, $A(x,u) \equiv Ax$,

$$(Ax - Ay,\ x - y) \geq \delta(\|x - y\|),\ \delta(1) > 0, \tag{1.9}$$

$$\|Ay\| \leq L\|y\| + c \tag{1.10}$$

There is an element $Nu \in C(u)$ such that

$$\|Nu\| \leq b\|u\| + c, \qquad b < \delta(1)/2L \tag{1.11}$$

The proof of this remark can be found in [4].

Next, for the solution of (1.2) we consider a Tikhonov regularization procedure. Let f be an admissible functional defined on U, $e_n > 0$, $e_n \to 0$. We ask for $[x_n, u_n] \in G(S)$ with the property

$$\|x_n - u_n\|^2 + e_n f(u_n) = \inf_{[y,v]\in G(S)} \{\|y - v\|^2 + e_n f(v)\} \tag{1.12}$$

where f is to be such that the solution of (1.12) is unique. However, the determination of such an f may be very difficult.

Theorem 1.2. Let f be w.-l.s.c. on U, bounded from below, and let $u_i \rightharpoonup u$, $\overline{\lim}\ f(u_i) \leq f(u)$ imply $u_i \to u$ (e.g., $f(v) = \|v - v_0\|^2$, Y uniformly convex). Then (1.12) has for every n at least one solution $[x_n, u_n]$. The sequence $[x_n, u_n]$ is (w, s)-compact in $Y \times Y$ and every accumulation point is a solution of (1.2).

The proof can be easily extracted from [7, p. 223].

At last, we come to an approximation of the solution of (1.2) by problems

$$\|x - u\|^2 = \inf_{[y,v]\in G(S_e)} \|y - v\|^2 \tag{1.13}$$

where S_e is single-valued. For this reason we "regularize" the parametric v.i. (1.1) with the duality mapping J.

Theorem 1.3. Under the assumptions

$$Y \text{ and } Y^* \text{ are both strictly convex} \tag{1.14}$$

$$\begin{cases} A(\cdot,v) \text{ is single-valued, monotone, hemicontinuous, } D(A(\cdot,v)) = Y \\ h(\cdot,v) \text{ is convex, l.s.c.} \end{cases} \tag{1.15}$$

the quasivariational inequality

$$\begin{cases} u \in C(u) \\ (A(u,u) - y^*,\ z - u) \geq h(u,u) - h(z,u) \qquad \forall z \in C(u) \end{cases} \tag{1.16}$$

is equivalent to the fixed point problem

$$u = S_e u \tag{1.17}$$

where S_e is the single-valued solution operator of the parametric variational inequality

$$(A(y,v) + eJ(y - v) - y^*,\ z - y) \geq h(y,v) - h(z,v) \qquad \forall z \in C(v) \tag{1.18}$$

and $e > 0$ is fixed.

Proof. First of all we point out that

$$(Jx - Jy, x - y) \geq (\|x\| - \|y\|)^2$$

holds, consequently J is coercive and under the assumption (1.14) J is single-valued, strictly monotoneous and demicontinuous (cf. [7]). Besides, it is clear that $J(0) = 0$ holds.

Then, by a standard theorem of monotonie-theory (1.18) is uniquely solvable, i.e., S_e is a unique operator on U. Because of $J(0) = 0$, (1.16) and (1.17) correspond to each other and have the same solutions.

Remarks.

(i) Theorem 1.3 can be generalized to multivalued, monotoneous, hemi-upper-semi-continuous operators $A(\cdot, v)$ (cf. papers of H. Brezis and R. T. Rockafellar on this subject).

(ii) Based on Theorem 1.3 other types of existence theorems can be proved.

(iii) As to an approximation procedure for (1.2) by problems (1.13) in the case where $\inf_{[y,v] \in G(S)} \|y - v\|^2 \neq 0$ cf. [1].

2. APPROXIMATION BY FINITE-DIMENSIONAL UNCONSTRAINED OPTIMIZATION PROBLEMS AND THEIR ITERATION

Let here H be a real Hilbert space with the scalar product $(\cdot, \cdot)$ and the norm $\|\cdot\|$, $H_n \subset H$ finite-dimensional subspaces, P_n operators of orthogonal projection on H_n with $\|P_n y - y\| \to 0$ for all $y \in H$. Let further w-closed nonempty sets $U \subset H$, $U_n \subset H_n$ be given with the property

$$\text{w-}\overline{\text{Lim}}\ U_n \subset U \subset \text{s-}\underline{\text{Lim}}\ U_n \tag{2.1}$$

and let for every $u \in U$ a nonempty, closed, convex set $C(u) \subset H$ be given such that

$$C(u) = \{y \in H,\ B(y, u) = 0\} \tag{2.2}$$

where the "penalty"-operator B is defined on the whole of $H \times U$, monotoneous in the first argument, demicontinuous, and is approximated (in a certain sense, cf. (2.8) below) by continuous and in the first argument monotoneous operators

$$B_n : H_n \times U_n \to H_n$$

Let finally A be a bounded, demicontinuous, single-valued, and everywhere defined operator from $H \times U$ into H and y^* a fixed element of H.

Our aim is to approximate the problem (1.2), where $h \equiv 0$, by the unconstrained optimization problems

$$K_n(x_n, u_n) = \inf_{[y,v] \in H_n \times U_n} K_n(y, v) \tag{2.3}$$

where

$$K_n(y, v) = \|y - v\|^2 + \frac{1}{e_n} \|\Lambda_n(y, v) - P_n y^*\|^2 \quad (e_n > 0,\ e_n \to 0) \tag{2.4}$$

$$\Lambda_n(y, v) = P_n A(y, v) + \frac{1}{\lambda_n} B_n(y, v) \qquad (\lambda_n > 0,\ \lambda_n \to 0) \tag{2.5}$$

Besides (1.5) and (1.6) we need the following assumptions:

$$\begin{cases} (A(x, v) - A(y, v), x - y) \geq \delta(\|x - y\|) \qquad \forall v \in U \\ \text{where } \delta(r) \geq 0 \text{ is independent of } v \text{ and } \delta(r) \to 0 \text{ implies } r \to 0 \end{cases} \tag{2.6}$$

$$\begin{cases} \forall u \in U \quad \exists\, Nu \in C(u) \quad \text{such that} \\ \|Nu\| \le b_0 \|u\| + c, \quad b_0 < 1 \\ \|A(Nu,u)\| \le d\|Nu - u\| + c, \quad d < \delta(1)/2 \end{cases} \tag{2.7}$$

$$\begin{cases} \text{For every } n \text{ there is a mapping } N_n : U_n \to H_n \text{ with} \\ \|N_n v\| \le b\|v\| + c, \quad b < 1 \\ \|\Lambda_n(N_n v, v)\| \le d\|N_n v - v\| + c, \quad d < \delta(1)/2 \end{cases} \tag{2.8}$$

$$\begin{cases} \text{If } v_j \in U_j,\ y_j \in H_j,\ w_j \in H_j,\ \{j\} \subseteq \{n\} \text{ with} \\ \quad v_j \rightharpoonup v,\ y_j \rightharpoonup y,\ w_j \to w \qquad \text{we have} \\ \underline{\lim}\,(B_j(w_j, v_j), y_j - w_j) \ge (B(w,v), y - w) \end{cases} \tag{2.9}$$

$$\begin{cases} \text{Let sequences } v_j \in U_j,\ v_j \rightharpoonup v, \text{ and } y_j \in H_j,\ y_j \rightharpoonup y \\ \text{and an element } y_0 \in C(v) \text{ be given.} \\ \text{Then there is a sequence } z_j \in H_j \text{ with } z_j \to y_0,\ A(z_j, v_j) \to A(y_0, v) \\ \text{and } \underline{\lim}\{(1/\lambda_j)(B_j(z_j, v_j), y_j - z_j)\} \ge 0. \end{cases} \tag{2.10}$$

Before we formulate our approximation theorem we are going to give some comments concerning these assumptions.

Assumption (2.6). This assumption implies

$$(A(x,v) - A(y,v),\ x - y) \ge \frac{\delta(1)}{2}\|x - y\|^2 \tag{2.11}$$

if $\|x - y\| \ge 2$ (cf. [2, Lemma 1]). Hence $A(\cdot, v)$ is coercive (independent of v).

Assumption (2.7). By (2.7) (together with (2.6)) (1.8) follows: We have to show that

$$(A(y,u),\ y - Nu)/\|y\| \to \infty \text{ if } \|y\| \to \infty,\ \|y - u\| \le c_1$$

Indeed, $\|y - Nu\| \ge \|y\| - \|Nu\| \ge (1 - b_0)\|y\| + c \to \infty$ if $\|y\| \to \infty$. Because of (2.6)

$$\begin{aligned} (A(y,u),\ y - Nu)/\|y\| &\ge (A(Nu,u),\ y - Nu)/\|y\| + \delta(\|y - Nu\|)/\|y\| \\ &\ge -\|A(Nu,u)\|\,\|y - Nu\|/\|y\| + \delta(\|y - Nu\|)/\|y\| \\ &\ge \left(-d\|y - Nu\|^2 + \frac{\delta(1)}{2}\|y - Nu\|^2\right)/\|y\| + c \to \infty \text{ if } \|y\| \to \infty \end{aligned}$$

Assumption (2.8) is in a similar way needed for the existence proof in the finite-dimensional space H_n.

Assumption (2.9) is a generalization of the lower semicontinuous operator-approximation in [7]. In the sequel we give a general example for operators B and B_n that fulfill (2.9) and (2.10) (cf. [2]). Besides the mapping $C: U \to 2^H$ with the property (1.5) let for every n a mapping

$$C_n : U_n \to 2^{H_n}, \quad C_n(v) \ne \phi, \text{ convex, closed}$$

be given with the property

$$u_r \in U_r,\ u_r \to u \Rightarrow \text{w-}\overline{\text{Lim}}\, C_r(u_r) \subset C(u) \subset \text{s-}\underline{\text{Lim}}\, C_r(u_r) \tag{2.12}$$

and A is to have the property

$$A(z_j, v_j) \to A(y_0, v) \text{ if } v_j \rightharpoonup v, \text{ and } z_j \to y_0 \text{ in the sense of (2.12)}$$

We show

$$(1.5) \Rightarrow P_{C(v_j)} w_j \to P_{C(v)} w \text{ if } v_j \rightharpoonup v, \ w_j \to w$$

$$(2.12) \Rightarrow P_{C_j(v_j)} w_j \to P_{C(v)} w \text{ if } v_j, w_j \in H_j, \ v_j \rightharpoonup v, \ w_j \to w$$

Indeed, let be $\zeta \in C(v)$, $\zeta_j \in C(v_j)$ with $\zeta_j \to \zeta$, $d_j = P_{C(v_j)} w_j$. We have $\|w_j - d_j\| \leq \|w_j - \zeta_j\| \leq c$ implying $d_i \rightharpoonup d$, $\{i\} \subseteq \{j\}$, $d \in C(v)$. Then $\|w - d\| \leq \underline{\lim} \|w_i - d_i\| \leq \overline{\lim} \|w_i - \zeta_i\| = \|w - \zeta\|$, i.e., $d = P_{C(v)} w$, $d_j \rightharpoonup P_{C(v)} w$. Let be $\xi_j \to d$, $\xi_j \in C(v_j)$, then $\|w - d\| \leq \underline{\lim} \|w_j - d_j\| \leq \overline{\lim} \|w_j - \xi_j\| = \|w - d\|$, i.e., $\|w_j - d_j\| \to \|w - d\|$ implying $d_j \to d$. The second assertion follows in the same way.

We now define

$$B(x, u) = (I - P_{C(u)}) x$$

$$B_n(x, u) = (I - P_{C_n(u)}) x$$

where I is the identity, and prove (2.9) and (2.10).

Let be v_j, y_j, w_j given as in (2.9). Then

$$(B_j(w_j, v_j), y_j - w_j) = (w_j - P_{C_j(v_j)} w_j, y_j - w_j) \to (w - P_{C(v)} w, y - w) = (B(w, v), y - w)$$

Let now v_j, v, y_j, y, y_0 be as in (2.10). Take $z_j \in C_j(v_j)$ such that $z_j \to y_0$ (cf. (2.12)). Then $B_j(z_j, v_j) = 0$ and (2.10) follows immediately.

Theorem 2.1. Under our assumptions ((1.5), (1.6), (2.1), (2.2), (2.6), (2.7), (2.8), (2.9), (2.10)) the problems (1.2) and (2.3) are solvable. If $[x_n, u_n]$ is a solution of (2.3) then the sequence $\{[x_n, u_n]\}$ is s-compact in $H \times H$ and each of its accumulation points is a solution of (1.2).

Proof. Since for every $v \in U$ $A(\cdot, v)$ is assumed to be strictly monotoneous, hemicontinuous and coercive and $D(A(\cdot, v)) = H$ the problem (1.1) has for every $v \in U$ exactly one solution Sv. By the same reasons

$$P_n A(y, v) + \frac{1}{\lambda_n} B_n(y, v) = P_n y^* \tag{2.13}$$

has exactly one solution $R_n v \in H_n$. The problem (1.2) is also solvable (cf. Theorem 1.1 and the comment concerning assumption (2.7)). The solvability of (2.3) follows from the continuity of K_n, the solvability of (2.13) and the boundedness of the (closed) set

$$\{[y, v] \in H_n \times U_n, \ K_n(y, v) \leq K_n(y_0, v_0), \ y_0 = R_n v_0, \ v_0 \text{ fixed}\}$$

that is implied by (2.6) and (2.8).

Let $[x_n, u_n]$ be a solution of (2.3). The proof of the boundedness of the sequence again uses (2.6) and (2.8); besides

$$v_n \rightharpoonup v, \ v_n \in U_n \text{ implies } R_n v_n \to Sv \tag{2.14}$$

We compare with a sequence $[R_n v_n, v_n]$ because for that sequence the functional (2.4) is very simple. The proof of (2.14) uses the boundedness of the sequence $R_n v_n$ (with the help of (2.6) and (2.8)), the fact that the weak accumulation points lie in C(v) (by (2.9)) and then the strong convergency of the whole sequence follows by (2.10) and (2.6).

By the same means finally the assertion follows where the strong convergency is implied by the weak one and by the convergency of the norms. As to a detailed proof cf. [3].

In the rest of this section we wish to approximate the problem (2.3), for n fixed, by an iteration procedure.

Let H be a finite-dimensional Hilbert-space, $U \subset H$ a closed set (of parameters), T an operator

$$T: H \times U \rightarrow H$$

and $y^* \in H$ a fixed element. We consider the problem

$$k(y, v) = \frac{1}{2}\|y - v\|^2 + \frac{1}{2}\|T(y, v) - y^*\|^2 \rightarrow \mathrm{Min}_{H\times U}! \tag{2.15}$$

(where T can be interpreted as $T = \sqrt{1/e_n}\,(P_n A + (1/\lambda_n)B_n))$.

We want to use a gradient method and assume that T is continuous and the partial derivatives T_y and T_v may exist. Then

$$k_y(y, v) = y - v + [T_y(y, v)]^*(T(y, v) - y^*) \tag{2.16}$$

$$k_v(y, v) = v - y + [T_v(y, v)]^*(T(y, v) - y^*) \tag{2.17}$$

where $[T_y(y, v)]^*$ is the adjoint of the linear operator $T_y(y, v)$. We construct an iteration sequence with the property that each of its accumulation points fulfills the equation $k' = [k_y, k_v] = 0$, a necessary condition for a minimum of (2.15).

Let h(y, v) be a functional on $H \times U$ with the gradient $h' = [h_y, h_v]$. To approximate its minimum we consider the iteration sequence

$$\begin{cases} x_0 \in H, \; u_0 \in U \\ x_{n+1} = x_n - a_n h_y(x_n, u_n) \\ u_{n+1} = u_n - b_n h_v(x_{n+1}, u_n) \end{cases} \tag{2.18}$$

where a_n, b_n are positive numbers that will be specified later. For the sequel we need the notion of a local Lipschitz-continuity (l. L.) of an operator of two arguments.

Definition 2.3. An operator S in two arguments

$$S: H_1 \times H_2 \rightarrow H_3$$

is called locally Lipschitzian (l. L.) in the first argument if there exists a bounded operator

$$L_1: H_1 \times H_2 \times R^+ \rightarrow R^+$$

such that for $\|y_1 - y_1'\| \leq r$

$$\|S(y_1, y_2) - S(y_1', y_2)\| \leq L_1(y_1, y_2, r)\,\|y_1 - y_1'\| \tag{2.19}$$

holds. S is called l. L. in the second argument if there exists a bounded operator

$$L_2: H_1 \times H_2 \times R^+ \rightarrow R^+$$

such that for $\|y_2 - y_2'\| \leq s$

$$\|S(y_1, y_2) - S(y_1, y_2')\| \leq L_2(y_1, y_2, s)\,\|y_2 - y_2'\| \tag{2.20}$$

holds.

If S is l. L. then S is Lipschitz continuous on bounded sets.

Without proof we quote from [3] the

Lemma 2.4. Let the functional $h(y, v)$ be nonnegative, and let h_y and h_v exist and be l.L. in the first resp. second argument. Let further $d_0 > 0$, $d > 0$ be fixed.

$$0 < a_n \leq \text{Min}\,\{1/(d_0 + \frac{1}{2} L_1(x_n, u_n, r_n)),\ d\} \tag{2.21}$$

$$0 < b_n \leq \text{Min}\,\{1/(d_0 + \frac{1}{2} L_2(x_{n+1}, u_n, s_n)),\ d\} \tag{2.22}$$

where $r_n = d\,\|h_y(x_n, u_n)\|$, $s_n = d\,\|h_v(x_{n+1}, u_n)\|$ and L_1 is the L.-function of h_y, L_2 is the function of h_v. Then for the iteration procedure (2.18) the following assertion is true: $h(x_n, u_n)$ is decreasing and $\|x_{n+1} - x_n\| + \|u_{n+1} - u_n\| \to 0$.

With the help of Lemma 2.4 we prove the

Theorem 2.5. Let the operator T have the following properties:

T is continuous and bounded and l.L. in both arguments. (2.23)

T_y and T_v exist and are continuous and bounded over $H \times U$; T_y is l.L. in the first, T_v is l.L. in the second argument (T_y and T_v are considered as mappings from $H \times H$ into the space of linear operators of H into H). (2.24)

$(T(x_1, v) - T(x_2, v),\ x_1 - x_2) \geq \delta(\|x_1 - x_2\|)$, $\delta(r) \geq 0$, $\delta(r) \to 0$ then $r \to 0$ (cf. (2.6)) (2.25)

$\forall\, u \in U\ \exists\, Nu \in H$ such that $\|Nu\| \leq b\|u\| + c$, $b < 1$, and $\|T(Nu, u)\| \leq b_1\|Nu - u\| + c$, $b_1 < \delta(1)/2$ (cf. (2.8)) (2.26)

Then we obtain:

(i) The functional k (cf. (2.15)) satisfies the assumptions of Lemma 2.4.
(ii) If a_n and b_n are chosen in accordance with (2.21) and (2.22) then the sequence (2.18) is bounded for $h = k$.
(iii) It is possible to choose a_n and b_n in such a way that $a_n > c_0$, $b_n > c_0 > 0$ and (2.21) and (2.22) hold.
(iv) If a_n and b_n are as under (iii) then every accumulation point of (2.18) satisfies the equation $k' = [k_y, k_v] = 0$.

Proof. (i) An easy calculation shows that for $\|y_1 - y_2\| \leq r$ and all $w \in H$

$$([T_y(y_1, v)]^* T(y_1, v) - [T_y(y_2, v)]^* T(y_2, w),\ w) \leq L(y_1, v, r)\|y_1 - y_2\|\ \|w\| \tag{2.27}$$

holds where L is composed of the L.-functions of T and T_y and the bounded terms $\|T(y_1, v)\|$ and $\|T_y(y_2, v)\|$ (cf. Lemma 3 in [3]). An analogous expression will hold when T_y is replaced by T_v. Because of (2.16) we get immediately from (2.27) for $\|y_1 - y_2\| \leq r$

$$(k_y(y_1, v) - k_y(y_2, v),\ w) \leq L'(y_1, v, r)\,\|y_1 - y_2\|\ \|w\| \tag{2.28}$$

(As to k_v we proceed in a similar way.) In Lemma 2.4 l.L.-property of k_y can be replaced by the (weaker) property (2.28).

(ii) By Lemma 2.4, $k(x_n, u_n)$ is decreasing. Then

$$\|x_n - u_n\|^2 \leq 2k(x_n, u_n) \leq \text{const}$$

i.e., x_n is bounded if and only if u_n is bounded. We show

$$\frac{1}{2}\|T(y, v) - y^*\|^2 \to \infty \text{ if } \|y\| \to \infty \text{ and } \|y - v\| \leq \text{const} \tag{2.29}$$

Indeed, because of (2.11) we have for big $\|y - Nv\|$

$$\frac{\delta(1)}{2}\|y - Nv\|^2 \le (T(y,v) - T(Nv,v),\ y - Nv)$$

hence

$$\|T(y,v) - y^*\| \ge \frac{\delta(1)}{2}\|y - Nv\| + (T(Nv,v) - y^*,\ y - Nv)/\|y - Nv\|$$

From (2.26) and $\|y - v\| \le c$ we get further

$$|(T(Nv,v) - y^*,\ y - Nv)|/\|y - Nv\| \le \|T(Nv,v)\| + \text{const} \le b_1\|Nv - v\| + \text{const}$$
$$\le b_1\|y - Nv\| + \text{const},\ b_1 < \frac{\delta(1)}{2}$$

Using again (2.26) we have

$$\|T(y,v) - y^*\| \ge (\delta(1)/2 - b_1)\|y - Nv\| + \text{const} \to \infty \text{ if } \|y\| \to \infty$$

Therefore, x_n and u_n have to be bounded.

(iii) Let us choose

$$\lambda L_{1,n} \le a_n \le L_{1,n},\quad \lambda L_{2,n} \le b_n \le L_{2,n}$$

where

$$L_{1,n} = \text{Min}\{1/(d_0 + \frac{1}{2}L_1(x_n, u_n, r_n)), d\},\ L_{2,n} = \text{Min}\{1/(d_0 + \frac{1}{2}L_2(x_{n+1}, u_n, s_n)), d\}$$

and λ with $0 < \lambda \le 1$ is arbitrary fixed. Then—as we have seen—x_n and u_n are bounded. Consequently, $r_n = d\|k_y(x_n, u_n)\|$ and $s_n = d\|k_v(x_{n+1}, u_n)\|$ are bounded (cf. (2.24)), hence $L_1(x_n, u_n, r_n)$ and $L_2(x_{n+1}, u_n, s_n)$ are bounded too. We then have

$$a_n \ge \lambda\cdot L_{1,n} \ge \lambda c_1 > 0,\quad b_n \ge \lambda\cdot L_{2,n} \ge \lambda c_1 > 0$$

(iv) We have (because of Lemma 2.4, (2.18), and (iii))

$$k_y(x_n, u_n) = (x_n - x_{n+1})/a_n \to 0$$

$$k_v(x_{n+1}, u_n) = (u_n - u_{n+1})/b_n \to 0$$

Let $[x_0, u_0]$ be an accumulation point of the iteration sequence $[x_n, u_n]$, i.e., $x_j \to x_0$, $u_j \to u_0$ for a partial sequence $\{j\} \subseteq \{n\}$. Since k_y and k_v are continuous we have $k_y(x_j, u_j) \to k_y(x_0, u_0)$, $k_v(x_{j+1}, u_j) \to k_v(x_0, u_0)$ ($x_j - x_{j+1} \to 0$ implies $x_{j+1} \to x_0$). Hence $k_y(x_0, u_0) = k_v(x_0, u_0) = 0$, i.e., $[x_0, u_0]$ solves the equation $k' = 0$.

3. AN ABSTRACT OBSTACLE-PROBLEM

Let H, U, H_n, U_n, P_n be as in the beginning of Sec. 2. Here we specialize our set $C(u)$ as

$$C(u) = Mu + C \tag{3.1}$$

where $C \subset H$ is nonempty, closed, and convex and M is a mapping from U into H. Let further be $h \equiv 0$, $A(x,u) \equiv Ax$, A single-valued, continuous, and bounded, $D(A) = H$, $f \in H$. That means, we consider here the quasivariational inequality

$$\begin{cases} u \in Mu + C \\ (Au - f,\ v - u) \ge 0 \quad \forall v \in Mu + C \end{cases} \tag{3.2}$$

A special case of (3.2) is the "obstacle-problem" of the following kind: We ask for $u \in U$ with the property

$$\begin{cases} u \leq Mu \\ (Au - f,\ v - u) \geq 0 \qquad \forall\, v \leq Mu \end{cases} \tag{3.3}$$

where H (a Sobolev space) is a lattice with respect to the partial ordering $\leq$ induced by a cone $K \subset H$ of nonnegative functions. Then clearly, the set $C(u) = \{v \in H,\ v \leq Mu\}$ can be written as (3.1), where $C = -K$ is a closed, convex subset of H.

Let more general C(u) be given as

$$C(u) = \{v \in H,\ v \geq \Psi(u) \text{ in } \Omega,\ D_i v|_\Gamma \geq v_i,\ D_j v|_\Gamma \leq v_j,\ i \in I,\ j \in J\} \tag{3.4}$$

where $H = W_2^m(\Omega)$, $\Omega \subset R^n$ a domain with the boundary Γ, Ψ a mapping from H into H, D_i, D_j linear operators on H with

$$D_i \Psi(u)|_\Gamma = w_i,\quad D_j \Psi(u)|_\Gamma = w_j \tag{3.5}$$

v_i, v_j, w_i, w_j fixed functions on Γ, and I and J finite sets of natural numbers. Then C(u) is of the form (3.1) with $Mu = \Psi u$ and

$$C = \{v \in H,\ v \geq 0 \text{ in } \Omega,\ D_i v|_\Gamma \geq v_i - w_i,\ D_j v|_\Gamma \leq v_j - w_j,\ i \in I,\ j \in J\} \tag{3.6}$$

Special cases are (3.3) where

$$C(u) = \{v \in H,\ v \geq \Psi(u) \text{ a.e.}\},\quad I \cup J = \phi,$$

and

$$C(u) = \{v \in H,\ v \geq \Psi(u) \text{ a.e.},\ \Psi(u)|_\Gamma = u_0,\ v|_\Gamma = v_1\}$$

where $I = J = \{1\}$, D_1 = identity.

(3.4) describes a general case where the solution depending obstacle lies within the domain. We want to look at another case where the obstacle is located on the boundary. We consider the set

$$K(u) = \{v \in H,\ \left.\frac{\partial^i v}{\partial \gamma^i}\right|_\Gamma \geq h_i(u),\ \frac{\partial^j v}{\partial \gamma^i} \leq h_j(u),\ i \in I,\ j \in J\} \tag{3.7}$$

where $\partial v/\partial \gamma$ is the normal derivative, $h_j: H^m(\Omega) \to H^{m-j-1/2}(\Gamma)$ and I, $J \subseteq \{0, 1, 2, \ldots, m\}$. Then $K(u) = K + Mu$, where

$$\left.\frac{\partial^k Mu}{\partial \gamma^k}\right|_\Gamma = h_k(u),\qquad k \in I \cup J \tag{3.8}$$

and

$$K = \{v \in H,\ \left.\frac{\partial^i v}{\partial \gamma^i}\right|_\Gamma \geq 0,\ \left.\frac{\partial^j v}{\partial \gamma^j}\right|_\Gamma \leq 0,\ i \in I,\ j \in J\}$$

The existence of $Mu \in H$ with the property (3.8) results from the trace-theorem (cf. [9]). Special cases are

$$K(u) = \{v \in H,\ v|_\Gamma \geq h_0(u)\},\ \text{where } I = \{0\},\ J = \phi$$

and

$$K(u) = \{v \in H,\ v|_\Gamma = h_0(u),\ \left.\frac{\partial v}{\partial \gamma}\right|_\Gamma \leq h_1(u)\},\ \text{where } I = \{0\},\ J = \{0, 1\}$$

Now, we go back to the general problem (3.2). A corresponding parametric variational inequality is

$$\begin{cases} y \in Mv + C \\ (Ay - f, z - y) \geq 0 \quad \forall z \in Mv + C \end{cases} \tag{3.9}$$

Let closed, convex, nonempty sets $C_n \subset H_n$ be given with the property

$$\text{w-}\overline{\text{Lim}}\, C_n \subset C \subset \text{s-}\underline{\text{Lim}}\, C_n \tag{3.10}$$

and monotoneous, everywhere defined, hemicontinuous operators $S: H \to H$ and $S_n: H_n \to H_n$ with

$$C = \{y \in H,\ Sy = 0\}, \quad C_n = \{y \in H_n,\ S_n y = 0\} \tag{3.11}$$

Let us put

$$B(x, v) = S(x - Mv), \quad B_n(x, v) = S_n(x - P_n Mv) \tag{3.12}$$

In the case $S = I - P_C$, $S_n = I - P_{C_n}$ we would have $B(x, v) = x - Mv - P_C(x - Mv)$, $B_n(x, v) = x - P_n Mv - P_{C_n}(x - P_n Mv)$ (cf. [3]). Then we obtain the

Theorem 3.1. Let the following conditions be fulfilled:

$$A \text{ is continuous, } \|Ax\| \leq L\|x\| + \text{const},\ L \text{ fixed} \tag{3.13}$$

$$(Ax - Ay, x - y) \geq \delta(\|x - y\|);\ \delta(r) \to 0 \Rightarrow r \to 0 \tag{3.14}$$

$$M \text{ is increasing continuous (i.e., } u_r \to u \Rightarrow Mu_r \to Mu) \tag{3.15}$$

$$\|Mu\| \leq b\|u\| + \text{const},\ b < \delta(1)/2(L + \delta(1)/2) \tag{3.16}$$

$$\underline{\lim}\, (S_j w_j, y_j - w_j) \geq (Sw, y - w) \text{ for } w_j \in H_j,\ w_j \rightharpoonup w,\ y_j \in H_j,\ y_j \to y,\ \{j\} \subseteq \{n\} \tag{3.17}$$

Then the problems

$$\|y - v\|^2 \to \text{Min! where } y \text{ solves (3.9) with the parameter } v \in U \tag{3.18}$$

and

$$\|y_n - v_n\|^2 + \frac{1}{e_n}\|P_n Ay_n + \frac{1}{\lambda_n} B_n(y_n, v_n) - P_n f\|^2 \to \text{Min}_{H_n \times U_n}! \tag{3.19}$$

are solvable. If $[x_n, u_n]$ is a solution of (3.19) then the sequence $\{[x_n, u_n]\}$ is s-compact in $H \times H$ and each of its accumulation points is a solution of (3.18).

Proof. Putting $C(u) = C + Mu$, $C_n(u) = C_n + P_n Mu$ we have

$$C(u) = \{y \in H,\ B(y, u) = 0\}, \quad C_n(u) = \{y \in H_n,\ B_n(y, u) = 0\}$$

B and B_n are monotoneous in the first argument, demicontinuous and everywhere defined and (1.5) and (2.12) hold. We show that also the other assumptions of Theorem 2.1 are fulfilled. (2.1) is clear as A is continuous, (2.6) is (3.14). To prove (2.7) we take $Nu = c_0 + Mu$, $c_0 \in C$ fixed. Then $\|Nu\| \leq \|Mu\| + \text{const} \leq b\|u\| + \text{const}$,

$$\|A(Nu)\| \leq L\|Mu\| + \text{const} \leq L\frac{b}{1 - b}\|Mu - u\| + \text{const} \tag{3.20}$$

since $\|Mu\| \leq b\|u\| + \text{const} \leq b\|u - Mu\| + b\|Mu\| + \text{const}$. But $b/(1 - b) < \delta(1)/2(L + \delta(1)/2)(1 - \delta(1)/2(L + \delta(1)/2)) = \delta(1)/2L$ and (3.20) gives (2.7). For the proof of (2.8) we define

$$N_n v = c_n + P_n Mv$$

and proceed in the same way as above. We come to (2.9). For $v_j \in H_j$, $y_j \in H_j$, $w_j \in H_j$, $\{j\} \subseteq \{n\}$ with $v_j \rightharpoonup v$, $y_j \rightharpoonup y$, $w_j \to w$ we have

$$\underline{\lim}\,(B_j(w_j, v_j), y_j - w_j) = \underline{\lim}\,(S_j(w_j - P_j Mv_j), y_j - w_j)$$

$$\geq (S(w - Mv), y - w) = (B(w,v), y - w)$$

because of (3.17) and $w_j - P_j Mv_j \to w - Mv$. Finally, we consider (2.10). For $v_j \in U_j$, $v_j \rightharpoonup v$, $y_j \in H_j$, $y_j \rightharpoonup y$, $\{j\} \subseteq \{n\}$, $y_0 \in C(v)$ we take according to (2.12) a sequence $z_j \in C_j(v_j) \subset H_j$ with the property $z_j \to y_0$, then also $Az_j \to Ay_0$ and $\underline{\lim}\{(1/\lambda_j)(B_j(z_j, v_j), y_j - z_j)\} = 0$ since $B_j(z_j, v_j) = 0$. Now, the assertion follows from Theorem 2.1.

Remark. In the case where $H = W_2^m(\Omega)$, $\Omega \subset R^n$, $n < 2m$,

$$C = \{v \in H,\ v \geq v_0 \text{ in } \Omega,\ D_i v|_\Gamma \geq v_i,\ i = 1, 2, \ldots, k\}$$

D_i linear operators with $D_i r = 0$ if r is a constant function, $D_i(Mu)|_\Gamma = w_i$ for all u; v_i, w_i given functions, $S = I - P_C$, $S_n = I - P_{C_n}$, $C_n \subset H_n$ have the property (3.10)—as a consequence of the Sobolev imbedding theorem—in Theorem 3.1 assumption (3.17) is not needed and (3.5) can be replaced by "M is w-closed" (i.e., $u_r \rightharpoonup u$, $Mu_r \rightharpoonup x \Rightarrow x = Mu$) (cf. [2]).

As an example we specify H, H_n, U, U_n, S, S_n in the following way and then verify (3.10), (3.11), and (3.17). Let be

$$H = W_2^1(0,1),\ U = H,\ K = \{y \in H,\ y \geq 0 \text{ a.e.}\},\ H_n = \langle \omega_0^n, \ldots, \omega_{n-1}^n \rangle,\ U_n = H_n$$

where ω_k^n is the linear spline function

$$\omega_k^n(t) = \begin{cases} 0 \text{ for } t \leq \dfrac{k-1}{n-1},\ t \geq \dfrac{k+1}{n-1} \\ 1 \text{ for } r = \dfrac{k}{n-1} \end{cases} \qquad k = 0, \ldots, n-1$$

$$K_n = \{y \in H_n,\ y \geq 0 \text{ a.e.}\} = \left\{ \sum_{0 \leq i \leq n-1} \alpha_i \omega_i^n,\ \alpha_i \geq 0 \right\}$$

Let Π_n be the n-th interpolation operator, i.e., if $y \in C(0,1)$ then

$$\Pi_n y = \sum_{0 \leq i \leq n-1} y\left(\frac{i}{n-1}\right) \omega_i^n$$

Π_n is a linear, continuous operator from $C(0,1)$ (and as $W_2^1 \subset C$ also from H) into H:

$$\|\Pi_n y\|_H \leq \sum \left| y\left(\frac{i}{n-1}\right) \right| \|\omega_i^n\|_H \leq \left(\sum \|\omega_i^n\|_H\right) \|y\|_C \leq C_n \|y\|_H$$

Besides, we have

$$|(\Pi_n y)(t)| \leq \max_t |y(t)|, \text{ i.e., } \|\Pi_n y\|_C \leq \|y\|_C$$

Lemma 3.2. For $y \in H$ we obtain

$$\Pi_n y \to y \quad \text{in } H \tag{3.21}$$

Proof. Let $y \in H$ and $\epsilon > 0$ be given. Since W_2^2 is dense in W_2^1 there is a $z \in W_2^2$ with the property

$$\|y - z\|_H \leq \epsilon$$

Besides, by a well-known formula (cf. [5]) we have for $z \in W_2^2$

$$\|z - \Pi_n z\|_H \leq k \cdot h_n \|z\|_{W_2^2}, \quad h_n = 1\ (n - 1)$$

Hence, for $n > n(\epsilon)$ we have

$$\|z - \Pi_n z\|_H \leq \epsilon$$

and

$$\|y - \Pi_n y\|_C \leq \|y - z\|_C + \|z - \Pi_n z\|_C + \|\Pi_n z - \Pi_n y\|_C$$

$$\leq 2k_0 \|y - z\|_H + k_0 \|z - \Pi_n z\|_H \leq k_0 (2\epsilon + \epsilon)$$

This means $\|y - \Pi_n y\|_C \to 0$, i.e.,

$$\Pi_n y \to y \quad \text{in } H$$

Then $\|\Pi_n y\|_H < \infty$, and the Banach-Steinhaus theorem (cf., e.g., [6]) gives

$$\|\Pi_n\| \leq m_0 \tag{3.22}$$

where Π_n is considered as a linear mapping from H into H. Hence,

$$\|y - \Pi_n y\|_H \leq \|y - z\|_H + \|z - \Pi_n z\|_H + \|\Pi_n z - \Pi_n y\|_H$$

$$\leq (1 + m_0)\|y - z\|_H + \|z - \Pi_n z\|_H \leq (m_0 + 2)\epsilon$$

Q.E.D.

Since $K_n \subset K$, K is w-closed and $\Pi_n y \in K_n$ if $y \in K$, we get immediately

$$\text{w-}\overline{\text{Lim}}\, K_n \subset K \subset \text{s-}\underline{\text{Lim}}\, K_n \tag{3.23}$$

Now, we define S and S_n. We consider the real function

$$f(t) = \begin{cases} (1 - 1/t + 1)^3 & \text{for } t \geq 0 \\ 0 & \text{for } t < 0 \end{cases} \tag{3.24}$$

f is a nonnegative, twice continuous differentiable function with

$$|f(t)| \leq 1, \quad |f'(t)| \leq 3, \quad |f''(t)| \leq 6$$

By

$$a(u, v) := \int_0^1 -f(-u(t))v(t)\,dt = (-F(-u), v)_{L_2}$$

for every $u \in H$ a linear continuous functional on H is defined. Here F is the superposition operator

$$(Fx)(t) = f(x(t))$$

mapping L_2 into itself. Let the operator S then be defined by

$$(Su, v) = (-F(-u), v)_{L_2} \tag{3.25}$$

S is monotoneous and Lipschitz-continuous:

$$(Sx - Sy, x - y) = (-F(-x) + F(-y), x - y)_{L_2} \geq 0$$

as F is monotoneous since f is a monotoneous function.

$$\|Sx - Sy\|^2 = (Sx - Sy,\ Sx - Sy) = \|-F(-x) + F(-y)\|^2_{L_2}$$
$$= \int_0^1 |-f(-x(t)) + f(-y(t))|^2\, dt$$

The function f is Lipschitz-continuous:

$$|f(t_1) - f(t_2)| \leq f_1 |t_1 - t_2|$$

where f_1 is an upper bound of $f'(t)$. Hence

$$\int_0^1 |-f(-x(t)) + f(-y(t))|^2\, dt \leq f_1^2 \int_0^1 |x(t) - y(t)|^2\, dt$$

such that

$$\|Sx - Sy\| \leq f_1 \|x - y\|_{L_2} \leq f_1 \|x - y\| \tag{3.26}$$

Besides,

$$\{y \in H,\ y \geq 0 \text{ a.e.}\} \quad \{y \in H,\ Sy = 0\} \tag{3.27}$$

Indeed, let be $y(t) \geq 0$ a.e.; then $-f(-y(t)) = 0$ a.e., i.e., $-F(-y) = 0$, hence $(Sy, v) = 0$ for all $v \in H$, i.e., $Sy = 0$. If $Sy = 0$ then $(F(-y), v)_{L_2} = 0$ for all $v \in H$, hence for all $v \in L_2$ since H is dense in L_2, i.e., $F(-y) = 0$, i.e., $y(t) \geq 0$ a.e.

Further, by

$$(S_n u, v) = \frac{1}{n} \sum_{0 \leq j \leq n-1} -f(-u(t_j))v(t_j), \quad t_j = j/(n-1) \tag{3.28}$$

we define an operator S_n on H_n being obviously monotoneous and Lipschitz-continuous:

$$\|S_n x - S_n y\|^2 = \frac{1}{n} \sum (-f(-x(t_j)) + f(-y(t_j))^2$$
$$\leq \frac{1}{n} \sum f_1^2 (x(t_j) - y(t_j))^2 \leq f_1^2 \|x - y\|_C^2 \leq c_0^2 f_1^2 \|x - y\|^2$$

Besides,

$$\{y \in H_n,\ y \geq 0 \text{ a.e.}\} = \{y \in H_n,\ S_n y = 0\} \tag{3.29}$$

Indeed, if $y(t) \geq 0$ then $-f(-y(t_j)) = 0$ hence $(S_n y, v) = 0$ for all $v \in H_n$ hence $S_n y = 0$. If $S_n y = 0$ then $-f(-y(t_j)) = 0$ for all j hence $y(t_j) \geq 0$. Since $y \in H_n$ we have $y(t) \geq 0$.

At last, if $w_n, y_n \in H_n$, $w_n \to w$, $y_n \to y$ we have

$$\frac{1}{n} \sum -f(-w_n(t_j^n))(y_n(t_j^n) - w_n(t_j^n)) \to \int_0^1 -f(-w(t))(y(t) - w(t))\, dt \tag{3.30}$$

The left hand side can be interpreted as the integral over a step-function $z_n(t)$ converging a.e. to $z(t) = -f(-w(t))(y(t) - w(t))$. Besides, $|z_n(t)| \leq \|-f(-w_n))\|_C \|y_n - w_n\|_C \leq$ const since our assumptions imply convergence in the sense of C. The theorem of Lebesgue then gives (3.30). Consequently, $(S_n w_n, y_n - w_n) \to (Sw, y - w)$, this is (3.17). This way, if also A and M have the assumed properties all assumptions of Theorem 3.1 are fulfilled.

Now, we come to the application of the iteration procedure (Theorem 2.5) to the special case (3.1). In this case the operator T of Theorem 2.5 is of the kind

$$T(y, v) = c_1(P_N Ay + c_2 B_N(y, v)) \tag{3.31}$$

where P_N is the orthogonal projection of H onto H_N, $B_N(y, v) = S_N(y - P_N Mv)$, S_N is a monotoneous operator from H_N into H_N with the property (3.11), N is a fixed natural number.

In the sequel we write:

$$H_N = H_0,\ P_N A = A_0,\ P_N M = M_0,\ \Pi_N = \Pi_0,\ S_N = S_0,\ B_N = B_0,\ P_n f = f_0$$

$$\text{and}\quad T(y,v) = A_0 y + S_0(y - M_0 v) \tag{3.32}$$

Then we have

$$T_y(y,v) = A_0'(y) + S_0'(y - M_0 v) \tag{3.33}$$

$$T_v(y,v) = -S_0'(y - M_0 v)M_0'(v) \tag{3.34}$$

where A_0' and S_0' are the Gateau-derivatives of A_0 and S_0, M_0' is the Frechet-derivative of M_0. (The existence of these derivatives may be supposed.) We prove (3.34):

$$(S_0(y - M_0 v))_v w = \lim \frac{1}{\theta}\{S_0(y - M_0(v + \theta w)) - S_0(y - M_0 v)\}$$

$$M_0(v + \theta w) = M_0 v + M_0'(v)\theta w + \theta\|w\|\epsilon(v, \theta w),\quad \epsilon \to 0 \text{ if } \theta \to 0$$

Then

$$(S_0(y - M_0 v))_v w = \lim \frac{1}{\theta}\{S_0(y - M_0 v - \theta M_0'(v)w - \theta\|w\|\epsilon) - S_0(y - M_0 v)\}$$

$$= -S_0'(y - M_0 v)M_0'(v)w$$

Theorem 3.3. Besides of (3.13), (3.14), (3.16) let the following assumptions be fulfilled:

A_0 is l.L. (3.35)

A_0' exists, is l.L. and bounded (3.36)

S_0 is monotonous, l.L. and bounded (3.37)

S_0' exists, is l.L. and bounded (3.38)

The Frechet-derivative M_0' of M_0 exists, is l.L. and bounded (3.39)

If a_n and b_n are suitably chosen (cf. (2.21), (2.22), and Theorem 2.5(iii)) then every accumulation point of the iteration sequence

$$x_0 \in H,\quad u_0 \in U$$

$$x_{n+1} = x_n - a_n k_y(x_n, u_n)$$

$$u_{n+1} = u_n - b_n k_v(x_{n+1}, u_n)$$

fulfills the equation $[k_y, k_v] = 0$, where

$$k_y(y,v) = y - v + (A_0'(y) + S_0'(y - M_0 v))^*(A_0 y + S_0(y - M_0 v) - f_0) \tag{3.40}$$

$$k_v(y,v) = v - y + (-S_0'(y - M_0 v)M_0' v)^*(A_0 y + S_0(y - M_0 v) - f_0) \tag{3.41}$$

Proof. Similar to the proof of Theorem 2.5 (Lemma 3 in [3]) we conclude that the l.L.-property and boundedness of T_y and T_v can be "composed" of the corresponding properties of A_0', S_0', and M_0'. This way, T satisfies the assumptions (2.23), (2.24), and (2.25). (2.26) is a consequence of (3.13) and (3.16) (cf. the proof of Theorem 3.1). Applying Theorem 2.5 we get the assertion.

Now, let us continue with our example and show that S_0 fulfills (3.38) ((3.37) is obvious). We have

$$\left(\frac{1}{\theta}(S_0(v + \theta w) - S_0 v),\ x\right) = \frac{1}{N\cdot\theta} \sum_{0\le j\le N-1} [-f(-v(t_j) - \theta w(t_j)) + f(-v(t_j))]\,x(t_j)$$

$$\xrightarrow{\theta\to 0} \frac{1}{N} \sum f'(-v(t_j))w(t_j)x(t_j)$$

since $(1/\theta)(f(s_1 + \theta s_2) - f(s_1)) \to f'(s_1)s_2$.

Let us define an operator $S_0' : H_0 \to L(H_0 \to H_0)$ by

$$(S_0'(v)w,\ x) = \frac{1}{N} \sum f'(-v(t_j))w(t_j)x(t_j)$$

then

$$\lim \left(\frac{1}{\theta}(S_0(v + \theta w) - S_0 v) - S_0'(v)w,\ x\right) = 0$$

for all $x \in H_0$, i.e., $(1/\theta)(S_0(v + \theta w) - S_0 v) \to S_0'(v)w$ in H_0, hence S_0' is the Gateau-derivative of S_0.

At last, we show the Lipschitz-continuity of S_0'. We have

$$\|(S_0'(x) - S_0'(y))w\|^2 = \frac{1}{N} \sum (f'(-x(t_j)) - f'(-y(t_j)))^2 w(t_j)^2$$

$$\le \frac{1}{N} \sum f_2^2 |x(t_j) - y(t_j)|^2 |w(t_j)|^2$$

$$\le f_2^2 \|x - y\|_C^2 \|w\|_C^2 \le c_0^4 f_2^2 \|x - y\|^2 \|w\|^2$$

since $H \subset C$ is continuously imbedded and

$$|f'(t_1) - f'(t_2)| \le f_2 |t_1 - t_2|$$

where f_2 is the upper bound of $f''(t)$.

REFERENCES

1. Bruckner, G.: Math. Nachr. 104, 209-216 (1981).
2. Bruckner, G.: Math. Nachr. 105, 293-306 (1982).
3. Bruckner, G.: On abstract quasi-variational inequalities. Approximation of solutions III. Preprint AdN der DDR, P-Math–24/84.
4. Bruckner, G.: Z. Anal. Anwendungen 3, 81-86 (1984).
5. Ciarlet, P.: The Finite Element Method for Elliptic Problems. North-Holland (1978).
6. Kantorovitch, L. W.; Akilov, G. P.: Functional Analysis. Moscow (1977).
7. Kuge, R.: Nichtlineare Variationsungleichungen. Berlin (1979).
8. Kluge, R.: Zur approximativen Lösung nichtlinearer Variationsungleichungen. Diss. B, Berlin (1970).
9. Lions, J. L.; Magenes, E.: Problèmes aux limites non homogènes et applications I. Paris (1968).

10 HELMHOLTZ DECOMPOSITION OF $W^{p,s}$ VECTOR FIELDS*

Murray R. Cantor[†] and John C. Matovsky[‡] Department of Mathematics, University of Texas at Austin, Austin, Texas

I. INTRODUCTION

A well-known classical problem is decomposing a vector field into the sum of a gradient field and a divergence-free field—the important Helmholtz decomposition. This plays a fundamental role in solving the equations of fluid mechanics and electromagnetism (cf. Marsden [6], Ladyzhenskaya [4]); in the Euler equations for a perfect fluid, for example, it implies the existence of the pressure function.

To couch this decomposition problem in the appropriate abstract mathematical setting, we know from physical considerations that the total kinetic energy of fluid flow is finite. Hence, the velocity field is square integrable and the L^2 Sobolev theory of derivatives arises naturally. The Hilbert spaces $W^{2,s}$ in fact are structurally related by their norms to the energy of the fluid flow.

To obtain the Helmholtz decomposition of the $W^{2,s}$ vector fields on a smooth bounded domain in $\mathbb{R}^n$, a partial differential equation with a boundary condition has to be solved. This can be seen by taking a vector field X and writing

$$X = \nabla u + Y$$

where Y has zero divergence and its tangential to the boundary. (This latter condition would correspond in hydrodynamics to particles of fluid not moving across the boundary.) Hence, if ν is the unit outward normal to be boundary, $Y \cdot \nu = 0$. Therefore,

$$\Delta u = \operatorname{div}(X)$$

$$\frac{\partial u}{\partial \nu} = \nabla u \cdot \nu = X \cdot \nu$$

Thus, a Neumann problem arises naturally in connection with the Helmholtz decomposition. In the PDE literature, sufficient conditions for existence and uniqueness of a solution to a Neumann problem with data in L^2 are well-known [4]. Below we give a clear proof of this result; in particular, by first posing the Neumann problem in a weak or distributional form, the introduction of boundary terms when integrating by parts will be seen.

Now the question naturally arises: Can the hypotheses be weakened to allow for more general data in L^p, $1 < p < \infty$? Using modern elliptic regularity theory, the answer is yes, and the restriction to $p = 2$ is unnecessary. This result seems to be well known but we know of no published proof. Moreover, by enlarging the class of solutions to the Sobolev spaces and adopting the viewpoint of splitting spaces of vector fields rather than the individual vector fields themselves, we avail ourselves of the clean

---

*Partially supported by National Science Foundation research grant #PHY 7901801

†Current affiliation: Department of Geophysical Research, Shell Development Company, Houston, Texas

‡Current affiliation: Department of Mathematics and Statistics, Louisiana Tech University, Ruston, Louisiana

mathematical apparatus of functional analysis to obtain the Helmholtz decomposition. This is of particular significance in fluid mechanics and electromagnetism where the L^p Sobolev spaces $W^{p,s}$ are used to allow for discontinuities in the velocity field of the fluid or charge density. Though arising from physical considerations we also have the well-posedness of the Neumann problem with data in L^p which is of independent mathematical interest.

Throughout, we denote the usual $W^{p,s}$ norm by $\| \ \|_{p,s}$ and the trace space $W^{p,s-1/p}(\partial\Omega)$, for s a positive integer, is the set of all restrictions to the boundary of functions in the usual L^p Sobolev space $W^{p,s}(\Omega)$; for $\phi \in W^{p,s-1/p}(\partial\Omega)$, $\|\phi\|_{p,s-1/p} \equiv \inf\{\|v\|_{p,s}: \phi = v|_{\partial\Omega}\}$. The space $C^\infty(\partial\Omega)$ is the set of all restrictions to the boundary of functions in $C^\infty(\bar{\Omega})$.

Standard Sobolev results used below are the continuity of the trace operator $\gamma_0 : W^{p,1}(\Omega) \to L^p(\partial\Omega)$ with range $W^{p,1-1/p}(\partial\Omega)$ dense in $L^p(\partial\Omega)$ [1]; the Rellich-Kondrachov Compactness Theorem [1]:

> If Ω is a bounded domain with C^∞ boundary and $s < t$, then the inclusion map $W^{p,t}(\Omega) \hookrightarrow W^{p,s}(\Omega)$ is compact;

the Sobolev Embedding Theorem [1]:

> If Ω is a bounded domain with C^∞ boundary and $s > m + (n/p)$, then $W^{p,s}(\Omega) \hookrightarrow C^m(\bar{\Omega})$;

and that $C^\infty(\bar{\Omega})$ is dense in $W^{p,s}(\Omega)$ for Ω with smooth boundary [1].

II. THE NEUMANN PROBLEM WITH DATA IN L^p

<u>Definition</u>. Let $\Omega \subset \mathbb{R}^n$ be a bounded domain with (n - 1)-dimensional boundary $\partial\Omega$ of class C^∞. Given $f \in L^p(\Omega)$, $\phi \in W^{p,1-1/p}(\partial\Omega)$, $1 < p < \infty$ (so there is a $v \in W^{p,1}(\partial\Omega)$ so that $v|_{\partial\Omega} = \phi$), a function $u \in W^{p,1}(\Omega)$ is a <u>generalized solution of the Neumann boundary-value problem in</u> $L^p(\Omega)$ if for all $v \in W^{p',1}(\Omega)$ $(1/p + 1/p' = 1)$:

$$\int_\Omega \nabla u \cdot \nabla v = \int_{\partial\Omega} \phi\gamma_0(v) - \int_\Omega fv$$

Then $\Delta u = f$ a.e. (Ω) by restricting v to be in the dense subspace $C_0^\infty(\Omega)$ of $L^{p'}(\Omega)$, and consequently

$$\int_{\partial\Omega} \phi\gamma_0(v) = \int_\Omega (\nabla u \cdot \nabla v + \Delta uv) = \int_{\partial\Omega} \frac{\partial u}{\partial \nu}\gamma_0(v)$$

by Green's Theorem. Hence

$$\phi = \frac{\partial u}{\partial \nu} \text{ a.e. } (\partial\Omega)$$

since the trace operator has dense range in $L^p(\partial\Omega)$ [1].

We give two proofs of the existence of weak solutions for the Neumann problem in L^2, the first using Fredholm theory and the second a nonlinear one using ideas from the calculus of variations.

<u>Theorem</u> (Existence of Weak Solutions). Let $f \in L^2(\Omega)$ and $\phi \in W^{2,\frac{1}{2}}(\partial\Omega)$ where Ω is a bounded domain with C^∞ boundary. Then there is a $u \in W^{2,1}(\Omega)$ such that for every $v \in W^{2,1}(\Omega)$

$$\int_\Omega \nabla u \cdot \nabla v = \int_{\partial\Omega} \phi\gamma_0 v - \int_\Omega fv$$

if and only if $\int_{\partial\Omega} \phi = \int_\Omega f$.

First Proof. Define a continuous bilinear form $B: W^{2,1}(\Omega) \times W^{2,1}(\Omega) \to \mathbb{R}$ by

$$B(u,v) \equiv \int_\Omega \nabla u \cdot \nabla v$$

and a continuous linear functional $G: W^{2,1}(\Omega) \to \mathbb{R}$ by

$$G(v) \equiv \int_{\partial\Omega} \phi\gamma_0 v - \int_\Omega fv$$

continuity of G following from that of the trace operator

$$\gamma_0 : W^{2,1}(\Omega) \to L^2(\partial\Omega)$$

For $u, v \in W^{2,1}(\Omega)$, let $Lu(v) = B(u,v)$ and $Ju(v) = \int_\Omega uv$, so $L, J: W^{2,1}(\Omega) \to W^{2,1}(\Omega)$. Note that, since $(L+J)(u): v \to \langle u,v\rangle_{W^{2,1}(\Omega)}$, it follows that the operator $\tilde{L} \equiv L + J$ is the canonical Hilbert space isomorphism.

Claim 1. The operator $J: W^{2,1}(\Omega) \to W^{2,1}(\Omega)^*$ is compact: Let $(u_m) \subset W^{2,1}(\Omega)$ be a bounded sequence. Then, by the Rellich compactness lemma, there is a $u_0 \in L^2(\Omega)$ and a subsequence $(u_{m,i})$ so that $\|u_{m,i} - u_0\|_{L^2(\Omega)} \to 0$. Since $\langle u_0, \cdot\rangle_{L^2(\Omega)}|W^{2,1}(\Omega) \in W^{2,1}(\Omega)^*$,

$$\|Ju_{m,i} - \langle u_0, \cdot\rangle_{L^2(\Omega)}\|_{W^{2,1}(\Omega)^*} = \sup_{\|v\|_{W^{2,1}(\Omega)}=1} |Ju_{m,i}(v) - \int_\Omega u_0 v|$$

$$= \sup_{\|v\|_{W^{2,1}(\Omega)}=1} |\int_\Omega (u_{m,i} - u_0)v|$$

$$\leq \|u_{m,i} - u_0\|_{L^2(\Omega)} \to 0$$

Thus, $Ju_{m,i}(\cdot) \to \langle u_0, \cdot\rangle_{L^2(\Omega)}$ in $W^{2,1}(\Omega)^*$. Q.E.D.

Claim 2. $\operatorname{Ker}(L) = \{v \in W^{2,1}(\Omega): v \text{ is constant a.e. } (\Omega)\}$: Let $v \in \ker(L)$: then $Lv(v) = 0$. Consequently, $\int_\Omega |\nabla v|^2 = 0$, so $\partial_i v = 0$ a.e. (Ω), $i = 1, 2, \ldots, n$. Since Ω is connected, v is a constant function a.e. (Ω).

Therefore, given $G \in W^{2,1}(\Omega)^*$, in order to solve the weak Neumann problem $Lu = G$, note that we have the equivalent equations

$$\tilde{L}u - Ju = G$$

and

$$u - (\tilde{L}^{-1} \cdot J)u = \tilde{L}^{-1}(G)$$

By Fredholm theory, the last equation has a solution $u \in W^{2,1}(\Omega)$ if and only if

$$\tilde{L}^{-1}(G) \in \ker(\mathrm{id} - \tilde{L}^{-1} \cdot J)^\perp = \ker(L)^\perp$$

that is, if and only if

$$\langle \tilde{L}^{-1}(G),\ 1\rangle_{W^{2,1}(\Omega)} = 0 = \tilde{L}(\tilde{L}^{-1}(G))(1) = G(1)$$

Thus, we obtain the compatibility condition

$$G(1) = 0 = \int_{\partial\Omega} \phi - \int_{\Omega} f$$

Q.E.D.

Second Proof. Define $F: W^{2,1}(\Omega) \to \mathbb{R}$ by

$$F(u) = \frac{1}{2}\int_{\Omega} |\nabla u|^2 + \int_{\Omega} fu - \int_{\partial\Omega} \phi\gamma_0 u$$

Let $W \equiv \{u \in W^{2,1}(\Omega): \int_{\Omega} u = 0\}$ and note that $W \subset W^{2,1}(\Omega)$ is closed, hence a Banach space. For $u \in W$, we have a Poincaré inequality [7]:

$$\int_{\Omega} |u|^2 \leq C \int_{\Omega} |\nabla u|^2$$

and so $\langle u, w\rangle_W \equiv \int_{\Omega} \nabla u \cdot \nabla w$ determines an equivalent norm $\|\cdot\|_W$ on W, given by

$$\|u\|_W \equiv \|\nabla u\|_{L^2(\Omega)}$$

Defining $G: W^{2,1}(\Omega) \to \mathbb{R}$ by

$$G(v) = \int_{\partial\Omega} \phi\gamma_0 v - \int_{\Omega} fv$$

we have $G \in W^{2,1}(\Omega)^*$ as before by the continuity of the trace operator $\gamma_0: W^{2,1}(\Omega) \to L^2(\partial\Omega)$. Consequently, $G|_W$ is $\|\cdot\|_W$-continuous. Also, $F|_W$ is bounded below:

$$\begin{aligned} F(u) &\geq \frac{1}{2}\|u\|_W^2 - \|(G|_W)\|_{W^*}\|u\|_W \\ &= \frac{1}{2}(\|u\|_W - \|(G|_W)\|_{W^*})^2 - \frac{1}{2}\|(G|_W)\|_{W^*}^2 \\ &\geq -\frac{1}{2}\|(G|_W)\|_{W^*}^2 \quad \text{for all } u \in W \end{aligned}$$

Therefore, $d \equiv \inf\{F(u): u \in W\} > -\infty$. By the parallelogram identity, for $u, v \in W$:

$$\begin{aligned} \|u - v\|_W^2 &= 2\|u\|_W^2 + 2\|v\|_W^2 - 4\left\|\frac{u+v}{2}\right\|_W^2 \\ &= 4(F(u) + F(v)) - 8F\left(\frac{u+v}{2}\right) \\ &\leq 4(F(u) + F(v)) - 8d \end{aligned}$$

Choose $u_n \in W$ so that $F(u_n) < d + (1/n)$: then

$$\begin{aligned} \|u_n - u_m\|_W^2 &\leq 4(F(u_n) + F(u_m)) - 8d \\ &< 4\left(\frac{1}{n} + \frac{1}{m}\right) \end{aligned}$$

Thus, $(u_n) \subset W$ is $\|\cdot\|_W$-Cauchy; let $(u_n) \to u \in W$. Then $F(u_n) \to F(u) = d$. Hence, the Gateaux derivative of F at u is zero:

$$F'(u)w \equiv \lim_{t\to 0} \frac{1}{t}\{F(u + tw) - F(u)\} = 0 = \int_{\Omega} \nabla u \cdot \nabla w - G(w)$$

Since for $v \in W^{2,1}(\Omega)$, $v - (1/|\Omega|)\int_\Omega v \in W$, so putting $\bar{v} = (1/|\Omega|)\int_\Omega v$:

$$\int_\Omega \nabla u \cdot \nabla v = \int_\Omega \nabla u \cdot \nabla(v - \bar{v})$$
$$= G(v - \bar{v})$$
$$= G(v) - \bar{v}G(1)$$

Thus for every $v \in W^{2,1}(\Omega)$, we have

$$\int_\Omega \nabla u \cdot \nabla v = \int_{\partial\Omega} \phi \cdot \gamma_0(v) - \int_\Omega f \cdot v$$

if and only if

$$\int_{\partial\Omega} \phi = \int_\Omega f$$

Q.E.D.

To obtain regularity at the boundary and the existence of strong solutions of the Neumann problem with data in L^p, we cite:

Theorem (Agmon, Douglis, Nirenberg [2]). For $1 < p < \infty$ and any nonnegative integer k, if Ω is a bounded domain in $\mathbb{R}^n$ whose boundary is an $n - 1$ dimensional C^{k+2}-manifold, $f \in W^{p,k}(\Omega)$, $\phi \in W^{p,k+1-1/p}(\partial\Omega)$ and $u \in W^{p,2}(\Omega)$ satisfies $\Delta u = f$, $\partial u/\partial\nu = \phi$, then $u \in W^{p,k+2}(\Omega)$ and

$$\|u\|_{W^{p,k+2}(\Omega)} \le C(\|f\|_{W^{p,k}(\Omega)} + \|\phi\|_{W^{p,k+1-1/p}(\partial\Omega)} + \|u\|_{L^p(\Omega)})$$

Corollary. Suppose $u \in W^{2,1}(\Omega)$ is a weak solution of the L^2-Neumann boundary-value problem with data $f \in C^\infty(\bar{\Omega})$, $\phi \in C^\infty(\partial\Omega)$, over bounded domain Ω with C^∞ boundary. Then $u \in C^\infty(\bar{\Omega})$.

Proof. The hypotheses on the data f, ϕ and the boundary $\partial\Omega$ imply $f \in W^{2,k}(\Omega)$ for every $k \ge 0$ and ϕ has a global extension in $C^\infty(\bar{\Omega})$, so $\phi \in W^{2,k+\frac{1}{2}}(\partial\Omega)$, all $k \ge 0$. By the regularity of weak solutions of the L^2-Neumann problem [8], $u \in W^{2,2}(\Omega)$. Hence by the global regularity theorem above, $u \in W^{2,k+2}(\Omega)$, all $k \ge 0$. By the Sobolev embedding theorem, $u \in C^m(\bar{\Omega})$ for all $m \ge 0$. Q.E.D.

Using the L^2 regularity theory of the Neumann problem, as in the above corollary, and the following inference from the global regularity estimate which gives sufficient conditions for removing the L^p norm from the right-hand side of that inequality, we will conclude well-posedness of the Neumann problem in L^p if and only if the data are compatible and the solution has average value zero.

Lemma. If the bounded domain Ω is a C^s-manifold-with-boundary, $s \ge 2$, then there is a $C > 0$ so that for every $u \in W^{p,s}(\Omega)$ with $\int_\Omega u = 0$, $p > 1$,

$$\|u\|_{p,s} \le C\left(\|\Delta u\|_{p,s-2} + \left\|\frac{\partial u}{\partial\nu}\right\|_{p,s-1-1/p}\right)$$

Proof. Suppose there is a sequence $(u_m) \subset W^{p,s}(\Omega)$ with $\int_\Omega u_m = 0$ such that $\|u_m\|_{p,s} = 1$ and $\|\Delta u_m\|_{p,s} \to 0$, $\|\partial u_m/\partial v\|_{p,s-1-1/p} \to 0$. Passing to a subsequence of (u_m), we can assume $u_m \to u$ in $L^p(\Omega)$ by the Rellich-Kondrachov Compactness Theorem [1]. Then

$$\|u_m - u_n\|_{p,s} \le C\left(\|\Delta(u_m - u_n)\|_{p,s-2} + \left\|\frac{\partial}{\partial\nu}(u_m - u_n)\right\|_{p,s-1-\frac{1}{p}} + \|u_m - u_n\|_p\right)$$

so (u_m) is Cauchy in $W^{p,s}(\Omega)$, hence converges in $W^{p,s}(\Omega)$ to a function of norm one. By uniqueness of limits, $u \in W^{p,s}(\Omega)$ with $\|u\|_{p,s} = 1$ and $\left|\int_\Omega u\right| = \left|\int_\Omega u - \int_\Omega u_m\right| \le |\Omega|^{1/p'} \cdot \|u - u_m\|_p \to 0$, so $\int_\Omega u = 0$. But u is then the solution of the homogeneous Neumann problem $\Delta u = 0$, $\partial u/\partial\nu = 0$, so by the global regularity theorem and the Sobolev Embedding Theorem, $u \in C^\infty(\bar{\Omega})$, hence by the ordinary Green's Theorem, $u \equiv 0$. This contradicts the previous conclusion that u has $W^{p,s}(\Omega)$-norm equal to one and the lemma follows. Q.E.D.

<u>Theorem</u>. Let Ω be a bounded domain in $\mathbb{R}^n$ with C^∞ boundary $\partial\Omega$. Suppose $f \in L^p(\Omega)$, $\phi \in W^{p,1-1/p}(\partial\Omega)$, $p > 1$. Then there is a unique function $u \in W^{p,2}(\Omega)$ such that

$$\Delta u = f \quad \text{a.e. } (\Omega)$$

$$\frac{\partial u}{\partial\nu} = \phi \quad \text{a.e. } (\partial\Omega)$$

$$\int_\Omega u = 0$$

if and only if $\int_\Omega f = \int_{\partial\Omega} \phi$.

<u>Proof</u>. Suppose $\int_\Omega f = \int_{\partial\Omega} \phi$. Let $\phi = v|_{\partial\Omega}$ for some global extension $v \in W^{p,1}(\Omega)$. Find sequences (f_m), $(v_m) \subset C^\infty(\bar{\Omega})$ so that $\|f - f_m\|_{L^p(\Omega)} \to 0$ and $\|v - v_m\|_{W^{p,1}(\Omega)} \to 0$. Putting $\phi_m = v_m|_{\partial\Omega} \in C^\infty(\partial\Omega)$, we have:

$$\|\phi - \phi_m\|_{W^{p,1-1/p}(\partial\Omega)} \le \|v - v_m\|_{W^{p,1}(\Omega)} \to 0$$

and

$$\left|\int_{\partial\Omega} \phi - \int_{\partial\Omega} \phi_m\right| \le |\partial\Omega|^{1/p'} \|\gamma_0(v - v_m)\|_{L^p(\partial\Omega)}$$

$$\le C|\partial\Omega|^{1/p'} \|v - v_m\|_{W^{p,1}(\Omega)} \to 0$$

Further, note that:

$$\left|\int_{\partial\Omega} \phi - \int_\Omega f_m\right| = \left|\int_\Omega f - \int_\Omega f_m\right|$$

$$\le |\Omega|^{1/p'} \cdot \|f - f_m\|_{L^p(\Omega)}$$

so $\int_\Omega f_m \to \int_{\partial\Omega} \phi$; consequently, putting

$$\tilde{f}_m \equiv f_m - \frac{1}{|\Omega|}\int_\Omega f_m + \frac{1}{|\Omega|}\int_{\partial\Omega} \phi_m$$

see that

$$\int_\Omega \tilde{f}_m = \int_{\partial\Omega} \phi_m$$

and

$$\|f - \tilde{f}_m\|_{L^p(\Omega)} \leq \|f - f_m\|_{L^p(\Omega)} + \frac{1}{|\Omega|}\left|\int_\Omega f_m - \int_{\partial\Omega} \phi_m\right| \cdot |\Omega|^{1/p} \to 0$$

Thus, without loss of generality, we can assume $\int_\Omega f_m = \int_{\partial\Omega} \phi_m$.

Therefore, let $u_m \in C^\infty(\bar{\Omega})$ be the unique function satisfying

$$\Delta u_m = f_m \quad \text{in } \Omega$$

$$\frac{\partial u}{\partial \nu}m = \phi_m \quad \text{on } \partial\Omega$$

$$\int_\Omega u_m = 0$$

By the above lemma:

$$\|u_m - u_n\|_{p,2} \leq C(\|f_m - f_n\|_p + \|\phi_m - \phi_n\|_{p,1-1/p})$$

Hence $(u_m) \subset W^{p,2}(\Omega)$ is Cauchy, so $(u_m) \to u \in W^{p,2}(\Omega)$. Consequently, since convergence in L^p implies pointwise convergence a.e. of a subsequence [5], reindexing we have:

$$\begin{aligned} \Delta u &= \lim_{m\to\infty} \Delta u_m && \text{a.e.}(\Omega) \\ &= f && \text{a.e.}(\Omega) \\ \frac{\partial u}{\partial \nu} &= \lim_{m\to\infty} \frac{\partial u_m}{\partial \nu} && \text{a.e.}(\partial\Omega) \\ &= \phi && \text{a.e.}(\partial\Omega) \end{aligned}$$

Also:

$$\begin{aligned} \left|\int_\Omega u\right| &= \left|\int_\Omega u - \int_\Omega u_m\right| \\ &\leq \int_\Omega |u - u_m| \\ &\leq |\Omega|^{1/p'} \cdot \|u - u_m\|_{L^p(\Omega)} \to 0 \end{aligned}$$

and so $\int_\Omega u = 0$.

Conversely, if $u \in W^{p,2}(\Omega)$ satisfies $\Delta u = f$ a.e. (Ω) and $\frac{\partial u}{\partial \nu} = \phi$ a.e. $(\partial\Omega)$, then

$$\int_\Omega f = \int_\Omega \Delta u = \int_{\partial\Omega} \frac{\partial u}{\partial \nu} = \int_{\partial\Omega} \phi$$

Q.E.D.

From the main regularity theorem, we have the

Corollary. For a bounded domain Ω with smooth boundary, if $f \in W^{p,k}(\Omega)$, $\phi \in W^{p,k+1-1/p}(\partial\Omega)$ (so there is a $v \in W^{p,k+1}(\Omega)$ so that $v|_{\partial\Omega} = \phi$) such that $\int_\Omega f = \int_{\partial\Omega} \phi$, then there is a unique function $u \in W^{p,k+2}(\Omega)$ satisfying

$$\Delta u = f \quad \text{in } \Omega$$

$$\frac{\partial u}{\partial \nu} = \phi \quad \text{on } \partial\Omega$$

$$\int_\Omega u = 0$$

III. APPLICATION OF THE NEUMANN BOUNDARY-VALUE PROBLEM TO THE DECOMPOSITION OF VECTOR FIELDS

Theorem (Helmholtz Decomposition). Let $\Omega \subset \mathbb{R}^n$ be a bounded domain with boundary $\partial\Omega$ an $(n-1)$-dimensional C^∞-manifold. Then for $1 < p < \infty$ and positive integer $s \geq 1$, the Banach space as $W^{p,s}$-vector fields over Ω splits as the following topological direct sum with closed summands:

$$W^{p,s}(\Omega, \mathbb{R}^n) = \nabla(W^{p,s+1}(\Omega)) \oplus D_t^{p,s}$$

where

$$D_t^{p,s} = \{X \in W^{p,s}(\Omega, \mathbb{R}^n) : \operatorname{div}(X) = 0,\ X \cdot \nu = 0\}$$

the divergence-free tangential $\mathbb{R}^n$-valued $W^{p,s}$-vector fields over Ω.

Proof. Define the projection onto the unit outward normal ν.

$$P_\nu : W^{p,s}(\Omega, \mathbb{R}^n) \to W^{p,s-1/p}(\partial\Omega)$$

by $P_\nu(S) = X \cdot \nu$, and observe we have the following chain of maps:

$$W^{p,s+1}(\Omega) \xrightarrow{\nabla} W^{p,s}(\Omega, \mathbb{R}^n) \xrightarrow{\operatorname{div} \times P_\nu} W^{p,s-1}(\Omega) \times W^{p,s-1/p}(\partial\Omega)$$

To prove the Banach space decomposition, it is sufficient to prove the following two algebraic conditions [3]:

1. $\ker((\operatorname{div} \times P_\nu) \circ \nabla) = \ker(\nabla)$
2. $\operatorname{Rng}((\operatorname{div} \times P_\nu) \circ \nabla) = \operatorname{Rng}(\operatorname{div} \times P_\nu)$

To prove the kernels are equal, let $u \in \ker((\operatorname{div} \times P_\nu) \circ \nabla)$; then u satisfies the homogeneous Neumann problem $\Delta u = 0$, $\partial u/\partial \nu = 0$. Translating u by its average $\bar{u} \equiv (1/|\Omega|) \int_\Omega u$, the function $u - \bar{u}$ has integral 0 and also satisfies the homogeneous problem. Hence, $u - \bar{u} \equiv 0$ a.e. (Ω) and $u \in \ker(\nabla)$.

To prove the ranges are equal, let $(f, \phi) \in \operatorname{Rng}(\operatorname{div} \times P_\nu)$, so $f = \operatorname{div}(X)$ and $\phi = X \cdot \nu$ for some $\mathbb{R}^n$-valued vector field $X \in W^{p,s}(\Omega, \mathbb{R}^n)$, hence are compatible data by the divergence theorem. Thus, there is a function $u \in W^{p,s+1}(\Omega)$ such that $\Delta u = f$ and $\partial u/\partial \nu = \phi$; that is, $(f, \phi) = (\operatorname{div} \times P_\nu) \circ \nabla(u)$ and proves 2. Q.E.D.

REFERENCES

1. Adams, R.: Sobolev Spaces. Academic Press,(1975).
2. Agmon, S.; Douglis, A.; Nirenberg, L.: Estimates near the boundary for solutions of elliptic partial differential equations satisfying general boundary conditions I. Comm. Pure Appl. Math. 12, 623-727 (1959).
3. Cantor, M.: Elliptic operators and the decomposition of tensor fields. Bull. A.M.S. 5, 235-262 (1981).
4. Ladyzhenskaya, O. A.: The Mathematical Theory of Viscous Incompressible Fluid Flows. Gordon and Breach (1969).

5. Lang, S.: Real Analysis. Addison-Wesley (1969).
6. Marsden, J.: Applications of Global Analysis in Mathematical Physics. Publish or Perish, Inc., Boston (1974).
7. Morrey, C.: Multiple Integrals and the Calculus of Variations. Springer (1966).
8. Showalter, R.: Hilbert Space Methods for Partial Differential Equations. Pitman, London (1977).

11 NONLINEAR DISPERSIVE WAVES AND VARIATIONAL PRINCIPLES

Lokenath Debnath / Department of Mathematics, University of Central Florida, Orlando, Florida

1. INTRODUCTION

Almost all wave motions are essentially oscillations which propagate in space. Of all oscillatory motions in nature, water waves are the most common and notable examples of wave phenomena. It is well known that the theory of water waves provides an excellent example of linear and nonlinear dispersion. Indeed, most of the general ideas about dispersive wave systems originated from the classical problems of water waves. Historically, the classical works on water waves by Cauchy and Poisson, Stokes, Russell, Korteweg, and de Vries, Boussinesq, Airy and Rayleigh marked the beginning of the modern development of dispersive waves in general. In his famous paper on water waves in 1847, Stokes obtained two remarkable results that periodic wave trains are possible in nonlinear systems and the dispersion relation involves the amplitude. These results have served as basis for all recent developments in the theory and applications of nonlinear dispersive systems.

Despite the special attention given to the linearized theory of water waves for more than a century, the systematic study of nonlinear wave propagation in dispersive media appears to have been neglected until the 1960s. In a series of notable papers [1-6] Whitham has first successfully initiated the study of the theory of nonlinear dispersive waves and its applications to various wave phenomena. His pioneering works include the new method of averaged variational principles for the description of waves in general. Whitham developed many new ideas, methods, and results of the theory of nonlinear wave propagation with mathematical justification. His contributions have had a tremendous impact on nonlinear mechanics of waves, and led subsequent important developments in this subject during the past twenty years.

This paper gives a brief survey and development of the theory and applications of nonlinear dispersive waves with special reference to the Whitham averaged variational principles. Section 2 deals with the basic ideas and important examples of linear dispersive waves. Some discussion is made on the importance of the Fourier superposition principle in constructing the general solution of a linear system. In Sec. 3, an asymptotic analysis of an initial value problem is carried out to obtain the asymptotic solution of the problem for large time and distance. The asymptotic solution is examined with a view to extend the classical ideas of wave number, frequency, and amplitude. Section 4 is essentially concerned with the Whitham theory of slowly varying nonlinear wave motions and its application to instability of water waves. This section also includes some of the remarkable contributions to the instability theory of Stokes waves and resonance interactions by Lighthill, Benjamin and Feir, and Phillips. In order to describe the nonlinear evolution of wave trains, wave packets, and small disturbances in fluid flows, the nonlinear Schrödinger equation is discussed in Sec. 5. The Whitham averaged variational principles for nonlinear dispersive systems are presented in Sec. 6. The perturbation expansion and Hamiltonian formulation of the theory of nonlinear systems are included in Sec. 7. Section 8 deals with the variational formulation of the nonlinear Klein-Gordon, Boussinesq and K-dV equations, and Sec. 9 contains the variational principles for Rossby waves in a β-plane ocean, and Boussinesq equation for a stratified fluid flows. The final Sec. 10 is devoted to the study of nonlinear dispersive wave propagation in dissipative media and the averaged pseudovariational principle for dissipative wave trains.

2. LINEAR DISPERSIVE WAVES

We consider a dynamical problem so that a small disturbance $\phi(\underline{r}, t)$ with reference to the undisturbed stable state is governed by a linear partial differential equation with constant coefficients

$$P\left(\frac{\partial}{\partial t}, \frac{\partial}{\partial x_1}, \frac{\partial}{\partial x_2}, \frac{\partial}{\partial x_3}\right) \phi(\underline{r}, t) = F(\underline{r}, t) \tag{2.1}$$

where P is a polynomial, t and $\underline{r} = (x_1, x_2, x_3) \equiv (x, y, z)$ are independent time and space variables; $F(\underline{r}, t)$ represents the action of external forces on the dynamical system and is usually referred to a given forcing term.

We assume a plane wave solution of the homogeneous equation

$$P\left(\frac{\partial}{\partial t}, \frac{\partial}{\partial x_i}\right) \phi(\underline{r}, t) = 0 \tag{2.2}$$

in the form

$$\phi(\underline{r}, t) = A \exp[i(\underline{\kappa} \cdot \underline{r} - \omega t)] \tag{2.3}$$

where A is the amplitude, $\underline{\kappa} = (k_1, k_2, k_3) \equiv (k, l, m)$ is the wave number and ω is the frequency.

Such a solution exists provided the dispersion relation

$$P(-i\omega, ik_1, ik_2, ik_3) \equiv D(\omega, \underline{k}) = 0 \tag{2.4}$$

is satisfied.

Thus there exists a direct relationship between the governing equation and the dispersion relation through the obvious correspondence

$$\frac{\partial}{\partial t} \longleftrightarrow -i\omega, \quad \frac{\partial}{\partial x_j} \longleftrightarrow ik_j \tag{2.5ab}$$

This correspondence allows us to obtain the governing dispersion relation from Eq. (2.2), and conversely Eq. (2.2) can be constructed from the dispersion relation (2.4).

If, in particular, the dispersion relation (2.4) can explicitly be solved in terms of real roots given by

$$\omega = \Omega(\underline{\kappa}) \tag{2.6}$$

then there may be several possible roots for the frequency corresponding to different modes of propagation.

The phase velocity of the waves is the velocity at which a surface of constant phase moves, and is defined by the relation

$$\underline{c} = \frac{\omega}{\kappa} \hat{\kappa} \tag{2.7}$$

where $\hat{\kappa}$ is the unit vector in the direction of the wave vector $\underline{\kappa}$.

Thus, for any particular mode, $\omega = \Omega(\underline{\kappa})$, the phase velocity depends upon the wave numbers k_1, k_2, k_3. In other words, the different waves propagate with different phase velocities, and such waves are called dispersive provided the determinant $|\partial^2\Omega/\partial k_i \partial k_j| \neq 0$. On the other hand, waves are nondispersive if the phase velocity does not depend on the wave numbers. For one-dimensional dispersive waves, the determinant reduces to $\Omega''(k) \neq 0$. However, in general, the governing dispersion relation $\omega = \Omega(k)$ may give complex ω for a real k. In such a case, the phase velocity depends not only on the wave numbers, but also on the effective amplitude of the waves. This means that the amplitude will grow or decay in time according as $I(\omega) > $ or < 0. So the former case leads to instability.

The group velocity of the wave motion is defined by the result

$$\underline{C}(\kappa) = \nabla_{\kappa}\omega = \left(\frac{\partial\omega}{\partial k}, \frac{\partial\omega}{\partial l}, \frac{\partial\omega}{\partial m}\right) \tag{2.8}$$

For one-dimensional waves, the group velocity $C(k) = d\omega/dk$.

It is noted that the group velocity plays a fundamental role in the theory of dispersive waves, and is much more important than the phase velocity.

Examples of Linear Dispersive Waves

1. Korteweg-de Vries (KdV) equation for long water waves

$$\phi_t + c_0\phi_x + \alpha\phi_{xxx} = 0, \quad (c_0^2 = gh), \quad \omega = c_0 k - \alpha k^3 \tag{2.9ab}$$

2. Boussinesq equation for long water waves

$$\phi_{tt} - c_0^2\phi_{xx} - \beta^2\phi_{xxtt} = 0, \quad \omega^2 = c_0^2 k^2(1 + \beta^2 k^2)^{-1} \tag{2.10ab}$$

3. Benjamin, Bona, and Mahony (BBM) equation

$$\phi_t + c_0\phi_x - \alpha\phi_{xxt} = 0, \quad \omega = c_0 k(1 + \alpha k^2)^{-1} \tag{2.11ab}$$

4. Klein-Gordon equation

$$\phi_{tt} - c^2\nabla^2\phi + \beta^2\phi = 0, \quad \omega = \pm(c^2\kappa^2 + \beta^2)^{\frac{1}{2}} \tag{2.12ab}$$

5. Internal waves in a stratified ocean are governed by the equation

$$\left(\frac{\partial}{\partial t} + U\frac{\partial}{\partial x}\right)^2\left(\frac{\partial^2\psi}{\partial x^2} + \frac{\partial^2\psi}{\partial z^2}\right) + N^2\frac{\partial^2\psi}{\partial x^2} = 0, \quad \omega = Uk \pm \frac{Nk}{\sqrt{k^2 + m^2}} \tag{2.13ab}$$

6. Inertial waves in a rotating liquid are governed by the equation

$$\left(\frac{\partial}{\partial t} + U\frac{\partial}{\partial z}\right)\nabla^2\chi + f^2\frac{\partial^2\chi}{\partial z^2} = 0, \quad (f = 2\Omega), \quad \omega = Um \pm \frac{fm}{\kappa} \tag{2.14ab}$$

7. Internal-inertial waves in a rotating stratified ocean are governed by the equation

$$\left(\frac{\partial^2}{\partial t^2} + N^2\right)\nabla_h^2\chi + \left(\frac{\partial^2}{\partial t^2} + f^2\right)\chi_{zz} = 0, \quad \omega^2 = \frac{f^2m^2 + N^2(k^2 + \ell^2)}{\kappa^2} \tag{2.15ab}$$

where $\nabla_h^2 \equiv \partial^2/\partial x^2 + \partial^2/\partial y^2$.

8. Rossby waves in a β-plane ocean are governed by the equation

$$(\nabla_h^2 - a^2)\psi_t + \beta\psi_x = 0, \quad \left(a^2 = \frac{f^2}{c_0^2}\right), \quad \omega = -\frac{\beta k}{k^2 + l^2 + a^2} \tag{2.16ab}$$

9. The governing dispersion relation for surface waves in water of constant depth h

$$\omega^2 = gk \tanh kh \tag{2.17}$$

10. Dispersion relation for the Alfvén-gravity waves in an electrically conducting liquid of constant depth h

$$\omega^2 = (gk + a^2k^2)\tanh kh \tag{2.18}$$

where $a = B_0/\sqrt{4\pi\rho}$ is the Alfvén velocity.

11. Dispersion relation for electromagnetic waves in plasmas

$$\omega^2 = \omega_p^2 + c^2k^2 \tag{2.19}$$

where $\omega_p = (4\pi n\, e^2/m)^{\frac{1}{2}}$ is the plasma frequency, c is the velocity of light, n is the number of electrons (per unit volume) of mass m and charge e.

12. Schrödinger equation and de Broglie waves

$$i\hbar\psi_t = V\psi - \frac{\hbar^2}{2m}\nabla^2\psi, \quad \omega = \frac{\hbar\kappa^2}{2m} + \frac{V}{\hbar} \tag{2.20ab}$$

where $2\pi\hbar$ is the Planck constant and V is a constant potential.

The phase and the group velocities associated with these examples can readily be calculated. And for dispersive waves, the former is different from the latter.

The general solution for any dispersive system can then be obtained by the Fourier superposition of the plane wave solution (2.3) for different wave numbers $\underline{\kappa}$ with the corresponding frequencies ω to satisfy the dispersion law (2.6), and is represented by

$$\phi(\underline{r}, t) = \iiint_{-\infty}^{\infty} f(\underline{\kappa}) \exp[i\{\underline{\kappa}\cdot\underline{r} - \Omega(\underline{\kappa})t\}]\, d\underline{\kappa} \tag{2.21}$$

where the spectrum function $f(\underline{\kappa})$ is determined from appropriate initial or boundary conditions.

Although (2.21) is the exact solution, the principal features of the dispersive waves cannot be understood without an exact or approximate evaluation of the integral in (2.21). We shall discuss this point in the next section concerned with initial value problems.

It is relevant to mention here that, in general, the governing equations for dispersive waves are inherently nonlinear, and the corresponding dispersion relation is also nonlinear in the sense that the frequency ω is not only a function of $\underline{\kappa}$, but also of other parameters, such as amplitude, local properties of the medium. The theory of water waves provides an excellent example of nonlinear dispersion. In the study of the linear dispersive systems, the above parameters are assumed to be small, and the governing equations for such systems are obtained by linearizing the original nonlinear equations. The classical theory of water waves satisfying Laplace's equation and the free surface conditions can again be cited as an excellent example of the process of linearization. However, in his pioneering work based upon variational principles, Whitham gives a new description of nonlinear water waves, and a completely different approach to nonlinear dispersive waves in general. It has become clear from Whitham's theory that nonlinear dispersion is much more important than the corresponding linearized concept, and plays a significant role in the general theory of dispersive systems.

3. INITIAL VALUE PROBLEMS AND ASYMPTOTIC SOLUTIONS

We consider the initial value problem of wave propagation in which the disturbance $\phi(\underline{r}, t)$ is given by $F(\underline{r})$ at time $t = 0$. The general complex solution for any particular mode of propagation with known dispersion law $\omega = \Omega(\underline{\kappa})$ is given by the integral

$$\phi(\underline{r}, t) = \int_{-\infty}^{\infty} f(\underline{\kappa}) \exp[i\{\underline{\kappa}\cdot\underline{r} - \Omega(\underline{\kappa})t\}]\, d\underline{\kappa} \tag{3.1}$$

where $\int d\underline{\kappa}$ represents line, area, or volume integral according as the number of dimensions n = 1, 2, or 3. At t = 0, (3.1) gives

$$F(\underline{r}) = \phi(\underline{r}, 0) = \int_{-\infty}^{\infty} f(\underline{\kappa}) \exp[i(\underline{\kappa}\cdot\underline{r})]\, d\underline{\kappa} \tag{3.2}$$

where $f(\underline{\kappa})$ is the generalized Fourier transform of the initial disturbance $F(\underline{r})$ so that

$$f(\underline{\kappa}) = \frac{1}{(2\pi)^{n/2}} \int_{-\infty}^{\infty} F(\underline{r})\, e^{-i\underline{\kappa}\cdot\underline{r}}\, d\underline{r} \tag{3.3}$$

Thus the spectrum function $f(\underline{\kappa})$ can be determined from the given initial data. Whatever the particular initial value problem may be, the general solution can always be represented by the integral (3.1). The number of modes of the type (3.1) depends on the particular problem. So the complete solution will be the sum of terms like (3.1) with one integral for each of the solutions $\omega = \Omega(\underline{\kappa})$.

It is noted that physical interpretation of the integral solution (3.1) is not at all obvious from its present form. On the other hand, the exact evaluation of (3.1) is also a difficult task. In order to investigate the principal features of dispersive waves, it is important to consider the asymptotic evaluation of (3.1) for large r and t. It is then necessary to resort to asymptotic methods. We next turn our attention to asymptotic evaluation of (3.1) by using the method of stationary phase. According to this method (Jones [7]), the main contribution to the integral (3.1) for large t is found from the terms associated with the stationary points, if any, where the phase function $\underline{\kappa}\cdot\underline{r} - t\Omega(\underline{\kappa})$ is stationary, that is, where

$$\frac{\partial\Omega(\underline{\kappa})}{\partial k_i} = \frac{x_i}{t} \tag{3.4}$$

It can be shown that the dominant contribution to the integral (3.1) will come from those values of $k_i = k_i(x_i, t)$ which satisfy (3.4). Thus, the asymptotic solution from one such point, k_i is obtained in the standard form

$$\phi(\underline{r}, t) \sim \frac{(2\pi)^{n/2} f(ki)}{\left\{t^n \det\left|\frac{\partial^2\Omega}{\partial k_i \partial k_j}\right|\right\}^{\frac{1}{2}}} \exp\left[i\left\{\underline{k}\cdot\underline{r} - t\Omega(\underline{k}) - \frac{\pi}{4}\,\mathrm{Sgn}\,\frac{\partial^2\Omega}{\partial k_i^2}\right\}\right] \tag{3.5}$$

where the summation convention is used.

The complete asymptotic solution of $\phi(\underline{r}, t)$ will be the sum of terms like (3.5) with one term for each of the stationary points k_i of (3.4).

The asymptotic solution (3.5) can be expressed in the form of the elementary solution

$$\phi(\underline{r}, t) \sim R[A(x_i, t) \exp\{i\theta(x_i, t)\}] \tag{3.6}$$

where R denotes the real part, the complex amplitude $A(x_i, t)$ and the phase function $\theta(x_i, t)$ are given by

$$A(x_i, t) = \frac{(2\pi)^{n/2} f(k_i)}{\left[t^n \det\left|\frac{\partial^2\Omega}{\partial k_i \partial k_j}\right|\right]^{\frac{1}{2}}} \exp\left[-\frac{i\pi}{4}\,\mathrm{Sgn}\,\frac{\partial^2\Omega}{\partial k_i^2}\right] \tag{3.7}$$

$$\theta(x_i, t) = \underline{k}\cdot\underline{r} - t\Omega(\underline{k}) \tag{3.8}$$

with the fact that $k_i(x_i, t)$ is the solution of (3.4).

The asymptotic solution (3.6) has not only the same form as the elementary plane wave solution, but also it represents an oscillatory wave train. But, in contrast to the elementary solution, (3.6) is not a uniform wave train because k_i is a function of x_i and t. Moreover, the quantities A and $\omega = \Omega(k_i)$ involved in (3.6) are no longer constant, but they are functions of x_i and t.

It follows from (3.8) combined with (3.4) that

$$\frac{\partial \theta}{\partial x_i} = k_i + \left\{ x_i - t \frac{\partial \Omega}{\partial k_i} \right\} \frac{\partial k_i}{\partial x_i} = k_i(x_i, t) \tag{3.9}$$

$$-\frac{\partial \theta}{\partial t} = \Omega(k_i) - \left\{ x_i - t \frac{\partial \Omega}{\partial k_i} \right\} \frac{\partial k_i}{\partial t} = \Omega(k_i) = \omega(x_i, t) \tag{3.10}$$

This means that $\partial\theta/\partial x_i$ and $-\partial\theta/\partial t$ still have the significance of a wave number and a frequency. They are no longer constant, but functions of x_i and t. Thus $k_i = \partial\theta/\partial x_i$ and $\omega = -\partial\theta/\partial t$ represent the local wave number and the local frequency, and are still governed by the dispersion relation. The most remarkable difference between the asymptotic solution (3.6) and the elementary plane wave solution is that the former represents an oscillatory nonuniform wave train, but the parameters A, k_i, ω are no longer constant. Indeed, they are slowly varying functions of x_i and t in the sense that $\Delta A \ll A$, $\Delta \underline{\kappa} \ll \kappa$ and $\Delta\omega \ll \omega$ over the intervals κ^{-1} and ω^{-1}. In other words, the relative changes of these quantities in one wave length and in one period are small. This point can readily be verified by using (3.4) and (3.7).

Finally, the present asymptotic analysis reveals two remarkable consequences concerning the dual role of the group velocity. First, a careful consideration of result (3.4) reveals that the local wave number propagates with the group velocity $\partial\Omega/\partial k_i$. Second, expression (3.7) indicates that $|A|^2$ is an energy like quantity and propagates with the group velocity.

4. NONLINEAR DISPERSIVE WAVES

The theory of linear dispersive waves is essentially based upon the Fourier superposition principle. However, in nonlinear problems, the superposition principle is no longer applicable to construct more general solutions for the wave trains. It follows from the asymptotic solution (3.6) that the form of the elementary solution, $\phi = a \exp(i\theta)$, $\theta = kx - \omega t$ can still be used to describe nonlinear wave trains provided a, ω, and k are no longer constants but slowly varying functions of x and t corresponding to the slow modulation of the wave train. For slowly varying wave trains, there exists a phase function $\theta(x, t)$ so that the local frequency $\omega(x, t)$ and the local wave number $k(x, t)$ are defined in terms of θ by

$$\omega(x, t) = -\theta_t, \quad k(x, t) = \theta_x \tag{4.1ab}$$

In view of the slow variation of k and ω, it seems reasonable to assume the dispersion relation in the form

$$\omega = \Omega(k) \tag{4.2}$$

We note that, for nonlinear wave trains, Ω may also depend on amplitude.

Elimination of θ from (4.1ab) gives immediately

$$\frac{\partial k_i}{\partial t} + \frac{\partial \omega}{\partial x} = 0, \qquad \frac{\partial k_i}{\partial x_j} - \frac{\partial k_j}{\partial x_i} = 0 \tag{4.3ab}$$

where the former is a relationship between the wave numbers and frequency. This can be thought of as an equation of continuity for phase.

Substitution of (4.2) into (4.3a) and use of (4.3b) imply

$$\frac{\partial k_i}{\partial t} + C_j(k) \frac{\partial k_i}{\partial x_j} = -\frac{\partial \Omega}{\partial x_i} \tag{4.4}$$

where $C_j = C_j(k)$ is called the group velocity defined by

$$C_j = \frac{\partial \Omega}{\partial k_j} \tag{4.5}$$

This represents the propagation velocity for the wave number k_i.

An observer traveling with the local group velocity $C_j(k)$ of the wave train moves along a path in space-time known as a ray. From the existence of the phase function $\theta(x_i, t)$ and the dispersion relation (4.2), it turns out that changes along rays are given by

$$\frac{dk_i}{dt} = -\frac{\partial \Omega}{\partial x_i}, \quad \frac{d\omega}{dt} = \frac{\partial \Omega}{\partial t} \tag{4.6ab}$$

Equation (4.4) can also be written in the characteristic form

$$\frac{dk_i}{dt} = -\frac{\partial \Omega}{\partial x_i}, \quad \frac{dx_i}{dt} = \frac{\partial \Omega}{\partial k_i} = C_i \tag{4.7ab}$$

These are identical with Hamilton's equations of motion for a particle with position x_i and generalized momentum k_i where Ω plays the role of the Hamiltonian. If Ω is independent of time, that is, if $\omega = \Omega(k_i, x_i)$, (4.7ab) imply that the frequency ω remains constant for the wave packet. This is identical with the important fact that, for a dynamical system with a time-independent Hamiltonian, every motion of the system carries a constant value of the Hamiltonian which is equal to the total energy. Thus the frequency behaves like energy and the wave number like momentum. This remarkable analogy between the classical (particle) mechanics and the quantum (wave) mechanics leads to the well-known duality exploited in quantum physics.

In uniform media, $\partial\Omega/\partial x_i = 0$, Eq. (4.6a) or (4.7a) implies that the wave number k_i is constant on each characteristic, and the characteristics are straight lines in (x_i, t) space. And each k_i propagates along the characteristics with the constant group velocity C_i. Further, if the medium is independent of time t, $\partial\Omega/\partial t = 0$, then the frequency ω is constant on each characteristic. In nonuniform media, Eqs. (4.7ab) show significant differences. First, k_i is no longer constant, but varies at the rate $-\partial\Omega/\partial x_i$. Second, it propagates along the characteristics with the variable group velocity $C_i(k_i, x_i, t)$ and the characteristics are no longer straight lines.

It is interesting to observe that the local dispersion law (4.2) with (4.1ab) gives the partial differential equation for $\theta(x,t)$ in the form

$$\frac{\partial \theta}{\partial t} + \Omega\left(\frac{\partial \theta}{\partial x_i}, x_i, t\right) = 0 \tag{4.8}$$

This is the well-known Hamilton-Jacobi equation. The solution of this equation determines the geometry (or the kinematics) of the wave train.

The above kinematic theory indicates one important role of the group velocity. However, this theory provides no indication of changes in amplitude or energy of a wave train or wave packet. Whitham [6] initiated further study of the amplitude function (3.7) and the related energy density in order to derive a differential equation for the amplitude as well as to determine the second role of the group velocity.

For one-dimensional case, the energy density described by (3.5) is proportional to $|A|^2$ so that

$$A^2 = \frac{2\pi f(k) f^*(k)}{t|\Omega''(k)|} \tag{4.9}$$

where $f^*(k)$ is the complex conjugate of $f(k)$.

The amount of energy $Q(t)$ between two points x_1 and x_2 is given by

$$Q(t) = \int_{x_1}^{x_2} g(k) A^2 \, dx \tag{4.10}$$

where g(k) is an arbitrary proportionality factor associated with the square of the amplitude and energy.

In a new coordinate system moving with the group velocity, that is, along the rays $x = C(k)t$, (4.10) reduces to the form

$$Q(t) = 2\pi \int_{k_1}^{k_2} g(k)\, f(k)\, f^*(k)\, dk \tag{4.11}$$

where $\Omega''(k) > 0$, k_1 and k_2 are defined by $x_1 = C(k_1)t$ and $x_2 = C(k_2)t$, respectively.

Using the principle of conservation of energy, that is, the energy between the points x_1 and x_2 traveling with the group velocities $C(k_1)$ and $C(k_2)$ remains invariant, it turns out that

$$\frac{dQ}{dt} = \int_{x_1}^{x_2} \frac{\partial}{\partial t}\{g(k)A^2\}\, dx + g(k_2)C(k_2)A^2(x_2, t) - g(k_1)C(k_1)A^2(x_1, t) = 0$$

which is, in the limit $x_2 - x_1 \to 0$,

$$\frac{\partial}{\partial t}\{g(k)A^2\} + \frac{\partial}{\partial x}\{g(k)C(k)A^2\} = 0 \tag{4.13}$$

This may be treated as the energy equation. The quantities $g(k)A^2$ and $g(k)C(k)A^2$ represent the energy density and the energy flux so that they are proportional to $|A|^2$ and $C(k)|A|^2$, respectively. The flux of energy propagates with the group velocity C(k). Hence, the group velocity has the double role, and is the propagation velocity for the wave number k and for the energy $g(k)|A|^2$.

Thus (4.3a) and (4.13) are known as Whitham's conservation equations for nonlinear waves. The former represents the conservation of wave number k and the latter is the conservation of energy (or more generally, the conservation of wave action). It is important to observe that, even in uniform media $(\partial\Omega/\partial x_i = 0)$, (4.3a) or its equivalent form given by (4.4) is the most fundamental hyperbolic equation, even if the original equation for ϕ is linear or nonhyperbolic. The conservation equations can more rigorously be derived from a general and extremely powerful approach which is now known as Whitham's average variational principle.

Whitham used his conservation equations combined with the nonlinear dispersion relation $\omega = \omega(k, a)$ to classify the nature of the equations for a nonlinear system and to examine modulations of linear wave trains. For moderately small amplitude, Stokes' expansion of ω has the form

$$\omega = \omega_0(k) + \omega_2(k)a^2 + \cdots \tag{4.14}$$

We next substitute (4.14) into (4.3a) and (4.13), replace $\omega'(k)$ by its linear values, $\omega_0'(k)$ and retain the terms of order a^2 to obtain the differential equations for k and a^2 in the form

$$\frac{\partial k}{\partial t} + \{\omega_0'(k) + \omega_2'(k)a^2\}\frac{\partial k}{\partial x} + \omega_2(k)\frac{\partial a^2}{\partial x} = 0 \tag{4.15}$$

$$\frac{\partial a^2}{\partial t} + \omega_0'(k)\frac{\partial a^2}{\partial x} + \omega_0''(k)a^2\frac{\partial k}{\partial x} = 0 \tag{4.16}$$

These represent modulation equations of the linear wave trains and are coupled. The characteristic velocities for the system (4.15)-(4.16) are found to be

$$C(k) = C_0(k) \pm a\sqrt{\omega_2(k)\omega_0''(k)} + 0(a^2) \tag{4.17ab}$$

where $C_0(k) = \omega_0'(k)$ is the linear group velocity.

Equations (4.17ab) are hyperbolic or elliptic according as $\omega_2\omega_0'' > 0$ or < 0. In the former case, the double characteristic velocity of the linear theory splits into two separate velocities which are called nonlinear group velocities. The splitting into two different velocities is one of the most remarkable results of the modulation theory. This means that any initial disturbance of finite extent will eventually split into two separate disturbances. This conclusion is completely different from that of the linear theory where an initial disturbance may distort due to the dependence of $C_0(k) = \omega_0'(k)$ on k, but would never split.

Another significant consequence of nonlinearity in the hyperbolic case is that compressive modulations will eventually distort and steepen. This leads to the multiple-valued solutions and instability. This is completely different from the linear behavior where the question of instability does not arise at all.

In the elliptic case ($\omega_2\omega_0'' < 0$), the characteristics are complex. This leads to ill-posed problems in the theory of nonlinear waves. Any small sinusoidal disturbances in k and a can be represented by solutions of the form $\exp[i\alpha(x - Ct)]$ where C is given by (4.17ab) for the unperturbed values of k and a. Thus the modulations, when C is complex, grow exponentially with time t. Thus, the wavetrains are unstable. An application of this result to the Stokes waves in sufficiently deep water leads to the most remarkable evidence of the instability of the Stokes waves in the history of water wave theory. For the Stokes waves in deep water, the famous dispersion relation is

$$\omega = \sqrt{gk}\,(1 + \tfrac{1}{2}a^2k^2) + 0(a^4) \tag{4.18}$$

Hence, $\omega_0''\omega_2 < 0$, and this corresponds to the elliptic case so that the characteristic velocities are complex. This is a conclusive proof of the remarkable fact that Stokes waves in deep water are definitely unstable. The instability of deep water waves came as a surprise to researchers in the field in view of the very long history of the subject. The question of instability went unrecognized for a long period of time, even though special efforts have been made to prove the existence of permanent shape of periodic water waves for all amplitudes less than the critical value at which the waves assume a sharp-crested shape.

In addition to Whitham's theory of Stokes water waves, Lighthill [8-9] used Whitham's conservation equations to investigate the evolution of weakly nonlinear dispersive waves in deep water. He has proved an elegant and remarkably simple result determining whether, under certain conditions, very gradual variations in the properties of a wave train are governed by elliptic or hyperbolic differential equations. It was shown that instability is associated with the elliptic case. In particular, two sets of initial data were investigated—(i) a slightly modulated Stokes waves and (ii) a wave packet with Gaussian envelope. In case (i) the modulation is found to grow and the envelope developed into a cusped shape. In case (ii), the envelope developed a cusp at its peak within a finite period of time. It was established that smooth solutions do not exist beyond this stage of evolution, and the time taken to reach the singularities is inversely proportional to the wave steepness. These findings have been the first indications of modulational instabilities for weakly nonlinear water waves on deep oceans. It must be noted here that Lighthill's findings were in perfect agreement with the results of instability due to Whitham.

In addition to Whitham's and Lighthill's independent stability analyses, it is important to outline the remarkable theoretical and experimental investigation of Benjamin and Feir [10, 11] into the instability of periodic wave trains in nonlinear dispersive systems with special reference to Stokes waves in deep water. Benjamin and Feir first discussed the essential features leading to instability in general terms and then developed a detailed stability theory based on perturbation equations for the uniform wave motions in water of arbitrary depth h. They have shown that the Stokes waves in water of depth h is in fact unstable when the fundamental wave number k satisfies $kh > 1.363$, but are otherwise stable. Their mathematical and experimental investigation is now regarded as conclusive evidence of the instability and consequent disintegration of the Stokes water waves.

The analysis of the Benjamin and Feir instability is essentially a treatment of the third-order interactions of a tetrad with wave numbers k, k, $-k(1 \pm \epsilon)$ for $0 < \epsilon << 1$. Their frequencies are approximately ω, ω, $-\omega(1 - \frac{1}{2}\epsilon)$, and $-\omega(1 + \frac{1}{2}\epsilon)$ for deep water waves so that the dispersion relation $\omega^2 = g|k|$ is still satisfied. The quantities $-\omega(1 \pm \frac{1}{2}\epsilon)$ are called side-band frequencies of the fundamental frequency ω of the basic wave train. Thus the pair of wave modes with side-band frequencies and wave numbers fractionally different from the fundamental frequency ω and wave number k constitute a disturbance which can gain energy from the fundamental wave mode. In consequence of nonlinear effects on these modes, energy is then transferred from the basic wave mode to the side-band modes so that they grow exponentially in time or space. Consequently, the growth of the side-band modes leads to the instability of the fundamental wave mode. It is also shown that any small disturbances in the form of a pair of side-band modes undergo unbounded magnification if $0 < \epsilon < 2\sqrt{2}\,ka$ where k and a are the fundamental wave number and amplitude of the perturbed wave train. Benjamin and Feir also provided the most direct experimental evidence in support of their remarkable discovery concerning the instability of the classical Stokes water waves. These findings resolved the famous controversy about the existence of water waves of permanent form.

Among the various notable features of the Whitham general theory of nonlinear dispersive waves, perhaps the most remarkable one deals with the success of predicting instability of the uniform Stokes water waves. The Whitham theory implies that the modulation equations for the Stokes waves in water of depth h is governed by a hyperbolic or an elliptic equation according as $kh < 1.363$ or > 1.363. When the governing equation is elliptic, the problem of finding the motion is mathematically ill-posed in that it is set as a Cauchy problem and the solution breaks down after a finite time. Thus the findings of the Whitham theory and the predictions of the Lighthill theory are in excellent agreement with the results of the Benjamin and Feir instability theory. Recent investigations of Stuart and DiPrima [12] further recognized the importance of the side-band instability. Based upon the method of amplitude equation, they showed that the mechanism of the Benjamin-Feir instability of the Stokes water waves is essentially equivalent to that of a side-band instability theory discovered earlier by Eckhaus [13-14].

With a view to make an extension of the Benjamin and Feir instability theory, Phillips [15] made an interesting investigation of various aspects of the resonant interactions among waves in dispersive fluid systems. Unlike the Benjamin and Feir theory, Phillips' resonant interaction is not restricted to neighboring sets of wave numbers. Moreover, the instability of Stokes water waves can be described in terms of these general resonant interactions, and is not restricted to purely two-dimensional wave motions. Indeed, it was shown by Phillips [16] that the basic wave motion is unstable to disturbances containing wave number pairs which lie just inside the figure-of-eight loop. Thus the theory of Benjamin and Feir is a special case of the resonant interaction theory of Phillips. Several reliable experiments support the remarkable theoretical findings of the Phillips theory.

5. NONLINEAR SCHRÖDINGER EQUATION

With the discovery of the mechanism of instability of the two-dimensional Stokes water waves, it has widely been recognized the need for the governing equations for the unsteady time evolution of weakly nonlinear deep water wave trains and wave packets. Several authors including Zakharov [17], Hasimoto and Ono [18], Davey and Stewartson [19], and Yuen and Lake [20] have employed the method of multiple scales to show that the evolution of wave trains satisfies the nonlinear Schrödinger equation. It was first derived by Zakharov using the multiple scale method when the wave number variation is small. Yuen and Lake derived the nonlinear Schrödinger equation as a consequence of Whitham's conservation equations when second variations in the dispersion law are included. Dysthe [21] has investigated the connection between Whitham's method and the nonlinear Schrödinger equation approach. For weakly nonlinear deep water waves with a carrier wave number k_0 and a carrier frequency ω_0, several authors, under the same set of assumptions, derived the same nonlinear Schrödinger equation for the complex wave train envelope $A(x, t) = a(x,t) \exp[i\theta(x,t)]$ in the form

$$i\left(\frac{\partial A}{\partial t} + \frac{\omega_0}{2k_0}\frac{\partial A}{\partial x}\right) - \frac{\omega_0}{8k_0^2}\frac{\partial^2 A}{\partial x^2} - \frac{1}{2}\omega_0 k_0^2 |A|^2 A = 0 \tag{5.1}$$

where the perturbed wave number and frequency are $\partial\theta/\partial x$ and $-\partial\theta/\partial t$, respectively, and the free surface elevation is

$$\eta(x, t) = \mathrm{Re}[A(x, t) \exp\{i(k_0 x - \omega_0 t)\}] \tag{5.2}$$

Equation (5.1) was found to be extremely useful for describing a wide variety of physical problems, and was first studied by Benney and Newell [22] in the general context.

One of the most remarkable features of the nonlinear Schrödinger equation is that it can be solved exactly for initial conditions that decay sufficiently rapidly as $|x| \to \infty$. Zakharov and Shabat [23] obtained the exact solution by using the inverse scattering method. It has shown that any initial wave packet eventually evolves into a definite number of envelope solitons and an oscillatory dispersive tail. This is completely different from the linearized theory where deep-water wave systems have highly dispersive and uncoupled wave components. The existence of the envelope solitions was experimentally confirmed by Yuen and Lake. In a subsequent paper [24], Lake et al. reported further experimental and computational results concerning the evolution of a nonlinear continuous wave train in deep water. It was shown that the initial stage of evolution leads to exponential growth of a modulational instability as confirmed by the stability analysis of Benjamin and Feir. At later stages of the evolution process, the wave trains exhibit the Fermi-Pasta-Ulam recurrence phenomenon. Their results of computational solutions of the nonlinear Schrödinger equation were in excellent agreement with the experimental findings. Also, it follows from discussions of Benney and Newell, Zakharov, and Hasimoto and Ono that the results of the Benjamin-Feir instability mechanism can be recovered from a stability analysis of the nonlinear uniform wave train solution $A = a_0 \exp(-\frac{1}{2} i k_0^2 \omega_0 a^2 t)$ to the nonlinear Schrödinger equation. Thus, in the case of solitons and wave packets, both qualitative and quantitative description of nonlinear deep-water wave dynamics is provided by the nonlinear Schrödinger equation.

In addition to the context of nonlinear water waves, the derivation of the nonlinear Schrödinger equation has received much attention in various problems in plasma physics, fluid and solid mechanics. Taniuti and Washimi [25], Watanabe [26] have used the method of multiple scales to investigate the modulational instability of a small, but finite amplitude, dispersive hydromagnetic wave propagating in a cold plasma without dissipation. They showed that, in a frame of reference moving downstream with the group velocity, the slow evolution in the complex amplitude A of the wave satisfies the nonlinear Schrödinger equation in the form

$$i\frac{\partial A}{\partial t} + p\frac{\partial^2 A}{\partial x^2} = iqA + rA|A|^2 \tag{5.3}$$

where the space coordinate x is in the direction of propagation, p, q, and r are known constants with $q = 0$ or $q = i\epsilon$, ϵ real (no dissipation), p and r are real.

In order to describe the nonlinear evolution of Stokes waves in deep water, Chu and Mei [27-28] have employed the Whitham theory to derive the modulation equations of the wave amplitude a and the wave number k in the form

$$\frac{\partial k}{\partial t} + \frac{\partial \omega}{\partial x} = 0 \tag{5.4a}$$

$$\frac{\partial}{\partial t}\left(\frac{a^2}{\omega_0}\right) + \frac{\partial}{\partial x}\left(C\frac{a^2}{\omega_0}\right) = 0 \tag{5.4b}$$

where $\omega_0 = (gk)^{\frac{1}{2}}$ is the first order approximation of the wave frequency ω given by

$$\omega = \omega_0 \left[1 + \epsilon^2 (\tfrac{1}{2} a^2 k^2 + \left\{ \left(\frac{a}{\omega_0}\right)_{tt} \div 2a\omega_0 \right\} \right] \tag{5.5}$$

and $C = g/2\omega_0$ is the group velocity. They have transformed (5.4ab) in a frame of reference moving with the group velocity to obtain the following equations in nondimensional form

$$\frac{\partial W}{\partial t} + \frac{\partial}{\partial x}\left(-\frac{W^2}{4} + \frac{1}{4}A^2 + \frac{A_{xx}}{16A}\right) = 0$$

$$\frac{\partial A^2}{\partial t} + \frac{\partial}{\partial x}\left(-\frac{1}{2}WA^2\right) = 0 \tag{5.6b}$$

These equations play the same role as the K-dV equation in the theory of shallow water waves.

Introducing a transformation $W = -2\phi_x$, equations (5.6ab) become

$$-2\frac{\partial^2 \phi}{\partial x\, \partial t} + \frac{\partial}{\partial x}\left(-\phi_x^2 + \frac{1}{4}A^2 + \frac{A_{xx}}{16A}\right) = 0 \tag{5.7a}$$

$$\frac{\partial A^2}{\partial t} + \frac{\partial}{\partial x}(A^2 \phi_x) = 0 \tag{5.7b}$$

where A is a small nondimensional amplitude and ϕ is a small phase change. Integrating (5.7a) with respect to x and putting the constant of integration to be zero, (5.7a) assumes the form

$$\phi_t + \frac{1}{2}\phi_x^2 - \frac{1}{8}A^2 - \frac{A_{xx}}{32A} = 0 \tag{5.8}$$

A transformation $\Psi = A \exp(4i\phi)$ is used to simplify (5.7b) and (5.8) which are then found to be equivalent to the nonlinear Schrödinger equation

$$i\Psi_t + \frac{1}{8}\Psi_{xx} = -\frac{1}{2}\Psi|\Psi|^2 \tag{5.9}$$

This equation has also been derived and exploited by several authors including Benney and Roskes [29], Davey and Stewartson [19], and Stuart and DiPrima [12] to examine the instability of Stokes waves in water and the stability of periodic flows.

Johnson [30] used the multiple scale method to derive the modulation equation for the amplitude of the fundamental when $\sigma_0 = kh$ is near to $\sigma_0 \approx 1.363$. He showed that, for $\sigma = \sigma_0$, the amplitude satisfies a new nonlinear Schrödinger equation which reduces to the classical nonlinear Schrödinger equation in the appropriate limit. This new equation has been used to investigate the stability of the Stokes water waves, and the new inequality $[\sigma < 1.363 + 25.1(\epsilon A)^2,\ \epsilon \to 0]$ for stability is derived. It has been shown that this analysis is consistent with both Benjamin and Feir's instability criterion [11] for the Stokes waves and the stable soliton solution of finite amplitude which leads to the usual conjecture that modulational instability of the Stokes waves eventually produce a train of solitons.

With a view to incorporate dissipation effects, Davey [31] has used the small amplitude equations of the Whitham theory of slowly varying wave trains to derive the modulation equation for a wave function Ψ satisfies a dissipative form of the nonlinear Schrödinger equation

$$i\Psi_t + \tfrac{1}{2}\gamma\Psi_{xx} = (\delta_1 + \delta_2|\Psi|^2)\Psi \tag{5.10}$$

where γ, δ_1, and δ_2 are complex constants with imaginary part of γ is negative and the real part of δ_1 is zero. It has also been shown that the nonlinear Schrödinger equation

(5.10) reduces to the K-dV equation for purely dispersive long waves ($k \to 0$), and to the Burgers equation for dispersive and dissipative long waves. Davey's analysis further indicates that the steady-state solutions of the K-dV equation and of the Burgers equation are the exact steady-state solutions of the Schrödinger equation.

Using the method of multiple scales, Stewartson and Stuart [32] have investigated a nonlinear instability theory for the evolution of small two-dimensional disturbances in marginally unstable plane Poiseuille flow, where dissipation effects are important. They have shown that the modulation of their wave is governed by (5.10). DiPrima et al. [33] have utilized a complicated discrete modal analysis to investigate the nonlinear growth of a small disturbance in the vicinity of a marginally stable state for a more general class of fluid flows including the plane Poiseuille flow. They have also derived the nonlinear Schrödinger equation in the dissipative form (5.10).

In the context of wave propagation in solids and lattices, Tappert and Verma [34] have obtained a nonlinear Schrödinger equation similar to (5.3) with $q = 0$.

With a view to make a generalization of the one-dimensional nonlinear Schrödinger equation, attention has been given to the propagation of wave trains in two space dimensions. As derived by Zakharov [17], the equation for the complex envelope of a wave train propagating in two space dimensions is given by

$$i\left(\frac{\partial A}{\partial t} + \frac{\omega_0}{2k_0}\frac{\partial A}{\partial x}\right) - \frac{\omega_0}{8k_0^2}\frac{\partial^2 A}{\partial x^2} + \frac{\omega_0}{4k_0^2}\frac{\partial^2 A}{\partial y^2} - \frac{1}{2}\omega_0 k_0^2 |A|^2 A = 0 \tag{5.11}$$

where $A(x, y, t) = a(x, y, t) \exp[i\theta(x, y, t)]$, $\underline{k}_0 = (k_0, 0)$ is the carrier wave number, the perturbed wave number and frequency are $\underline{k} = (\partial\theta/\partial x, \partial\theta/\partial y)$ and $\omega = -(\partial\theta/\partial t)$, respectively.

Equation (5.11) is well known as the two-dimensional nonlinear Schrödinger equation. However, it is not elliptic, but a hyperbolic equation because of the opposite sign of the second x- and y-derivative terms involved in Eq. (5.11). It is noted that the carrier wave still propagates in the x-direction, but its envelope incorporates two-dimensional variations. If the wave envelope is independent of y, Eq. (5.11) reduces to the one-dimensional nonlinear Schrödinger equation.

One of the most striking features of the nonlinear Schrödinger equation in two space dimensions is that it admits two-dimensional plane envelope solitons provided the angle between the directions of the carrier wave and of the envelope is less than 35.26°. It has also an oblique plane solitons. These two-dimensional plane envelope solitons are infinite in extent and are found to be unstable to two-dimensional disturbances as suggested by Zakharov and Rubenchik [35] and Saffman and Yuen [36]. Equation (5.3) reproduces two-dimensional instability which reduces to the Benjamin-Feir result in the limit $K_y \to 0$. However, in contrast to one-dimensional result with a high-wave number cutoff, the two-dimensional instability does not have a high-wave number cutoff. This conclusion has important implications for the long-time evolution of an unstable wave mode.

The initial value investigation of Yuen and Ferguson [37] confirmed the existence of the long-time evolution process and of the Fermi-Pasta-Ulam recurrence phenomenon for wave trains in two space dimensions. For any fully two-dimensional unstable perturbation vector (K_x, K_y), there exists an infinite set of integers (m, n) such that (mK_x, nK_y) lie in the instability regime. Finally, the connection between initial conditions and long-term evolution for the two-dimensional Schrödinger equation indicates that the long-time evolution consists of a periodic return to the initial state together with a gradual transfer of energy to higher and higher modes.

The nonlinear Schrödinger equation for deep water waves was derived by several authors using perturbation analysis to $0(\epsilon^3)$ in the wave-steepness $\epsilon = ka$. The predictions of the equation in the parametric range $ka > 0.15$ are not in agreement with those of the Longuet-Higgins exact instability theory [38-39]. This means that the nonlinear Schrödinger equation has a limited range of validity. In order to eliminate the shortcomings of the Schrödinger equation, Dysthe [40] has recently carried out the perturbation analysis to the next order ϵ^4 and suggested a new modification to the nonlinear

Schrödinger equation for application to deep water waves. The modified nonlinear Schrödinger equation is found to have the nondimensional form

$$2i(A_t + \frac{1}{2}A_x) - \frac{1}{4}A_{xx} + \frac{1}{2}A_{yy} - A|A|^2$$
$$= A(\bar{\phi}_x - i\bar{\phi}_z) - \frac{i}{8}(6A_{xyy} - A_{xxx}) + \frac{3}{2}iA(AA^*_x - A^*A_x) + \frac{1}{2}i|A|^2A_x \tag{5.12}$$

where $\bar{\phi}$ is the slowly varying function of space and time representing the mean flow brought about by the radiation stress of the wave.

The entire right hand side of (5.12) arises from the perturbation analysis to $0(\epsilon^4)$, and disappears when all terms of order ϵ^4 are neglected. In other words, when terms of the order ϵ^4 are neglected, that is, when the entire right hand side of (5.12) is zero, (5.12) reduces to the nonlinear Schrödinger equation as obtained by earlier authors. The modified equation seems to be a significant improvement. The first two linear terms on the right hand side of (5.12) are merely corrections to the dispersive terms of the Schrödinger equation. Dysthe used the modified equation (5.12) to study stability of the finite amplitude wave train, and showed a significant improvement on the results of the stability analysis. The notable new effect introduced by the terms of the order ϵ^4 is the mean flow response to nonuniformities in the radiation stress caused by modulation of the finite amplitude wave. The horizontal component of this mean flow along the direction of wave propagation causes a slowly varying Doppler shift of the wave as represented by the term $A\bar{\phi}_x$ in the right hand side of (5.12). This Doppler shift then has a detuning effect on the modulational instability.

It is also relevant to mention other works concerning the derivation of the nonlinear Schrödinger equation in two space dimensions. Benney and Roskes [29] and Davey and Stewartson [19] have studied the evolution of a three-dimensional packet of surface waves of small amplitude propagating in one direction in water of finite depth. These authors were able to prove that the amplitude modulation of the wave packet can be described by a nonlinear Schrödinger equation coupled to a Poisson type differential equation for the mean part of the velocity potential of the flow. These equations are used to investigate the stability of the uniform Stokes water waves to small disturbances whose length scale is large compared to the wave length $2\pi/k$ of the wave. In a recent paper [41], Hui and Hamilton have obtained exact solutions of the nonlinear Schrödinger equation derived by Davey and Stewartson, and applied them to investigate the nonlinear evolution of Kelvin ship wave pattern generated by a moving ship on a sea. They proved that there always exist solutions for the group envelope of the elliptic cn and dn functions (Dutta and Debnath [42]). These solutions represent permanent wave groups whose envelope varies periodically in space and time. Their common limiting solution represents the solitary wave. Longuet-Higgins [43] has used the theory of Davey and Stewartson for the calculation of the nonlinear energy transfer in a narrow gravity wave spectrum of a random ocean. It has been shown that energy from an isolated peak in the wave spectrum tends to spread outwards along two characteristic lines in the wave number space, making angles $\Psi_c = \tan^{-1}(1/\sqrt{2}) = 35°$ or $-\Psi_c$ with the wave field.

Hocking et al. [44] have considered the nonlinear growth of a small three dimensional disturbance in a plane parallel flow, and showed that the modulated wave amplitude satisfies a nonlinear Schrödinger equation in two space dimensions.

Thus the nonlinear Schrödinger equation plays a significant role in theory of evolution of wave packet of waves of small amplitude or of small disturbances in a fluid flows where the dissipative effects are important, and is of remarkable importance in the nonlinear instability theory.

6. WHITHAM'S AVERAGED VARIATIONAL PRINCIPLE

In his pioneering work, Whitham first developed a general approach to linear and nonlinear dispersive waves using a Lagrangian. It is well known that most of the general ideas about dispersive waves originated from the classical problems of water waves. First, we consider a variational principle for water waves. The governing equations for water waves in an incompressible inviscid fluid (water) of depth h are the Euler equations for an irrotational flow of the fluid with a free surface at $z = \eta(x, y, t)$:

$$\nabla^2 \phi = 0, \quad -h < z < \eta(x, y, t) \tag{6.1}$$

$$\eta_t + \nabla\phi \cdot \nabla\eta - \frac{\partial \phi}{\partial z} = 0 \quad \text{on } z = \eta(x, y, t) \tag{6.2}$$

$$\phi_t + \frac{1}{2}(\nabla\phi)^2 + g\eta = -\frac{p}{\rho} \quad \text{on } z = \eta(x, y, t) \tag{6.3}$$

$$\phi_z = 0 \quad \text{on } z = -h \tag{6.4}$$

where $\phi = \phi(x, y, z; t)$ is the velocity potential, $\eta = \eta(x, y, t)$ is the free surface, g is the acceleration due to gravity, ρ is the density of water, p is the external surface pressure, the horizontal coordinates are (x, y) and the vertical coordinate z is positive upwards.

Luke [45] has first proved that the Euler equations for two-dimensional water waves can be derived directly from the Hamilton variational principle

$$\delta \iint_R L \, dx \, dt = 0 \tag{6.5}$$

where the Lagrangian L is assumed to be equal to the pressure and is given by

$$L = -\rho \int_{-h}^{\eta(x,t)} [\phi_t + \tfrac{1}{2}(\nabla\phi)^2 + gz] \, dz \tag{6.6}$$

where R is an arbitrary region in the (x, t) space.

Using the standard procedure in the calculus of variations and integrating by parts, (6.5) with (6.6) reduces to the form

$$0 = -\delta \iint_R \frac{L}{\rho} dx \, dt = \iint_R \left\{ [\tfrac{1}{2}(\nabla\phi)^2 + \phi_t + gz]_{z=\eta} \delta\eta + \int_{-h}^{\eta} (\nabla\phi \cdot \nabla\delta\phi + \delta\phi_t) \, dz \right\} dx \, dt$$

$$= \iint_R \left\{ [\phi_t + \tfrac{1}{2}(\nabla\phi)^2 + gz]_{z=\eta} \delta\eta + \left[\frac{\partial}{\partial t} \int_{-h}^{\eta} \delta\phi \, dz + \frac{\partial}{\partial x} \int_{-h}^{\eta} \phi \, \delta\phi \, dz \right] \right.$$

$$\left. - \int_{-h}^{\eta} (\phi_{xx} + \phi_{zz}) \delta\phi \, dz - [(\eta_t - \phi_z + \eta_x \phi_x)\delta\phi]_{z=\eta} + [\phi_z \, \delta\phi]_{z=-h} \right\} dx \, dt \tag{6.7}$$

The second term integrates out to the boundaries of the domain R, and is equal to zero if $\delta\phi$ is assumed to vanish on the boundaries of R. Next, we set $\delta\eta = 0$, $\delta\phi = 0$ at $z = -h$ and $z = \eta$ so that (6.7) gives the Laplace equation (6.1) for an arbitrary $\delta\phi$. Since $\delta\eta$ and $\delta\phi$ at $z = -h$, $z = \eta$ can take arbitrary values, (6.7) gives the water wave equations (6.1)-(6.4).

In all dynamical problems where the governing equations admit uniform periodic wave train solutions, it is generally true that the system can be described by a Hamilton variational principle

$$\delta \iint L(u_t, u_{x_i}, u) \, d\underline{x} \, dt = 0 \tag{6.8}$$

where the dependent variable $u = u(\underline{x}, t)$.

The Euler-Lagrange equation for (6.8) is

$$\frac{\partial L_1}{\partial t} + \frac{\partial L_2}{\partial x_i} - L_3 = 0 \tag{6.9}$$

where

$$L_1 = \frac{\partial L}{\partial u_t}, \quad L_2 = \frac{\partial L}{\partial u_{x_i}}, \quad L_3 = \frac{\partial L}{\partial u} \tag{6.10abc}$$

Equation (6.9) is a second order partial differential equation for $u(\underline{x}_i, t)$. We assume that this equation has periodic wave train solutions in the form

$$u = \Phi(\theta), \qquad \theta = \underline{\kappa} \cdot \underline{x} - \omega t \tag{6.11abc}$$

where $\underline{\kappa}$ and ω are constants and represent the wave number and frequency, respectively. Since (6.9) is a second-order equation, its solution depend on two arbitrary constants of integration. One is the amplitude a and the other is the phase-shift. Omitting the latter constant, it turns out that solution of (6.9) exists provided the three parameters ω, $\underline{\kappa}$, a are connected by a dispersion relation

$$D(\omega, \underline{\kappa}, a) = 0 \tag{6.12}$$

In linear problems with the wave train solution in the form $u = \Phi(\theta) = ae^{i\theta}$, the dispersion relation (6.12) does not involve the amplitude a.

For slowly varying dispersive wave trains, the solution maintains the elementary form $u = \Phi(\theta, a)$, but ω, $\underline{\kappa}$, and a are no longer constants so that θ is not a linear function of $\underline{x}$ and t. The local wave number and frequency are defined by

$$k_i = \frac{\partial \theta}{\partial x_i}, \qquad \omega = -\frac{\partial \theta}{\partial t} \tag{6.13ab}$$

The quantities ω, $\underline{\kappa}$, and a are slowly varying functions of $\underline{x}$ and t corresponding to the slow modulation of the wave train.

The Whitham averaged Lagrangian over the phase of the integral of L is defined by

$$\mathcal{L}(\omega, \underline{\kappa}, a, \underline{x}, t) = \frac{1}{2\pi} \int_0^{2\pi} L\, d\theta \tag{6.14}$$

and is calculated by putting the uniform periodic solution $u = \Phi(\theta, a)$ in L. Whitham first formulated the averaged variational principle in the form

$$\delta \iint \mathcal{L}\, d\underline{x}\, dt = 0 \tag{6.15}$$

to derive the equations for ω, $\underline{\kappa}$, and a.

It is noted that the dependence of $\mathcal{L}$ on $\underline{x}$ and t reflects possible nonuniformity of the medium supporting the wave motion. In a uniform medium, $\mathcal{L}$ is independent of $\underline{x}$ and t so that the Whitham function $\mathcal{L} = \mathcal{L}(\omega, \underline{\kappa}, a)$. However, in a nonuniform medium, some additional variables also appear only through their derivatives. They represent potentials whose derivatives are important physical quantities.

The Euler equations resulting from the independent variations of δa and $\delta\theta$ in (6.15) with $\mathcal{L} = \mathcal{L}(\omega, \underline{\kappa}, a)$ are

$$\delta a: \quad \mathcal{L}_a(\omega, \underline{\kappa}, a) = 0 \tag{6.16}$$

$$\delta\theta: \quad \frac{\partial}{\partial t}\mathcal{L}_\omega - \frac{\partial}{\partial x_i}\mathcal{L}_{k_i} = 0 \tag{6.17}$$

The θ-eliminant of (6.13ab) gives the consistency equations

$$\frac{\partial k_i}{\partial t} + \frac{\partial \omega}{\partial x_i} = 0, \qquad \frac{\partial k_i}{\partial x_j} - \frac{\partial k_j}{\partial x_i} = 0 \tag{6.18ab}$$

Thus (6.16)-(16.18ab) represent the Whitham equations for describing the slowly varying wave train in a uniform medium, and constitute a closed set from which the triad ω, $\underline{\kappa}$, a can be determined.

In linear problems, L is, in general, a quadratic in u and its derivatives. Hence, if $\Phi(\theta) = a \cos \theta$ is substituted in (6.14), $\mathcal{L}$ must always take the form

$$\mathcal{L}(\omega, \underline{\kappa}, a) = D(\omega, \underline{\kappa})a^2 \tag{6.19}$$

so that the dispersion relation must be

$$D(\omega, \underline{\kappa}) = 0 \tag{6.20}$$

We note that the stationary value of $\mathcal{L}$ is, in fact, zero for linear problems. In those cases, $\mathcal{L}$ is equal to the difference between kinetic and potential energy. This proves the well-known principle of equipartition of energy.

We next apply the Whitham averaged variational principle to the uniform and nonuniform Stokes water waves. For nonuniform water wave trains, the solution has the general form

$$\left.\begin{aligned} \eta &= N(\theta) \\ \phi &= \psi + \Phi(\theta, z) \end{aligned}\right\} \tag{6.21ab}$$

$$\left.\begin{aligned} \theta_x &= k, \quad \theta_t = -\omega \\ \psi_x &= \beta, \quad \psi_t = -\gamma \end{aligned}\right\} \tag{6.22ab}$$

where N and Φ are periodic functions of the phase function θ, and contain the amplitude a and the mean value b of wave height η. The quantities β and γ behave like a pseudo wave number and frequency in ϕ corresponding to k and ω in θ. Thus the solution of the nonuniform wave train depends on two triads of quantities (k, ω, a) and (β, γ, b) which are all slowly varying functions of x and t. The closed set of equations for these functions is derived from Whitham's averaged variational principle.

Whitham calculated the average Lagrangian $\mathcal{L}$ from the Stokes expansion of periodic wave trains in water of arbitrary depth in the form

$$\mathcal{L} = (\tfrac{1}{2}\beta^2 - \gamma)h + \tfrac{1}{2}gb^2 + \frac{E}{2}\left[1 - \frac{(\omega - \beta k)^2}{gk \tanh kh}\right] + \frac{k^2E^2D_0}{2g} + 0(E^3) \tag{6.23}$$

where the total mean depth $h = h_0 + b$ and

$$E = \tfrac{1}{2}ga^2, \quad D_0 = \frac{1}{8T_0^4}(9T_0^4 - 10T_0^2 + 9), \quad T_0 = \tanh kh_0 \tag{6.24abc}$$

It also follows from Whitham's averaged equations with $\mathcal{L}$ given by (6.23) that the evolution of a wave-packet is governed either by a hyperbolic or an elliptic equation and the transition from the former to the latter occurs as kh increased through the critical value 1.363. In the hyperbolic case, changes in the wave trains propagate according to the usual theory of characteristics and the characteristic velocities provide a generalization of the linear group velocity to nonlinear wave problems. On the other hand, the elliptic case leads to the instability of Stokes water wave trains.

In this connection, mention may be made of Lighthill's formulation where the nonlinear dispersion relation can be used to eliminate a from the Lagrangian density so that $\mathcal{L}$ can be treated as a function, $\mathcal{L} = \mathcal{L}(\omega, k)$. Following Whitham, a phase function $\theta(x,t)$ with the local wave number $k = \theta_x$ and the local frequency $\omega = -\theta_t$ can be used to describe a more general wave train. The Euler equation for the variational principle

$$\delta \iint \mathcal{L}(-\theta_t, \theta_x)\, dx\, dt = 0 \tag{6.25}$$

specifying the function $\theta(x,t)$ can be derived as

$$\frac{\partial}{\partial t}\left(\frac{\partial \mathcal{L}}{\partial \omega}\right) = \frac{\partial}{\partial x}\left(\frac{\partial \mathcal{L}}{\partial k}\right) \tag{6.26}$$

This is the Whitham equation (6.17) obtained earlier. Lighthill used the function $\mathcal{L}(\omega,k)$ to introduce the wave-action density $\mathcal{A} = \mathcal{L}_\omega$ which has an important implication in the theory of nonlinear dispersive waves. Thus the Whitham equation (6.25) can be interpreted directly as a conservation law for the wave-action

$$\frac{\partial \mathcal{A}}{\partial t} + \frac{\partial \mathcal{B}}{\partial x} = 0 \tag{6.27}$$

where $\mathcal{B} = -\mathcal{L}_k$ represents the wave-action flux.

It has also been shown by Lighthill that total wave energy density, $W = 2T - \mathcal{L}$, and $\mathcal{A} = \mathcal{L}_\omega = 2T/\omega$ where T is the average kinetic energy per unit distance. Thus it follows from the result $W = \omega\mathcal{L}_\omega - \mathcal{L}$ that

$$\begin{aligned}\frac{\partial W}{\partial t} &= \left(\omega \frac{\partial}{\partial t}\mathcal{L}_\omega + \omega_t \mathcal{L}_\omega\right) - (\omega_t \mathcal{L}_\omega + k_t \mathcal{L}_k) \\ &= \omega \frac{\partial}{\partial x}\mathcal{L}_k + \omega_x \mathcal{L}_k, \text{ by (6.17)-(6.18a)} \\ &= \frac{\partial}{\partial x}(\omega \mathcal{L}_k) = -\frac{\partial I}{\partial x}\end{aligned} \tag{6.28}$$

where $I = -\omega \mathcal{L}_k$ is the energy flux. Hence the wave-action conservation law leads to the energy conservation law

$$\frac{\partial W}{\partial t} + \frac{\partial I}{\partial x} = 0 \tag{6.29}$$

7. PERTURBATION EXPANSION AND HAMILTONIAN FORMULATION

Whitham [46] successfully justified his general variational approach to nonlinear systems by formal perturbation expansion and introduced the method of two time scales: the fast oscillations of the wave train and the slow variations of the quantities ω, $\underline{\kappa}$, and a. The method recognizes explicitly in the dependent variables that changes occur on the two time scales. The essence of the method is to define the fast variables Θ, X, T and slow variables θ, x, t by the relations

$$\theta = \epsilon^{-1}\Theta(X, T), \quad \underline{X} = \epsilon \underline{x}, \quad T = \epsilon t \tag{7.1abc}$$

where ϵ is a small parameter that measures the ratio of the fast and the slow time scales.

The function $\phi(\underline{x}, t)$ can be expressed in the form

$$\phi(\underline{x}, t) = \Phi(\theta, \underline{X}, T, \epsilon) \tag{7.2}$$

Consequently, the basic Euler equation (6.9) can be written in terms of the fast variables as

$$\frac{\partial}{\partial \theta}(\nu L_1 + kL_2) - L_3 + \epsilon\left(\frac{\partial L_1}{\partial T} + \frac{\partial L_2}{\partial X}\right) = 0 \tag{7.3}$$

where ν and $\underline{k}$ are defined by

$$\nu(\underline{X}, T) = -\omega(\underline{X}, T) = \Theta_T, \quad \underline{k}(X, T) = \Theta_X \tag{7.4ab}$$

and the new arguments of L_j are given by

$$L_j = L_j(\nu\Phi_\theta + \epsilon\Phi_T, k\Phi_\theta + \epsilon\Phi_X, \Phi) \tag{7.5}$$

The solution of the Euler equation (7.3) is obtained by a formal perturbation expansion in ϵ

$$\Phi(\theta, X, T, \epsilon) = \sum_{n=0}^{\infty} \epsilon^n \Phi_n(\theta, X, T) \tag{7.6}$$

Equation (7.3) can be written in the conservation form

$$\frac{\partial}{\partial\theta}\{(\nu L_1 + kL_2)\Phi_\theta - L\} + \epsilon\left[\frac{\partial}{\partial T}(L_1\Phi_\theta) + \frac{\partial}{\partial X}(L_2\Phi_\theta)\right] = 0 \tag{7.7}$$

Then the averaged variational principle

$$\delta\int\int \mathfrak{L}\ dX\, dT = 0 \tag{7.8}$$

where

$$\mathfrak{L} = \frac{1}{2\pi}\int_0^{2\pi} L(\nu\Phi_\theta + \epsilon\Phi_T,\ k\Phi_\theta + \epsilon\Phi_X,\ \Phi)\, d\theta \tag{7.9}$$

leads to the Euler equations

$$\frac{\partial}{\partial\theta}L_{\Phi_\theta} + \frac{\partial}{\partial T}L_{\Phi_T} + \frac{\partial}{\partial X}L_{\Phi_X} - L_\Phi = 0 \tag{7.10}$$

$$\frac{\partial}{\partial T}\mathfrak{L}_\nu + \frac{\partial}{\partial X}\mathfrak{L}_k = 0 \tag{7.11}$$

The former agrees with the basic equation (7.3).

Equating the terms of successive orders in ϵ to zero, it follows from (7.6) and (7.3), to the lowest order in ϵ that

$$\frac{\partial}{\partial\theta}\{\nu L_1^{(0)} + kL_2^{(0)}\} - L_3^{(0)} = 0 \tag{7.12}$$

where

$$L_j^{(0)} = L_j(\nu\Phi_{0\theta},\ k\Phi_{0\theta},\ \Phi_0) \tag{7.13}$$

The averaged variational principle

$$\delta\int\int \mathfrak{L}_0\ dX\, dT = 0 \tag{7.14}$$

with

$$\mathfrak{L}_0 = \frac{1}{2\pi}\int_0^{2\pi} L^{(0)}(\nu\Phi_{0\theta},\ k\Phi_{0\theta},\ \Phi_0)\, d\theta \tag{7.15}$$

gives two partial differential equations for $\Phi_0(\theta, X, T)$ and $\Theta(X, T)$. These equations dropping the subscript zero, have the form

$$\frac{\partial}{\partial\theta}(\nu L_1 + kL_2) - L_3 = 0 \tag{7.16}$$

$$\frac{\partial}{\partial T}\mathfrak{L}_\nu + \frac{\partial}{\partial X}\mathfrak{L}_k = 0 \tag{7.17}$$

where L_j and $\mathfrak{L}$ are given by (7.13) and (7.15).

The first integral of (7.16) is

$$\nu L_1 + kL_2 - L = A(X, T) \tag{7.18}$$

where A can be recognized as the amplitude as a function of the fast variables.

An extension of the above perturbation analysis for higher order in ϵ, and to more variables is given by Whitham [46] and Luke [47].

Finally, it turns out that the above method is essentially a Hamiltonian formulation of the equations. The basic idea is to eliminate Φ_θ in favor of $\partial L/\partial\Phi_\theta$ just as $\dot{q}$ is eliminated in favor of a generalized momentum $p = \partial L/\partial\dot{q}$ in classical mechanics.

We introduce a new variable Π defined by

$$\Pi = \frac{\partial L}{\partial \Phi} = \nu L \ + kL \tag{7.19}$$

and the Hamiltonian $H(\Pi, \Phi, \Theta)$ is defined by

$$H = \Phi_\theta \frac{\partial}{\partial\Phi} - L = \Phi_\theta(\nu L_1 + kL_2) - L \tag{7.20}$$

These Hamiltonian transformations (7.19)-(7.20) imply

$$\frac{\partial\Phi}{\partial\theta} = \frac{\partial H}{\partial\Pi} \tag{7.21}$$

and (7.16) gives

$$\frac{\partial\Pi}{\partial\theta} = -\frac{\partial H}{\partial\Phi} \tag{7.22}$$

These may be treated as the Hamiltonian equations for the system, and replace the second order Lagrangian (7.16) by two first order equations for Φ and Π. Consequently, the variational principle (7.14), dropping the subscript zero, can be rewritten in terms of Π and H as

$$\iint \mathcal{L}\, dX\, dT = 0 \tag{7.23}$$

with

$$\mathcal{L} = \frac{1}{2\pi}\int_0^{2\pi} (\Pi\Phi_\theta - H)\, d\theta \tag{7.24}$$

Moreover, (7.18) assumes the form

$$H(\Pi, \Phi, \Theta) = A(X, T) \tag{7.25}$$

which can be solved for the function Π

$$\Pi = \Pi(\Phi, A; \Theta) \tag{7.26}$$

These allow us to rewrite the variational principle (7.23)-(7.24) as

$$\iint \mathcal{L}(A, \Theta)\, dX\, dT = 0 \tag{7.27}$$

where

$$\mathcal{L}(A, \Theta) = \frac{1}{2\pi}\oint \Pi(\Phi, A; \Theta)\, d\Phi - A \tag{7.28}$$

This variational principle implies

$$\mathcal{L}_A = 0 \tag{7.29}$$

$$\frac{\partial}{\partial T}\mathcal{L}_\nu + \frac{\partial}{\partial X}\mathcal{L}_k = 0 \tag{7.30}$$

These equations are identically the same with those obtained earlier by the Lagrangian formulation where $\nu = -\omega$.

8. THE VARIATIONAL FORMULATION OF THE KLEIN-GORDON, BOUSSINESQ, AND KORTEWEG-DE VRIES EQUATIONS

It can easily be verified from (6.9) that the nonlinear Klein-Gordon equation

$$\phi_{tt} - \phi_{xx} + V'(\phi) = 0 \tag{8.1}$$

is the Euler-Lagrange equation for the variational principle

$$\delta \iint L \, dx \, dt = 0 \tag{8.2}$$

where the Lagrangian L is

$$L = \tfrac{1}{2}\phi_t^2 - \tfrac{1}{2}\phi_x^2 - V(\phi) \tag{8.3}$$

and $V(\phi)$ is the nonlinear potential energy.

It appears to be generally true that a nonlinear dispersive system has periodic wave solution in the form

$$\phi = \Phi(\theta), \qquad \theta = kx - \omega t \tag{8.4ab}$$

where Φ is a periodic function of phase θ. Thus the existence of periodic wave solution of (8.1) implies

$$(\omega^2 - k^2)\Phi_{\theta\theta} + V'(\Phi) = 0 \tag{8.5}$$

which has the first integral

$$\tfrac{1}{2}(\omega^2 - k^2)\Phi_\theta^2 + V(\Phi) = A \tag{8.6}$$

where A is a constant of integration and will be identified as the amplitude.

The solution of (8.4) can be written as

$$\theta = \{\tfrac{1}{2}(\omega^2 - k^2)\}^{\frac{1}{2}} \int \frac{d\Phi}{\sqrt{A - V(\Phi)}} \tag{8.7}$$

where $V(\Phi)$ is either a cubic, a quartic, or a trigonometric function so that $\Phi(\theta)$ can be represented in terms of standard elliptic functions (Dutta and Debnath [42]).

The periodic wave train solution of (8.1) is described by (8.6)-(8.7) and involve the parameters ω, k, and A. The Whitham averaged Lagrangian function

$$\mathcal{L} = \frac{1}{2\pi}\int_0^{2\pi} L \, d\theta \tag{8.8}$$

where L, given by (8.3), can be expressed in terms of $\Phi(\theta)$ as

$$L = \tfrac{1}{2}(\omega^2 - k^2)\Phi_\theta^2 - V(\Phi) \tag{8.9}$$

Using (8.6), the result (8.8) becomes

$$\begin{aligned}\mathcal{L}(\omega, k, A) &= \frac{1}{2\pi}\int_0^{2\pi} (\omega^2 - k^2)\Phi_\theta \, d\theta - A \\ &= \frac{1}{2\pi}\int_0^{2\pi} (\omega^2 - k^2)\Phi_\theta \, d\Phi - A \\ &= \frac{1}{2\pi}\{2(\omega^2 - k^2)\}^{\frac{1}{2}} \oint \sqrt{A - V(\Phi)} \, d\Phi - A\end{aligned} \tag{8.10}$$

where the loop integral is taken over a complete period of the integrand.

With an assumption that ω, k, and A are slowly varying functions of x and t, the Whitham averaged variational principle

$$\delta \iint \mathcal{L}\,(\omega,\ k,\ A)\ dx\ dt = 0 \tag{8.11}$$

gives

$$\mathcal{L}_A = \frac{1}{2\pi}\left\{\frac{1}{2}(\omega^2 - k^2)\right\}^{\frac{1}{2}} \oint \frac{d\Phi}{\sqrt{A - V(\Phi)}} - 1 = 0 \tag{8.12}$$

$$\frac{\partial}{\partial t}\mathcal{L}_\omega - \frac{\partial}{\partial x}\mathcal{L}_k = 0 \tag{8.13}$$

Equation (8.12) can be recognized as the correct nonlinear dispersion relation involving ω, k, and A.

In the linear case, $V(\Phi) = \frac{1}{2}\Phi^2$, and $\Phi = a \cos \theta$ with $A = \frac{1}{2}a^2$ so that $\mathcal{L}$ given by (8.10) has the form

$$\mathcal{L} = [(\omega^2 - k^2)^{\frac{1}{2}} - 1]\,A \tag{8.14}$$

and (8.12) gives the linear dispersion law

$$\omega^2 = k^2 + 1 \tag{8.15}$$

and no dependence on the amplitude A.

The Boussinesq equations for long water waves in a liquid of uniform initial depth h_0 are

$$\frac{\partial h}{\partial t} + \frac{\partial}{\partial x_i}\left(h\frac{\partial \phi}{\partial x_i}\right) = 0 \tag{8.16}$$

$$\frac{\partial \phi}{\partial t} + \frac{1}{2}\left(\frac{\partial \phi}{\partial x_i}\right)^2 + gh + \nu_0\frac{\partial^2 h}{\partial t^2} = 0 \tag{8.17}$$

where $h(\underline{x}, t)$ is the depth, $\phi(\underline{x}, t)$ is the velocity potential for the mean flow with velocity $u_i = \partial\phi/\partial x_i$, g is the acceleration due to gravity and $\nu_0 = (1/3)h_0$.

These equations follow from a variational principle

$$\delta \iint L\ d\underline{x}\ dt = 0 \tag{8.18}$$

where the Lagrangian is given by

$$L(h_t,\ h;\ \phi_t,\ \phi_{x_i}) = \tfrac{1}{2}\nu_0 h_t^2 - \tfrac{1}{2}gh^2 - h(\phi_t + \tfrac{1}{2}\phi_{x_i}^2) \tag{8.19}$$

The Euler equations are

$$\frac{\partial}{\partial t}\left(\frac{\partial L}{\partial \phi_t}\right) + \frac{\partial}{\partial x_i}\left(\frac{\partial L}{\partial \phi_{x_i}}\right) = 0 \tag{8.20}$$

$$\frac{\partial}{\partial t}\left(\frac{\partial L}{\partial h_t}\right) - \frac{\partial L}{\partial h} = 0 \tag{8.21}$$

These are equivalent to (8.16)-(8.17).

Whitham has shown that the Lagrangian L to the lowest order in ϵ has the form

$$L = \tfrac{1}{2}\nu h_\theta^2 - \tfrac{1}{2}gh^2 - h\{\gamma + \nu\Phi_\theta + \tfrac{1}{2}(\beta + k\Phi_\theta)^2\} \tag{8.22}$$

where the slowly varying wave trains are described by

$$\phi = \Psi + \Phi(\theta,\ X,\ T) \tag{8.23}$$

$$\theta = \epsilon^{-1}\Theta(X,\ T), \qquad \Psi = \epsilon^{-1}\Psi(X,\ T) \tag{8.24ab}$$

$$\nu = \Theta_T, \quad k = \Theta_X, \quad \gamma = \Psi_T \text{ and } \beta = \Psi_X \tag{8.25}$$

Using the Hamiltonian transformations

$$\Pi_1 = \frac{\partial L}{\partial h_\theta} = \nu^2 h_\theta, \quad \Pi_2 = \frac{\partial L}{\partial \Phi_\theta} = -h\{\nu + k(\beta + k\Phi_\theta)\} \tag{8.26ab}$$

$$H = \frac{1}{2\nu^2}\Pi_1^2 - \frac{1}{2hk^2}\{\Pi_2 + h(\nu + \beta k)^2\} + h\left(\gamma + \frac{1}{2}\beta^2\right) + \frac{1}{2}gh^2 \tag{8.27}$$

Whitham obtained the averaged Lagrangian function $\mathcal{L}$ in the form

$$\mathcal{L} = \frac{1}{2\pi}\oint \Pi_1\, dh - A$$

$$= \frac{1}{2\pi}\oint \nu\{k^{-2}h^{-1}(B + h\nu + h\beta k)^2 - h(2\gamma + \beta^2) - gh^2 + 2A\}^{\frac{1}{2}}\, dh - A \tag{8.28}$$

where $\Pi_2 = B(X, T)$.

The Korteweg-de Vries (KdV) equation is given by

$$\eta_t + 6\eta\eta + \eta_{xxx} = 0 \tag{8.29}$$

There is no variational principle for this form of (8.29). It is then necessary to transform (8.29) with $\eta = \phi_x$ in the form

$$\phi_{xt} + 6\phi_x\phi_{xx} + \phi_{xxxx} = 0 \tag{8.30}$$

This equation has the Lagrangian

$$L = -\tfrac{1}{2}\phi_t\phi_x - \phi_x^3 + \tfrac{1}{2}\phi_{xx}^2 \tag{8.31}$$

We assume the uniform wave train solution in the form

$$\phi = \Psi(\theta) + \psi, \qquad \theta = kx - \omega t, \qquad \psi = \beta x - \gamma t \tag{8.32abc}$$

so that

$$\eta = \beta + k\Phi_\theta \tag{8.33}$$

where β is the parameter representing the mean value of η.

Thus the uniform wave train solution of (8.29) has form

$$k^2\eta_{\theta\theta\theta} + 6\eta\eta_\theta - \omega k^{-1}\eta_\theta = 0 \tag{8.34}$$

The first and the second integrals of this equation are

$$k^2\eta_{\theta\theta} + 3\eta^2 - \omega k^{-1}\eta + B = 0 \tag{8.35}$$

$$k^2\eta_\theta^2 + 2\eta^3 - \omega k^{-1}\eta^2 + 2B\eta - 2A = 0 \tag{8.36}$$

where A and B are constants of integration.

The Lagrangian L given by (8.31) can then be expressed in terms of η in the form

$$L = \frac{1}{2}\left(\gamma - \frac{\omega}{k}\beta\right)\eta + \frac{1}{2}\frac{\omega}{k}\eta^2 - \eta^3 + \frac{1}{2}k^2\eta_\theta^2 \tag{8.37}$$

which, in view of (8.36), assumes the form

$$L = k^2\eta_\theta^2 + \left\{B + \frac{1}{2}\left(\gamma - \frac{\omega}{k}\right)\beta\right\}\eta - A \tag{8.38}$$

Finally, the averaged Lagrangian $\mathcal{L}$ is given by

$$\mathcal{L} = \frac{1}{2\pi}\int_0^{2\pi} L\, d\theta \tag{8.39}$$

Using (8.33) and (8.36), it turns out that

$$\mathcal{L} = k\Upsilon(A, B, C) + \beta B + \tfrac{1}{2}\beta\gamma - \tfrac{1}{2}C\beta^2 - A \tag{8.40}$$

where

$$\Upsilon(A, B, C) = \frac{1}{2\pi} \oint \{2A - 2B\eta + C\eta^2 - 2\eta^3\}^{\frac{1}{2}} \tag{8.41}$$

and the nonlinear phase velocity is $c = \omega/k$.

9. VARIATIONAL PRINCIPLES FOR ROSSBY WAVES AND BOUSSINESQ EQUATION FOR A STRATIFIED FLUID

The linearized Rossby wave equation for rotating fluid flows in the ocean on the surface of the earth with the β-plane approximation is (Debnath [48])

$$\nabla^2\Psi_t + \beta\Psi_x = 0 \tag{9.1}$$

where $\Psi(x, y, t)$ is the stream function defined by $u = \Psi_y$ and $u = -\Psi_x$, $\beta = df/dy$, $f = f(y)$, the x-axis is eastward and y-axis is northward.

Equation (9.1) follows from the variational principle

$$\delta \iiint \{\tfrac{1}{2}(\nabla X_t)^2 - \tfrac{1}{2}\beta X_x X_t\}\, dx\, dy\, dt = 0 \tag{9.2}$$

where $\Psi = X_t$.

However, the nonlinear Rossby wave equation is

$$\frac{\partial}{\partial t}(\nabla^2\Psi) + \beta\Psi_x = J(\Psi, \nabla^2\Psi) \tag{9.3}$$

where J is the Jacobian.

No variational principle for (9.3) is known.

The Boussinesq equation of motion for the stream function Ψ in a stratified liquid with $\rho_0(y)$ as the density distribution is given by

$$\rho_0(y)\Psi_{xxtt} + [\rho_0(y)\Psi_{yt}]_{yt} - g\rho_0'(y)\Psi_{xx} = 0 \tag{9.4}$$

This equation follows from the variational principle

$$\delta \iiint [\tfrac{1}{2}\rho_0(y)(\nabla\Psi_t)^2 + \tfrac{1}{2}g\rho_0'(y)\Psi_x^2]\, dx\, dy\, dt = 0 \tag{9.5}$$

10. NONLINEAR WAVES IN A DISSIPATIVE SYSTEM AND AVERAGED PSEUDO-VARIATIONAL PRINCIPLE

Jimenez and Whitham [49] considered a dispersive and dissipative system governed by the differential equation

$$N(\phi, \epsilon) = 0 \tag{10.1}$$

where $N(\phi, 0)$ is conservative and can be derived from a variational principle with a classical Lagrangian, and $N(\phi, \epsilon) - N(\phi\ 0) = o(1)$. The pseudo-Lagrangian for this system can be written as

$$F(\phi', \phi) = L(\phi') + M(\phi', \phi, \epsilon) \tag{10.2}$$

where $M(\phi', \phi, \epsilon) = o(1)$ and represents the reversible part of the system.

We consider a slightly damped harmonic oscillator governed by the equation

$$\phi_{tt} + \epsilon\phi_t + \alpha^2\phi = 0 \tag{10.3}$$

where ϵ is a small parameter and the main part of the system is reversible. Then the pseudo-Lagrangian for the system (10.3) can be expressed in the form (10.2) and is given by

$$F(\phi', \phi) = \tfrac{1}{2}(\phi_t'^2 - \alpha^2\phi'^2) - \epsilon\phi'\phi_t \tag{10.4}$$

In the absence of dissipation ($\epsilon = 0$), equation (10.3) admits a slowly varying wave train in sinusoidal form

$$\phi = a \sin\theta, \qquad \theta = kx - \omega t \tag{10.5ab}$$

We assume that ϕ' has the same form so that $\phi' = a' \sin\theta'$. Putting these results into (10.4) and taking an average over a few cycles, it turns out that the averaged pseudo-Lagrangian $\mathcal{F}$ is given by

$$\mathcal{F} = \frac{1}{4}a'^2(\omega'^2 - \alpha^2) - \frac{1}{2}\epsilon a'a\omega \sin(\theta' - \theta) \tag{10.6}$$

The Euler equations for the variations of a' and θ' with $a' = a$ and $\theta' = \theta$ are

$$\omega = \alpha, \qquad \frac{d}{dt}(a^2\omega) + \epsilon a^2\omega = 0 \tag{10.7ab}$$

The former one represents the dispersion relation and the latter is the amplitude equation describing the decay of amplitude. In the limit $\epsilon \to 0$, the dispersion relation remains unchanged, but the amplitude equation (10.7b) reduces to the corresponding result for the linear harmonic oscillator without dissipation. In fact, (10.7b) gives $a^2\omega$ = constant which corresponds to the result for the adiabatic invariant of the harmonic oscillator. Also, the Whitham averaged Lagrangian $\mathcal{L}$ follows directly from (10.6) in the limit of $\epsilon \to 0$ and given by

$$\mathcal{L} = \frac{1}{4}a^2(\omega^2 - \alpha^2)$$

We now consider a nonlinear oscillator with dissipation governed by the equation

$$\phi_{tt} + V_\phi(\phi) + \epsilon f(\phi)\phi_t = 0 \tag{10.9}$$

The pseudo-Lagrangian for this equation is

$$F(\phi', \phi) = \tfrac{1}{2}\phi_t'^2 - V(\phi') - \epsilon\phi' f(\phi)\phi_t \tag{10.10}$$

In the limit $\epsilon \to 0$, equation (10.9) admits periodic solution $\Phi(\theta)$ which satisfies

$$\omega^2\Phi_{\theta\theta} + V_\phi(\Phi) = 0, \qquad \omega = -\theta_t \tag{10.11ab}$$

The first integral of this equation is

$$\tfrac{1}{2}\omega^2\Phi_\theta^2 + V(\Phi) = A \tag{10.12}$$

where A is a constant of integration.

We assume the form of $\phi' = \Phi'(\theta')$ with parameters ω', A' in (10.12) to express F in terms of Φ' and Φ so that (10.10) has the form

$$F(\Phi', \Phi) = \tfrac{1}{2}\omega'^2\Phi_{\theta'}'^2 - V(\Phi') - \epsilon\omega\Phi'(\theta')f(\Phi)\Phi_\theta \tag{10.13}$$

Replacing $\Phi'(\theta')$ with its approximate form

$$\Phi'(\theta') \sim \Phi(\theta) + (\theta' - \theta)\Phi_\theta(\theta) \tag{10.14}$$

We use the energy relation (10.12) combined with (10.14) to obtain the final form of (10.13) as

$$F(\Phi', \Phi) = \omega'^2\Phi_{\theta'}'^2 - A' - \epsilon\omega f(\Phi)\Phi\Phi_\theta - \epsilon(\theta' - \theta)\omega f(\Phi)\Phi_\theta^2 \tag{10.15}$$

Noting that the third term of (10.15) does not contribute to the variational equations, and using Whitham's ideas, we derive the averaged pseudo-Lagrangian in the form

$$\mathcal{F} = \langle F(\Phi', \Phi)\rangle = \langle \omega'^2\Phi^2_{\theta'} - A'\rangle - \epsilon\langle \omega\Phi\Phi_\theta f(\Phi)\rangle - \epsilon\langle \theta' - \theta)\langle \omega f(\Phi)\Phi_\theta\rangle \tag{10.16}$$

where $\langle\ \rangle$ denotes the average value over the phases θ and θ'. Thus, it turns out that

$$\mathcal{F} = \frac{\omega'}{2\pi}\oint (2A' - 2V)^{\frac{1}{2}} d\Phi' - A' - \frac{\epsilon(\theta' - \theta)}{2\pi}\oint f(\Phi)(2A - 2V)^{\frac{1}{2}} d\Phi \tag{10.17}$$

where the loop integrals are functions of A' and a, respectively.

For a reversible system, the average Lagrangian $\mathcal{L} = \mathcal{L}(\omega, k, A)$. According to the Whitham averaged variational principle (6.15), the fundamental equations for ω, k, and A are

$$\mathcal{L}_A = 0, \quad \frac{\partial}{\partial t}\mathcal{L}_\omega - \frac{\partial}{\partial x}\mathcal{L}_k = 0, \quad \frac{\partial k}{\partial t} + \frac{\partial \omega}{\partial x} = 0 \tag{10.18abc}$$

When dissipation is included in the system, these equations remain unaltered. However, the averaged pseudo-Lagrangian $\mathcal{F}$ can be expressed in the form similar to (10.2) so that

$$\mathcal{F} = \mathcal{L}(\omega', k', A') + \epsilon(\theta' - \theta)M(\omega, k, A) \tag{10.19}$$

The Euler equation corresponding to the variation of θ can be derived from the variational principle and has the form

$$\frac{\partial}{\partial t}\mathcal{L}_\omega - \frac{\partial}{\partial x}\mathcal{L}_k + \epsilon M = 0 \tag{10.20}$$

This is the Whitham fundamental equation for nonlinear dispersive and dissipative wave trains and the third term on the left hand side of (10.20) represents the effect of dissipation.

Equation (10.20) has been used to examine the effect of dissipation on a linear system. For a linear system, the averaged pseudo-Lagrangian $\mathcal{F}$ is linear in amplitude A and can be written as

$$\mathcal{F} = A'D(\omega', k') + \epsilon(\theta' - \theta)H(\omega, k) \tag{10.21}$$

And the dispersion relation $D(\omega, k) = 0$ is independent of amplitude A so that we can express $\omega = \omega(k)$. Thus (10.18c) gives the group velocity $C = d\omega/dk = -D_k/D_\omega$. In view of these results, Eq. (10.20) can be simplified to obtain

$$A_t + (CA)_x + \epsilon\mu(k)A = 0, \qquad \mu(k) = -\frac{H}{D_\omega} \tag{10.22ab}$$

For a fixed wave number k, the solution for A is given by $\exp(-\epsilon\mu t)F(x - Ct)$ where $\mu(k)$ represents the rate of decay.

REFERENCES

1. Whitham, G. B.: Comm. Pure Appl. Math. 14, 675-691 (1961).
2. Whitham, G. B.: Proc. Roy. Soc. A 283, 238-261 (1965).
3. Whitham, G. B.: J. Fluid Mech. 22, 273-283 (1965).
4. Whitham, G. B.: J. Fluid Mech. 27, 399-412 (1967).
5. Whitham, G. B.: Proc. Roy. Soc. A 299, 6-25 (1967).
6. Whitham, G. B.: Linear and Nonlinear Waves, Wiley (1974).
7. Jones, D. S.: Generalized Functions, McGraw-Hill (1966).
8. Lighthill, M. J.: J. Inst. Math. Applic. 1, 269-306 (1966).
9. Lighthill, M. J.: Proc. Roy. Soc. A 299, 28-53 (1967).
10. Benjamin, T. B.; Feir, J. E.: J. Fluid Mech. 27, 417-430 (1967).

11. Benjamin, T. B.: Proc. Roy. Soc. A 299, 59-75 (1967).
12. Stuart, J. T.; DiPrima, R. C.: Proc. Roy. Soc. A 362, 27-41 (1978).
13. Eckhaus, W.: J. Mech. II, 153-172 (1963).
14. Eckhaus, W.: Studies in Nonlinear Stability Theory, vol. 6. Springer (1965).
15. Phillips, O. M.: Proc. Roy. Soc. A 299, 104-119 (1967).
16. Phillips, O. M.: J. Fluid Mech. 9, 437-456 (1960).
17. Zakharov, V. E.: J. Appl. Mech. Tech. Phys. 2, 190-194 (1968).
18. Hasimoto, H.; Ono, H.: J. Phys. Soc. Jpn. 33, 805-811 (1972).
19. Davey, A.; Stewartson, K.: Proc. Roy. Soc. A 338, 101-110 (1974).
20. Yuen, H. C.; Lake, B. M.: Phys. Fluids 18, 956-960 (1975).
21. Dysthe, K. B.: J. Plasma Phys. 11, 63-76 (1974).
22. Benney, D. J.; Newell, A. C.: J. Math. Phys. 46, 133-139 (1967).
23. Zakharov, V. E.; Shabat, A. B.: Sov. Phys. JETP 34, 62-69 (1971).
24. Lake, B. M.; Yuen, H. C.; Rungaldier, H.; Ferguson, W. E.: J. Fluid Mech. 83, 49-74 (1977).
25. Taniuti, T.; Washmi, H.: Phys. Rev. Lett. 21, 209-212 (1968).
26. Watanabe, T.: J. Phys. Soc. Jpn. 27, 1341-1350 (1969).
27. Chu, V. H.; Mei, C. C.: J. Fluid Mech. 41, 873-887 (1970).
28. Chu, V. H.; Mei, C. C.: J. Fluid Mech. 47, 337-351 (1981).
29. Benney, D. J.; Roskes, G.: Stud. Appl. Math. 48, 377-385 (1969).
30. Johnson, R. S.: Proc. Roy. Soc. A 357, 131-141 (1977).
31. Davey, A.: J. Fluid Mech. 53, 769-781 (1972).
32. Stewartson, K.; Stuart, J. T.: J. Fluid Mech. 48, 529-545 (1971).
33. DiPrima, R. C.; Eckhaus, W.; Segel, L. A.: J. Fluid Mech. 49, 705-744 (1971).
34. Tappert, F. D.; Varma, C. M.: Phys. Rev. Lett. 25, 1108-1111 (1970).
35. Zakharov, V. E.; Rubenchik, A. M.: Soviet Phys. JETP 38, 494-500 (1974).
36. Saffman, P. G.; Yuen, H. C.: Phys. Fluids 21, 1450-1451 (1978).
37. Yuen, H. C.; Ferguson, W. E.: Phys. Fluids 21, 1275-1278 (1978).
38. Longuet-Higgins, M. S.: Proc. Roy. Soc. A 360, 471-488 (1978).
39. Longuet-Higgins, M. S.: Proc. Roy. Soc. A 360, 489-505 (1978).
40. Dysthe, K. B.: Proc. Roy. Soc. A 369, 105-116 (1979).
41. Hui, W. H.; Hamilton, J.: J. Fluid Mech. 93, 117-133 (1979).
42. Dutta, M.; Debnath, L.: Elements of the Theory of Elliptic and Associated Functions with Applications. World Press Pvt. Ltd. (1965).
43. Longuet-Higgins, M. S.: Proc. Roy. Soc. A 347, 311-328 (1976).
44. Hocking, L. M.; Stewartson, K.; Stuart, J. T.: J. Fluid Mech. 51, 705-735 (1972).
45. Luke, J. C.: J. Fluid Mech. 27, 395-397 (1967).
46. Whitham, G. B.: J. Fluid Mech. 44 (1970).
47. Luke, J. C.: Proc. Roy. Soc. A 292, 403-412 (1966).
48. Debnath, L.: Nat. Acad. Sci., India, Professor P. L. Bhatnagar Commemoration Volume, 169-184 (1979).
49. Jimenez, J.; Whitham, G. B.: Proc. Roy. Soc. A 349, 277-287 (1976).
50. Debnath, L.: Plasma Phys. 19, 263-275 (1977).
51. Debnath, L. Lectures on Dynamics of Oceans, University of Calcutta (1976).

12 A PRIORI GROWTH AND HÖLDER ESTIMATES FOR HARMONIC MAPPINGS

Halldór I. Elíasson / Faculty of Engineering and Science, University of Iceland, Reykjavík, Iceland

INTRODUCTION

Let M and N be C^∞ Riemannian manifolds, with M compact and of dimension $n \geq 2$. A natural extension of the Dirichlet problem to mappings $f: M \to N$ consists in finding extremals to the variation integral given by the so-called energy

$$E(f) = \frac{1}{2} \int_M \|df\|^2$$

The corresponding Euler-Lagrange equation can be written $\Delta f = 0$, with Δ a natural extension of the Laplace or Beltrami operator to such mappings [2]. The problem of existence of extremals, the harmonic mappings, has been an outstanding problem in nonlinear global analysis for many years [1] and typical for many truly nonlinear geometric variational problems, where the highest order linear elliptic part of the Euler-Lagrange operator does not necessarily dominate the nonlinear terms of lower order. The classical method to obtain existence of extremals of variation integrals on linear function spaces is to establish the existence first in a complete function space with topology weaker than the uniform topology. Such topologies are difficult and unnatural when we are mapping into a nonlinear space. Moreover, we want to establish existence in some homotopy class and should then use a topology not weaker than the uniform one. Thus, if we want to use the direct method of the calculus of variations we should find a minimizing sequence of mappings, which is equicontinuous. This is of course not possible except we assume something more than boundedness of E(f). It has been suggested to request the integral

$$H_p(f) = \int_M \|\Delta f\|^p, \quad p > \frac{n}{2}$$

bounded [3, 6], which should not be serious since we want $\Delta f = 0$. The space of mappings can then be taken as the Sobolev space $L_2^p(M, N)$, which is a Banach manifold [2, 3]. Of course, we must then discuss whether $H_p(f)$ does not take its upper bounds when we minimize E(f) with $H_p(f)$ bounded, or if we minimize a linear combination of E and H_p as in [3, 7], we get an Euler-Lagrange equation which may have more solutions than the harmonic maps. Things like that cannot be avoided, since there are examples where there is no harmonic map in a homotopy class [1]. Our hope would be to get solutions to a problem more restricted than the nonlinear Dirichlet problem and then search for harmonic maps in the larger solution space.

The main object of this paper is to provide the a priori Hölder estimates which are needed if we are to attack problems of the kind we have described. Applications of these estimates, we believe, could extend to a variety of nonlinear problems and their importance is not restricted to the case of mappings between manifolds. The precision that must repeatedly be used in order to get to useful growth estimates is not at all due to the presence of curved space. The curvature does not more than add slightly to the complexity of the formulas. However, the curving of large portions of either manifold has quite an effect. This does not matter very much in the case of the domain, since we do not need to consider very distant points there, but the effects in the case of the range are serious. This is felt in our main result, which we choose to formulate at this point of our discussion as follows.

Theorem. Let M be a C^∞ Riemannian manifold of dimension $n \geq 2$ and with bounded sectional curvatures. Let N be a C^∞ Riemannian manifold with sectional curvatures bounded from above by λ^2 with $\lambda \geq 0$. Let μ, ρ, and p be real parameters satisfying:

$$0 < \mu \leq \frac{3}{8}, \quad \lambda\rho < \frac{\pi}{2}, \quad p > n\left(2 - \frac{11\mu}{3}\right)^{-1}$$

and define

$$\kappa = \mu(\lambda\rho \cot \lambda\rho)^2$$

Given $r_0 > 0$, define R(x) for x in M, such that R(x) equals the radius of convexity of M at x if it is less than r_0 and else let $R(x) = r_0$. Define a Hölder constant of a map f from M into N by

$$L_\kappa(f, r) = \sup\left\{ d(f(x), f(y))\, d(x, y)^{-\kappa} : x \neq y \text{ in M s.t. } d(x, y)^2 < r^2 < \frac{1}{2} R(x)\, R(y),\ \frac{1}{2} < \frac{R(x)}{R(y)} < 2 \right\}$$

Then for r_0 chosen sufficiently small, the Hölder constant $L_\kappa(f, r)$ is bounded on a set of C^∞ maps $f: M \to N$, such that f maps any convex ball B in M of radius $\leq r_0$ into a ball in N of radius ρ and the integrals:

$$\int_B \|df\|^2, \quad \int_B \|\Delta f\|^p$$

over all such balls B are bounded.

Remark. Just how small r_0 must be taken depends on bounds for the sectional curvatures of M and estimates can be found in the propositions to follow. L_κ is defined only if r is small.

Observe that we do not request either manifold complete. In particular, M could be the interior of a manifold with boundary and the mentioned family of maps would be equicontinuous on any compact subset of M. However, study of boundary behavior is not included in our work here, although that could be done by similar methods and some additional work.

It is interesting to compare our theorem with the result of Hildebrandt, Kaul, and Widman [4] on the existence of a solution to the Dirichlet problem for mapping a Riemannian manifold with boundary, fixing the boundary map, into a ball of radius less than $\pi/2\lambda$ in another Riemannian manifold. There they could use classical methods, establishing first the existence of a weak solution. Their assumption, like a global counterpart to our mapping condition, is very strong and excludes in fact most examples of interest to the geometrician. Then, on the other hand, it remains to be seen how useful our theorem really is in establishing existence. However, we can point out that the mapping condition should not be too hostile to applications in variational problems, as is in fact demonstrated in the case of the similar global condition in ([4], Lemma 4).

In order to present some geometric application that does not take too much space, we mention the following evident consequence of the theorem, which is new in the case of positive curvature.

Corollary. Let M be compact and N complete. Then any closed set of harmonic mappings from M into N, with bounded total image in N, is compact if $\lambda = 0$ and locally compact if $\lambda > 0$, in the uniform topology.

Remark. The one-parameter family of harmonic mappings of the two sphere onto itself, which is provided by carrying the transformations $x \to tx$ of three space, parametrized by t in $0 < t < \infty$, to the sphere by stereographic projections, is an interesting example of a noncompact family of harmonic maps, in the same homotopy class and with bounded image. Of course, this family violates the mapping condition in our theorem, but E is constant and $H_p = 0$.

We divide the paper into the following sections:

1. Geometric inequalities
   1.1 Metric inequalities
   1.2 Hölder estimates
   1.3 Local energy inequalities
2. Comparison and growth estimates
   2.1 Weak estimates
   2.2 Strong estimates
3. Equicontinuity

The core of our technique consists in integration of some differential identities or inequalities of divergence type for a map f over a ball and derive ordinary differential inequalities for relevant emerging integrals, called local energies, as functions of the radius of the ball. These local energies and the basic inequalities are introduced in 1.3. In the second section we derive estimates on the growth of the local energies as the radius increases, so that we can apply an estimate on the Hölder constant given in 1.2. We are able to do all this work in an intrinsic manner, without ever referring to a coordinate system for either manifold, due to the metric inequalities presented in 1.1. In fact, the Riemannian metric is visible only through the appearance of trigonometric functions in place of the usual powers of the radius. In case of negative curvature, the trigonometric functions transform into their hyperbolic counterparts. In Sec. 3 we give the proof of the theorem and discuss some alternative applications of our main estimate given in Prop. 7.

Added in Proof. A referee has pointed out that estimates equivalent to some of the estimates here when applied to the case of harmonic maps, i.e., if $H_p(f) = 0$, have been published [8]. In particular, the results in [8] could be used to obtain the corollary above.

1. GEOMETRIC INEQUALITIES

M shall denote a connected Riemannian manifold of dimension $n \geq 2$ and of class C^∞. We denote by d(x, y) the distance between two points x and y in M. We shall make extensive use of the function $r = d(x, x_0)$ measuring the distance r of x from a fixed point of reference x_0. As a preparation, we estimate the eigenvalues of the Hessian of this function r in terms of sectional curvatures of M in our first section. Equivalent information should be available in the literature, but possibly not in this setting, and are included for the sake of completeness. In Sec. 1.1 we describe a method to estimate the Hölder constant of a map of M into some other Riemannian manifold. This is in fact an old trick, but here we must take extra care since we want to prepare an implicit estimate on the Hölder constant. Such an estimation is carried out in Sec. 2 and is in terms of determinants of the map, which we call local energies and correspond to local norms in the euclidean case. We introduce these local energies in Sec. 1.3 and establish the basic differential inequalities behind our comparison and growth estimates.

1.1 Metric Inequalities

Let X_0 and X_1 be some unit tangent vectors to the manifold M at the point x_0 of reference. The geodesic

$$c(s) = \exp(sX_0)$$

is well defined on $0 \leq s < \epsilon$, if ϵ is sufficiently small. We define a variation of c by

$$b(s, t) = \exp(sX_0 + stX_1), \quad \text{so} \quad \frac{\partial b}{\partial t}(0, 0) = X_1$$

This is a variation through geodesics so that the distance r of b from x_0 is given by

$$r^2(b(s, t)) = s^2 \left\|\frac{\partial b}{\partial s}\right\|^2 = s^2 \|X_0 + tX_1\|^2$$

Taking the derivative with respect to s and also t, we obtain:

$$(dr^2 \cdot b)\frac{\partial b}{\partial s} = 2s\left\|\frac{\partial b}{\partial s}\right\|^2$$

$$(dr^2 \cdot b)\frac{\partial b}{\partial t} = 2s^2\left\langle\frac{\partial b}{\partial s}, \nabla_t \frac{\partial b}{\partial s}\right\rangle$$

∇_t denotes covariant derivation with respect to t. The vectorfields $\partial b/\partial s$ and

$$\nabla_t\left(\frac{\partial b}{\partial s}\right) = \nabla_s\left(\frac{\partial b}{\partial t}\right)$$

are orthogonal for all s, if they are so for $s = 0$, so the gradient of r^2 is given along b by

$$\nabla r^2 \cdot b = 2s\frac{\partial b}{\partial s}$$

Differentiating covariantly on both sides with respect to t we obtain

$$(\nabla^2 r^2 \cdot b)\frac{\partial b}{\partial t} = 2s\nabla_s\left(\frac{\partial b}{\partial t}\right)$$

Now we put $Y = \partial b/\partial t$ at $t = 0$. Y is known as a Jacobi field along c(s). Putting $t = 0$ in the equation above, we get

$$(\nabla^2 r^2 \cdot c)Y = 2s\nabla Y$$

This equation holds for all Jacobi fields Y of the type we can construct with such a variation, but they are determined by the following second order equation and initial condition

$$\nabla^2 Y + R(Y, c')c' = 0, \qquad Y(0) = 0$$

We are thinking of the covariant derivative $\nabla^2 r^2$ of the gradient field ∇r^2 as a linear operator on the tangent space of M at the base point. Its equation can also be written in terms of the covariant derivative of dr^2, which is a symmetric bilinear form at each point:

$$(\nabla dr^2 \cdot c)(Y_1, Y_2) = 2s\langle\nabla Y_1, Y_2\rangle$$

with at least Y_1 a Jacobi field vanishing at $s = 0$. We shall take both Y_1 and Y_2 equal Y in order to estimate the eigenvalues. In the case that Y is tangent to c' at some point, then Y is proportional to sc' and

$$s\langle Y, \nabla Y\rangle = \|Y\|^2$$

In the case that Y is orthogonal to c', we apply a known estimate of $\langle Y, \nabla Y\rangle$, see [5]. The result is given in the following proposition.

Proposition 1. Let M be a C^∞ Riemannian manifold and x_0 some fixed point in M. Let s denote the distance to x_0 as a function on the ball of some radius r less than the injectivity radius of the exponential map at x_0. Then the covariant derivative $\nabla^2 s^2$ of the gradient ∇s^2 is a C^∞ section on the ball which values are symmetric operators on the tangent spaces of M. Moreover, ∇s is an eigenvector of $\nabla^2 s^2$ with eigenvalue 2 and if X is a unit tangent vector orthogonal to ∇s, then we have the estimate:

$$2bs \cot(bs) \leq \langle \nabla^2 s^2 X, X \rangle \leq 2as \cot(as), \qquad 0 \leq s \leq r,$$

where a and b are real or pure imaginary numbers, such that:

$$a^2 \leq K \leq b^2$$

holds for all sectional curvatures K of tangent planes at points in distance $s \leq r$ from x_0. In particular, we have for the Laplacian of s^2:

$$2(1 + (n - 1)bs \cot bs) \leq \Delta s^2 \leq 2(1 + (n - 1)as \cot as)$$

if $s \leq r$. Moreover, the volume of the sphere of radius s, denoted by $\omega(s)$, satisfies:

$$\left(\frac{1}{b} \sin bs\right)^{n-1} \omega_n \leq \omega(s) \leq \left(\frac{1}{a} \sin as\right)^{n-1} \omega_n$$

where ω_n denotes the volume of the unit n - 1 sphere in the euclidean n space.

The injectivity radius at a point is the maximal distance, such that the exponential map is well defined and injective on the open ball of that radius in the tangent space. We see from the estimate above that $\nabla^2 s^2$ is a strictly positive operator at all points in the ball, if b^2 is negative or $br < \pi/2$, in case of positive curvature. This corresponds to a known estimate for the so-called radius of convexity, which takes this upper bounds for r, if the exponential map is well defined for such a large radius. Recall that we do not assume the manifold M to be complete.

The volume estimate can also be obtained from the formulas in [5], but we give a different proof here. With r and $s \leq r$ measuring the distance, we have:

$$\omega(r) = \int_{\partial B_r} 1 = \frac{1}{2r} \int_{\partial B_r} \langle \nabla s^2, \nabla s \rangle = \frac{1}{2r} \int_{B_r} \Delta s^2$$

with B_r and ∂B_r denoting the ball and its boundary. If we multiply by r and differentiate both sides, then we get half the integral of Δs^2 over the r-sphere, which is estimated by the inequality in Prop. 1 to give:

$$\omega(n - 1)br \cot br \leq \cot br \leq r\omega' \leq \omega(n - 1)ar \cot ar$$

Each inequality is then integrated over some interval from $s > 0$ and to r. The result is:

$$\left(\frac{\sin bs}{\sin br}\right)^{1-n} \omega(s) \leq \omega(r) \leq \left(\frac{\sin as}{\sin ar}\right)^{1-n} \omega(s)$$

and the limit as $s \to 0$ is easily obtained, proving our claim.

1.2 Hölder Estimates

We shall assume there is given a function R on the Riemannian manifold M with the following properties: Any ball $B_r(x)$, centered at x in M and of radius $r < R(x)$, is strongly convex, i.e., there is a unique minimal geodesic connecting any two points in $B_r(x)$ and that geodesic lies entirely in the ball. Furthermore, $R(x) > 0$ for any x in M and:

$$\left|1 - \frac{R(y)}{R(x)}\right| \leq \frac{d(x, y)}{R(x)}$$

for all y in distance less than R(x) from x. The convexity radius [6] is an example of such a function and it is either finite at all points or always infinite. However, we expect it might be convenient to choose some lower bounds for the convexity radius for R.

We shall denote by W_r the set of all (x, y) in $M \times M$, such that:

$$d(x, y)^2 < r^2 < \frac{1}{2}R(x)R(y), \quad \frac{1}{2} < \frac{R(x)}{R(y)} < 2$$

W_r is then an open neighborhood of a given compact subset of the diagonal, if r is sufficiently small. For (x_1, x_2) in W_r, put:

$$r_i = rR_i(r + R_i)^{-1}, \quad R_i = R(x_i), \quad B_i = B_{r_i}(x_i)$$

for i = 1, 2. Then r_i is less than both r and R_i and

$$r_1 + r_2 - r \geq \frac{r_1 r_2}{2r} > 0$$

The intersection of the balls B_1 and B_2 is then nonempty and contains the ball

$$B_0 = B_{r_0}(x_0), \quad r_0 = \frac{1}{2}(r_1 + r_2 - r)$$

with center x_0 on the minimal geodesic segment midway between x_1 and x_2.

Lemma 1. Given x_1 in M and $r < R_1 = R(x_1)$, then (x_1, x) belongs to W_{r_1} for all x in the ball B_1 constructed above.

Proof. We have:

$$R(x) \geq R_1 - d(x, x_1) > R_1 - r_1 = R_1^2(r + R_1)^{-1}$$

Thus it suffices to show:

$$r^2 R_1^2 (r + R_1)^{-2} \leq \frac{1}{2} R_1^3 (r + R_1)^{-1}$$

in order to obtain the first condition:

$$r_1^2 < \frac{R(x)R_1}{2}$$

But that inequality is of second order in r:

$$r^2 - \frac{1}{2}R_1 r - \frac{1}{2}R_1^2 < 0$$

and the roots are $-R_1/2$ and R_1, with r in between, so it holds. The second condition is a consequence of:

$$\left|1 - \frac{R(x)}{R_1}\right| \leq \frac{d(x, x_1)}{R_1} < \left(\frac{R(x)}{2R_1}\right)^{\frac{1}{2}}$$

which implies:

$$\frac{1}{2} < \frac{R(x)}{R_1} < 2$$

this being precisely the solution interval.

Now let f: M → N be a map into a second Riemannian manifold N. We define a Hölder constant of f for an exponent μ, $0 < \mu \leq 1$, and depending on upper bounds for the distance $r > 0$, by:

$$L_\mu(f, r) = \sup\{d(f(x), f(y))d(x, y)^{-\mu} : (x, y) \in W_r, \ x \neq y\}$$

The map f is Hölder continuous if $L_\mu(f, r)$ is finite for some $r > 0$, since W_r is an open neighborhood of the diagonal. We aim to estimate the Hölder constant in terms of the local energies:

$$w(x_0, f, r) = \int_B \rho^2, \qquad B = B_r(x_0), \qquad \rho(x) = d(f(x_0), f(x))$$

or rather in terms of some upper bounds independent of the base point x_0. However, we assume that the estimate is obtained with the aid of a local Hölder constant:

$$L_\mu(x_0, f, r) = \sup\{d(f(x_0), f(x))\, d(x_0, x)^{-\mu}: 0 < d(x_0, x) < r\}$$

Our method is described more precisely in the following:

Proposition 2. Let M be a connected Riemannian manifold with a given convexity function as above. Let N be a Riemannian manifold and A a set of maps $f: M \to N$. Suppose $0 \le \mu \le 1$ and $\psi(r, L)$, $\phi(t)$ are increasing positive functions of each real variable such that:

$$w(x, f, s) \le s^{2\mu+n}\psi(r, L), \text{ on } 0 < s < r, \text{ with } L = L_\mu(x, f, r)$$

for all f in A, x in M, and $r > 0$ satisfying:

$$r < R(x), \quad r < \phi\left(\frac{1}{L}\right)$$

Then the inequality

$$L_\mu(f, r)^2 \le n\, 4^{n+1} \frac{\psi(r, L_\mu(f, r))}{\omega_n \min\{1, \Omega_n(br)\}}$$

holds for any $r > 0$ and f in A that satisfy:

$$r < \phi\left(\frac{1}{L_\mu(f, r)}\right) \quad \text{and} \quad \Omega_n(c) = n\int_0^1 \left(\frac{1}{c}\sin ct\right)^{n-1} dt$$

Proof. Suppose (x_1, x_2) is in W_r and that r, f satisfy the condition in the conclusion. We prepare our estimation of the Hölder constant by integration of both sides of the triangle inequality:

$$d(f(x_1), f(x_2)) \le d(f(x_1), f(x)) + d(f(x), f(x_2))$$

as functions of x over the ball B_0 that we constructed before together with balls B_1 and B_2 centered at x_1 and x_2 and containing B_0. With ρ_1 and ρ_2 denoting the distances on the right hand side, we get by means of Hölder inequality:

$$d(f(x_1), f(x_2)) \le \mathrm{Vol}(B_0)^{-1}\left(\int_{B_0}\rho_1 + \int_{B_0}\rho_2\right)$$

$$\le \mathrm{Vol}(B_0)^{-\frac{1}{2}}\left(w(x_1, f, r_1)^{\frac{1}{2}} + w(x_2, f, r_2)^{\frac{1}{2}}\right)$$

This is in fact an old trick to obtain a uniformity estimate, but we have to check on the conditions for the estimate on w in the hypothesis in our proposition. First we have $r < R(x_i)$ for both $i = 1, 2$ by definition of W_r. Next we have with L_i denoting the local Hölder constant at x_i:

$$L_i \le L_\mu(f, r_i) \le L_\mu(f, r)$$

since (x, x_i) belongs to W_r for all x in B_i by Lemma 1 and since the radius r_i of B_i is less than r by definition of the balls B_i. Then:

$$r < \phi\left(\frac{1}{L_\mu(f, r)}\right) \le \phi\left(\frac{1}{L_i}\right)$$

first by assumption on r and then since ϕ is increasing. Then we have established the conditions and get:

$$d(f(x_1), f(x_2)) \le \left(\frac{\psi(r, L)}{Vol(B_0)}\right)^{\frac{1}{2}} (r_1^\sigma + r_2^\sigma)$$

with $\sigma = \mu + n/2$ and $L = L_\mu(f, r)$. We can estimate the volume of B_0 by means of Prop. 1 from which we get the meaning of the constants b and ω_n:

$$Vol(B_0) \ge \int_0^{r_0} \omega_n \left(\frac{1}{b} \sin bs\right)^{n-1} \ge \frac{1}{n} \omega_n r_0^n \min\{1, \Omega_n(br)\}$$

We estimate the r_i separately less than r in the factors r_i^μ and then we use:

$$\frac{r_1}{r_0} = R_1(r + R_2)(R_1 R_2 - r^2)^{-1} \le 4$$

since R_1 and R_2 are restricted to the domain:

$$R_1 R_2 > 2r^2, \quad \frac{1}{2} < \frac{R_1}{R_2} < 2$$

We have the same estimate in case of i = 2 and then we have shown that

$$(d(f(x_1), f(x_2))r^{-\mu})^2$$

is bounded by the bounds we claim we have for the Hölder constant. But we can replace r in the whole discussion above by any s in $d(x_1, x_2) < s \le r$, since (x_1, x_2) is then also in W_s. Then we can replace r by $d(x_1, x_2)$ and take the supremum over all (x_1, x_2) in W_r. This proves the proposition.

1.3 Local Energy Inequalities

From now on we shall fix a base point x_0 in M and a C^∞ map f: M → N. We use mainly the two independent variables r and $s \le r$ to measure distances to x_0 in M and the variable ρ to measure distances to $f(x_0)$ in N. We shall discuss properties of the following functions:

$$u(r) = \int_B \|df\|^2, \quad v(r) = \int_B \|df\, \nabla s\|^2$$

$$w(r) = \int_B \rho^2 \circ f, \quad H_p(r) = \left(\int_B \|\Delta f\|^p\right)^{2/p}, \qquad 2 \le p < \infty$$

where the integration is over the ball $B = B_r$ of radius r and center x_0. ∇s is the gradient of s as a function on B_r. Our main concern is to estimate w in terms of u and H_p, but the simplest estimate of w turns out to be in terms of the local energy v, which then must be compared with u. The derivatives of the functions u, v, w are given by the corresponding integrals over the boundary ∂B_r of the ball and we shall compute their second derivatives. We shall denote by div X the divergence of a tangent vector field X on M, which is defined by contracting the covariant derivative ∇X with some orthonormal bases. Then we have Green's formula:

$$\int_{B_r} \operatorname{div} X = \int_{\partial B_r} \langle X, \nabla s\rangle - \frac{d}{dr}\int_{B_r} \langle X, \nabla s\rangle$$

This is then our tool to compute derivatives of second order of local energies, since the first expression is easily differentiated to provide the second derivative of the local energy with density h, if we put $X = h\nabla s$.

We start by observing that

$$\operatorname{div}(\rho^2\circ f\,\nabla s^2) = \langle \nabla\rho^2\circ f,\ df\,\nabla s^2\rangle + \rho^2\circ f\,\Delta s^2$$
$$\leq 4s\,\rho\circ f\,\|df\,\nabla s\| + 2(1 + (n-1)\,as\cot as)\rho^2\circ f$$

where we have estimated Δs^2 by means of Prop. 1. Then we get:

$$(rw')' = \frac{1}{2}\frac{d}{dr}\Big(\int_{\partial B_r} \rho^2\circ f\,\langle \nabla r^2,\ \nabla r\rangle\Big)$$
$$= \frac{1}{2}\int_{\partial B_r} \operatorname{div}(\rho^2\circ f\,\nabla s^2)$$
$$\leq 2r(w'v')^{\frac{1}{2}} + (1 + (n-1)\,ar\cot ar)w'$$

This gives us the differential inequality:

$$\frac{d}{dr}(\theta w')^{\frac{1}{2}} \leq (\theta v')^{\frac{1}{2}}, \qquad \theta(r) = \Big(\frac{1}{a}\sin ar\Big)^{1-n}$$

which we can integrate immediately to obtain our first basic comparison inequality:

$$\theta(r)w'(r) \leq \Big(\int_0^r (\theta(s)v'(s))^{\frac{1}{2}}\,ds\Big)^2$$

Of course, w' tends to zero asymptotically like s^{n+1} as s tends to zero, since f is assumed C^∞ and v' like s^{n-1}, but we do not know how. The trivial comparison $v' \leq u'$ shall be improved, but let us first compare u with w by means of:

$$\operatorname{div}(d^*f\,\nabla\rho^2\circ f) = \langle(\nabla^2\rho^2\circ f)\,df,\ df\rangle + \langle\nabla\rho^2\circ f,\ \Delta f\rangle$$

Here d^*f denotes the adjoint linear operator to the differential df, so we are pulling back the gradient of ρ^2 to a vector field on M and taking its divergence. We shall estimate the first term on the right carefully. For that purpose we denote by $\partial_i f$ the partial derivatives of f in the directions given by some orthonormal basis at the relevant point in M. Then we introduce the vectors:

$$Y_i = \partial_i f - \langle X,\ \partial_i f\rangle X, \qquad X = (\nabla\rho)\circ f$$

which span the tangent space of the image point in N, such that the Y_i are orthogonal to X. Then by Prop. 1:

$$\langle\nabla^2\rho^2\ X,\ Y_i\rangle = 0$$
$$\langle\nabla^2\rho^2\ X,\ X\rangle = 2$$
$$\langle\nabla^2\rho^2\ Y_i,\ Y_i\rangle \geq 2\lambda\rho\cot(\lambda\rho)\,\|Y_i\|^2$$
$$= 2\lambda\rho\cot(\lambda\rho)(\|\partial_i f\|^2 - \langle X,\ \partial_i f\rangle^2)$$

Then we get:

$$\langle(\nabla^2\rho^2\circ f)\,df,\ df\rangle = \Sigma_i\langle(\nabla^2\rho^2\circ f)\,\partial_i f,\ \partial_i f\rangle$$
$$= \Sigma_i(\langle\nabla^2\rho^2 Y_i,\ Y_i\rangle + \langle X,\ \partial_i f\rangle^2\langle\nabla^2\rho^2 X,\ X\rangle)$$
$$\geq 2\lambda\rho\cot(\lambda\rho)\|df\|^2 + 2(1 - \lambda\rho\cot\lambda\rho)\|d^*f\,\nabla\rho\circ f\|^2$$

Now f is mapping the ball B_r into some ball of radius ρ_r, let's say, which tends to zero with r. We shall denote by ρ_s the supremum of $\rho\circ f$ on the ball of radius s and we choose

$\lambda \geq 0$, such that λ^2 is an upper bound for all sectional curvatures of N on the image by f of the ball B_r of radius $r \geq s$. So λ replaces b in Prop. 1 then applied to N at $f(x_0)$ in place of M at x_0. This means that we must request ρ_r within the normal range of the exponential map of N at $f(x_0)$, or less than the injectivity radius at $f(x_0)$, as is done in Prop. 1. Thus we must be able to control the size ρ_r of the image. This can be done either by an a priori hypothesis or by use of:

$$\rho_s \leq s^{\mu} L_{\mu}(x_0, f, r), \quad \text{if } 0 \leq s \leq r$$

If we take $\mu > 0$, then we are able to estimate ρ_s uniformly as a function on $0 \leq s \leq r$ vanishing at $s = 0$ and we are prepared by Prop. 2 to let our estimates depend implicitly on the Hölder constant L_μ. However, we shall leave this choice open and use:

$$\rho_s = L_0(x_0, f, s)$$

to estimate $\rho \circ f$. The function x cot x is decreasing on $0 \leq x < \pi$, so that we get at a point x in distance s from x_0:

$$\operatorname{div}(d^*f \nabla\rho^2 \circ f) \geq 2(1 - \Lambda_r(s)) \|df\|^2 + 2\Lambda_r(s) \|d^*f \nabla\rho \circ f\|^2 - 2\rho \circ f \|\Delta f\|$$

for $s \leq r$ and provided $\lambda\rho_r < \pi$, where we have introduced for abbreviation:

$$\Lambda_r(s) = 1 - x \cot x, \quad x = \lambda\rho_s$$

We shall proceed by integration and we are then led to a new local energy:

$$k(r) = \int_{B_r} \|d^*f \nabla\rho \circ f\|^2$$

Then with Λ to be chosen as some function of s with values in the interval $0 \leq \Lambda \leq 1$, we continue by estimating:

$$\begin{aligned}
\int_0^r ((1 - \Lambda_r(s))u' + \Lambda_r(s)k')\, ds &\leq \frac{1}{2} \int_{\partial B_r} \langle d^*f \nabla\rho^2 \circ f, \nabla s \rangle + \int_0^r \int_{\partial B_r} \rho \circ f \|\Delta f\|\, ds \\
&\leq \int_{\partial B_r} (\Lambda\rho \circ f \|d^*f \nabla\rho \;\; f\| + (1 - \Lambda)\rho \circ f \|df \nabla s\|) \\
&\quad + \int_0^r (w'H_2')^{\frac{1}{2}}\, ds \\
&\leq \Lambda(w'k')^{\frac{1}{2}} + (1 - \Lambda)(w'v')^{\frac{1}{2}} + \int_0^r (w'H_2')^{\frac{1}{2}}\, ds
\end{aligned}$$

We have $0 \leq k' \leq u'$, so that we can get rid of the k terms by choosing $\Lambda = 0$. However, we intend to make a better choice reflecting the weights the corresponding terms carry.

We have mentioned that we need a precise comparison of u and v and that is offered by utilization of the important identity:

$$\operatorname{div}(d^*f\, df - \frac{1}{2}\|df\|^2 I) = d^*f\, \Delta f$$

Here I denotes the identity operator on the tangent spaces of M and div the divergence of operators of that kind, by means of covariant derivation and contraction. In fact:

$$\operatorname{div}(d^*f\, df) = d^*f\, \Delta f + \Sigma_i \nabla_i d^*f\, \partial_i f$$

$$\langle \nabla \|df\|^2, X \rangle = 2\Sigma_i \langle \nabla df(X, e_i), df e_i \rangle$$

$$= 2\Sigma_i \langle X, \nabla d^*f(e_i) df e_i \rangle$$

since ∇df is a symmetric bilinear map [2]. Of course, we have been using some orthonormal frame $\{e_i\}$. We have:

$$2(rv')' - (ru')' = \int_{\partial B_r} \operatorname{div}(\langle d^*f df, \nabla s^2 \rangle - \|df\|^2 \nabla s^2/2)$$

$$= \int_{\partial B_r} (\langle \operatorname{Div}(d^*f df - \tfrac{1}{2}\|df\|^2 I), \nabla s^2 \rangle + \langle d^*f df - \tfrac{1}{2}\|df\|^2 I, \nabla^2 s^2 \rangle)$$

We use our identity on the first term and estimate the second by means of Prop. 1 as follows:

$$\langle d^*f df, \nabla^2 s^2 \rangle \leq 2as \cot(as)(\|df\|^2 - \|df \nabla s\|^2) + 2\|df \nabla s\|^2$$

$$\langle I, \nabla^2 s^2 \rangle = \Delta s^2 \geq 2(1 + (n-1) bs \cot bs)$$

with b and a depending on bounds for the sectional curvatures of M on the ball B_r. Then we get our third basic energy inequality:

$$2rv'' - ru'' \leq (2ar \cot ar - (n-1) br \cot br)u' - 2ar \cot ar\, v' + 2r(v'H_2')^{\frac{1}{2}}$$

We shall collect our results so far in a proposition, for the sake of precision and later reference. There we have rewritten the third inequality in a rather obvious way after use of:

$$(v'H_2')^{\frac{1}{2}} \leq (\xi'/\xi)v' + (\xi/4\xi')H_2'$$

Proposition 3. Let M be a C^∞ Riemannian manifold of dimension $n \geq 2$ and let $f: M \to N$ be a C^∞ map into a second C^∞ Riemannian manifold. Let $r > 0$ be chosen so small that estimates as in Prop. 1 hold at a point x_0 in M and also at the point $f(x_0)$ in N. Then the following inequalities hold for the local energies we have associated with f and x_0:

(1) With a chosen as in Prop. 1 and with the abbreviation

$$\theta(s) = \left(\frac{1}{a} \sin as\right)^{1-n}$$

we have:

$$\theta(s)w'(s) \leq \int_0^s (\theta(t)v'(t))^{\frac{1}{2}} dt \qquad \text{for } 0 \leq s \leq r$$

(2) With $\lambda \geq 0$ so that λ^2 bounds the sectional curvatures on the ball in N of radius $\rho_r = L_0(x_0, f, r)$ from above, assuming $\lambda\rho_r < \pi$, with:

$$\Lambda_r(s) = 1 - \lambda\rho_s \cot(\lambda\rho_s)$$

and with $\Lambda(s)$ some function taking values in the unit interval, we have on $0 \leq s \leq r$:

$$\int_0^s ((1 - \Lambda_s(t))u'(t) + \Lambda_s(t)k'(t))\, dt \leq \Lambda(w'k')^{\frac{1}{2}} + (1 - \Lambda)(w'v')^{\frac{1}{2}} + \int_0^s (w'H_2')^{\frac{1}{2}} dt$$

(3) With r, a, and b satisfying the conditions in Prop. 1 and with $\xi(s)$ some positive function with positive derivative and with the abbreviation

$$P(s) = \left(\frac{1}{a}\sin as\right)^2\left(\frac{1}{b}\sin bs\right)^{1-n}$$

we have on $0 < s < r$:

$$\frac{d}{ds}\left(\sin as\,\frac{v'}{\xi a}\right) \le \frac{\sin as}{2\xi Pa}\,\frac{d}{ds}(Pu') + \frac{\sin as}{4a\xi'}\,H_2'$$

Let us observe how far this third inequality alone brings us in estimating the growth of u. We have:

Proposition 4. Let r, a, and b satisfy the conditions in Prop. 1 and put:

$$\eta(s) = \frac{a}{2}(\sin as)^{-1}\left(\frac{1}{b}\sin bs\right)^{n-1}\exp\left(-\alpha(s/r)^{\nu}\right)$$

and α and ν chosen such that:

$$0 < \nu < 2, \qquad 0 < \alpha < \frac{1}{\nu}(n - 2 + \nu)$$

Assume further in the case $b^2 > 0$ that:

$$(n - 1)\,br\cot br > 1 - \nu + \alpha\nu$$

$$(2 - n)\,a^2r^2 \le \alpha\nu((n - 1)\,br\cot br + \nu - 1 - \alpha\nu)$$

Then we have for $0 < s < r$:

$$\frac{d}{ds}\frac{u}{\eta} \ge \frac{2}{\eta}v' - \frac{1}{P\eta^2}\int_0^s \frac{\sin as}{4a\xi'}H_2'\,ds, \qquad \xi(s) = e^{\alpha(s/r)^{\nu}}$$

Moreover, assuming further in the case $a^2 > 0$:

$$ar\cot ar \ge \nu - 1$$

then we have for $0 \le t \le s \le r$:

$$u(t) \le \eta(t)\left(\frac{u}{\eta}(s) + \int_0^s \frac{\tau\xi(\tau)}{2\alpha\nu\eta(\tau)}\left(\frac{\tau}{r}\right)^{-\nu} H_2(\tau)\,d\tau\right)$$

Proof. Let us first observe that η satisfies:

$$(P\eta')' = \frac{\sin as}{2a}\left(\left(\frac{1}{\xi}\right)'' + \frac{a^2}{\xi}(2 - n) + (n - 1)\,b\cot bs\left(\frac{1}{\xi}\right)'\right)$$

$$+ \frac{n - 1}{2a\xi}\sin as\left(a\cot as\; b\cot bs - \left(\frac{b}{\sin bs}\right)^2 + a^2\right)$$

$$\le \frac{\sin as}{2a\xi}\left((2 - n)a^2 - \alpha\nu\left((n - 1)\,bs\cot bs + \nu - 1 + \alpha\nu\left(\frac{s}{r}\right)^{\nu}\right)\frac{s^{\nu-2}}{r^{\nu}}\right) \le 0$$

Here we have estimated the first term just by using the definition of ξ and the conditions on the parameters in the final estimate. The second term is negative for $0 \le s \le r$ and all a and b. This is not obvious, but the trigonometric function in the bracket there has the derivative:

$$-b\cot bs\left(\frac{a}{\sin as}\right)^2 - a\cot as\left(\frac{b}{\sin bs}\right)^2 + 2b\cot bs\left(\frac{b}{\sin bs}\right)^2 \le 0$$

The verification of those inequalities is not too difficult, the first is a consequence of the obvious fact that

$$b^3\cot bs\,(\sin bs)^{-2}$$

is a decreasing function for all real and pure imaginary b. Now we get by use of (3) in Prop. 3:

$$P\eta^2 \frac{d}{ds}\frac{u}{\eta} = \int_0^s (\eta(Pu')' - u(P\eta')')\,ds$$

$$\geq \int_0^s \frac{2aP\xi\eta}{\sin as}\Big(\frac{d}{ds}\Big(\sin as\,\frac{v'}{a\xi}\Big) - \frac{\sin as}{4a\xi'}H_2'\Big)\,ds$$

$$= 2\eta Pv'\Big|_0^s - \int_0^s \frac{\sin as}{4a\xi}H_2'\,ds$$

since $2aP\xi\eta/\sin as = 1$. This proves our first assertion and the second follows by integration following the use of $v' \geq 0$. The additional assumption is only used to simplify the double integral, which we get rid of by partial integration and drop the derivative of $s^{1-\nu}\sin as$ in the process.

The result of our efforts is that u(s) is uniformly bounded by s in power n - 2 on a set of maps f with u(r) and $H_p(r)$, $p > n/2$, bounded. This just falls short of proving that such a set is equicontinuous, which is just as well since we have counterexamples to such a statement in this generality. In fact, if u(s) tends to zero with s in power $n - 2 + \epsilon$, for some $\epsilon > 0$, then we can resolve the integral in (1) Prop. 3, using only that v is less than u, and we could then apply Prop. 2. It is noteworthy that properties of the second manifold N are quite irrelevant in Prop. 4. Our hope is then placed on (2) in Prop. 3.

2. COMPARISON AND GROWTH ESTIMATES

Let's look at our first basic inequality in Prop. 3 and see how far it leads us. By Hölder inequality, we get for any positive function q(s):

$$\theta(r)w'(r) \leq \int_0^r \frac{ds}{q}\int_0^r q\theta v'\,ds$$

We shall have to estimate an integral involving v' on several occasions. Therefore, we state for reference:

<u>Lemma 2</u>. Let $\xi(s)$ be a positive strictly increasing C^∞ function on $0 \leq s < r$. Let q(s) be a positive function on $0 < s < r$, such that sq tends to zero as $s \to 0$. With θ and P as in Prop. 3, define:

$$Q(s) = \int_s^r \frac{a\theta\xi q}{\sin as}\,ds\,, \qquad Q_1(s) = \frac{\sin as}{4a\xi'}Q\,,$$

$$Q_2(s) = \frac{d}{ds}\Big(P\frac{d}{ds}\Big(\frac{\sin as}{2aP\xi}Q\Big)\Big)$$

Then with the same assumptions as in Prop. 3, we have:

$$\int_0^r q\theta v'\,ds \leq \frac{1}{2}(q\theta u)(r) + \int_0^r (Q_2 u + Q_1 H_2')\,ds$$

<u>Proof</u>. This follows immediately from Prop. 3 (3) and by application of partial integration, since:

$$\int_0^r q\theta v'\,ds = \int_0^r Q\left(\frac{\sin as}{a\xi}v'\right)'ds$$

by definition of Q. Q.E.D.

To continue our discussion, note that if q behaves like s in power $1 - \nu$, say, with $0 < \nu < 2$, then Q_2 behaves like s in power $1 - n - \nu$, so in our first attempt to estimate w' we are not able to get rid of the weak expression of u, involving the integral of $Q_2 u$. We shall in fact observe that we cannot choose the weight functions in such a way that this weak expression of u loses strength. Our next task, and the most difficult one, is therefore to obtain a weak comparison inequality where singularities can be compared. In fact, we shall see the strong one cancel.

2.1 Weak Estimates

Our intention now is to rewrite inequality (2) in Prop. 3 in a weak form with the aim of eliminating v and w by means of the other inequalities (1) and (3) and dropping k as well after some benefits, in order to get an estimate for a weak expression of u, modulo H_2. Then our plan is to go back to the estimate of w' we just started on. We shall proceed in this section discovering gradually the properties we need to have for our weight functions instead of stating the results beforehand. Hopefully, this form of exposition can benefit a reader who would like to see possibilities for improvements.

We may replace the s in Λ_s in inequality (2) by the larger r and then we multiply both sides with θq_0, where q_0 is some positive function left to be chosen, and integrate over the interval $0 < s < r$. Using the abbreviation:

$$Q_3(s) = \frac{1}{\theta}\int_s^r \theta q_0\,ds$$

we obtain:

$$\int_0^r \theta Q_3((1 - \Lambda_r)u' + \Lambda_r k')\,ds \le I_1 + I_2 + I_3$$

with:

$$I_1 = \int_0^r (1 - \Lambda)\theta q_0 (w'v')^{\frac{1}{2}}\,ds$$

$$I_3 = \int_0^r Q_3\,\theta(w'H_2')\,ds, \qquad I_2 = \int_0^r \Lambda\theta q_0 (w'k')^{\frac{1}{2}}\,ds$$

We shall start estimating the integrals I_j by the same method, applying inequality (1) or Prop. 3 to obtain, with e_j as any one of the three relevant energy densities:

$$\int_0^r \theta q(w'e_j')^{\frac{1}{2}}\,ds \le \int_0^r q(\theta e_j')^{\frac{1}{2}}\left(\int_0^s (\theta v')^{\frac{1}{2}}\,ds\right)ds$$

$$= \int_0^r \left(\int_s^r q(\theta e_j')^{\frac{1}{2}}\,ds\right)(\theta v')^{\frac{1}{2}}\,ds$$

$$\le \int_0^r \left(\int_s^r \left(\frac{\theta q^2}{q_3}\right)ds \int_s^r q_3 e_j'\,ds\right)^{\frac{1}{2}}(\theta v')^{\frac{1}{2}}\,ds$$

$$\le \frac{1}{2}\int_0^r \theta q_2 v'\,ds + \frac{1}{2}\int_0^r \theta q_4 e_j'\,ds$$

with $\theta q_4 = q_3 \int_0^s \frac{1}{q_2}\left(\int_t^r \left(\frac{\theta q^2}{q_3}\right) ds\right) dt$. The q in these estimates is to be replaced by the corresponding weight function in one of the integrals I_j and we shall denote it by q_1 in that context. We shall replace the weight function q_3 by a new variable depending on s:

$$y = \int_s^r \left(\frac{\theta q_1^2}{q_3}\right) ds$$

We intend to replace the integral equation defining q_4 by a more transparent differential equation. Now we have:

$$\frac{y}{q_2} = \frac{d}{ds}\left(\frac{\theta q_4}{q_3}\right) = -\frac{d}{ds}\left(\left(\frac{q_4}{q_1^2}\right)\frac{dy}{ds}\right)$$

and q_4/q_3 must vanish at $s = 0$. Then we introduce a new variable z by:

$$dz = -\left(\frac{q_1^2}{q_4}\right) ds$$

Then z is a strictly decreasing function of s and we intend to let $z \to \infty$ as $s \to 0$, in accordance with the requested singular behavior of the weight functions at $s = 0$. Then we can just as well consider y as a function of z and in these variables our equation becomes:

$$\frac{d^2y}{dz^2} = -\frac{y}{q^2}, \quad \text{and} \quad \frac{dy}{dz} \to 0 \text{ as } z \to \infty$$

with $q = q_1(q_2/q_4)^{\frac{1}{2}}$. The boundary condition corresponds to q_4/q_3 being zero at $s = 0$. Furthermore we must request $y = 0$ for $z = z_0$, if z_0 is to correspond to the value r of s and y must then be a strictly increasing function of z on $z_0 \leq z < \infty$, such that y turns out to be a strictly decreasing function of s. Now we may choose q and y as some functions of z satisfying those conditions and then choose z as some function of s, decreasing from ∞ at $s = 0$ to the zero point z_0 of y at $s = r$. Then the weight functions are uniquely determined and have the requested properties, if we suppose q_1 is given. In fact:

$$q_4 = -q_1^2\left(\frac{dz}{ds}\right)^{-1}, \quad q_2 = q_4\left(\frac{q}{q_1}\right)^2$$

and q_3 is determined by the derivative of y with respect to s, but does not interest us further. We shall later on in the estimate of I_1 choose $q_4 = q_2$, since $e_1 = v$, and then $q_1 = q$, which determines $(1 - \Lambda)q_0$. Then we may consider the other q_1's in the definition of I_2 and I_3 as determined except for a still free choice of Λ.

Our next step in the estimation of the I_j's is to apply Lemma 2, so that:

$$\int_0^r \theta q_2 v' \, ds \leq \frac{1}{2}(\theta q_2 u)(r) + \int_0^r (Q_2 u + Q_1 H_2')\, ds$$

of course with q_2 replacing q elsewhere. Here we shall also introduce new variables. First of all we shall introduce a new function U by:

$$U = \frac{Q}{\xi}\Phi, \quad \text{with} \quad \Phi = \left(\frac{1}{b}\sin bs\right)^{(n-1)/2}$$

Then U is mapped to Q_2 by a known second order operator according to the definition of Q_2:

$$Q_2 = \frac{\sin as}{2a\Phi}\left(U'' - \left(\frac{G_0}{s^2}\right)U\right)$$

with

$$G_0(s) = -a^2s^2 + \left(\frac{n-1}{2}bs\cot bs\right)^2 + \frac{n-1}{2}\left(\frac{bs}{\sin bs}\right)^2 - (n-1)as\cot as\, bs\cot bs$$

The function U must have a zero at s = r just as Q and is then determined up to a multiplication by constant by Q_2, if we like. Computing the derivative of Q from its defining equation and replacing ξ there by its equal from the definition of U:

$$\xi = \frac{\Phi Q}{U}$$

we get the equation:

$$-\frac{d}{ds}\log Q = \frac{a\theta\Phi}{\sin as}\left(\frac{q_2}{U}\right), \qquad Q(r) = 0$$

which may serve to define Q, after q_2 and U have been determined. The main condition in Lemma 2 we still have to worry about is that ξ should have positive derivative, which we must bound away from zero in order to estimate Q_1. Now:

$$\frac{d}{ds}\log \xi = h_2 + \frac{d}{ds}\log Q$$

with the abbreviation:

$$h_2 = \frac{d}{ds}\log\left(\frac{\Phi}{U}\right) = h + \frac{n-1}{2}b\cot bs$$

with

$$h = -\frac{U'}{U}$$

Then we may write:

$$\frac{d}{ds}\log\xi = h_2q_2\left(\frac{1}{q_2} - \frac{\theta a\Phi}{\sin as}(Uh_2)^{-1}\right)$$

which must be positive. Then h_2 must be positive, which is no problem with U singular at s = 0, and $1/q_2$ must bound an expression entirely depending on U. So let's see how we can get q_2 small. We may study q_2 by means of:

$$q_2 = -q^2\left(\frac{dz}{ds}\right)^{-1} = \left(\frac{d}{ds}\frac{1}{q}\right)^{-1}\frac{dq}{dz}$$

Recall that we can choose q as a function of z independently of its dependence on s. Thus it is natural to choose first dq/dz as small as possible and let then the choice of q as a function of s undergo the restriction:

$$\frac{d}{ds}\frac{1}{q} > \frac{a\theta\Phi}{\sin as}(Uh_2)^{-1}\frac{dq}{dz}$$

Most important, and critical in our estimate of I_1 as we shall see shortly, is to get dq/dz small at the point $z = \infty$, corresponding to s = 0.

The Story of dq/dz at ∞. We have a severe restriction on q by the differential equation for y and other requirements on y previously explained. By the change $z = 1/\zeta$ of variable the differential equation is transformed to:

$$\zeta^2\frac{d^2y}{d\zeta^2} + 2\zeta\frac{dy}{d\zeta} + \left(\frac{z}{q}\right)^2 y = 0$$

It is in accordance with our request for small dq/dz to assume $\zeta = 0$ to be a regular singularity for the equation, i.e., z/q analytic at $\zeta = 0$. Then we have:

$$\lim\frac{z}{q} = \left(\lim\frac{dq}{dz}\right)^{-1} \qquad \text{as} \qquad z \to \infty$$

and we denote this number by m. The roots of the characteristic equation corresponding to the regular singularity are:

$$\frac{1}{2}(-1 \pm (1 - 4m^2)^{\frac{1}{2}})$$

This means that the solution oscillates if $m > 1/2$ and y has then zeros arbitrarily close to $z = \infty$, which contradicts our request that y should be increasing on some interval from a zero point z_0 to ∞. Thus we must have $m \leq 1/2$, which gives us 2 as a lower limit for dq/dz. If we choose q equal to 2z, then y is to satisfy an Euler equation with $\sqrt{z}$ and $\sqrt{z} \log z$ as two independent solutions. If we choose the first one for y, then we would have to take $z_0 = 0$ and q would be zero at $s = r$ as well as y. We want to avoid this and prefer:

$$y = \sqrt{z} \log\left(\frac{z}{z_0}\right) \qquad \text{for some } z_0 > 0$$

It may seem tempting to get dq/dz below 2 for finite values of z by a different choice of q and y. However, it is doubtful whether the gain is worth the effort, since we are still free to choose q as a function of s. Anyhow, we shall stick to the definition:

$$q = 2z$$

We find it convenient to interpret U as a solution of the second order equation:

$$s^2 U'' = G_0(s)U + G(s)U \qquad \text{on } 0 < s \leq r, \quad U(r) = 0$$

In fact, U is determined up to multiplication by some constant, after some choice of G. Of course, we want to choose G as small as possible.

Since we want $1/q$ to vanish at $s = 0$, we can integrate our previous condition on q to get:

$$\frac{1}{q} > \int_0^s \frac{a\theta\Phi}{\sin as}\left(\frac{2}{Uh_2}\right) ds$$

In particular, the implied integrability condition shows that U/s must tend to ∞ stronger than s in power $1 - n/2$ as $s \to 0$. This means that G must take a positive value in $s = 0$, since with

$$\sigma = \left(\left(\frac{n}{2} - 1\right)^2 + G(0)\right)^{\frac{1}{2}} - \frac{1}{2}$$

$-\sigma$ is the smaller root of the characteristic equation of the differential equation for U. Then U is going to be asymptotic to $s^{-\sigma}$ as $s \to 0$ and stay positive until it reaches zero at $s = r$. That does not mean that q has to be zero at $s = r$, since h_2 has a simple pole there if U has a simple zero.

Looking back at the logarithmic derivative of ξ, we do not think there is any good choice for ξ, and we shall interpret it as causing a deflection from the best possible course and we shall want to keep that deflection small. We put

$$h_3 = h_2 - \eta \qquad \text{with} \qquad \eta = \frac{d}{ds} \log \xi$$

Then q_2 is defined in terms of U and h_3 (or η) by:

$$q_2 = \frac{\sin as}{a\theta\Phi} Uh_3$$

From q_2 we can compute q and we shall also represent it by another function V, such that:

$$\frac{1}{q} = \int_0^s \left(\frac{2}{q_2}\right) ds = \left(2 \frac{\sin as}{aq_2}\right) V$$

Writing up the integral for V, we have:

$$V(s) = \int_0^s \left(\frac{\sin at}{\sin as}\right)^{1-n} \left(\frac{\sin bt}{\sin bs}\right)^{\frac{n-1}{2}} \left(\frac{U(s)h_3(s)}{U(t)\,h_3(t)}\right) \frac{adt}{\sin at}$$

We see easily, since U behaves like $s^{-\sigma}$, that:

$$V(0) = \frac{1}{\nu} \qquad \text{with} \qquad \nu = \sigma - \frac{n-3}{2}$$

But this looks serious. We have:

$$Q_2 = g\theta q \qquad \text{with} \qquad g = VG\frac{\sin as}{as^2 h_3}$$

$$h_3(s)s \to \sigma + \frac{n-1}{2} = \frac{G(0)}{\nu} \qquad \text{as} \quad s \to 0$$

and this means $g = 1$ in $s = 0$. This is the largest value of g we can accept, if our estimation is to be successful, since q must be equal to q_0 at $s = 0$ in the I_1 estimate as we shall see later and the integral of $q_0 u$ is just what we have started with. However, we might stay in business if we manage to keep $1 - g$ positive for $s > 0$ and approach zero as s in some power μ less than ν, so that $(1 - g)q$ is still singular at $s = 0$. In fact it is evident from the equations above that q_2/s and q both tend to infinity as $s^{-\nu}$ as $s \to 0$. The only restriction on ν (see Lemma 2) is: $0 < \nu < 2$.

Let us then take a closer look at the function V. It can also be defined by:

$$\frac{\sin as}{a}\frac{dV}{ds} = 1 - h_1 V, \qquad V(0) = \frac{1}{\nu}$$

with

$$h_1 = \frac{\sin as}{a}\left(h - \frac{h_3'}{h_3} - (n-1)\,a\cot as + \frac{n-1}{2}\,b\cot bs\right)$$

$$h = -\frac{U'}{U}$$

From the differential equation for U we get immediately:

$$h' = h^2 - (G + G_0)s^{-2}$$

Using this formula we can compute the derivative of h and get to the remarkable result:

$$h_3' = \left(h - \frac{n-1}{2}b\cot bs + \eta\right)h_3 - \frac{G}{s^2} + \eta(\eta - (n-1)\,b\cot bs)$$

$$- \eta' + (n-1)\left(a\cot as\; b\cot bs - \left(\frac{b}{\sin bs}\right)^2\right) + a^2$$

That means:

$$h_1 = G\frac{\sin as}{as^2 h_3} - \frac{\Theta}{a}\sin as$$

with

$$\Theta = \eta + (n-1)(a\cot as - b\cot bs)$$
$$+ \frac{\eta(\eta - (n-1)\,b\cot bs - \eta' + (n-1)(a\cot as\; b\cot bs - (b/\sin bs)^2) + a^2}{h_3}$$

The remarkable thing is that the first term in h_1 multiplied by V gives us g and also that the influence of η is dominating in Θ for small values of s, if η/s is unbounded as s tends to zero. That is easily taken care of, we have only to be sure that $s\eta$ tends to zero with s as we did in computing the value of g at $s = 0$. Observe also that the curvature terms in Θ drop out if M is of constant curvature, i.e., if $b = a$. The dependence of Θ

on h_3 only helps in keeping Θ small, since sh_3 takes a large value at $s = 0$ and tends to infinity as $s \to r$ because of the zero of U in $s = r$. This means that we can interpret Θ as a small perturbation, requesting η small and r small, depending on the curvature of M. However, someone might like to discuss the influence of the curvature on Θ in order to keep r large, so we keep Θ as we go along.

We have now a good reason to rewrite the differential equation for V as an equation to determine g:

$$g = 1 - \frac{1}{a} \sin as\,(V' - \Theta V)$$

Then g is essentially determined by the derivative of V, except for the perturbation of Θ. We shall find its convenient to choose q first and then to obtain V from:

$$V = \frac{a}{\sin as}\left(-\frac{d}{ds}\log q\right)^{-1}$$

which follows from our first definition of V and the connection between q and q_2. We have only to choose ν as the order of singularity for q, so that V takes the value $1/\nu$ in $s = 0$. Then we get q_2 from the simple equation:

$$q_2 = qV\frac{2}{a}\sin as$$

So far ξ has not had any influence except on the perturbation Θ. Moreover its influence on U, which we can take from the following formula, is mild:

$$\frac{Q\Phi}{\xi} = U = \frac{\Phi}{\xi}\int_s^r 2\xi\theta qV\,ds$$

Actually, we do not have to know much about U, except possibly in order to estimate Q_1, which enters into the H_2 terms from the application of Lemma 2 and we can compute from:

$$Q_1 = \frac{\sin as}{4a\xi'}\int_s^r 2\xi\theta qV\,ds$$

We shall simplify the H_2 term by assuming $\sin as/a\xi'$ increasing, which is an acceptable condition on ξ, and neglect its derivative in performing a partial integration. That gives:

$$\int_0^r Q_1 H_2'\,ds \le \int_0^r \frac{\theta}{2\eta}\left(\left(\frac{1}{q}\right)'\right)^{-1} H_2\,ds$$

This is then our general framework in estimating the I_j's. Let's take a look at the expression we are estimating by their sum before we turn to the individual estimates. By partial integration, we get:

$$\int_0^r \theta Q_3((1 - \Lambda_r)u' + \Lambda_r k')\,ds \ge \int_0^r \theta q_0((1 - \Lambda_r)u + \Lambda_r k)\,ds$$

since Λ_r is increasing (see Prop. 3) and $k \le u$.

The Estimate for I_1. Here we have:

$$q_4 = q_2 \quad \text{and} \quad q_1 = q = (1 - \Lambda)q_0$$

the first equality by choice and the next as a consequence. We shall wait with our choice of q and ξ, but then V, q_2, g, and Q_1 are determined by the general framework above and

$$I_1 \le \frac{1}{2}(\theta q_2 u)(r) + \int_0^r \theta g q u\,ds + \int_0^r Q_1 H_2'\,ds$$

The Estimate for I_2. Here we have $q_1 = \Lambda q_0$ and q_0 has been fixed in the I_1 estimate, but we have not decided what to do about Λ. Our main concern is that the singularity of q does not exceed the singularity of what remains on the left after I_1 has been subtracted. If Λ must vanish at s = 0 we are having a drop in singularity due to g = 1 at s = 0, but we do not want the singularity to vanish. We are therefore led to define the q for the I_2 estimate by:

$$q_{12} = \gamma q^m, \qquad 0 < m < 1$$

with some positive constant γ and where q is the same as in the I_1 estimate. This implies for the V, g, q_2 functions we must use in the I_2 estimate, marked by the suffix 2:

$$V_2 = \frac{1}{m}V, \qquad V = V_1$$

$$g_2 = 1 - \frac{1}{ma}\sin as\,(V - \Theta_2 V)$$

$$q_{22} = \frac{2\gamma}{ma}\sin as\; Vq^m$$

Of course we are allowed to choose ξ differently here which would result in a different Θ. We get q_4 from the general formula:

$$q_4 = q_2\left(\frac{q_1}{q}\right)^2 \qquad \text{and now} \qquad q_1 = \frac{\Lambda q}{1 - \Lambda}$$

then we get:

$$q_{42} = \frac{2\sin as}{\gamma ma}\left(\frac{\Lambda}{1-\Lambda}\right)^2 Vq^{2-m}$$

Now let's look back at the estimate we had for I_2 together with the k-term from the left hand side and continue:

$$I_2 - \int_0^r \theta q_0 \Lambda_r k\, ds \le \int_0^r \left(\frac{1}{2}\theta q_2 v' + \frac{1}{2}\theta q_4 k' - \theta q_0 \Lambda_r k\right) ds$$

$$\le \int_0^r \left(-\frac{1}{2}(\theta q_{42})' - \theta q_0 \Lambda_r\right) k\, ds + \frac{1}{2}(\theta q_{42} k)(r)$$

$$+ \frac{1}{4}(\theta q_{22} u)(r) + \frac{1}{2}\int_0^r (Q_{22} u + Q_{12} H_2')\, ds$$

Let's look at the first term. We request:

$$\Lambda_r \le B\Lambda$$

for some positive function B. Then:

$$q_0 \Lambda_r \le \theta q \frac{B\Lambda}{1-\Lambda} = -\frac{1}{2}(\theta q_{42})'$$

if we define Λ through this equation. This is in fact a differential equation for Λ, which we can easily solve:

$$\frac{\Lambda}{1-\Lambda} = -\left(\frac{\sin as}{\gamma ma} V\theta q^{2-m}\right)^{-\frac{1}{2}} \int \left(\frac{\sin as}{\gamma ma} Vq^{2-m}\right)^{-\frac{1}{2}} \theta q B\, ds$$

where the constant of integration is to be chosen such that the right hand side is positive, but then Λ is uniquely defined with values in the unit interval. The object of this maneuver is to get the factor of k in the integral in our last estimate positive, so that we can estimate k away by $k \le u$, getting rid of k for good. Then we get:

$$I_2 - \int_0^r \theta q_0 \Lambda_r k \, ds \le \int_0^r \theta\Big(q_0(B\Lambda - \Lambda_r) + \frac{\gamma}{2} g_2 q^m\Big) u \, ds$$
$$+ \frac{\sin ar}{ma}\Big(\frac{\gamma}{2} + \frac{1}{\gamma}\Big(\frac{\Lambda}{1 - \Lambda}\Big)^2 q^{2-2m}\Big) V q^m \theta u \Big|_{s=r}$$
$$+ \int_0^r \theta\Big(\frac{2}{\gamma}\eta_2 (q^{-m})'\Big)^{-1} H_2 \, ds$$

We have used our general estimate for the H_2 term and must then respect the corresponding condition on the $\xi = \xi_2$ we want to use here, but η_2 denotes the logarithmic derivative of ξ_2.

The Estimate for I_3. Now we have $q_1 = Q_3$, but we choose V, q, and q_2 as in the I_2 estimate, except that we replace γ by another constant ϵ, which we intend to take so small that the terms coming from the application of Lemma 2 can be considered as small perturbations of the corresponding terms from the I_2 estimate. We had:

$$I_3 \le \frac{1}{2}\int_0^r \theta q_{23} v' \, ds + \frac{1}{2}\int_0^r \theta q_{43} H_2' \, ds$$

and the first term is estimated as in the case of I_2 with γ replaced by ϵ. Then we get a new H_2 term where:

$$q_{43} = \frac{2\epsilon}{ma} \sin as \, Vq^m\Big(\frac{Q_3}{\epsilon q^m}\Big)^2 = \frac{2}{\epsilon}(-(q^m)')^{-1}\Big(\frac{1}{\theta}\int_s^r \frac{\theta q}{1 - \Lambda} ds\Big)^2$$

Then we get:

$$I_3 \le \frac{\sin ar}{2ma} Vq^m \theta u\Big|_{s=r} + \int_0^r \frac{\epsilon}{2} g_2 q^m \theta u \, ds + \int_0^r \epsilon\theta(2\eta_2(q^{-m})')^{-1} H_2 \, ds$$
$$- \int_0^r \frac{2}{\epsilon}\frac{d}{ds}\Big(((q^m)'\theta)^{-1}\Big(\int_s^r \frac{\theta q}{1 - \Lambda} ds\Big)^2\Big) H_2 \, ds$$

The Final Estimate. Collecting terms, we get

$$\int_0^r X\theta u \, ds \le Y\theta u\Big|_{s=r} + \int_0^r Z H_2 \, ds$$

with

$$Z = \frac{\theta}{2}\Big(\eta\Big(\frac{1}{q}\Big)'\Big)^{-1} + (\gamma + \epsilon)(2\eta_2(q^{-m})')^{-1} - \frac{2}{\epsilon}\frac{d}{ds}\Big((\theta(q^m)')^{-1}\Big(\int_s^r \frac{\theta q}{1 - \Lambda} ds\Big)^2\Big)$$

$$Y = \Big(1 + \frac{\gamma + \epsilon}{2m} q^{m-1} + \frac{1}{m\gamma}\Big(\frac{\Lambda}{1 - \Lambda}\Big)^2 q^{1-m}\Big)\Big(\Big(\frac{1}{q}\Big)'\Big)^{-1}\Big|_{s=r}$$

$$X = q_0(1 - \Lambda_r - g(1 - \Lambda) - (B\Lambda - \Lambda_r)) - \frac{\gamma + \epsilon}{2} g_2 q^m$$
$$= (1 - B)\frac{\Lambda}{1 - \Lambda} q + \frac{\sin as}{a}\Big(qV' + \frac{\gamma + \epsilon}{2m}(q^m V)'\Big) - \Big(\Theta + \frac{\gamma + \epsilon}{2m} q^{m-1}\Theta_2\Big)\Big(\Big(\frac{1}{q}\Big)'\Big)^{-1}$$

The function Λ is defined by:

$$\frac{\Lambda}{1-\Lambda} = \frac{\gamma}{q}\left(-\frac{(q^m)'}{\theta}\right)^{\frac{1}{2}}\left(B_0 + \int_s^r (-\theta(q^m)')^{\frac{1}{2}} B\, ds\right)$$

where $B_0 \geq 0$ in some constant. The function Λ_r does not enter the estimate except through the condition:

$$\Lambda_r \leq B\Lambda$$

We shall keep B strictly positive on $s \geq 0$ and with values less than 1, such that the first term in X is positive and it behaves then like q^m as s tends to zero, but Λ is asymptotically proportional to q^{m-1} as $s \to 0$. This means that we can just as well use the first term in X to dominate the last one, instead of using the second term for that purpose. We have written the second term in X in a form that resembles the following form of the weight function in our reformulation of Lemma 2:

$$qg = \frac{\sin as}{a}(-qV)', \qquad V = \frac{a}{\sin as}\frac{-q}{q'}$$

except that the first term qV' is not a derivative, it is what is left of the cancellation of the left side against the u term from I_1 essentially, and the second is a derivative but with the wrong sign. We want to put this right, since we intend to apply our estimate following the use of Lemma 2. Thus we put:

$$qV' = (-yV)' \qquad \text{so} \qquad d \log V = \frac{-dy}{q+y}$$

Then of course we want y to dominate q^m and it is obviously preferable to define y as a function of q. Let us however first look for other properties we would like that function to have.

We would like Y to be small and then it seems desirable that q' tends to $-\infty$ as s tends to r with q positive at r. From the formula for Λ we see that Y has then the limit:

$$Y \to \frac{\gamma B_0}{\theta(r)} \qquad \text{if} \quad s \to r,\ q' \to -\infty,\ q(r) > 0$$

This looks quite nice although we certainly must have B_0 positive now, since else we would have $\Lambda = 1$ in $s = r$, which is good since we use Λ as upper bounds not only for Λ_r but also for the Θ terms in X. In fact it follows from our intention above that the second term in X is most likely negative for s close to r, so that the first term must be used to dominate the third one. We shall look into that later and also at Z, but we can see at a glance that $q' \to -\infty$ forces Z to zero at $s = r$, which is good.

Now, let us turn to the definition of q. We denote the value of q at $s = r$ by $q_r > 0$ and enforce a singularity at $s = r$ by defining:

$$V = \frac{1}{\nu}\left(1 - \left(\frac{q}{q_r}\right)^{-\beta}\right)^{\frac{1}{\beta}-\alpha-1}\left(1 + \frac{\alpha\beta}{1-\alpha\beta}\left(\frac{q}{q_r}\right)^{-\beta}\right)^{\alpha-1}$$

The parameters α and β are restricted by:

$$\alpha < \frac{1}{\beta} < 2\alpha$$

and V is so chosen that:

$$d(\log V) = \frac{-dy}{q+y}$$

with

$$y = \frac{2\alpha\beta - 1}{1 - \alpha\beta} q \left(\frac{q}{q_r}\right)^{-\beta} - \frac{\alpha\beta}{1 - \alpha\beta} q \left(\frac{q}{q_r}\right)^{-2\beta}$$

The restriction on the parameters implies that the coefficient of the first term is positive and it dominates q^m if $1 - \beta \geq m$, with γ and ϵ sufficiently small in case of equality. In order to compute q itself, we have:

$$d\left(\log \operatorname{tg}\left(\frac{as}{2}\right)\right) = \frac{a\, ds}{\sin as} = -V \frac{dq}{q}$$

so that we get by integration:

$$\log \left(\frac{\operatorname{tg}(as/2)}{\operatorname{tg}(ar/2)}\right)^{-\nu} = \int_{q_r}^{q} \nu V \frac{dq}{q}$$

$$= \frac{1}{\beta} \int_z^1 x^{-1}(1 - x)^{\frac{1}{\beta} - \alpha - 1} \left(1 + \frac{\alpha\beta}{1 - \alpha\beta} x\right)^{\alpha - 1} dx, \qquad z = \left(\frac{q}{q_r}\right)^{-\beta}$$

Then we have the following asymptotic formulas:

$$\frac{q}{q_r} \sim \left(\frac{\operatorname{tg}(as/2)}{\operatorname{tg}(ar/2)}\right)^{-\nu} \qquad \text{as } s \to 0$$

$$\left(\beta\left(\frac{q}{q_r} - 1\right)\right)^{\frac{1}{\beta} - \alpha} \sim (1 - \alpha\beta)^{\alpha} \frac{\nu a}{\sin ar}(r - s) \qquad \text{as } s \to r$$

We must request:

$$\left(\frac{1}{\beta} - \alpha\right)^{-1} < 1 \quad \text{or} \quad \frac{1}{\beta} > 1 + \alpha$$

since we want q' to tend to $-\infty$ as $s \to r$. This additional restriction implies $\alpha > 1$ due to our earlier request. The integral in the definition of Λ converges since $1/\beta$ is larger than α and the integral containing the inverse of $1 - \Lambda$ in the last term of the H_2 factor Z converges to the same order.

The function V takes its maximum in the interior and is bounded by $2^{\alpha - 1}/\nu$. This gives us the estimate:

$$(-q') \geq \frac{aq\nu}{\sin as} 2^{1-\alpha}$$

which suffices to estimate the factor Z of H_2. There we are mainly concerned with the behavior at $s = 0$. We get:

$$Z \leq 2^{\alpha - 1}\left(\frac{\sin as}{a}\right)^{2-n}\left(\frac{q}{\nu\eta} + q^m \frac{\gamma + \epsilon}{\nu m \eta_2}\right) + \frac{C}{\epsilon}\left(\frac{\sin as}{a}\right)^{3-n}\left(\frac{\sin as}{\sin ar}\right)^{\nu m - 2\nu}$$

where C is some constant we do not bother with. The last term is independent of η and if we do not want the first term to show worse behavior we should let η behave like s in power $2\mu - 1$, say, with $2\mu \leq \nu - \nu m$. On the other hand we have the following asymptotic behavior:

$$q\Lambda\left(\left(\frac{1}{q}\right)'\right) \sim \frac{2\gamma m \nu^2}{n - 2 + \nu m} B(0) s^{(1-m)\nu - 1} \qquad \text{as } s \to 0$$

so that if this is to dominate Θ, i.e., the first term in X to dominate the last one, we must have $\nu(1 - m) \leq 2\mu$. In fact we are quite happy to take $2\mu = (1 - m)\nu$. We may add as an explanation that Λ behaves now like s in power 2μ, which is the same behavior as of Λ_r, if μ becomes the meaning of a Hölder exponent for the map f (see Prop. 3).

We should now be sufficiently motivated to estimate Θ and choose η, but that estimate is not the only influencing factor on our choice of η, as the discussion above shows. We give our result in the following lemma and take also the opportunity to present our reformulation of Lemma 2.

Lemma 3. Let $0 < \mu < 1$ and define:

$$\eta = \delta\left(\frac{1}{b}\sin bs\right)^{2\mu-1} + \delta_1 \int_0^s \left(\frac{\sin bs}{\sin bt}\right)^{2\mu-1} dt$$

Suppose δ_1, $\delta > 0$ and $r > 0$ are chosen such that

$$\eta - (\delta_1 + \delta_0)\frac{1}{\eta} \leq (n - 2 + 2\mu)\, b \cot bs \qquad \text{and} \qquad \eta > 0$$

holds on $0 < s \leq r$, where we use for abbreviation:

$$\delta_0 = \frac{2}{3}(n - 1)b^2 + \frac{n - 4}{3}a^2$$

and a^2 and b^2 denote the bounds on the sectional curvatures of M on the ball of radius r (Prop. 3). Then:

$$\Theta \leq \Theta_\mu = \eta + (n - 1)(a \cot as - b \cot bs)$$

holds on $0 < s \leq r$. If we suppose further that:

$$\eta + \delta_1 \frac{1}{\eta} \leq a \cot as - (2\mu - 1)\, b \cot bs$$

holds on $0 < s \leq r$ and that p is some positive differentiable function on $0 < s < r$ with $1/p$ integrable on that interval, then:

$$\int_0^r \theta p v' \, ds \leq \frac{1}{2}\theta p u\Big|_{s=r} + \int_0^r \theta\left(\frac{\sin as}{a}\frac{d}{ds}\left(-\frac{ap}{2\sin as}\right) + \frac{p}{2}\Theta_\mu\right) u\, ds + \int_0^r \frac{\theta p}{4\eta} H_2\, ds$$

Remark. The second condition on η is not essential, but allows a simplification of the H_2 term. However, it is automatically satisfied for s sufficiently small, since the right hand side is asymptotic to $(2 - 2\mu)/s$ and dominates the left hand side.

Proof. We use again, as in Prop. 4, that the function

$$(n - 1)\left(a \cot as\, b \cot bs - \left(\frac{b}{\sin bs}\right)^2\right) + a^2$$

is decreasing and tends to the value $-\delta_0$ as s tends to 0. Moreover, we use:

$$\eta' = \eta(2\mu - 1)\, b \cot bs + \delta_1$$

Then the first condition on η in our lemma obviously implies that the term depending on h_3 in our definition of Θ is negative and can be dropped. Our estimate for Θ is precisely what remains. The second condition on η is precisely that $\sin as/a\xi'$ should be increasing, which was used to estimate the H_2 term. It is clear that integration of $\eta = \xi'/\xi$ gives us an admissible function ξ, which does not bother us further. Then we have already established the final inequality, we have only to notice that p is in the role of q_2 and q is obtained by integration of $2/p$ and inversion.

Let us finally formulate the results we have so far, collecting all the assumptions in the following weak estimate for the local energy u, where of course several choices are left open for the sake of applications.

Proposition 5. Let M be a C^∞ Riemannian manifold of dimension $n \geq 2$ and let $f: M \to N$ be a C^∞ map into a second C^∞ Riemannian manifold. Let r measure the distance from some fixed point x_0 in M and let a^2 and b^2 denote lower and upper bounds for the sectional curvatures of M on the ball of radius r, with r chosen less than the radius of injectivity for the exponential map at x_0 and also sufficiently small that the conditions for η in Lemma 3 are satisfied, depending on some admissible choice of μ, δ, and δ_1. In particular let br be less than $\pi/2$ in case b^2 is positive. Let λ^2 denote upper bounds for the sectional curvatures of N on the image of the r-ball. Let γ, ϵ, B_0, q_r be positive constants and μ, ν, α, and β be other parameters restricted by:

$$0 < 2\mu < \nu < 2, \qquad 1 + \alpha < \frac{1}{\beta} < 2\alpha$$

Define the functions q, Q, and Λ as functions of the radius s on $0 < s < r$, by:

$$\log\left(\frac{tg(as/2)}{tg(ar/2)}\right)^{-\nu} = \int_a^1 \frac{1}{\beta} x^{-1}(1-x)^{\frac{1}{\beta}-\alpha-1}\left(1 + \frac{\alpha\beta x}{1-\alpha\beta}\right)^{\alpha-1} dx$$

with $z = (q/q_r)^{-\beta}$ and then with $m = 1 - 2\mu/\nu$:

$$Q = -2\frac{q}{q'}\left(\frac{2\alpha\beta - 1}{1-\alpha\beta} qz - \frac{\alpha\beta}{1-\alpha\beta} qz^2 - \frac{\gamma+\epsilon}{2m} q^m\right)$$

$$\frac{\Lambda}{1-\Lambda} = \frac{\gamma}{q}\left(-\frac{(q^m)'}{\theta}\right)^{\frac{1}{2}}\left(B_0 + \int_s^r (-\theta(q^m)')^{\frac{1}{2}} B\, ds\right)$$

where B is some continuous function on $0 \leq s \leq r$ with values on $0 < B \leq 1$. Then there is a constant C, such that:

$$\int_0^r \theta\left(\frac{\sin as}{a}\frac{d}{ds}\left(-\frac{aQ}{2\sin as}\right) + (1-B)\frac{\Lambda q}{1-\Lambda} - \left(q + \frac{\gamma+\epsilon}{2m} q^m\right)\frac{q}{q'}\Theta_\mu\right) u\, ds$$

$$\leq \gamma B_0 u(r) + C\int_0^r \left(\frac{\sin as}{a}\right)^{3-n-\nu-2\mu} H_2(s)\, ds \left(\frac{\sin ar}{a}\right)^{\nu+2\mu}$$

where Θ_μ is defined in Lemma 3, provided we have on $0 \leq s \leq r$:

$$1 - \lambda\rho_s \cot \lambda\rho_s \leq B\Lambda \quad \text{and} \quad \lambda\rho_r < \pi$$

2.2 Strong Estimates

We can now continue with our comparison of w with u, initiated in the beginning of this second section. Our first step is to prove:

Proposition 6. Let M be a C^∞ Riemannian manifold of dimension $n \geq 2$ and let $f: M \to N$ be a C^∞ map into a second Riemannian manifold N. Let r measure the distance from some fixed point x_0 in M and let a^2 and b^2 denote lower and upper bounds for the sectional curvatures of M on the ball of radius r, with r chosen less than the radius of injectivity for the exponential map at x_0 and also sufficiently small that the conditions for η in Lemma 3 are satisfied, depending on some admissible choice of μ, δ, and δ_1. In any case, let $br < \pi/2$ if b can be taken positive and let $ar \leq \pi$ if a can be taken positive. Restrict r further if necessary such that:

$$\lambda\rho_r \leq \frac{\pi}{2}$$

where λ^2 denotes upper bounds for the sectional curvatures of N on the image of the r-ball and where λ is requested nonnegative. Let the following parameters be chosen within the stated bounds:

$$0 < 4\mu < \nu \leq 2, \quad \epsilon > 0$$

Then there is a constant C, such that

$$w'(r) \leq \frac{\omega}{\mu} \frac{\sin ar}{a} u(r) + C \int_0^r \left(\frac{\sin as}{\sin ar}\right)^{1-n-\nu-2\mu} \left(\frac{\sin as}{a}\right)^2 H_2 \, ds$$

where ω is defined by:

$$\omega = \frac{1}{4}(y+1)^{-1}\left(\frac{y}{y+1} - \Delta\right)^{-y-1} (1 + y - y\Delta)^y, \quad y = \frac{\nu}{2\mu} - 2$$

$$\Delta = \frac{\delta + \epsilon}{\nu - 2\mu}\left(1 + \frac{n-2}{\nu - 2\mu}\right)\left(\frac{2}{a} \operatorname{tg}\left(\frac{ar}{2}\right)\right)^{2\mu}$$

Remark. We shall find it useful to get ω below 1, but we have $\omega = 1$ if $y = 1$ and $\Delta = 0$. Moreover, ω is less than 1 in case $\Delta = 0$, if $2/3 \leq y < 1$ and can therefore be taken less than 1 for y in this interval by making Δ sufficiently small by choice of δ and ϵ or r.

Proof. We apply Lemma 3, with the choice:

$$p = Q - 2 \frac{q}{q'} \frac{P}{V} q_r = Q + 2 \frac{\sin as}{a} Pq_r$$

$$= -2 \frac{q}{q'}\left(y - \frac{\gamma + \epsilon}{2m} q^m + \frac{Pq_r}{V}\right), \quad \text{y as before}$$

with q and V as in Prop. 5 and P some constant chosen sufficiently large such that p is positive on $0 < s < r$. Then we can replace the Q in the inequality in Prop. 5 by p and get:

$$\theta(r)w'(r) \leq \int_0^r \frac{ds}{p} \int_0^r p\theta v' \, ds$$

$$\leq \int_0^r \frac{ds}{p}\left(\gamma B_0 u(r) + \theta(r) \frac{\sin ar}{a} Pq_r u(r) + \int_0^r \left(C\left(\frac{\sin as}{a}\right)^{3-n-\nu-2\mu} + \frac{\theta p}{4\eta}\right) H_2 \, ds\right)$$

provided the conditions in Prop. 5 are satisfied and also:

$$(1 - B)\frac{\Lambda}{1 - \Lambda} \geq (V_1 + Pz^{\frac{1}{\beta}}) \frac{\sin as}{a} \Theta_\mu$$

with

$$V_1 = \left(1 + \frac{y}{q}\right) V = \frac{1}{\nu}(1 - z)^{\frac{1}{\beta} - \alpha}\left(1 + \frac{\alpha\beta z}{1 - \alpha\beta}\right)^\alpha$$

The additional H_2 term is of the same type as before, so that we have only to increase the constant C. Then we have only left to discuss the Λ condition and the coefficients of u(r) in the boundary terms, but that is some work. Let us look first at the integral:

$$\int_0^r \frac{ds}{p} = \int_0^1 \frac{q}{-\beta q' p} \frac{dz}{z} = \int_0^1 \left(2\beta q_r\left(c_0 - c_1 z + \left(\frac{P}{V}\right) z^{\frac{1}{\beta} - 1}\right)\right)^{-1} z^{\frac{1}{\beta} - 2} \, dz$$

with

$$c_0 = \frac{2\alpha\beta - 1}{1 - \alpha\beta} - \frac{\gamma + \epsilon}{2m} q_r^{m-1}, \quad c_1 = \frac{\alpha\beta}{1 - \alpha\beta}$$

Here we have chosen $\beta = 2\mu/\nu$ or $m = 1 - \beta$. That means that we must restrict μ further by $4\mu < \nu$, which implies $\mu < 1/2$. We could avoid this by taking $\beta = \mu/\nu$ and include the perturbation in c_1, but then we are not able to get ω below 1. Next we shall estimate the integrand by use of the elementary:

$$z \leq \frac{1 - 2\beta}{1 - \beta} \epsilon_1 + \epsilon_1^{2-\frac{1}{\beta}} \frac{\beta}{1 - \beta} z^{\frac{1}{\beta}-1}$$

with

$$\epsilon_1 = (\tau(1 - z))^{\kappa}\left(\tau\left(1 + \frac{\alpha\beta z}{1 - \alpha\beta}\right)\right)^{1-\kappa} \leq \tau - \left(\kappa - \frac{\alpha\beta}{1 - \alpha\beta}(1 - \kappa)\right)\tau z$$

$$\kappa = \left(\frac{1}{\beta} - \alpha - 1\right)\left(\frac{1}{\beta} - 2\right)^{-1}$$

and τ some positive constant, sufficiently small that P defined by

$$\nu P = \frac{\beta}{1 - \beta}\left(1 - \frac{2\alpha\beta - 1}{1 - \alpha\beta}\tau\right)^{-1} \tau^{2-\frac{1}{\beta}} c_1$$

is positive and then we obtain:

$$c_0 - c_1 z + \left(\frac{P}{V}\right) z^{\frac{1}{\beta}-1} \geq c_2 = c_0 - c_1 \frac{1 - 2\beta}{1 - \beta}\tau\left(1 - \frac{2\alpha\beta - 1}{1 - \alpha\beta}\tau\right)^{-1}$$

Then it is easy to evaluate the integral:

$$Pq_r \int_0^r \frac{ds}{p} \leq P(2(1 - \beta)c_2)^{-1}$$

This is what we have as a factor in one of the boundary terms in our estimate of w' and we shall choose τ such that it takes its minimum, which happens for:

$$\frac{1}{\tau} = \frac{c_1}{c_0} + \frac{2\alpha\beta - 1}{1 - \alpha\beta}\,\frac{1 - \beta}{1 - 2\beta}$$

where

$$\frac{P}{c_2} = \left(\frac{c_1}{\nu c_0}\right)\tau^{2-\frac{1}{\beta}}$$

We obtain the boundary term in the inequality in Prop. 6, if we let α take its lower limit $1/\beta - 1$ and put

$$\omega = \frac{\beta}{4(1 - \beta)}\left(\frac{c_1}{c_0}\right)\tau^{2-\frac{1}{\beta}}$$

We shall see that this is permissible, but first we must study the Λ condition.

Let us use the abbreviations:

$$F = (-\theta(q^m)')^{\frac{1}{2}} = \left(\frac{\sin as}{a}\right)^{-\frac{n}{2}} q^{\frac{m}{2}} \left(\frac{m}{V}\right)^{\frac{1}{2}}$$

$$\Theta_\mu^0 = (V_1 + Pz^{\frac{1}{\beta}})\, z^{-2} \frac{\sin as}{a}\, \Theta_\mu$$

From the definition of Θ_μ in Lemma 3 and q in Prop. 5, we see that Θ_μ^0 has a positive limit as $s \to 0$:

$$\Theta_\mu^0(0) = \frac{\delta}{\nu}\left(\frac{2}{a}\,\mathrm{tg}\left(\frac{ar}{2}\right)\right)^{2\mu}$$

The equation defining Λ can now be written:

$$\frac{\Lambda}{1-\Lambda} = \frac{\gamma F}{\theta q}\left(B_0 + \int_s^r FB\,ds\right)$$

and the new Λ condition is

$$\frac{\Lambda}{1-\Lambda} \geq (1-B)^{-1} z^2 \Theta_\mu^0$$

and holds, if we request B to satisfy the differential inequality:

$$FB \geq -\frac{d}{ds}\left((1-B)^{-1}\left(\frac{\theta q^m}{\gamma F q_r^{m-1}}\right)\Theta_\mu^0\right)$$

and

$$(1-B)^{-1}\left(\frac{\theta q^m}{\gamma F q_r^{m-1}}\right)\Theta_\mu^0 = B_0 \qquad \text{in } s = r$$

Actually, we can just define B_0 by this expression. We find it more convenient to replace B by:

$$g = (1-B)^{-1}\left(\frac{V}{m}\right)^{\frac{1}{2}}\Theta_\mu^0$$

and then the differential inequality reads:

$$1 \geq \frac{1}{g}\left(\frac{V}{m}\right)^{\frac{1}{2}}\Theta_\mu^0 + \frac{g}{2\Gamma}(mV)^{-\frac{1}{2}}\left(m + (n-2)V\cos as - 2\frac{\beta}{g}z\frac{dg}{dz}\right)$$

with $\Gamma = \gamma q_r^{m-1}$. We could define g by replacing the inequality by an equality, but we find that nonlinear differential equation too difficult to handle, so that we choose to minimize the right hand side as a function of g with g'/g fixed, by the choice:

$$g^{-2}\Theta_\mu^0 = \frac{1}{2\Gamma}\left(m + (n-2)V\cos as - 2\frac{\beta}{g}z\frac{dg}{ds}\right)$$

In fact, this is the best thing to do at $z = 0$ and now we have a linear equation for g^2, which we can solve:

$$g^2 = z^{\frac{m}{\beta}}\left(\frac{\sin as}{a}\right)^{n-2}\left(d + \int_z^1 \frac{2}{\beta}\Gamma\Theta_\mu^0\left(\frac{\sin as}{a}\right)^{2-n} z^{-\frac{m}{\beta}}\frac{dz}{z}\right)$$

The inequality for g is now:

$$\frac{2}{g}\left(\frac{V}{m}\right)^{\frac{1}{2}}\Theta_\mu^0 \leq 1$$

and this holds, if Γ and the constant of integration d is sufficiently large. Now let us observe that we have estimated one of the two boundary terms as independent of q_r and the other depends on q_r as:

$$\left(\frac{1}{q_r}\right)\gamma B_0 = \sqrt{d}\, q_r^{-\frac{m}{2}}$$

and tends to zero as q_r tends to infinity. Moreover, we can go right to the limit, since the H_2 terms are easily seen to be bounded, and this means that the mentioned boundary term drops out. Then let us look at the remaining condition in Prop. 5:

$$1 - \lambda\rho_s \cot \lambda\rho_s \leq B\Lambda$$

We have chosen B (or g) such that:

$$\Lambda \geq g\left(\frac{m}{V}\right)^{\frac{1}{2}} z^2\left(1 + g\left(\frac{m}{V}\right)^{\frac{1}{2}} z^2\right)^{-1}$$

so that the condition is implied by:

$$\lambda\rho_s \cot \lambda\rho_s \geq (1 + z^2\Theta_\mu^0)\left(1 + gz^2\sqrt{\frac{m}{V}}\right)^{-1}, \qquad 0 \leq s \leq r$$

This condition holds at s = 0 and the right hand side tends to zero uniformly on $\epsilon \leq s \leq r$ for any $\epsilon > 0$, if d tends to ∞. Moreover, since the right hand side is zero at s = r, it suffices to request $\lambda\rho_r \leq \pi/2$ as is done in Prop. 6.

Now, in our formula for c_1 in the boundary-term estimate, we must replace γq_r^{m-1} by Γ and similarly we must replace ϵq_r^{m-1} by another constant independent of q_r, which we shall still denote by ϵ. We cannot take Γ smaller than the value we get by letting the equality hold in our inequality for g and that means:

$$\Gamma = \frac{2}{\nu}\left(1 + \frac{n-2}{\nu - 2\mu}\right)\delta\left(\frac{2}{a}\,\mathrm{tg}\left(\frac{ar}{2}\right)\right)^{2\mu}$$

So we must either take δ or r small, if we want Γ small. We have redefined ϵ as a perturbation of δ in Prop. 6, but of course we cannot go to 0 with either ϵ or δ, since the H_2 terms would then tend to infinity. On the other hand, there is nothing now that prevents us from going to the limit $1/\beta - 1$ with α and that completes our proof of Prop. 6. It is perhaps surprising that we did not need a uniform estimate for ρ_s here.

We are now able to improve the growth estimate for u in Prop. 4 and then to reach our goal to estimate the growth of w.

<u>Proposition 7</u>. Let the assumptions be as in Prop. 6 with the additional restriction:

$$\gamma = 2\frac{\mu}{\omega}(1 - \epsilon)^2 \leq \nu$$

Let $\alpha > 0$ be chosen and r restricted such that the conditions in Prop. 4 are satisfied with the ν replaced by 2μ. Define a function Φ by:

$$\Phi(s) = \exp\left(-\gamma \int_s^r (\lambda\rho_t \cot \lambda\rho_t)^2 \frac{a\, dt}{\sin at}\right)$$

and choose a new parameter p such that:

$$4 - \nu - 2\mu > \frac{2n}{p}$$

Then there is a constant C, such that

$$u(s) \leq \eta(s)\Phi(s)\left(\frac{u(r)}{\eta(r)} + Cr^{4-\frac{2n}{p}} H_p(r)\right) \quad \text{on } 0 \leq s \leq r$$

with η defined as in Prop. 4. Moreover, if r is further restricted if necessary such that:

$$(n-1)(bs \cot bs - 1) - 2\mu\alpha\left(\frac{s}{r}\right)^2 \leq 0 \quad \text{on } 0 < s \leq r$$

then there is a constant C such that

$$w(s) \leq \frac{\omega}{2\mu n} s^n \Phi(s)\left(\frac{u(r)}{\eta(r)} + Cr^{4-\frac{2n}{p}} H_p(r)\right)$$

on $0 \leq s \leq r$.

Proof. We shall apply Prop. 3(2) again, but now in a simplified version where we put $\Lambda = 0$ and drop the k terms. Then:

$$(cs^{\mu} \cot cs^{\mu})u \leq \frac{A}{2} w' + \frac{1}{2A} v' + Bw + \frac{1}{4B} H_2 \qquad \text{at } s$$

where A and B can be any positive functions of s. Then we apply Prop. 4 with the ν there replaced by 2μ and get:

$$v' \leq \frac{\eta}{2} \frac{d}{ds} \frac{u}{\eta} + \frac{1}{2\eta P} \int_0^s \frac{\sin as}{4a\xi'} H_2' \, ds$$

with

$$\eta = \frac{a}{2\xi} (\sin as)^{-1} \left(\frac{1}{b} \sin bs\right)^{n-1}, \qquad \xi = \exp\left(\alpha\left(\frac{s}{r}\right)^{2\mu}\right)$$

$$P\eta = \frac{\sin as}{2a\xi}$$

We use Prop. 6 to estimate w' in terms of u and H_2 and integrate that inequality to get an estimate for w. Then we get:

$$\frac{d}{ds}\frac{u}{\eta} \geq 4A\left(\lambda\rho_s \cot \lambda\rho_s - \frac{A}{2\mu} \frac{\sin as}{a} - B\frac{\omega}{\mu} 2\left(\frac{1}{a} \sin \frac{as}{2}\right)^2\right)\frac{u}{\eta} - Z$$

$$Z = \frac{s}{4\mu\alpha\eta}\left(\frac{s}{r}\right)^{-2\mu} H_2 + \frac{2}{\eta} A^2 C Z_1 + \frac{4}{\eta} ABCZ_0 + \frac{A}{B\eta} H_2$$

$$Z_j = \int_0^s \left(\frac{\sin at}{\sin as}\right)^{j-n-\nu-2\mu} \left(\frac{\sin at}{a}\right)^{3-j} H_2 \, dt$$

Now we choose:

$$B = \frac{\mu}{2\omega} \epsilon \left(\frac{a}{\sin as/2}\right)^2 X, \qquad A = \frac{\mu a(1-\epsilon)}{\omega \sin as} X, \qquad X = \lambda\rho_s \cot \lambda\rho_s$$

for some $\epsilon > 0$, which we for convenience take as the same ϵ as in Prop. 6. Then we obtain the differential inequality:

$$\frac{d}{ds}\frac{u}{\eta} \geq (\lambda\rho_s \cot \lambda\rho_s)^2 \frac{\gamma a}{\sin as} \frac{u}{\eta} - Z$$

with γ as defined in Prop. 7. Integration over an interval from s to r gives:

$$\frac{u}{\eta}(s) \leq \Phi(s)\left(\frac{u}{\eta}(r) + \int_s^r \Phi(t)^{-1} Z(t) \, dt\right)$$

and we have only left to estimate the remaining integral to obtain the inequality for u in Prop. 7. Here we can be brief, since we do not bother to know the constant C. We may observe that

$$\Phi(s) \geq \left(\frac{\operatorname{tg} as/2}{\operatorname{tg} ar/2}\right)^{\gamma}$$

which makes it possible to estimate the remaining integral independent of $\lambda\rho_s$. Moreover, the trigonometric functions there can be replaced by powers and we use:

$$\int_s^r s^{\kappa-n} H_2 \, ds \leq C\left(\int_s^r (s^{\kappa-n+1})^{\frac{p}{p-2}} s^{n-1} ds\right)^{1-\frac{2}{p}} \left(\int_{B_r} \|\Delta f\|^p\right)^{\frac{2}{p}}$$

$$\leq Cr^{1+\kappa-\frac{2n}{p}} H_p(r) \quad \text{if } 1+\kappa > \frac{2n}{p}$$

An inspection of the exponents shows that κ is the sum of 3, 2μ, and either ν or γ, hence the condition $\gamma \leq \nu$ and then the condition for p in Prop. 7 is just what is needed.

The inequality for w is now obtained by integration from the inequality for w' in Prop. 6 replacing there r by s and estimating u(s) in terms of u(r) by the growth estimate for u we just established. The additional condition in Prop. 7 is there only because we have estimated Φ by its maximum and the remaining factor by a power in the integration. This condition is in fact satisfied if $b \geq 0$. The H_2 term is treated as before, which finishes our proof.

3. EQUICONTINUITY

We shall first discuss the application of Prop. 7, where Φ is estimated by use of the local Hölder constant on the basis of

$$\rho_s \leq L_\kappa(x_0, f, r) s^\kappa$$

and where the condition on the range of f in Prop. 6 is taken care of by

$$c = \lambda L_\kappa(x_0, f, r) \leq \frac{\pi}{2} r^{-\kappa}$$

This is of course a stronger condition, but the object would be to get a single inequality for $L_\kappa(f, r)$ by means of Prop. 2 which would be satisfied, hopefully, when the Hölder constant is in some bounded interval corresponding to a component of the set of mappings f under consideration.

Using the estimates:

$$\frac{1}{a}\sin as \leq \frac{s}{d}, \qquad d = \begin{cases} ar/\sin ar, & \text{if } a < 0 \\ 1 \quad, & \text{if } a \geq 0 \end{cases}$$

$$\log\left(\frac{\sin cr^\kappa}{\sin cs^\kappa}\right)^{\frac{1}{\kappa}} \leq \int_s^r (ct^\kappa \cot ct^\kappa) \frac{dt}{t}$$

$$\leq \left(\log\frac{r}{s}\right)^{\frac{1}{2}} \left(\int_s^r (ct^\kappa \cot ct^\kappa)^2 \frac{dt}{t}\right)^{\frac{1}{2}}$$

$$\leq \frac{1}{2}\log\frac{r}{s} + \frac{1}{2}\int_s^r (ct^\kappa \cot ct^\kappa)^2 \frac{dt}{t}$$

we obtain:

$$\Phi(s) \leq \left(\frac{\sin cs^{\kappa}}{\sin cr^{\kappa}}\right)^{\frac{2d\gamma}{\kappa}} \left(\frac{r}{s}\right)^{d\gamma} \leq \left(\frac{s}{r}\right)^{d\gamma} \left(\frac{cr^{\kappa}}{\sin cr^{\kappa}}\right)^{\frac{2d\gamma}{\kappa}}$$

Then it is clear that we can apply Props. 2 and 7 to get an inequality satisfied by the Hölder constant. We get:

$$\Theta^2 \leq \omega 4^{n+1} (2\mu\omega_n \min\{1, \Omega_n(br)\})^{-1} \left(\frac{\lambda\Theta}{\sin\lambda\Theta}\right)^{\frac{2d\gamma}{\kappa}} \left(\frac{u(r)}{\eta(r)} + Cr^{4-\frac{2n}{p}} H_p(r)\right)$$

for $\Theta = L_\kappa(f, r)r^\kappa$, if $2\kappa \leq d\gamma$, provided we assume that all sectional curvatures of M are bounded within the interval from a^2 to b^2 and all sectional curvatures of N are bounded from above by λ^2, and other conditions on r and the parameters in Prop. 7 are satisfied.

Now, we can make the coefficient of H_p as small as we please by making r small, but u/η is only bounded when r gets small, due to Prop. 4, that is if we do not use the growth estimate for u in Prop. 7, which would imply restrictions for the Hölder constant for a radius exceeding r. We would need to be able to conclude from the above inequality that $\lambda\Theta \leq \pi/2$ would imply that the strict inequality must hold. That is then not possible except we have $u(r)r^{2-n}$ sufficiently small. It may be possible to handle an a priori condition of that kind, but the condition in our theorem for ρ_r is probably preferable. Thus we let this discussion rest and turn to the proof of our theorem in the Introduction.

Proof of Theorem. With $0 < \mu < 3/8$ and $\nu = 16\mu/3$ we have $\omega < 1$ according to our remark to Prop. 6 and then we are able to choose the ϵ in the definition of γ in Prop. 7 such that $\gamma = 2\mu$. Moreover, our condition for p in the theorem implies the condition for p in Prop. 7. Choosing a^2 and b^2 as bounds for all sectional curvatures of M and r sufficiently small, all the assumptions in Prop. 7 hold, at an arbitrary point x_0. Let's say that the conditions on r are satisfied for $r \leq r_0$. Then we can define the function R(x) used in the construction of W_r in 1.2 as r_0 or the convexity radius of M at x, if it is smaller, and the estimates in Prop. 7 hold at any point x_0, if

$$r < R(x) \quad \text{and} \quad L_0(x_0, f, r) = \rho_r \leq \rho$$

with ρ restricted as in the theorem. We observe that Prop. 2 still holds if we replace the condition containing Φ there by $\rho_r \leq \rho$, but take $L_\mu = L_\kappa$ in the final conclusion, with κ defined as in the theorem. Estimating $\rho_s \leq \rho$ in the definition of $\Phi(s)$ in Prop. 7, we get

$$\Phi(s) \leq \left(\frac{\sin as}{\sin ar}\right)^{2\kappa}$$

Then we can proceed as in our first alternative and obtain:

$$L_\kappa(f, r)^2 \leq \omega 4^{n+1} (2\mu\omega_n \min\{1, \Omega_n(br)\})^{-1} \left(\frac{a/d}{\sin ar}\right)^2 \left(\frac{u(r)}{\eta(r)} + Cr^{4-\frac{2n}{p}} H_p(r)\right)$$

Here, d is as above and we must interpret u(r) and $H_p(r)$ as made independent of the center x_0 of the ball by taking the supremum over all the admissible balls. This concludes our proof of the theorem. Observe that $L_\kappa(f, r)$ is defined only if r is sufficiently small and the conclusion of the theorem is not void for any such r.

REFERENCES

1. Eells, J.; Lemaire, L.: Bull. London Math. Soc. 10, 1-68 (1978).
2. Elíasson, H. I.: J. Diff. Geom. 1, 169-194 (1967).
3. Elíasson, H. I.: Variation integrals in fibre bundles, in Proc. Symp. in Pure Math., vol. 16. Amer. Math. Soc. (1970), pp. 67-89.
4. Hildebrandt, S.; Kaul, H.; Widman, K.-O.: Acta Math. 138, 1-16 (1977).
5. Im Hof, H.-C.; Ruh, E. A.: Comm. Math. Helv. 50, 389-401 (1975).
6. Klingenberg, W.; Gromoll, D.; Meyer, W.: Riemannsche Geometry im Grossen. Springer (1968).
7. Uhlenbeck, K.: Bull. Am. Math. Soc. 76, 1082-1087 (1970).
8. Hildebrandt, S.; Jost, J.; Widman, K.-O.: Inventiones Math. 62, 269-298 (1980).

13 CONSERVATION LAWS IN GAUGE FIELD THEORIES*

Arthur E. Fischer / Department of Mathematics, University of California, Santa Cruz, California

INTRODUCTION

Since the time of Emmy Noether, the relationships between differential identities, conservation laws, and symmetry groups in theoretical physics have been one of the most profoundly studied and written about topics in modern physics. Noether's original ideas have been applied to virtually every area of theoretical physics, from classical mechanics to quantum chromodynamics.

The recent emergence of gauge field theories as being of fundamental importance in elementary particle physics dictates that Noether's Theorem can be profitably applied to the study of these theories. Importantly, in Noether's original work [14], the generality is already present for the study of conservation laws in gauge theories. Noether's work has currently been applied by many authors, among whom are Trautman [17], Sternberg [16], Moncrief [13], Jackiw and Manton [10], Jackiw [9], and Forgacs and Manton [8]. These authors have currently clarified our understanding of conservation laws in gauge field theories.

For the most part, the above authors are concerned with fields over a spacetime M. One of our goals is to consider conservation laws in the more general context of principal fiber bundles. As is well-known, a principal fiber bundle over a spacetime M is the natural space on which to consider gauge field theories.

Working on a principal fiber bundle, we are able to present a unified picture of the differential identities and conservation laws associated with a generally covariant Lagrangian field theory. We are particularly interested in studying the relationship between these identities and their corresponding conservation laws with the gauge and symmetry groups of the field theory.

Let

$$\pi: P(M,G) \to M$$

be a principal fiber bundle with structure group G over the spacetime M, an n-dimensional connected oriented manifold without boundary. The spaces that play the central role in our work are

$\mathcal{M}$, the space of pseudo-riemannian metrics on M with fixed signature;
$\mathcal{C}$, the space of connections on P;
$\mathcal{H}$, the space of k-form Higgs fields on P; and
AUT(P), the group of automorphisms of P.

Let

$$\mathcal{L}: \mathcal{M} \times \mathcal{C} \times \mathcal{H} \to C_d^\infty(M)$$

*This research was partially supported by the National Science Foundation of the United States, and by the Max-Planck-Institut für Physik und Astrophysik, Garching bei München, Federal Republic of Germany.

denote a Lagrangian density on $\mathcal{M} \times \mathcal{C} \times \mathcal{H}$, where $C_d^\infty(M) = C^\infty(\Lambda^n(M))$ is the space of densities, or n-forms, on M. This Lagrangian density describes the interaction between gravity, gauge fields, and matter fields. However, all of these fields need not be taken as dynamical fields simultaneously. Indeed, the specification of some of these fields as nondynamical, or parameter, fields, will be important in our classication of differential identities and conservation laws.

The group AUT(P) naturally unifies the concept of spacetime, or coordinate transformation, with that of internal, or gauge transformation. We define the Lagrangian density $\mathcal{L}$ to be generally covariant with respect to AUT(P) if

$$\mathcal{L}(f^*g,\ F^*\omega,\ F^*\varphi) = f^*(\mathcal{L}(g,\omega,\varphi))$$

for all $(g,\omega,\varphi) \in \mathcal{M} \times \mathcal{C} \times \mathcal{H}$ and $F \in AUT(P)$, where f is the diffeomorphism of M covered by F. Here "*" represents pullback of g and $\mathcal{L}(g,\omega,\varphi)$ by f, and ω and φ by F.

By basing our concept of general covariance on the group AUT(P), we are able to give a unified treatment of coordinate and gauge transformations which minimizes the complications arising from the mixing of spacetime and internal degrees of freedom.

If the Lagrangian density given above is generally covariant, then by using modern methods of global analysis and differential geometry, we are able to quickly and efficiently derive two general covariance differential identities, or master equations, for $\mathcal{L}$. One of these differential identities reflects the infinitesimal version of covariance of $\mathcal{L}$ in the vertical or internal direction, and the other in the horizontal or spacetime direction. These general covariance differential identities can be used to derive the gauge differential identities and the local and global conservation laws.

To see our method and to describe our terminology, let ξ represent a fixed parameter field (or fields), and let $AUT_\xi(P)$ denote the subgroup (possibly trivial) of AUT(P) such that $\mathcal{L}$ is covariant with respect to AUT_ξ acting on the dynamical fields only. $AUT_\xi(P)$ is then the covariance group of $\mathcal{L}$ at ξ. If $AUT_\xi(P)$ is infinite-dimensional, it is a gauge group for $\mathcal{L}$ at ξ. If it is finite-dimensional it is a symmetry group for $\mathcal{L}$ at ξ.

In the presence of a gauge group, the master equations lead to gauge differential identities, whereas in the presence of a symmetry group, they generate local conservation laws when the dynamical field equations are satisfied. Moreover, in this latter case, there is a global conservation law for each dimension of $AUT_\xi(P)$.

The master equations can be interpreted geometrically as follows. Let

$$J:\ T^*(\mathcal{M} \times \mathcal{C} \times \mathcal{H})_M \to \mathcal{X}^*_G(P)$$

denote the momentum map associated with the action of AUT(P) on the natural cotangent bundle $T^*(\mathcal{M} \times \mathcal{C} \times \mathcal{H})_M$ of $\mathcal{M} \times \mathcal{C} \times \mathcal{H}$. Here $\mathcal{X}^*_G(P)$ is the natural dual of $\mathcal{X}_G(P)$, the Lie algebra of G-invariant vector fields on P. The master equations for a generally covariant Lagrangian density can then be expressed using the momentum map J and the Lagrangian 1-form of $\mathcal{L}$,

$$\Psi_{\mathcal{L}} : \mathcal{M} \times \mathcal{C} \times \mathcal{H} \to T^*(\mathcal{M} \times \mathcal{C} \times \mathcal{H})_M$$

$$(g,\omega,\varphi) \longmapsto \left(g,\ \omega,\ \varphi,\ \frac{\delta\mathcal{L}}{\delta g}(g,\omega,\varphi),\ \frac{\delta\mathcal{L}}{\delta\omega}(g,\omega,\varphi),\ \frac{\delta\mathcal{L}}{\delta\varphi}(g,\omega,\varphi)\right)$$

The master equations are then very simply

$$J \circ \Psi_{\mathcal{L}} = 0$$

or

$$J\left(g,\ \omega,\ \varphi,\ \frac{\delta\mathcal{L}}{\delta g}(g,\omega,\varphi),\ \frac{\delta\mathcal{L}}{\delta\omega}(g,\omega,\varphi),\ \frac{\delta\mathcal{L}}{\delta\varphi}(g,\omega,\varphi)\right) = 0$$

for all $(g, \omega, \varphi) \in \mathcal{M} \times \mathcal{C} \times \mathcal{H}$, a fact we shall refer to as the vanishing of the momentum theorem.

There are several features of our treatment of differential identities and conservation laws that we wish to point out. Firstly, our expressions involving the Lie derivatives of connections and tensorial quantities on P are straightforwardly derived and formulated in a manner that is easily applied to covariant Lagrangian field theories. Secondly, we use the concepts of natural dual, natural cotangent bundle, and natural L_2-adjoint. These natural objects have several advantages over their usual counterparts. For example, we do not need to introduce a riemannian structure on M or on the vector bundles involved. For another, we are able to give a compact and convenient definition of the functional derivative of $\mathscr{L}$ which is valid for all finite order Lagrangians, and which bypasses the rather cumbersome pointwise expressions usually involved.

Thirdly, we consider the global conservation laws for $\mathscr{L}$ when the dynamical fields are derivatively coupled to gravity or to the gauge field. The usual expressions for conserved currents for Yang-Mills or Higgs systems are for Lagrangians that are nonderivatively coupled to gravity and to the gauge field. If derivative coupling is present, then the currents must be modified by a term that reflects this coupling. Our definition of infinitesimal symmetry for $\mathscr{L}$ automatically makes this modification in the master equations.

Finally, we clarify the relation between the symmetry groups of $\mathscr{L}$ and the structure group G of the principal fiber bundle P(M, G). These are often taken to be equal, but this is not the general case. For example, for the Higgs system discussed in Sec. 3.2, the symmetry group $AUT_{(g,\omega)}(P)$ of $\mathscr{L}_{Higgs}$ at $(g, \omega) \in \mathcal{M} \times \mathcal{C}$ may be properly larger than G for two reasons. Firstly, the internal symmetry group $Aut_\omega(P) \subseteq AUT_{(g,\omega)}(P)$ of $\mathscr{L}$ may be properly larger than $I_\omega^{int}(P)$, the internal symmetry group of ω. Secondly, $I_\omega^{int}(P)$ itself need not be equal to G, unless M is simply connected and ω is flat.

Throughout this work, many people were consulted. I would especially like to thank Professors R. Abraham, D. Christodoulou, J. Ehlers, D. Fried, R. Jantzen, J. Marsden, R. Palais, B. Schmidt, and M. Walker for many valuable conversations and helpful remarks. I would also like to thank Professor J. Ehlers and the Max-Planck-Institut für Physik und Astrophysik, Garching bei München, F.R.G., for their kind invitation, gracious hospitality, and generous support in the Fall of 1980, during which part of this work was prepared.

1. PRINCIPAL FIBER BUNDLES

In this section we summarize the results from the theory of principal fiber bundles and connections that we shall need. The standard reference for this material is [12].

1.1 The Group of Automorphisms of a Principal Fiber Bundle

Let $\pi: P(M, G) \to M$ be a C^∞ principal fiber bundle over a C^∞ n-dimensional manifold M, with structure group G, a Lie group acting on the right with group action

$$\Phi: P \times G \to G \qquad (p, a) \longmapsto \Phi(p, a) = p \cdot a$$

and with projection $\pi: P \to M = P/G$. For $a \in G$, let

$$R_a: P \to P, \quad p \longmapsto p \cdot a$$

denote the diffeomorphism of P induced by $a \in G$ and the action Φ, and for $p \in P$, let

$$\Phi_p: G \to P, \quad a \longmapsto p \cdot a$$

Let e be the identity element of G, and let $\mathfrak{G} = T_eG$ denote the Lie algebra of G. Thus

$$T_e\Phi_p : \mathfrak{G} \to T_pP$$

If $A \in \mathfrak{G}$, we let $A^*: P \to TP$ denote the fundamental vector field induced by A, defined by

$$A^*(p) = T_e\Phi_p \cdot A \in T_pP$$

Thus the fundamental vector fields are the infinitesimal generators of the action of G on P.

An automorphism $F: P \to P$ is a C^∞ diffeomorphism that is equivariant with respect to the action of G on P, i.e., such that $F(p \cdot a) = F(p) \cdot a$ for all $a \in G$. Thus for $a \in G$, the following diagram commutes:

$$\begin{array}{ccc} P & \xrightarrow{F} & P \\ R_a \downarrow & & \downarrow R_a \\ P & \xrightarrow[F]{} & P \end{array}$$

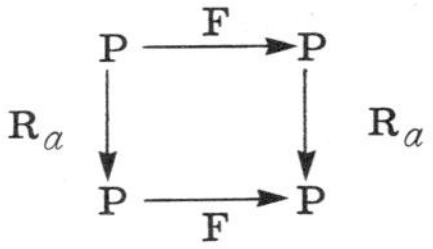

An automorphism $F: P \to P$ maps fibers to fibers, and induces a diffeomorphism of the base manifold M. Let $\hat{\pi}(F) = f: M \to M$ denote this induced diffeomorphism. Thus $\pi \circ F = f \circ \pi$, or the following diagram commutes:

$$\begin{array}{ccc} P & \xrightarrow{F} & P \\ \pi \downarrow & & \downarrow \pi \\ M & \xrightarrow[f]{} & M \end{array}$$

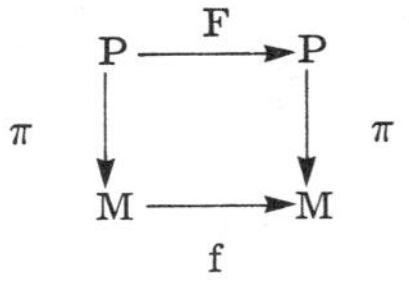

We denote the various infinite dimensional groups that we shall be considering as follows:

$\mathcal{D}(P) = \mathrm{Diff}(P) = \{F: P \to P \mid F \text{ is a } C^\infty \text{ diffeomorphism of } P\}$, the group of diffeomorphisms of P;

$\mathrm{AUT}(P) = \{F \in \mathrm{Diff}(P) \mid F(p \cdot a) = F(p) \cdot a\}$, the group of automorphisms of P;

$\mathrm{Aut}(P) = \{F \in \mathrm{AUT}(P) \mid \hat{\pi}(F) = \mathrm{id}_M\}$, the group of automorphisms of P that cover the identity diffeomorphism id_M of M;

$\mathcal{D}(M) = \mathrm{Diff}(M) = \{f: M \to M \mid f \text{ is a } C^\infty \text{ diffeomorphism}\}$, the group of diffeomorphisms of M.

Let

$$\hat{\pi}: \mathrm{AUT}(P) \to \mathrm{Diff}(M); \qquad F \longmapsto \hat{\pi}(F) = f \in \mathrm{Diff}(M)$$

be the map that takes an automorphism of P to its induced diffeomorphism of M. $\hat{\pi}$ is then a homomorphism whose kernel is Aut(P), a normal subgroup of AUT(P). The quotient group AUT(P)/Aut(P) is then naturally isomorphic to a subgroup of Diff(M), containing the connected component of the identity, by the map

$$[F] = F \circ \mathrm{Aut}(P) \longmapsto \hat{\pi}(F \circ \mathrm{Aut}(P)) = \hat{\pi}(F) = f$$

We shall also be considering the following infinite dimensional Lie algebras:

$\mathscr{X}(P) = \{Z \mid Z \text{ is a } C^\infty \text{ vector field on } P\}$, the smooth vector fields on P;

$\mathscr{X}_G(P) = \{Z \in \mathscr{X}(P) \mid (R_a)_* Z = Z \text{ for all } a \in G\}$, the space of smooth G-invariant vector fields on P, where $(R_a)_* Z = TR_a \circ Z \circ R_a^{-1}$ denotes the push forward of the vector field Z by the diffeomorphism R_a;

$\mathscr{X}_G^{\mathrm{vert}}(P) = \{Z \in \mathscr{X}_G(P) \mid T\pi \circ Z = 0\}$, the space of G-invariant vertical vector fields on P;

$\mathscr{X}(M) = \{X \mid X \text{ is a } C^\infty \text{ vector field on } M\}$, the smooth vector fields on M.

These Lie algebras are taken with the usual bracket of vector fields as their Lie algebra structure. When M is compact, AUT(P), Aut(P), and $\mathscr{D}$(M) are infinite dimensional Lie groups. The brackets on their corresponding Lie algebras are then the right Lie algebra structures of their corresponding Lie groups.

The map $\pi: P \to M$ induces a Lie algebra homomorphism $\pi_*: \mathscr{X}_G(P) \to \mathscr{X}(M)$, $Z \longmapsto \pi_*(Z)$, defined by $\pi_*(Z)(x) = T_p\pi \cdot Z(p)$ where $x \in M$, $p \in \pi^{-1}(x)$ is arbitrary, and $T_p\pi: T_pP \to T_xM$. That this definition is independent of the choice of P follows from the G-invariance of Z. The kernel of π_* is then $\mathscr{X}_G^{vert}(P)$.

When M is a non-compact manifold, an important question is in what sense can the above groups be made into infinite dimensional Lie groups (see [11], p. 23, for a discussion on this point). One of the main difficulties is in finding the proper Lie algebra. For example, the Lie algebra $\mathscr{X}(M)$ is too large to be the Lie algebra of $\mathscr{D}$(M) since the noncomplete vector fields cannot be integrated globally. On the other hand, the space of complete vector fields is not closed under addition.

Although the ultimate answer to the above question will be of importance in Lagrangian field theory, we can bypass this question by working with diffeomorphisms of M that are the identity outside of compact subsets of M and with automorphisms of P that project onto such diffeomorphisms of M. However, for ease of discussion, and so as not to lose sight of the pertinent geometry, we shall assume, beginning in Sec. 2, that M is compact.

1.2 Vector-Valued k-Forms

Let $\pi: P(M, G) \to M$ be a principal fiber bundle over M, as above. Let E be a real finite-dimensional vector space, and let GL(E) be the general linear group of E. Let $g\ell(E)$ denote the Lie algebra of GL(E), the vector space of endomorphisms of E with the usual Lie algebra structure, $[A, B] = A \circ B - B \circ A$.

Let $\rho: G \to GL(E)$ be a continuous homomorphism of Lie groups. Then G acts on $P \times E$ on the right,

$$(P \times E) \times G \to P \times E, \qquad ((p, e), a) \longmapsto (p \cdot a, \rho(a^{-1}) \cdot e)$$

and the resulting quotient space $P \times_G E$ is the total space of the vector bundle $E_\rho(M) = E(M)$ over M associated with P(M, G), E, and ρ.

Similarly let $E^* = L(E, R)$ denote the dual of E. Then ρ induces a representation of G on E^*,

$$\rho^*: G \to GL(E^*)$$

defined by

$$\rho^*(a) = (\rho(a^{-1}))^*: E^* \to E^*$$

where $(\rho(a^{-1}))^*$ denotes the adjoint of $\rho(a^{-1})$: $E \to E$. We let $E_\rho^*(M) = E^*(M)$ denote the associated vector bundle associated with E^* and the representation ρ^*, and shall refer to it as the vector bundle dual to $E_\rho(M)$.

For k a nonnegative integer, let $C^\infty_{\rho,hor}(\Lambda^k(P) \otimes E)$ denote the horizontal E-valued k-forms on P of type ρ; i.e., if $\varphi \in C^\infty_{\rho,hor}(\Lambda^k(P) \otimes E)$, then φ satisfies

1. $R_a^*\varphi = \rho(a^{-1}) \cdot \varphi$ for all $a \in G$,
2. $\varphi(Z_1, \ldots, Z_k) = 0$ if any Z_i, $i = 1, \ldots, k$, is a vertical vector.

Note that the concept of horizontal E-valued k-form does not need a connection on P since vertical vectors are defined without a connection. Kobayashi and Nomzu [12] refer to such objects as tensorial. The word horizontal or tensorial is motivated by the following basic theorem, whose proof is straightforward (see for example [12], p. 116, or [4]).

1.2.1 Theorem. There is a natural isomorphism (i.e., linear bijection) between $C^\infty_{\rho,hor}(\Lambda^k(P) \otimes E)$ and $C^\infty(\Lambda^k(M) \otimes E_\rho(M))$, the space of C^∞ sections of the tensor product bundle $\Lambda^k(M) \otimes E_\rho(M) \to M$. We denote this isomorphism by $\varphi \longleftrightarrow \varphi_M$, where $\varphi \in C^\infty_{\rho,\mathrm{hor}}(\Lambda^k(P) \otimes E)$ and $\varphi_M \in C^\infty(\Lambda^k(M) \otimes E_\rho(M))$.

Remarks. 1. E-valued 0-forms on P of type ρ are always horizontal, so we suppress the horizontal designation in this case. Thus

$$C^\infty_\rho(\Lambda^0(P) \otimes E) = C^\infty_\rho(P, E) \approx C^\infty(E_\rho(M))$$

where $C^\infty_\rho(P, E) = \{\varphi : P \to E \mid R^*_a \varphi = \varphi \circ R_a = \rho(a^{-1}) \cdot \varphi$ for all $a \in G\}$, so that for $\varphi \in C^\infty_\rho(P;E)$, $\varphi(p \cdot a) = \rho(a^{-1}) \cdot \varphi(p)$. Thus the G-equivariant maps of P to E are isomorphic to the space of sections of $E_\rho(M)$.

2. For $k > n$, $C^\infty_{\rho,\mathrm{hor}}(\Lambda^n(P) \otimes E) \approx C^\infty(\Lambda^k(M) \otimes E_\rho(M)) = 0$, so we shall restrict k to $0 \leq k \leq n$.

1.2.2 Example. Let Ad: $G \to GL(\mathfrak{G})$ denote the adjoint representation of the Lie group G on its Lie algebra $\mathfrak{G}$. Then the vector bundle associated to P(M, G) and Ad is the adjoint bundle $\mathfrak{G}_{Ad}(M) \to M$, a bundle of Lie algebras over M. From Theorem 1.2.1, the space of sections $C^\infty(\mathfrak{G}_{Ad}(M))$ is naturally isomorphic with $C^\infty_{Ad}(P, \mathfrak{G}) = \{\phi : P \to \mathfrak{G} \mid R^*_a \phi = Ad(a^{-1}) \cdot \phi\}$. Moreover, we have the following additional isomorphism, again easily proved.

1.2.3 Proposition. The spaces $\mathscr{X}^{\mathrm{vert}}_G(P)$, $C^\infty_{Ad}(P, \mathfrak{G})$, and $C^\infty(\mathfrak{G}_{Ad}(M))$ are all naturally isomorphic. We denote these isomorphisms by

$$T \longleftrightarrow \phi_T \longleftrightarrow \phi_M$$

where $T \in \mathscr{X}^{\mathrm{vert}}_G(P)$, $\phi_T \in C^\infty_{Ad}(P, \mathfrak{G})$, and $\phi_M \in C^\infty(\mathfrak{G}_{Ad}(M))$.

Note that the above isomorphisms are natural and do not depend on any connection on P.

On E-valued k-forms of type ρ, one can define exterior differentiation

$$d: C^\infty_\rho(\Lambda^k(P) \otimes E) \to C^\infty_\rho(\Lambda^{k+1}(P) \otimes E)$$

and interior multiplication, i.e., contraction with a vector field $Z \in \mathscr{X}_G(P)$,

$$\varphi \longmapsto i_Z\varphi, \qquad i_Z\varphi \cdot (Z_1, \ldots, Z_{k-1}) = \varphi(Z, Z_1, \ldots, Z_{k-1})$$

and Lie differentiation with respect to $Z \in \mathscr{X}_G(P)$,

$$L_Z: C^\infty_\rho(\Lambda^k(P) \otimes E) \to C^\infty_\rho(\Lambda^k(P) \otimes E), \qquad \varphi \longmapsto L_Z\varphi$$

in the usual manner (see for example, [15], p. 255). If $Z \in \mathscr{X}_G(P)$, these operations respect the type ρ. Moreover, Lie differentiation respects horizontal forms on P (see below). However, to define exterior differentiation that respects horizontal forms on P will require a connection (see Sec. 1.3 below).

The Lie derivative $L_Z\varphi$ for $\varphi \in C^\infty_\rho(\Lambda^k(P) \otimes E)$ and $Z \in \mathscr{X}_G(P)$ is defined as follows. Let $F_\lambda \in AUT_{loc}(P)$, $\lambda \in (-\delta, \delta)$, $\delta > 0$ be a local flow for Z, defined on neighborhoods of the form $\pi^{-1}(U)$, U open in M, for each $p \in P$. Thus $F_0 = id_P$, $(dF_\lambda/d\lambda)|_{\lambda=0} = Z$, and $F^*_\lambda\varphi$, $\lambda \in (-\delta, \delta)$, is a curve in $C^\infty_\rho(\Lambda^k(P) \otimes E)$ in each such neighborhood. The Lie derivative is defined locally in each such neighborhood by

$$L_Z\varphi \equiv \frac{d}{d\lambda}(F_\lambda^*\varphi)\big|_{\lambda=0} \in C_\rho^\infty(\Lambda^k(P) \otimes E)$$

As for real valued k-forms, one easily establishes that

$$L_Z\varphi = i_Z d\varphi + di_Z\varphi$$

(see for example, [1], p. 115).

The following proposition will be useful (see for example [2], p. 47, or [4]).

1.2.4 Proposition. Let $\varphi \in C^\infty_{\rho,hor}(\Lambda^k(P) \otimes E)$ and $F \in AUT(P)$. Then $F^*\varphi \in C^\infty_{\rho,hor}(\Lambda^k(P) \otimes E)$. Also, if $Z \in \mathscr{X}_G(P)$, then

$$L_Z: C^\infty_{\rho,hor}(\Lambda^k(P) \otimes E) \to C^\infty_{\rho,hor}(\Lambda^k(P) \otimes E)$$

Remark. That $L_Z\varphi$ is horizontal if φ is horizontal and $Z \in \mathscr{X}_G(P)$ is the infinitesimal version of the fact that $F^*\varphi$ is horizontal if φ is horizontal and $F \in AUT(P)$.

In general there is no wedge product between E-valued k-forms on P. However, let

$$B: E_1 \times E_2 \to E_3$$

be a bilinear map. Then B induces a map

$$\wedge_B: C^\infty(\Lambda^k(P) \otimes E_1) \times C^\infty(\Lambda^\ell(P) \otimes E_2) \to C^\infty(\Lambda^{k+\ell}(P) \otimes E_3)$$

$$(\varphi_1, \varphi_2) \longmapsto \varphi_1 \wedge_B \varphi_2$$

defined by

$$\varphi_1 \wedge_B \varphi_2 \cdot (Z_1, \ldots, Z_{k+\ell}) =$$

$$\frac{1}{k!\,\ell!} \sum_{\sigma \in S_{(k+\ell)}} (\text{sign}\,\sigma)\, B(\varphi_1(Z_{\sigma(1)}, Z_{\sigma(2)}, \ldots, Z_{\sigma(k)}), \varphi_2(Z_{\sigma(k+1)}, \ldots, Z_{\sigma(k+\ell)}))$$

where Σ is taken over the permutation group $S_{(k+\ell)}$ on $k + \ell$ elements, a group of order $(k + \ell)!$. Our convention on the wedge product is that of [1], p. 102.

Exterior differentiation and interior multiplication behave as skew derivations with respect to this generalized wedge product. Thus if $\varphi_1 \in C^\infty(\Lambda^k(P) \otimes E\ )$, $\varphi_2 \in C^\infty(\Lambda^\ell(P) \otimes E_2)$, then

$$d(\varphi_1 \wedge_B \varphi_2) = d\varphi_1 \wedge_B \varphi_2 + (-1)^k \varphi_1 \wedge_B d\varphi_2$$

and

$$i_Z(\varphi_1 \wedge_B \varphi_2) = i_Z\varphi_1 \wedge_B \varphi_2 + (-1)^k \varphi_1 \wedge_B i_Z\varphi_2$$

Lie differentiation on the other hand behaves as a "Leibnitz" derivation

$$L_Z(\varphi_1 \wedge_B \varphi_2) = L_Z\varphi_1 \wedge_B \varphi_2 + \varphi_1 \wedge_B L_Z\varphi_2$$

An important example of the generalized wedge product is the case when $E_1 = E_2 = E_3 = \mathfrak{G}$ is a Lie algebra with Lie bracket [,]. We then denote the induced pairing

$$C^\infty(\Lambda^k(P) \otimes \mathfrak{G}) \times C^\infty(\Lambda^\ell(P) \otimes \mathfrak{G}) \to C^\infty(\Lambda^{k+\ell}(P) \otimes \mathfrak{G})$$

by $(\varphi_1, \varphi_2) \longmapsto [\varphi_1 \wedge \varphi_2]$, the wedge bracket of Lie algebra valued forms. The wedge bracket operation is thus a hybrid of the wedge product and the Lie bracket.

As another example of importance, let $E_2 = E_3 = E$, and let $E_1 = g\ell(E)$, the Lie algebra of linear maps on E. Then there is a wedge product

$$C^\infty(\Lambda^k(P) \otimes g\ell(E)) \times C^\infty(\Lambda^\ell(P) \otimes E) \to C^\infty(\Lambda^{k+\ell}(P) \otimes E)$$

given by $(\varphi_1, \varphi_2) \longmapsto \varphi_1 \wedge \varphi_2$, where the necessary "multiplication" of φ_1 and φ_2 is the natural bilinear pairing of $g\ell(E) \times E \to E$; $(A, e) \longmapsto A \cdot e$. Thus there is a natural wedge product between a $g\ell(E)$-valued form and an E-valued form, which we shall denote by $\varphi_1 \wedge \varphi_2 \in C^\infty(\Lambda^{k+\ell}(P) \otimes E)$, where $\varphi_1 \in C^\infty(\Lambda^k(P) \otimes g\ell(E))$ and $\varphi_2 \in C^\infty(\Lambda^\ell(P) \otimes E)$.

1.3 The Space of Connections on a Principal Fiber Bundle

In this section, we study the space of connections on a principal fiber bundle, and properties of the gauge covariant exterior derivative.

Let Γ be a smooth connection in a principal fiber bundle $P(M, G)$, and let ω denote its associated Lie algebra valued connection 1-form on P (see [12] for the basic definitions). Thus ω satisfies $R_a^*\omega = \mathrm{Ad}(a^{-1}) \cdot \omega$ for all $a \in G$, and $\omega(A^*) = A$ for all fundamental vector fields A^*. These conditions characterize a connection 1-form. We shall identify a connection with its connection 1-form. Thus we let

$$\mathscr{C} = \mathscr{C}(P) = \{\omega \in C^\infty_{\mathrm{Ad}}(\Lambda^1(P) \otimes \mathfrak{G}) \mid \omega(A^*) = A \text{ for all } A \in \mathfrak{G}\}$$

denote the space of connections of P.

The structure of $\mathscr{C}(P)$ is easily described. We set $A^1(P) = C^\infty_{\mathrm{Ad,hor}}(\Lambda^1(P) \otimes \mathfrak{G})$ and $A^1(M) = C^\infty(\Lambda^1(M) \otimes \mathfrak{G}_{\mathrm{Ad}}(M))$.

1.3.1 Theorem. Let $\omega_0 \in \mathscr{C}(P)$. Then

$$\mathscr{C}(P) = \{\omega_0\} + A^1(P)$$

Thus the tangent space to $\mathscr{C}(P)$ at $\omega \in \mathscr{C}(P)$ is

$$T_\omega \mathscr{C}(P) = \{\omega\} \times A^1(P) \approx \{\omega\} \times A^1(M)$$

Proof. One shows that the difference of two connections $\omega - \omega_0 \in A^1(P)$. Conversely, if $\varphi \in A^1(P)$, then $\omega_0 + \varphi \in \mathscr{C}(P)$.

Q.E.D.

Remark. Thus $\mathscr{C}(P)$ can be thought of as the space $C^\infty_{\mathrm{Ad,hor}}(\Lambda^1(P) \otimes \mathfrak{G})$ up to a choice of origin. In other words, $\mathscr{C}(P)$ is the affine space associated with the vector space $A^1(P)$.

A connection 1-form $\omega \in \mathscr{C}(P)$ defines a horizontal subspace H_p at each $p \in P$ by $H_p = \ker \omega(p) = \{Z_p \in T_pP \mid \omega(p) \cdot Z_p = 0\}$. A vector field $Z \in \mathscr{X}_G(P)$ is horizontal (with respect to ω) if $\omega(Z) = 0$. Let

$$\mathscr{X}_G^{\mathrm{hor}}(P) = \{Z \in \mathscr{X}_G(P) \mid Z \text{ is a horizontal vector field}\}$$

denote the G-invariant horizontal vector fields on P. For $X \in \mathscr{X}(M)$, X has a unique horizontal lift to P, denoted $\bar{X}$, and such that $\bar{X} \in \mathscr{X}_G^{\mathrm{hor}}(P)$. By this association, $\mathscr{X}(M)$ is isomorphic to $\mathscr{X}_G^{\mathrm{hor}}(P)$.

For $Z \in \mathscr{X}_G(P)$, let $Z_{\mathrm{vert}} \in \mathscr{X}_G^{\mathrm{vert}}(P)$ be defined by $Z_{\mathrm{vert}}(P) = T_e\Phi_p \cdot (\omega(p) \cdot Z(p))$, and $Z_{hor} = \overline{(\pi_* Z)} = \overline{X}$ where $X = \pi_* Z$. Thus every vector field $Z \in \mathscr{X}_G(P)$ has a unique decomposition

$$Z = Z_{vert} + Z_{hor} = T + \overline{X} \text{ where } T = Z_{vert} \text{ and } \overline{X} = \overline{(\pi_* Z)}$$

which gives the following vector space direct sum decomposition,

$$\mathscr{X}_G(P) = \mathscr{X}_G^{vert}(P) \oplus \mathscr{X}_G^{hor}(P) \approx C^\infty(\mathfrak{G}_{Ad}(M)) \oplus \mathscr{X}(M)$$

We shall return to this decomposition of $\mathscr{X}_G(P)$ in Sec. 2.3.

If $\varphi \in C_\rho^\infty(\Lambda^k(P) \otimes E)$, then $\varphi_{hor} \in C_{\rho,hor}(\Lambda^k(P) \otimes E)$ is defined by

$$\varphi_{hor}(p) \cdot (Z_1, \ldots, Z_k) = \varphi(p) \cdot ((Z_1)_{hor}, \ldots, (Z_k)_{hor})$$

where $Z_1, \ldots, Z_k \in T_pP$, and $(Z_i)_{hor}$ denotes the projection of Z_i, $i = 1, \ldots, k$, into the horizontal subspace at p. The gauge covariant exterior derivative of $\varphi \in C_\rho^\infty(\Lambda^k(P) \otimes E)$ is defined by

$$D\varphi = (d\varphi)_{hor} \in C_{\rho,hor}^\infty(\Lambda^{k+1}(P) \otimes E)$$

The gauge covariant exterior derivative thus gives rise to a sequence of first order linear differential operators on P,

$$C_\rho^\infty(P;E) \xrightarrow{D} C_{\rho,hor}^\infty(\Lambda^1(P) \otimes E) \xrightarrow{D} \cdots \xrightarrow{D} C_{\rho,hor}^\infty(\Lambda^{n-1}(P) \otimes E) \xrightarrow{D} C_{\rho,hor}^\infty(\Lambda^n(P) \otimes E)$$

For $\varphi_M \in C^\infty(\Lambda^k(M) \otimes E_\rho(M))$, we define

$$D_M\varphi_M \in C^\infty(\Lambda^{k+1}(M) \otimes E_\rho(M))$$

by $D_M\varphi_M \equiv (D\varphi)_M$, where $\varphi \in C_{\rho,hor}^\infty(\Lambda^k(P) \otimes E)$ corresponds to φ_M by Theorem 1.2.1. Then the above sequence descends to M as a sequence of first order linear differential operators on M,

$$C^\infty(E_\rho(M)) \xrightarrow{D_M} C^\infty(\Lambda^1(M) \otimes E_\rho(M)) \xrightarrow{D_M} \cdots \xrightarrow{D_M} C^\infty(\Lambda^{n-1} \otimes E_\rho(M))$$

$$\xrightarrow{D_M} C^\infty(\Lambda^n(M) \otimes E(M))$$

For a connection 1-form $\omega \in C_{Ad}^\infty(\Lambda^1(P) \otimes \mathfrak{G})$, its curvature form Ω is a Lie algebra valued 2-form, $\Omega \in C_{Ad,hor}^\infty(\Lambda^2(P) \otimes \mathfrak{G})$, defined by $\Omega = D\omega = (d\omega)_{hor}$. We let $\Omega_M \in C^\infty(\Lambda^2(M) \otimes \mathfrak{G}_{Ad}(M))$ denote the curvature projected down to M.

Given a continuous (and hence smooth) homomorphism $\rho: G \to GL(E)$, let $\rho' = T_e\rho: \mathfrak{G} \to g\ell(E)$ be the induced homomorphism of Lie algebras. Let $\rho' \circ \omega \in C^\infty(\Lambda^1(P) \otimes g\ell(E))$ denote the composition, defined by $(\rho' \circ \omega)(p) \cdot Z_p = \rho'(\omega(p) \cdot Z_p) \in g\ell(E)$. Then if $\varphi \in C_\rho^\infty(\Lambda^k(P) \otimes E)$, the wedge product $(\rho' \circ \omega) \wedge \varphi$ is defined, since $\rho' \circ \omega$ takes values in a vector space of linear maps on E. Similarly, $\rho' \circ \Omega \in C^\infty(\Lambda^2(P) \otimes g\ell(E))$, so that $(\rho' \circ \Omega) \wedge \varphi$ is defined.

The following formulas concerning the gauge covariant derivative are basic (see for example [2], [4], or [12]).

1.3.2 Theorem. Let ω be a connection on P with curvature $\Omega = D\omega$. Then

(1) $\Omega = D\omega = d\omega + \frac{1}{2}[\omega \wedge \omega]$, the structure equation for $D\omega$

(2) $D\Omega = 0$, Bianchi's identity.

If $\varphi \in C^{\infty}_{\rho,\mathrm{hor}}(\Lambda^k(P) \otimes E)$, then

(3) $D\varphi = d\varphi + (\rho' \circ \omega) \wedge \varphi$, the structure equation for $D\varphi$

(4) $D^2\varphi = D(D\varphi) = (\rho' \circ \Omega) \wedge \varphi$, Ricci's identity.

Note that $D\Omega = D(D\omega) = D^2\omega$. Thus the first two equations give formulas for $D\omega$ and $D^2\omega$, and the second two equations give formulas for $D\varphi$ and $D^2\varphi$.

1.4 The Lie Derivative of Vector Valued k-Forms

In Sec. 1.2 we defined $L_Z\varphi$ for $\varphi \in C^{\infty}_{\rho}(\Lambda^k(P) \otimes E)$ and $Z \in \mathscr{X}_G(P)$. We saw that $L_Z\varphi \in C^{\infty}_{\rho}(\Lambda^k(P) \otimes E)$, and that $L_Z\varphi = di_Z\varphi + i_Z d\varphi$. We wish to specialize this formula to the case that φ is a connection 1-form, and to the case that φ is horizontal.

The following formulas concerning Lie differentiation are fundamental.

<u>1.4.1 Theorem</u>. Let $\omega \in \mathscr{C}(P)$ be a connection on P with curvature form Ω. Let $T \in \mathscr{X}^{\mathrm{vert}}_G(P)$, $X \in \mathscr{X}(M)$, and let $\bar{X} \in \mathscr{X}^{\mathrm{hor}}_G(P)$ be its horizontal lift. Let $\varphi \in C^{\infty}_{\rho,\mathrm{hor}}(\Lambda^k(P) \otimes E)$. Then

$$L_T\omega = D\phi_T \in C^{\infty}_{\mathrm{Ad,hor}}(\Lambda^1(P) \otimes \mathfrak{G}), \text{ where } \phi_T = \omega(T) \in C^{\infty}_{\mathrm{Ad}}(P; \mathfrak{G}) \tag{1}$$

$$L_{\bar{X}}\omega = i_{\bar{X}}\Omega \in C^{\infty}_{\mathrm{Ad,hor}}(\Lambda^1(P) \otimes \mathfrak{G}) \tag{2}$$

$$L_T\varphi = -\rho'(\phi_T) \cdot \varphi \in C^{\infty}_{\rho,\mathrm{hor}}(\Lambda^k(P) \otimes E), \text{ where "}\cdot\text{" denotes the multiplication of } g\ell(E) \text{ on } E \tag{3}$$

$$L_{\bar{X}}\varphi = i_{\bar{X}}D\varphi + Di_{\bar{X}}\varphi \in C^{\infty}_{\rho,\mathrm{hor}}(\Lambda^k(P) \otimes E) \tag{4}$$

Thus if $Z \in \mathscr{X}_G(P)$, and $Z = T + \bar{X}$ is its canonical decomposition with respect to $\omega \in \mathscr{C}(P)$, then

$$L_Z\omega = L_T\omega + L_{\bar{X}}\omega = D\varphi_T + i_{\bar{X}}\Omega \in C^{\infty}_{\mathrm{Ad,hor}}(\Lambda^1(P) \otimes \mathfrak{G}) \tag{5}$$

$$L_Z\varphi = L_T\varphi + L_X\varphi = -\rho'(\phi_T) \cdot \varphi + i_{\bar{X}}D\varphi + Di_{\bar{X}}\varphi \in C^{\infty}_{\rho,\mathrm{hor}}(\Lambda^k(P) \otimes E) \tag{6}$$

<u>Proof</u>. (1) $L_T\omega = i_Td\omega + di_T\omega = i_Td\omega + d\phi_T$ since $i_T\omega = \omega(T) = \phi_T$. Since $D\omega$ is horizontal and T is vertical, $i_TD\omega = 0$. Thus using the structure equation for $D\omega$,

$$\begin{aligned} 0 = i_TD\omega &= i_Td\omega + \frac{1}{2}i_T[\omega \wedge \omega] \\ &= i_Td\omega + \frac{1}{2}([\omega(T) \wedge \omega] - [\omega \wedge \omega(T)]) \\ &= i_Td\omega - [\omega \wedge \phi_T] \end{aligned}$$

so that $i_Td\omega = [\omega \wedge \phi_T]$. Thus $L_T\omega = [\omega \wedge \phi_T] + d\phi_T = D\phi_T$.

(2) $L_{\bar{X}}\omega = i_{\bar{X}}d\omega + di_{\bar{X}}\omega = i_{\bar{X}}d\omega$ since $\omega(\bar{X}) = 0$. But $i_{\bar{X}}\Omega = i_{\bar{X}}D\omega = i_{\bar{X}}d\omega + \frac{1}{2}i_{\bar{X}}[\omega \wedge \omega] = i_{\bar{X}}d\omega$. Thus $L_{\bar{X}}\omega = i_{\bar{X}}\Omega$.

(3) Since φ and $D\varphi$ are horizontal, $i_T\varphi = 0$ and $i_T D\varphi = 0$. Thus $L_T\varphi = i_T d\varphi + di_T\varphi = i_T d\varphi$, and

$$0 = i_T D\varphi = i_T(d\varphi + \rho'\circ\omega \wedge \varphi)$$
$$= i_T d\varphi + \rho'\circ\omega(T) \wedge \varphi \qquad \text{(using } i_T\varphi = 0 \text{ again)}$$
$$= i_T d\varphi + \rho'(\phi_T) \wedge \varphi$$

so that $i_T d\varphi = -\rho'(\phi_T) \wedge \varphi = -\rho'(\phi_T)\cdot\varphi$ where the dot replaces the wedge since $\rho'(\phi_T)$ is a $g\ell(E)$-valued 0-form. Thus $L_T\varphi = i_T d\varphi = -\rho'(\phi_T)\cdot\varphi$.

(4)

$$i_{\bar{X}}D\varphi + Di_{\bar{X}}\varphi = i_{\bar{X}}(d\varphi + (\rho'\circ\omega)\wedge\varphi) + d(i_{\bar{X}}\varphi) = (\rho'\circ\omega)\wedge i_{\bar{X}}\varphi$$
$$= i_{\bar{X}}d\varphi - (\rho'\circ\omega)\wedge i_{\bar{X}}\varphi + di_{\bar{X}}\varphi + (\rho'\circ\omega)\wedge i_{\bar{X}}\varphi$$
$$= i_{\bar{X}}d\varphi + di_{\bar{X}}\varphi = L_{\bar{X}}\varphi$$

Equations (5) and (6) follow from Eqs. 1-4. Q.E.D.

Remarks. 1. Note that $L_Z\omega \in C^\infty_{Ad,hor}(\Lambda^1(P)\otimes \mathfrak{G})$ is horizontal, even though ω is not. This fact is the infinitesimal version of the fact that the difference of two connections is horizontal (see Theorem 1.3.1).

2. Equation (3) for $L_T\varphi$ does not require a connection on P, but Eq. (4) for $L_{\bar{X}}\varphi$ does require one.

Equations 1-6 can be projected down to M, to give

$$(L_T\omega)_M = (D\phi_T)_M = D_M\phi_M \quad \text{where } \phi_M = (\phi_T)_M \in C^\infty(\mathfrak{G}_{Ad}(M)) \tag{1}$$

$$(L_{\bar{X}}\omega)_M = (i_{\bar{X}}\Omega)_M = i_X\Omega_M \in C^\infty(\Lambda^1(M)\otimes \mathfrak{G}_{Ad}(M)) \tag{2}$$

$$(L_T\varphi)_M = -\left(\rho'(\phi_T)\cdot\varphi\right)_M = -\rho'_M(\phi_M)\cdot\varphi_M \in C^\infty(\Lambda^k(M)\otimes E_\rho(M)) \tag{3}$$

where $\rho'_M: \mathfrak{G}_{Ad}(M) \to g\ell(E_\rho(M))$ is the vector bundle morphism induced by the map $\rho': \mathfrak{G} \to g\ell(E)$, and $g\ell(E_\rho(M)) = \bigcup_{x\in M} g\ell(E_x)$ is the Lie algebra bundle of linear maps of $E_x = (E(M))_x$ to itself.

$$(L_{\bar{X}}\varphi)_M = (i_{\bar{X}}D\varphi + Di_{\bar{X}}\varphi)_M$$
$$= i_X(D\varphi)_M + D_M(i_X\varphi_M)$$
$$= i_X(D_M\varphi_M) + D_M(i_X\varphi_M) \in C^\infty(\Lambda^k(M)\otimes E_\rho(M)) \tag{4}$$

For $\varphi_M \in C^\infty(\Lambda^k(M)\otimes E_\rho(M))$, $X \in \mathscr{X}(M)$, and $\omega \in \mathscr{C}(P)$, we define the Lie derivative $L_X\varphi_M \in C^\infty(\Lambda^k(M)\otimes E_\rho(M))$ by $L_X\varphi_M = (L_{\bar{X}}\varphi)_M = i_X(D_M\varphi_M) + (D_M i_X\varphi_M)$. Note that unlike for a real valued k-form on M, a connection is needed to define the Lie derivative of a more general vector-bundle valued k-form φ_M on M with respect to a vector field X on M (see Remark 2 above).

Finally, projecting Eqs. (5) and (6) to M gives

$$(L_Z\omega)_M = D_M\phi_M + i_X\Omega_M, \text{ where } \phi_M = (\omega(Z))_M \text{ and } X = \pi_* Z \tag{5}$$

$$(L_Z\varphi)_M = -\rho'_M(\phi_M)\cdot\varphi_M + L_X\varphi_M \tag{6}$$

which summarizes Eqs. 1-4 above.

The above equations will play an important role in our later derivations.

2. COVARIANT LAGRANGIAN GAUGE FIELD THEORIES

We now give a detailed study of gauge field theories described by a Lagrangian density of the form

$$\mathscr{L} : \mathscr{M} \times \mathscr{C} \times \mathscr{H} \to C_d^\infty(M)$$

where $\mathscr{M}$ is the space of smooth pseudo-riemannian metrics with fixed signature on the spacetime manifold M, $\mathscr{C}$ is the space of connections on the principal fiber bundle $P(M, G) \to M$, and $\mathscr{H} = C^\infty_{\rho,\mathrm{hor}}(\Lambda^k(P) \otimes E)$ is the space of k-form Higgs fields on P of type (ρ, E), and $C_d^\infty(M)$ is the space of densities, or n-forms, on M. For ease of discussion, we shall assume that M is compact, although with suitable modifications, many of our results are valid when M is noncompact.

When $\mathscr{L}$ is generally covariant, we derive two differential identities, or master equations, for $\mathscr{L}$. We then give a geometric interpretation of these equations using the momentum mapping of the action of AUT(P) on the natural cotangent bundle of $\mathscr{M} \times \mathscr{C} \times \mathscr{H}$.

In this section, we consider all field variables on an equal footing, making no distinction between dynamical and parameter fields. Thus we consider the Lagrangian density $\mathscr{L}$ above as a "raw" Lagrangian. When in the applications we specify which fields are dynamical fields and which are parameter fields, the master equations will generate the specific gauge differential identities, local conservation laws, and global conservation laws of the Lagrangian field theory.

2.1 The Lagrangian Density for Gauge Field Theories

As in Sec. 1, we let $\pi: P(M, G) \to M$ be a principal fiber bundle over M with structure group G, E a finite-dimensional vector space, and $\rho: G \to GL(E)$ a representation of G on E.

For gauge field theories, a wide class of Lagrangian densities are of the form

$$\mathscr{L} : \mathscr{M} \times \mathscr{C} \times \mathscr{H} \to C_d^\infty(M), \qquad (g, \omega, \varphi) \longmapsto (g, \omega, \varphi)$$

Other structures, such as nondegenerate bilinear forms on $\mathfrak{G}$ and E may be involved in constructing $\mathscr{L}$. These additional structures are considered as part of the Lagrangian and not as additional fields. Such additional structures can in fact be generalized to fields; we shall pursue this point of view elsewhere.

For the Lagrangian density

$$\mathscr{L} : \mathscr{M} \times \mathscr{C} \times \mathscr{H} \to C_d^\infty$$

let

$$T\mathscr{L} : T(\mathscr{M} \times \mathscr{C} \times \mathscr{H}) \to TC_d^\infty$$

denote the tangent of $\mathscr{L}$. Let $S_2(M)$ denote the space of smooth 2-covariant symmetric tensor fields on M. Then $\mathscr{M}$ is open in $S_2(M)$, so for $g \in \mathscr{M}$, the tangent space of $\mathscr{M}$ at g is given by

$$T_g\mathscr{M} = \{g\} \times S_2(M)$$

Similarly, for $\omega \in \mathscr{C}$,

$$T_\omega \mathscr{C} = \{\omega\} \times A^1(P)$$

(see Theorem 1.3.1). Since $\mathscr{H}$ and $C_d^\infty(M)$ are linear spaces, for $\varphi \in \mathscr{H}$,

$$T_\varphi \mathscr{H} = \{\varphi\} \times \mathscr{H}$$

and for $\nu \in C_d^\infty(M)$,

$$T_\nu C_d^\infty(M) = \{\nu\} \times C_d^\infty(M)$$

Let

$$D\mathscr{L}(g,\omega,\varphi): S_2(M) \times A(P) \times \mathscr{H}(P) \to C_d^\infty$$

denote the principal part of the map $T\mathscr{L}$. Thus the following diagram commutes:

$$\begin{array}{ccc}
T_{(g,\omega,\varphi)}(\mathscr{M} \times \mathscr{C} \times \mathscr{H}) & \xrightarrow{T_{(g,\omega,\varphi)}\mathscr{L}} & T_{\mathscr{L}(g,\omega,\varphi)} C_d^\infty \\
\downarrow P_2 & & \downarrow P_2' \\
S_2(M) \times A^1(P) \times \mathscr{H}(P) & \xrightarrow[D\mathscr{L}(g,\omega,\varphi)]{} & C_d^\infty
\end{array}$$

where P_2 and P_2' are the obvious projections onto the second factors.

Let

$$D_g \mathscr{L}(g,\omega,\varphi): S_2(M) \to C_d^\infty(M)$$

$$D_\omega \mathscr{L}(g,\omega,\varphi): A^1(P) \to C_d^\infty(M)$$

and

$$D_\varphi \mathscr{L}(g,\omega,\varphi): \mathscr{H}(P) \to C_d^\infty(M)$$

denote the associated partial derivatives of $\mathscr{L}$ at (g,ω,φ) with respect to the variables g, ω, and φ, respectively. Thus if $(h,\sigma,\tau) \in S_2(M) \times A^1(P) \times \mathscr{H}(P)$,

$$D\mathscr{L}(g,\omega,\varphi) \cdot (h,\sigma,\tau) = D_g\mathscr{L}(g,\omega,\varphi) \cdot h + D_\omega \mathscr{L}(g,\omega,\varphi) \cdot \sigma + D_\varphi \mathscr{L}(g,\omega,\varphi) \cdot \tau$$

Let $A^1(M) = C^\infty(\Lambda^1(M) \otimes \mathfrak{G}_{Ad}(M))$ and let $\mathscr{H}(M) = C^\infty(\Lambda^k(M) \otimes E_\rho(M))$. Let $(h, \sigma_M, \tau_M) \in S_2(M) \times A^1(M) \times \mathscr{H}(M)$ correspond to $(h,\sigma,\tau) \in S_2(M) \times A^1(P) \times \mathscr{H}(P)$ by the isomorphism of Theorem 1.2.1. Then we define the projected derivatives of $D\mathscr{L}$, $D_\omega \mathscr{L}$, and $D_\varphi \mathscr{L}$ as follows:

$$(D\mathscr{L}(g,\omega,\varphi))_M: S_2(M) \times A^1(M) \times \mathscr{H}(M) \to C_d^\infty(M); \qquad (h, \omega_M, \tau_M) \longmapsto D\mathscr{L}(g,\omega,\varphi) \cdot (h,\sigma,\tau)$$

$$(D_\omega \mathscr{L}(g,\omega,\varphi))_M: A^1(M) \to C_d^\infty(M); \qquad \sigma_M \longmapsto (D_\omega \mathscr{L}(g,\omega,\varphi)) \cdot \sigma$$

and

$$(D_\varphi \mathscr{L}(g,\omega,\varphi))_M: \mathscr{H}(M) \to C_d^\infty(M); \qquad \tau_M \longmapsto (D_\varphi \mathscr{L}(g,\omega,\varphi)) \cdot \tau$$

Let $A^*(M) = C^\infty((TM)_d \otimes \mathfrak{G}^*_{Ad}(M))$ denote the space of $\mathfrak{G}^*_{Ad}(M)$-valued vector densities on M, let $\mathscr{H}^*(M) = C^\infty(\Lambda_k(M)_d \otimes E^*_\rho(M))$ denote the space of $E^*_\rho(M)$-valued contravariant k-form densities on M, and let $S_2^* = S_d^2(M)$ denote the space of smooth symmetric 2-contravariant tensor densities on M. There are natural bilinear pairings between $S_2(M)$ and $S_2^*(M)$, between $A^1(M)$ and $A^*(M)$, between $\mathscr{H}(M)$ and $\mathscr{H}^*(M)$, and between $C_d^\infty(M)$ and $C^\infty(M;R)$ (the space of smooth real-valued functions on M) which yield densities (n-forms) on M. With respect to these pairings we can define the natural L_2-adjoints of the differential operators above (see [7] for more information regarding the natural L_2-adjoints). The natural L_2-adjoints of these differential operators then map as follows:

$$(D\mathscr{L}(g,\omega,\varphi))_M^*: C^\infty(M;R) \to S_2^*(M) \times A^*(M) \times \mathscr{H}^*(M)$$

$$(D_g\mathscr{L}(g,\omega,\varphi))^*: C^\infty(M;R) \to S_2^*(M) = S_d^*(M)$$

$$(D_\omega\mathscr{L}(g,\omega,\varphi))_M^*: C^\infty(M;R) \to A^*(M) = C^\infty((\Lambda^1(M) \otimes \mathfrak{G}_{Ad}(M))_d^*)$$

$$(D_\varphi\mathscr{L}(g,\omega,\varphi))_M^*: C^\infty(M;R) \to \mathscr{H}^*(M) = C^\infty((\Lambda^k(M) \otimes E_\rho(M))_d^*)$$

The functional derivatives of $\mathscr{L}$ are then defined in terms of these natural L_2-adjoints as follows:

2.1.1 Definition. Let

$$\mathscr{L}: \mathscr{M} \times \mathscr{C} \times \mathscr{H} \to C_d^\infty$$

be a Lagrangian density. The (partial) functional derivatives of $\mathscr{L}$ with respect to the variables g, ω, and φ are, respectively, the maps

$$\frac{\delta\mathscr{L}}{\delta g}: \mathscr{M} \times \mathscr{C} \times \mathscr{H} \to S_2^*(M); \quad (g,\omega,\varphi) \longmapsto (D_g\mathscr{L}(g,\omega,\varphi))_M^* \cdot 1$$

$$\frac{\delta\mathscr{L}}{\delta\omega}: \mathscr{M} \times \mathscr{C} \times \mathscr{H} \to A^*(M); \quad (g,\omega,\varphi) \longmapsto (D_\omega\mathscr{L}(g,\omega,\varphi))_M^* \cdot 1, \text{ and}$$

$$\frac{\delta\mathscr{L}}{\delta\varphi}: \mathscr{M} \times \mathscr{C} \times \mathscr{H} \to \mathscr{H}^*(M); \quad (g,\omega,\varphi) \longmapsto (D_\varphi\mathscr{L}(g,\omega,\varphi))_M^* \cdot 1$$

where $1 \in C^\infty(M,R)$ is the constant function one.

The functional derivative of $\mathscr{L}$ is the map

$$\left(\frac{\delta\mathscr{L}}{\delta g}, \frac{\delta\mathscr{L}}{\delta\omega}, \frac{\delta\mathscr{L}}{\delta\varphi}\right): \mathscr{M} \times \mathscr{C} \times \mathscr{H} \to S_2^*(M) \times A^*(M) \times \mathscr{H}^*(M)$$

$$(g,\omega,\varphi) \longmapsto (D\mathscr{L}(g,\omega,\varphi))^* \cdot 1 = \left(\frac{\delta\mathscr{L}}{\delta g}(g,\omega,\varphi), \frac{\delta\mathscr{L}}{\delta\omega}(g,\omega,\varphi), \frac{\delta\mathscr{L}}{\delta\varphi}(g,\omega,\varphi)\right)$$

The functional derivative of $\mathscr{L}$ also defines the Lagrangian 1-form

$$\Psi_{\mathscr{L}}: \mathscr{M} \times \mathscr{C} \times \mathscr{H} \to T^*(\mathscr{M} \times \mathscr{C} \times \mathscr{H})_M = (\mathscr{M} \times \mathscr{C} \times \mathscr{H}) \times (S_2^*(M) \times A^*(M) \times \mathscr{H}^*(M))$$

$$(g,\omega,\varphi) \longmapsto \Psi_{\mathscr{L}}(g,\omega,\varphi) = (g,\omega,\varphi;(D\mathscr{L}(g,\omega,\varphi))_M^* \cdot 1) =$$

$$\left(g, \omega, \varphi, \frac{\delta\mathscr{L}}{\delta g}(g,\omega,\varphi), \frac{\delta\mathscr{L}}{\delta\omega}(g,\omega,\varphi), \frac{\delta\mathscr{L}}{\delta\varphi}(g,\omega,\varphi)\right)$$

Note that we are defining the natural cotangent bundle of $\mathscr{M} \times \mathscr{C} \times \mathscr{H}$ as $T^*(\mathscr{M} \times \mathscr{C} \times \mathscr{H})_M = (\mathscr{M} \times \mathscr{C} \times \mathscr{H}) \times (S_2^*(M) \times A^*(M) \times \mathscr{H}^*(M))$. With this definition there is a natural bilinear pairing between $T(\mathscr{M} \times \mathscr{C} \times \mathscr{H}) = (\mathscr{M} \times \mathscr{C} \times \mathscr{H}) \times (S_2(M) \times A^1(P) \times \mathscr{H}(P))$ and $T^*(\mathscr{M} \times \mathscr{C} \times \mathscr{H})_M$, given by

$$T^*(\mathscr{M} \times \mathscr{C} \times \mathscr{H})_M \times T(\mathscr{M} \times \mathscr{C} \times \mathscr{H}) \to C_d^\infty(M)$$

$$((g,\omega,\varphi;k,\alpha_M,\beta_M), (g,\omega,\varphi;h,\sigma,\tau)) \longmapsto k \cdot h + \alpha_M \cdot \sigma_M + \beta_M \cdot \tau_M$$

In particular, for $(h,\sigma,\tau) \in S_2(M) \times A^1(P) \times \mathscr{H}(P)$,

$$\Psi\mathscr{L}(g,\omega,\varphi) \cdot (g,\omega,\varphi;\, h,\sigma,\tau) = ((D\mathscr{L}(g,\omega,\varphi))^*_M \cdot 1) \cdot (h,\sigma_M,\tau_M)$$

$$= \frac{\delta\mathscr{L}}{\delta g}(g,\omega,\varphi) \cdot h + \frac{\delta\mathscr{L}}{\delta\omega}(g,\omega,\varphi) \cdot \sigma_M + \frac{\delta\mathscr{L}}{\delta\varphi}(g,\omega,\varphi) \cdot \tau_M$$

2.2 The Master Equation for Covariant Lagrangian Densities

In this section we derive the master equations for generally covariant gauge theory Lagrangian densities of the form

$$\mathscr{L}\colon \mathscr{M} \times \mathscr{C} \times \mathscr{H} \to C^\infty_d(M)$$

On $\mathscr{M}\times\mathscr{C}\times\mathscr{H}$, consider the right action by AUT(P)

$$\Phi_1 : (\mathscr{M}\times \mathscr{C} \times \mathscr{H}) \times \mathrm{AUT}(P) \to \mathscr{M} \times \mathscr{C} \times \mathscr{H};\ ((g,\omega,\varphi),F) \longmapsto (f^*g,\ F^*\omega,\ F^*\varphi)$$

where $f = \hat{\pi}(F) \in \mathrm{Diff}(M)$. Also, Diff(M) acts on $C^\infty_d(M)$ by pull-back,

$$\Phi_2 : C^\infty_d(M) \times \mathrm{Diff}(M) \to C^\infty_d(M);\quad (\nu, f) \longmapsto f^*\nu$$

With respect to these actions, we make the following definition.

2.2.1 Definition. Let

$$\mathscr{L}\colon \mathscr{M} \times \mathscr{C} \times \mathscr{H} \to C^\infty_d(M)$$

be a Lagrangian density. Then $\mathscr{L}$ is covariant with respect to AUT(P) if for $(g,\omega,\varphi) \in \mathscr{M}\times\mathscr{C} \times\mathscr{H}$, $F \in \mathrm{AUT}(P)$, and $f = \hat{\pi}(F) \in \mathrm{Diff}(M)$,

$$\mathscr{L}(f^*g,\ F^*\omega,\ F^*\varphi) = f^*(\mathscr{L}(g,\omega,\varphi))$$

In other words, $\mathscr{L}$ is covariant with respect to AUT(P) if $\mathscr{L}$ is equivariant with respect to the two actions given above, and the homomorphism

$$\hat{\pi}\colon \mathrm{AUT}(P) \to \mathrm{Diff}(M)$$

or equivalently, if the following diagram commutes:

$$\begin{array}{ccc}
(\mathscr{M}\times\mathscr{C}\times\mathscr{H}) \times \mathrm{AUT}(P) & \xrightarrow{\ \Phi_1\ } & \mathscr{M}\times\mathscr{C}\times\mathscr{H} \\
\Big\downarrow {\scriptstyle \mathscr{L}\times\hat{\pi}} & & \Big\downarrow {\scriptstyle \mathscr{L}} \\
C^\infty_d(M) \times \mathrm{Diff}(M) & \xrightarrow[\ \Phi_2\ ]{} & C^\infty_d(M)
\end{array}$$

If $\mathscr{L}$ is covariant with respect to AUT(P), then we shall say that $\mathscr{L}$ is generally covariant and AUT(P) is the general covariance group for $\mathscr{L}$.

Now let

$$\mathscr{L}\colon \mathscr{M} \times \mathscr{C} \times \mathscr{H} \to C^\infty_d(M)$$

be a Lagrangian density that is covariant with respect to AUT(P). We then have the following infinitesimal version of general covariance with respect to AUT(P).

2.2.1 Theorem. Let

$$\mathscr{L} : \mathscr{M} \times \mathscr{C} \times \mathscr{H} \to C^\infty_d(M)$$

be a Lagrangian density that is covariant with respect to AUT(P). Let $(g, \omega, \varphi) \in \mathcal{M} \times \mathcal{C} \times \mathcal{H}$. Then for $Z \in \mathcal{X}_G(P)$ with $X = \pi_* Z$,

$$D_g \mathcal{L}(g,\omega,\varphi) \cdot L_X g + D_\omega \mathcal{L}(g,\omega,\varphi) \cdot L_Z \omega + D_\varphi \mathcal{L}(g,\omega,\varphi) \cdot L_Z \varphi = \mathrm{div}_{nat}(X \mathcal{L}(g,\omega,\varphi)) \quad \text{(a)}$$

For $\phi \in C^\infty_{Ad}(P; \mathfrak{G})$,

$$D_\omega \mathcal{L}(g,\omega,\varphi) \cdot D\phi - D_\varphi \mathcal{L}(g,\omega,\varphi) \cdot (\rho'(\phi) \cdot \varphi) = 0 \quad \text{(b)}$$

and for $X \in \mathcal{X}(M)$,

$$D_g \mathcal{L}(g,\omega,\varphi) \cdot L_X g + D_\omega \mathcal{L}(g,\omega,\varphi) \cdot i_{\bar{X}} \Omega + D_\varphi \mathcal{L}(g,\omega,\varphi) \cdot L_{\bar{X}} \varphi = \mathrm{div}_{nat}(X \mathcal{L}(g,\omega,\varphi)) \quad \text{(c)}$$

where $\bar{X}$ is the horizontal lift of X with respect to the connection ω.

Projected to M, these equations are

$$D_g \mathcal{L}(g,\omega,\varphi) \cdot L_X g + (D_\omega \mathcal{L}(g,\omega,\varphi))_M \cdot (L_Z \omega)_M + (D_\varphi \mathcal{L}(g,\omega,\varphi))_M \cdot (L_Z \varphi)_M = \mathrm{div}_{nat}(X \mathcal{L}(g,\omega,\varphi)) \quad \text{(a')}$$

$$(D_\omega \mathcal{L}(g,\omega,\varphi))_M \cdot D_M \varphi_M - (D_\varphi \mathcal{L}(g,\omega,\varphi))_M \cdot (\rho'_M(\phi_M) \cdot \varphi_M) = 0 \quad \text{(b')}$$

where $\rho'_M \in C^\infty(\mathfrak{G}^*_{Ad}(M) \otimes g\ell(E_\rho(M)))$ (see Sec. 1.3); and

$$D_g \mathcal{L}(g,\omega,\varphi) \cdot L_X g + (D_\omega \mathcal{L}(g,\omega,\varphi))_M \cdot i_{\bar{X}} \Omega_M + (D_\varphi \mathcal{L}(g,\omega,\varphi))_M \cdot L_X \varphi_M = \mathrm{div}_{nat}(X \mathcal{L}(g,\omega,\varphi)) \quad \text{(c')}$$

<u>Note</u>. Let $\mathcal{X}_d(M) = C^\infty((TM)_d)$ denote the space of vector densities on M. Then if $Y_d \in \mathcal{X}_d(M)$, the natural divergence $\mathrm{div}_{nat}\, Y_d \in C^\infty_d(M)$ is defined in local coordinates by $\mathrm{div}_{nat}\, Y_d = Y^\alpha_{d,\alpha}$, where the comma denotes ordinary differentiation. Since Y_d is a vector density, $\mathrm{div}_{nat}\, Y_d$ is a density. Note that the operator div_{nat} on vector densities is defined without a riemannian metric on M.

<u>Proof</u>. Let F_λ be the flow of Z. Then $\hat{\pi}(F_\lambda) = f_\lambda$ is the flow of $\pi_* Z = X$. Since $\mathcal{L}$ is covariant with respect to AUT(P), for all $\lambda \in R$,

$$\mathcal{L}(f^*_\lambda g, F^*_\lambda \omega, F^*_\lambda \varphi) = f^*_\lambda(\mathcal{L}(g,\omega,\varphi))$$

Differentiating with respect to λ and evaluating at $\lambda = 0$ gives Eq. (a)

$$D_g \mathcal{L} \cdot L_X g + D_\omega \mathcal{L} \cdot L_Z \omega + D_\varphi \mathcal{L} \cdot L_Z \varphi = \mathrm{div}_{nat}(X \mathcal{L})$$

Equation (b) then follows from Eq. (a) by letting $Z = T$ correspond to $\phi \in C^\infty_{Ad}(P, \mathfrak{G})$ as per Proposition 1.2.3. Then $\phi = \omega(T)$, and by Theorem 1.4.1, $L_T \omega = D\phi$, $L_T \varphi = -\rho'(\phi) \cdot \varphi$, $\pi_* T = 0$, and so Eq. (b) follows.

Equation (c) follows from (a) by letting $Z = \bar{X}$. Then $L_{\bar{X}} \omega = i_{\bar{X}} \Omega$, and $\pi_* \bar{X} = X$.

Q.E.D.

Using the above equations, we now derive our first master equation. We shall need the following operators. For $\omega \in \mathcal{C}$, let

$$D_M : C^\infty(\mathfrak{G}_{Ad}(M)) \to C^\infty(\Lambda^1(M) \otimes \mathfrak{G}_{Ad}(M))$$

denote the gauge covariant derivative. Then its natural L_2-adjoint is

$$D^*_M : C^\infty((TM)_d \otimes \mathfrak{G}^*_{Ad}(M)) \to C^\infty(\mathfrak{G}^*_{Ad}(M)_d)$$

Let $\mathrm{DIV} = -D^*_M$ and $A^*(M) = C^\infty((TM)_d \otimes \mathfrak{G}^*_{Ad}(M))$.

For $\varphi_M \in \mathscr{H}(M)$, let

$$\alpha_{\varphi_M} : \mathscr{X}(M) \to \mathscr{H}(M); \quad X \longmapsto \alpha_{\varphi_M}(X) = L_X \varphi_M$$

and let

$$\alpha^*_{\varphi_M} : \mathscr{H}^*(M) \to C^\infty(\Lambda^1_d(M))$$

denote its natural L_2-adjoint.

2.2.2 Theorem. Let

$$\mathscr{L} : \mathscr{M} \times \mathscr{C} \times \mathscr{H} \to C^\infty_d(M)$$

be covariant with respect to AUT(P). Then for $(g, \omega, \varphi) \in \mathscr{M} \times \mathscr{C} \times \mathscr{H}$

$$\mathrm{DIV}\left(\frac{\delta \mathscr{L}}{\delta \omega}(g, \omega, \varphi)\right) + \left(\frac{\delta \mathscr{L}}{\delta \varphi}(g, \omega, \varphi)\right) \cdot (\rho'_M \cdot \varphi_M) = 0 \tag{a}$$

and

$$-2\left(\mathrm{div}_g \frac{\delta \mathscr{L}}{\delta g}(g, \omega, \varphi)\right)^\flat + \frac{\delta \mathscr{L}}{\delta \omega}(g, \omega, \varphi) \cdot \Omega_M + \alpha^*_{\varphi_M}\left(\frac{\delta \mathscr{L}}{\delta \varphi}(g, \omega, \varphi)\right) = 0 \tag{b}$$

where

$$\mathrm{DIV}: C^\infty((\Lambda^1(M) \otimes \mathfrak{G}_{Ad}(M))^*_d) \to C^\infty(\mathfrak{G}^*_{Ad}(M)_d)$$

is the divergence with respect to the connection ω,

$$\mathrm{div}_g : S^2_d \to \mathscr{X}_d(M)$$

is the metric divergence, and

$$\alpha^*_{\varphi_M} : \mathscr{H}^*(M) = C^\infty((\Lambda^k(M) \otimes E_\rho(M))^*_d) \to C^\infty_d(\Lambda^1(M))$$

is the natural L_2-adjoint of α_{φ_M}.

In local coordinates, these equations are

$$\left(\frac{\delta \mathscr{L}}{\delta \omega}\right)^\alpha_{a:\alpha} + \left(\frac{\delta \mathscr{L}}{\delta \varphi}\right)^{\alpha_1 \cdots \alpha_k}_A (\rho'_M)^A_{aB}(\varphi_M)^B_{\alpha_1 \cdots \alpha_k} = 0$$

$$-2\left(\frac{\delta \mathscr{L}}{\delta g}\right)^\beta_{\alpha;\beta} + \left(\frac{\delta \mathscr{L}}{\delta \omega}\right)^\beta_a (\Omega_M)^a_{\alpha\beta} + \alpha^*_{\varphi_M}\left(\frac{\delta \mathscr{L}}{\delta \varphi}\right) = 0$$

Note. The flat sign "$\flat$" lowers the contravariant index using g,

$$\flat : \mathscr{X}_d \to C^\infty(\Lambda^1_d(M)) \qquad X_d \longmapsto X^\flat_d$$

which in local coordinates is $(X_d)^\alpha \longmapsto g_{\alpha\beta}(X_d)^\beta$.

Proof. From Eqs. (b') and (c') of Proposition 2.2.1,

$$(D_\omega \mathscr{L}(g, \omega, \varphi))_M \cdot D_M \phi_M - (D_\varphi \mathscr{L}(g, \omega, \varphi))_M \cdot (\rho'_M(\phi_M) \cdot \varphi_M) = 0 \tag{b'}$$

and

$$D_g\mathscr{L}(g,\omega,\varphi)\cdot L_X g + (D_\omega \mathscr{L}(g,\omega,\varphi))_M \cdot i_X\Omega_M + (D_\varphi\mathscr{L}(g,\omega,\varphi))_M \cdot L_X\varphi_M = \mathrm{div}_{nat}(X\mathscr{L}(g,\omega,\varphi)) \qquad (c')$$

for $(g,\omega,\varphi) \in \mathscr{M}\times\mathscr{C}\times\mathscr{H}$, and for $(\phi_M, X) \in C^\infty(\mathfrak{G}_{Ad}(M)) \times \mathscr{X}(M)$. Thus if $\phi_M \in C^\infty(\mathfrak{G}_{Ad}(M))$,

$$\int (D_\omega \mathscr{L}(g,\omega,\varphi))_M \cdot D_M\phi_M - \int (D_\varphi\mathscr{L}(g,\omega,\varphi))_M \cdot (\rho'_M(\phi_M)\cdot\varphi_M)$$
$$= \int ((D_\omega\mathscr{L}(g,\omega,\varphi)^*_M \cdot 1)\cdot D_M\phi_M - \int ((D_\varphi\mathscr{L}(g,\omega,\varphi)^*_M\cdot 1)\cdot(\rho'_M(\phi_M)\cdot\varphi_M)$$
$$= -\int \left(\mathrm{DIV}\left(\frac{\delta\mathscr{L}}{\delta\omega}(g,\omega,\varphi)\right)\right)\cdot\phi_M - \int\left(\frac{\delta\mathscr{L}}{\delta\varphi}(g,\omega,\varphi)\right)\cdot(\rho'_M(\phi_M)\cdot\varphi_M) = 0$$

where we have integrated by parts freely (since M is compact) to get the functional derivatives of $\mathscr{L}$ and the natural L_2 adjoint of D_M. Since this equation is true for all $\phi_M \in C^\infty(\mathfrak{G}_{Ad}(M))$,

$$\mathrm{DIV}\left(\frac{\delta\mathscr{L}}{\delta\omega}(g,\omega,\varphi)\right) - \frac{\delta\mathscr{L}}{\delta\varphi}(g,\omega,\varphi)\cdot(\rho'_M\cdot\varphi_M) = 0$$

Similarly, for $X \in \mathscr{X}(M)$, integrating Eq. (c') over M gives

$$\int D_g\mathscr{L}(g,\omega,\varphi)\cdot L_X g + \int (D_\omega\mathscr{L}(g,\omega,\varphi))_M\cdot i_X\Omega_M + \int(D_\varphi\mathscr{L}(g,\omega,\varphi))_M\cdot L_X\varphi_M$$
$$= \int \mathrm{div}_{nat}(X\mathscr{L}(g,\omega,\varphi)) = 0$$

since the last integral is the divergence of a vector density on compact manifold. Integrating the left hand side by parts then gives

$$\int((D_g\mathscr{L}(g,\omega,\varphi))^*\cdot 1)\cdot L_X g + \int((D_\omega\mathscr{L}(g,\omega,\varphi))^*_M\cdot 1)\cdot i_X\Omega_M + \int((D_\varphi\mathscr{L}(g,\omega,\varphi))^*_M\cdot 1)\cdot L_X\varphi_M$$
$$= -2\int\left(\mathrm{div}_g\left(\frac{\delta\mathscr{L}}{\delta g}(g,\omega,\varphi)\right)\right)\cdot X + \int\left(\frac{\delta\mathscr{L}}{\delta\omega}(g,\omega,\varphi)\right)\cdot i_X\Omega_M + \int\left(\alpha^*_{\varphi_M}\left(\frac{\delta\mathscr{L}}{\delta\varphi}(g,\omega,\varphi)\right)\right)\cdot X = 0$$

Since this equation is true for all $X \in \mathscr{X}(M)$, it follows that

$$-2\left(\mathrm{div}_g\left(\frac{\delta\mathscr{L}}{\delta g}(g,\omega,\varphi)\right)\right) + \left(\frac{\delta\mathscr{L}}{\delta\omega}(g,\omega,\varphi)\right)\cdot\Omega_M + \alpha^*_{\varphi_M}\left(\frac{\delta\mathscr{L}}{\delta\varphi}(g,\omega,\varphi)\right) = 0$$

For the local coordinate expressions, note that

$$\frac{\delta\mathscr{L}}{\delta\omega}\cdot i_X\Omega_M = \left(\frac{\delta\mathscr{L}}{\delta\omega}\right)_a^\beta X^\alpha(\Omega_M)^a_{\alpha\beta}$$

so

$$\left(\frac{\delta\mathscr{L}}{\delta\omega}\cdot\Omega_M\right)_\alpha = \left(\frac{\delta\mathscr{L}}{\delta\omega}\right)_a^\beta(\Omega_M)^a_{\alpha\beta}$$

Similarly,

$$\frac{\delta\mathscr{L}}{\delta\varphi}\cdot(\rho'_M(\phi_M)\cdot\varphi_M) = \left(\frac{\delta\mathscr{L}}{\delta\varphi}\right)_A^{\alpha_1\cdots\alpha_k}(\rho')^A_{aB}(\phi_M)^a\varphi^B_{\alpha_1\cdots\alpha_k}$$

$$\left(\frac{\delta\mathscr{L}}{\delta\varphi}\cdot(\rho'_M\cdot\varphi_M)\right)_a = \left(\frac{\delta\mathscr{L}}{\delta\varphi}\right)_A^{\alpha_1\cdots\alpha_k}(\rho')^A_{aB}\varphi^B_{\alpha_1\cdots\alpha_k}$$

The local coordinate expressions for the other terms follow similarly. Q.E.D.

Remarks. 1. The equations (a) and (b) are true for any covariant Lagrangian density, and so are the general covariance differential identities for $\mathscr{L}$. These equations depend on the functional derivatives of $\mathscr{L}$ with respect to all three variables g, ω, and φ.

2. No hypotheses other than general covariance are placed on $\mathscr{L}$. For example, it is not required that $\mathscr{L}$ be of first order, or that the fields be minimally or non-derivatively coupled to each other.

3. The direct sum of Eqs. (a) and (b) lives in the space

$$C^\infty(\mathfrak{G}^*_{Ad}(M)_d) \oplus C^\infty(\Lambda^1_d(M)) \approx \mathscr{X}^*_G(P)$$

We shall pursue this interpretation further in the next section.

Our second set of master equations will be our main tool in the study of global conservation laws (Sec. 3).

2.2.3 Theorem. Let

$$\mathscr{L} : \mathscr{M} \times \mathscr{C} \times \mathscr{H} \to C^\infty_d(M)$$

be a generally covariant Lagrangian density where $\mathscr{H} = C^\infty_{\rho,hor}(\Lambda^k(P) \otimes E)$, $0 \le k \le n$. Then for $(g,\omega,\varphi) \in \mathscr{M} \times \mathscr{C} \times \mathscr{H}$ and $Z \in \mathscr{X}_G(P)$, and $1 \le k \le n$,

$$\operatorname{div}_{nat}\left(2X^\flat \cdot \frac{\delta\mathscr{L}}{\delta g}(g,\omega,\varphi) + \phi_M \cdot \frac{\delta\mathscr{L}}{\delta\omega}(g,\omega,\varphi) + \frac{(-1)^{k-1}}{(k-1)!}\, i_X\varphi_M \cdot \frac{\delta\mathscr{L}}{\delta\varphi}(g,\omega,\varphi)\right)$$

$$= L_Xg \cdot \frac{\delta\mathscr{L}}{\delta g}(g,\omega,\varphi) + (L_Z\omega)_M \cdot \frac{\delta\mathscr{L}}{\delta\omega}(g,\omega,\varphi) + (L_Z\varphi)_M \cdot \frac{\delta\mathscr{L}}{\delta\varphi}(g,\omega,\varphi) \tag{a}$$

where

$$X = \pi_* Z \in \mathscr{X}(M) \quad \text{and} \quad \phi_M = (\omega(Z))_M \in C^\infty(\mathfrak{G}_{Ad}(M))$$

If k = 0, Eq. (a) remains true with the last term on the left hand side deleted ($i_X\varphi_M = 0$).

In local coordinates, Eq. (a) is

$$\left(2g_{\beta\gamma}X^\gamma\left(\frac{\delta\mathscr{L}}{\delta g}\right)^{\alpha\beta} + \phi^a_M\left(\frac{\delta\mathscr{L}}{\delta\omega}\right)^\alpha_a + \frac{(-1)^{k-1}}{(k-1)!}X^\beta(\varphi_M)^A_{\beta\alpha_1\cdots\alpha_{k-1}}\left(\frac{\delta\mathscr{L}}{\delta\varphi}\right)^{\alpha_1\cdots\alpha_{k-1}\alpha}_A\right)_{,\alpha}$$

$$= (L_Xg)_{\alpha\beta}\left(\frac{\delta\mathscr{L}}{\delta g}\right)^{\alpha\beta} + ((L_Z\omega)_M)^a_\alpha\left(\frac{\delta\mathscr{L}}{\delta\omega}\right)^\alpha_a + ((L_Z\varphi)_M)^A_{\alpha_1\cdots\alpha_k}\left(\frac{\delta\mathscr{L}}{\delta\varphi}\right)^{\alpha_1\cdots\alpha_k}_A$$

where we have suppressed the variables (g, ω, φ) in the functional derivative.

Note. $X^\flat \cdot (\delta\mathscr{L}/\delta g)(g,\omega,\varphi) \in \mathscr{X}_d(M)$ is the natural contraction of the 1-form $X^\flat \in C^\infty(\Lambda^1(M))$ with $(\delta\mathscr{L}/\delta g)(g,\omega,\varphi) \in S^2_d(M)$ to give a vector density.

Proof. Using the equations in Theorem 2.2.2 and suppressing the variables in the functional derivatives, we compute as follows:

$$\operatorname{div}_{nat}\left(2X^\flat \cdot \frac{\delta\mathscr{L}}{\delta g}\right) = L_Xg \cdot \frac{\delta\mathscr{L}}{\delta g} + 2X^\flat \cdot \operatorname{div}_g\left(\frac{\delta\mathscr{L}}{\delta g}\right)$$

$$= L_Xg \cdot \frac{\delta\mathscr{L}}{\delta g} + i_X\left(\frac{\delta\mathscr{L}}{\delta\omega} \cdot \Omega_M\right) + i_X\left(\alpha^*_{\varphi_M}\left(\frac{\delta\mathscr{L}}{\delta\varphi}\right)\right)$$

$$= L_Xg \cdot \frac{\delta\mathscr{L}}{\delta g} + \frac{\delta\mathscr{L}}{\delta\omega} \cdot i_X\Omega_M + i_X\alpha^*_{\varphi_M}\left(\frac{\delta\mathscr{L}}{\delta\varphi}\right)$$

and

$$\mathrm{div}_{\mathrm{nat}}\left(\phi_M \cdot \frac{\delta\mathscr{L}}{\delta\omega}\right) = D_M\phi_M \cdot \frac{\delta\mathscr{L}}{\delta\omega} + \phi_M \cdot \mathrm{DIV}\left(\frac{\delta\mathscr{L}}{\delta\omega}\right) = D_M\phi_M \cdot \frac{\delta\mathscr{L}}{\delta\omega} - \frac{\delta\mathscr{L}}{\delta\varphi} \cdot (\rho'_M(\phi_M) \cdot \varphi_M)$$

Moreover, a computation shows that for $k \geq 1$,

$$\frac{(-1)^{k-1}}{(k-1)!}\,\mathrm{div}_{\mathrm{nat}}\left(i_X\varphi_M \cdot \frac{\delta\mathscr{L}}{\delta\varphi}\right) = \frac{\delta\mathscr{L}}{\delta\varphi} \cdot L_X\varphi_M - i_X\left(\alpha^*_{\varphi_M}\left(\frac{\delta\mathscr{L}}{\delta\varphi}\right)\right)$$

Thus

$$\mathrm{div}_{\mathrm{nat}}\left(2X^\flat \cdot \frac{\delta\mathscr{L}}{\delta g} + \phi_M \cdot \frac{\delta\mathscr{L}}{\delta\omega} + \frac{(-1)^{k-1}}{(k-1)!}\, i_X\varphi_M \cdot \frac{\delta\mathscr{L}}{\delta\varphi}\right)$$

$$= L_X g \cdot \frac{\delta\mathscr{L}}{\delta g} + \frac{\delta\mathscr{L}}{\delta\omega} \cdot i_X\Omega_M + i_X\left(\alpha^*_{\varphi_M}\left(\frac{\delta\mathscr{L}}{\delta\varphi}\right)\right) + D_M\varphi_M \cdot \frac{\delta\mathscr{L}}{\delta\omega}$$

$$- \frac{\delta\mathscr{L}}{\delta\varphi} \cdot (\rho'_M(\phi_M) \cdot \varphi_M) + L_X\varphi_M \cdot \frac{\delta\mathscr{L}}{\delta\varphi} - i_X\left(\alpha^*_{\varphi_M}\left(\frac{\delta\mathscr{L}}{\delta\varphi}\right)\right)$$

$$= L_X g \cdot \frac{\delta\mathscr{L}}{\delta g} + (D_M\varphi_M + i_X\Omega_M) \cdot \frac{\delta\mathscr{L}}{\delta\omega} + (-\rho'_M(\phi_M) \cdot \varphi_M + L_X\varphi_M) \cdot \frac{\delta\mathscr{L}}{\delta\varphi}$$

$$= L_X g \cdot \frac{\delta\mathscr{L}}{\delta g} + (L_Z\omega)_M \cdot \frac{\delta\mathscr{L}}{\delta\omega} + (L_Z\varphi)_M \cdot \frac{\delta\mathscr{L}}{\delta\varphi}$$

If $k = 0$, $i_X\varphi_M = 0$, $(L_Z\varphi)_M = (i_Z D\varphi)_M = -\rho'(\phi_M) \cdot \varphi_M + i_X D_M\varphi_M$, and $\alpha^*_{\varphi_M}(\delta\mathscr{L}/\delta\varphi) = D_M\varphi_M \cdot (\delta\mathscr{L}/\delta\varphi)$. Thus

$$\mathrm{div}_{\mathrm{nat}}\left(2X^\flat \cdot \frac{\delta\mathscr{L}}{\delta g} + \phi_M \cdot \frac{\delta\mathscr{L}}{\delta\omega}\right) = L_X g \cdot \frac{\delta\mathscr{L}}{\delta g} + (D_M\phi_M + i_X\Omega_M) \cdot \frac{\delta\mathscr{L}}{\delta\omega}$$

$$+ i_X\left(\alpha^*_{\varphi_M}\left(\frac{\delta\mathscr{L}}{\delta\varphi}\right)\right) - \frac{\delta\mathscr{L}}{\delta\varphi} \cdot (\rho'_M(\phi_M) \cdot \varphi_M)$$

$$= L_X g \cdot \frac{\delta\mathscr{L}}{\delta g} + (L_Z\omega)_M \cdot \frac{\delta\mathscr{L}}{\delta\omega} + i_X\left(D_M\varphi_M \cdot \frac{\delta\mathscr{L}}{\delta\varphi}\right) - \frac{\delta\mathscr{L}}{\delta\varphi} \cdot (\rho'_M(\phi_M) \cdot \varphi_M)$$

$$= L_X g \cdot \frac{\delta\mathscr{L}}{\delta g} + (L_Z\omega)_M \cdot \frac{\delta\mathscr{L}}{\delta\omega} + (i_Z D\varphi)_M \cdot \frac{\delta\mathscr{L}}{\delta\varphi}$$

since $\alpha^*_{\varphi_M}(\delta\mathscr{L}/\delta\varphi) = D_M\varphi_M \cdot (\delta\mathscr{L}/\delta\varphi)$. Q.E.D.

Remark. For $1 \leq k \leq n$, consider the vector density

$$Y_d = 2X^\flat \cdot \frac{\delta\mathscr{L}}{\delta g} + \phi_M \cdot \frac{\delta\mathscr{L}}{\delta\omega} + \frac{(-1)^{k-1}}{(k-1)!}\, i_X\varphi_M \cdot \frac{\delta\mathscr{L}}{\delta\varphi} \in \mathscr{X}_d(M)$$

and for $k = 0$, delete the last term. Then in local coordinates,

$$\mathrm{div}_{\mathrm{nat}}\, Y_d = \mathrm{div}_\alpha (Y_d)^\alpha = (Y_d)^\alpha_{,\alpha}$$

is a metric-independent expression (see Note following Theorem 2.2.1). Using Y_d, Theorem 2.2.3 can be summarized by the very compact expression

$$\mathrm{div}_{\mathrm{nat}}\, Y_d = L_X g \cdot \frac{\delta\mathscr{L}}{\delta g} + (L_Z\omega)_M \cdot \frac{\delta\mathscr{L}}{\delta\omega} + (L_Z\varphi)_M \cdot \frac{\delta\mathscr{L}}{\delta\varphi}$$

This equation is of importance inasmuch as it expresses the geometrical quantity on the right hand side as the natural divergence of a vector density. We shall study the geometrical interpretation of this equation in the next section.

2.3 The Vanishing of the Momentum Map

In this section we consider an interesting interaction between covariant Lagrangian densities and the momentum map associated with the cotangent action of AUT(P) on $T^*(\mathcal{M}\times\mathcal{C}\times\mathcal{H})_M$. This interaction involves the momentum mapping

$$J: T^*(\mathcal{M}\times\mathcal{C}\times\mathcal{H})_M \to \mathcal{X}^*_G(P)$$

for the cotangent action and the Lagrangian 1-form

$$\Psi_{\mathcal{L}}: \mathcal{M}\times\mathcal{C}\times\mathcal{H} \to T^*(\mathcal{M}\times\mathcal{C}\times\mathcal{H})_M$$

Our main theorem, the vanishing of the momentum theorem states that if $\mathcal{L}$ is covariant with respect to AUT(P), then

$$J \circ \Psi_{\mathcal{L}} = 0$$

We shall see that this theorem is the geometrical formulation of the master equations of Theorem 2.2.2 and 2.2.3. For more information regarding the momentum mapping of a cotangent action see [1], pp. 276 and 283.

As we have seen, the direct sum of Eqs. (a) and (b) in Theorem 2.2.2 lives in the space $C^\infty(\mathfrak{G}^*_{Ad}(M)_d) \oplus C^\infty(\Lambda^1_d(M))$. This direct sum in turn is the natural dual of the space

$$C^\infty(\mathfrak{G}_{Ad}(M)) \oplus \mathcal{X}(M)$$

which, if a connection ω is given, is isomorphic to the direct sum

$$\mathcal{X}^{vert}_G(P) \oplus \mathcal{X}^{hor}_G(P) = \mathcal{X}_G(P)$$

the Lie algebra of AUT(P). Thus using the connection ω, the natural dual $\mathcal{X}^*_G(P)$ of the Lie algebra $\mathcal{X}_G(P)$ is given by

$$\mathcal{X}^*_G(P) \approx C^\infty(\mathfrak{G}^*_{Ad}(M)_d) \oplus C^\infty(\Lambda^1_d(M))$$

Now consider again the right action of AUT(P) on $\mathcal{M}\times\mathcal{C}\times\mathcal{H}$,

$$\Phi: (\mathcal{M}\times\mathcal{C}\times\mathcal{H}) \times AUT(P) \to \mathcal{M}\times\mathcal{C}\times\mathcal{H}\,; \quad ((g,\omega,\varphi), F) \longmapsto (f^*g, F^*\omega, F^*\varphi)$$

where $f = \hat{\pi}(F) \in \mathcal{D}(M)$. For $(g,\omega,\varphi) \in \mathcal{M}\times\mathcal{C}\times\mathcal{H}$, let

$$\Phi_{(g,\omega,\varphi)}: AUT(P) \to \mathcal{M}\times\mathcal{C}\times\mathcal{H}, \quad F \longmapsto \Phi_{(g,\omega,\varphi)}(F) = \Phi(g,\omega,\varphi;F) = (f^*g, F^*\omega, F^*\varphi)$$

denote the orbit map at (g,ω,φ). Then, letting $\mathcal{X}(\mathcal{M}\times\mathcal{C}\times\mathcal{H}) = C^\infty(T(\mathcal{M}\times\mathcal{C}\times\mathcal{H}))$ denote the smooth vector fields on $\mathcal{M}\times\mathcal{C}\times\mathcal{H}$, the infinitesimal generator of the action Φ is

$$\Phi': \mathcal{X}_G(P) \to \mathcal{X}(\mathcal{M}\times\mathcal{C}\times\mathcal{H})$$

where

$$Z \to \Phi'(Z) = Z^*: \mathcal{M}\times\mathcal{C}\times\mathcal{H} \to T(\mathcal{M}\times\mathcal{C}\times\mathcal{H})$$

is defined by

$$Z^*(g,\omega,\varphi) = T_{id}\Phi_{(g,\omega,\varphi)}(Z) = (g,\omega,\varphi, L_Xg, L_Z\omega, L_Z\varphi) \in T_{(g,\omega,\varphi)}(\mathcal{M}\times\mathcal{C}\times\mathcal{H})$$

where $X = \pi_* Z \in \mathcal{X}(M)$.

We now describe the tangent and cotangent action associated with the right action Φ. For $F \in AUT(P)$, let

$$\Phi_F: \mathcal{M}\times\mathcal{C}\times\mathcal{H} \to \mathcal{M}\times\mathcal{C}\times\mathcal{H}, \qquad (g,\omega,\varphi) \longmapsto (f^*g, F^*\omega, F^*\varphi)$$

denote the diffeomorphism of $\mathcal{M} \times \mathcal{C} \times \mathcal{H}$ induced by F. Then Φ_F lifts to a vector bundle isomorphism of the tangent bundle,

$$T\Phi_F: T(\mathcal{M}\times\mathcal{C}\times\mathcal{H}) \to T(\mathcal{M}\times\mathcal{C}\times\mathcal{H}); \quad (g,\omega,\varphi;h,\sigma,\tau) \longmapsto (f^*g,\ F^*\omega,\ F^*\varphi;\ f^*h,\ F^*\sigma,\ F^*\tau)$$

Also, Φ_F lifts to a vector bundle isomorphism of the cotangent bundle

$$T^*\Phi_F: T^*(\mathcal{M}\times\mathcal{C}\times\mathcal{H})_M \to T^*(\mathcal{M}\times\mathcal{C}\times\mathcal{H})_M$$

where for $(g,\omega,\varphi;k,\alpha_M,\beta_M) \in T^*_{(g,\omega,\varphi)}(\mathcal{M}\times\mathcal{C}\times\mathcal{H})_M$

$$T^*\Phi_F(g,\omega,\varphi;k,\alpha_M,\beta_M) \in T^*_{\Phi_{F^{-1}}(g,\omega,\varphi)}(\mathcal{M}\times\mathcal{C}\times\mathcal{H})_M$$

$$= T^*_{((f^{-1})^*g,\ (F^{-1})^*\omega,\ (F^{-1})^*\varphi)}(\mathcal{M}\times\mathcal{C}\times\mathcal{H})_M$$

is defined by requiring that for all $((f^{-1})^*g,\ (F^{-1})^*\omega,\ (F^{-1})^*\varphi;\ h,\sigma,\tau) \in T_{((f^{-1})^*g,\ (F^{-1})^*\omega,\ (F^{-1})^*\varphi)}(\mathcal{M}\times\mathcal{C}\times\mathcal{H})$,

$$\int T^*\Phi_F(g,\omega,\varphi;k,\alpha_M,\beta_M)\cdot((f^{-1})^*g,\ (F^{-1})^*\omega,\ (F^{-1})^*\varphi;\ h,\ \sigma_M,\ \tau_M)$$

$$= \int (g,\omega,\varphi;\ k,\ \alpha_M,\ \beta_M)\cdot(T\Phi_F((f^{-1})^*g,\ (F^{-1})^*\omega,\ (F^{-1})^*\varphi;\ h,\sigma,\tau))_M$$

$$= \int (k\cdot f^*h + \alpha_M\cdot(F^*\sigma)_M + \beta_M\cdot(F^*\tau)_M)$$

The right action Φ then lifts to a right tangent and a right cotangent action as follows:

The tangent action

$$\Phi^T: T(\mathcal{M}\times\mathcal{C}\times\mathcal{H}) \times \mathrm{AUT}(P) \to T(\mathcal{M}\times\mathcal{C}\times\mathcal{H})$$

is defined by

$$((g,\omega,\varphi;h,\sigma,\tau),\ F) \longmapsto \Phi^T((g,\omega,\varphi;h,\sigma,\tau),\ F) = \Phi^T_F(g,\omega,\varphi;h,\sigma,\tau)$$

$$= (f^*g,\ F^*\omega,\ F^*\varphi;\ f^*h,\ F^*\sigma,\ F^*\tau)$$

where $\Phi^T_F = T\Phi_F$.

The cotangent action

$$\Phi^{T^*}: T^*(\mathcal{M}\times\mathcal{C}\times\mathcal{H})_M \times \mathrm{AUT}(P) \to T^*(\mathcal{M}\times\mathcal{C}\times\mathcal{H})_M$$

is defined by

$$((g,\omega,\varphi;k,\ \alpha_M,\ \beta_M),\ F) \longmapsto \Phi^{T^*}((g,\omega,\varphi;k,\ \alpha_M,\ \beta_M),\ F) = \Phi^{T^*}_F(g,\omega,\varphi;\ k,\ \alpha_M,\ \beta_M)$$

where $\Phi^{T^*}_F = T^*\Phi_{F^{-1}}$. Thus for $(g,\omega,\varphi;k,\ \alpha_M,\ \beta_M) \in T^*_{(g,\omega,\varphi)}(\mathcal{M}\times\mathcal{C}\times\mathcal{H})_M$, and $(f^*g,\ F^*\omega,\ F^*\varphi;\ h,\sigma,\tau) \in T_{(f^*g,\ F^*\omega,\ F^*\varphi)}(\mathcal{M}\times\mathcal{C}\times\mathcal{H})$

$$\int \Phi^{T^*}((g,\omega,\varphi;k,\ \alpha_M,\ \beta_M),\ F)\cdot(f^*g,\ F^*\omega,\ F^*\varphi;\ h,\ \sigma_M,\ \tau_M)$$

$$= \int (g,\omega,\varphi;k,\ \alpha_M,\ \beta_M)\cdot(\Phi^T_{F^{-1}}(f^*g,\ F^*\omega,\ F^*\varphi;h,\sigma,\tau))$$

$$= \int (k\cdot(f^{-1})^*h + \alpha_M\cdot((F^{-1})^*\sigma)_M + \beta_M\cdot((F^{-1})^*\tau)_M)$$

Since Φ is a right action, both the tangent and the cotangent actions are right actions. Moreover, the cotangent action is a symplectic action with respect to the canonical symplectic structure of $T^*(\mathcal{M}\times\mathcal{C}\times\mathcal{H})_M$, a symplectic structure which generalizes the symplectic structure of $T^*\mathcal{M}$ (see [7], p. 333).

Associated with every cotangent action is a momentum mapping (see [1], p. 283). In our case the mapping

$$J: T^*(\mathcal{M}\times\mathcal{C}\times\mathcal{H})_M \to \mathcal{X}^*_G(P);\ \mathcal{X}(g,\omega,\varphi;k,\alpha_M,\beta_M) \longmapsto J(g,\omega,\varphi;k,\alpha_M,\beta_M)$$

is defined by requiring that for all $Z\in\mathcal{X}_G(P)$,

$$\int J(g,\omega,\varphi;k,\alpha_M,\beta_M)\cdot Z = \int (g,\omega,\varphi;k,\alpha_M,\beta_M)\cdot (Z^*(g,\omega,\varphi))_M$$

$$= \int (k\cdot L_X g + \alpha_M\cdot (L_Z\omega)_M + \beta_M\cdot (L_Z\varphi)_M)$$

Note that $J(g,\omega,\varphi;k,\alpha_M,\beta_M)\in\mathcal{X}^*_G(P)$ is defined by using the fact that an element $\mathcal{Y}\in\mathcal{X}^*_G(P)$ is defined if we know its action on $\mathcal{X}_G(P)$ with respect to the L_2-pairing

$$\mathcal{X}^*_G(P)\times\mathcal{X}_G(P)\to\mathbb{R},\quad (\mathcal{Y},Z)\longmapsto\int\mathcal{Y}\cdot Z$$

Also, using the (varying) connection $\omega\in\mathcal{C}$, we identify $\mathcal{X}_G(P)$ with $C^\infty(\mathfrak{G}_{Ad}(M))\oplus\mathcal{X}(M)$, and $\mathcal{X}^*_G(P)$ with $C^\infty(\mathfrak{G}^*_{Ad}(M)_d)\oplus C^\infty(\Lambda^1_d(M))$, so that with this identification, $J(g,\omega,\varphi;k,\alpha_M,\beta_M)\in C^\infty(\mathfrak{G}^*_{Ad}(M)_d)\oplus C^\infty(\Lambda^1_d(M))$.

For $Z\in\mathcal{X}_G(P)$, we also define

$$J_Z: T^*(\mathcal{M}\times\mathcal{C}\times\mathcal{H})_M\to C_d(M)$$

by

$$(g,\omega,\varphi;k,\alpha_M,\beta_M)\longmapsto J_Z(g,\omega,\varphi;k,\alpha_M,\beta_M) = J(g,\omega,\varphi;k,\alpha_M,\beta_M)\cdot Z$$

where the dot "$\cdot$" here indicates the pointwise pairing

$$\mathcal{X}^*_G(P)\times\mathcal{X}_G(P)\to C^\infty_d(M)$$

Our main theorem, the vanishing of the momentum theorem for generally covariant Lagrangian densities, is then the following (see also [5]):

<u>2.3.1 Theorem</u>. Let

$$\mathcal{L}:\mathcal{M}\times\mathcal{C}\times\mathcal{H}\to C^\infty_d$$

be a generally covariant Lagrangian density. Let

$$\Psi_{\mathcal{L}}:\mathcal{M}\times\mathcal{C}\times\mathcal{H}\to T^*(\mathcal{M}\times\mathcal{C}\times\mathcal{H})_M$$

$$(g,\phi\ \varphi)\longmapsto\left(g,\omega,\varphi;\frac{\delta\mathcal{L}}{\delta g}(g,\omega,\varphi),\ \frac{\delta\mathcal{L}}{\delta\omega}(g,\omega,\varphi),\ \frac{\delta\mathcal{L}}{\delta\varphi}(g,\omega,\varphi)\right)$$

be the Lagrangian 1-form (see Definition 2.1.1), and let

$$J: T^*(\mathcal{M}\times\mathcal{C}\times\mathcal{H})_M\to\mathcal{X}^*_G(P)$$

be the momentum mapping for the cotangent action of AUT(P) on $T(\mathcal{M}\times\mathcal{C}\times\mathcal{H})$. Then

$$J\circ\Psi_{\mathcal{L}}:\mathcal{M}\times\mathcal{C}\times\mathcal{H}\to\mathcal{X}^*_G(P)$$

is identically zero,

$$J\circ\Psi_{\mathcal{L}} = 0$$

Thus for $(g,\omega,\varphi) \in \mathcal{M} \times \mathscr{C} \times \mathcal{H}$,

$$J\left(g,\ \omega,\ \varphi,\ \frac{\delta\mathscr{L}}{\delta g}(g,\omega,\varphi),\ \frac{\delta\mathscr{L}}{\delta\omega}(g,\omega,\varphi),\ \frac{\delta\mathscr{L}}{\delta\varphi}(g,\omega,\varphi)\right) = 0$$

Proof. From the definition for J, for $Z \in \mathscr{X}_G(P)$, and from Theorem 2.2.3, we have for $k \geq 1$

$$\int J \circ \Psi_{\mathscr{L}}(g,\omega,\varphi) \cdot Z = \int \Psi_{\mathscr{L}}(g,\omega,\varphi) \cdot (Z^*(g,\omega,\varphi))_M$$

$$= \int\left(\frac{\delta\mathscr{L}}{\delta g}(g,\omega,\varphi) \cdot L_X g + \frac{\delta\mathscr{L}}{\delta\omega}(g,\omega,\varphi) \cdot (L_Z\omega)_M + \frac{\delta\mathscr{L}}{\delta\varphi} \cdot (L_Z\varphi)_M\right)$$

$$= \int \mathrm{div}_{nat}\left(2X^\flat \cdot \frac{\delta\mathscr{L}}{\delta g}(g,\omega,\varphi) + \phi_M \cdot \frac{\delta\mathscr{L}}{\delta\omega}(g,\omega,\varphi) + \frac{(-1)^{k-1}}{(k-1)!} i_X\varphi_M \cdot \frac{\delta\mathscr{L}}{\delta\varphi}(g,\omega,\varphi)\right)$$

$$= \int \mathrm{div}_{nat} Y_d = 0$$

where $X = \pi_* Z$ and $\phi_M = (\omega(Z))_M$. Since the above equations are true for all $Z \in \mathscr{X}_G(P)$, it follows that $J \circ \Psi_{\mathscr{L}} \equiv 0$. For $k = 0$, the above equations are true with the $i_M\varphi_M$ term deleted.
Q.E.D.

The proof directly reflects the fact that the vanishing of the momentum theorem is the geometrical formulation of the master equations. Interestingly, in the above theorem, we did not actually have to calculate J. This however is easily done using a computation similar to the one in Theorem 2.2.2.

2.3.2 Proposition. Let

$$J:\ T^*(\mathcal{M} \times \mathscr{C} \times \mathcal{H}) \rightarrow \mathscr{X}^*_G(P)$$

be the momentum map associated with the symplectic cotangent action of AUT(P) on $T^*(\mathcal{M} \times \mathscr{C} \times \mathcal{H})_M$. Then for $(g,\omega,\varphi;\ k,\ \alpha_M,\ \beta_M) \in T^*(\mathcal{M} \times \mathscr{C} \times \mathcal{H})_M$,

$$J(g,\omega,\varphi;\ k,\ \alpha_M,\ \beta_M)$$

$$= -(\mathrm{DIV}\ \alpha_M + \beta_M \cdot (\rho'_M \cdot \varphi_M)) + (-2(\mathrm{div}_g\ k)^\flat + \alpha_M \cdot \Omega_M + \alpha^*_{\varphi_M}(\beta_M))$$

$$\in C^\infty(\mathfrak{G}^*_{Ad}(M)_d) \oplus C^\infty(\Lambda^1_d(M)) \approx \mathscr{X}^*_G(P)$$

where $\mathscr{X}^*_G(P)$ is identified with the direct sum by using the connection $\omega \in \mathscr{C}$.

Note. See Theorem 2.2.2 for notation and for a coordinate expression for each of the above terms.

Proof. For $(g,\omega,\varphi;\ k,\ \alpha_M,\ \beta_M) \in T^*(\mathcal{M} \times \mathscr{C} \times \mathcal{H})_M$, $Z \in \mathscr{X}_G(P)$, $X = \pi_* Z$, and $\phi_M = (\omega(T))_M$, we have

$$\int J(g,\omega,\varphi;\ k,\ \alpha_M,\ \beta_M) \cdot Z = \int (k \cdot L_X g + \alpha_M \cdot (L_Z\omega)_M + \beta_M \cdot (L_Z\varphi)_M)$$

$$= -2\int (\mathrm{div}_g\ k)^\flat \cdot X + \int \alpha_M \cdot (D_M\phi_M + i_X\Omega_M) + \int \beta_M \cdot (-\rho'_M(\phi_M) \cdot \varphi_M + L_X\varphi_M)$$

$$= -2\int (\mathrm{div}_g\ k)^\flat \cdot X - \int \mathrm{DIV}\ \alpha_M \cdot \phi_M + \int \alpha_M \cdot i_X\Omega_M$$

$$- \int \beta_M \cdot (\rho'_M(\phi_M) \cdot \varphi_M) + \int \alpha^*_{\varphi_M}(\beta_M) \cdot X$$

$$= -\int \phi_M \cdot (\mathrm{DIV}\, \alpha_M + \beta_M \cdot (\rho'_M \cdot \varphi_M)) + \int X \cdot (-2(\mathrm{div}_g k)^\flat + \alpha_M \cdot \Omega_M + \alpha^*_{\varphi_M}(\beta_M))$$

Since these equations are true for all $Z \in \mathscr{X}_G(P)$, it follows that

$$J(g, \omega, \varphi; k, \alpha_M, \beta_M) = -(\mathrm{DIV}\, \alpha_M + \beta_M \cdot (\rho'_M \cdot \varphi_M)) + (-2(\mathrm{div}_g k)^\flat + \alpha_M \cdot \Omega_M + \alpha^*_{\varphi_M}(\beta_M))$$

Q.E.D.

Note that the formula for J together with Theorem 2.2.2 gives

$$J \circ \Psi_{\mathscr{L}}(g, \omega, \varphi) = J\left(g, \omega, \varphi; \frac{\delta \mathscr{L}}{\delta g}(g, \omega, \varphi), \frac{\delta \mathscr{L}}{\delta \omega}(g, \omega, \varphi), \frac{\delta \mathscr{L}}{\delta \varphi}(g, \omega, \varphi)\right)$$

$$= -\mathrm{DIV}\left(\frac{\delta \mathscr{L}}{\delta \omega}(g, \omega, \varphi)\right) - \frac{\delta \mathscr{L}}{\delta \varphi}(g, \omega, \varphi) \cdot (\rho'_M \cdot \varphi_M)$$

$$-2\left(\mathrm{div}_g\left(\frac{\delta \mathscr{L}}{\delta g}(g, \omega, \varphi)\right)\right)^\flat + \frac{\delta \mathscr{L}}{\delta \omega}(g, \omega, \varphi) \cdot \Omega_M + \alpha^*_{\varphi_M}\left(\frac{\delta \mathscr{L}}{\delta \varphi}(g, \omega, \varphi)\right) = 0$$

thereby giving a direct (but equivalent) proof of Theorem 2.3.1.

3. APPLICATIONS

We are now ready to apply the master equations of Sec. 2.2 to covariant gauge field theories. We study a general Yang-Mills and a general Higgs Lagrangian field theory. In each system, the parameter and dynamical fields are specified. The differential identities and conservation laws generated by the master equations for these systems are then derived.

3.1 The Yang-Mills System

By a Yang-Mills gauge field theory, we mean a Lagrangian field theory

$$\mathscr{L}_{YM}: \mathscr{M} \times \mathscr{C} \to C^\infty_d(M)$$

where the dynamical fields are the connections $\mathscr{C}$ on a principal fiber bundle P, and the parameter fields are the pseudo-riemannian metrics $\mathscr{M}$ on M with fixed signature. The Yang-Mills theory is generally covariant if $\mathscr{L}_{YM}$ is covariant with respect to AUT(P). The dynamical field equations are

$$\frac{\delta \mathscr{L}_{YM}}{\delta \omega}(g, \omega) = 0$$

The Yang-Mills stress energy tensor density at $(g, \omega) \in \mathscr{M} \times \mathscr{C}$ is

$$\mathscr{T}_{YM}(g, \omega) = 2\frac{\delta \mathscr{L}_{YM}}{\delta g}(g, \omega) \in S^2_d(M)$$

Fix a parameter field $g \in \mathscr{M}$. Then

$$\mathrm{AUT}_g(P) = \{F \in \mathrm{AUT}(P) \mid \mathscr{L}_{YM}(g, F^*\omega) = f^*(\mathscr{L}_{YM}(g, \omega))$$
$$\text{for all } \omega \in \mathscr{C}, \text{ where } f = \hat{\pi}(F) \in \mathscr{D}(M)\}$$

is the covariance group of $\mathscr{L}_{YM}$ at g, $\mathrm{Aut}_g(P) = \mathrm{AUT}_g(P) \cap \mathrm{Aut}(P)$ is the internal covariance group of $\mathscr{L}_{YM}$ at g, and $\mathscr{D}_g(M) = \hat{\pi}(\mathrm{AUT}_g(P))$ is the spacetime covariance group of $\mathscr{L}_{YM}$ at g. Also, $I_g(M) = \{f \in \mathscr{D}(M) \mid f^*g = g\}$ is the isometry group of g, and $I_g(P) =$

$\{F \in AUT(P) \mid \hat{\pi}(F) \in I_g(M)\}$ is the <u>group of automorphisms of</u> P <u>covering</u> $I_g(M)$. Note that $Aut(P) \subseteq I_g(P)$, so that $I_g^{int}(P) = I_g(P) \cap Aut(P) = Aut(P)$.

Now suppose $\mathscr{L}_{YM}$ is generally covariant. Then $I_g(P) \subseteq AUT_g(P)$, and so $Aut(P) = I_g^{int} \subseteq Aut_g(P) \subseteq Aut(P)$. Thus $Aut_g(P) = Aut(P)$ is an infinite-dimensional group, the <u>internal gauge group of</u> $\mathscr{L}_{YM}$ <u>at</u> g. Thus $AUT_g \supseteq Aut_g(P)$ is also infinite-dimensional, the <u>gauge group for</u> $\mathscr{L}_{YM}$ <u>at</u> g. Also, $I_g(M) \subseteq \mathscr{D}_g(M)$. Moreover, we shall assume that $\mathscr{D}_g(M)$ is a finite-dimensional group, and we thus define it as the <u>symmetry group for</u> $\mathscr{L}_{YM}$ <u>at</u> g. For "physical" Lagrangians, this assumption is realistic.

Continuing with our assumption that $\mathscr{L}_{YM}$ is generally covariant, the following characterizations for the Lie algebras of <u>infinitesimal covariance transformations</u>, <u>infinitesimal internal covariance transformations</u>, and <u>infinitesimal spacetime covariance transformations</u>, respectively, <u>of</u> $\mathscr{L}_{YM}$ <u>at</u> g, are easily derived:

$$\mathscr{X}_{G,g}(P) = \{Z \in \mathscr{X}_G(P) \mid D_g\mathscr{L}_{YM}(g,\omega)\cdot L_Xg = 0,\ X = \pi_*Z,\ \text{for all } \omega \in \mathscr{C}\} \supseteq \mathscr{I}_g(P)$$

$$\mathscr{X}_{G,g}^{vert}(P) = \mathscr{X}_{G,g}(P) \cap \mathscr{X}_G^{vert}(P) = \{T \in \mathscr{X}_G^{vert}(P) \mid D_\omega\mathscr{L}_{YM}(g,\omega)\cdot L_T\omega = 0$$

$$\text{for all } \omega \in \mathscr{C}\} = \mathscr{X}_G^{vert}(P)$$

and

$$\mathscr{X}_g(M) = \pi_*(\mathscr{X}_{G,g}(P)) = \{X \in \mathscr{X}(M) \mid D_g\mathscr{L}_{YM}(g,\omega)\cdot L_Xg = 0 \text{ for all } \omega \in \mathscr{C}\} \supseteq \mathscr{I}_g(M)$$

where $\mathscr{I}_g(M) = \{X \in \mathscr{X}(M) \mid L_Xg = 0\}$ is the Lie algebra of <u>Killing vector fields for</u> g, and $\mathscr{I}_g(P) = \{Z \in \mathscr{X}_G(P) \mid \pi_*Z \in \mathscr{I}_g(M)\}$ is the Lie algebra of G-<u>invariant vector fields on</u> P <u>covering</u> $\mathscr{I}_g(M)$.

To preview the expected results regarding $\mathscr{L}_{YM}$, we have the following:

If $\mathscr{L}_{YM}$ is generally covariant with respect to AUT(P), there will be two general covariance differential identities (Theorem 2.2.2). If $g \in \mathscr{M}$ is a fixed parameter field, then $Aut_g(P) = Aut(P)$ is an infinite-dimensional gauge group. One of the covariance differential identities will be a gauge differential identity, i.e., will depend on g, ω, and the functional derivative $(\delta\mathscr{L}_{YM}/\delta\omega)(g,\omega)$ only. Assume that $\mathscr{D}_g(M)$ is a finite-dimensional group, i.e., a symmetry group for $\mathscr{L}_{YM}$ at g. Then the other covariance identity involves both functional derivatives and thus generates a local conservation law when the dynamical field equations are satisfied. Moreover, for each dimension of $\mathscr{D}_g(M)$, there will be a continuity equation, or global conservation law.

The details of these results are contained in the next two theorems.

<u>3.1.1 Theorem</u>. Let

$$\mathscr{L}_{YM}: \mathscr{M} \times \mathscr{C} \to C_d^\infty(M)$$

be covariant with respect to AUT(P). Then for $(g,\omega) \in \mathscr{M}\times\mathscr{C}$

$$DIV\Big(\frac{\delta\mathscr{L}_{YM}}{\delta\omega}(g,\omega)\Big) = 0 \tag{a}$$

$$-(\operatorname{div}_g \mathscr{T}_{YM}(g,\omega))^\flat + \frac{\delta\mathscr{L}_{YM}}{\delta\omega}(g,\omega)\cdot\Omega_M = 0 \tag{b}$$

If in addition $(g,\omega) \in \mathscr{M}\times\mathscr{C}$ satisfies the dynamical field equations

$$\frac{\delta\mathscr{L}_{YM}}{\delta\omega}(g,\omega) = 0$$

then

$$\operatorname{div}_g(\mathscr{T}_{YM}(g,\omega)) = 0 \tag{c}$$

Proof. These equations follow directly from Theorem 2.2.2 and the definition $\mathscr{T}_{YM}(g,\omega) = 2(\delta\mathscr{L}_{YM}/\delta g)(g,\omega)$. Q.E.D.

Remark. Equations (a) and (b) are the covariance differential identities. They hold independently of the field equations, and are functions of g, ω, $(\delta\mathscr{L}_{YM}/\delta g)(g,\omega)$, and $(\delta\mathscr{L}_{YM}/\delta\omega)(g,\omega)$. Because of the gauge group $\operatorname{Aut}_g(P) = \operatorname{Aut}(P)$, the internal differential identity (a) is a gauge differential identity involving only the functional derivative $(\delta\mathscr{L}_{YM}/\delta\omega)(g,\omega)$. The spacetime differential identity (b) involves both functional derivatives and generates the local conservation law (c) when the dynamical equations hold. The local conservation law involves only the functional derivative $(\delta\mathscr{L}_{YM}/\delta g)(g,\omega)$.

Global conservation laws for $\mathscr{L}_{YM}$ arise as follows:

3.1.2 Theorem. Let

$$\mathscr{L}_{YM}: \mathscr{M}\times\mathscr{C} \to C_d^\infty(M)$$

be covariant with respect to AUT(P). Then for $(g,\omega) \in \mathscr{M}\times\mathscr{C}$ and $Z \in \mathscr{X}_G(P)$,

$$\operatorname{div}_{nat}\left(X^\flat\cdot\mathscr{T}_{YM}(g,\omega) + \phi_M\cdot\frac{\delta\mathscr{L}_{YM}}{\delta\omega}\cdot(g,\omega)\right)$$
$$= \frac{1}{2}L_X g\cdot\mathscr{T}_{YM}(g,\omega) + (L_Z\omega)_M\cdot\left(\frac{\delta\mathscr{L}_{YM}}{\delta\omega}(g,\omega)\right) \tag{a}$$

where $X = \pi_* Z$ and $\phi_M = (\omega(Z))_M$.

If for fixed $g \in \mathscr{M}$, $\omega \in \mathscr{C}$ satisfies the dynamical field equations

$$\frac{\delta\mathscr{L}_{YM}}{\delta\omega}(g,\omega) = 0$$

and $X \in \mathscr{X}(M)$ satisfies condition C'

$$L_X g\cdot\mathscr{T}_{YM}(g,\omega) = 0 \tag{C'}$$

then the vector density $X^\flat\cdot\mathscr{T}_{YM}(g,\omega) \in \mathscr{X}_d(M)$ satisfies

$$\operatorname{div}_{nat}(X^\flat\cdot\mathscr{T}_{YM}(g,\omega)) = 0 \tag{b}$$

Proof. Equation (a) is a direct consequence of Theorem 2.2.3. Equation (b) then follows from (a), the field equations $(\delta\mathscr{L}_{YM}/\delta\omega)(g,\omega) = 0$, and condition (C'). Q.E.D.

Remarks. 1. Equation (b) can also be proven from the expression

$$\operatorname{div}_{nat}(X^\flat\cdot\mathscr{T}_{YM}(g,\omega)) = \frac{1}{2}L_X g\cdot\mathscr{T}_{YM}(g,\omega) = X^\flat\cdot\operatorname{div}_g(\mathscr{T}_{YM}(g,\omega))$$

Then if $(\delta\mathscr{L}_{YM}/\delta\omega)(g,\omega) = 0$, from Theorem 3.1.1(c), $\operatorname{div}_g(\mathscr{T}_{YM}(g,\omega)) = 0$. Condition C' then implies the result.

2. The vector density $X^\flat\cdot\mathscr{T}_{YM}(g,\omega) \in \mathscr{X}_d(M)$ is also referred to as a current density for $\mathscr{L}_{YM}$.

We now make some further remarks comparing condition C' that appears in the above theorem, rewritten as

$$\frac{1}{2}\mathcal{T}_{YM}(g,\omega)\cdot L_X g = \frac{\delta\mathcal{L}_{YM}}{\delta g}(g,\omega)\cdot L_X g = 0 \tag{C'}$$

with the condition C that occurs when $X \in \mathscr{X}_g(M)$, namely

$$D_g\mathcal{L}_{YM}(g,\omega)\cdot L_X g = 0 \tag{C}$$

These conditions are related but not equivalent. In particular, $X \in \mathscr{X}_g(M)$ need not in general satisfy condition C'. Note that condition C is required to hold for all $\omega \in \mathscr{C}$, whereas condition C' is taken only for solutions to the field equations. At this point, we are not concerned with this distinction (but see Remark 6 following Prop. 3.1.5 below).

Conditions C and C' are related by the concept of derivative coupling.

3.1.3 Definition. A Lagrangian field theory

$$\mathcal{L}_{YM}: \mathcal{M} \times \mathscr{C} \to C_d^\infty$$

is k-th order covariantly coupled to gravity if, in local coordinates,

(1) $\mathcal{L}_{YM}$ is a function of the k-th order partial derivatives of g, and is not a function of any higher order partial derivative of g,
(2) these partial derivatives only appear in the Christoffel symbols and their partial derivatives (of order $\leq k - 1$).

Thus, in particular,

$$D_g\mathcal{L}_{YM}(g,\omega): S_2(M) \to C_d^\infty(M)$$

is a linear k-th order differential operator.

If $k = 0$ so that $\mathcal{L}_{YM}$ does not contain any derivatives of g, then $\mathcal{L}_{YM}$ is non-derivatively coupled to gravity; if $k \geq 1$, then $\mathcal{L}_{YM}$ is derivatively coupled to gravity.

3.1.4 Proposition. Let

$$\mathcal{L}_{YM}: \mathcal{M} \times \quad \to C_d^\infty$$

be a Yang-Mills Lagrangian field theory that is first-order covariantly coupled to gravity. Then for all $(g,\omega) \in \mathcal{M}\times\mathscr{C}$ and $h \in S_2(M)$,

$$D_g\mathcal{L}_{YM}(g,\omega)\cdot h = \frac{\delta\mathcal{L}_{YM}}{\delta g}(g,\omega)\cdot h + \operatorname{div}_{nat}(P(g,\omega)\cdot h)$$

where $P(g,\omega) \in C^\infty(((TM \otimes_{sym} TM) \otimes TM)_d)$ is given in local coordinates by

$$P^{\mu\nu\lambda} = \frac{1}{2}\partial_{\Gamma^\gamma_{\alpha\beta}}\mathcal{L}_{YM}(g^{\gamma(\mu}\delta_\beta^{\nu)}\delta_\alpha^\lambda + g^{\gamma(\nu}\delta_\alpha^{\mu)}\delta_\beta^\lambda - g^{\gamma\lambda}\delta_\alpha^{(\mu}\delta_\beta^{\nu)}$$

where $(\mu\nu) = (1/2)(\mu\nu + \nu\mu)$, δ_α^β is the Kronecker delta, and where $P(g,\omega)\cdot h = P^{\mu\nu\lambda}\cdot h_{\mu\nu}$.

Proof. In order that the partial derivatives of $\mathcal{L}_{YM}$ be tensor differential operators, it is necessary to consider the undifferentiated metric coefficients that appear in the Christoffel symbols separately from the other differentiated metric coefficients (see [7], p. 387). Thus we write

$$\mathcal{L}_{YM}(g,\omega) = \mathcal{L}_{YM}(g,\Gamma,\omega)$$

where $\mathscr{L}_{YM}(g, \Gamma, \omega)$ depends, respectively, on the undifferentiated metric coefficients g that do not appear in the Christoffel symbols, on the undifferentiated Christoffel symbols (since k = 1), and on the connection coefficients ω, both differentiated and undifferentiated.

Let $\partial_g \mathscr{L}_{YM} = \partial_g \mathscr{L}_{YM}(g, \Gamma, \omega)$ denote the derivative of $\mathscr{L}_{YM}$ with respect to the undifferentiated g's that do not appear in the Christoffel symbols. Using this notation, we have

$$
\begin{aligned}
D_g \mathscr{L}_{YM}(g, \omega) \cdot h &= D_g \mathscr{L}_{YM}(g, \Gamma, \omega) \cdot h \\
&= \partial_g \mathscr{L}_{YM}(g, \Gamma, \omega) \cdot h + \partial_\Gamma \mathscr{L}_{YM}(g, \Gamma, \omega) \cdot (D\Gamma \cdot h) \\
&= \partial_g \mathscr{L}_{YM} \cdot h + \frac{1}{2} \partial_{\Gamma^\gamma_{\alpha\beta}} \mathscr{L}_{YM} \cdot (h^\gamma{}_{\beta;\alpha} + h_\alpha{}^\gamma{}_{;\beta} - h_{\alpha\beta}{}^{;\gamma}) \\
&= \partial_g \mathscr{L}_{YM} \cdot h + \frac{1}{2} \partial_{\Gamma^\gamma_{\alpha\beta}} \mathscr{L}_{YM} \cdot (g^{\gamma\mu} \delta^\nu_\beta \delta^\lambda_\alpha + g^{\gamma\nu} \delta^\mu_\alpha \delta^\lambda_\beta - g^{\gamma\lambda} \delta^\mu_\alpha \delta^\nu) h_{\mu\nu;\lambda} \\
&= \partial_g \mathscr{L}_{YM} \cdot h + \frac{1}{2} \partial_{\Gamma^\gamma_{\alpha\beta}} \mathscr{L}_{YM} (g^{\gamma(\mu} \delta^{\nu)}_\beta \delta^\lambda_\alpha + g^{\gamma(\nu} \delta^{\mu)}_\beta \delta^\lambda_\beta - g^{\gamma\lambda} \delta_\alpha{}^{(\mu} \delta^{\nu)}_\beta) h_{\mu\nu;\lambda} \\
&= \partial_g \mathscr{L}_{YM} \cdot h + P^{\mu\nu\lambda} h_{\mu\nu;\lambda}
\end{aligned}
$$

Now, for all $h \in S_2(M)$,

$$
\begin{aligned}
\int \frac{\delta \mathscr{L}_{YM}}{\delta g} \cdot h &= \int D_g \mathscr{L}_{YM} \cdot h \\
&= \int \partial_g \mathscr{L}_{YM} \cdot h + \int P^{\mu\nu\lambda} h_{\mu\nu;\lambda} \\
&= \int \partial_g \mathscr{L}_{YM} \cdot h - \int P^{\mu\nu\lambda}{}_{;\lambda} h_{\mu\nu}
\end{aligned}
$$

Thus

$$
\frac{\delta \mathscr{L}_{YM}}{\delta g} = \partial_g \mathscr{L}_{YM} - P^{\mu\nu\lambda}{}_{;\lambda} h_{\mu\nu}
$$

and so

$$
\begin{aligned}
D_g \mathscr{L}_{YM}(g, \omega) \cdot h &= \partial_g \mathscr{L}_{YM} \cdot h + P^{\mu\nu\lambda} h_{\mu\nu;\lambda} \\
&= \frac{\delta \mathscr{L}_{YM}}{\delta g} \cdot h + P^{\mu\nu\lambda}{}_{;\lambda} h_{\mu\nu} + P^{\mu\nu\lambda} h_{\mu\nu;\lambda} \\
&= \frac{\delta \mathscr{L}_{YM}}{\delta g} \cdot h + (P^{\mu\nu\lambda} h_{\mu\nu})_{;\lambda} \\
&= \frac{\delta \mathscr{L}_{YM}}{\delta g} \cdot h + \operatorname{div}_{nat}(P \cdot h)
\end{aligned}
$$

where the last divergence can be taken to be natural since $P \cdot h = P^{\mu\nu\lambda} \cdot h_{\mu\nu}$ is a vector density.

Q.E.D.

Thus the tensor $P(g,\omega)$ is a measure of the derivative coupling. Indeed, with a little calculation, one can demonstrate that

$$P(g,\omega) = 0 \text{ iff } \partial_\Gamma \mathscr{L}_{YM} = 0 \text{ iff } k = 0$$

In the latter case, $D_g \mathscr{L}_{YM} \cdot L_X g = (\delta \mathscr{L}_{YM}/\delta g) \cdot L_X g$, so that conditions C and C' are equivalent.

If $\mathscr{L}_{YM}$ is derivatively coupled to gravity ($k \geq 1$), then conditions C and C' are not equivalent. Since an infinitesimal spacetime symmetry transformation satisfies condition C, and not, in general, condition C', condition C is, presumably, the more natural condition to require in the presence of derivative coupling. Since Theorem 3.1.2 gives global conservation laws when C' is satisfied, we modify this theorem in the presence of first order derivative coupling as follows:

3.1.5 Proposition. Let

$$\mathscr{L}_{YM}: \mathscr{M} \times \mathscr{C} \to C_d^\infty$$

be a covariant Yang-Mills Lagrangian field theory that is first order covariantly coupled to gravity. For fixed $g \in \mathscr{M}$, let $\omega \in \mathscr{C}$ satisfy the field equations

$$\frac{\delta \mathscr{L}_{YM}}{\delta\omega}(g,\omega) = 0$$

and let $X \in \mathscr{X}(M)$ satisfy condition C

$$D_g \mathscr{L}_{YM}(g,\omega) \cdot L_X g = 0 \tag{C}$$

Then

$$\operatorname{div}_{nat}(X^\flat \cdot \mathscr{T}_{YM}(g,\omega) + P(g,\omega) \cdot L_X g) = 0$$

Proof. From Eq. 3.1.2(a), if $(\delta \mathscr{L}_{YM}/\delta\omega)(g,\omega) = 0$, then

$$\operatorname{div}_{nat}(X^\flat \cdot \mathscr{T}_{YM}(g,\omega)) = \frac{1}{2}\mathscr{T}_{YM}(g,\omega) \cdot L_X g$$

and from Prop. 3.1.4, if $D_g \mathscr{L}_{YM} \cdot L_X g = 0$, then

$$\frac{\delta \mathscr{L}_{YM}}{\delta g} \cdot L_X g = \frac{1}{2}\mathscr{T}_{YM} \cdot L_X g = -\operatorname{div}_{nat}(P \cdot L_X g)$$

Thus

$$\operatorname{div}_{nat}(X^\flat \cdot \mathscr{T}_{YM} + P \cdot L_X g) = 0 \qquad \text{Q.E.D.}$$

Note. Similar but more complicated expressions are available for higher order coupling.

Remarks. 1. If $\mathscr{L}_{YM}$ is nonderivatively coupled to gravity, so that $P(g,\omega) = 0$, conditions C and C' are equivalent, and Theorem 3.1.2 and Prop. 3.1.5 give the same continuity equation. In the derivatively coupled case, the results of 3.1.2 and 3.1.5 can be summarized by the following:

$$C' \Longleftrightarrow \mathscr{T}_{YM}(g,\omega) \cdot L_X g = 0 \Longleftrightarrow \operatorname{div}_{nat}(X^\flat \cdot \mathscr{T}_{YM}(g,\omega)) = 0$$

$$C \Longleftrightarrow D_g \mathscr{L}_{YM}(g,\omega) \cdot L_X g = 0 \Longleftrightarrow \operatorname{div}_{nat}(X^\flat \cdot \mathscr{T}_{YM}(g,\omega) + P(g,\omega) \cdot L_X g) = 0$$

Although these implications are equivalent for first order covariantly coupled Lagrangians, and although the first implications do not require $\mathscr{L}_{YM}$ to be either first order or covariantly coupled to gravity, condition C is the more natural condition in the presence of derivative coupling since $X \in \mathscr{X}_g(M)$ satisfies condition C (see p. 236) and $\mathscr{X}_g(M)$ is naturally the space of infinitesimal spacetime symmetries of $\mathscr{L}_{YM}$ at g. Thus we take the second set of implications as generating the set of global conservation laws for $\mathscr{L}_{YM}$ at g. In current practice, most Yang-Mills Lagrangian densities are nonderivatively coupled to gravity, and so this rather unfortunate complication does not occur.

2. It is convenient to speak of the spacetime symmetry group $\mathscr{D}_g(M)$ as the group generating the global conservation laws, although strictly speaking it is the Lie algebra $\mathscr{X}_g(M)$ that generates the global conservation laws. For example, a discrete group does not generate any global conservation laws.

3. An interesting question is what conditions on $\mathscr{L}_{YM}$ guarantee that the Lie algebra homomorphism

$$\mathscr{X}_g(M) \to \mathscr{X}_d(M), \qquad X \longmapsto X^\flat \cdot \mathscr{T}_{YM}(g,\omega) + P(g,\omega) \cdot L_X g$$

is an injection? Presumably this map is injective for general physical Lagrangians. In that case, dim $\mathscr{X}_g(M)$ would count the number of linearly independent global conservation laws for the Yang-Mills system at g.

4. Note that since $\mathscr{I}_g(M) \subseteq \mathscr{X}_g(M)$, $\mathscr{I}_g(M)$ is a minimal Lie algebra that always generates global conservation laws for $\mathscr{L}_{YM}$. The Lie group $I_g(M)$ is then a minimal Lie group that generates global conservation laws for $\mathscr{L}_{YM}$ at g, subject to the qualifications of Remark 2.

5. The Lie algebra $\mathscr{I}_g(M)$ of Killing vector fields is taken by some authors to be the Lie algebra $\mathscr{X}_g(M)$ of infinitesimal spacetime symmetries of $\mathscr{L}_{YM}$ at g. However, in general, $\mathscr{I}_g(M)$ is properly contained in $\mathscr{X}_g(M)$, and so there could be unaccounted for global conservation laws if $\mathscr{X}_g(M)$ were taken as equal to $\mathscr{I}_g(M)$. An interesting question is what conditions on $\mathscr{L}_{YM}$ imply that $\mathscr{I}_g(M) = \mathscr{X}_g(M)$?

6. If $X \in \mathscr{X}_g(M)$, then X satisfies condition C for all $\omega \in \mathscr{C}$. The global conservation law of the proposition then follows for any solution to the dynamical field equations. However, the proposition does not require that X satisfy condition C for all ω, or even for <u>all</u> solutions. It is sufficient that X satisfy condition C for a <u>particular</u> solution, and then there will be a global conservation law for that solution. In general such an X need not satisfy condition C for another solution. Such an X should be called a <u>sporadic</u> (or <u>spontaneous</u>) <u>infinitesimal spacetime symmetry of</u> $\mathscr{L}_{YM}$ <u>at</u> (g,ω). In general, field theorists are interested in infinitesimal symmetries that generate global conservation laws for any solution to the field equations, so sporadic infinitesimal symmetries are of less interest. However, they remain an interesting theoretical possibility.

These are several popular conditions on vector fields that make them satisfy condition C (or C'), and which thus lead to global conservation laws. Since $\mathscr{I}_g(M) \subseteq \mathscr{X}_g(M)$, the most obvious of course is that X be a Killing vector field for g, $L_X g = 0$. The associated global conservation law is then

$$\operatorname{div}_{nat}(X^\flat \cdot \mathscr{T}_{YM}(g,\omega)) = 0$$

Note that in this case the distinction between derivative and nonderivative coupling is immaterial, although $P(g,\omega)$ need not be zero, inasmuch as $P(g,\omega) \cdot L_X g = 0$, so that both conditions C and C' are satisfied.

Another popular condition arises when $\mathscr{L}_{YM}$ is <u>conformally invariant</u>, i.e., when

$$\mathscr{L}_{YM}(pg,\omega) = \mathscr{L}_{YM}(g,\omega)$$

for $(g,\omega) \in \mathscr{M} \times \mathscr{C}$ and for $p \in \mathrm{Pos}(M) = \{p \in C^\infty(M;\mathbb{R}) \mid p(x) > 0,\ x \in M\}$, the smooth positive functions on M.

For $g \in \mathcal{M}$, let

$$C_g(M) = \{f \in \mathcal{D}(M) \mid f^*g = pg \text{ for some } p \in \mathrm{Pos}(M)\}$$

denote the conformal group of g, and let

$$c_g(M) = \{X \in \mathcal{X}(M) \mid L_X g = \sigma g \text{ for some } \sigma \in C^\infty(M;\mathbb{R})\}$$

denote the Lie algebra of conformal Killing vector fields.

3.1.6 Theorem. Let

$$\mathcal{L}_{YM}: \mathcal{M} \times \mathcal{C} \to C_d^\infty$$

be a conformally invariant Yang-Mills Lagrangian density. If $(g,\omega) \in \mathcal{M} \times \mathcal{C}$ and $\lambda \in C^\infty(M,\mathbb{R})$, then

$$D_g \mathcal{L}_{YM}(g,\omega) \cdot \lambda g = 0$$

and

$$\mathrm{tr}_g(\mathcal{T}_{YM}(g,\omega)) = 0$$

In particular, if X is a conformal Killing vector field,

$L_X g = \sigma g$, $\sigma \in C^\infty(M,\mathbb{R})$, then for all $(g,\omega) \in \mathcal{M} \times \mathcal{C}$,

$$D_g \mathcal{L}_{YM}(g,\omega) \cdot L_X g = 0$$

and

$$\mathcal{T}_{YM}(g,\omega) \cdot L_X g = 0$$

Proof. For $\lambda \in C^\infty(M;\mathbb{R})$, let $p(s) = 1 + s\lambda$. Then for $\delta > 0$ sufficiently small, $p(s) \in \mathrm{Pos}(M)$ for $s \in (-\delta,\delta)$. Since $\mathcal{L}_{YM}$ is conformally invariant, for $(g,\omega) \in \mathcal{M} \times \mathcal{C}$ and $s \in (-\delta,\delta)$,

$$\mathcal{L}_{YM}(p(s)g,\omega) = \mathcal{L}_{YM}(g,\omega)$$

Differentiating with respect to s and evaluating at s = 0 gives

$$D_g \mathcal{L}_{YM}(g,\omega) \cdot \lambda g = 0$$

Integrating this equation over M then gives

$$\begin{aligned} 0 &= \int D_g \mathcal{L}_{YM}(g,\omega) \cdot \lambda g \\ &= \int ((D_g \mathcal{L}_{YM}(g,\omega))^* \cdot 1) \cdot \lambda g \\ &= \int \frac{\delta \mathcal{L}_{YM}}{\delta g}(g,\omega) \cdot \lambda g \\ &= \int \lambda \, \mathrm{tr}_g\Big(\frac{\delta \mathcal{L}_{YM}}{\delta g}(g,\omega)\Big) = 0 \end{aligned}$$

Since this is true for all $\lambda \in C^\infty(M;\mathbb{R})$, $\mathrm{tr}_g((\delta \mathcal{L}_{YM}/\delta g)(g,\omega)) = 0$.

If X is a conformal Killing vector field, $L_X g = \sigma g$, then

$$D_g \mathscr{L}_{YM}(g,\omega) \cdot L_X g = D_g \mathscr{L}_{YM}(g,\omega) \cdot \sigma g = 0$$

and

$$\mathscr{T}_{YM}(g,\omega) \cdot L_X g = \mathscr{T}_{YM}(g,\omega) \cdot \sigma g = \sigma\, \mathrm{tr}_g(\mathscr{T}_{YM}(g,\omega)) = 0 \qquad \text{Q.E.D.}$$

Thus if $\mathscr{L}_{YM}$ is conformally invariant, $g \in \mathscr{M}$, and $X \in c_g(M)$ is a conformal Killing vector field, then X satisfies conditions C and C' for all $\omega \in \mathscr{C}$. In particular, $X \in \mathscr{X}_g(M)$, so $c_g(M) \subseteq \mathscr{X}_g(M)$. Thus as an immediate corollary we have

3.1.7 Corollary. Let

$$\mathscr{L}_{YM}: \mathscr{M} \times \mathscr{C} \to C_d^{\infty}$$

be a conformally invariant generally covariant Yang-Mills Lagrangian density. For $g \in \mathscr{M}$, let $X \in \mathscr{X}_g(M)$. Then if $\omega \in \mathscr{C}$ satisfies the field equations

$$\frac{\delta \mathscr{L}_{YM}}{\delta \omega}(g,\omega) = 0$$

then

$$\mathrm{div}_{nat}(X^{\flat} \cdot \mathscr{T}_{YM}(g,\omega)) = 0$$

Thus $C_g(M)$ is a minimal symmetry group for conformally invariant generally covariant $\mathscr{L}_{YM}$ at g.

From Theorem 3.1.6, $\mathrm{tr}_g\, \mathscr{T}_{YM}(g,\omega) = 0$ for conformally invariant $\mathscr{L}_{YM}$. For $\mathscr{L}_{YM}$ first order covariantly coupled to gravity, we then have the following:

3.1.8 Corollary. Let $\mathscr{L}_{YM}$ be conformally invariant and first order covariantly coupled to gravity. Then

$$\mathrm{tr}_g\, P(g,\omega) = 0$$

where $\mathrm{tr}_g\, P(g,\omega) \in \mathscr{X}_d(M)$ is a vector density on M, given in local coordinates by

$$(\mathrm{tr}_g\, P(g,\omega))^{\lambda} = g_{\mu\nu} P^{\mu\nu\lambda}$$

Proof. From Prop. 3.1.4 for $(g,\omega) \in \mathscr{M} \times \mathscr{C}$ and $h \in S_2(M)$,

$$D_g \mathscr{L}_{YM} \cdot h = \frac{\delta \mathscr{L}_{YM}}{\delta g} \cdot h + \mathrm{div}_{nat}(P \cdot h)$$

Letting $h = \lambda g$, from Corollary 3.1.7, $D_g \mathscr{L}_{YM} \cdot \lambda g = 0$ and $(\delta \mathscr{L}_{YM}/\delta g) \cdot \lambda g = 0$, and so

$$\mathrm{div}_{nat}(P \cdot \lambda g) = \mathrm{div}_{nat}(\lambda \mathrm{tr}_g P) = 0$$

for all $\lambda \in C^{\infty}(M;\mathbb{R})$. Taking $\lambda = 1$ gives $\mathrm{div}_{nat}(\mathrm{tr}_g P) = 0$. Then expanding

$$\mathrm{div}_{nat}(\lambda \mathrm{tr}_g P) = d\lambda \cdot \mathrm{tr}_g P + \lambda\, \mathrm{div}_{nat}(\mathrm{tr}_g P) = d\lambda \cdot \mathrm{tr}_g P = 0$$

for all $\lambda \in C^{\infty}(M;\mathbb{R})$, and hence $\mathrm{tr}_g P = 0$. Q.E.D.

We summarize our discussion of global conservation laws for the Yang-Mills system as follows:

3.1.9 Theorem. Let

$$\mathscr{L}_{YM}: \mathscr{M} \times \mathscr{C} \to C_d^{\infty}$$

be a generally covariant Yang-Mills density. For $g \in \mathcal{M}$, let $\omega \in \mathcal{C}$ satisfy the dynamical field equations

$$\frac{\delta \mathcal{L}_{YM}}{\delta\omega}(g,\omega) = 0$$

Let $X \in \mathcal{X}(M)$ satisfy any of the following conditions:

(1) $L_X g = 0$

(2) $\mathcal{T}_{YM}(g,\omega) \cdot L_X g = 0$

(3) $\mathcal{L}_{YM}$ is conformally invariant and $X \in c_g(M)$

(4) $\mathcal{L}_{YM}$ is nonderivatively coupled to gravity and $D_g \mathcal{L}_{YM}(g,\omega) \cdot L_X g = 0$ (so (2) is satisfied).

Then

$$\mathrm{div}_{nat}(X^\flat \cdot \mathcal{T}_{YM}(g,\omega)) = 0$$

(5) If $\mathcal{L}_{YM}$ is first-order covariantly coupled to gravity, and $D_g \mathcal{L}_{YM}(g,\omega) \cdot L_X g = 0$, then

$$\mathrm{div}_{nat}(X^\flat \cdot \mathcal{T}_{YM}(g,\omega) + P_{YM}(g,\omega) \cdot L_X g) = 0$$

3.2 The Higgs System

By a Higgs field theory we mean a Lagrangian field theory

$$\mathcal{L}_{Higgs}: \mathcal{M} \times \mathcal{C} \times \mathcal{H} \to C^\infty_d$$

where the dynamical fields are $\mathcal{H} = C^\infty_{\rho,hor}(\Lambda^k(P) \otimes E)$, $0 \le k \le n = \dim M$, the Higgs k-form fields of type (ρ, E), and where the parameter fields are $\mathcal{M} \times \mathcal{C}$.

There are many similarities of a Higgs field theory with a Yang-Mills field theory. On the other hand, there are some interesting and important differences. For example, both a generally covariant Yang-Mills and Higgs field theory have two general covariance differential identities. However, unlike a Yang-Mills field theory, a Higgs theory does not have any gauge group, or gauge differential identities. Moreover, a Higgs field theory has both an internal and spacetime local conservation law, whereas a Yang-Mills field theory has only a spacetime local conservation law. Also, consequently, a Higgs field theory may have an internal global conservation law.

We now explore further these similarities and differences.

For fixed parameter fields $(g,\omega) \in \mathcal{M} \times \mathcal{C}$, the dynamical field equation for $\mathcal{L}_{Higgs}$ is

$$\frac{\delta \mathcal{L}_{Higgs}}{\delta\varphi}(g,\omega,\varphi) = 0$$

the Higgs stress energy tensor density at $(g,\omega,\varphi) \in \mathcal{M} \times \mathcal{C} \times \mathcal{H}$ is

$$\mathcal{T}_{Higgs}(g,\omega,\varphi) = 2\frac{\delta \mathcal{L}_{Higgs}}{\delta g}(g,\omega,\varphi) \in S^2_d(M)$$

and the Higgs current density at (g,ω,φ) is

$$\mathcal{J}_{Higgs}(g,\omega,\varphi) = \frac{\delta \mathcal{L}_{Higgs}}{\delta\omega}(g,\omega,\varphi) \in A^*(M)$$

Also let

$$\mathrm{AUT}_{(g,\omega)}(P) = \{F \in \mathrm{AUT}(P) \mid \mathscr{L}_{\mathrm{Higgs}}(g, \omega, F^*\varphi) = f^*(\mathscr{L}_{\mathrm{Higgs}}(g, \omega, \varphi)),$$
$$\text{where } f = \hat{\pi}(F) \in \mathscr{D}(M), \text{ for all } \varphi \in \mathscr{H}\}$$

denote the <u>covariance group of</u> $\mathscr{L}_{\mathrm{Higgs}}$ <u>at</u> (g,ω), let

$$\mathrm{Aut}_{(g,\omega)}(P) = \mathrm{AUT}_{(g,\omega)}(P) \cap \mathrm{Aut}(P)$$

denote the <u>internal covariance group of</u> $\mathscr{L}_{\mathrm{Higgs}}$ <u>at</u> (g,ω), and let

$$\mathscr{D}_{(g,\omega)}(M) = \hat{\pi}(\mathrm{AUT}_{(g,\omega)}(P))$$

denote the <u>spacetime covariance group of</u> $\mathscr{L}_{\mathrm{Higgs}}$ <u>at</u> (g,ω).

Now assume $\mathscr{L}_{\mathrm{Higgs}}$ is generally covariant (i.e., covariant with respect to AUT(P)). Then the following characterizations for the Lie algebras of <u>infinitesimal covariance transformations</u>, <u>infinitesimal internal covariance transformations</u>, and <u>infinitesimal spacetime covariance transformations</u>, respectively, <u>of</u> $\mathscr{L}_{\mathrm{Higgs}}$ <u>at</u> (g,ω) are easily derived:

$$\mathscr{X}_{G,(g,\omega)}(P) = \{Z \in \mathscr{X}_G(P) \mid D_g\mathscr{L}_{\mathrm{Higgs}}(g,\omega,\varphi)\cdot L_X g + D_\omega \mathscr{L}_{\mathrm{Higgs}}(g,\omega,\varphi)\cdot L_Z\omega = 0,$$
$$X = \pi_* Z, \text{ for all } \varphi \in \mathscr{H}\}$$

$$\mathscr{X}^{\mathrm{vert}}_{G,(g,\omega)}(P) = \mathscr{X}_{G,(g,\omega)}(P) \cap \mathscr{X}^{\mathrm{vert}}_G(P)$$
$$= \{T \in \mathscr{X}^{\mathrm{vert}}_G(P) \mid D_\omega \mathscr{L}_{\mathrm{Higgs}}(g,\omega,\varphi)\cdot L_T\omega = 0, \text{ for all } \varphi \in \mathscr{H}\}$$

and

$$\mathscr{X}_{(g,\omega)}(M) = \pi_*(\mathscr{X}_{G,(g,\omega)}(P))$$
$$= \{X \in \mathscr{X}(M) \mid \text{there exists } \phi \in C^\infty_{\mathrm{Ad}}(P;\mathfrak{G}) \text{ such that}$$
$$D_g \mathscr{L}_{\mathrm{Higgs}}(g,\omega,\varphi)\cdot L_X g + D_\omega \mathscr{L}_{\mathrm{Higgs}}(g,\omega,\varphi)\cdot L_{\bar{X}}\omega$$
$$+ D_\omega \mathscr{L}_{\mathrm{Higgs}}(g,\omega,\varphi)\cdot D\phi = 0 \text{ for all } \varphi \in \mathscr{H}\}$$

where $\bar{X}$ is the horizontal lift of X using ω.

Moreover, for $(g,\omega) \in \mathscr{M} \times \mathscr{C}$, let

$$I_\omega(P) = \{F \in \mathrm{AUT}(P) \mid F^*\omega = \omega\}$$

denote the <u>automorphism group of</u> ω;

$$I^{\mathrm{int}}_\omega(P) = I_\omega(P) \cap \mathrm{Aut}(P)$$

the <u>internal automorphism group of</u> ω (i.e., those automorphisms that cover the identity), and

$$I_{(g,\omega)}(P) = \{F \in I_\omega(P) \mid \hat{\pi}(F) \in I_g(M)\}$$

the automorphism group of ω that covers the isometry group of g, or the <u>simultaneous automorphism</u>, or <u>symmetry group of</u> (g,ω). Let

$$I^{\mathrm{int}}_{(g,\omega)}(P) = I_{(g,\omega)}(P) \cap \mathrm{Aut}(P)$$

and note that $I^{\mathrm{int}}_{(g,\omega)}(P) = I^{\mathrm{int}}_\omega(P)$. Let

$$I_{(g,\omega)}(M) = \hat{\pi}(I_{(g,\omega)}(P)) = \{f \in \mathscr{D}(M) \mid \text{there exists an } F \in I_\omega(P) \text{ such that } \hat{\pi}(F) = f \in I_g(M)\} \subseteq I_g(M)$$

Note that $\hat{\pi}(I_{(g,\omega)}(P))$ need not equal $I_g(M)$, as there may be isometries of g that cannot be lifted to automorphisms of ω. $I_{(g,\omega)}$ is the simultaneous spacetime symmetry group of (g,ω). Let

$$\mathscr{I}_\omega(P) = \{Z \in \mathscr{X}_G(P) \mid L_Z\omega = 0\}$$

denote the Lie algebra of infinitesimal automorphisms, or symmetries, of ω,

$$\mathscr{I}_\omega^{int}(P) = \mathscr{I}_\omega(P) \cap \mathscr{X}_G^{vert}(P)$$

$$= \{T \in \mathscr{X}_G^{vert}(P) \mid L_T\omega = D\phi_T = 0, \ \phi_T = \omega(T)\}$$

the Lie algebra of vertical infinitesimal automorphisms of ω, and let

$$\mathscr{I}_{(g,\omega)}(P) = \{Z \in \mathscr{X}_G(P) \mid L_Z\omega = 0 \text{ and } L_Xg = 0, \ X = \pi_*Z\}$$

denote the Lie algebra of simultaneous infinitesimal symmetries of (g,ω). Then

$$\mathscr{I}^{vert}_{(g,\omega)}(P) = \mathscr{I}_{(g,\omega)}(P) \cap \mathscr{X}_G^{vert}(P) = \mathscr{I}_\omega^{vert}(P)$$

Also, let

$$\mathscr{I}_{(g,\omega)}(M) = \pi_*(\mathscr{I}_{(g,\omega)}(P))$$

$$= \{X \in \mathscr{X}(M) \mid L_Xg = 0 \text{ and } D_M\phi_M + i_X\Omega_M = 0 \text{ for some } \phi_M \in C^\infty(\mathfrak{G}_{Ad}(M))\} \subseteq \mathscr{I}_g(M)$$

so $\mathscr{I}_{(g,\omega)}(M)$ is the Lie algebra of infinitesimal simultaneous spacetime symmetries of (g,ω).

If $\mathscr{L}_{Higgs}$ is generally covariant, then

$$I_{(g,\omega)}(P) \subseteq AUT_{(g,\omega)}(P) \text{ and } \mathscr{I}_{(g,\omega)}(P) \subseteq \mathscr{X}_{G,(g,\omega)}(P)$$

$$I^{int}_{(g,\omega)}(P) = I^{int}_\omega(P) \subseteq Aut_{(g,\omega)}(P) \text{ and } \mathscr{I}^{int}_{(g,\omega)}(P) = \mathscr{I}^{int}_\omega(P) \subseteq \mathscr{X}_{G,(g,\omega)}(P)$$

and

$$\hat{\pi}(I_{(g,\omega)}(P)) \subseteq \mathscr{D}_{(g,\omega)}(M), \text{ and } \pi_*(\mathscr{I}_{(g,\omega)}(P)) \subseteq \mathscr{X}_{(g,\omega)}(M)$$

The differential identities and local conservation laws for a Higgs field theory follow immediately from Theorem 2.2.2.

3.2.1 Theorem. Let

$$\mathscr{L}_{Higgs}: \mathscr{M} \times \mathscr{C} \times \mathscr{H} \to C_d^\infty$$

be a generally covariant Higgs field theory. Then for $(g,\omega,\varphi) \in \mathscr{M} \times \mathscr{C} \times \mathscr{H}$,

$$DIV(\mathscr{J}_{Higgs}(g,\omega,\varphi)) + \frac{\delta\mathscr{L}_{Higgs}}{\delta\varphi}(g,\omega,\varphi)\cdot(\rho_M'\cdot\varphi_M) = 0 \qquad \text{(a)}$$

$$-(\mathrm{div}_g\, \mathcal{T}_{\mathrm{Higgs}}(g,\omega,\varphi))^{\flat} + \mathcal{J}_{\mathrm{Higgs}}(g,\omega,\varphi)\cdot\Omega_M + \alpha^*_{\varphi_M}\Big(\frac{\delta\mathcal{L}_{\mathrm{Higgs}}}{\delta\varphi}(g,\omega,\varphi)\Big) = 0 \tag{b}$$

Let $(g,\omega) \in \mathcal{M}\times\mathcal{C}$, and let $\varphi \in \mathcal{H}$ satisfy the dynamical field equations

$$\frac{\delta\mathcal{L}_{\mathrm{Higgs}}}{\delta\varphi}(g,\omega,\varphi) = 0$$

Then

$$\mathrm{DIV}\,(\mathcal{J}_{\mathrm{Higgs}}(g,\omega,\varphi)) = 0 \tag{c}$$

$$-(\mathrm{div}_g\, \mathcal{T}_{\mathrm{Higgs}}(g,\omega,\varphi))^{\flat} + \mathcal{J}_{\mathrm{Higgs}}(g,\omega,\varphi)\cdot\Omega_M = 0 \tag{d}$$

Equations (a) and (b) are the general covariance differential identities. They are functions of the fields and of the functional derivatives of $\mathcal{L}_{\mathrm{Higgs}}$ with respect to both the dynamical and parameter fields.

For $(g,\omega) \in \mathcal{M}\times\mathcal{C}$, the covariance group $\mathrm{AUT}_{(g,\omega)}(P)$, for $\mathcal{L}_{\mathrm{Higgs}}$ at (g,ω) is for most "physical" Higgs Lagrangian densities a finite-dimensional group, or a symmetry group. Thus $\mathrm{Aut}_{(g,\omega)}(P)$ and $\mathcal{D}_{(g,\omega)}(M)$ are also symmetry groups, so in this case $\mathcal{L}_{\mathrm{Higgs}}$ does not have any gauge group. Thus neither of the covariance differential identities are gauge differential identities, i.e., neither depend on only the functional derivative $(\delta\mathcal{L}_{\mathrm{Higgs}}/\delta\varphi)$. Thus when the field equations are satisfied, the general covariance identities generate equations (c) and (d), which are local conservation laws for $\mathcal{L}_{\mathrm{Higgs}}$ at (g,ω) when $\mathrm{AUT}_{(g,\omega)}(P)$ is a finite-dimensional group.

Global conservation laws are given by the following (see also Theorem 3.2.5 below and the remarks thereafter):

3.2.2 Theorem. Let

$$\mathcal{L}_{\mathrm{Higgs}}: \mathcal{M}\times\mathcal{C}\times\mathcal{H} \to C^\infty_d$$

be a generally covariant Higgs field theory. Then for

$$(g,\omega,\varphi) \in \mathcal{M}\times\mathcal{C}\times\mathcal{H},\ Z \in \mathcal{X}_G(P),\ \text{and } k \geq 1$$

$$\mathrm{div}_{\mathrm{nat}}\Big(X^{\flat}\cdot\mathcal{T}_{\mathrm{Higgs}}(g,\omega,\varphi) + \phi_M\cdot\mathcal{J}_{\mathrm{Higgs}}(g,\omega,\varphi) + \frac{(-1)^{k-1}}{(k-1)!}\,i_X\varphi_M\cdot\frac{\delta\mathcal{L}_{\mathrm{Higgs}}}{\delta\varphi}(g,\omega,\varphi)\Big)$$

$$= \frac{1}{2}L_Xg\cdot\mathcal{T}_{\mathrm{Higgs}}(g,\omega,\varphi) + (L_Z\omega)_M\cdot\mathcal{J}_{\mathrm{Higgs}}(g,\omega,\varphi) + (L_Z\varphi)_M\cdot\frac{\delta\mathcal{L}_{\mathrm{Higgs}}}{\delta\varphi}(g,\omega,\varphi)$$

where $X = \pi_* Z$ and $\phi_M = (\omega(Z))_M$. If $k = 0$, the last term on the left hand side is deleted.

For $(g,\omega) \in \mathcal{M}\times\mathcal{C}$, let $\varphi \in \mathcal{H}$ satisfy the dynamical field equations

$$\frac{\delta\mathcal{L}_{\mathrm{Higgs}}}{\delta\varphi}(g,\omega,\varphi) = 0$$

and let $Z \in \mathcal{X}_G(P)$ satisfy

$$\frac{1}{2}L_Xg\cdot\mathcal{T}_{\mathrm{Higgs}}(g,\omega,\varphi) + (L_Z\omega)_M\cdot\mathcal{J}_{\mathrm{Higgs}}(g,\omega,\varphi) = 0 \tag{C'}$$

where $X = \pi_* Z$. Then

$$\operatorname{div}_{\text{nat}}(X^\flat \cdot \mathcal{T}_{\text{Higgs}}(g,\omega,\varphi) + \phi_M \cdot \mathcal{J}_{\text{Higgs}}(g,\omega,\varphi)) = 0$$

$\phi_M = (\omega(Z))_M$. If $T \in \mathcal{X}_G^{\text{vert}}(P)$ satisfies

$$(L_T\omega)_M \cdot \mathcal{J}_{\text{Higgs}}(g,\omega,\varphi) = D_M\phi_M \cdot \mathcal{J}_{\text{Higgs}}(g,\omega,\varphi) = 0 \qquad (C'_{\text{vert}})$$

then

$$\operatorname{div}_{\text{nat}}(\phi_M \cdot \mathcal{J}_{\text{Higgs}}(g,\omega,\varphi)) = 0$$

If $X \in \mathcal{X}(M)$ satisfies

$$\frac{1}{2}L_X g \cdot \mathcal{T}_{\text{Higgs}}(g,\omega,\varphi) + i_X\Omega_M \cdot \mathcal{J}_{\text{Higgs}}(g,\omega,\varphi) = 0 \qquad (C'_{\text{space}})$$

then

$$\operatorname{div}_{\text{nat}}(X^\flat \cdot \mathcal{T}_{\text{Higgs}}(g,\omega,\varphi)) = 0$$

Proof. The proof follows directly from Theorem 2.2.3. Q.E.D.

Remark. The quantities $X^\flat \cdot \mathcal{T}_{\text{Higgs}}$ and $\phi_M \cdot \mathcal{J}_{\text{Higgs}}$ are both vector densities, so the natural divergence operator can be applied.

As in the Yang-Mills system, the fields $\varphi \in \mathcal{H}$ may be derivatively coupled to gravity. The Higgs field theory is further complicated by the fact that the fields φ may be derivatively coupled to the connection field as well. Since the above theorem does not refer to any derivative coupling, it remains valid in the derivatively coupled case. However, in the presence of derivative coupling, the conditions C', C'_{vert}, and C'_{space} of the theorem are not equivalent to the conditions satisfied by $Z \in \mathcal{X}_{G,(g,\omega)}(P)$, $T \in \mathcal{X}^{\text{vert}}_{G,(g,\omega)}(P)$, and $X \in \mathcal{X}_{(g,\omega)}(M)$, respectively; namely, that for all $\varphi \in \mathcal{H}$,

$$D_g\mathcal{L}_{\text{Higgs}}(g,\omega,\varphi) \cdot L_X g + D_\omega\mathcal{L}_{\text{Higgs}}(g,\omega,\varphi) \cdot L_Z\omega = 0, \quad X = \pi_* Z \qquad (C)$$

$$D_\omega\mathcal{L}_{\text{Higgs}}(g,\omega,\varphi) \cdot L_T\omega = 0 \qquad (C_{\text{vert}})$$

and

$$D_g\mathcal{L}_{\text{Higgs}}(g,\omega,\varphi) \cdot L_X g + D_\omega\mathcal{L}_{\text{Higgs}}(g,\omega,\varphi) \cdot i_{\overline{X}}\Omega = 0 \qquad (C_{\text{space}})$$

Since we wish to consider the consequences of infinitesimal symmetries Z, T, and X of $\mathcal{L}_{\text{Higgs}}$ at (g,ω) that satisfy conditions C, C_{vert}, and C_{space}, respectively, we modify the global conservation laws of the above theorem in the presence of derivative coupling. The derivative coupling to gravity for $\mathcal{L}_{\text{Higgs}}$ is essentially the same as for $\mathcal{L}_{\text{YM}}$, so we concentrate on the derivative coupling to the connection.

3.2.3 Definition. A Higgs field theory

$$\mathcal{L}_{\text{Higgs}}: \mathcal{M} \times \mathcal{C} \times \mathcal{H} \to C_d^\infty$$

is ℓ-th order covariantly coupled to the connection field ω if in local coordinates

(1) $\mathcal{L}_{\text{Higgs}}$ is a function of the ℓ-th order partial derivatives of ω,

(2) these partial derivatives only appear as the curvature Ω and its gauge-covariant derivatives (not gauge covariant exterior derivative) of order $\leq \ell - 1$.

If $\ell = 0$, $\mathscr{L}_{\text{Higgs}}$ is <u>nonderivatively coupled to</u> ω, and if $\ell \geq 1$, $\mathscr{L}_{\text{Higgs}}$ is <u>derivatively coupled to</u> ω.

Note that since $D\Omega = 0$, ℓ can be ≥ 2 only if $\mathscr{L}_{\text{Higgs}}$ depends on the (non-exterior) gauge covariant derivatives $\Omega^a_{\alpha\beta:\gamma}$, $\Omega_{\alpha\beta:\gamma:\delta}$, etc.

Analogous to Prop. 3.1.4, we then have the following:

3.2.4 Proposition. Let $\mathscr{L}_{\text{Higgs}}: \mathscr{M} \times \mathscr{C} \times \mathscr{H} \to C^\infty_d$ be first order covariantly coupled to ω. Then for $(g, \omega, \varphi) \in \mathscr{M} \times \mathscr{C} \times \mathscr{H}$ and $\sigma \in A^1(P)$

$$D_\omega \mathscr{L}_{\text{Higgs}}(g, \omega, \varphi) \cdot \sigma = \frac{\delta \mathscr{L}_{\text{Higgs}}}{\delta\omega}(g, \omega, \varphi) \cdot \sigma_M + 2 \operatorname{div}_{\text{nat}}(\partial_\Omega \mathscr{L}_{\text{Higgs}}(g, \omega, \varphi) \cdot \sigma)_M$$

Proof. Since $\mathscr{L}_{\text{Higgs}}$ is first order covariantly coupled to ω, $\mathscr{L}_{\text{Higgs}}$ depends on the connection and its derivatives only through an algebraic dependence on ω and Ω. To display this dependence, write $\mathscr{L}_{\text{Higgs}}(g, \omega, \Omega, \varphi)$. Then

$$\begin{aligned}
D_\omega \mathscr{L}_{\text{Higgs}}(g, \omega, \Omega, \varphi) \cdot \sigma &= \partial_\omega \mathscr{L}_{\text{Higgs}} \cdot \sigma + \partial_\Omega \mathscr{L}_{\text{Higgs}} \cdot (D_\omega \Omega \cdot \sigma) \\
&= \partial_\omega \mathscr{L}_{\text{Higgs}} \cdot \sigma + \partial_\Omega \mathscr{L}_{\text{Higgs}} \cdot D\sigma \\
&= (\partial_\omega \mathscr{L}_{\text{Higgs}})_M \cdot \sigma_M + (\partial_\Omega \mathscr{L}_{\text{Higgs}})_M \cdot D_M \sigma_M \\
&= \partial_{\omega^a_\alpha} \mathscr{L}_{\text{Higgs}} \cdot \sigma^a_\alpha + \partial_{\Omega^a_{\alpha\beta}} \mathscr{L}_{\text{Higgs}} (D\sigma)^a_{\alpha\beta} \quad \text{(suppressing the subscript M)} \\
&= \partial_{\omega^a_\alpha} \mathscr{L}_{\text{Higgs}} \cdot \sigma^a_\alpha - 2(\partial_{\Omega^a_{\alpha\beta}} \mathscr{L}_{\text{Higgs}})_{:\alpha} \sigma^a_\beta + 2(\partial_{\Omega^a_{\alpha\beta}} \mathscr{L}_{\text{Higgs}} \cdot \sigma^a_\beta)_{;\alpha} \\
&= \frac{\delta \mathscr{L}_{\text{Higgs}}}{\delta\omega} \cdot \sigma_M + 2 \operatorname{div}_{\text{nat}}(\partial_\Omega \mathscr{L}_{\text{Higgs}} \cdot \sigma)_M
\end{aligned}$$

where we have used $(D\sigma)^a_{\alpha\beta} = \sigma^a_{\beta:\alpha} - \sigma^a_{\alpha:\beta}$, and

$$\frac{\delta \mathscr{L}_{\text{Higgs}}}{\delta\omega} = \partial_\omega \mathscr{L}_{\text{Higgs}} - 2 \operatorname{DIV}((\partial_\Omega \mathscr{L}_{\text{Higgs}})_M)$$

Q.E.D.

Remark. The quantity $\partial_\Omega \mathscr{L}_{\text{Higgs}}$ is a measure of the derivative coupling to ω, and is the analog of the quantity P_{Higgs} defined as in Prop. 3.1.4 (with $\mathscr{L}_{\text{Higgs}}$ replacing $\mathscr{L}_{\text{YM}}$), which measures the derivative coupling to g. In fact, $\partial_\Omega \mathscr{L}_{\text{Higgs}} = 0$ if and only if $\mathscr{L}_{\text{Higgs}}$ is nonderivatively coupled to ω ($\ell = 0$), in which case

$$(D_\omega \mathscr{L}_{\text{Higgs}})_M = (\partial_\omega \mathscr{L}_{\text{Higgs}})_M = \frac{\delta \mathscr{L}_{\text{Higgs}}}{\delta\omega}$$

When $\mathscr{L}_{\text{Higgs}}$ is first order covariantly coupled to g and ω, we modify Theorem 3.2.2 as follows:

3.2.5 Proposition. Let

$$\mathscr{L}_{\text{Higgs}}: \mathscr{M} \times \mathscr{C} \times \mathscr{H} \to C^\infty_d$$

be a generally covariant Higgs field theory that is first order covariantly coupled to g and ω.

For $(g,\omega) \in \mathcal{M} \times \mathcal{C}$, let $\varphi \in \mathcal{H}$ satisfy the dynamical field equations

$$\frac{\delta \mathcal{L}_{Higgs}}{\delta \varphi}(g,\omega,\varphi) = 0$$

and let $Z \in \mathcal{X}_G(P)$ satisfy

$$D_g \mathcal{L}_{Higgs}(g,\omega,\varphi) \cdot L_X g + D_\omega \mathcal{L}_{Higgs}(g,\omega,\varphi) \cdot L_Z \omega = 0, \qquad X = \pi_* Z \tag{C}$$

Then

$$\operatorname{div}_{nat}(X^\flat \cdot \mathcal{T}_{Higgs}(g,\omega,\varphi) + \phi_M \cdot \mathcal{J}_{Higgs}(g,\omega,\varphi)$$
$$+ P_{Higgs}(g,\omega,\varphi) \cdot L_X g + 2\partial_\Omega \mathcal{L}_{Higgs}(g,\omega,\varphi) \cdot L_Z \omega) = 0$$

where $\phi_M = (\omega(T))_M$ and $P_{Higgs}(g,\omega,\varphi)$ is defined as in Prop. 3.1.4.

If $T \in \mathcal{X}_G^{vert}(P)$ satisfies

$$D_\omega \mathcal{L}_{Higgs}(g,\omega,\varphi) \cdot L_T \omega = 0 \tag{C_{vert}}$$

then

$$\operatorname{div}_{nat}(\phi_M \cdot \mathcal{J}_{Higgs}(g,\omega,\varphi) + 2\partial_\Omega \mathcal{L}_{Higgs}(g,\omega,\varphi) \cdot D\phi) = 0$$

If $X \in \mathcal{X}(M)$ satisfies

$$D_g \mathcal{L}_{Higgs}(g,\omega,\varphi) \cdot L_X g + D_\omega \mathcal{L}_{Higgs}(g,\omega,\varphi) \cdot i_{\bar{X}} \Omega = 0 \tag{C_{space}}$$

then

$$\operatorname{div}_{nat}(X^\flat \cdot \mathcal{T}_{Higgs}(g,\omega,\varphi) + P_{Higgs}(g,\omega,\varphi) \cdot L_X g) = 0$$

<u>Proof</u>. If the dynamical field equations are satisfied, then from Theorem 3.2.2, for $(g,\omega,\varphi) \in \mathcal{M} \times \mathcal{C} \times \mathcal{H}$, and $Z \in \mathcal{X}_G(P)$,

$$\operatorname{div}_{nat}(X^\flat \cdot \mathcal{T}_{Higgs} + \phi_M \cdot \mathcal{J}_{Higgs}) = \frac{1}{2} L_X g \cdot \mathcal{T}_{Higgs} + (L_Z \omega)_M \cdot \mathcal{J}_{Higgs}$$

Since $\mathcal{L}_{Higgs}$ is first order covariantly coupled to g and ω, from Props. 3.1.4 and 3.2.4

$$D_g \mathcal{L}_{Higgs} \cdot L_X g = \frac{\delta \mathcal{L}_{Higgs}}{\delta g} \cdot L_X g + \operatorname{div}_{nat}(P_{Higgs} \cdot L_X g)$$

and

$$D_\omega \mathcal{L}_{Higgs} \cdot L_Z \omega = \frac{\delta \mathcal{L}_{Higgs}}{\delta \omega} \cdot L_Z \omega + 2 \operatorname{div}_{nat}(\partial_\Omega \mathcal{L}_{Higgs} \cdot L_Z \omega)$$

Thus if

$$D_g \mathcal{L}_{Higgs} \cdot L_X g + D_\omega \mathcal{L}_{Higgs} \cdot L_Z \omega = 0$$

then

$$\frac{\delta \mathcal{L}_{Higgs}}{\delta g} \cdot L_X g + \frac{\delta \mathcal{L}_{Higgs}}{\delta \omega} \cdot (L_Z \omega)_M = -\operatorname{div}_{nat}(P_{Higgs} \cdot L_X g + 2\partial_\Omega \mathcal{L}_{Higgs} \cdot L_Z \omega)$$
$$= \frac{1}{2} \mathcal{T}_{Higgs} \cdot L_X g + \mathcal{J}_{Higgs} \cdot L_Z \omega$$

Thus

$$\mathrm{div}_{\mathrm{nat}}(X^{\flat}\cdot\mathscr{T}_{\mathrm{Higgs}} + \phi_M \cdot \mathscr{J}_{\mathrm{Higgs}}) = -\mathrm{div}_{\mathrm{nat}}(P_{\mathrm{Higgs}}\cdot L_X g + 2\partial_\Omega \mathscr{L}_{\mathrm{Higgs}}\cdot L_Z\omega)$$

The remaining two conclusions follow directly. Q.E.D.

Remarks. 1. In the nonderivatively coupled case, $P_{\mathrm{Higgs}}(g,\omega,\varphi) = 0$ and $\partial_\Omega \mathscr{L}_{\mathrm{Higgs}}(g,\omega,\varphi) = 0$, so the conditions and conclusions of the above theorem are the same as in Theorem 3.2.2. In the derivatively coupled case, the above global conservation laws arise more naturally than those in Theorem 3.2.2. For example, if $T \in \mathscr{X}^{\mathrm{vert}}_{G,(g,\omega)}(P)$, then condition C_{vert} is satisfied (but not C'_{vert}), and so

$$\mathrm{div}_{\mathrm{nat}}(\phi_M \cdot \mathscr{J}_{\mathrm{Higgs}}(g,\omega,\varphi) + 2\partial_\Omega \mathscr{L}_{\mathrm{Higgs}}(g,\omega,\varphi)\cdot D\phi) = 0$$

Similarly, if $X \in \mathscr{X}_{(g,\omega)}(M)$, then C_{space} is satisfied (but not C'_{space}), and so

$$\mathrm{div}_{\mathrm{nat}}(X^{\flat}\cdot\mathscr{T}_{\mathrm{Higgs}}(g,\omega,\varphi) + P_{\mathrm{Higgs}}(g,\omega,\varphi)\cdot L_X g) = 0$$

Inasmuch as $\mathscr{X}^{\mathrm{vert}}_{G,(g,\omega)}(P)$ and $\mathscr{X}_{(g,\omega)}(M)$ measure the infinitesimal symmetries of $\mathscr{L}_{\mathrm{Higgs}}$ at (g,ω), the above global conservation laws are the appropriate ones for $\mathscr{L}_{\mathrm{Higgs}}$.

We also remark that as for $\mathscr{L}_{\mathrm{YM}}$, most $\mathscr{L}_{\mathrm{Higgs}}$ are nonderivatively coupled to both g and ω, and so the above modifications are rarely necessary. However, they do remain an interesting theoretical possibility.

2. As in Remark 5 after Prop. 3.1.5, there exists the possibility of sporadic infinitesimal symmetries of $\mathscr{L}_{\mathrm{Higgs}}$ at (g,ω), i.e., $Z \in \mathscr{X}_G(P)$ that satisfy condition C of the above proposition for a particular solution $\varphi \in \mathscr{H}$ but not for all φ, or even for all solutions. Such a Z then would generate a global conservation law for that particular solution, but not for all solutions.

Since $\mathscr{I}_{(g,\omega)}(P) \subseteq \mathscr{X}_{G,(g,\omega)}(P)$, $\mathscr{I}_{(g,\omega)}(P)$ is a minimal Lie algebra that will generate global conservation laws for $\mathscr{L}_{\mathrm{Higgs}}$ at (g,ω). The Lie group $I_{(g,\omega)}(P)$ is also said to generate global conservation laws for $\mathscr{L}_{\mathrm{Higgs}}$ at (g,ω); however, see Remarks 2 and 4 following Prop. 3.1.5 which also apply to $\mathscr{L}_{\mathrm{Higgs}}$.

Although in general $I_{(g,\omega)}(P) \subseteq \mathrm{AUT}_{(g,\omega)}(P)$, the symmetry group $\mathrm{AUT}_{(g,\omega)}(P)$ for $\mathscr{L}_{\mathrm{Higgs}}$ at (g,ω) depends on $\mathscr{L}_{\mathrm{Higgs}}$, whereas $I_{(g,\omega)}$ will be a minimal Lie group of symmetries for all generally covariant $\mathscr{L}_{\mathrm{Higgs}}$. Thus it is useful to study the group $I_{(g,\omega)}(P)$ since the Lie algebras $\mathscr{I}_{(g,\omega)}(P)$, $\mathscr{I}^{\mathrm{vert}}_{(g,\omega)}(P) = \mathscr{I}^{\mathrm{vert}}_{\omega}(P)$, and $\mathscr{I}_{(g,\omega)}(M) = \pi_*(\mathscr{I}_{(g,\omega)}(P))$ are minimal Lie algebras which generate global conservation laws for all $\mathscr{L}_{\mathrm{Higgs}}$ at (g,ω). Thus if

$$T \in \mathscr{I}^{\mathrm{vert}}_{\omega}(P) = \{T \in \mathscr{X}^{\mathrm{vert}}_{G}(P) \mid L_T\omega = D\phi_T = 0,\quad \phi_T = \omega(T)\}$$

then T satisfies condition C_{vert} and C'_{vert} and

$$\mathrm{div}_{\mathrm{nat}}(\phi_M \cdot \mathscr{J}_{\mathrm{Higgs}}(g,\omega,\varphi)) = 0, \qquad \phi_M = (\omega(T))_M$$

for solutions φ of the field equations.

Similarly, if $X \in \mathscr{I}_{(g,\omega)}(M) = \{X \in \mathscr{X}(M) \mid L_X g = 0$, and there exists $\phi_M \in C^\infty(\mathfrak{G}_{Ad}(M))$ such that $D_M \phi_M + i_X \Omega_M = 0\}$, where D_M and Ω_M are formed using ω, then $Z = T + \bar{X}$ satisfies $L_Z \omega = 0$, where $(\omega(T))_M = \phi_M$, and $\bar{X}$ is the horizontal lift of X using ω. Also, $X = \pi_* Z$ satisfies $L_X g$. Thus $Z \in \mathscr{I}_{(g,\omega)}(P)$, and so Z satisfies conditions C and C', and so from either Theorem 4.2.2 or Prop. 4.2.5,

$$\mathrm{div}_{nat}(X^\flat \cdot \mathscr{T}_{Higgs}(g,\omega,\varphi) + \phi_M \cdot \mathscr{J}_{Higgs}(g,\omega,\varphi)) = 0$$

for all solutions $\varphi \in \mathscr{H}$ of the field equations.

Another important feature of $\mathscr{I}_{(g,\omega)}(P)$ is that its elements satisfy both conditions C and C' and thus generate global conservation laws which do not involve the derivative coupling terms P_{Higgs} or $\partial_\Omega \mathscr{L}_{Higgs}$, even for derivatively coupled $\mathscr{L}_{Higgs}$. This is completely analogous to the role $\mathscr{I}_g(M)$ plays in Yang-Mills field theories (see Remark 4 following Prop. 3.1.5).

A natural question about the group $I_{(g,\omega)}(P)$ is how it is related to the structure group G of the bundle. This question in turn can be reduced to a study of the group of internal symmetries

$$I_\omega^{int}(P) = \{F \in \mathrm{Aut}(P) \quad F^*\omega = \omega\}$$

of the connection ω (see [6] for more details regarding the group of internal symmetries of a connection ω).

Let $\mathrm{Hol}(\omega) \subseteq G$ be the holonomy group of the connection ω, and assume that $\mathrm{Hol}(\omega)$ is a closed subgroup of G. Then $I_\omega^{int}(P)$ can be shown to be isomorphic to the centralizer of $\mathrm{Hol}(\omega)$ in G, namely,

$$C_G(\mathrm{Hol}(\omega)) = \{a \in G \mid ab = ba \text{ for all } b \in \mathrm{Hol}(\omega)\}$$

In elementary particle theories, M is usually taken to be R^4 or S^4, and ω is taken to be flat (see [3] and the references therein). In this case (since M is simply connected), $\mathrm{Hol}(\omega) = \{id\}$ (see [12], p. 92) and so

$$I_\omega^{int}(P) \approx C_G(\{id\}) = G$$

Thus in this classical case, the minimal internal symmetry group $I_\omega^{int}(P)$ is isomorphic to the structure group of the bundle. If, however, the connection is nonflat, then the internal symmetry group is reduced to the smaller group $I_\omega^{int}(P)$. Thus a nonflat gauge potential provides a natural mechanism for spontaneous symmetry breaking. We are currently studying the importance of this phenomenon in specific elementary particle theories.

There is an interesting analogy to the fact that a simply connected manifold M and a flat connection ω imply $I_\omega^{int}(P) \approx G$. Consider a tensor field theory over a spacetime M with Lorentz metric g. Then the isometry group $I_g(M)$ is a minimal symmetry group for the theory. But $I_g(M)$ need not be isomorphic to the Poincaré group unless M is simply connected and g is complete and flat ([18], p. 68). Turning this around, a complete nonflat g causes a symmetry breaking from the Poincaré group to $I_g(M)$, just as a nonflat connection ω causes a symmetry breaking from the structure group G of the bundle to $I_\omega^{int}(P)$.

Finally, we remark that the master equations of Theorems 2.2.2 and 2.2.3 are very general and have other applications. We have seen how they can be used to deduce the differential identities and conservation laws of a generally covariant Lagrangian field theory. It is not unreasonable to expect that these master equations may contain other information regarding Lagrangian field theories. We are currently pursuing this possibility (see [4]).

We hope that in this paper we have been able to justify our contention that even though the study of symmetry groups and conservation laws is one of the oldest studies in theoretical physics, it is still today on the forefront of research, and indeed, holds a promising future in helping to unravel the deep mysteries of general relativity, gauge theory, and elementary particle physics.

REFERENCES

1. Abraham, R.; Marsden, J. E.: Foundations of Mechanics, second edition, Benjamin/Cummings (1978).
2. Bleecker, D.: Gauge Theory and Variational Principles, Addison-Wesley (1981).
3. Fischer, A.: Gen. Relativity Gravit. 14, 683-689 (1982).
4. Fischer, A.: Lecture Notes on Gauge Field Theories, to appear, 1985.
5. Fischer, A.: The vanishing of the momentum theorem for generally covariant Lagrangian gauge field theories, to appear, 1985.
6. Fischer, A.: The internal symmetry group of a connection on a principal fiber bundle with applications to gauge field theories, to appear, 1985.
7. Fischer, A.; Marsden, J.: Topics in the dynamics of general relativity, in Isolated Gravitating Systems in General Relativity, J. Ehlers, ed. North-Holland (1979).
8. Forgacs, P.; Manton, N. S.: Commun. Math. Phys. 72, 15-35 (1980).
9. Jackiw, R.: Invariance, symmetry and periodicity in gauge theories, NFS ITP preprint 80-15 (1980).
10. Jackiw, R.; Manton, N. S.: Nucl. Phys. B 158, 141 (1979).
11. Kobayashi, S.: Transformation Groups in Differential Geometry, Springer (1972).
12. Kobayashi, S.; Nomizu, K.: Foundations of Differential Geometry, Vol. I. Interscience (1963).
13. Moncrief, V.: Ann. Phys. 108, 387-400 (1977).
14. Noether, E.: Nachrichten Gesell. Wissenschaft. Gottingen, Math.-Phys. 2, 235-257 (1918).
15. Sternberg, S.: Lectures on Differential Geometry, Prentice-Hall (1964).
16. Sternberg, S.: On the role of field theories in our physical conception of geometry, in Differential Geometrical Methods in Mathematical Physics II, K. Bleuler, H. R. Petry, and A. Reetz, eds. Springer (1978).
17. Trautman, A.: On groups of gauge transformations, in Geometrical and Topological Methods in Gauge Theories, J. P. Harnad and S. Shnider, eds. Springer (1980).
18. Wolf, J. A.: Spaces of Constant Curvature. McGraw-Hill (1967).

14 THE BOREL SPECTRAL SEQUENCE: SOME REMARKS AND APPLICATIONS

Hans R. Fischer and Floyd L. Williams / Department of Mathematics and Statistics, University of Massachusetts at Amherst, Amherst, Massachusetts

1. INTRODUCTION

1.1. The Borel spectral sequence of a holomorphic fibration $X \to Y$ relates the Dolbeault cohomology of X (with vector bundle coefficients) with that of Y and the fibre. In his original treatment [12] Borel assumed the coefficient bundle, say $\underset{\sim}{E}$, to be a pull-back of a bundle over Y. In a later treatment [13, 14] Le Potier weakened the latter assumption and required only that $\underset{\sim}{E}$ should be "locally trivial over Y" in the sense of Sec. 2.1b below. In the present paper we maintain Le Potier's local triviality assumption and comment on the computation of the E_2-term. The latter term involves an important "cohomology bundle" over Y. We compute this bundle when X, Y are homogeneous (showing that it too is homogeneous) and thereby derive results of Botts [5] and Ahiezer [1, 2]. We also outline a construction of cohomology bundles in the general context of open real orbits in complex flag manifolds, and point out their application to mathematical physics in connection with the so-called Penrose transform [17, 18, 20]; see Corollary 2 and the example in Sec. 3.2. Finally, in this paper, we outline a theory of Dolbeault cohomology-valued characteristic classes of holomorphic vector bundles over compact complex manifolds (here too Borel's spectral sequence is employed). These classes are shown to satisfy the Grothendieck axioms of [11]. Moreover they may be obtained by a construction analogous to the classical one of Chern and Weil.

1.2. Notation and Conventions. In this note all manifolds are assumed to be smooth (i.e., C^∞) and countable at infinity, and therefore are paracompact. If X is such a manifold, A(X) or $A^\cdot(X)$ denotes the module of smooth differential forms on X. For a smooth vector bundle $\underset{\sim}{E} \to X$, $A(X, \underset{\sim}{E}) = \bigoplus_{K \geq 0} A^K(X, \underset{\sim}{E})$ is the module of $\underset{\sim}{E}$-valued forms on X. If X is a complex manifold, we sometimes use Λ^{pq} to denote the bundle of scalar (p, q)-forms on denote by $A^{pq}(X, \underset{\sim}{E})$ the $\underset{\sim}{E}$-valued (p, q)-forms on X when $\underset{\sim}{E}$ is a holomorphic vector bundle. The smooth sections of $\underset{\sim}{E}$ will be denoted by $\Gamma(\underset{\sim}{E})$ and when X is compact, $\Gamma(\underset{\sim}{E})$ will carry its usual C^∞-topology which makes it a Fréchet space. One notes by the closed graph theorem this is the unique Fréchet topology on $\Gamma(\underset{\sim}{E})$ which is finer than the product topology. $\underline{O}_X(\underline{O}_X^*)$, Ω_X will denote the sheaves of (non-vanishing) holomorphic functions and holomorphic differential forms, respectively, on X when X is a complex manifold. Further conventions will be introduced as they are needed.

2. THE BOREL SPECTRAL SEQUENCE

2.1. In this section, we are going to establish a spectral sequence for the Dolbeault cohomology of a holomorphic fibre bundle, due originally to Borel and later generalized by Le Potier; cf. [12-14]. We shall be dealing with a diagram

$$\begin{array}{ccc} \underset{\sim}{E} & \to & X \\ & & \downarrow \pi \\ & & Y \end{array} \tag{2.1}$$

with the following data: $X \xrightarrow{\pi} Y$ is a holomorphic fibre bundle with fibre F and group G,

$\underset{\sim}{E} \to X$ is a holomorphic vector bundle with fibre E. The fibre $\pi^{-1}(y)$ will also be denoted by X_y and the induced bundle $\underset{\sim}{E}|X_y$ by $\underset{\sim}{E}_y$. Moreover, we impose the following conditions on the data:

a) The fibre F is compact.
b) The bundle $\underset{\sim}{E}$ is "locally trivial over Y" in the following sense: There is a holomorphic vector bundle $\underset{\sim}{E}_0$ over F with the property that every $y \in Y$ has an open neighborhood U such that $\pi^{-1}(U)$ admits a trivialization $\phi_U: \pi^{-1}(U) \simeq U \times F$ and a holomorphic isomorphism $\psi_U: \underset{\sim}{E}|\pi^{-1}(U) \simeq U \times \underset{\sim}{E}_0$ over ϕ_U; the following diagram therefore commutes:

$$\begin{array}{ccc} \underset{\sim}{E}|\pi^{-1}(U) & \xrightarrow{\psi_U} & U \times \underset{\sim}{E}_0 \\ \downarrow & & \downarrow \\ \pi^{-1}(U) & \xrightarrow{\phi_U} & U \times F \\ & \searrow^{\pi} \quad \swarrow & \\ & U & \end{array} \tag{2.2}$$

In particular, for each $y \in U$, there is the isomorphism $\psi_U(y)$ of $\underset{\sim}{E}_y$ onto $\underset{\sim}{E}_0$ over $\phi_U(y): X_y \simeq F$, and $\psi_U(y)$ "depends holomorphically on y."

Note that the condition b) is satisfied, in particular, whenever $\underset{\sim}{E}$ is the pullback to X of a holomorphic vector bundle on Y, the case considered in [12].

2.2. The fibration $X \xrightarrow{\pi} Y$ leads to a decreasing filtration of the module $A(X, \underset{\sim}{E})$ of $\underset{\sim}{E}$-valued forms on Y, compatible with $\bar{\partial}$, in the following manner:

For given (p, q) and $r \geq 0$, let $F^r A^{p,q}(X, \underset{\sim}{E})$ be the $C^\infty(X)$-submodule of $A^{p,q}(X, \underset{\sim}{E})$ generated by all forms ω which may be written as

$$\omega = \pi^*\alpha \wedge \beta \tag{2.3}$$

with $\alpha \in A^{a,b}(Y)$, $\beta \in A^{c,d}(X, \underset{\sim}{E})$ such that $a + c = p$, $b + d = q$ and $a + b \geq r$. This construction yields the filtration of $A^{p,q}(X, \underset{\sim}{E})$ "by base degree." It is clear that $\bar{\partial}(F^r A^{p,q}(X, \underset{\sim}{E})) \subset F^r A^{p,q+1}(X, \underset{\sim}{E})$ and, moreover, that $F^r A^{p,q}(X, \underset{\sim}{E}) = 0$ for $r > p + q$ $F^0 A^{p,q}(X, \underset{\sim}{E}) = A^{p,q}(X, \underset{\sim}{E})$. $A^{p,\cdot}(X, \underset{\sim}{E}) = \oplus A^{p,q}(X, \underset{\sim}{E})$ now is filtered as usual and this filtration will be compatible with $\bar{\partial}$.

Thus, for each fixed $p \geq 0$, there is the standard spectral sequence $({}^pE_r^{s,t})$ of the filtered graded module $A^{p,\cdot}(X, \underset{\sim}{E})$ which converges to the cohomology $H^{p,\cdot}(X, \underset{\sim}{E})$ of X. The family of spectral sequences defined in this manner constitutes what is known as the Borel spectral sequence converging to $H_{\bar{\partial}}(X, \underset{\sim}{E})$.

We shall add some remarks on the E_0-, E_1-, and E_2-terms of this spectral sequence below, but state the main result before any further details:

Theorem [12-14]. For each pair (c, d), the vector fibration

$$\underset{\sim}{H}^{c,d}(\underset{\sim}{E}) = \bigcup_y H^{c,d}(X_y, \underset{\sim}{E}_y) \tag{2.4}$$

of fibre cohomologies is a holomorphic vector bundle over Y with fibre $H^{c,d}(F, \underset{\sim}{E}_0)$ (the (c, d)-cohomology bundle) and with this, the E_2-term ${}^pE_2^{s,t}$ has the following realization:

$${}^pE_2^{s,t} \simeq \bigoplus_i H^{i,s-i}(Y, \underset{\sim}{H}^{p-i,t+i}(\underset{\sim}{E})) \tag{2.5}$$

Remark. If $\underset{\sim}{E} = \pi^*\underset{\sim}{G}$, $\underset{\sim}{G} \to Y$ a holomorphic vector bundle, then one can easily show that $\underset{\sim}{H}^{c,d}(\underset{\sim}{E}) \simeq \underset{\sim}{H}^{c,d} \otimes \underset{\sim}{G}$ where $\underset{\sim}{H}^{c,d}$ is the bundle of scalar cohomologies. This yields the E_2-term in the form it has in [12]. In many practical cases, the scalar cohomology bundles will be trivial and in such a case, the last observation together with (2.5) shows that the E_2-term then is expressed directly in terms of the scalar fibre cohomology $H_{\bar\partial}(F)$ and the base cohomology $H_{\bar\partial}(Y, \underset{\sim}{G})$. One also observes that for $p = 0$, one again obtains the Leray spectral sequence for the sheaf $\underset{\sim}{O}(\underset{\sim}{E})$ of holomorphic sections of $\underset{\sim}{E}$; cf. [14] for this and further remarks.

In order to establish the claim that the cohomology bundles are holomorphic, one may proceed as follows:

Firstly, let $\underline{\underline{GL}}(\underset{\sim}{E}_0)$ be the group of all holomorphic automorphisms of $\underset{\sim}{E}_0$, i.e., of all biholomorphic maps $\phi: \underset{\sim}{E}_0 \to \underset{\sim}{E}_0$ which are fibrewise linear. By results of [15] and [16], this will be a (finite-dimensional) complex Lie group since F is compact. Its Lie algebra $\underline{\underline{L}}(\underset{\sim}{E}_0)$ may be realized as the algebra of holomorphic first-order differential operators L on E_0 whose symbol $\sigma(L)$ is a holomorphic vector field X on F (i.e., a holomorphic section of T(F)); cf. [13] and also [9]. Any $\phi \in \underline{\underline{GL}}(\underset{\sim}{E}_0)$ induces a biholomorphic map $f: F \to F$ and this construction yields a holomorphic homomorphism $\underline{\underline{GL}}(\underset{\sim}{E}_0) \to \mathrm{Aut}(F)$ whose derivative maps $L \in \underline{\underline{L}}(\underset{\sim}{E}_0)$ to $\sigma(L) = X$. Using f, the automorphism ϕ then may be interpreted as a specific isomorphism $\alpha: \underset{\sim}{E}_0 \simeq f^*\underset{\sim}{E}_0$; in this sense, $\phi = (f, \alpha)$. The action of $\underline{\underline{GL}}(\underset{\sim}{E}_0)$ on $\Gamma(\underset{\sim}{E}_0)$ which leads to the realization $L(\underset{\sim}{E}_0)$ of the algebra is given by

$$R_\phi s = \alpha^{-1} f^* s \tag{2.6}$$

and lifts to $\underset{\sim}{E}_0$-valued differential forms in the obvious manner: given the decomposable $\underset{\sim}{E}_0$-valued (p, q)-form $\omega \otimes s$ $(\omega \in A^{p,q}(F),\ s \in \Gamma(\underset{\sim}{E}_0))$, one sets

$$R_\phi^{p,q}(\omega \otimes s) = f^*\omega \otimes R_\phi s \tag{2.7}$$

The action is smooth in the sense that $\phi \to R_\phi^{p,q}\beta$ is a differentiable map of $GL(\underset{\sim}{E}_0)$ into the Fréchet space $A^{p,q}(F, \underset{\sim}{E}_0)$. Its derivative at the identity is the map $L \to L^{p,q}\beta$,

$$L^{p,q}(\omega \otimes x) = \theta(X)\omega \otimes s + \omega \otimes Ls \tag{2.8}$$

$X = \sigma(L)$, $\theta(X)$ the Lie derivative along X. Clearly, this is $\mathbb{C}$-linear in L and so the action of $\underline{\underline{GL}}(\underset{\sim}{E}_0)$ is actually holomorphic. Both actions commute with $\bar\partial$ and thus induce actions on the cohomology $H^{p,q}(F, \underset{\sim}{E}_0)$; we set $\pi^{p,q}(\phi)$ for the automorphism induced by $R_\phi^{p,q}$. The crucial observation now is that $\phi \to \pi^{p,q}(\phi)$ is <u>holomorphic</u>, as follows from (the details of) the above considerations. One also notes that

$$0 \to \mathrm{im}(\bar\partial) \to \ker(\bar\partial) \to H(F, \underset{\sim}{E}_0) \to 0 \tag{2.9}$$

actually is an exact sequence of $\underline{\underline{GL}}(\underset{\sim}{E}_0)$-<u>modules</u>, a fact used below.

Suppose next that (U_α) is an open covering of Y such that over each U_α, there is a "trivialization" $(\phi_\alpha, \psi_\alpha)$ of $\underset{\sim}{E}$ in the sense of (2.2). Setting $\psi_{\alpha\beta} = \psi_\alpha \cdot \psi_\beta^{-1}$, one obtains an automorphism of $(U_\alpha \cap U_\beta) \times \underset{\sim}{E}_0$ of the form $(y, e) \to (y, \psi_{\alpha\beta}(y)e)$ where now $\psi_{\alpha\beta}(y)$ lies in $\underline{\underline{GL}}(\underset{\sim}{E}_0)$ and covers (the automorphism of F induced by) $g_{\alpha\beta}(y) = \phi_\alpha \circ \phi_\beta^{-1} | X_y \in G$. Setting

$$h_{\alpha\beta}^{p,q} = \pi^{p,q}(\psi_{\beta\alpha}) \tag{2.10}$$

one obtains a holomorphic cocycle with values in $\underline{\underline{GL}}(H^{p,q}(F, \underset{\sim}{E}_0))$ and therefore a holomorphic vector bundle $\underset{\sim}{H}^{p,q}(\underset{\sim}{E})$ over Y with fibre $H^{p,q}(F, \underset{\sim}{E}_0)$. It is easy to see that this bundle may be identified with the fibration $\bigcup_y H^{p,q}(X_y, \underset{\sim}{E}_y)$ in a canonical manner, establishing the first part of the theorem.

2.3. In order to arrive at (2.5), we need to find a more explicit description of the E_0-terms and the differential d_0. To this end, let $V(X) = \ker(T\pi)$ be the vertical tangent bundle of $X \to Y$. The complex structure "along the fibres" of X yields a splitting

$$V(X) = \underset{\sim}{V} \oplus \bar{\underset{\sim}{V}} \tag{2.11}$$

into anti-isomorphic complex subbundles and we define the module of <u>vertical k-forms</u> ("fibre forms" of [12], "relative forms" of [13, 14]) to be $A_V^k(X, \underset{\sim}{E}) = \Gamma(\Lambda^k V^*(X)^{\mathbb{C}} \otimes \underset{\sim}{E}))$; (2.11) then yields the double grading of $A_V^k(X, \underset{\sim}{E})$ by the submodules

$$A_V^{c,d}(X, \underset{\sim}{E}) = \Gamma(\Lambda^c \underset{\sim}{V}^* \otimes \Lambda^d \bar{\underset{\sim}{V}}^* \otimes \underset{\sim}{E}) \tag{2.12}$$

of "vertical (c,d)-forms" on X with values in $\underset{\sim}{E}$ (the "fibre $\underset{\sim}{E}$-(c,d)-forms" of [12]); in the notations of [14], these are the modules $A^{c,d}(X/Y, \underset{\sim}{E})$. Moreover, essentially following the definitions of [7], one introduces a differential

$$\bar{\partial}_V : A_V^{c,d}(X, \underset{\sim}{E}) \to A_V^{c,d+1}(X, \underset{\sim}{E}) \tag{2.13}$$

with $\bar{\partial}_V^2 = 0$; this amounts to using the standard formula for $\bar{\partial}$ using only vertical vector fields since, indeed, the terms $Z_i\omega(Z_1, \ldots, \hat{Z}_i, \ldots, Z_k)$ and $\omega([Z_i, Z_j], Z_1, \ldots, \hat{Z}_i, \ldots, \hat{Z}_j, \ldots, Z_k)$ are well-defined as long as all Z_i are vertical. In the notations of [7], one chooses $\underset{\sim}{F} = \underset{\sim}{V} \oplus \bar{\underset{\sim}{V}}$ and then splits d_F into two conjugate differentials; cf. [9] for details. The cohomology given by this situation is denoted by

$$H_V^{c,d}(X, \underset{\sim}{E}) = \ker(\bar{\partial}_V \,|\, A_V^{c,d}(X, \underset{\sim}{E}))/\bar{\partial}_V(A_V^{c,d-1}(X, \underset{\sim}{E})) \tag{2.14}$$

and is called the <u>vertical</u> cohomology (relative cohomology in [13, 14]).

The reason for the introduction of all this is the following: the vertical forms on X can be realized as differential forms on X, e.g., by the choice of a splitting

$$T(X) = W(X) \oplus V(X) \tag{2.15}$$

into complex subbundles, i.e., by the choice of a "horizontal" bundle W(X). Then $W(X)^{\mathbb{C}} = \underset{\sim}{W} \oplus \bar{\underset{\sim}{W}}$ shows that there is a quadruple grading of the differential forms; in particular,

$$A^{p,q}(X, \underset{\sim}{E}) = \oplus\, \Gamma(\Lambda^a \underset{\sim}{W}^* \otimes \Lambda^b \bar{\underset{\sim}{W}}^* \otimes \Lambda^c \underset{\sim}{V}^* \otimes \Lambda^d \bar{\underset{\sim}{V}}^* \otimes \underset{\sim}{E})$$

the sum being extended over all (a,b,c,d) with a + c = p, b + d = q; cf. also [12, 13]. Since $W(X)^* \simeq \pi^* T^*(Y)$, this leads to the following explicit description of the E_0-term:

$$\mathrm{gr}^r(FA^{p,q}(X, \underset{\sim}{E})) \simeq \bigoplus_i A^{i,s-i}(Y) \underset{C^\infty(Y)}{\otimes} A_V^{p-i,t+i}(X, \underset{\sim}{E}) \tag{2.16}$$

and d_0 corresponds to $\epsilon \otimes \bar{\partial}_V$ under this isomorphism, $\epsilon \,|\, A^{a,b}(Y) = (-1)^{a+b}$.

Next, $\bar{\partial}_V$ is $C^\infty(Y)$-linear and one therefore has the short exact sequence of <u>$C^\infty(Y)$-modules</u> $0 \to \mathrm{im}(\bar{\partial}_V) \to \ker(\bar{\partial}_V) \to H_V(X, \underset{\sim}{E}) \to 0$. Since for each (a,b), $A^{a,b}(Y)$ is a projective $C^\infty(Y)$-module, this readily implies that

$${}^pE_1^{s,t} \simeq \bigoplus_i A^{i,s-i}(Y) \underset{C^\infty(Y)}{\otimes} H_V^{p-i,t+i}(X, \underset{\sim}{E}) \tag{2.17}$$

cf. [13, 14].

The final step now consists of a suitable re-interpretation of these terms which will also lead to a more concrete description of the image of d_1.

For this purpose, we introduce certain Fréchet bundles over Y with infinite-dimensional fibres, also considered in [12, 14], beginning with the fibrations

$$\underset{\sim}{A}^{c,d}(\underset{\sim}{E}) = \bigcup_y A^{c,d}(X_y, \underset{\sim}{E}_y) \tag{2.18}$$

whose fibres are the modules $A^{c,d}(F, \underset{\sim}{E}_0)$ (which are infinite-dimensional Fréchet spaces). The 1-cocycle defining such a bundle for an open cover (U_α) such as was used in the construction of the cohomology bundles is, of course, given by $r^{p,q}_{\alpha\beta} = R^{p,q}_{\psi_{\beta\alpha}}$. For each α, $\underset{\sim}{A}^{c,d}(\underset{\sim}{E}) | U_\alpha \simeq U_\alpha \times A^{c,d}(F, \underset{\sim}{E}_0)$ and this makes it possible to define smooth (holomorphic) sections of $\underset{\sim}{A}^{c,d}(\underset{\sim}{E})$ in a standard manner by using their local representatives which are maps $U_\alpha \to A^{c,d}(F, \underset{\sim}{E}_0)$; cf. [9] for more details. One easily argues that under the trivializations $(\phi_\alpha, \psi_\alpha)$, the vertical bundles $\Lambda^c \underset{\sim}{V}^* \otimes \Lambda^d \bar{\underset{\sim}{V}}^*$ over $\pi^{-1}(U_\alpha)$ map to $\Lambda^{c,d}$ on F and with this, one obtains a natural map

$$\rho: A_V^{c,d}(X, \underset{\sim}{E}) \to \Gamma(A^{c,d}(\underset{\sim}{E})) \tag{2.19}$$

which is readily seen to be an isomorphism of $C^\infty(Y)$-modules.

Moreover, the $\bar{\partial}$-operator of F induces bundle maps

$$D: \underset{\sim}{A}^{c,d}(\underset{\sim}{E}) \to \underset{\sim}{A}^{c,d+1}(\underset{\sim}{E}), \quad D^2 = 0 \tag{2.20}$$

and the cohomology of the bundle complex thus obtained is given by the bundles $\underset{\sim}{H}^{c,d}(\underset{\sim}{E})$ introduced earlier. The isomorphism ρ satisfies $D \circ \rho = \rho \circ \bar{\partial}_V$ and consequently, it induces isomorphisms $\ker(\bar{\partial}_V) \simeq \Gamma(\ker(D))$, $\operatorname{im}(\bar{\partial}_V) \simeq \Gamma(\operatorname{im}(D))$. Note here that the compactness of F implies that im(D) also is a Fréchet bundle and that there is a fibrewise splitting short exact sequence

$$0 \to \operatorname{im}(D) \to \ker(D) \to \underset{\sim}{H}^{\cdot\cdot}(\underset{\sim}{E}) \to 0 \tag{2.21}$$

of Fréchet bundles. Since the section sheaves of these bundles are modules over the sheaf of smooth functions on Y, they are fine and thus there is the corresponding short exact sequence

$$0 \to \Gamma(\operatorname{im}(D)) \to \Gamma(\ker(D)) \to \Gamma(\underset{\sim}{H}^{\cdot\cdot}(\underset{\sim}{E})) \to 0 \tag{2.22}$$

of $C^\infty(Y)$-modules. One finally obtains the commutative diagram

$$\begin{array}{ccccccccc} 0 \to & \bar{\partial}_V(A_V^{c,d-1}(X, \underset{\sim}{E})) & \to & \ker(\bar{\partial}_V | A^{c,d}(X, \underset{\sim}{E})) & \to & H_V^{c,d}(X, \underset{\sim}{E}) & \to 0 \\ & \rho\downarrow & & \rho\downarrow & & \swarrow & \\ 0 \to & \Gamma(\operatorname{im}(D^{c,d-1})) & \to & \Gamma(\ker(D^{c,d})) & \to & \Gamma(\underset{\sim}{H}^{c,d}(\underset{\sim}{E})) & \to 0 \end{array}$$

which shows that the map ρ induces natural isomorphisms

$$H_V^{c,d}(X, \underset{\sim}{E}) \simeq \Gamma(\underset{\sim}{H}^{c,d}(\underset{\sim}{E})) \tag{2.23}$$

Accordingly, the E_1-terms of the Borel spectral sequence now may be written in the form

$$^pE_1^{s,t} \simeq \bigoplus_i A^{i,s-i}(Y, \underset{\sim}{H}^{p-i,t+i}(\underset{\sim}{E}))$$

A last step, involving a reduction to the globally trivial case in the sense of (2.2), then shows that under these isomorphisms, d_1 corresponds to the standard $\bar{\partial}$-operator on Y for bundle-valued forms and this concludes the proof of the theorem.

3. SOME APPLICATIONS

3.1. Generalizations of a Theorem of Bott. We introduce here a large class of holomorphic bundles $X \to Y$ and vector bundles $\underset{\sim}{E} \to X$ which satisfy the conditions of Sec. 2.1, so that the Borel spectral sequence applies in these cases. It should be noted that the results of Sec. 2 can be verified directly in the present situation; we limit the exposition to a brief outline with some emphasis on the naturality of some of the earlier constructions.

The current context is the following: Y denotes a complex manifold, H_1 a complex Lie group and $Z \xrightarrow{\pi_1} Y$ a holomorphic principal H_1-bundle. Next, let $H_2 \subset H_1$ be a closed complex Lie subgroup such that H_1/H_2 is compact. Finally, $X = Z/H_2$ is the quotient of Z modulo the natural right action of H_2.

X is a complex manifold such that $Z \xrightarrow{\pi_2} X$ is a holomorphic principal H_2-bundle and, since $H_2 \subset H_1$, there is a natural map $\pi: X \to Y$ such that

$$\begin{array}{ccc} Z & \xrightarrow{\pi_2} & X \\ & {\scriptstyle\pi_1}\searrow & \downarrow{\scriptstyle\pi} \\ & & Y \end{array} \tag{3.1}$$

commutes. Using the bundle structures of Z over X and Y, one shows that $X \xrightarrow{\pi} Y$ is a holomorphic fibre bundle with fibre H_1/H_2.

Next, let $\lambda: H_2 \to GL(V)$ be a holomorphic finite-dimensional representation of H_2. We denote by $\underset{\sim}{E} = \underset{\sim}{E}_\lambda$ the holomorphic vector bundle over X, with fibre V, associated to the principal bundle $Z \to X$ by means of λ: $\underset{\sim}{E} = Z \times_{H_2} V \to X$. Thus, we have the situation

$$\begin{array}{ccc} \underset{\sim}{E} & \to & X \\ & & \downarrow{\scriptstyle\pi} \\ & & Y \end{array}$$

considered in Sec. 2.1. Furthermore, there also is the holomorphic vector bundle $\underset{\sim}{E}_0 = H_1 \times_{H_2} V \to H_1/H_2$ associated to the principal bundle $H_1 \to H_1/H_2$. The construction of "local trivializations" of $\underset{\sim}{E}$ over Y in the sense of Sec. 2.1 here proceeds as follows:

Since $Z \to Y$ is a holomorphic principal H_1-bundle, one can choose a holomorphic section $\delta_1: U \to Z$ of π_1 over the sufficiently small open set $U \subset Y$. Then, if $y \in U$ and $h \in H_1$,

$$\phi_U^{-1}(y, hH_2) = \pi_2(\delta_1(y)h) \tag{3.2}$$

defines a holomorphic trivialization of $\pi^{-1}(U) \subset X$.

The (holomorphic) isomorphism ψ_U over ϕ_U required in (2.2) is obtained as follows: Given $y \in U$ and $[h, v] \in \underset{\sim}{E}_0$ (where $[h, v]$ denotes the equivalence class of $(h, v) \in H_1 \times V$), set

$$\psi_U^{-1}(y, [h, v]) = (\phi_U^{-1}(y, hH_2), [\delta_1(y)h, v]) \tag{3.3}$$

where the right-hand side is in $\underset{\sim}{E} | \pi^{-1}(U)$ (= pull-back of $\underset{\sim}{E}$ to $\pi^{-1}(U) \subset X$). In particular, for $y \in U$, there is the map $\psi_U(y): \underset{\sim}{E}_y \to \underset{\sim}{E}_0$ which is a bundle isomorphism over $\phi_U(y): X_y \simeq H_1/H_2$; thus, for $e \in \underset{\sim}{E}_0$, $\psi_U^{-1}(y)(e) = \psi_U^{-1}(y, e)$. Moreover, $\phi_U(y)$ and $\psi_U(y)$ induce isomorphisms $\psi_U(y)^*: H^{p,q}(H_1/H_2, \underset{\sim}{E}_0) \simeq H^{p,q}(X_y, \underset{\sim}{E}_y)$.

Suppose now that $\epsilon_1: W \to Z$ is a second holomorphic section of π_1, $U \cap W \neq \emptyset$, and let $\phi_W: \pi^{-1}(W) \to W \times H_1/H_2$, $\psi_W: \underset{\sim}{E} | \pi^{-1}(W) \to W \times \underset{\sim}{E}_0$ the corresponding induced trivializations; in particular, there also are the isomorphisms $\psi_W(y)^*$ of the (p, q)-cohomologies. Following (2.10), we then have the map $h_{UW}: U \cap W \to GL(H^{p,q}(H_1/H_2, \underset{\sim}{E}_0))$ with the property that $h_{UW}(y) = \psi_U(y)^{*-1}\psi_W(y)^*$ for $y \in U \cap W$, $= \psi_{WU}(y)^*$ in the notations of (2.10). We know that h_{UW} is holomorphic. In the current situation, this can be seen directly as follows:

Let $\pi^{p,q}$ be the standard representation of H_1 on $V^{p,q} = H^{p,q}(H_1/H_2, \underset{\sim}{E}_0)$. $\pi^{p,q}$ is induced by the following action of H_1 on $A^{p,q}(H_1/H_2, {}_0)$: For a form ω and $h \in H_1$, define $h \cdot \omega$ by

$$(h \cdot \omega)(x)(L_1, \ldots, L_k) = h \cdot \omega(h^{-1} \cdot x)(d\ell_{h^{-1}}(x)L_1, \ldots, d\ell_{h^{-1}}(x)L_k) \tag{3.4}$$

where $k = p + q$, $x \in H_1/H_2$, $L_i \in T_x(H_1/H_2)^{\mathbb{C}}$ and where $\ell_h: H_1/H_2 \to H_1/H_2$ is left translation by $h \in H_1$.

Then $h_{UW}(y) = \pi^{p,q}(\psi_{WU}(y))$ in accordance with (2.10). Since $\pi^{p,q}$ and ψ_{WU} are holomorphic, we are done.

One concludes that if one sets $\underset{\sim}{H}^{p,q}(\underset{\sim}{E}) = \bigcup_y H^{p,q}(X_y, \underset{\sim}{E}_y)$, then $\underset{\sim}{H}^{p,q}(\underset{\sim}{E})$ has the structure of a holomorphic vector bundle over Y with the h_{UW} as its transition maps and, of course, that $\underset{\sim}{H}^{p,q}(\underset{\sim}{E})|U \simeq U \times H^{p,q}(H_1/H_2, \underset{\sim}{E}_0)$.

On the other hand, since $Z \to Y$ is a principal H_1-bundle, we can form the associated holomorphic vector bundle

$$\underset{\sim}{K}^{p,q} = Z \times_{H_1} V^{p,q} \tag{3.5}$$

over Y for the action of H_1 defined above. By general principles, the transition functions of this bundle are precisely the h_{UW} considered so far. Therefore, we have the

Proposition 1. The bundles $\underset{\sim}{H}^{p,q}(\underset{\sim}{E})$ and $\underset{\sim}{K}^{p,q}$ are canonically isomorphic holomorphic vector bundles.

In particular, there exist canonical isomorphisms of the $H^{a,b}(Y, \underset{\sim}{H}^{c,d}(\underset{\sim}{E}))$ onto the spaces $H^{a,b}(Y, Z \times_{H_1} H^{c,d}(H_1/H_2, \underset{\sim}{E}_0))$.

The main theorem of Sec. 2 therefore leads to the following generalization of a theorem of Bott ([5], theorem VI):

Theorem 1. For every $p \geq 0$, there is a spectral sequence ${}^pE_t^{s,t}$ which converges to $H^{p,\cdot}(X, \underset{\sim}{E}_\lambda)$, whose E_2-term is given by

$${}^pE_2^{s,t} \simeq \bigoplus_i H^{i,s-i}(Y, Z \times_{H_1} H^{p-i,t+i}(H_1/H_2, \underset{\sim}{E}_0)) \tag{3.6}$$

In addition, we also obtain the following more special result:

Theorem 2. Under the conditions of this section, assume moreover that the representation $\lambda: H_2 \to GL(V)$ extends to a holomorphic representation $\tilde{\lambda}: H_1 \to GL(V)$. Then ${}_0 = H_1 \times_{H_2} V$ is holomorphically trivial: $E_0 = H_1/H_2 \times V$. Therefore,

$$H^{p,q}(H_1/H_2, H_1 \times_{H_2} V) \simeq H^{p,q}(H_1/H_2) \otimes V \tag{3.7}$$

If $\pi^{p,q}$ denotes the representations of H_1 on the cohomologies $H^{p,q}(H_1/H_2, {}_0)$ and $H^{p,q}(H_1/H_2)$, respectively, then the isomorphism (3.7) maps $\pi^{p,q}$ to $\pi^{p,q} \otimes \tilde{\lambda}$.

Under the conditions of theorem 2, one concludes that $\underset{\sim}{H}^{p,q}(\underset{\sim}{E}) \simeq \underset{\sim}{H}^{p,q} \otimes \underset{\sim}{E}_{\tilde{\lambda}}$, and if the fibre H_1/H_2 is Kähler, H_1 connected, then the bundles $\underset{\sim}{H}^{p,q}$ are trivial; these remarks imply part of Ahiezer's main result; cf. theorem 1 of [1, 2].

3.2. Open Real Orbits in a Complex Flag Manifold. It is evident that the considerations of Sec. 3.1 apply, in particular, when $Z = G$ is a complex Lie group and $H_2 \subset H_1$ are closed complex Lie subgroups, in which case both $X = G/H_2$ and $Y = G/H_1$ are homogeneous spaces. There are, however, more general situations of practical interest where at least part of the apparatus of Sec. 2 and of Sec. 3.1 still can be established; this is true, e.g., of the cohomology bundles $\underset{\sim}{H}^{c,d}(\underset{\sim}{E})$ for the case of open real orbits in complex flag manifolds under suitable conditions on the groups involved. We present a brief outline of the results and refer to [8] for all details.

The situation we shall deal with is the following: G is a connected, semisimple complex Lie group, $P \subset G$ is a parabolic subgroup and $G_0 \subset G$ is a noncompact real form of G. We assume once and for all that

$$H = G_0 \cap P \tag{3.8}$$

is compact. One chooses maximal compact subgroups M, K of G, G_0, respectively, such that $H \subset K \subset M$; one has $H = K \cap P = M \cap P$ and G_0/H has the complex manifold structure of the open real orbit $G_0 \cdot 0$ in the complex flag manifold G/P; here $0 = 1 \cdot P \in G/P$. Moreover, there are the following basic facts; cf. [10, 19].

(i) M acts transitively on G/P with isotropy $M \cap P = H$, so that M/H inherits the complex structure of G/P;

(ii) K acts transitively on $K^{\mathbb{C}}/K^{\mathbb{C}} \cap P$ with isotropy $K \cap P = H$, so that K/H inherits the complex structure of $K^{\mathbb{C}}/K^{\mathbb{C}} \cap P$;

(iii) Let $\lambda: H \to GL(V)$ be an irreducible unitary representation of H on the complex vector space V. Then λ extends uniquely to a holomorphic irreducible representation of P.

In particular, the vector bundles $G_0 \times_H V \to G_0/H$ and $K \times_H V \to K/H$ possess holomorphic structures as pull-backs of holomorphic vector bundles over $G_0 \cdot 0$ and $K^{\mathbb{C}}/K^{\mathbb{C}} \cap P$, respectively.

Assume now that λ is given as in (iii). By (ii), the injection map $i: K \to K^{\mathbb{C}}$ induces an isomorphism of holomorphic vector bundles:

$$\begin{array}{ccc} K \times_H V & \xrightarrow{\ i\ } & K^{\mathbb{C}} \times_{K^{\mathbb{C}} \cap P} V \\ \downarrow & & \downarrow \\ K/H & \xrightarrow{\ i\ } & K^{\mathbb{C}}/K^{\mathbb{C}} \cap P \end{array} \tag{3.9}$$

Thus, there are induced isomorphisms

$$i^*: H^{p,q}(K^{\mathbb{C}}/K^{\mathbb{C}} \cap P, K^{\mathbb{C}} \times_{K^{\mathbb{C}} \cap P} V) \simeq H^{p,q}(K/H, K \times_H V) \tag{3.10}$$

For the sake of convenience, we write $S = K^{\mathbb{C}}/K^{\mathbb{C}} \cap P$, $\underset{\sim}{E} = K^{\mathbb{C}} \times_{K^{\mathbb{C}} \cap P} V$, $\underset{\sim}{E}_0 = K \times_H V$. The isomorphism i^* of (3.10) is induced by a map i^* at the level of differential forms. By an easy computation, one shows:

Proposition 2. Let $\pi^{p,q}$ be the standard holomorphic representation of $K^{\mathbb{C}}$ on $H^{p,q}(S, \underset{\sim}{E})$. At the level of differential forms, the isomorphism i^* intertwines $\pi^{p,q}|K$ with the standard representation $\pi_0^{p,q}$ of K on $A^{p,q}(K/H, \underset{\sim}{E}_0)$.

Corollary 1. For $k \in K$, $\pi_0^{p,q}(k)$ commutes with $\bar{\partial}$. Hence there is an induced representation $\pi_0^{p,q}$ of K on $H^{p,q}(K/H, E_0)$ and $\pi_0^{p,q}$ coincides with $i^* \circ \pi^{p,q}|K \circ i^{*-1}$. In particular, $\pi_0^{p,q}$ is smooth and we can form the induced C^∞ vector bundle $\underset{\sim}{H}_0^{p,q} = G_0 \times_K H^{p,q}(K/H, \underset{\sim}{E}_0)$ over G_0/K.

As before, we may consider the situation

$$\begin{array}{ccc} G_0 \times_H V \to & G_0/H = X \\ & \downarrow \pi \\ & G_0/K = Y \end{array} \tag{3.11}$$

where π is just a C^∞ fibration; similarly, $G_0 \to G_0/K$ is a C^∞ principal K-bundle (in the earlier notations, G_0, K and H correspond to Z, H_1 and H_2); in other words: even though $G_0 \times_H V$ and G_0/H have holomorphic structures, we are now dealing with a case where the data are just C^∞.

With these notations, one obtains the following main result after some work which we omit here (cf. [8]):

Theorem 3. Using the holomorphic structures introduced above,

(i) the "local triviality condition" (2.2) holds for the bundle $G_0 \times_H V$ where ψ_U now is just a diffeomorphism, but each $\psi_U(y)$, $y \in U$, actually is biholomorphic.

(ii) $\bigcup_y H^{p,q}(\pi^{-1}(y), (G_0 \times_H V)|\pi^{-1}(y))$ is a C^∞ vector bundle and as such it is isomorphic to the bundle $\mathbb{H}_0^{p,q}$ of Corollary 1.

Corollary 2. If G_0/K admits a G_0-invariant complex structure, then the bundle of theorem 3 (ii) admits a holomorphic structure.

Proof of the Corollary. $\mathbb{H}_0^{p,q}$ is induced by the finite-dimensional representation $\pi_0^{p,q}$ of K. If P^+ is the abelian Lie subgroup of G corresponding to the algebra p^+ of holomorphic tangent vectors at $0 \in G_0/K$, then $K^{\mathbb{C}}P^+ \subset G$ is a parabolic subgroup and G_0/K has the complex structure of the real open orbit $G_0 \cdot 1K^{\mathbb{C}}P^+$ in $G/K^{\mathbb{C}}P^+$. $\pi_0^{p,q}$ extends uniquely to a holomorphic representation of $K^{\mathbb{C}}P^+$ (such that $\pi_0^{p,q}(P^+) = 1$) and hence $\mathbb{H}_0^{p,q}$ is the pull-back of a holomorphic vector bundle over the above open orbit.

Remark. Corollary 2 serves as a basis for the construction of a "Penrose transform" in the general context of the present section.

We conclude this brief outline with the following

Example. We choose $G = SL(4, \mathbb{C})$, $G_0 = SU(2,2)$, P as the group of matrices

$$\begin{bmatrix} * & * & * & * \\ 0 & * & * & * \\ 0 & 0 & * & * \\ 0 & 0 & * & * \end{bmatrix}$$

so that $H = G_0 \cap P$ consists of the matrices

$$\begin{bmatrix} a & 0 & 0 & 0 \\ 0 & e & 0 & 0 \\ 0 & 0 & \multicolumn{2}{c}{\multirow{2}{*}{σ}} \\ 0 & 0 & & \end{bmatrix}$$

of determinant 1, where $|a|^2 = |e|^2 = 1$, $\sigma \in U(2)$; in other words, $H = S(U(1) \times U(1) \times U(2))$. In addition, $K = S(U(2) \times U(2))$. In this case, the fibration $G_0/H \to G_0/K$ of (3.11) actually is holomorphic and is denoted by $F_{12}^+ \to M^+$ in [17, 18]. In fact, one has a double fibration

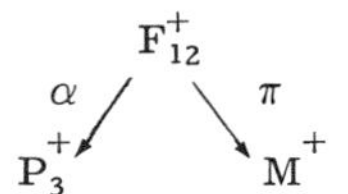

where P_3^+ is projective twistor space. In Singer's "integral geometry approach" to the Penrose transform, it is crucial to know that the set $\bigcup_y H^{0,1}(\pi^{-1}(y), \alpha^*H^{-n-2})$, denoted by V_n in [17, 18], where H is the hyperplane bundle, has the structure of a holomorphic vector bundle. Wells establishes this result by entirely different methods (direct image sheaves, etc.), but we note that the result is immediate from Corollary 2.

3.3. Characteristic Classes in $H_{\bar\partial}(X)$. In [11], Grothendieck pointed out that a theory of characteristic classes of holomorphic vector bundles (over compact complex manifolds) where the classes lie in the Dolbeault cohomology $H^\cdot(X,\Omega^\cdot) = H_{\bar\partial}(X)$ should be possible, using his axiomatic approach to such theories, and that the theory should be closely related to that of the Atiyah classes of holomorphic principal bundles (cf. [3]).

We present here a brief outline of such a theory and refer for all details to [9].

We propose to deal with the category $\underline{V}$ of compact complex manifolds and holomorphic maps; in this case, if $\underset{\sim}{E} \to X$ is a holomorphic bundle, the associated projective bundle $P(\underset{\sim}{E})$ is in $\underline{V}$, so that the first requirement of [11] is satisfied. The cohomology theory A of [11] to be used here is, of course,

$$A(X) = H^\cdot(X,\Omega^\cdot) \simeq H_{\bar\partial}(X) \tag{3.13}$$

the Dolbeault cohomology.

If $\underline{P}(X)$ denotes the Picard group of X, then $\underline{P}(X) \simeq H^1(X, \underline{O}_X^*)$ and the (functorial) homomorphism p: $\underline{P} \to A$ of [11] is obtained as follows: There is a natural map

$$\frac{1}{2\pi i}\, d \log : \underline{O}_X^* \to \Omega_X^1 \tag{3.14}$$

and we define p to be the composition p: $\underline{P}(X) \to H^1(X, \underline{O}_X^*) \to H^1(X, \Omega_X^1)$; with this, given the holomorphic line bundle $\underset{\sim}{L}$, we define

$$\alpha_1(\underset{\sim}{L}) = p([\underset{\sim}{L}]) \tag{3.15}$$

$[\underset{\sim}{L}]$ the isomorphism class of $\underset{\sim}{L}$ in $\underline{P}(X)$. This provides the normalization for the theory. One notes that $\alpha_1(\underset{\sim}{L})$ coincides with the Atiyah class of the associated principal $\mathbb{C}^*$-bundle (frame bundle of $\underset{\sim}{L}$).

In the approach of [11] (which avoids the use of classifying spaces), a crucial role is played by the cohomology structure of $P(\underset{\sim}{E})$, $\underset{\sim}{E}$ a holomorphic vector bundle. It is at this point that the Borel spectral sequence is used and we sketch the argument here:

$P(\underset{\sim}{E})$ carries a standard line bundle $\underset{\sim}{L}_E$, determined by the requirement that for each $x \in X$, $\underset{\sim}{L}_E \mid P(\underset{\sim}{E})_x$ should be the "universal line bundle" of $P(\underset{\sim}{E})_x = P(E_x)$. Let $\ell_{\underset{\sim}{E}} = \underset{\sim}{L}_E^*$ be the "bundle of linear forms" of $P(\underset{\sim}{E})$. If f: $P(\underset{\sim}{E}) \to X$ is the projection, the claim now is that

> under f^*: $H_{\bar\partial}(X) \to H_{\bar\partial}(P(\underset{\sim}{E}))$, the latter is a free $H_{\bar\partial}(X)$-module with basis 1, $\xi_{\underset{\sim}{E}}$, ..., ξ^{r-1}, r = rank($\underset{\sim}{E}$), $\underset{\sim}{\xi}_{\underset{\sim}{E}} = p([\underset{\sim}{L}_{\underset{\sim}{E}}]) \in H^{1,1}(P(\underset{\sim}{E}))$

One notes, first of all, that the assertion is true when X is a point; in other words: if E is a complex vector space, P(E) its projective space, ℓ_E the dual of the universal bundle, then the powers ξ_E^i, $0 \le i \le \dim(E) - 1$, form a basis of $H_{\bar\partial}(P(E))$: the easy way to see this is to use the fact that it is clear from the construction that under the Frölicher spectral sequence relating the $\bar\partial$-cohomology to the de Rham cohomology $H(P(E), \mathbb{C})$, ξ_E will correspond to the complex Chern class $c_1(\ell_E)$. Since P(E) is compact Kähler, the spectral sequence degenerates ($d_1 = 0$: this yields the "Hodge decomposition theorem") and one obtains the result.

For the global version, we use the Borel spectral sequence whose E_2-term will be

$$^pE_2^{s,t} = \bigoplus_i H^{i,s-i}(X, \underset{\sim}{H}^{p-i,t+i}), \text{ by (2.5)}$$

The fibre of $P(\underset{\sim}{E})$ is $P(E)$, again a compact Kähler manifold, and the group of $P(\underset{\sim}{E})$ is $GL(E)$ (or $PGL(E)$), a connected Lie group. The isomorphism $\bigoplus_{p+q=k} H^{p,q}(P(E)) \simeq H^k(P(E), \mathbb{C})$ of the hodge decomposition is equivariant for the actions of $GL(E)$ and by the homotopy axiom, the action is trivial on the de Rham cohomology, hence now also on $H_{\bar\partial}(P(E))$. As a consequence, the bundles $\underset{\sim}{H}^{c,d}$ over X are holomorphically trivial and we see that

$$ {}^pE_2^{s,t} = \bigoplus_i H^{i,s-i}(X) \otimes H^{p-i,t+i}(P(E)) \tag{3.16} $$

Next, there is a natural map $\Theta: H_{\bar\partial}(P(E)) \to H_{\bar\partial}(P(\underset{\sim}{E}))$, obtained by linear extension of $\xi_E^i \to \xi_{\underset{\sim}{E}}^i$. Using this, one shows that for each x, the inclusion i_x induces a surjective map $i_x^*: H_{\bar\partial}(P(\underset{\sim}{E})) \to H_{\bar\partial}(P(\underset{\sim}{E})_x) = H_{\bar\partial}(P(E_x))$, in fact: that $i_x^*\Theta$ is an isomorphism for each x. Accordingly, the Borel spectral sequence degenerates ($d_2 = 0$) and one finally concludes from (3.16) that

$$ H_{\bar\partial}(X) \otimes H_{\bar\partial}(P(E)) \simeq H_{\bar\partial}(P(\underset{\sim}{E})) \tag{3.17} $$

under the map $(a,b) \to f^*a \cdot \Theta b$ ("Leray-Hirsch theorem"!). This proves the claim.

Lastly, if $Y \subset X$ is a closed complex submanifold of codimension p, then the Serre duality theorem can be used to construct a group homomorphism $i_*: H_{\bar\partial}(Y) \to H_{\bar\partial}(X)$ of degree 2p which will have the properties required in [11]. With this, Chern classes of holomorphic vector bundles are now defined in $H_{\bar\partial}(X)$ as usual: In $H(P(\underset{\sim}{E}))$, there is the relation $\Sigma_0^r \alpha_i(\underset{\sim}{E})\xi_{\underset{\sim}{E}}^{r-i} = 0$ and the total Chern class $\alpha(\underset{\sim}{E}) \in H_{\bar\partial}(X)$ of $\underset{\sim}{E}$ is defined to be

$$ \alpha(\underset{\sim}{E}) = \sum_i \alpha_i(\underset{\sim}{E}) \tag{3.18} $$

($\alpha_0 = 1$, $\alpha_i = 0$ for $i > r$). One verifies the Hirzebruch axioms (in suitable form) and with this, the theory is established.

Using hermitian connections (whose curvature forms are of type (1,1)), one constructs a Chern-Weil homomorphism following the usual pattern, a process which leads to the classes $\beta(\underset{\sim}{E}) \in H_{\bar\partial}(X)$. Some of the techniques of [6] then can be used to show that β satisfies the Hirzebruch axioms and accordingly <u>coincides with</u> α. Among other things, this shows that in all generality, under the Frölicher spectral sequence $H_{\bar\partial}(X) \Longrightarrow H(X, \mathbb{C})$, $\alpha(\underset{\sim}{E})$ corresponds to the total Chern class $c(\underset{\sim}{E})$. It can also be shown that the theory of Chern classes in $H_{\bar\partial}$ is properly finer than the one of complex classes; cf. [9] for all details.

The preceding considerations may be summarized in the following

<u>Theorem</u>. On the category of compact complex manifolds and holomorphic maps, there exists a unique theory of characteristic classes of holomorphic vector bundles with values in the Dolbeault cohomology $H_{\bar\partial}$ which satisfies the axioms of Grothendieck-Hirzebruch. Moreover, this theory may also be obtained by means of a Chern-Weil homomorphism which, in turn, establishes the relations between the $H_{\bar\partial}$-Chern classes and both the complex Chern classes and the Atiyah classes for the holomorphic vector bundles.

REFERENCES

1. Ahiezer, D.: Mat. Sb. 84, 290-300 (1971); Math. USSR Sb. 13, 285-296 (1971).
2. Ahiezer, D.: Mat. Sb. 87, 607-614 (1972).
3. Atiyah, M. F.: Trans. Am. Math. Soc. 85, 181-207 (1957).
4. Auer, W.: Pac. J. Math. 44, 33-43 (1973).

5. Bott, R.: Ann. Math. 66, 203-248 (1957).
6. Bott, R.; Chern, S. S.: Acta Math. 114, 71-112 (1965).
7. Fischer, H. R.; Williams, F. L.: Trans. Am. Math, Soc. 252, 163-195 (1979).
8. Fischer, H. R.; Williams, F. L.: Construction of cohomology bundles in the case of open real orbits in complex flag manifolds, in preparation.
9. Fischer, H. R.; Williams, F. L.: Characteristic classes in the Dolbeault cohomology, in preparation.
10. Griffiths, P.; Schmid, W.: Acta Math. 123, 253-302 (1969).
11. Grothendieck, A.: Bull. Soc. Math. France 86, 137-158 (1958).
12. Hirzebruch, F.: Topological Methods in Algebraic Geometry. Springer (1966), Appendix II (A. Borel).
13. Le Potier, J.: Bull. Soc. Math. France 38, 107-119 (1974).
14. Le Potier, J.: Math. Ann. 218, 35-53 (1975).
15. Matsushima, Y.: Nagoya J. Math. 14, 1-24 (1959).
16. Morimoto, A.: Nagoya J. Math. 13, 157-168 (1958).
17. Wells, R., Jr.: Bull. Am. Math. Soc. 1, 296-336 (1979).
18. Wells, R., Jr.: Cohomology and the Penrose transform, in: Complex Manifold Techniques in Theoretical Physics, D. E. Lerner and P. D. Sommers, eds. Pitman, San Francisco (1979).
19. Wolf, J.: Bull. Amer. Math. Soc. 75, 1121-1237 (1969), added in proof.
20. Eastwood, M.; Penrose, R.; Wells, R., Jr.: Commun. Math. Phys. 78, 305-351 (1981).

15 NEWTON, EULER, AND POE IN THE CALCULUS OF VARIATIONS

S. H. Gould / Department of Mathematics, Brown University, Providence, Rhode Island

1. Newton's Problem. The earliest[1] problem in the history of the calculus of variations is found in a scholium—a comment or short note—in the second book of Newton's *Principia*; namely, to find the generating curve for a solid of revolution that will encounter the least resistance to motion in a resistive medium in the direction of its axis. The complete solution of this problem has required the efforts of Newton, Euler, Lagrange, Legendre, Jacobi, Weierstrass, and many other eminent mathematicians.

2. A Preliminary Theorem Proved by the Methods of the Ancient Greeks. The scholium is preceded by a theorem which could have been omitted but forms a useful introduction to the general ideas. In Motte's translation (1729) from the Latin of the third edition (1726), the theorem (Proposition XXXIV, 3rd Edit.) is stated as follows:

> If in a rare medium, consisting of equal particles freely disposed at equal distances from each other, a sphere[2] and a cylinder described on equal diameters move with equal velocities in the direction of the axis of the cylinder, the resistance of the sphere will be but half so great as that of the cylinder.

Although Newton obtained his results by means of his newly invented infinitesimal calculus, the proof[3] of this theorem, as of all theorems in the *Principia*, proceeds in the ancient geometric manner of Euclid, Archimedes and Apollonius, as follows (see Fig. 1).

In Fig. 1 let IA = 2r be the axis of revolution, so that the cylinder is represented by the square RGNO and the sphere by the circle IKAT of radius r, and let the motion be in the direction from I to A. Let the vector FB represent the force with which a particle impinges directly on an element of surface of the cylinder at the point b, so that the resistance to the entire cylinder is proportional to the area πr^2 of its leading face. To find the resistance on an element of surface at the corresponding point B of the sphere, produce FB to cut KC at E and mark the point H with $bH = BE^2/bE$ $(= BE^2/CB)$. Then Newton asserts that the effect on the sphere at B will be to the effect on the cylinder at b as bH is to bE.

For let FP and BP be the components of FB tangential and normal to the sphere at B. Then the tangential component FP will have no effect on the motion of the sphere,

---

[1] In 1685 Newton solved what can be viewed as the first really deep problem in the calculus of variations, Goldstine, p. xiii.

[2] Motte writes "globe" for "sphere," and there are a few similar changes in the quotations from him below. Newton also usually writes "*globus*" for "*sphaera*."

[3] The present account of Newton's Problem and of Euler's contributions to it is based on Volume VI of Whiteside's *Newton's Papers* and Goldstine's *History of the Calculus of Variations* . . . . Both books are extremely valuable for the information they contain on Newton's Problem and on a wide range of other mathematical questions.

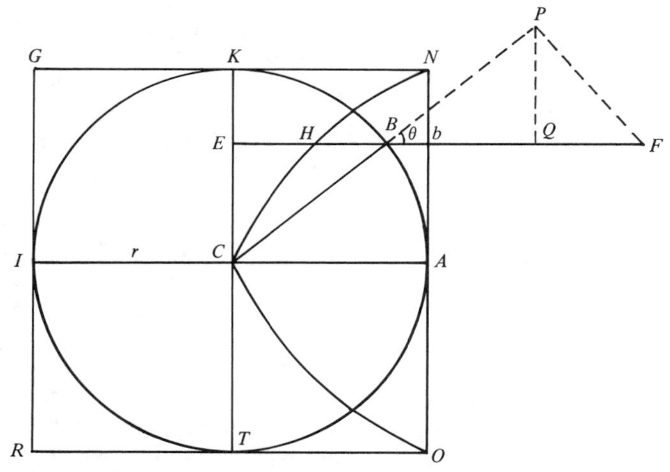

Fig. 1. From the Principia, Proposition XXXIV in 3rd edit., slightly modified.

since Newton's description of the medium implies absence of friction; and as for the normal force BP, its component PQ perpendicular to the direction of motion will be negated by a second particle impinging on the sphere at a point symmetric to B with respect to the axis IA. So the effect on the sphere at the point B will be to the effect on the cylinder at b as $BQ = FB \cos^2 \theta$ is to FB, where $\theta = FBP$ is the angle between the direction of motion and the normal to the surface. Newton then summarily concludes his proof with the words:

> and therefore the solid which is formed by all the right lines bH will be to the solid formed by all the right lines bE as the effect of all the particles upon the sphere to the effect of all the particles upon the cylinder. But the former of these solids is a paraboloid . . . and the latter solid is a cylinder circumscribed about the paraboloid, and it is known that a paraboloid is half its circumscribed cylinder.

So we see that while Newton did not expect his readers to know anything about the new infinitesimal calculus, he did expect them to be familiar with classical Greek geometry. The curve OCHN is a parabola by Theorem 11 in Book I of Apollonius, On Conic Sections, since

$$HE = bE - bH = CB - \frac{BE^2}{CB} = \frac{CB^2 - BE^2}{CB} = \frac{1}{r} \cdot CE^2$$

or $y = px^2$ with $y = HE$, $x = CE$, and $p = 1/r$; and the paraboloid is half of the circumscribed cylinder because it is 3/2 of the circumscribed cone by Proposition 21 in Archimedes' On Conoids and Spheroids, while the cone is 1/3 of the cylinder by Proposition 10 in Book 12 of Euclid's Elements.

3. Proof of the Theorem by Means of Calculus. If Newton had been willing to use his infinitesimal calculus, he could have proved the theorem as follows (see Fig. 2).

Let CA be taken for x-axis and CK for y-axis, and let a solid of revolution be formed by revolving about the x-axis an arbitrarily given curve joining the point $P_0(0, y_0)$ to the point $P_1(x_1, y_1)$; for example, for the sphere in Fig. 1 the curve $P_0 P_1$ is a quarter-circle with $y_0 = x_1 = r$ and $y_1 = 0$. Then an arbitrary point B (x, y) will trace out a circle of circumference $2\pi y$ and the number of particles striking an element of arc-length ds at the point B will be proportional to $ds \cdot \cos \theta$, so that the number striking the circular band obtained by revolving ds will be proportional to $2\pi y \cdot \cos \theta \, ds$. But we have seen that the effect of each such particle is proportional to $\cos^2 \theta$, so that the effect on the band is proportional to $2\pi y \cos^3 \theta \, ds$. With suitable choice of the units of measurement, the effect R on the entire solid may therefore be written in the form

$$R = 2\pi \int_{P_0}^{P_1} y \cos^3 \theta \, ds = 2\pi \int_{y_1}^{y_0} y \cos^2 \theta \, dy$$

since $dy = -\cos \theta \, ds$. But then, writing q for $dx/dy = -\tan \theta$, we have

$$\cos^2 \theta = \frac{dy^2}{ds^2} = \frac{dy^2}{dx^2 + dy^2} = \frac{1}{1 + q^2}$$

and therefore,

$$R = 2\pi \int_{y_1}^{y_0} \frac{y \, dy}{1 + q^2}$$

or, writing p for $dy/dx = 1/q = -\cot \theta$, we get

$$R = 2\pi \int_{0}^{x_1} \frac{yp^3 \, dx}{1 + p^2}$$

So for the sphere in Fig. 1, or in Fig. 3, with $\sin \theta = y/r$ and therefore $\cos^2 \theta = 1 - y^2/r^2$, we have

$$R = 2\pi \int_{0}^{r} \left(1 - \frac{y^2}{r^2}\right) y \, dy = \frac{\pi r^2}{2}$$

which is one-half of the effect πr^2 on the circumscribed cylinder, as was asserted by Newton.

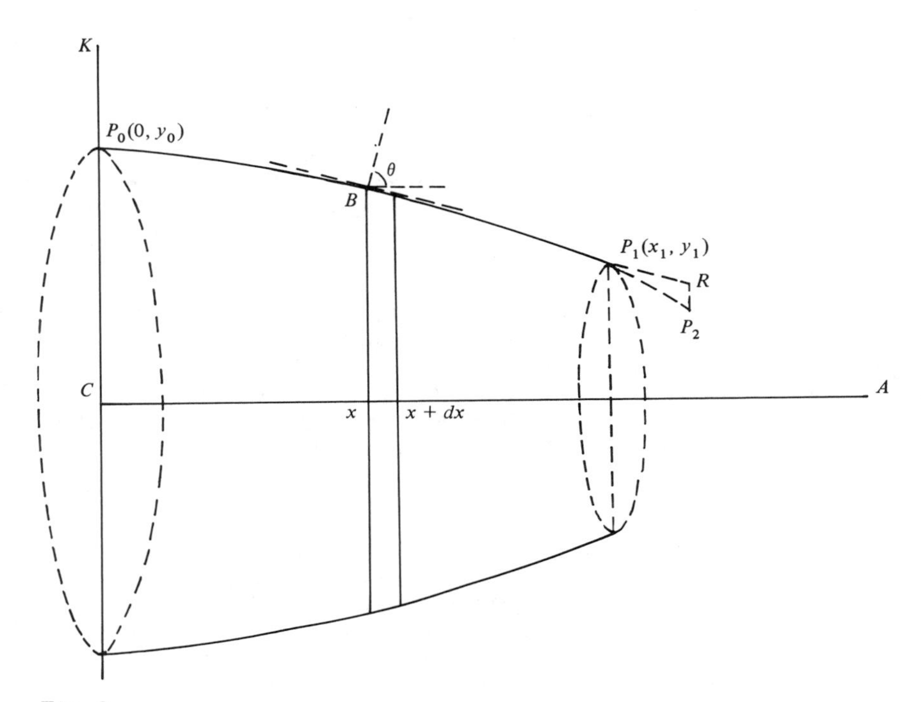

Fig. 2

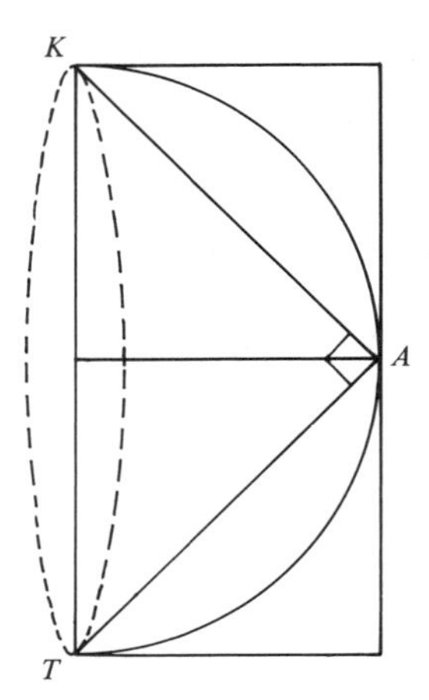

Fig. 3

Similarly, for the inscribed cone KAT in Fig. 3, with constant $\theta = 45° = \pi/4$ and therefore $\cos^2 \theta = 1/2$, we have

$$R = 2\pi \int_0^r \frac{y}{2} dy = \frac{\pi r^2}{2}$$

so that the resistance on this cone is the same as on the sphere.

More generally, for a cone with arbitrary base-radius r and base-angle θ we have

$$R = 2\pi \cos^2 \theta \int_0^r y\, dy = \pi r^2 \cos^2 \theta = \frac{\pi r^2}{1 + q^2}$$

4. The Three Statements in the Scholium. This preliminary theorem is followed by a scholium, consisting of three statements.

Statement I. Among all frusta of cones with the same breadth 2r (the diameter of the base ECHB in Fig. 4) and the same altitude OD = 2a, the vertex S of the cone that provides the frustum of least resistance is determined by bisecting OD at Q and setting QS = QC.

Statement II. If the solid (see Fig. 5) be generated by the revolution of an elliptical or oval figure ADBE about its axis AB and the generating figure be touched by three right lines FG, GH, HI at the points F, B and I, so that GH is perpendicular to the axis in the point of contact B and FG, HI are inclined to GH in the angles FGB, BHI of 135°, the solid arising from the revolution of the figure ADFGBHIE will be less resisted than the former solid . . . which proposition I conceive may be of use in the building of ships.

Statement III. If DNFG (see Fig. 6) be such a curve that, if from any point N thereof the perpendicular NM be let fall on the axis CB and from the given point G there be drawn the right line parallel to the right line touching the figure in N and cutting the axis produced in R the line MN becomes to GR as GR^3 to $4BR \cdot GB^2$, then the solid described by the revolution of this figure about its axis CB will be less resisted than any other circular solid whatsoever, described of the same length and breadth.

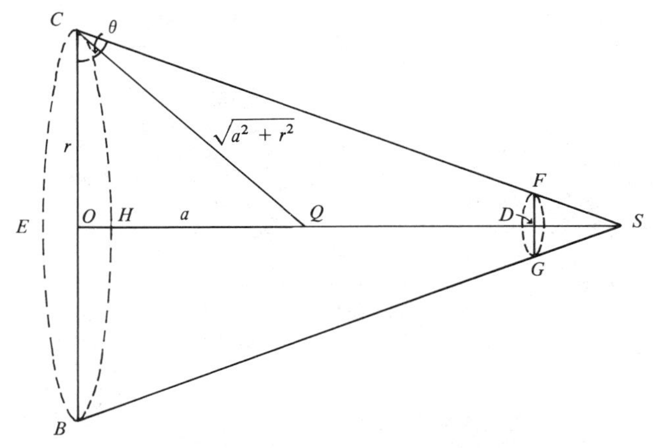

Fig. 4. First figure in the Scholium to Proposition XXXIV.

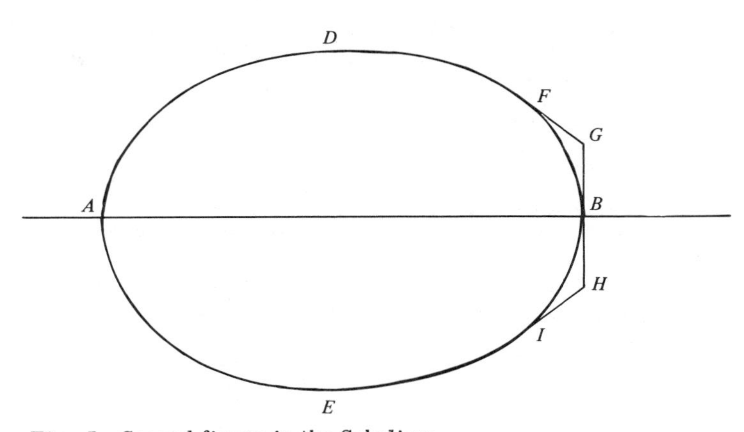

Fig. 5. Second figure in the Scholium.

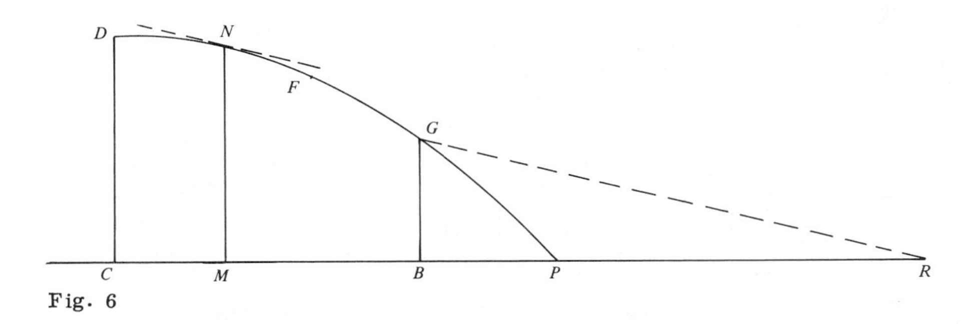

Fig. 6

Nowhere in the Principia does Newton give any hint of a proof for these three statements, which baffled his contemporaries, including even Leibniz. As Whiteside says (Vol. VI, p. 456),

> the printed Principia does scant justice to the complexity and sophistication of Newton's preliminary analysis as we may now know it from his autograph worksheets and drafts [only recently published in full by Whiteside]. This is most notably true of his determination of the solid of revolution of least resistance to rectilinear motion, whose geometrically framed defining differential equation is merely enunciated in the Principia without any explanation. . . . Newton's reticence here caused his contemporaries and eighteenth-century successors a good deal of effort and frustration when they sought to replicate his calculations, and the question as to exactly how he achieved his stated end has remained to plague historians down to the present day.

In fact, Newton prepared a considerably augmented revision of the scholium for incorporation in the second edition of the Principia (1713), but like many others of his proposed changes it remained unprinted. The proofs given below are intended to be as much like Newton's as possible with the use of the differential notation invented by Leibniz.

5. Proof of Statement I. The resistance, call it R_1, on the larger cone BCS in Fig. 4, with base-radius r and base-angle θ, is given by $R_1 = \pi r^2 \cos^2 \theta = \pi r^2/(1 + q^2)$ with $q = -\tan \theta$, as was shown at the end of Sec. 3. Also, the base-radius of the smaller cone GFS is $FD = DS \cot \theta = -DS/q$, where $-DS = 2a - r \tan \theta = 2a + rq$, so that $FD = (2a + rq)/q$. Thus the resistance on the smaller cone is

$$R_2 = \pi \cdot FD^2 \cdot \cos^2 \theta = \frac{(2a + rq)^2}{q^2(1 + q^2)}$$

Finally, the resistance R_3 on the leading face of the frustum is

$$R_3 = \pi FD^2 = \frac{\pi(2a + rq)^2}{q^2}$$

and the entire resistance R on the frustum is

$$R = R_1 - R_2 + R_3 = \frac{\pi r^2}{1 + q^2} - \frac{\pi(2a + rq)^2}{q^2(1 + q^2)} + \frac{\pi(2a + rq)^2}{q^2}$$

which reduces to

$$\frac{1}{\pi} \cdot R = \frac{r^2 + (2a + rq)^2}{1 + q^2}$$

Differentiating R with respect to q gives

$$\frac{1}{4a\pi} \cdot \frac{dR}{dq} = \frac{-rq^2 - 2aq + r}{(1 + q^2)^2}$$

so that for a minimum value with $dR/dq = 0$ we must have

$$q = \frac{-a - \sqrt{a^2 + r^2}}{r} \qquad \text{or} \qquad r \tan \theta = a + \sqrt{a^2 + r^2}$$

in conformity with Newton's statement that OS = OQ + QC. In the proof of Statement II this result will be needed in the following form. If $R(r, \varphi, a)$ denotes the resistance on a frustum with breadth 2r, altitude 2a, and base-angle $\varphi < \pi/4$, then

$$R(r, \varphi, a) > R(r, \pi/4, a)$$

which is at once implied by the argument just given, since for $q = -\tan \varphi$ and therefore $-1 < q < 0$ the derivative

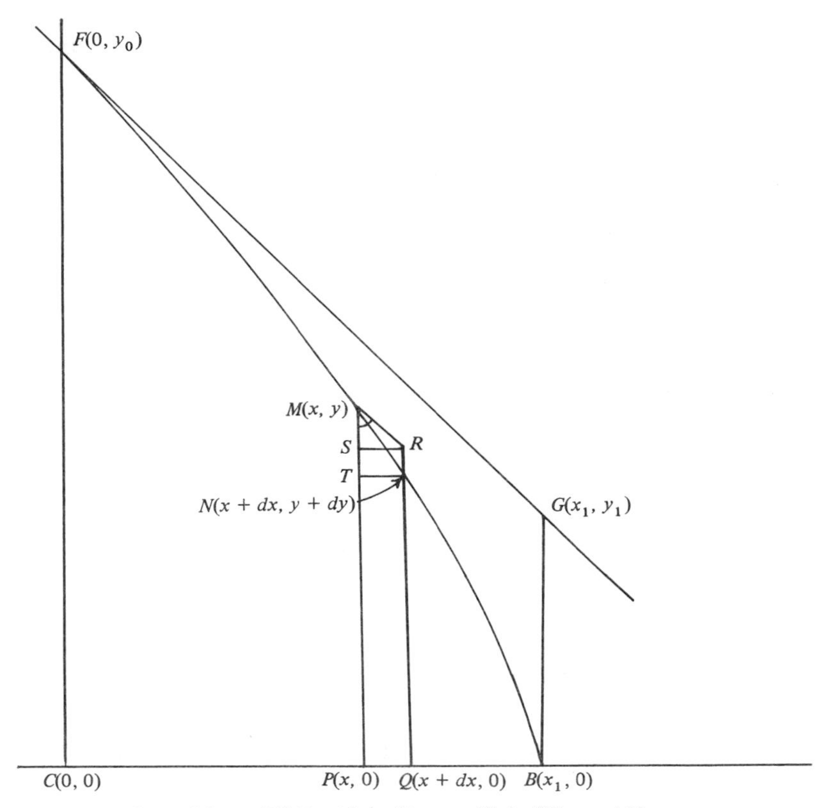

Fig. 7. Adapted from Whiteside's figure, Vol. VI, p. 462.

$$\frac{dR}{d\varphi} = \frac{dR}{dq}\frac{dq}{d\varphi} = \frac{-rq^2 - 2aq + r}{(1 + q^2)^2} \cdot (-1 - q^2) = \frac{r(1 - q^2) - 2aq}{1 + q^2}$$

is positive.

6. Proof of Statement II. Newton's proof of Statement II has not been found among his papers but, as suggested by Whiteside, it must have gone about as follows (see Fig. 7).

In Fig. 7, which is the right-hand part of Fig. 5 on a larger scale, let M(x,y) be any point on the curve FMB, and draw the ordinate PM. Let N have the coordinates (x + dx, y + dy) with $dx > 0$ and therefore $dy < 0$, and draw the ordinate QN. Draw MR parallel to FG (and thus inclined at 135° to the x-axis) to meet QN at R. Then $\angle PMR = \pi/4$ and $\angle PMN < \pi/4$, so that by the inequality at the end of Sec. 5 the resistance on the body generated by revolving the trapezoid PMNQ is greater[4] than the corre-

[4]In his tentative reconstruction of Newton's lost proof of this inequality Whiteside proceeds, in Vol. VI, p. 462, as follows (his figure is slightly different from our Fig. 7 because he takes $dx < 0$ and therefore $dy > 0$, whereas we take $dx > 0$ and therefore $dy < 0$):

> where the height OD of the cone frustum CFGB in the previous paragraph [i.e., Statement I; cf. Fig. 4] is infinitesimally small, it will be clear that (CQ≈)CO and (QS≈)OS come to be equal, so that CF is in this limiting case inclined at 45° to the axis OD. Whence, if N, n are two indefinitely close points [points M and R in

sponding resistance for the trapezoid PMRQ. Subtracting the resistance on the circle generated by the line NQ, which is common to these two solids, we obtain

$$\{MN\} > \{MR\} + \{RN\}$$

where $\{MN\}$ is the resistance on the slant surface generated by revolving the chord MN and similarly for $\{MR\}$, and $\{RN\} = 2\pi \cdot NQ \cdot RN$ is the resistance on the annulus generated by revolving the line RN.

But (see Sec. 3) for the line MR with $\theta = \pi/4$, we have

$$\{MR\} = 2\pi y \cos^3 \theta \, ds = \pi y \, dx$$

since $dx = -dy = \cos \theta \, ds$ and $\cos^2 = 1/2$. Also, since $MS = SR = PQ = dx$, we see that

$$RN = ST = MT - MS = -dy - dx$$

which is positive, since dy is negative with $|dy| > dx$, and we also have $NQ = y + dy$.

Thus $\{RN\} = 2\pi \cdot NQ \cdot RN = -2\pi(y + dy)(dy + dx) = -2\pi y(dy + dx)$ up to infinitesimals of second order.

Consequently,

$$\{MN\} > -2\pi y \, dy - \pi y \, dx$$

Since $\{MN\}$ is the resistance for the arc MN, integration over the curve gives

$$R_c > -2\pi \int_{y_0}^{0} y \, dy - \pi \int_{0}^{x_1} y \, dx = \pi(y_0^2 - Q)$$

where R_c denotes the resistance corresponding to the entire curve and Q is the area under the curve.

On the other hand, the cone frustum generated by the line FG will encounter the resistance πy_1^2 on its leading face, and on its slant surface the resistance will be

$$2\pi \int_{y_0}^{y_1} y \cos^3 \theta \, ds = -\pi \int_{y_0}^{y_1} y \, dy = \frac{\pi}{2}(y_0^2 - y_1^2)$$

So the total resistance on the frustum, call it R_f, will be

$$R_f = \pi y_1^2 + \frac{\pi}{2}(y_0^2 - y_1^2) = \pi\left[y_0^2 - \frac{1}{2}(y_0^2 - y_1^2)\right] = \pi[y_0^2 - A]$$

---

our Fig. 7] in the arc BF wholly contained by the tangent FG (inclined at 45° to CB) and the vertical BG at its respective end-points F and B, on constructing the (perpendicular ordinates MN, mn [our MP, RQ] and drawing no [our MR] parallel to FG till it meets MN [our QN] in o [our R] it at once follows that the resistance on the surface formed by rotating MN + Nn [our PM + MN] is greater than that on the cone frustum generated by likewise rotating Mo + on [our PM + MR].

Since the desired inequality holds for all possible values of the height, and not just for sufficiently small values, it seems desirable to give a proof that avoids passage to the limit as the height approaches zero. How closely our proof resembles Newton's lost proof remains, of course, an open question.

There seems to be a slip in the printing of Whiteside's proof. The first summand $2\pi \cdot MN \cdot \nu n$ in the last equality on p. 462 should read $2\pi MN(\nu n - \omega n)$.

In Goldstine's (p. 14) account of Whiteside's proof, the sentence "the cone frustum having this slope [tan 135°] for its generator experiences minimal resistance among cone frustums with common bases" seems hard to understand, since for any cone frustum with base-angle $\pi/4$—namely, for any possible value of the height—we can find a cone frustum with the same base and same height but with different base-angle ($> \pi/4$) and with smaller resistance.

where $A = \frac{1}{2}(y_0^2 - y_1^2)$ is the trapezoidal area CFGB under the line FG (note that $\angle CFG = 45°$).

But $A > Q$, and consequently

$$R_c - R_f = A - Q > 0$$

as stated by Newton.

The purpose of this Statement II is to show that the desired minimizing curve, say P_0P_1 in Fig. 2, can go no farther than the point P_1 where its slope is equal to -1. For if it went farther, say to a point P_2, the present argument would show that the broken line $P_0P_1RP_2$ engenders less resistance than the curve $P_0P_1P_2$.

7. The Euler Differential Equation in the Calculus of Variations. The first two statements in the scholium are in the nature of lemmas to Statement III, the only one that actually involves the calculus of variations. Newton's manuscript of 1685 includes only an abbreviated discussion along the same lines as the methods later used by the Bernoullis and Euler. So we now reproduce Euler's proof, in order to give the reader at least some idea of the remarkable flavor of Euler's masterpiece A Method of Finding Curves with a Maximum or Minimum Property (1744). Commentators enthusiastically agree that any mathematician who dips into this book will enjoy the experience.

To minimize the integral

$$W = \int Z(x,y,p)\,dx$$

where $p = dy/dx$ and Newton's Problem is the particular case with $Z = yp^3/(1 + p^2)$, Euler proceeds as follows (see Fig. 8).

Let the abscissa AZ (see Fig. 8) be divided into infinitely many infinitesimal and equal parts, such as IK, KL, LM, etc. Let an arbitrary part AM be called x. Let the value of Z at M be denoted by Z, at N by Z', at O by Z'', at P by Z''', etc.; and at L by $Z_{,}$, at K by $Z_{,,}$, at I by $Z_{,,,}$, etc. Also let the ordinate at M be denoted by y, at N by y', etc.; and at L by $y_{,}$, at K by $y_{,,}$, etc., and set

$$p = \frac{y' - y}{dx}, \quad p' = \frac{y'' - y'}{dx}, \quad p'' = \frac{y''' - y''}{dx}, \quad \text{etc.}$$

[Note that the letter Z is here used in two different senses; to denote the integrand and also the right-hand endpoint of the range of integration. Note also that the primes, superscript and subscript, are merely indices, so that in contrast to modern notation the superscripts do not indicate differentiation. And finally, while we would regard the expressions $p = (y' - y)/dx$, etc. as approximations to derivatives by means of finite differences, for Euler they are the actual derivatives, since y' is "infinitely close" to y.]

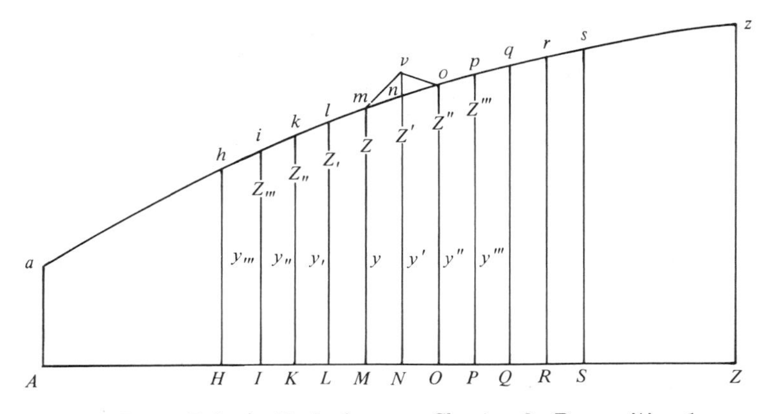

Fig. 8. From Euler's Method . . ., Chapter 2, Proposition 1.

Now let $\int Z\,dx$ denote the integral over the abscissa AM. [Note this third sense for the letter Z.] Then the integral over AZ will have the value

$$W = \int Z\,dx + Z\,dx + Z'\,dx + Z''\,dx + Z'''\,dx, \text{ etc.}$$

in infinitum up to the last point Z. [These are precisely Euler's words. Note that for him $\int Z\,dx$ is actually equal to Z dx, again because N is infinitely close to M.]

If $Nn = y'$ is now given a small increment $n\nu = dy'$ as in Fig. 8, we will have

$$dp = \frac{n\nu}{dx} \quad \text{and} \quad dp' = -\frac{n\nu}{dx}$$

while the values of x and all other values of y and p will remain unaffected, so that in particular $dx = dy = 0$. Thus among the summands in the expression for W only Z dx and Z' dx will change, and therefore

$$dW = dZ + dZ'$$

Since in this case Z is a function of x, y, p only, the differential dZ can be written in the form

$$dZ = M\,dx + N\,dy + P\,dp$$

where M, N, P are functions of x, y, p. [Note that Euler uses the same letters M, N, P in two unrelated senses: for these functions and for certain points on the x-axis.]

Thus for an extreme value of W we must have

$$dW = dZ + dZ' = (M\,dx + N\,dy + P\,dp) + M'\,dx + N'\,dy' + P'\,dp'$$

$$= n\nu(P + N'\,dx - P') = 0$$

But $P' - P = dP$ and in place of N' we may substitute N. [Euler writes these words without comment; again, N and M are "infinitely close" together. Note the tacit assumption of continuity for the function N.]

Then division by $n\nu\,dx$ gives

$$N - \frac{dP}{dx} = 0$$

which is the first appearance of the famous Euler differential equation in the calculus of variations.

It will be seen that Euler's method does not meet all the modern requirements for rigor, and in his earlier days it led him into mistakes. But as Carathéodory says in his Introduction to Euler's Method . . . , "from his calculations he gradually deduced a correct theory and thereby elevated the calculus of variations, which up to then had suffered a precarious existence on the outskirts of analysis, into a special branch of mathematics." Euler's predecessors dealt with separate problems, each with its own special construction, so that even simple problems led to complicated calculations. But Euler deals with many problems at once, each of them according to a fixed pattern of the utmost clarity and conciseness. It is true that, with his own active cooperation, his method was soon replaced by the superior one suggested to him by Lagrange in the famous letter (Oeuvres, Vol. XIV, pp. 138-144) of August 12, 1755, written when Lagrange was nineteen years old. As a result Euler's method was neglected until the end of the nineteenth century, when it was rigorously transformed into the "Euler method of finite differences," which now takes its place among the most useful of present-day direct methods, along with the Rayleigh-Ritz method[5] and others.

---

[5]See, e.g., Courant-Hilbert, pp. 176-177. It is difficult to see what is meant by the statement on p. 177 that "the Euler method may be regarded as a special case of the Ritz method, with suitable piecewise linear coordinate functions." The two methods would seem to be independent, since the Euler method does not consider linear manifolds of gradually increasing dimension.

8. Solution of Newton's Problem by Euler's Method. In more modern notation Euler's differential expression

$$dZ = M\,dx + N\,dy + P\,dz$$

becomes

$$dZ = \frac{\partial Z}{\partial x}dx + \frac{\partial Z}{\partial y}dy + \frac{\partial Z}{\partial p}dp$$

so that his equation takes the familiar form

$$\frac{\partial Z}{\partial y} - \frac{d}{dx}\frac{\partial Z}{\partial p} = 0$$

or with y for the independent variable

$$\frac{\partial Z}{\partial x} - \frac{d}{dy}\frac{\partial Z}{\partial q} = 0 \quad \text{with} \quad q = \frac{dx}{dy} = \frac{1}{p}$$

If we apply this result to the integral

$$R = 2\pi \int \frac{y}{1 + q^2}\,dy$$

in Newton's problem, we have the result

$$\frac{\partial Z}{\partial x} = 0 \quad \text{and therefore also} \quad \frac{d}{dy}\frac{\partial Z}{\partial q} = 0$$

since Z does not contain x. Thus

$$\frac{\partial Z}{\partial q} = \frac{-2yq}{(1 + q^2)^2} = \text{const}$$

or

$$\frac{-yq}{(1 + q^2)^2} = \text{const} = \frac{y_1}{4}$$

since $q = -1$ at the right-hand end of the minimizing curve (cf. Fig. 2).

In his Example V, in Chapter II of the Method, Euler now remarks (though he speaks in terms of $p = 1/q$) that this first-order equation is not integrable. As he puts it, "the further development of this equation cannot be carried out in such a way as to eliminate q and so it is appropriate to express the two coordinates x and y in terms of the same variable q."

Setting $-y_1/4 = a$ for brevity, we have

$$y = \frac{a(1 + q^2)^2}{q} = a(q^{-1} + 2q + q^3)$$

and

$$\frac{dx}{dq} = q \cdot \frac{dy}{dq} = a(-q^{-1} + 2q + 3q^3)$$

so that integration with respect to q gives

$$x = a\left(q^2 + \frac{3}{4}q^4 - \log|q|\right) + b$$

where the role of the two constants of integration is clear from the final words of Statement III, to the effect that the solids to be compared must all be of the same length and breadth, which means that the minimizing curve P_0P_1 in Fig. 2 must pass through the point P_0 (the condition of constant breadth 2r) and must have slope equal to -1 at the point P_1 (the condition of constant length x_1).

9. Proof of Newton's Statement III. Finally, we can now show at once that the proportion stated by Newton—namely, MN is to GR as GR^3 is to $4BR \cdot GB^2$—is merely a geometric statement of the analytic description

$$\frac{-yq}{(1+q^2)^2} = \frac{y_1}{4}$$

of the minimizing curve. For, $p = 1/q$ denotes the slope of the tangent at any point N of the curve and GR is drawn parallel to the tangent, so that $BR^2/BG^2 = q^2$. Also $MN = y$ and $BG = y_1$. Thus

$$(1+q^2) = 1 + \frac{BR^2}{GB^2} = \frac{GB^2 + BR^2}{GB^2} = \frac{GR^2}{GB^2}$$

so that

$$\frac{-yq}{(1+q^2)^2} = \frac{y_1}{4} \quad \text{implies} \quad \frac{MN \cdot BR}{GB} \Big/ \frac{GR^4}{GB^4} = \frac{MN \cdot BR \cdot BG^3}{GR^4} = \frac{GB}{4}$$

and multiplication on both sides by $GR^3/BR \cdot GB^3$ gives

$$\frac{MN}{GR} = \frac{GR^3}{4BR \cdot GB^2}$$

for every point N, as stated by Newton.

10. Poe's Application of Newton's Ideas in His Story "A Descent into the Maelstrom." In our treatment of Newton's ideas we have left many questions unanswered. For example, does the curve we have found provide a minimum or a maximum (Legendre 1786); does it actually run as far as the point where its slope is equal to -1 (Weierstrass-Erdmann corner condition, 1865 and 1877); and what about sufficient conditions for a minimum (Jacobi, 1836; Weierstrass, 1865-1890, etc.)? But for our study of Poe's ideas it is sufficient to examine just the Principia itself.

In his story "A Descent into the Maelstrom," first published in Graham's Magazine in 1841, Poe prepares us for the subsequent adventure by quoting a supposedly matter-of fact description[6] from a certain Jonas Ramus [though in fact Poe takes it without acknowledgment from the Third Edition of the Encyclopedia Britannica (1799)]:

> . . . It likewise happens frequently that whales come too near the stream, and are overpowered by its violence; and then it is impossible to describe their howlings and bellowings in their fruitless struggles to disengage themselves. A bear once, attempting to swim from Lofoden to Moskoe [here Poe omits the picturesque detail that the bear was enticed into the water by the sight of sheep grazing on the other side], was caught by the stream and borne down, while he roared terribly, so as to be heard on shore. Large stocks of firs and pine trees, after being absorbed into the current, rise again broken and torn to such a degree as if bristles grew upon them. This plainly shows the bottom to consist of craggy rocks, among which they are whirled to and fro. This stream is regulated by the flux and reflux of the sea—it being constantly high and low water every six hours. . . . But the intervals of

[6] This fact was first pointed out in Stedman and Woodberry, vol. IV, pp. 289-294 (cf. Campbell, p. 173). The third edition of Encyclopedia Britannica (1799) took the information, without acknowledgment, from Pontoppidan, who mentions Jonas Ramus. But then the ninth edition assigned it to Poe (who had got it from the third), as follows: "the various reports of travellers . . . were faithfully collated and thrown into stereoscopic relief by Edgar Allan Poe in his celebrated story A Descent into the Maelstrom." But by the time of the Eleventh Edition (1911) the encyclopedia had become more sensible. It simply says: "the tales of ships being swallowed in this whirpool are fables."

tranquillity are only at the turn of the ebb and flood, and in calm weather, and last but a quarter of an hour, the violence gradually returning.

Then Poe tells us how he himself, in company with a Norwegian fisherman, observed the stream from the top of Mount Helseggen. At the moment of its maximum violence it suddenly assumed the shape of a "circle more than a mile in diameter, the mouth of a terrific funnel whose interior . . . was a smooth, shining, and jet-black wall of water, inclined to the horizon at an angle of some forty-five degrees, speedily dizzily round and round with a swaying and sweltering motion."

At this point the story is taken up by the fisherman, who explains that "in all violent eddies at sea there is good fishing, at proper opportunities, if one has only the courage to attempt it." So he and his two brothers

> kept our smack in a cove about five miles higher up the coast; and it was our practice, in fine weather, to take advantage of the fifteen minutes' slack to push across the main channel of the Moskoe-strom, far above the pool, where the eddies are not so violent. Here we used to remain until nearly time for slack water again, when we weighed and made for home.

But on the tenth of July 18—, the three fishermen were overtaken by a hurricane which "at its first puff carried both masts overboard as if they had been sawed off—the mainmast taking with it my youngest brother, who had lashed himself to it for safety." The two older brothers were then carried down into the maelstrom at the height of its violence.

> Our first slide into the abyss itself . . . had carried us a great distance down the slope; but our farther descent was by no means proportionate. Round and round we swept—not with any uniform movement—but in dizzying swings and jerks, that sent us sometimes only a few hundred yards—sometimes nearly the complete circuit of the whirl. Our progress downward, at each revolution, was slow, but very perceptible.

[At this point we learn that a full moon, nearly overhead, suddenly lit up the scene through a rift in the jet-black clouds.]

> Looking about me upon the wide waste of liquid ebony on which we were thus borne, I perceived that our boat was not the only object in the embrace of the whirl. Both above and below us were visible fragments of vessels, large masses of building timber and trunks of trees, with many smaller articles, such as pieces of house furniture, broken boxes, barrels and staves. . . . I soon made three important observations. The first was that, as a general rule, the larger the bodies were, the more rapid their descent—the second, that between two masses of equal extent, the one spherical, and the other of any other shape, the superiority in speed of descent was with the sphere—the third, that between two masses of equal size, the one cylindrical and the other of any other shape, the cylinder was absorbed the more slowly. Since my escape, I have had several conversations on this subject with an old school-master of the district; and it was from him that I learned the use of the words "cylinder" and "sphere." He explained to me—although I have forgotten the explanation—how what I observed was, in fact, the natural consequence of the forms of the floating fragments—and showed me how it happened that a cylinder, swimming in a vortex, offered more resistance to its suction, and was drawn in with greater difficulty than any equally bulky body, of any form whatever.
>
> I no longer hesitated what to do. I resolved to lash myself securely to the water cask upon which I now held, to cut it loose from the counter, and to throw myself with it into the water. I attracted my brother's attention by signs . . . and did everything in my power to make him understand what I was about to do. . . . But he shook his head despairingly, and refused to move from his station by the ring-bolt. . . .

So with a bitter struggle, I resigned him to his fate; fastened myself to the cask by means of the lashings which secured it to the counter, and precipitated myself with it into the sea.

The result was precisely what I had hoped it might be. . . . It may have been an hour, or thereabout, after my quitting the smack, when, having descended to a vast distance beneath me, it made three or four wild gyrations in rapid succession, and, bearing my beloved brother with it, plunged headlong, at once and forever, into the chaos of foam below. The barrel to which I was attached sunk very little farther than half the distance between the bottom of the gulf and the spot at which I leaped overboard, before a great change took place in the character of the whirlpool. The slope of the sides of the vast funnel became momently less and less steep. The gyrations of the whirl grew gradually less and less violent. By degrees . . . the bottom of the gulf seemed slowly to uprise. The sky was clear, the winds had gone down, and the full moon was setting radiantly in the west, when I found myself on the surface of the ocean, in full view of the shores of Lofoden, and above the spot where the pool of the Moskoe-strom *had been*. It was the hour of the slack—but the sea still heaved in mountainous waves from the effects of the hurricane. I was borne violently into the channel of the Strom, and in a few minutes was hurried down the coast into the "grounds" of the fishermen. A boat picked me up, exhausted from fatigue and, now that the danger was removed, speechless from the memory of its horror.

11. *Poe's Method of Composing His Stories*. In his *Marginalia* Poe tells us how he kept notebooks for jottings from his reading in science, literature or history, with a view to incorporating suitable details into his own work. The initial inspiration for his "Descent into the Maelstrom" seems to have come from an article in the Philadelphia newspaper *Alexander's Messenger*[7] on October tenth 1838, describing whirlpools in the ebb and flow of the fjord at Trondheim, halfway up the coast of Norway. It would then be natural for Poe to turn to the Encyclopedia Britannica, where he found the details for his story about the Lofoden maelstrom, three hundred miles farther north.

From many passages in his writings, especially in the philosophical essay *Eureka*, it is clear that Poe also spent considerable time on Newton's *Principia*; and although his attention would naturally be concentrated on the third book *System of the World*, we may suppose that he at least glanced at Book II with the title *On the Motion of Bodies in Resisting Media*. And he then creates his story by combining his two sources in characteristic fashion. His geographical details are only remotely similar to the actual environment; he makes the astronomical error[8] of putting the moon nearly overhead at a place within the Arctic circle; he states that the sphere offers the minimum resistance, although Newton had proved otherwise; and he applies the Newtonian theory to bodies swept along in a current, although it is in fact applicable only to bodies moving through a motionless fluid, like barges towed in a canal.[9] There has never been a maelstrom like the one described by Poe; and if one did exist, it would offer no special advantage to spheres over cylinders.

12. *Poe's Misattribution of Newton's Ideas to Archimedes*. In the second edition (1845) of "Descent into the Maelstrom" Poe adds a footnote: "See Archimedes *De Incidentibus in Fluido*, lib. 2." He gives a Latin title for the two books *On Floating Bodies* because the Greek text was lost from 1550 to 1899, although there were several translations into Latin—some of them intended as improvements on others—with various

[7]See Campbell, p. 166.

[8]In fact, a full moon on July 10 might not rise at all at 68 degrees of latitude and could in no case rise more than ten degrees above the horizon.

[9]Here it would be possible to quote numerous mathematical articles; e.g., Francis, which deals with "the speed of free floating bodies in the surface of a water stream in a sloping channel."

renderings of the title; de insidentibus aquae, de iis quae in humido vehuntur and so on. And finally de insidentibus in humido is the title used by Peyrard, whose Oeuvres Complètes d'Archimède (1807) is the first translation of Archimedes into a modern language. So it seems likely that Poe's title is taken from the lengthy account of Peyrard's translation in Volume 3 of the Quarterly Review (1812), since the journal was well known to Poe, and his choice of a title is closest to Peyrard's, although again we must note that Poe has been careless; his one-letter change of insidentibus to incidentibus converts floating bodies into falling bodies. Moreover, the footnote is doubly misleading. Not only is the title wrong, but Poe is wrong in referring to Archimedes at all, since the two books On Floating Bodies deal only with positions of stable equilibrium in still water. It seems that Poe had forgotten where he saw the sentences from the Principia and now assigned them to Archimedes because of the famous stories concerning his leap from the bathtub and the engraving of a sphere and cylinder on his tomb.

In any case, Poe was quite unhampered by any sense of scholarly responsibility. Campbell, who seems to be the only commentator to have looked up the reference to Archimedes, writes (in Modern Language Notes):

> Poe appears to have indulged in something of fiction in the interest of verisimilitude—in which, I feel, he was quite within his rights as a creative artist. . . . His inaccurate citation of the title of his alleged authority is entirely in character.

13. Poe's Effectiveness. But whatever may be thought of Poe's methods, there can be no doubt about their effectiveness. Quinn (p. 313) expresses his admiration of "A Descent into the Maelstrom" in these words:

> the source of the details is of slight importance. What makes the story impressive is the . . . succession of moods. The decrease of terror when hope is abandoned—the interest taken in the race of the different objects toward "the horrible inner edge" of the maelstrom, the "new hope that made the heart beat more heavily" all identify the reader's own feelings with the most universal of all emotions, the struggle for self-preservation.

This tale holds the attention of every reader. Its opening paragraph sets the scene sharply before our eyes; it continues with realistic touches that lead us on from the probable to the impossible; and every sentence follows from the one before with such apparent logic that we never stop to analyze Poe's assertions or the state of our own emotions. And later on, when we can examine the story more calmly, we readily forgive the garbled science because we realize that it has made an important contribution to the total effect.

REFERENCES

1. Apollonius Pergaeus: On Conic Sections, T. L. Heath, ed. Cambridge, England (1896).
2. Archimedes: The Works of Archimedes, T. L. Heath, ed. Cambridge University Press (1897, 1912).
3. Archimedes: Oeuvres Complètes d'Archimède, Francois Peyrard, tr. Paris (1807).
4. Campbell, K.: The Mind of Poe, Cambridge, Mass. (1933).
5. Campbell, K.: Modern Language Notes XLII, No. 8 (1927).
6. Courant, R.; Hilbert, D.: Methods of Mathematical Physics, vol. I. Interscience (1953).
7. Euclid: The Elements, T. L. Heath, tr. Cambridge (1908).
8. Euler, L.: Leonhardi Euleri Opera Omnia, Series I, Opera Mathematica. I. XXIV: Methodus Inveniendi Lineas Curvas Maximi Minimive Proprietate Gaudentes sive Solutio Problematis Isoperimetrici Latissimo Sensu Accepti, C. Carathédory, ed. Bern (1952).
9. Francis, J. R. D.: J. Fluid Mech. 1, 517-560 (1956).

10. Goldstine, H. H.: A History of the Calculus of Variations from the 17th through the 19th Century. Springer (1980).
11. Lagrange, J. L.: Oeuvres, 14 vols. J. A. Serret and G. Darboux, eds. Paris (1867-1892).
12. Newton, I.: Philosophiae Naturalis Principia Mathematica. Many editions. First edition, London, 1687. Most recent edition, A. Koyré and I. B. Cohen, eds. Harvard University Press (1972). English translation by A. Motte: Sir Isaac Newton's Principles of Natural Philosophy . . . . London (1729), revised by F. Cajori, Berkeley, Calif. (1934).
13. Newton, I.: The Mathematical Papers of Isaac Newton, 7 vols. D. T. Whiteside, ed. Cambridge (1967-1976).
14. Pontoppidan, E.: Natural History of Norway. London (1755).
15. Quinn, A. H.: Edgar Allan Poe, A Critical Biography. Appleton-Century (1942).
16. Stedman, E. C.; Woodberry, G.E., eds.: The Works of E. A. Poe, 10 vols. New York (1903).

16 STABILITY OF MINIMUM POINTS FOR PROBLEMS WITH CONSTRAINTS

Donald H. Hyers / Department of Mathematics, University of Southern California, Los Angeles, California

1. INTRODUCTION

The stability of stationary points for functions of class C^2 defined on finite dimensional spaces was studied by Ulam and Hyers [12] and by Hyers [6]. In [7] some sufficient conditions for the stability of minimum points were established for functions defined on Banach spaces and for simple problems of the calculus of variations.

In the present paper, Sec. 2 contains a basic lemma, while Sec. 3 is devoted to a stability theorem for a minimum problem for smooth functions on R^n subject to equality and inequality constraints.

Minimum problems for smooth functions defined on a Banach space with equality constraints are discussed in Sec. 4. Theorem 4.1 provides a generalization to Banach spaces of the converse part of theorem 9.2 in Hestenes [5]. This result is essentially known (see, e.g., Ioffe [8]). It is useful in application of the stability theorem of Sec. 5 for isoperimetric problems.

2. PRELIMINARY CONSIDERATIONS

The following lemma was established in Hyers [7] and is reproduced here for the convenience of the reader.

<u>Lemma 2.1</u>. Let ρ be a real valued strictly increasing function defined on $[0, \infty)$ with $\rho(0) = 0$. Let E be a subset of a normed vector space X, let x_0 be a point of E, and let f be a real valued function defined on E such that when $x \in E$,

$$f(x) - f(x_0) \geq \rho(\|x - x_0\|) \tag{2.1}$$

For a given $\epsilon > 0$, let $B(x_0, \epsilon)$ denote the closed ball centered at x_0 with radius ϵ. Put $\delta = \rho(\epsilon)/2$, and let g be another real valued function defined on E such that when $x \in E \cap B(x_0, \epsilon)$,

$$|g(x) - f(x)| < \delta \tag{2.2}$$

Then $g(x_0) < \inf\{g(x) : x \in E, \|x - x_0\| = \epsilon\}$ whenever the indicated set is not empty.

<u>Proof</u>. Assume that the intersection of E and the boundary of $B(x_0, \epsilon)$ is not empty. Put

$$h = \inf\{f(x) - f(x_0) : x \in E, \|x - x_0\| = \epsilon\}$$

By (2.1), $h \geq \rho(\epsilon)$ so that $\delta \leq h/2$. When $x \in E$ and $\|x - x_0\| = \epsilon$ we have by (2.2) that $g(x) - f(x_0) > f(x) - f(x_0) - \delta \geq h/2$, so that

$$\inf\{g(x) - f(x_0) : x \in E, \|x - x_0\| = \epsilon\} \geq h/2 \tag{2.3}$$

Again, by (2.2) and (2.3),

$$\begin{aligned} g(x_0) &< f(x_0) + \delta \leq f(x_0) + h/2 \\ &\leq \inf\{g(x) : x \in E, \|x - x_0\| = \epsilon\} \end{aligned}$$

Remark. The lemma shows that if g has a minimum on the set $E \cap B(x_0, \epsilon)$, it must lie in the interior of the ball $B(x_0, \epsilon)$.

On the basis of this lemma, the following stability theorem is easily proved.

Theorem 2.1. Let the conditions of lemma 2.1 be satisfied and assume that the set $E \cap B(x_0, \epsilon)$ is a weakly sequentially compact subset of X. If the function g of the lemma is lower semicontinuous with respect to weakly convergent sequences in $E \cap B(x_0, \epsilon)$, then g has a minimum value on this set which is taken on at an interior point $\bar{x}$ of $B(x_0, \epsilon)$.

Of course not all minimum problems have stable minimum points. For example, let $f: R^3 \to R$ be defined by $f(x,y,z) = (x - y)^2 + z^2$ and $\phi(x,y,z) = z - 1$. Then the point (0, 0, 1) is a minimum point for the function f subject to the constraint $\phi = 0$. However, if we define $g(x,y,z) = (x - y)^2 + k(x + y) + z^2$ we find that g has no minimum points subject to the constraint $\phi = 0$ when $k \neq 0$, so the minimum point (0, 0, 1) is unstable.

3. A FINITE DIMENSIONAL PROBLEM

We denote the gradient of a function $\phi: R^n \to R$ by ϕ_x and its second differential with increment h by $\phi_{xx}(x)(h, h)$. With f and f_i $(i = 1, \ldots, p)$ defined on R^n, consider the problem of minimizing f subject to:

$$f_i(x) \leq 0, \quad 1 \leq i \leq m; \quad f_i(x) = 0, \quad m < i \leq p \tag{3.1}$$

Let S be the set of points in R^n which satisfy the constraints (3.1) and let x_0 be a point of S. We assume that the functions f, $f_1, \ldots, f_p$ belong to the class C^2 in the neighborhood of x_0. Following Hestenes [5] we define the tangent cone of S at x_0 as follows. A sequence $\{x_q\}$ of points of S will be said to converge to x_0 in the direction h if h is a unit vector, $x_q \neq x_0$ and $\lim |x_q - x_0| = 0$, while $\lim (x_q - x_0)/|x_q - x_0| = h$. The tangent cone of S at x_0 is the cone determined by all the unit vectors h for which there is a sequence in S converging to x_0 in the direction h. According to the theorem 10.3 on page 37 of Hestenes [5] we have the following sufficient condition for a solution to the above minimum problem:

Theorem H. Assume that there are multipliers $\lambda_1, \ldots, \lambda_p$ such that $\lambda_i \geq 0$ $(1 < i \leq m)$ and $L_x(x_0, \lambda_1, \ldots, \lambda_p) = 0$, where $L(x, \lambda_1, \ldots, \lambda_p) = f(x) + \Sigma_{j=1}^{p} \lambda_j f_j(x)$ is the Lagrangian function of the problem. Let Γ be the indices $\gamma \leq m$ such that $\lambda_\gamma > 0$. Suppose that the quadratic form $L_{xx}(x_0, \lambda_1, \ldots, \lambda_p)(h, h)$ is positive for all unit vectors h in the tangent cone of S at x_0 such that $f'_\gamma(x_0, h) = 0$ for $\gamma \in \Gamma$ (f' denotes differential of f).
Then there is a neighborhood N of x_0 and a constant $k > 0$ such that

$$f(x) - f(x_0) \geq k|x - x_0|^2$$

for all $x \in N \cap S$.

On the basis of this result together with those of Sec. 2 we easily obtain the following:

Theorem 3.1. Let the functions f and $f_1, \ldots, f_p$ belonging to class C^2 in the neighborhood of the point $x_0 \in X$ satisfy the conditions of theorem H. Given $\epsilon > 0$, with $B(x_0, \epsilon) \subset N$, let $g: S \to R$ be a lower semicontinuous function satisfying the inequality

$$|g(x) - f(x)| < \frac{k\epsilon^2}{2}$$

for $x \in B(x_0, \epsilon) \cap S$. Then g has a minimum value in $B(x_0, \epsilon) \cap S$ which is taken on at an interior point $\bar{x}$ of $B(x_0, \epsilon)$.

Proof. Since the functions f_i ($i = 1, \ldots, n$) are continuous in N, the set S is closed, while $B(x_0, \epsilon)$ is closed by definition, so g has a minimum point $\bar{x}$ in $B(x_0, \epsilon) \cap S$. By lemma 2.1, with $\rho(s) = ks^2$, $\bar{x}$ lies in the interior of $B(x_0, \epsilon)$.

4. MINIMUM PROBLEMS FOR SMOOTH FUNCTIONS ON BANACH SPACES WITH EQUALITY CONSTRAINTS

Let X and Y be Banach spaces and consider a function $f : X \to R$ and a function $F : X \to Y$. We shall deal with the problem of minimizing f subject to the constraint $F(x) = 0$.

Put $Q = \{x \in X : F(x) = 0\}$. The tangent cone to Q at a point $x_0 \in Q$ may be defined as it was in Sec. 3. An equivalent characterization of this tangent cone, which is more convenient for our present purposes, is given by I. V. Girsanov on page 40 of [4], as follows. A vector h is called a tangent direction to Q at x_0 if for any ϵ with $0 < \epsilon < \epsilon_0$ there exists a point $x(\epsilon) = x_0 + \epsilon h + r(\epsilon)$ in Q such that $\lim_{\epsilon \to 0} r(\epsilon)/\epsilon = 0$. The tangent cone to Q at x_0 is then the set of all tangent directions to Q at x_0.

The following theorem of L. A. Liusternik (see p. 61 of [4], p. 30 of [9], and Sec. 46 of [10]) often allows us to find the tangent cone.

Liusternik's Theorem. Let $F : X \to Y$ be continuously differentiable in the neighborhood of x_0 and let its derivative $F_x(x_0)$ be surjective (i.e., map X onto the whole space Y). Then the tangent cone to Q at $x_0 \in Q$ is the subspace $\{h : F_x(x_0)h = 0\}$.

Liusternik (theorem 2, p. 209 of [10]) also proved the following necessary condition: Let f and F be continuously differentiable in the neighborhood of the point $x_0 \in Q$ and let $F_x(x_0)$ be surjective. If f has a local minimum in Q at x_0, then there exists a continuous linear functional y^* defined on Y such that the Lagrangian functional

$$L(x, y^*) = f(x) + \langle y^*, F(x) \rangle \tag{4.1}$$

satisfies the equation $L_x(x_0, y^*) = 0$.

We next turn to a second order sufficient condition, analogous to theorem H, for the case of equality constraints.

Theorem 4.1. Assume that f and F are twice continuously differentiable in the neighborhood of $x_0 \in Q$, and that $F_x(x_0)$ is surjective. Suppose there exists a linear functional y^* in the conjugate space Y^* such that the Lagrangian functional (4.1) satisfies $L_x(x_0, y^*) = 0$ and

$$L_{xx}(x_0, y^*)(h, h) \geq k\|h\|^2 \tag{4.2}$$

for all h in the tangent cone to Q at x_0, and for some constant $k > 0$.

Then there exists a positive number α and a neighborhood N of x_0 such that

$$f(x) - f(x_0) \geq \alpha\|x - x_0\|^2$$

for all $x \in N \cap Q$.

Proof. By Taylor's formula, with $x \in Q$,

$$\begin{aligned} L(x, y^*) - L(x_0, y^*) &= L_x(x_0, y^*)(x - x_0) + \frac{1}{2} L_{xx}(x_0, y^*)(x - x_0, x - x_0) + o(\|x - x_0\|^2) \\ &= \frac{1}{2} L_{xx}(x_0, y^*)(x - x_0, x - x_0) + o(\|x - x_0\|^2) \end{aligned}$$

since $L_x(x_0, y^*) = 0$ by hypothesis.

Using Liusternik's theorem 2, p. 204 of [10], we can associate with each point x of the set Q a point $x_0 + h$ of the tangent manifold at x_0 (i.e., h lies in the tangent cone) such that $z = o(\|x - x_0\|)$, where $z = x_0 + h - x$, or $h = x - x_0 + o(\|x - x_0\|)$. Thus from (4.2) we obtain the inequality:

$$L(x, y^*) - L(x_0, y^*) \geq \frac{1}{2}k\|x - x_0\|^2 + o(\|x - x_0\|^2)$$

so that, since both x and x_0 belong to Q, we have:

$$f(x) - f(x_0) = L(x, y^*) - L(x_0, y^*) \geq \alpha\|x - x_0\|^2 \quad \text{for } x \in N \cap Q$$

where $0 < \alpha < \frac{1}{2}k$ and N is a sufficiently small neighborhood of x_0.

5. AN ISOPERIMETRIC PROBLEM

Consider the problem of minimizing

$$I[x] = \int_0^1 f(t, x(t), \dot{x}(t))\, dt \tag{5.1}$$

with fixed end conditions:

$$x(0) = a_0, \quad x(1) = a_1 \tag{5.2}$$

subject to constraints of the form:

$$I_\gamma[x] = \int_0^1 \{b_\gamma(t, x) + \ell_\gamma(t, x)\,\dot{x}(t)\}\, dt = 0, \qquad \gamma = 1, \ldots, m \tag{5.3}$$

where the dot denotes differentiation with respect to t. We assume that each of the functions $f(t, x, y)$, $b_\gamma(t, x)$, and $\ell_\gamma(t, x)$ is continuous for $0 \leq t \leq 1$ and all $x \in R^n$ and $y \in R^n$. In this section $C[0, 1]$ will denote the Banach space of n dimensional vector functions x(t) which are continuous on [0, 1], and the notation $L_p(0, 1)$ will be used similarly.

Let Q denote the class of all absolutely continuous arcs satisfying the fixed end continuous arcs satisfying the fixed end conditions (5.2) and the constraints (5.3) such that $\dot{x}(t) \in L_p(0, 1)$ with $p > 1$.

Hypothesis. There exists an $x_0(t) \in Q$ and a positive number α such that

$$I[x] - I(x_0] \geq \alpha \int_0^1 |\dot{x} - \dot{x}_0|^p\, dt \quad \text{for all } x(t) \in Q \tag{5.4}$$

Put $u(t) = x(t) - x_0(t)$ and assume that (5.4) holds. Then $u(0) = 0$ and $\dot{u} \in L_p(0, 1)$ with $p > 1$. By Holder's inequality, since u is absolutely continuous when $0 \leq t \leq 1$,

$$|u(t)| = |u(t) - u(0)| = \left|\int_0^t \dot{u}(s)\, ds\right| \leq \left(\int_0^1 |\dot{u}|^p dt\right)^{1/p}$$

so that

$$\|u\| = \sup_{0 \leq t \leq 1} |u(t)| \leq \left(\int_0^1 |\dot{u}|^p dt\right)^{1/p}$$

Hence for $x(t) \in Q$ we have

$$I[x] - I[x_0] \geq \alpha\|x - x_0\|^p \tag{5.5}$$

where the norm is that for $C[0, 1]$. Choose $\epsilon > 0$ and let $B(x_0, \epsilon)$ denote the closed ball in this space with center x_0 and radius ϵ.

Theorem 5.1. Let the functional $I[x]$ given by (5.1) satisfy the inequality (5.4) for all $x \in Q \cap B(x_0, \epsilon)$ so that x_0 furnishes a minimum for I on this set. Let $J[x] = \int_0^1 g(t, x(t), \dot{x}(t))\,dt$ be another functional satisfying the following conditions: For $0 \leq t \leq 1$, $|x - x_0(t)| \leq \epsilon$ and all y

(a) $g(t,x,y)$ is continuous and g is convex in y for fixed (t,x);

(b) $g(t,x,y) \geq \Psi(t) - \beta |y|^p$, where Ψ is summable on $(0,1)$ and β is a constant.

(c) $|J[x] - I[x]| < \delta$ for all $x(t) \in Q \cap B(x_0, \epsilon)$, where $\delta = \alpha\epsilon p/2$.

Then there exists a function $\bar{x} \in Q$ with $\|\bar{x} - x_0\| < \epsilon$ such that $J[\bar{x}]$ is an absolute minimum for J on the set $Q \cap B(x_0, \epsilon)$.

Proof. Put $\mu = \inf\{J[x] : x \in Q \cap B(x_0, \epsilon)\}$ and let $\{x_n\}$ be a minimizing sequence for J in $Q \cap B(x_0, \epsilon)$ such that $J[x_n] \leq J[x_1]$, $n = 1, 2, 3, \ldots$, and

$$\lim_{n\to\infty} J[x_n] = \mu \tag{5.6}$$

From inequalities (5.4) and condition (c) of the theorem it follows that

$$\begin{aligned} \alpha \int_0^1 |\dot{x}_n - \dot{x}_0|^p\,dt &\leq J[x_n] + \delta - I[x_0] \\ &\leq J[x_1] + \delta - I[x_0] \end{aligned}$$

so that the sequence $\{\dot{x}_n\}$ of derivatives form a bounded set in the L_p norm:

$$\left(\int_0^1 |\dot{x}_n|^p\,dt\right)^{1/p} \leq K \tag{5.7}$$

Thus the set $\{x_n(t)\}$ of functions is equicontinuous and uniformly bounded on the closed interval $[0,1]$. By standard arguments (see Secs. 28 and 29 of Akhiezer [1]) it follows that there is a subsequence (also to be denoted by $x_n(t)$) which converges uniformly to an absolutely continuous limit function $\bar{x}(t)$ whose derivative $\dot{\bar{x}}(t)$ belongs to $L_p(0,1)$, and finally that $\dot{x}_n(t)$ converges weakly to $\dot{\bar{x}}(t)$ in $L_p(0,1)$.

It remains to prove that the function $\bar{x}(t)$ satisfies the constraints (5.3), and that it furnishes an absolute minimum for J on $Q \cap B(x_0, \epsilon)$.

We have for $\gamma = 1, \ldots, m$:

$$\left| \int_0^1 \{b_\gamma(t, x_n) + \ell_\gamma(t, x_n)\dot{x}_n\}\,dt - \int_0^1 \{b_\gamma(t, \bar{x}) + \ell_\gamma(t, \bar{x})\,\bar{x}\}\,dt \right|$$

$$\leq \int_0^1 |b_\gamma(t, x_n) - b_\gamma(t, \bar{x})|\,dt + \left| \int_0^1 \{\ell_\gamma(t, x_n) - \ell_\gamma(t, \bar{x})\}\dot{x}_n\,dt \right| + \left| \int_0^1 \{\ell_\gamma(t, \bar{x})(\dot{x}_n - \dot{\bar{x}})\,dt \right|$$

Since b_γ, ℓ_γ, and $\bar{x}$ are continuous, and since x_n converges uniformly to $\bar{x}$ and $\dot{x}_n$ converges weakly to $\dot{\bar{x}}$ in $L_p(0,1)$, it follows that the first and last integrals approach zero as $n \to \infty$. As to the second integral, it is bounded by

$$\left(\int_0^1 |\ell_\gamma(t, x_n) - \ell_\gamma(t, x)|^q\right)^{1/q} \left(\int_0^1 |\dot{x}_n|^p\,dt\right)^{1/p}$$

where $q = p/(p-1)$, by use of the Holder inequality. Hence in view of the inequality (5.7), the second integral also tends to zero as $n \to \infty$. Since each $x_n(t)$ satisfies (5.3), it follows that $\bar{x}(t)$ does also. Thus $\bar{x} \in Q \cap B(x, \epsilon)$.

Since $x_n(t) \to \bar{x}(t)$ uniformly (and hence strongly in L_p) and $\dot{x}_n(t) \to \dot{\bar{x}}(t)$ weakly in L_p, we can apply theorem 1 of Berkovitz [3] (together with remark 1 following that theorem) to conclude that $\mu = \lim_{n\to\infty} J[x_n] > J[\bar{x}]$. Therefore $J[\bar{x}]$ is an absolute minimum on the set $Q \cap B(x, \epsilon)$. Finally, from inequality (5.5) and lemma 2.1 with $\rho(s) = \alpha s^p$ we conclude that $\|\bar{x} - x_0\| < \epsilon$.

Example 5.1. As an example we look at the problem of minimizing

$$I[x] = \int_0^1 \dot{x}^2 \, dt$$

with the end conditions $x(0) = x(1) = 0$ subject to the constraint

$$H[x] \equiv \int_0^1 x^2 \, dt = 1$$

In order to apply theorem 4.1 to this example we will let X be the space consisting of all functions $x(t)$ which are absolutely continuous on $[0, 1]$, which satisfy the end conditions, with $\dot{x}(t) \in L_2(0, 1)$. This space is a Banach space with the norm

$$\|x\|_2 = \sqrt{\int_0^1 |\dot{x}|^2 \, dt}$$

In this case we have $F(x) = H(x) - 1$, so that F maps X into $Y = R$. It is easily seen that its Frechet differential is

$$F_x(x)h = 2\int_0^1 xh \, dt$$

We take the Lagrangian functional for this problem to be of the form:

$$L(x, \lambda) = I[x] - \lambda(H[x] - 1)$$

so that

$$L_x(x, \lambda)(h) = 2\int_0^1 (\dot{x}\dot{h} - \lambda xh) \, dt$$

The solutions to the equation $L_x(x, \lambda) = 0$ satisfying the end conditions and the constraint $H[x] = 1$ are the eigenfunctions $x_n = \sqrt{2} \sin n\pi t$ corresponding to the eigenvalues $\lambda_n = n^2\pi^2$ where $n = 1, 2, 3, \ldots$. The minimum is attained for $x_1 = \sqrt{2} \sin \pi t$, with $\lambda_1 = \pi^2$.

The second Frechet differential is $L_{xx}(x, \lambda)(h, h) = 2\int_0^1 (\dot{h}^2 - \lambda h^2) \, dt$. In this example the set $Q = \{x \in X : H[x] = 1\}$, and the Frechet derivative $F_x(x)$ is easily seen to be continuous at $x = x_1$. Also

$$F_x(x_1) = 2\int_0^1 (\sqrt{2} \sin \pi t) \, h(t) \, dt$$

clearly maps the space X onto $Y = R$. Thus the conditions of Liusternik's theorem are satisfied and thus the tangent cone C to Q at X_1 is the subspace $\{h \in X : F_x(x_1)h = 0\}$. That is, C consists of the functions $h \in X$ which are orthogonal to x_1. It remains to show that the inequality (4.2) holds for some $k > 0$ and all $h \in C$.

We expand $h(t) \in C$ in a series of orthonormal eigenfunctions, $h(t) = \sum_{n=1}^{\infty} b_n x_n(t)$, with $b_n = \sqrt{2}\int_0^1 h(t) \sin n\pi t\, dt$. Since h is orthogonal to x_1, $b_1 = 0$. We have:

$$h(t) = \sum_{n=2}^{\infty} b_n \sqrt{2} \sin n\pi t$$

$$\dot{h}(t) = \sum_{n=2}^{\infty} n\pi b_n \sqrt{2} \cos n\pi t$$

$$L_{xx}(x_1, \lambda_1)(h,h) = 2\int_0^1 (\dot{h}^2 - \pi^2 h^2)\, dt$$

$$= 2\sum_{n=2}^{\infty} (n^2\pi^2 b_n^2 - \pi^2 b_n^2)$$

$$= 2\sum_{n=2}^{\infty} (n^2 - 1)\pi^2 b_n^2$$

$$\geq \frac{3}{2}\sum_{n=2}^{\infty} n^2\pi^2 b_n^2 = \frac{3}{2}\int_0^1 \dot{h}^2\, dt$$

Consequently, inequality (4.2) is satisfied with x_0 replaced by x_1, y^* by $-\lambda_1$ and $k = 3/2$, with the norm

$$\|h\|_2 = \left(\int_0^1 \dot{h}^2\, dt\right)^{\frac{1}{2}}$$

By theorem 4.1 it follows that $I[x] - I[x_1] \geq \alpha\int_0^1 |\dot{x} - \dot{x}_1|^2\, dt$ for some $\alpha > 0$ and $x \in Q$.

We can now apply theorem 5.1 to example 5.1, since the crucial inequality (5.4) has been demonstrated with $p = 2$. For example we may take $g(t,x,y) = y^2 + q(t)x^2$, where $q(t)$ is continuous and $\sup\{|q(t)|: 0 \leq t \leq 1\} < \delta = \alpha\epsilon$.

REFERENCES

1. Akhiezer, N. I.: The Calculus of Variations. Blaisdell, New York (1962).
2. Berkovitz, L. D.: Optimal Control Theory. Springer (1974).
3. Berkovitz, L. D.: Trans. Am. Math. Soc. 192, 51-57 (1974).
4. Girsanov, I. V.: Lectures on Mathematical Theory of Extremum Problems. Springer (1972).
5. Hestenes, M. R.: Calculus of Variations and Optimal Control Theory. Wiley (1966).
6. Hyers, D. H.: J. Math. Anal. Appl. 36, 622-626 (1971).
7. Hyers, D. H.: J. Math. Anal. Appl. 62, 530-537 (1978).
8. Ioffe, A. D.: SIAM J. Control Optimization 17, 266-288 (1979).
9. Ioffe, A. D.; Tihomirov, V. M.: Theory of Extremal Problems. North-Holland (1979).
10. Liusternik, L. A.; Sobolev, V. J.: Elements of Functional Analysis. F. Ungar (1961).
11. Smith, D. R.: Variational Methods in Optimization. Prentice-Hall (1974).
12. Ulam, S. M.; Hyers, D. H.: Math. Magazine 28, 59-64 (1954).

17 LÉONHARD EULER: MATHEMATICAL MODELLER AND MODEL FOR MATHEMATICIANS

Richard D. Järvinen* / Department of Mathematics, College of St. Benedict's/ St. John's University, St. Joseph, Minnesota

The occasion of the two hundredth anniversary since the passing of Léonhard Euler (1707-1783) provides an opportunity to applaud his greatness. Historians have in fact called him the greatest mathematician of all time, and all who have seriously pursued mathematical invention with an historical background are convinced that his accomplishments are monumental. His work, exposed in some eighty volumes, with all of its variety, quality, and quantity speaks for itself.

But here, instead of focusing upon one of his mathematical accomplishments from which there are so many options for selection, we first look to some of the attitudes of this man and touch upon some of his personality traits. It is easy to overlook the fact that the personality of an individual has as much to do with that individual's achievements as do special skills and aptitudes. We make the point that Euler's outlook was strongly a part of his overall success.

Secondly, and finally, we turn to a problem of current significance. It is one that has been approached through a special blend of mathematical and statistical subjects. We are inclined to think that this would have been the kind of problem Euler would have delighted in mathematizing: creating mathematics appropriate for the circumstances and writing it into the mathematical model that solves the problem. It is interesting to conjecture that had he engaged this problem that he would have invented mathematics never known.

Euler was a battler. He had zeal for what he undertook to do. He was ambitious and a willing participant in a wide range of endeavors. He was excited by mathematical thought and its problem solving impact in the world of problems at large. But it is the opinion of this author that his greatest asset was his openmindedness. He welcomed the challenge of new problems in entirely new settings, settings that oftentimes were at once mathematical and oftentimes required mathematical handles to be built. In these contexts he was inspired to create new mathematics, mathematics that was applicable and frequently developed into full-blown theories.

In addition to the mathematics in his life he was at minimum conversant and at maximum an expert among the intellectuals of his time in matters which include physics, chemistry, astronomy, geography, philosophy, politics, literature, history, weather, and music. Knowledgeable in these areas and a significant contributor to most of them, we must mention that his positions on politics and philosophy were often strongly disputed. Specific occupations and activities of Euler include that of textbook expositor, commissioner of weights and measurements, number theory, problem solving for prizes and contests, artillery, northern lights, sound, the tides, navigation, shipbuilding, hydrodynamics, magnetism, light, telescopic design, canal construction, annuities, compilation of the formalisms of classic analysis, calculus of variations, mechanics, and combinatorial topology. He was often called upon to offer advice on a

*Current affiliation: Department of Mathematics and Statistics, St. Mary's College, Winona, Minnesota

variety of problematical matters related to special projects of short duration. See [8]. Euler, we should add, was also an effective long range planner of his own career.

Euler spent the bulk of his professional life in association with one or another academy and in company of many great minds. His first academic position was at St. Petersburg, an appointment arranged with the help of Nicolas and Daniel Bernoulli. He accepted an offer to take a position in physiology for which he was obliged to learn medicine and anatomy as preparation ([9], pp. 61ff). Euler was not long into this position before he became primarily occupied with mathematics.

Euler became well versed in medicine. He also wrote mathematics for lotteries. These two endeavors of his connect us to a problem we next discuss, although we do so only briefly and in a summarizing fashion. We think it is appropriate to call this problem an "Euler-type problem" for reasons yet to be discussed. Mathematical analysis has a role in what we do in the form of the so-called logistic function. Probability and statistics form the primary setting in which this problem is framed. But perhaps the single most contributing factor in bringing the problem to a successful solution is attitudinally based, resting on the spirit, tenacity, creative drive, and willingness to reach out to deal with the intricacies of what we call today an applied problem. In the arena of this problem there has proved to be and continues to be space for mathematical invention.

During 1982 and more briefly in 1983 while on sabbatical leave the author was a research associate in the Department of Medical Research Statistics at the Mayo Clinic in Rochester, Minnesota, United States of America. He worked on a project entitled the Health Status of American Men which was organized and sponsored by the Contraceptive Evaluation Branch of the Center for Population Research, National Institute of Child Health and Human Development. Four centers were established in a cooperative effort to uncover possible side effects of the sterilization procedure known as vasectomy. The four centers were the University of California at Los Angeles, the University of Southern California, the University of Minnesota, and the Mayo Clinic.

No long term study examining possible side effects of vasectomy had previously been conducted. Work with vasectomized monkeys elsewhere indicated that atherosclerosis could perhaps be a side effect of the procedure in man. The scenario for the development of atherosclerosis in the experimental monkeys gives an autoimmunological response as the cause of the disease. The HSAM study was designed to have a broad base, to include approximately fifty-five diseases each of which might be a side effect of vasectomy and most of which were thought to be autoimmunological-response related [10].

The study is considered to be of great importance, for it has been estimated that over forty million men throughout the world have been vasectomized during the last twenty years with over six million of them citizens of the United States of America. Approximately ten thousand matched pairs of American men were included in the HSAM study. An enormous amount of time and effort was consumed collecting elaborate health information for each man in the study using interviews and medical center health records.

To each man in the study an n-dimensional vector $\vec{x} = (x_1, x_2, \ldots, x_n)$ is associated

f : [stick figure] $\longrightarrow \vec{x}$

Here each x_i designates a characteristic of interest to the researcher for each man in the study. For example, the i-th component might indicate numerically to what extent the man was a smoker, or to what extent he was a drinker, or yet whether or he had been vasectomized. Males were paired if they were similar with respect to certain major characteristics [10], except that one had a vasectomy—the case—and his match had not—the control. Each pair was followed over a designed interval of time.

Let d denote a disease variable, e.g., d might refer to atherosclerosis. Then d is assigned the value 1 if the disease was developed in a man in question and is given the value 0 otherwise. Let V indicate vasectomy and ~V that no vasectomy occurred. Let

$$p_1 = P(d = 1 | V)$$

meaning that p_1 gives the probability that the disease developed under the condition that the man had been vasectomized. Similarly define

$$q_1 = P(d = 1 | \sim V)$$

$$p_2 = P(d = 0 | V)$$

$$q_2 = P(d = 0 | \sim V)$$

The value

$$R = p_1/q_1$$

is called the relative risk of disease. Using the usual definition of odds, the odds of getting a disease d is given by

$$P(d = 1)/P(d = 0)$$

independently of vasectomy.

We form

$$\psi = [p_1/q_1]/[p_2/q_2]$$

where p_1 and q_1 are the probabilities of disease and not disease, resp., granted exposure to vasectomy, and p_2 and q_2 are the probabilities of disease and not disease, resp., granted no exposure. We call ψ the odds ratio ([12], p. 33). If a disease is fairly rare, then

$$R \doteq \psi$$

An early use of the logistic curve was made by the two statisticians R. Pearl and L. Reed. In the 1920s they used the curve to analyze population and biological growth. But as Yamane ([15], p. 1057) writes, the use of the logistic curve in statistical settings did not turn out to be very productive. Not until the 1960s, we add. In the 1960s and '70s the productivity of the logistic model was reversed. It came to play an entirely significant role in the analysis of individual and joint effects of a set of variables on the risk of disease [4, 5, 13]. Today the model is used in both cohort (prospective) and case-control (retrospective) studies [11] in a most ingenious fashion, indeed a highly creative act.

The logistic model specifies that the probability of disease, $p_{\vec{x}}$, for a given man depends upon his vector of characteristics $\vec{x} = (x_1, x_2, \ldots, x_n)$ as follows

$$p_{\vec{x}} = P(d = 1 | \vec{x})$$

$$= [1 + e^{-(\beta_0 + \vec{\beta} \cdot \vec{x})}]^{-1}$$

where the x_i are variables and the β_i parameters. The β_i represent the effect of the x_i on the probability of disease.

We let

$$q_{\vec{x}} = 1 - p_{\vec{x}}$$

$$= P(d = 0 | \vec{x})$$

Then

$$p_{\vec{x}}/q_{\vec{x}} = e^{\beta_0 + \beta \cdot \vec{x}}$$

Suppose vector $\vec{x}$ is associated with one man, e.g., a man with a vasectomy, and vector $\vec{x}^*$ with another, e.g., a nonvasectomized match of the first man. If the two men are the same in all ways except the first has had a vasectomy and the second has not, the following odds ratio is obtained

$$[p_{\vec{x}}/q_{\vec{x}}]/[p_{\vec{x}^*}/q_{\vec{x}^*}] = e^{\beta_j}$$

where $x_j = 1$ and $x_j^* = 0$ with the j-th component giving the vasectomy information and $x_k^* = x_k$ for all $k \neq j$.

Using the relative risk information or its approximation, the odds ratio above, a client with given characteristics and not vasectomized could be told what his risk of getting a certain disease, e.g., atherosclerosis, would be if he undertook to have the procedure performed. The risk varies from man to man as the vector $\vec{x}$ of characteristics changes. The fact that the parameters of the logistic model are so easily interpretable in terms of relative risk is one of the main reasons for using the model.

We next briefly address the question of how the β_i are determined. Until the advent of the modern computer and some clever algorithms, the β_i were virtually impossible to compute using maximum likelihood strategy. Even Euler, famous for making incredibly difficult calculations in mind alone, would have welcomed the power of the computer. The β_i can be computed by fitting, so to speak, the "best" logistic curve through all of the data in the study using the maximum likelihood procedure. A recent and powerful methodology for accomplishing the latter is found in [1] and is discussed in context in ([3], p. 201).

The applications of the logistic model, especially in medical research statistics, have been richly developed during the last two decades. Much exciting research using the model is presently taking place. The important books on this subject by Breslow and Day [3] and Schlesselman [12] bring organization and focus to this effort. Both books contain excellent bibliographies. Let us also mention that as of this writing the HSAM project is still in progress under the direction of Massey at the University of California at Los Angeles.

The ingenuity to take a rather ordinary looking function from analysis, namely, the logistic function, and join it as has been done to probabilistic and statistical elements to produce the results that have appeared in both pure and applied contexts is truly a great achievement. This is the kind of mathematical involvement upon which Euler seems to have thrived. The author is inclined to call the problematic aspects in the HSAM project an Euler-type problem, one that requires the blending of mathematics _per se_ with the problems of the world as they are observed for the purpose of resolution. In this process of problem solving new mathematics is often necessary to create.

The great mathematician Euler might mathematically be modelled as the resultant of the vectors pure-mathematician-Euler and applied-mathematician-Euler. His magnitude is maximal; his direction that of pathfinder; and his force is with us.

REFERENCES

1. Baker, R.; Nelder, J.: The GLIM System, Release 3, Numerical Algorithms Group, Royal Statistical Society (1978).
2. Bochner, S.: The Role of Mathematics in the Rise of Science. Princeton University Press (1960).
3. Breslow, N.; Day, N.: Statistical Methods in Cancer Research, Volume 1, The Analysis of Case-Control Studies. International Agency for Research on Cancer (1980).
4. Cornfield, J.: Fed. Proc. 21, 58-61 (1962).
5. Cornfield, J.; Gordon, T.; Smith, W.: Bull. Int. Statist. Inst. 38, 97-115 (1961).
6. Cox, D.: The Analysis of Binary Data. Methuen (1970).
7. Day, N.; Kerridge, D.: Biometrics 23, 313-323 (1967).

8. Kramer, E.: The Nature and Growth of Modern Mathematics. Princeton University Press (1981).
9. Langer, R.: Scripta Math. 3, 61ff (1935).
10. Massey, F.; O'Fallon, W.; et al.: Collaborative Study of the Health Status of American Men (project in progress at the University of California at Los Angeles) (to appear).
11. Prentice, R.; Pyke, R.: Biometrika 66, No. 3, 403-411 (1979).
12. Schlesselman, J.: Case-Control Studies. Oxford University Press (1982).
13. Truett, J.; Cornfield, J.; Kannel, W.: J. Chronic Dis. 20, 511-524 (1967).
14. Walker, S.; Duncan, D.: Biometrika 54, 167-179 (1967).
15. Yamane, T.: Statistics. Harper and Row (1973).

18 NONCOMMUTATIVE CALCULUS OF VARIATIONS

Boris A. Kupershmidt* / Department of Mathematics, The University of Michigan, Ann Arbor, Michigan

1. INTRODUCTION

Factor-objects appear in classical mechanics wherever there is a common set of symmetries for the class of dynamical systems under consideration. The most important case among such situations occurs when the configuration space is itself a Lie group, say G. Then Hamiltonian formalism on the cotangent bundle T*G is invariant under the lift into T*G of the left action of G on itself. Taking the quotient with respect to this action results in the usual Poisson structure on the dual space $\mathfrak{G}^*$ of the Lie algebra $\mathfrak{G}$ of G.

In the calculus of variations, no close analog of the above-mentioned reduction $T^*G \to \mathfrak{G}^*$ is known, and it is the purpose of this paper to give an exposition of the appropriate formalism. The presentation is algebraic since it requires far less preparatory development than the corresponding geometric versions. However, the idea of the construction is most easily seen geometrically, as follows: Let $\pi: E \to G$ be a trivial bundle over a Lie group G. Then the left action of G on itself can be naturally lifted into the infinite jet bundle $J^\infty\pi$. Considering only the left-invariant part of the usual Lagrangian formalism on $J^\infty\pi$ (see [1], ch. II), results in the reduction of $J^\infty\pi$ to its fiber over the identity element $e \in G$.

Since G need not be commutative, the basic derivations of the ring $K = C^\infty(G)$ also do not commute, and this is where noncommutativity must enter into the usual algebraic constructions of the calculus of variations (see, e.g., [4], ch. I).

The plan of the paper is as follows. In Sec. 2 we construct the Euler-Lagrange operator δ in a purely algebraic fashion and describe how this construction is related to the formula for the first variation. In the next section we give an exact description of the image of δ by constructing an analog of the geometric resolvent of δ. In Sec. 4 a few remarks are made concerning more general situations when both derivations and automorphisms are present.

The idea that a noncommutative differential calculus of variations should exist was suggested to me by Joseph Bernstein. I am very much indebted to him for the inspiration. I am also obliged to Arthur Greenspoon for suggesting various improvements in the manuscript.

2. CONSTRUCTION OF THE EULER-LAGRANGE OPERATOR δ

Let k be a commutative ring and K a commutative k-algebra. Let $\mathfrak{G}$ be a finite-dimensional Lie algebra over k which acts by derivations on K/k. We fix a basis $(\partial_1, \ldots, \partial_m)$ in $\mathfrak{G}$. Thus we obtain a basis $\{\partial_1^{\sigma_1} \cdots \partial_m^{\sigma_m} \mid \sigma = (\sigma_1, \ldots, \sigma_m) \in \mathbb{Z}_+^m\}$ in the universal

*Current affiliation: The University of Tennessee Space Institute, Tullahoma, Tennessee

enveloping algebra $U(\mathfrak{G})$, which we consider as a subalgebra in the algebra Diff K/k of differential operators on K/k.

For any $\sigma, \mu \in \mathbb{Z}_+^m$, we have

$$\partial^\sigma \partial^\mu = \partial^{\sigma+\mu} + c^\tau_{\sigma,\mu} \partial^\tau \tag{2.1}$$

for some $c^\tau_{\sigma,\mu} \in k$ (summation is always taken over repeated indices). The quantities $c^\tau_{\sigma,\mu}$ satisfy the equations

$$c^\tau_{\sigma,\mu} = 0, \quad \text{for } |\tau| \geq |\sigma| + |\mu| \tag{2.2}$$

where $|\sigma| := \sigma_1 + \cdots + \sigma_m$. For $|\sigma| = |\mu| = |\tau| = 1$, $c^\tau_{\sigma,\mu}$ are the structure constants of $\mathfrak{G}$. Also,

$$\partial^\sigma \partial^0 = \partial^\sigma \tag{2.3}$$

(Informally: for $k = \mathbb{R}$, $K = C^\infty(G)$, $\partial_1, \ldots, \partial_m$—left-invariant vector fields on G.)

Let J be a finite set,

$$|J| < \infty \tag{2.4}$$

Define $C := K[q_j^{(\sigma)}]$, $j \in J$, $\sigma \in \mathbb{Z}_+^m$. We extend the derivations $\partial_1, \ldots, \partial_m$ to C by the formulae

$$\partial^\sigma(q_j^{(\mu)}) = q_j^{(\sigma+\mu)} + c^\tau_{\sigma,\mu} q_j^{(\tau)} \tag{2.5}$$

Denoting

$$q_j = q_j^{(0)}$$

we obtain from (2.3), (2.5)

$$\partial^\sigma(q_j) = q_j^{(\sigma)} \tag{2.6}$$

(Informally: C is the ring of functions on the jet bundle of the bundle $\mathbb{R}^{|J|} \times G \to G$.)

Let $\Omega^1(C) = \{f_j^\sigma \, dq_j^{(\sigma)} \mid f_j^\sigma \in C, \text{ finite sums}\}$ be the C-module of 1-forms with the usual derivation $d : C \to \Omega^1(C)$ over K acting as $d : q_j^{(\sigma)} \mapsto dq_j^{(\sigma)}$. We extend the action of ∂_i's on $\Omega^1(C)$ such that the following diagram commutes:

$$d \circ \partial^\sigma = \partial^\sigma \circ d, \quad \forall \sigma \in \mathbb{Z}_+^m \tag{2.7}$$

Define $\operatorname{Im} \mathscr{D} := \Sigma_i \operatorname{Im} \partial_i$ (in C and in $\Omega^1(C)$). From (2.7) we obtain

$$d(\operatorname{Im} \mathscr{D}) \subset \operatorname{Im} \mathscr{D} \tag{2.8}$$

We shall write $a \sim b$ if $(a - b) \in \operatorname{Im} \mathscr{D}$. Elements of $\operatorname{Im} \mathscr{D}$ are called <u>trivial</u>.

Let $\Omega_0^1(C) = \{f_j dq_j \mid f_j \in C\}$ be the module of reduced forms. We define the projection $\hat{\delta} : \Omega^1(C) \to \Omega_0^1(C)$ by

$$\hat{\delta}(f \, dq_j^{(\sigma)}) = dq_j (-\partial)^{\bar{\sigma}}(f) \tag{2.9}$$

where

$$(-\partial)^{\bar{\sigma}} := (-\partial_m)^{\sigma_m} \cdots (-\partial_1)^{\sigma_1} \tag{2.10}$$

Obviously,

$$\hat{\delta}\omega \sim \omega, \quad \forall \omega \in \Omega^1(C) \tag{2.11}$$

Lemma 2.12. $(-\partial)^{\bar{\mu}}(-\partial)^{\bar{\sigma}} = (-\partial)^{\overline{\mu+\sigma}} + c^{\tau}_{\sigma,\mu}(-\partial)^{\bar{\tau}}$.

Proof. Consider $U(\mathfrak{G})$ as the algebra of right-invariant differential operators on G, rather than left-invariant ones. This amounts to reading (2.1) mirror-fashion after the change $\partial_i \mapsto -\partial_i$, which gives (2.12). Q.E.D.

Corollary 2.13. $\hat{\delta}[f\, d(\partial^{\sigma}\partial^{\mu}(q_j)] = dq_j(-\partial)^{\bar{\mu}}(-\partial)^{\bar{\sigma}}(f)$.

Proof. $\hat{\delta}[f\, d(\partial^{\sigma}\partial^{\mu}(q_j))] = \hat{\delta}[f\, d(q_j^{(\sigma+\mu)} + c^{\tau}_{\sigma,\mu}q^{(\tau)})]$

$$= dq_j[(-\partial)^{\overline{\sigma+\mu}}(f) + c^{\tau}_{\sigma,\mu}(-\partial)^{\bar{\tau}}(f)] = [\text{by } (2.12)]$$

$$= dq_j(-\partial)^{\bar{\mu}}(-\partial)^{\bar{\sigma}}(f)$$ Q.E.D.

Lemma 2.14. $\hat{\delta}(\operatorname{Im}\mathcal{D}) = 0$

Proof. $\hat{\delta}[\partial_i(f\, dq_j^{(\sigma)})] = \hat{\delta}[\partial_i(f)\, dq_j^{(\sigma)} + f\, d(\partial_i\partial^{\sigma}(q_j))] = [\text{by } (2.13)]$

$$= dq_j\{(-\partial)^{\bar{\sigma}}\partial_i(f) + (-\partial)^{\bar{\sigma}}(-\partial_i)(f)\} = 0$$ Q.E.D.

Definition 2.15. (i) The Euler-Lagrange operator δ is $\delta = \hat{\delta}d: C \to \Omega^1_0(C)$.

(ii) The functional derivatives $\delta/\delta q_j: C \to C$ are given by $\delta/\delta q_j = (-\partial)^{\bar{\sigma}}(\partial/\partial q_j^{(\sigma)})$.

Corollary 2.16. For any $H \in C$, $\delta H = dq_j(\delta H/\delta q_j)$.

Corollary 2.17. $(\delta/\delta q_j)(\operatorname{Im}\mathcal{D}) = 0$.

Proof. For $H \in \operatorname{Im}\mathcal{D}$, $dH \in \operatorname{Im}\mathcal{D}$ by (2.8), thus $dq_j(\delta H/\delta q_j) = \delta H = \hat{\delta}(dH) = 0$ by (2.14). Q.E.D.

Lemma 2.18. If $f \in C$ and $fC \sim 0$, then $f = 0$.

Proof. Choose n such that $\partial f/\partial q_j^{(\sigma)} = 0$, $\forall\sigma$ with $|\sigma| \geq n$. For any $N > n$, consider $g = f(q_j^{(N,0,\dots,0)})^2$. Since $g \sim 0$, $\delta g/\delta q_j = 0$ by (2.17). Thus (no sum over j)

$$0 = \frac{\partial}{\partial q_j^{(2N,0,\dots,0)}}\left(\frac{\delta g}{\delta q_j}\right) = \frac{\partial}{\partial q_j^{(2N,\dots,0)}}\left\{(-\partial)^{\bar{\tau}}\left[\frac{\partial f}{\partial q_j^{(\tau)}}(q_j^{(N,\dots,0)})^2\right] + (-\partial_1)^N[2f q_j^{(N,\dots,0)}]\right\} = (-1)^N 2f.$$ Q.E.D.

Let Der (C) denote the set of all derivations of C over K: $\operatorname{Der}(C) = \{X = X_j^{\sigma}(\partial/\partial q_j^{(\sigma)}) \mid X_j^{\sigma} \in C$, infinite sums are allowed$\}$. Let $D^{ev}(C)$ be the subalgebra of derivations in Der (C) which commute with $\mathfrak{G}$: $D^{ev}(C) = \{X = \partial^{\sigma}(X_j)\cdot\partial/\partial q_j^{(\sigma)} \mid X_j = X(q_j) \in C\}$.

Lemma 2.19. If $\omega \in \Omega^1_0(C)$ and $\omega(D^{ev}(C)) \sim 0$ then $\omega = 0$.

Proof. Let $\omega = f_j dq_j$ and suppose that not all f_j vanish; say, $f_1 \neq 0$. Then $\omega(\partial^{\sigma}(g)(\partial/\partial q_1^{(\sigma)})) = f_1 g \sim 0$. Hence $f_1 = 0$ by Lemma 2.18, a contradiction. Q.E.D.

Theorem 2.20. a) $\Omega_0^1(C) \cap \operatorname{Im} \mathscr{D} = (0)$; b) for $\omega \in \Omega^1(C)$, $\omega \sim 0$ iff $\omega(D^{ev}(C)) \sim 0$; c) $\hat{\delta} : \Omega^1(C) \to \Omega_0^1(C)$ is uniquely defined by the property $(\hat{\delta}\omega)(X) \sim \omega(X)$, $\forall X \in D^{ev}(C)$.

Proof. For any $\omega \in \Omega^1(C)$, $X \in D^{ev}(C)$,

$$[(\hat{\delta} - 1)\omega](X) \sim 0 \tag{2.21}$$

Indeed, $(\hat{\delta} - 1)[f\, dq_j^{(\sigma)}] = dq_j(-\partial)^{\bar{\sigma}}(f) - f\, dq_j^{(\sigma)}$, so $[(\hat{\delta} - 1)\omega](X) = X_j(-\partial)^{\bar{\sigma}}(f) - f\, \partial^{\sigma}(X_j) \sim 0$.

(b) $\Rightarrow$: Since $\omega \sim 0$, $\hat{\delta}\omega = 0$ by (2.14), thus $\omega = -(\hat{\delta} - 1)\omega$ and we apply (2.21);

(a) If $\omega \in \operatorname{Im} \mathscr{D}$, then $\omega(D^{ev}(C)) \sim 0$ by b)$\Rightarrow$, and by lemma (2.19) $\omega = 0$;

(c) Uniqueness follows from lemma 2.19, and existence follows from (2.11) and (b)$\Rightarrow$;

(b) $\Leftarrow$: Since $\omega \sim \hat{\delta}\omega$ and $\omega(D^{ev}(C)) \sim 0$, then $(\hat{\delta}\omega)(D^{ev}(C)) \sim 0$ by (b)$\Rightarrow$. Thus $\hat{\delta}\omega = 0$ by lemma 2.19. Hence $\omega \sim \hat{\delta}\omega = 0$. Q.E.D.

Remark 2.22. Part (c) of the above theorem provides a connection with the traditional approach to the calculus of variations, where one has real "variations." To see this notice that $\forall H \in C$, $\forall X \in D^{ev}(C)$,

$$X(H) = (dH)(X) \sim (\delta H)(X) = \bar{X}^t \frac{\delta H}{\delta \bar{q}} \tag{2.23}$$

where $\bar{X}$ is the vector column $\{X_j\}$, "t" stands for "transpose," and $\delta H/\delta \bar{q}$ is the vector $\{\delta H/\delta q_j\}$. Thus we obtain in (2.23) the (local part of the) formula for the first variation.

3. FIRST COMPLEX FOR THE OPERATOR δ

It is easy to show that $\operatorname{Ker} \delta = \operatorname{Im} \mathscr{D} + K$. The classical problem of the calculus of variations is to describe $\operatorname{Im} \delta$.

Let $\bar{\mathfrak{G}} = \mathfrak{G} \oplus k$ be the one-dimensional trivial central extension of $\mathfrak{G}$. We denote the new basis vector in $\bar{\mathfrak{G}}$ by ∂_{m+1} and we let ∂_{m+1} act trivially on K. Let us write multiindices in $\mathbb{Z}_+^{m+1}$ as pairs $(\sigma|p)$, $\sigma \in \mathbb{Z}_+^m$, $p \in \mathbb{Z}_+$. Denote $\bar{C} = K[q_j^{(\sigma|p)}]$, and repeat all the constructions of the preceding section with the ring $\bar{C}$. The new Euler-Lagrange operator will be denoted by δ^1.

Notice that we have a natural imbedding $\bar{\tau} : C \hookrightarrow \bar{C}$ defined by $\bar{\tau}(q_j^{(\sigma)}) = q_j^{(\sigma|0)}$. Let us extend $\bar{\tau}$ to the map of the C-module $\Omega^1(C)$ into $\bar{C}$ by

$$\bar{\tau}(dq_j^{(\sigma)}) = q_j^{(\sigma|1)} \tag{3.1}$$

Thus $\bar{\tau}$ commutes with ∂^{σ}, $\sigma \in \mathbb{Z}_+^m$, and hence

$$\bar{\tau}(\operatorname{Im} \mathscr{D}) \subset \operatorname{Im} \mathscr{D} \tag{3.2}$$

Lemma 3.3. $\forall H \in C$, $\partial_{m+1}(\bar{\tau}H) = \bar{\tau}(dH)$.

Proof. $\bar{\tau}(dH) = \bar{\tau}((\partial H/\partial q_j^{(\sigma)})dq_j^{(\sigma)}) = (\partial(\bar{\tau}H)/\partial q_j^{(\sigma|0)})q_j^{(\sigma|1)}$. Q.E.D.

Theorem 3.4. $\delta^1 \bar{\tau} \delta = 0$.

Proof. $\forall H \in C$, $\delta H = \hat{\delta}\, dH \sim dH$, thus $\bar{\tau}\delta H \sim \bar{\tau}\, dH$ by (3.2), hence $\delta^1 \bar{\tau}\, dH = \delta^1 \partial_{m+1} \bar{\tau} H = 0$ by lemmas 3.3 and 2.14. Q.E.D.

Thus we obtain the complex $C \xrightarrow{\delta} \Omega_0^1(C) \xrightarrow{\delta^1 \bar{\tau}} \Omega_0^1(\bar{C})$. We are now going to show that this is an exact sequence.

For any $f \in C$, define $D_j(f) := (\partial f/\partial q_j^{(\sigma)})\partial^\sigma$. If $\bar{R}$ is a (column) vector with elements in C, we denote by $D(\bar{R})$ the matrix differential operator with entries

$$[D(\bar{R})]_{ij} = D_j(R_i) \tag{3.5}$$

The matrix $D(\bar{R})$ is called the Fréchet derivative of $\bar{R}$. It can be defined by the property

$$D(\bar{R})\bar{X} = X(\bar{R}), \quad \forall\, X \in D^{ev}(C) \tag{3.6}$$

Recall that if $F: C^N \to C^M$ is a matrix differential operator, its adjoint $F^*: C^M \to C^N$ is defined by the property

$$u^t F v \sim (F^* u)^t v, \quad \forall u \in C^N, \ \forall v \in C^M \tag{3.7}$$

and we treat all vectors as columns unless specified otherwise. If $F = (F_{ij})$, then $(F^*)_{ij} = (F_{ji})^*$; also, $(f\partial^\sigma)^* = (-\partial)^{\bar{\sigma}} f$, $f \in C$.

Theorem 3.8. For any $H \in C$, the operator $D(\delta H/\delta \bar{q})$ is symmetric.

Proof. By theorem 3.4, $0 = \delta^1 \bar{\tau}\delta H = \delta^1 \bar{\tau}(dq_j(\delta H/\delta q_j)) = \delta^1[\bar{\tau}(\delta H/\delta q_j)q_j^{(0|1)}] = dq_i\{(-\partial)^{\bar{\sigma}}[(\partial/\partial q_i^{(\sigma|0)})(\bar{\tau}(\delta H/\delta q_j)) \cdot q_j^{(0|1)}] - \partial_{m+1}\bar{\tau}(\delta H/\delta q_i)\}$, hence

$$(-\partial)^{\bar{\sigma}}\left[\bar{\tau}\frac{\partial}{\partial q_i^{(\sigma)}}\left(\frac{\delta H}{\delta q_j}\right)\cdot q_j^{(0|1)}\right] = \bar{\tau}\frac{\partial}{\partial q_j^{(\sigma)}}\left(\frac{\delta H}{\delta q_i}\right)\cdot \partial^\sigma q_j^{(0|1)} \tag{3.9}$$

Since $q_j^{(0|1)}$ are free independent variables and $\bar{\tau}$ is injective, (3.9) implies

$$(-\partial)^{\bar{\sigma}}\cdot\frac{\partial}{\partial q_i^{(\sigma)}}\left(\frac{\delta H}{\delta q_j}\right) = \frac{\partial}{\partial q_j^{(\sigma)}}\left(\frac{\delta H}{\delta q_i}\right)\partial^\sigma$$

or

$$\left[\frac{\partial}{\partial q_i^{(\sigma)}}\left(\frac{\delta H}{\delta q_j}\right)\partial^\sigma\right]^* = D_j\left(\frac{\delta H}{\delta q_i}\right)$$

or

$$\left[D_i\left(\frac{\delta H}{\delta q_j}\right)\right]^* = \left(\left[D\left(\frac{\delta H}{\delta \bar{q}}\right)\right]_{ji}\right)^* = \left[D\left(\frac{\delta H}{\delta \bar{q}}\right)^*\right]_{ij} = \left[D\left(\frac{\delta H}{\delta \bar{q}}\right)\right]_{ij} \qquad \text{Q.E.D.}$$

Theorem 3.10. Let $\bar{R} \in C^N$, $N = |J|$. If $D(\bar{R})$ is symmetric then $\exists H \in C$ such that $\bar{R} = \delta H/\delta\bar{q}$.

Proof. Define a map $A_t: C \to C[t]$ by $A_t(f) = t^{\deg f} f$ for homogeneous f; here $\deg(\bar{k}q_{j_1}^{(s_1)} \cdots q_{j_n}^{(s_n)}) = n$, $\bar{k} \in K$. Now set

$$H = q_i \int_0^1 A_t(R_i)\, dt \tag{3.11}$$

We want to show that $\delta H/\delta\bar{q} = \bar{R}$. By theorem 2.20(c) in the form (2.23), this is equivalent to $X(H) \sim \bar{X}^t\bar{R}$, $\forall X \in D^{ev}(C)$. We have

$$X(H) = X_i \int_0^1 A_t(R_i)\, dt + q_i \int_0^1 X(A_t(R_i))\, dt \tag{3.12}$$

Notice that if $D(\bar{R})$ is symmetric, then so is $D(A_t(\bar{R}))$. Hence we can transform the second summand in (3.12) as

$$\int_0^1 q_i D_j(A_t(R_i)) X_j \, dt = \int_0^1 \bar{q}^t [D(A_t(\bar{R})] \bar{X} \, dt \sim \int_0^1 \bar{X}^t [D(A_t(\bar{R}))] \bar{q} \, dt$$

$$= X_i \int_0^1 D_j(A_t(R_i)) q_j \, dt$$

Thus

$$X(H) \sim X_i \left\{ \int_0^1 dt [A_t(R_i) + D_j(A_t(R_i)) q_j] \right\} \tag{3.13}$$

Now we notice that

$$\frac{\partial}{\partial q_j^{(\sigma)}} A_t = t A_t \frac{\partial}{\partial q_j^{(\sigma)}} \tag{3.14}$$

Thus $D_j(A_t(R_i))q_j = (\partial/\partial q_j^{(\sigma)})[A_t(R_i)] \cdot \partial^{\sigma}(q_j) = tA_t(\partial R_i/\partial q_j^{(\sigma)})q_j^{(\sigma)} = A_t((\partial R_i/\partial q_j^{(\sigma)})q_j^{(\sigma)})$. Hence (3.13) becomes

$$X(H) \sim X_i \int_0^1 dt \, A_t \left[R_i + \frac{\partial R_i}{\partial q_j^{(\sigma)}} q_j^{(\sigma)} \right] \tag{3.15}$$

If f is a homogeneous polynomial in C of degree n, then $A_t(f + (\partial f/\partial q_j^{(\sigma)})q_j^{(\sigma)}) =$ $A_t(n+1)f = t^n(n+1)f$, thus $\int_0^1 dt \, A_t(\mathbb{1} + q_j^{(\sigma)}(\partial/\partial q_j^{(\sigma)})) = \mathbb{1}$, and therefore (3.15) yields $X(H) \sim X_i R_i$. Q.E.D.

<u>Remark 3.16</u>. For $|J| = \infty$ theorem 3.10 fails. In addition to $D(\bar{R})$ being symmetric, one also needs some growth conditions on the R_i's.

4. GENERALIZATIONS

The idea of noncommutativity in the calculus of variations can be arrived at from a different perspective, namely from the discrete calculus of variations [2].

Let $\hat{G}$ be a countable group which acts by automorphisms $\{\hat{g}\}$ on K/k, $\hat{g} \in \hat{G}$. Set $C = K[q_j^{(g)}]$, $j \in J$, $\hat{g} \in G$, extend $\hat{G}$ to act on C, $\Omega^1(C)$, etc. Define Im $\mathscr{D}$ = $\Sigma_{\hat{g} \in \hat{G}} \operatorname{Im}(\hat{g} - 1)$. Once again one defines $\hat{\delta} : \Omega^1(C) \to \Omega_0^1(C) \approx \Omega^1(C)/\operatorname{Im} \mathscr{D}$ by the formula $\hat{\delta}(f \, dq_j^{(g)}) = dq_j \hat{g}^{-1}(f)$. This results in the formulae for the functional derivatives

$$\frac{\delta}{\delta q_j} = \sum_{\hat{g} \in \hat{G}} \hat{g}^{-1} \frac{\partial}{\partial q_j^{(g)}} \tag{4.1}$$

When $\hat{G}$ is abelian, this case was considered in [2]. However, one can go one step further. As was noticed in ([3], ch. VII) (for the abelian case), one can have both differential and discrete degrees of freedom.

The appropriate set-up can be briefly outlined as follows. Let $\mathfrak{G}$ act by derivations on K/k and $\hat{G}$ act by automorphisms. Assume that those actions are compatible in the sense that the ring of operators generated by all these actions can be generated, as a k-vector space, by the operators of the form $U(\mathfrak{G}) \circ \hat{G}$. For example one can have the following possibilities:

1) $\partial_i \circ \hat{g} = \hat{g} \circ \partial_i, \quad \forall \hat{g} \in \hat{G}, \ \forall \partial_i \in \mathfrak{G}$ (4.2)

2) Let $\sigma: \hat{G} \to \mathrm{Aut}(\mathfrak{G})$ be an antihomomorphism, and

$$\partial_i \circ \hat{g} = \hat{g} \circ (\sigma(\hat{g})(\partial_i)), \quad \forall \hat{g} \in \hat{G}, \ \forall \partial_i \in \mathfrak{G} \tag{4.3}$$

3) Let $\mathfrak{G}$ be commutative let $\theta: \hat{G} \to \hat{G}$ be an automorphism, and

$$\partial_i \circ \hat{g} = \theta(\hat{g}) \circ \partial_i \tag{4.4}$$

Defining $C = K[q_j^{(\sigma|g)}]$, $\sigma \in \mathbb{Z}_+^m$, $\hat{g} \in \hat{G}$, $\Omega^1(C)$, etc., one arrives at the following formulae for the functional derivatives

$$\frac{\delta}{\delta q_j} = \sum \hat{g}^{-1}(-\partial)^{\bar{\sigma}} \frac{\partial}{\partial q_j^{(\sigma|g)}} \tag{4.5}$$

I shall not dwell on the details. The results of Sections 2 and 3 hold true after the obvious modifications are made. From the point of view of applications, the main results about Lie algebras over rings with calculus are the same as in the abelian case ([3], ch. VIII): (a) one-to-one correspondence between Lie algebras and linear Hamiltonian operators; (b) one-to-one correspondence between generalized 2-cocycles on Lie algebras and affine Hamiltonian operators; (c) generation of canonical quadratic maps associated with representations of Lie algebras.

In conclusion, let me mention that if we start off with a nontrivial bundle $\pi: E \to G$ on which G still acts, it is again possible to reduce the calculus of variations on $J^\infty\pi$ onto the fiber over $e \in G$, resulting in a sort of Yang-Mills-Lagrangian formalism. But I will leave this to another occasion.

REFERENCES

1. Kupershmidt, B. A.: Geometry of jet bundles and the structure of Lagrangian and Hamiltonian formalisms, in Lect. Notes Math. No. 775. Springer (1980), pp. 162-218.
2. Kupershmidt, B. A.: Lett. Math. Phys. 6, 85-89 (1982).
3. Kupershmidt, B. A.: Discrete Lax Equations and Differential-Difference Calculus, E.N.S. Lecture Notes, Paris (1982).
4. Manin, Yu. I.: J. Sov. Math. 11, 1-152 (1979).

19 ON THE ROLE OF RECIPROCITY CONDITIONS IN THE FORMULATION OF CONSERVATION LAWS AND VARIATIONAL PRINCIPLES

Ernesto A. Lacomba / Department of Mathematics, Universidad Autónoma Metropolitana—Iztapalapa, México, D. F., Mexico

Diego Bricio Hernández* / Department of Process Engineering and Hydraulics, Universidad Autónoma Metropolitana—Iztapalapa, México, D. F., Mexico

1. INTRODUCTION

This paper is intended to explore the various interconnections between concepts such as reciprocity, variational principle, and conservation law (all of them full of physical content) as well as the important mathematical notion of selfadjointness of linear operators. The main discussion is preceded by preliminaries such as exterior calculus on Banach manifolds (Sec. 3), incorporating an elementary account of symplectic structures and Lagrangian submanifolds which does not appear to be available elsewhere. This is done in Sec. 4 (Geometry and Reciprocity in Physics), where a physical formulation involving a closed differential 1-form is seen to imply reciprocity (a local notion, physically meaningful) as well as the Lagrangian character of a submanifold annihilating a suitable symplectic form, a mathematical notion providing a global counterpart of reciprocity. This part follows the work of Tulczyjew et al. [19, 31] rather closely, the motivations being afforded by a presentation of thermostatics inspired in Callen [7].

Section 5 (Reciprocity and Conservation Laws) is largely devoted to showing that conservation appears whenever reciprocity and a bit more can be asserted, namely existence of an integrating factor guaranteeing complete integrability of a suitable differential 1-form. Reciprocity follows naturally in this case from the ensuing closedness of the form once it is multiplied by such factor. The relevant mathematical tool is the Frobenius Integration Theory, which is presented specifically for 1-forms together with Caratheodory's Inaccessibility Theorem. The whole machinery is put to work in giving a formulation of thermostatics in the spirit of the late J. M. Jauch (Geneve), a problem whose solution is interwoven with the mathematical developments along the section.

The paper is concluded in Sec. 7 with some comments on the complementarity of variational principles, a natural question to consider in connection with the inverse problem of the calculus of variations. These principles were previously introduced in Sec. 6, special care being afforded to the role played by selfadjointness in establishing the existence of a variational principle associated with a given differential equation. One of the main theses of this work being that both variational principles and conservation laws are concrete instances of the mathematical concept of closedness of a differential 1-form, special care is placed here on clarifying this idea.

The key notion appearing at the root of so many physical formulations is that of reciprocity, embodying a certain equilibrium and complementarity in Nature that must find its counterpart in any mathematical formulation of physical phenomena. This ubiquitous notion is given ample coverage in Sec. 2 (Reciprocity in Physics) where its connections with the thermodynamics of irreversible processes are expounded at large, illustrating the general discussion with examples taken from various fields of physics and chemistry. In turn, this important physical notion is globalized into the mathematical notion of Lagrangian submanifold, this being done in Sec. 4, as already indicated. This is one of the main points of this paper.

---

*Current affiliation: Department of Mathematics

The treatment is reasonably selfcontained, but an acquaintance with basic concepts about differentiable manifolds is expected on the part of the reader.

2. RECIPROCITY PRINCIPLES IN PHYSICS

Physics and chemistry provide a large number of examples of systems in which various phenomena occur concurrently and simultaneously. Typical examples in abundance are provided by the physical chemistry of multicomponent mixtures subject to both temperature and concentration gradients. Their behavior is given ample treatment in the chemical engineering literature where substantial application is made of them. See, e.g., Bird et al. [2]. Energy (and matter) flow in the direction of decreasing temperature (resp. concentration), but interaction between these two phenomena do occur. These are the Dufour and Soret effects, namely energy flow due to concentration gradients and mass flow due to temperature gradients—in addition to the main effects, molecular diffusion and heat conduction ([2], p. 564).

A number of thermoelectric and thermomagnetic effects are described in most thermodynamics texts (see, e.g., [7], Chap. 17). For instance, the Seebeck effect refers to the production of an electromotive force in a thermocouple. Conversely, the Peltier effect refers to the evolution of heat accompanying the flow of an electric current across an isothermal junction of two materials. In addition there is the Thompson effect, consisting in the evolution of heat as an electric current traverses a temperature gradient in a material. In addition to these, Callen mentions a host of thermomagnetic effects, referring to heat flowing in the presence of a magnetic field in spite of the material being isothermal, as well as magnetic fields being induced by temperature gradients.

The first thermodynamics treatment of the various thermoelectric effects was given by Lord Kelvin in a series of papers [28]. Lord Kelvin based his analysis upon the two basic laws of thermodynamics plus an additional assumption concerning the "reversible contributions to the process." Attempts to establish this additional hypothesis have been unsuccessful and in fact it is known now that this hypothesis cannot be justified [11]. Nevertheless, L. Onsager succeeded in 1931 in giving a correct proof of Lord Kelvin's results. Onsager based his own analysis upon the assumption of microscopic reversibility, itself the cornerstone of the modern discipline of thermodynamics of irreversible processes [25]. This is a very general theory encompassing all the various phenomena mentioned thus far and many others.

Briefly, this discipline starts out by assuming the existence of a function giving the entropy of the system in terms of the extensive parameters characterizing the latter, say n of them. Thus, the state space of the system of interest is an open subset of $\mathbb{R}^n$, to be denoted in what follows by Σ, coordinates of states $\sigma \in \Sigma$ being the aforementioned extensive parameters. In addition, there is a C^∞ real function on Σ called S, the entropy, the extra assumption being made that this function is identical with entropy of equilibrium states. This last assumption is normally referred to as that of local equilibrium.

Letting the coordinates in Σ be $X_1, \ldots, X_n$, the intensive parameters $\phi_1, \ldots, \phi_n$ are defined by

$$\phi_k = \frac{\partial S}{\partial X_k} \qquad k = 1, \ldots, n$$

By the local equilibrium assumption, the functional relations giving the intensive parameters for every $\sigma \in \Sigma$ is identical with the corresponding equilibrium relations.

A <u>process</u> in Σ is a C^∞-map $\gamma : [0, \infty) \to \Sigma$, each $\gamma(t)$ being "the state of the system at time t during process γ." The instantaneous rate of entropy production during process γ is then given by

$$\dot{S}_\gamma(t) = \sum_{k=1}^{n} \phi_k \dot{X}_k(t) \tag{2-1}$$

Associated with each intensive parameter X_k there is a C^∞ three dimensional time varying vector field on an open set $\Omega \subset \mathbb{R}^3$, the space representation for the system under study. Let $j_k : [0,\infty) \times \Omega \to \mathbb{R}^3$ be such vector field, subject to the interpretation $j_k(t,p) :=$ vector flux of the k-th extensive parameter per unit time and unit area, measured at instant t and location p.

At this point it is convenient to invoke the assumption of local equilibrium in the following form: each $p \in \Omega$ acts as an independent system with the same state space Σ but undergoing its own individual process γ_p, with $(t,p) \longmapsto \gamma_p(t)$ a C^∞ map. Then entropy is defined on a pointwise basis, thus defining a function $S: \Omega \times \Sigma \to \mathbb{R}$, also assumed to be of class C^∞. In addition, the local equilibrium assumption requires that

$$S(p,\sigma) = S(q,\sigma) \qquad \forall\, \sigma \in \Sigma$$

for every choice of p, $q \in \Omega$.

The rate of entropy production at point p is then

$$\frac{\partial S}{\partial t}(p, \gamma_p(t)) = \sum_{k=1}^{n} \phi_k(p, X_1(t,p), \ldots, X_n(t,p)) \frac{\partial X_k}{\partial t}(t,p) \tag{2-2}$$

Now the various extensive parameters can be neither produced nor destroyed, a remark translating into the corresponding continuity equation [2]

$$\frac{\partial X_k}{\partial t} + \operatorname{div} j_k = 0 \tag{2-3}$$

Equation (2-1) suggests a definition of entropy current density as the time dependent vector field on $\Omega \times \Sigma$ given by

$$j_s = \sum_{k=1}^{n} \phi_k j_k \tag{2-4}$$

An entropy balance similar to that giving rise to the continuity equation (2-3) would yield

$$\frac{\partial S}{\partial t} + \operatorname{div} j_s = \dot{S} \tag{2-5}$$

the right hand side denoting the local rate of entropy production during process γ, a C^∞ function on $[0,\infty) \times \Omega$.

By combining the last four equations one readily gets

$$\dot{S}_\gamma(t,p) = \sum_{k=1}^{n} \nabla\phi_k(t, X_1(t,p), \ldots, X_n(t,p)) \cdot j_k(t,p)$$

each $\nabla\phi_k$ being termed the corresponding affinity, hereafter denoted by α_k, thus giving the local rate entropy production during process γ by

$$\dot{S}_\gamma = \sum_{k=1}^{n} \alpha_k \cdot j_k \tag{2-6}$$

Now, for a large number of materials, the various fluxes are related to the state and affinity variables by a function

$$j_k = J_k(X_1, \ldots, X_n, \alpha_1, \ldots, \alpha_n) \tag{2-7}$$

A natural requirement on J_k is that it should satisfy

$$J_k(X_1, \ldots, X_n, 0, \ldots, 0) = 0$$

hence

$$j_k = \sum_{\ell=1}^{n} \frac{\partial J_k}{\partial \alpha_\ell} \alpha_\ell + \cdots \tag{2-8}$$

The partial derivatives in (2-8) are evaluated at points $(X_1, \ldots, X_n, 0, \ldots, 0)$, the terms omitted being those of second and higher order.

Let

$$L_{ij}(X_1, \ldots, X_n) = \frac{\partial J_i}{\partial \alpha_j}(X_1, \ldots, X_n, 0, \ldots, 0)$$

denote the so called kinetic coefficient. Thus, up to first order terms

$$j_k = \sum_{m=1}^{n} L_{km} \alpha_m \tag{2-9}$$

The local rate of entropy production is thus given by the quadratic form

$$\dot{S}_\gamma = \sum_{i,j=1}^{n} \alpha_i L_{ij} \alpha_j \tag{2-10}$$

Note that subscript γ in $\dot{S}_\gamma$ justified, insofar as each term in the right-hand side depends on the extensive parameters traversed by the system during the process. Irreversible thermodynamics then proceeds by choosing a process γ^* satisfying

$$\dot{S}_{\gamma^*} = \sup \dot{S}_\gamma \tag{2-11}$$

as the one followed by the system. Maximization in (2-11) is carried out over the class of all processes satisfying all external constraints upon the system [7].

Let's go back to Eq. (2-9), the so-called phenomenological equation. By applying the concept of microscopic reversibility to them—a statistical mechanical notion related to symmetry of physical laws under time reversal—Onsager proved that

$$L_{km} = L^t_{mk} \tag*{(2-12)'}$$

that is,

$$\frac{\partial J_k}{\partial \alpha_m} = \frac{\partial J_m}{\partial \alpha_k} \tag*{(2-12)''}$$

from which Lord Kelvin's relations for the various thermoelectric phenomena can be proved to follow [11]. Analogously, equations expressing symmetry between the Dufour and Soret effects can be given suitably specializing Eq. (2-11)—see p. 567 of [2].

The reciprocity embodied in Onsager's relations (2-12) occur in many fields distinct from thermodynamics of irreversible processes but akin to it, such as the following ones:

a) Thermostatics, in the form of Maxwell's relations

$$\left(\frac{\partial V}{\partial S}\right)_p = \left(\frac{\partial T}{\partial p}\right)_S, \quad \left(\frac{\partial S}{\partial V}\right)_T = \left(\frac{\partial p}{\partial T}\right)_V, \quad \left(\frac{\partial S}{\partial p}\right)_T = \left(\frac{\partial V}{\partial T}\right)_p$$

[7], each of them being equivalent to the existence of a thermodynamic potential.

b) Particle mechanics, say the n dimensional dynamical system

$$m\ddot{x} + F(x) = 0 \tag{2-13}$$

the relevant reciprocity condition being

$$\frac{\partial F_i}{\partial x_j} = \frac{\partial F_j}{\partial x_i}$$

a condition known to be equivalent to the existence of a potential energy function V such that

$$F(x) = \text{grad } V(x)$$

In turn, this is equivalent to total energy E being an invariant of the motion, i.e.,

$$E = \frac{1}{2} m|\dot{x}|^2 + V(x) \tag{2-14}$$

is constant along the trajectories of (2-13).

c) Hamiltonian systems, described in terms of canonical coordinates p, q. For these systems, Lagrangian and Hamiltonian functions of the coordinates can be constructed such that

$$H + L = p \cdot \dot{q}$$

along any trajectory $t \mapsto (p(t), q(t))$. For the Lagrangian

$$dL = \dot{p} \cdot dq + p \cdot d\dot{q}$$

whereas the Hamiltonian satisfies

$$dH = -\dot{p} \cdot dq + \dot{q} \cdot dp$$

The corresponding reciprocity equations are then

$$\frac{\partial \dot{p}}{\partial \dot{q}} = \left(\frac{\partial p}{\partial q}\right)^t, \qquad \frac{\partial \dot{p}}{\partial p} = -\left(\frac{\partial \dot{q}}{\partial q}\right)^t$$

respectively [5].

d) Electric circuits. Let an electric network consist of n independent meshes, each an LRC circuit. Assume isothermal operation, i.e., all Joule effect heat is removed as soon as it is generated. Let the total voltage impressed on the meshes be $U_1, \ldots, U_n$. By Kirchoff's laws, each of them can be related to the meshes parameters (inductance, resistance, and capacitance), the corresponding relation being

$$U_i(t) = \sum_{k=1}^{n} \left[\frac{1}{C_{ik}} q_k(t) + R_{ik}\dot{q}_k(t) + L_{ik}\ddot{q}(t)\right] \tag{2-15}$$

Here $q_k(t)$ = charge that has traversed mesh k of network up to time t assuming $q_k(-\infty) = 0$. Classical circuit analysis [17] then shows that

$$C_{ik} = C_{ki}, \quad R_{ik} = R_{ki}, \quad L_{ik} = L_{ki} \tag{2-16}$$

Such symmetry conditions are shown to follow from a suitably generalized version of Eqs. (2-12) by Meixner in [23].

Thus we see that reciprocity relations are rather frequent in physics, this bearing witness to symmetry in nature. The next sections will be devoted to showing some mathematical features underlying reciprocity. For instance, reciprocity will be related to the fact that Σ can be lifted to a Lagrangian submanifold of a suitable symplectic space.

3. DIFFERENTIAL FORMS AND EXTERIOR CALCULUS

This section is devoted to developing the notions of differential forms on Banach manifolds and exterior calculus on them. Only 0, 1, and 2-forms will be examined, focusing our attention upon their exactness and closedness properties, i.e., Poincaré's Lemma and its converse. Most of these properties being local, often there will be no loss of generality in working on a vector (Banach) space, so we shall follow Cartan [8] rather closely.

For, let B denote a real Banach space equipped with a continuous symmetric bilinear map $(\ , \ ): B \times B \to \mathbb{R}$. In addition, this map will be assumed to be weakly nondegenerate in the following sense: For each $u \in B$, $(u,v) = 0 \ \forall \ v \in B \Rightarrow u = 0$. Equivalently, the induced continuous linear map $B \to B^*$ given by $u \mapsto u^*$ —with $u^*(v) = (u,v)$ for $v \in B$—is injective. Alternatively $(\ , \ )$ is said to be a weakly nondegenerate inner product if $(u,u) > 0$ for every $u \neq 0$.

Let M be a Banach manifold. A weak Riemannian structure can be induced on M by means of a smooth assignment $x \mapsto (\ , \ )_x$ on M, the image of every $x \in M$ being a weakly nondegenerate inner product on the tangent space T_xM. Incidentally, note that the map $T_xM \mapsto (T_xM)^*$ given by $u \to (u, \ )$ need not be onto: the usual L_2 inner product

$$(u,v) = \int_0^1 u(x)\, v(x)\, dx$$

is a weak Riemannian metric on $C([0,1], \mathbb{R}) = :B$ with the supremum norm, yet the induced linear map $u \mapsto (u, \cdot)$ maps B onto an isomorphic copy of itself, a proper subset of B^* in view of Riesz's representation theorem [21]. A continuous symmetric, bilinear map $(\cdot, \cdot)$ on a Banach space B is nondegenerate if the above linear map is indeed onto. There is no distinction in finite dimension.

A differential 0-form on a Banach manifold M is a C^∞ real function on M. Since the Banach manifold M is modeled on a Banach space B, the charts on M will be ordered pairs (U, ϕ), with U a neighborhood in M and $\phi: U \to B$, the system of charts satisfying the usual compatibility and smoothness assumptions [3]. Then, "$f: M \to \mathbb{R}$ is a 0-form on M" means that $f \circ \phi^{-1}$ is C^∞ on $\phi(U)$, an open subset of B.

Given a 0-form f on M, its differential df,

$$df_x: T_xM \to T_{f(x)}\mathbb{R} = \mathbb{R}, \qquad x \in M$$

is defined for each $v \in T_xM$ and for each chart (U, ϕ) at x by

$$df_x(v) = f'_\phi(y)\, \phi_* v \tag{3-1}$$

In (3-1), f_ϕ is the local representative of f, given by $f_\phi : = f \circ \phi^{-1}$, taken with respect to chart (U, ϕ) with $x \in U$. In addition, $y = \phi(x) \in B$, the derivative being understood in the sense of Fréchet [21]. Moreover, $\phi_* v$ stands for $(\phi \circ \gamma)'(0) \in B$, where $\gamma: \mathbb{R} \to M$ is any curve whose velocity vector at 0 is v. Note that each df_x is a linear map on the corresponding T_xM, so that df is a smooth assignment $x \mapsto df_x$ of a linear map on T_xM to every $x \in M$.

There are many other instances of maps ω on M, assigning a continuous linear form on T_xM to each $x \in M$ in a smooth fashion. Each such map will be called a 1-form on M, its value on an element $x \in M$ being denoted by $\langle \omega(x), \cdot \rangle$. With respect to the local chart (U, ϕ) at x, a 1-form ω has a local representative ω_ϕ, its value $\omega_\phi(y)$ at $y = \phi(x) \in \phi(U)$ being defined by the equation

$$\langle \omega(x), u \rangle = \langle \omega_\phi(y), \ \phi_* u \rangle \tag{3-2}$$

The following example of a 1-form will be very important in the sequel. First a few definitions.

A manifold domain of a Banach manifold M is a dense subset D of M which is a Banach manifold on its own right, such that the inclusion map $i: D \to M$ is C^∞ and its differential has a dense range everywhere [22].

A densely defined operator on M is a vector field F on D, that is, a map $F: D \to TM$ such that $F(x) \in T_xM$ for every $x \in D$.

Example. Let F be a densely defined operator on a Banach manifold M equipped with a weak Riemannian structure. Then F induces a 1-form ω_F on D, the domain of F, given by

$$\langle \omega_F(x), v \rangle = (F(x), v)_x, \qquad v \in T_xD, \quad x \in D \tag{3-3}$$

The concept of a 2-form on a manifold M can be defined quite analogously: just think of assigning to every $x \in M$ a continuous alternating bilinear form on T_xM and doing this smoothly. Thus, the value of a 2-form ω at $x \in M$ is a bilinear form, its value at $(u, v) \in T_xM \times T_xM$ being denoted by

$$\langle u, \omega(x)v \rangle$$

Moreover, each bilinear form $\omega(x)$ must be alternating, in the sense that

$$\langle u, \omega(x)v \rangle + \langle v, \omega(x)u \rangle = 0 \tag{3-4}$$

for every $(u, v) \in T_xM \times T_xM$.

In addition, local representatives for 2-forms can be defined in much the same way as for 1-forms: it suffices to require that

$$\langle u, \omega(x)v \rangle = \langle \phi_* u, \omega_\phi(y)\phi_*(v) \rangle \tag{3-5}$$

for every pair (u, v) and every local chart (U, ϕ) at x, with $y = \phi(x)$.

The concept of a p-form ω on M can be defined in an entirely analogous fashion for $p > 2$. For it suffices to regard each ω as a smooth assignment of a continuous p-linear alternating form $\omega(x)$ on T_xM to every $x \in M$. The value of such a multilinear form at $(v_1, \ldots, v_p) \in (T_xM)^p$ can be most aptly denoted by

$$\omega(x)(v_1, \ldots, v_p)$$

In terms of this notation the alternancy condition (3-4) can be expressed as

$$\omega(x)(v_1, \ldots, v_p) + \omega(x)(v_{\sigma 1}, \ldots, v_{\sigma p}) = 0 \tag{3-6}$$

σ denoting any permutation of $\{1, \ldots, p\}$ exchanging just two elements and leaving the remaining ones fixed. As to the definition of local representative ω_ϕ of such a p-form, (3-5) generalizes to

$$\omega(x)(v_1, \ldots, v_p) = \omega_\phi(y)(\phi_* v_1, \ldots, \phi_* v_p)$$

all symbols maintaining their usual meaning. Let $\Lambda^p(M)$ denote the space of all p-forms on M.

Note that, if M is n-dimensional, then all p-forms on M with $p > n$ are identically zero. This follows from the fact that

$$\omega(x)(v_1, \ldots, v_p) = 0$$

if any two v_i's coincide, as (3-6) shows. Thus we see that $\Lambda^p(M) = 0$ for $p > n$ while $\Lambda^n(M)$ is one dimensional.

In the sequel, most instances of differential forms considered will be either 0-, 1-, or 2-forms. Let's now define the exterior, wedge, or Grassmann product as well as exterior differentiation, two basic operations enabling us to construct new forms from old ones. In the finite dimensional case, the exterior product will allow us to give local representations of 1- and 2-forms.

The exterior product $\wedge$ of differential forms on a Banach manifold M is a map $\Lambda^p(M) \times \Lambda^q(M) \to \Lambda^{p+q}(M)$, defined for p, $q \geq 0$ arbitrary. We only need the precise definitions for the following two cases:

1) If $f \in \Lambda^0(M)$, $\alpha \in \Lambda^1(M)$, then $f \wedge \alpha \in \Lambda^1(M)$ is given by

$$\langle f \wedge \alpha(x), v \rangle = f(x) \langle \alpha(x), v \rangle \tag{3-7}$$

if $v \in T_x M$, $x \in M$.

2) If $\alpha, \beta \in \Lambda^1(M)$, then $\alpha \wedge \beta \in \Lambda^2(M)$ is given by

$$\langle u, (\alpha \wedge \beta)(x) v \rangle = \langle \alpha(x), u \rangle \langle \beta(x), v \rangle - \langle \alpha(x), v \rangle \langle \beta(x), u \rangle \tag{3-8}$$

if u, $v \in T_x M$, $x \in M$.

Note the antisymmetry of $\wedge$ in case 2, namely

$$\alpha \wedge \beta = -\beta \wedge \alpha \tag{3-9}$$

apparent from (3-8).

If M is finite (say n) dimensional, then it is modeled on $\mathbb{R}^n$, whose points will be denoted by the letter y. Letting x_i denote the i-th coordinate projection, $x_i(y) = y_i$, then $dx_i \in \Lambda^1(\mathbb{R}^n)$, for $i = 1, \ldots, n$. In fact, $dx_1, \ldots, dx_n$ constitute a basis of $(\mathbb{R}^n)^*$ at each point $y \in \mathbb{R}^n$, so that each $\omega \in \Lambda^1(M)$ admits a local representation of the form

$$\sum_{i=1}^{n} f_i(y)\, dx_i \tag{3-10}$$

Likewise, $dx_i \wedge dx_j$ for $i < j$ is a basis for 2-forms at each point of $\mathbb{R}^n$, hence any $\omega \in \Lambda^2(M)$ can be locally written as

$$\sum_{i<j} a_{ij}(y)\, dx_i \wedge dx_j \tag{3-11}$$

Local representations of 1- and 2-forms such as (3-10) and (3-11) will appear quite frequently in the applications discussed in Sec. 4 and elsewhere.

Exterior differentiation d operates linearly on $\Lambda^p(M)$ to produce p + 1-forms. Thus, if $f \in \Lambda^0(M)$, df stands for its differential defined in (3-1). If $\omega \in \Lambda^1(M)$, then $d\omega \in \Lambda^2(M)$ is given on local chart (U, ϕ) by

$$\langle u, d\omega_\phi(y) v \rangle = \langle u, \omega'_\phi(y) v \rangle - \langle v, \omega'_\phi(y) u \rangle \tag{3-12}$$

where $y \in \phi(U)$, u, $v \in B$. ω_ϕ stands for the local representative defined in (3-2), primes denoting Fréchet differentiation, so that $\omega'_\phi(y)$ is a continuous bilinear map from B to B^*.

In local coordinates, $(df)_\phi = f'_\phi$, hence by (3-12)

$$\langle u, (d^2 f)_\phi(y) v \rangle = \langle u, f''_\phi(y) v \rangle - \langle v, f''_\phi(y) u \rangle = 0$$

in view of the symmetry of f'' [12]. Thus

$$d^2 f = 0 \quad \text{for any } f \in \Lambda^0(M)$$

It is true in general that

$$d^2\omega = 0 \quad \text{for any } \omega \in \Lambda^p(M) \tag{3-13}$$

a result that follows readily along the foregoing lines once a local representation similar to (3-10) or (3-11) is given for ω. See Cartan [8] for a proof.

Remark. Note from (3-12) that the statement

$$d\omega = 0 \tag{3-14}$$

is equivalent to the symmetry of $\omega'_\phi(y)$ for each $y = \phi(x) \in B$, given any local chart (U, ϕ). A differential form satisfying (3-14) is said to be closed.

An an example, consider the 1-form ω_F given in (3-3). An easy calculation—see below—shows that in terms of coordinates (U, ϕ), with $y \in \phi(U)$

$$\langle u, (\omega_F)'_\phi(y)\, v \rangle = (u, F'_\phi(y)\, v) \tag{3-15}$$

for $u, v \in B$, F_ϕ being the local expression of operator F.

Lemma 1. For a densely defined operator F on a Banach manifold M and for each chart (u, ϕ) at $x \in M$, $y = \phi(x)$,

$$\langle u, (d\omega_F)_\phi(y)\, v \rangle = (F'_\phi(y)\, u, v)_y - (F'_\phi(y)\, v, u)_y$$

for $u, v \in B$.

Proof. In view of (3-12), it suffices to verify (3-15). Indeed, we have from (3-3) that

$$\langle (\omega_F)_\phi(y + u) - (\omega_F)_\phi(y), v \rangle = (F_\phi(y + u) - F_\phi(y), v)_y$$

and the assertion follows from the definition of the Fréchet derivatives. Q.E.D.

It will suffice in most of the remaining part of this section to work locally on a Banach space B, and we will do so unless otherwise stated.

Note that a Banach space being locally convex [34], in particular it is locally star-shaped.

Observation. It follows from the foregoing lemma that—in local coordinates—$d\omega_F = 0$ is equivalent to saying that $F'(u)$ is self-adjoint for every u.

Let $U \subset B$ be an open, star shaped neighborhood of the origin, a point hereafter assumed to be a center of the star. The following result will prove to be very useful below and it follows readily from the chain rule for derivatives:

Lemma 2. Let ω be a 1-form defined on star shaped open $U \subset B$, centered at the origin. Let $h \in B$. Then, $t \mapsto \langle t\omega(tu), h \rangle$ is differentiable and

$$\frac{d}{dt} \langle t\omega(tu), h \rangle = \langle \omega(tu) + \omega'(tu)\, u, h \rangle$$

Recall (3-13), according to which an exact differential 1-form is necessarily closed. By exactness of a 1-form ω it is meant the existence of a 0-form f such that

$$\omega = df$$

Such an f is said to be a potential for ω.

An important result termed Poincaré's Lemma says that this is true together with its converse. An important ingredient in the proof of that result is the concept of line integral of a differential form λ on a manifold M: for a smooth curve $\gamma: [0, 1] \to M$, define the line integral of ω along the path γ by

$$\int_\gamma \omega = \int_0^1 \langle \omega(\gamma(t)), \gamma'(t) \rangle dt$$

where $\gamma'(t) \in T_{\gamma(t)}M$ for each $t \in [0, 1]$ are the derivative (tangent) vectors to γ.

Poincaré's Lemma. Let B be a Banach space and let $U \subset B$ be open and star shaped, centered at x_0. Let $\omega \in \Lambda^1(U)$. Then ω is exact if and only if it is closed. If this is the case, a potential for ω is

$$V_\omega(x) = \int_{\overline{x_0 x}} \omega$$

where $\overline{x_0 x}$ denotes the path given by $t \longmapsto x_0 + t(x - x_0)$.

Proof. Without any loss of generality assume $x_0 = 0$. Then let's put

$$V(x) = \int_0^1 \langle \omega(tx), x \rangle \, dt$$

the goal is to prove that

$$\langle V'(x), h \rangle = \langle \omega(x), h \rangle$$

or, equivalently [21]

$$\lim_{\alpha \to 0} \frac{V(x + \alpha h) - V(x)}{\alpha}$$

For, note that

$$\begin{aligned} \langle \omega(t(x + \alpha h)), x + \alpha h \rangle - \langle \omega(tx), x \rangle &= \langle \omega(tx + t\alpha h) - \omega(tx), x \rangle + \alpha \langle \omega(tx + t\alpha h), h \rangle \\ &= \alpha t \langle h, \omega'(tx) x \rangle + \alpha \langle \omega(tx), h \rangle + o(\alpha) \\ &= \alpha t \langle x, \omega'(tx) h \rangle + \alpha \langle \omega(tx), h \rangle + o(\alpha) \end{aligned}$$

where use has been made of both Lemma 1 and the assumption of closedness of ω as applied to (3-12). Thus

$$\begin{aligned} \langle \omega(t(x + \alpha h)), x + \alpha h \rangle - \langle \omega(tx), x \rangle &= \alpha \langle \omega(tx) + t\omega'(tx) x, h \rangle + o(\alpha) \\ &= \alpha \frac{d}{dt} \langle t\omega(tx), h \rangle + o(\alpha) \end{aligned}$$

in view of Lemma 2. Therefore

$$(V(x + \alpha h) - V(x))/\alpha = \int_0^1 \frac{d}{dt} \langle t\omega(tx), h \rangle dt + o(\alpha)/\alpha = \langle \omega(x), h \rangle + o(\alpha)/\alpha$$

from which the result follows upon letting $\alpha \to 0$. The converse assertion is a direct consequence of (3-13). Q.E.D.

We state without proof the most general global version of Poincaré's Lemma for 1-forms [8, 1.3].

Theorem. Let ω be a closed differential 1-form on a Banach manifold M. If M is simply connected then there is a potential $V \in \Lambda^0(M)$ for ω, i.e., $\omega = dV$.

An important special case is obtained if we take a weakly nondegenerate inner product on B and then $\omega = \omega_F (p = 1)$ for some nonlinear operator F on B, differentiable on a star shaped subset U of B. By Lemma 1, closedness of ω_F amounts to $F'(x)$ being selfadjoint at each x with respect to the weakly nondegenerate inner product (,) on B. By Poincaré's Lemma,

$$dV = \omega_F$$

for a suitable potential V.

By a further abuse of notation one sometimes writes $V' = F$. In fact, in the finite dimensional case F is a vector field, closedness of $\omega_F := f_1(y)\,dx_1 + \cdots + f_n(y)\,dx_n$ translates into each f_i being the partial derivative $\partial V/\partial x_i$, i.e., $F = \text{grad } V$. This is so if $B = \mathbb{R}^n$, case in which $F' = V''$, a symmetric matrix. This takes us back to the reciprocity conditions of Sec. 2 (Example b).

4. GEOMETRY AND RECIPROCITY IN PHYSICS

We would like to probe a bit further into the ubiquitous reciprocity of physical systems, such as previously presented in Sec. 2. The main aim of this presentation is to exhibit some geometrical features common to the various physical systems considered here, aspects that become apparent when described in terms of differentiable manifolds.

As an illustration let's consider thermostatics, where reciprocity appeared in the guise of Maxwell relations, displayed in example a) of Sec. 2.

Following Gibbs and in analogy with the discussion on irreversible thermodynamics given in Sec. 2, the <u>state</u> of a pure substance is defined in terms of its extensive parameters, namely entropy S and volume V. The state space Σ is now the positive quadrant of the (S, V) plane, an open set. Experience leads us to postulate the existence of a C^∞ function $U: \Sigma \to \mathbb{R}$ called the <u>internal energy</u> of the system, subject to the following interpretation:

Let $\sigma_1, \sigma_2 \in \Sigma$ and let the system be taken from σ_1 to σ_2 while enclosed in adiabatic walls, by merely applying mechanical work on it, say W. Then

$$\Delta U := U(\sigma_2) - U(\sigma_1) = W$$

The foregoing equality fails to hold if heat is allowed to escape from the system. If such is the case then

$$\Delta U < W$$

and, in fact (see Callen [7], Sec. 1.6, 1.7),

$$Q := W - \Delta U$$

is the heat released. Therefore

$$W = \Delta U + Q \tag{4-1}$$

gives the right energy balance.

Note that both Q and W are thought of as being positive, whereas ΔU can have any sign. In fact, both Q and W can turn out to be negative, negative Q meaning that heat is absorbed instead of being released, with the analogous sign convention for W. If the system is a gas and only compression work is involved, elementary mechanical considerations show that the work supplied to the system is

$$W = -\int_\gamma p\,dV \tag{4-2}$$

In this equation γ is a path in Σ connecting σ_1 and σ_2, its points being the states traversed by the system when undergoing a (<u>quasistatic</u>) process $\sigma_1 \to \sigma_2$. As to p, it is <u>pressure</u>, always a positive quantity. More precisely, $p: \Sigma \to \mathbb{R}$ is defined by

$$p := \frac{\partial U}{\partial V} \tag{4-3)'$$

Analogously, absolute temperature $T: \Sigma \to \mathbb{R}$ is defined by

$$T := \frac{\partial U}{\partial S} \tag{4-3)''$$

and it is postulated to take only nonnegative values.

In virtue of Eqs. (4-3)

$$dU = T\,dS - p\,dV \tag{4-4}$$

which can be integrated along γ to yield

$$\Delta U = \int_\gamma T\,dS + W$$

An immediate comparison with (4-1) shows that the heat released by the system is precisely

$$Q = -\int_\gamma T\,dS$$

Moreover, it is clear from (4-4) that

$$\frac{\partial T}{\partial V} = -\frac{\partial p}{\partial S} \tag{4-5}$$

one of the Maxwell relations embodying reciprocity in this case. Eq. (4-5) amounts to the necessary equality of the mixed second partial derivatives of U.

Equation (4-5) is a direct consequence of the existence of the potential U for the 1-form

$$T\,dS - p\,dV$$

By means of Legendre transformations the extensive variables S, V can be partially or totally exchanged with their conjugates, T and -p, respectively, thus giving rise to other potentials for appropriately chosen 1-forms. For instance, a potential for

$$-S\,dT + V\,dp$$

is given by Gibbs' free energy function G, defined by

$$G: = U - TS + pV$$

This choice of potential gives rise to a new reciprocity relation, namely

$$\frac{\partial V}{\partial T} = -\frac{\partial S}{\partial p}$$

See Chap. 5 of [7] for the details of this and related constructions. Let's now go back to the original formulation in terms of extensive parameters leading to (4-5). Let $M = \Sigma \times \mathbb{R}^2$ be the set of all extended states (S, V, T, -p), an open subset of $\mathbb{R}^4$. Let L denote the set

$$L: = \left\{(S, V, T, -p) \in M: T = \frac{\partial U}{\partial S}(S, V),\ -p = \frac{\partial U}{\partial V}(S, V)\right\}$$

whose points are the extended states that can in fact be occupied by the system, in view of (4-3).

Both M and L are differentiable manifolds: M has dimension 4 while the dimension of L is 2, as can be easily verified.

Noting that $d^2U = 0$, in view of (3-13), it follows from (4-4) that the 2-form

$$dT \wedge dS - dp \wedge dV \tag{4-6}$$

vanishes on L. Note also that the 1-form in the right side of (4-4) can be thought of as a map $dU: \Sigma \to \mathbb{R}^{2*}$, whose graph is exactly L (if $\mathbb{R}^{2*}$ is identified with $\mathbb{R}^2$).

M with the 2-form (4-6) is an example of a symplectic manifold. In turn, L is a submanifold of M which is Lagrangian, in view of its annihilating the symplectic form (4-6) and its dimension being half the dimension of M. These terms will be clarified below, following Abraham-Marsden [1]. See Sec. 1.1 of [19] for a discussion of thermostatics in connection with symplectic structures and Lagrangian submanifolds.

It is no accident that a symplectic manifold with a Lagrangian submanifold could be associated with this physical situation. In fact, quoting from p. 402 of [1]: "Virtually every physical system has a symplectic manifold associated with it, and the behavior of the system may be described in terms of Lagrangian submanifolds." It is no accident either that a reciprocity condition such as Maxwell's relation (4-5) holds in this case. In fact, in this context Tulczyjew et al. [19, 30, 31] have studied reciprocity as determined by a Lagrangian submanifold of a symplectic manifold. Let's devote the rest of this section to clarifying these concepts and then applying them to the remaining examples presented in Sec. 2.

Definition (Marsden [22]). Let M be a manifold modeled on a Banach space B. $\omega \in \Lambda^2(M)$ is said to be symplectic if

i) ω is closed, i.e., $d\omega = 0$
ii) For every $x \in M$, $\omega(x)$ is nondegenerate as bilinear from on T_xM.

The form is said to be weakly symplectic if the bilinear form on T_xM in ii) is merely weakly nondegenerate. Then, (M, ω) will be said to be a symplectic manifold (resp. a weakly symplectic manifold). In finite dimensions a weakly nondegenerate bilinear map is necessarily nondegenerate, hence the term symplectic manifold can be used without any further qualifications.

There is a nice characterization of nondegeneracy for 2-forms on a finite dimensional space ([1], Prop. 3.1.5): $\omega \in \Lambda^2(M)$ is nondegenerate if and only if M is even dimensional and $\omega^n = \omega \wedge \ldots \wedge \omega$ is a volume element on M, where dim M = 2n. Recall that $\Lambda^{2n}(M)$ is one dimensional on each T_xM, so ω^n a volume element means that it never vanishes ([1], p. 105). Thus there are only even dimensional symplectic manifolds in the finite dimensional case. Moreover, the following characterization of the symplectic character of a nondegenerate form can be given (Darboux's theorem, see p. 175 of [1]):

A nondegenerate form ω on a 2n-dimensional manifold M is symplectic if and only if for each $x \in M$ there is a coordinate neighborhood (U, ϕ) such that, in terms of local coordinates, $\phi(u) = (x_1, \ldots, x_n, y_1, \ldots, y_n)$, $u \in U$,

$$\omega = \sum_{i=1}^{n} dy_i \wedge dx_i \quad \text{on } U \tag{4-7}$$

By direct calculation in terms of these local coordinates, it follows that the volume element ω^n is given by ([14], Sec. 2.3)

$$\omega^n = (-1)^{n(n-1)/2} n! \, dy_1 \wedge \ldots \wedge dy_n \wedge dx_1 \wedge \ldots \wedge dx_n \tag{4-8}$$

The normalized volume element

$$dy_1 \wedge \ldots \wedge dy_n \wedge dx_1 \wedge \ldots \wedge dx_n$$

is the classical phase density of Hamiltonian mechanics (see, e.g., Sec. 10.3 of [14]).

Given a Banach manifold N, the cotangent bundle T^*N of N can be given a canonical symplectic structure. For, let $\tau: T^*N \to N$ be the projection

$$z \mapsto x \quad (x \in M, \; z \in T_x^*N)$$

and define the canonical 1-form $\Theta \in \Lambda^1(T^*N)$ by

$$\langle \Theta(z), \xi \rangle := \langle z, T\tau(\xi) \rangle$$

where $z \in T^*N$, $\xi \in T_z(T^*N)$, and $T\tau(\xi) \in TN$ since $T\tau$ is the tangent map of τ at z [1, p. 45]. In local coordinates, this is the same as saying that

$$\langle \Theta(u, \alpha), (e, \beta) \rangle = \langle \alpha, e \rangle, \quad \text{for } u, e \in B, \; \alpha, \beta \in B^*$$

If N is finite (say 2n-) dimensional, we have

$$\Theta = \sum_{i=1}^{n} y_i \, dx_i$$

where $(x_1, \ldots, x_n, y_1, \ldots, y_n)$ are natural chart coordinates [1, p. 57] for T*N.

The canonical weak symplectic form ω on T*N is defined as $\omega = d\Theta$. Locally, using (3-12) we have

$$\langle (e_1, \alpha_1), \omega(u, \alpha)(e_2, \alpha_2) \rangle = \langle \alpha_2, e_1 \rangle - \langle \alpha_1, e_2 \rangle \tag{4-9}$$

and in the finite dimensional case we get just (4-7).

In the infinite dimensional case, ω given by the above construction is symplectic if and only if B is a reflexive Banach space. See [22].

Let's illustrate these concepts in terms of the thermostatics example given above. The 2-form of (4-6) is clearly nondegenerate, hence symplectic in view of Darboux's theorem. As to the phase density, it coincides with the four dimensional Lebesgue measure on the open set M, $-dS \wedge dV \wedge dT \wedge dp$. There it was mentioned that the 2-dimensional manifold L was Lagrangian. Let's now explore this notion, following [1].

Let M be a Banach manifold with a weak symplectic 2-form ω on it. Let $L \subset M$ be an immersed submanifold of M, i.e., such that the inclusion map $i: L \to M$ has a Fréchet derivative injective at every point of L.

Definition [1]. An (immersed) submanifold L of a symplectic manifold (M, ω) is Lagrangian if (i) For every $x \in L$, and every $u, v \in T_xL$, $\langle u, \omega(x)v \rangle = 0$. (ii) L is maximal among those submanifolds L' of M satisfying i), meaning that $x \in L \cap L' \Rightarrow T_xL' \subset T_xL$.

Property (i) is sometimes referred to as isotropicity of L as a submanifold of symplectic M. It is normally paraphrased by saying that "L annihilates the symplectic 2-form ω." Thus, a Lagrangian submanifold of the symplectic manifold (M, ω) is one whose tangent spaces are maximal among the isotropic submanifolds of M.

Let N be a Banach manifold (not necessarily weakly symplectic) and pick $\alpha \in \Lambda^1(N)$. Let the image of α be denoted by $\alpha(N)$, a subset of the cotangent bundle T*N—whose symplectic structure is defined in (4-9)—given by

$$\alpha(N) := \{\alpha(x) \in T_x^*N : x \in N\}$$

The inclusion map $i: \alpha(N) \to T^*N$ has everywhere an injective Fréchet derivative, hence $\alpha(N)$ is a submanifold. It is interesting to note that $\alpha(N)$ is Lagrangian if and only if α is a closed 1-form. See pp. 4-10 of [1] for a proof.

Referring back to the thermostatics example given above, the 1-form dU of (4-4) is obviously closed, hence its image is a two-dimensional submanifold of the four dimensional $\Sigma \times \mathbb{R}^2$. Noting that in manifold terminology

$$dU(\Sigma) = \left\{(S, V, T, -p) : T = \frac{\partial U}{\partial S}, \; -p = \frac{\partial U}{\partial V}\right\}$$

we get the Lagrangian character of the submanifold L defined there in connection with the 1-form (4-4). Here $\Sigma \times \mathbb{R}^2$ are local coordinates for $T^*\Sigma$.

In general, the notion of image of a 1-form on a manifold corresponds to its graph in local coordinates, provided we forget about base points.

Note that, in general, $\dim(T^*N) = 2n$ if $\dim N = n$. Also note that $\dim \alpha(N) = n$ if $\alpha \in \Lambda^1(N)$ in this case. No wonder that in the thermostatics example

$$\dim dU(\Sigma) = \frac{1}{2} \dim (\Sigma \times \mathbb{R}^2)$$

in view of the following characterization of the Lagrangian character of a submanifold in the finite dimensional case ([1], Prop. 5-3 14):

Let L be a submanifold of M annihilating symplectic $\omega \in \Lambda^1(M)$. Then L is Lagrangian if and only if dim M = 2 dim L.

Again, noting that

$$dT \wedge dS - dp \wedge dV = 0 \quad \text{on } dU(\Sigma)$$

with dU as in (4-4), the foregoing characterization yields the Lagrangian character of $dU(\Sigma)$.

In general, let M be an m-dimensional manifold and let θ be a closed 1-form on M. Then L: = θ(M) is a Lagrangian submanifold of T*M whose symplectic structure is given in (4-9). On the other hand, by Poincaré's Lemma, presented in Sec. 3 above, $\theta = dV$ locally, with V a C^∞ function on an open subset of M—a potential. Letting $(x_1, \ldots, x_m)$ be local coordinates for M, so that

$$dV = \sum_{i=1}^{m} f_i(x_1, \ldots, x_m)dx_i$$

(again locally), we get the reciprocity conditions

$$\frac{\partial f_i}{\partial x_j} = \frac{\partial f_j}{\partial x_i} \qquad i, j = 1, \ldots, m \tag{4-10}$$

Thus, a closed form $\theta \in \Lambda^1(M)$ implies both reciprocity (a local notion) and the existence of a Lagrangian submanifold θ(M) of T*M (a global notion). Note that the Lagrangian character of this submanifold is the global counterpart of the local, rich in physical content notion of reciprocity.

Regarding Lagrangian submanifolds in general, we know now that any closed 1-form on a manifold M gives rise to a Lagrangian submanifold L in T*M, with the canonical symplectic structure defined there. It is true that not every Lagrangian submanifold is generated in this fashion. However, that a partial converse is locally valid follows from a theorem due to Weinstein ([1], p. 411): given any Lagrangian submanifold $L \subset T^*M$, there exists a local diffeomorphism taking L into the zero section of T*L and preserving the symplectic structures. The null 1-form being trivially closed, all this contributes to establishing the close correspondence between closedness of 1-forms (i.e., reciprocity) and the Lagrangian character of a certain submanifold.

To illustrate the foregoing concepts let's go back to the remaining examples presented at the end of Sec. 2 above.

Consider the equations of particle mechanics (2-13), total energy being defined as in (2-14). The relevant manifold is $M = \mathbb{R}^{6n}$, whose cotangent space T*M can be identified with $\mathbb{R}^{12n}$, with symplectic structure given by

$$\omega = \sum_{i=1}^{3n} dF_i \wedge dx_i + \sum_{i=1}^{3n} dp_i \wedge dx_i$$

Here n denotes the number of particles, hence $m_1 = m_2 = m_3$, $m_4 = m_5 = m_6$, etc. in the expression for ω.

Let θ be the 1-form given by

$$\theta = \sum_{i=1}^{3n} F_i(x)\, dx_i + \sum_{i=1}^{3n} m_i \dot{x}_i\, d\dot{x}_i$$

which is a closed form if F = grad V: if such is the case then $\theta = dE$, with

$$E = V(x_1, \ldots, x_{3n}) + \frac{1}{2}\sum_{i=1}^{3n} m_i \dot{x}_i^2$$

Then L: = θ(M) is a Lagrangian submanifold of T*M. The ensuing reciprocity relations (4-10) specialize to

$$\frac{\partial F_i}{\partial x_j} = \frac{\partial F_j}{\partial x_i} \tag{4-11}$$

the remaining ones being trivial.

More generally, consider a classical Hamiltonian system on $T^*\mathbb{R}^n = \mathbb{R}^n \times \mathbb{R}^{n*}$ with coordinates (q, p) there. Assume the associated Legendre Transformation is a global diffeomorphism, so that the corresponding Lagrange formulation is well defined. Thus, both Lagrangian L and Hamiltonian H are defined, with

$$H + L = p \cdot \dot{q} \tag{4-12}$$

along the system's trajectories.

A symplectic structure is defined on $T\,T^*\mathbb{R}^n$ with coordinates $(q, p, \dot{p}, \dot{q})$ by the 2-form.

$$\omega = d\dot{p} \wedge dq + dp \wedge d\dot{q}$$

In these coordinates, consider the 1-form

$$\theta_L: = \dot{p} \cdot dq + p \cdot d\dot{q}$$

its image I_L consisting of all those points of $T\,T^*\mathbb{R}^n$ determined by Lagrange's equations

$$\dot{p} = \frac{\partial L}{\partial q}, \quad p = \frac{\partial L}{\partial \dot{q}} \tag{4-13)'$$

i.e.,

$$\frac{d}{dt}\left(\frac{\partial L}{\partial \dot{q}}\right) = \frac{\partial L}{\partial q} \tag{4-13)''$$

Then I_L is a Lagrangian submanifold of $T\,T^*\mathbb{R}^n$, thus giving rise to reciprocity relations

$$\frac{\partial \dot{p}}{\partial \dot{q}} = \left(\frac{\partial p}{\partial q}\right)^t \tag{4-14}$$

in the spirit of (4-10).

Correspondingly, let I_H be the image of closed 1-form

$$\theta_H: = \dot{p} \cdot dq - \dot{q} \cdot dp$$

again a Lagrangian submanifold of $T\,T^*\mathbb{R}^n$.

As in (4-10), there result reciprocity relations in the form of

$$\frac{\partial \dot{p}}{\partial p} = -\left(\frac{\partial \dot{q}}{\partial q}\right)^t \tag{4-15}$$

It can be shown [31] that both I_L and I_H are in fact the same Lagrangian submanifold of $T\,T^*\mathbb{R}^n$, by taking into account relation (4-12). In some sense, this Lagrangian submanifold can be thought of as the image of either dL or -dH, with $L = L(q, \dot{q})$, $H = H(p, q)$: it suffices to invoke Lagrange's equations (4-13) as well as Hamilton's equations

$$\dot{q} = \frac{\partial H}{\partial p}, \quad \dot{p} = -\frac{\partial H}{\partial q}$$

These two representations can be related via the Legendre transformation

$$(q, p) \mapsto (q, \dot{q}) \tag{4-16}$$

defined by (4-12). Incidentally, note that this Legendre transformation is partial, since only p is replaced by its conjugate variable $\dot{q}$. This circumstance explains the minus sign [5, Th2] in (4-15), a feature not present in (4-14): θ_H gets one minus sign, because (4-12) has to be satisfied.

The configuration space of a Hamiltonian system can be a general manifold M and not necessarily $\mathbb{R}^n$. In this more general case L will be defined on TM, H on T*M. The symplectic form ω on T T*M can be globally defined from the canonical 2-form on T*M. This globalization is not trivial though [30].

As a final example consider the electric circuits presented at the end of Sec. 2. For simplicity suppose all resistances are zero and write $\Gamma_{ik} = 1/C_{ik}$. Then (2-15) can be written in vector matrix form as

$$L\ddot{q} + \Gamma q = u \tag{4-17}$$

with the obvious identifications. The underlying manifold is

$$M = \mathbb{R}^{2n}$$

with the symplectic structure on $T^*M = \mathbb{R}^{4n}$ given by

$$\omega = \sum_i d\phi_i \wedge d\dot{q}_i + \sum_i dv_i \wedge dq_i$$

where ϕ_i, v_i are magnetic flux and capacitor voltage drop. Consider the 1-form

$$\theta = \sum_{i,k} L_{ik}\dot{q}_k d\dot{q}_i + \sum_{i,k} \Gamma_{ik} q_k dq_i \tag{4-18}$$

which in fact is the differential of the total stored energy

$$E = \frac{1}{2}\sum_{i,k} L_{ik}\dot{q}_i\dot{q}_k + \frac{1}{2}\sum_{i,k} \Gamma_{ik} q_i q_k$$

Then θ is closed, hence $\theta(M)$ is a Lagrangian submanifold of T*M and reciprocity relations are to be expected, in the form of (4-10). In fact, these relations specialized to (4-18) turn out to be

$$\frac{\partial}{\partial \dot{q}_j}\Big(\sum_k L_{ik}\dot{q}_k\Big) = \frac{\partial}{\partial \dot{q}_i}\Big(\sum_k L_{jk}\dot{q}_k\Big), \quad \frac{\partial}{\partial q_j}\Big(\sum_k \Gamma_{ik} q_k\Big) = \frac{\partial}{\partial q_i}\Big(\sum_k \Gamma_{jk} q_k\Big)$$

plus two trivial ones, thus resulting in Eqs. (2-16).

5. RECIPROCITY AND CONSERVATION LAWS

The important role played by reciprocity has already been made clear in the previous sections. In particular, it has been shown to be always related to the existence of a closed differential 1-form ω on a certain manifold, and accompanied by the presence of a Lagrangian submanifold of a given symplectic structure. Moreover, we saw in Sec. 3 that closedness of ω_F (1-form constructed in terms of a densely defined nonlinear operator F) is locally equivalent to self-adjointness of F' at every point. In the finite dimensional case, this last condition is interpreted as the symmetry of a matrix, the Hessian matrix of the local potential, i.e., as reciprocity. We would like now to investigate reciprocity in connection with the integrability of certain Pfaffian equations.

For the sake of motivation, let's go back to our presentation of thermostatics given early in Sec. 4, with the state of one mole of pure substance defined in terms of its extensive parameters. Only now our choice of these parameters will be different, more in accord with experiment: internal energy U and volume V. The state space Σ is now the upper half of the euclidean plane, its points being labeled now as (U, V). The physical arguments leading to Eq. (4-1) remain valid. Together with (4-2), this allows us to define the heat generated during process γ as

$$Q := -\Delta U - \int_\gamma p\, dV$$

if only mechanical work is involved. Alternatively,

$$Q = \int_{\gamma} q$$

with $q \in \Lambda^1(\Sigma)$ given by

$$q: = -dU - p\,dV \tag{5-1}$$

Once entropy is shown to exist in this formalism also, pressure p will again be given by (4-3)'. In fact, by the implicit function theorem, pressure can be obtained from entropy as

$$p = \left(\frac{\partial S}{\partial V}\right)_U \Big/ \left(\frac{\partial S}{\partial U}\right)_V \tag{5-2}$$

For the time being, it will be necessary to assume the existence of such a C^∞ function $p: \Sigma \to \mathbb{R}$ representing pressure, assumption that can be removed through use of (5-2) once the existence of entropy has been established.

In this context, a process will be said to be adiabatic if $\int_{\gamma} q = 0$, and in fact $\int_{\gamma_1} q = 0$ for any subprocess γ_1 of γ. NB γ_1 is said to be a subprocess of γ if there is a subinterval $[a, b]$ of $[0, 1]$ and a bijection $s: [a, b] \to [0, 1]$ such that $\gamma_1 \circ s = \gamma|_{[a,b]}$. Thus, an adiabatic process satisfies the condition.

$$-U'(t) - p(\gamma(t))V'(t) = 0, \quad 0 \leq t \leq 1 \tag{5-3)'$$

Alternatively, such γ is said to be an integral curve of the Pfaffian equation

$$-dU - p\,dV = 0 \tag{5-3)''$$

Thus, adiabatic processes are those annihilating the 1-form q. This means (5-3) and no more than that.

Note that

$$dq = -\left(\frac{\partial p}{\partial U}\right)_V dU \wedge dV \tag{5-4}$$

so that q is not closed, hence it is not exact either.

The Second Law of Thermodynamics asserts the existence of an integrating factor μ for q though: there is a never vanishing, C^∞ function μ on Σ such that

$$d(\mu q) = d\mu \wedge q + \mu \wedge dq = 0$$

hence

$$dq = -\frac{d\mu}{\mu} \wedge q$$

Physical arguments are given by Fermi [13] to define absolute temperature T as the inverse of one of those integrating factors, a positive C^∞ function T on Σ, so that $\mu = 1/T$ and

$$dq = \frac{dT}{T} \wedge q \tag{5-5}$$

An obvious substitution of (5-1) and (5-4) into (5-5) yields

$$p\frac{\partial T}{\partial U} - \frac{\partial T}{\partial V} = T\frac{\partial p}{\partial U} \tag{5-6}$$

thus characterizing T as a solution of a hyperbolic partial differential equation of the first order.

By Poincaré's Lemma (Sec. 3), q/T is exact, say equal to -dS, for a suitable C^∞ function $S: \Sigma \to \mathbb{R}$. Moreover, S is given up to a constant by

$$-S(U,V) = \int_{(U_0,V_0)}^{(U,V)} q/T = -\int_{(U_0,V_0)}^{(U,V)} \frac{1}{T}\,dU + \frac{p}{T}\,dV$$

with $(U_0, V_0) \in \Sigma$, otherwise arbitrary. The path of integration is any process joining (U_0, V_0) with (U, V). Thus

$$dS = \frac{1}{T}\,dU + \frac{p}{T}\,dV \tag{5-7}$$

which shows that

$$\frac{1}{T} = \left(\frac{\partial S}{\partial U}\right)_V, \quad \frac{p}{T} = \left(\frac{\partial S}{\partial V}\right)_U$$

thus validating (5-2).

Equation (5-7) is normally written as (4-4):

$$dU = T\,dS - p\,dV$$

a relation implied by Eqs. (4-3), with the <u>entropy</u> S defined by

$$S(\sigma) = -\int_{\overline{\sigma_0\sigma}} \frac{q}{T}, \quad \sigma \in \Sigma \tag{5-8}$$

in the notation of Poincaré's Lemma. Note that q is thought of here as heat <u>produced</u>, hence the discrepancies in sign between (5-8) and most other presentations.

Note that (5-3) and (5-7) imply that, for an adiabatic process γ

$$\frac{d}{dt} S(\gamma(t)) = 0, \quad 0 \leq t \leq 1 \tag{5-9)'}$$

so that adiabatic processes are necessarily <u>isentropic</u>, i.e., entropy is conserved along them. Just as with (5-3), the foregoing condition can be rewritten as the Pfaffian equation

$$\frac{1}{T}\,dU + \frac{p}{T}\,dV = 0 \tag{5-9)''}$$

Thus, every solution of (5-3) satisfies (5-9). We would like to make sure that adiabatic processes, i.e., integral curves of (5-3), do exist. Let's investigate this and other related questions on 1-forms ω given in local coordinates y (in finite dimension) by expressions such as

$$\sum_{i=1}^{n} f_i(y)\,dx_i$$

In local coordinates, the corresponding Pfaffian equation looks like

$$\sum_{i=1}^{n} f_i(y)\,dx_i = 0 \tag{5-10)'}$$

the integral curves of (5-10)' being those satisfying

$$\sum_{i=1}^{n} f_i(\gamma(t))x_i'(t) = 0, \quad 0 \leq t \leq 1 \tag{5-10)''}$$

where $\gamma(t) = (x_1(t), \ldots, x_n(t))$, $0 \leq t \leq 1$.

Let's now proceed to express these concepts in a sufficiently ample setting, congruent with the treatment given so far. For, let M be a Banach manifold and pick $\omega \in \Lambda^1(M)$ as in Sec. 3. The <u>Pfaffian equation</u> associated with ω (with (5-3)'' and (5-9)'' as particular instances) is denoted by

$$\langle \omega(x),\ x' \rangle = 0 \tag{5-11}$$

its solutions being curves γ on M such that $\gamma_*\omega = 0$. Recall that $\gamma_*\omega$ is the 1-dimensional 1-form given by

$$\langle \gamma_*\omega(t),\ h \rangle := \langle \omega(\gamma(t)),\ \gamma'(t) \rangle h \qquad t \in [0,1],\quad h \in \mathbb{R}$$

Thus, solutions of (5-11) are curves on M which annihilate ω.

Assume that ω is everywhere nonzero on M. If that were not the case, we can always restrict our attention to an open submanifold of M where ω does not vanish. Note that for each $x \in M$,

$$\ker \omega(x) := \{v \in T_xM : \langle \omega(x), v \rangle = 0\}$$

is a co-dimension one vector subspace of T_xM.

Thus setting Pfaffian equation such as (5-11) amounts to assigning a codimension one Banach subspace of T_xM for every $x \in M$ and doing this smoothly. In turn, solving one such equation amounts to connecting some of those codimension one subspaces in a smooth fashion. This point of view leads to the general idea of a distribution Δ on M, namely a smooth assignment of a subspace $\Delta(x)$ of T_xM for every $x \in M$. A requirement is that each two of these subspaces must be isomorphic. Thus, $\ker \omega$ is a distribution on M and a solution γ of (5-11) is such that $T_{\gamma(t)}\gamma \subset \ker \omega(\gamma(t))$. See Boothby [3] for a more systematic treatment of distributions on a manifold.

If N is a codimension one Banach submanifold of M such that $T_xN = \ker \omega(x)$ for each $x \in N$, then N is said to be an integral submanifold for ω. In this case ω vanishes on N, hence so does $d\omega$. This fact motivates the following definition: A form $\omega \in \Lambda^1(M)$ is completely integrable if $\ker \omega$ annihilates $d\omega$.

This condition means that for every $x \in M$,

$$u, v \in \ker \omega(x) \Longrightarrow \langle u,\ d\omega(x)v \rangle = 0 \tag{5-12}$$

Thus complete integrability of ω is necessary for the existence of integral submanifolds passing through every point of M. Let's now investigate whether the heat form q is completely integrable.

From (5-4), dq is a nonvanishing constant times the alternating bilinear form with matrix

$$\begin{pmatrix} 0 & 1 \\ -1 & 0 \end{pmatrix}$$

while q is minus the linear form

$$(a,b) \longmapsto a + bp$$

Hence (5-12) is easily seen to reduce to the statement that "if the system $ax + by = 0$, $cx + dy = 0$ is satisfied by $(1,p)$, then $ad - bc = 0$," which is certainly true. Hence q is completely integrable.

That complete integrability of q suffices to ensure the existence of adiabatic processes, follows from the very important Frobenius' Theorem (global version [33]):

Let ω be a 1-form vanishing nowhere on a Banach manifold M. There is an integral submanifold of ω through any point of M if and only if ω is completely integrable.

However, it is not clear yet that adiabatic processes are necessarily isentropic. In general, it is required to find out when an integral submanifold of a given form $\omega \in \Lambda^1(M)$ is the level set of a C^∞ function V on M, at least locally. Just as condition (5-5) was seen to play a very important role in establishing the existence of entropy, the relation

$$d\omega = \theta \wedge \omega \tag{5-13}$$

can be similarly shown to be crucial in proving the potentialness of ω in the general case.

Let's begin by proving that (5-13) guarantees that ω has an integrating factor whenever θ is exact, say $\theta = dU$. For, letting $\mu = e^{-U}$ it follows that

$$d(\mu\omega) = -e^{-U} dU \wedge \omega + e^{-U} d\omega = 0$$

as asserted. Thus, $\mu\omega$ has a local potential V by Poincaré's Lemma and therefore ω has a local conservation law.

Clearly, (5-13) implies the integrability condition

$$\omega \wedge d\omega = 0 \tag{5-14}$$

which in turn is equivalent to the complete integrability (5-12) of ω [33], and is a more handy condition. Incidentally, (5-13) can be directly shown to imply (5-12), since (3-8) gives

$$\langle u, d\omega(x)v \rangle = \langle \theta(x), u \rangle \langle \omega(x), v \rangle - \langle \omega(x), u \rangle \langle \theta(x), v \rangle \tag{5-15}$$

A local converse is given by Frobenius' Theorem (local version [33]; [4], p. 33):

Let ω be a 1-form vanishing nowhere on a Banach manifold M. If ω is completely integrable, then any $x \in M$ has a neighborhood U and there is a C^∞ function ϕ defined on U such that

$$d\omega = d\phi \wedge \omega$$

Summing up, under the assumption of a 1-form never being zero, complete integrability is more general than reciprocity (closedness), but they are locally equivalent modulo an integrating factor.

The precise form of the relationship between reciprocity (selfadjointness) and conservation laws for (5-1) is obtained when $\omega = \omega_F$, F being a densely defined vector field on M as in (3-3). The following statement is a consequence of the foregoing developments:

Proposition. Let M be a Banach manifold and F a densely defined operator on it with $F(x) \in T_xM$ for each $x \in M$. Construct ω_F as in (3-3) and assume it never vanishes. Then the following conditions are equivalent:

1) The Pfaffian equation $\langle \omega_F(x), x' \rangle = 0$ has a local conservation law.
2) There is a nonvanishing C^∞ function μ on M such that (in local coordinates) $(\mu F)'$ is self-adjoint.
3) There is a nonvanishing C^∞ function μ on M such that $\mu\omega_F$ is closed.

Notice that the existence of an integrating factor μ as in 3) for ω guarantees (5-13) with $\theta = -d(\log \mu)$, hence complete integrability, this being the main ingredient in the proof of 1) via the local version of Frobenius' Theorem.

To supplement our study of Pfaffian equations, let's go back to the problem of constructing adiabatic processes in the thermostatics setting we have been considering in this section. So far we have dealt with the initial value problem for the Pfaffian equation (5-3), whose solution can be established via Frobenius Integrability Theorem, thus leading to the existence of adiabatic processes passing through any prespecified state, in particular. The question suggests itself of considering a two-point boundary value problem associated with (5-11):

Given $x, y \in M$, $x \neq y$, find a solution curve of (5-11) passing through x and y.

In the thermostatics approximation with $\omega = q$, this question amounts to the requirement of finding an adiabatic process connecting any two prespecified states of a given system. A moment's reflection shows that requiring this problem to have a solution for any two prespecified states is equivalent to the construction of a perpetual motion machine [6], which is impossible as asserted by the second law of thermodynamics. Thus this principle can be given an alternate statement, due to Carathéodory, which goes as follows:

In any neighborhood of every equilibrium state $\sigma_0 \in \Sigma$, there are states which cannot be reached along solutions of $q = 0$.

A formulation of thermodynamics based entirely on this form of the second law of thermodynamics is given by Buchdahl in [6]. Its equivalence with the previously given formulation of this law was proved by C. Carathéodory in 1909 and can be summarized as follows:

Inaccessibility Theorem [33]. Let M be an n-dimensional manifold with $\partial M = \phi$ and let $\omega \in \Lambda^1(M)$ be nowhere vanishing. Then, ω admits an integrating factor if and only if for every $x \in M$, there are points of M arbitrarily close to x that cannot be joined to x along integral curves of $\langle \omega(x), x' \rangle = 0$.

As a final application of Frobenius' theory, let's consider the problem of determining absolute temperature for a given thermodynamic system, say a perfect crystalline solid. For such a system, the so-called third law of thermodynamics [7] asserts that entropy vanishes precisely at zero absolute temperature. When combined with the second law in any of its equivalent forms, this condition requires that

$$T = 0 \quad \text{on } \Sigma_0 \tag{5-16}$$

with

$$\Sigma_0 := \{(U, V) \in \Sigma : S(U, V) = 0\}$$

Σ standing for the upper half of the (U, V)-plane. This gives a natural way of fixing the additive constant involved in S.

Thus, determining absolute temperature amounts to solving the first order partial differential equation (5-6) with the initial condition (5-16), assuming p is known as a function of (U, V).

For, we assume that

$$p \frac{\partial S}{\partial U} - \frac{\partial S}{\partial V} \neq 0 \quad \text{on } \Sigma_0 \tag{5-17}$$

meaning that the "initial manifold" Σ_0 is not characteristic for (5-6), i.e., the vector field (p, -1) is not tangent to Σ_0. Then ([33], p. 248), there is a neighborhood G of Σ_0 and a unique differentiable function T on G which obeys both (5-6) and (5-16), thus proving independently that absolute temperature is determined by the laws of thermodynamics.

6. VARIATIONAL PRINCIPLES AND SELF-ADJOINTNESS

As is well known and much appreciated in physics [20], the variational formulation of field theories allows for a degree of unification absent from their versions in terms of differential equations. In fact, a single operation (extremizing a functional) enables the analyst to obtain in every case the equations governing the behavior of the system under study. Moreover, from the more practical viewpoint of "actually computing the solution" it often happens that solving the variational problem independently is more expedient than solving the associated differential equations, at least if approximate solutions are acceptable (see, e.g., [27]). It is therefore important to characterize those cases in which a variational formulation can be given [32].

This section is devoted to examining this circle of ideas from the more abstract viewpoint afforded by the existence of a closed 1-form ω_F associated with a given nonlinear system of differential equations $F(u) = 0$. This same 1-form leads to the symplectic framework of reciprocity in field theories, as already indication in Sec. 4 above.

For, let's consider a Lagrangian $L: U \times \mathbb{R}^n \to \mathbb{R}$, with U open in $\mathbb{R}^n$, hereafter assumed to be C^4. Let $\mathcal{U}$ consist of all U-valued C^2 functions on an interval [a, b], satisfying there the boundary conditions

$$u(a) = \alpha, \quad u(b) = \beta \tag{6-1}$$

for given $\alpha, \beta \in U$. (The restriction to functions of one independent variable in all that follows is not essential: minor changes in this treatment allow us to deal with the case of several independent variables.) Let $\mathscr{H}$ stand for the space of all C^2 functions $h: [a, b] \to \mathbb{R}^n$, equipped with the norm

$$\|h\| = \sup_{a \leq t \leq b} |h(t)| + \sup_{a \leq t \leq b} \left|\frac{dh}{dt}\right|$$

where $| \ |$ stands for the euclidean norm in $\mathbb{R}^n$. See Cartan ([8], 11.1) for a proof that $(\mathscr{H}, \| \ \|)$ is a Banach space. Then $\mathscr{U}$ is an open subset ([8], 11.1) in the relative topology of the affine subspace $\mathscr{M}$ of $\mathscr{H}$ consisting of all functions $u \in \mathscr{H}$ satisfying (6-1). Let $I: \mathscr{U} \to \mathbb{R}$ be given by

$$I(u) = \int_a^b L(u(t), \dot{u}(t))\, dt \tag{6-2}$$

Finally, let $\mathscr{S}$ stand for the closed subspace of $\mathscr{H}$ which is parallel to $\mathscr{M}$, i.e.,

$$\mathscr{S} = \{u - u_0 : u \in \mathscr{M}\}$$

for any fixed $u_0 \in \mathscr{M}$, so that $\mathscr{M} = u_0 + \mathscr{S}$. Note that $\mathscr{S}$ is isomorphic to $\mathscr{M}$, which in turn can be modeled after $\mathscr{S}$ and thus be made into a Banach manifold in its own right.

This is the setting adopted here in order to pose the problem of extremizing I, i.e., of finding those $u \in \mathscr{U}$ such that $I'(u) = 0$, where I' stands for the Fréchet derivative of I (N.B. The theory of extrema developed below being essentially local, there is no harm in restricting ourselves to local coordinates in this section. In general, the Lagrangian function is defined on a finite dimensional manifold M, whereas the associated functional I has as domain a Banach manifold of curves joining two prescribed points of M ([1], Sec. 3.8).)

It is easily verified that

$$\langle I'(u), h \rangle = \int_a^b v(t) \cdot h(t)\, dt \qquad (u \in \mathscr{U},\ h \in \mathscr{H}) \tag{6-3}$$

where $\cdot$ denotes the usual dot product in $\mathbb{R}^n$ and $v = E(u)$, with

$$v_i(t) = L_{u_i}(u(t), \dot{u}(t)) - \frac{d}{dt} L_{\dot{u}_i}(u(t), \dot{u}(t)) \tag{6-4}$$

for $i = 1, \ldots, n$. The right hand side E_i of (6-4) is often referred to as the i-th component of the Euler expression for the Lagrangian L, E denoting the Euler operator associated with L.

Suppose $u \in \mathscr{U}$ extremizes I. Then

$$\langle I'(u), h \rangle = 0 \quad \text{for every } h \in \mathscr{S}$$

implies that ([8], 11.1)

$$L_{u_i} - \frac{d}{dt} L_{\dot{u}_i} = 0, \qquad i = 1, \ldots, n \tag{6-5)'$$

i.e.,

$$E(u) = 0 \tag{6-5)''$$

Note that (6-3) can be rewritten following (3-3) as

$$\langle I'(u), h \rangle = (E(u), h) = \langle \omega_E(u), h \rangle$$

in terms of the weak inner product $(\ , \ )$ on $\mathscr{H}$ defined by

$$(v, h) = \int_a^b v(t) \cdot h(t)\, dt$$

In general, let F be any second order differential operator defined on $\mathcal{U}$ like E, F(u) being a continuous $\mathbb{R}^n$-valued function on [a,b], with components $F_1(u), \ldots, F_n(u)$. Consider the associated 1-form ω_F on $\mathcal{U}$ defined as in (3-3) by

$$\langle \omega_F(u), v \rangle = (F(u), v)$$

This 1-form will be very useful in the following investigation on whether F is the Euler operator associated with a suitable Lagrangian L_F. Suppose this is indeed the case and let L be the corresponding Lagrangian, then form I as in (6-2). Then

$$\bar{u} \text{ extremizes } I \Longleftrightarrow F(\bar{u}) = 0$$

and whether an extremizer of I gives a maximum, a minimum or something else results from considering the second derivative of I, say I''. It can be shown ([16], p. 118) that

$$\langle v, I''(u)\, v \rangle = \int_a^b J(u(t); v(t), \dot{v}(t))\, dt \quad (u \in \mathcal{U},\ v \in \mathcal{S})$$

where J stands for the Jacobi operator given by

$$J(u; v, h) = \frac{1}{2} \sum_{i,k=1}^{n} (L_{u_i u_k} v_i v_k + 2L_{\dot{u}_i u_k} h_i v_k + L_{\dot{u}_i \dot{u}_k} h_i h_k)$$

a second order homogeneous function in the variables v, h.

Incidentally, a continuous second derivative is necessarily symmetric [12], that is

$$\langle v, I''(u)\, w \rangle = \langle w, I''(u)\, v \rangle \qquad \text{for each } u \in \mathcal{U};\ v, w \in \mathcal{S}$$

a fact that will prove to be very useful in the sequel, once we have proved that

$$\langle v, I''(u)\, w \rangle = (v, F'(u)\, w) \tag{6-6}$$

for all values of $u \in \mathcal{U}$, $v, w \in \mathcal{S}$. Assuming (6-6) has been established, it will follow that

$$(v, F'(u)\, w) = (w, F'(u)\, v) \tag{6-7}$$

thus proving that self-adjointness of F'(u) as a linear operator on $\mathcal{S}$ is necessary for the existence of Lagrangian L_F associated with the nonlinear operator F.

To prove (6-6) recall the homogeneity properties of $(v, h) \mapsto J(u; v, h)$ on $\mathbb{R}^{2n}$ and invoke Euler's theorem on homogeneous functions ([9], p. 109), from which

$$\sum_{i=1}^{n} (v_i J_{v_i} + \dot{v}_i J_{\dot{v}_i}) = 2J$$

Hence, for $u \in \mathcal{U}$, $v \in \mathcal{S}$

$$\langle v, I''(u)\, v \rangle = \frac{1}{2} \sum_{i=1}^{n} \int_a^b \left(v_i J_{v_i} + \dot{v}_i J_{\dot{v}_i} \right) dt$$

$$= \frac{1}{2} \sum_{i=1}^{n} \int_a^b v_i \left(J_{v_i} - \frac{d}{dt} J_{\dot{v}_i} \right) dt$$

upon integrating by parts and recalling that $v \in \mathcal{S}$ implies v(a) = v(b) = 0. Now, it suffices to consider (6-4) and differentiate in order to obtain

$$[F'(u)\, v]_i = J_{v_i} - \frac{d}{dt} J_{\dot{v}_i}$$

thus leading to

$$\langle v, I''(u)v\rangle = \frac{1}{2}\sum_{i=1}^{n}\int_a^b v_i[F'(u)v]_i\,dt = (v, F'(u)v)$$

Finally, letting $v, w \in \mathcal{S}$ observe that

$$\langle v, I''(u)w\rangle = \frac{1}{2}\{\langle v+w, I''(u)(v+w)\rangle - \langle v, I''(u)v\rangle - \langle w, I''(u)w\rangle\}$$

$$= \frac{1}{2}\{(v+w, F'(u)(v+w)) - (v, F'(u)v) - (w, F'(u)w)\} = (v, F'(u)w)$$

which proves (6-6). Summing up, only nonlinear operators F with $F'(u)$ everywhere self-adjoint on $\mathcal{S}$ are candidates for being the Euler expression of a Lagrangian.

Recall, however, that $F'(u)$ being self-adjoint for each u is equivalent to ω_F being closed, as pointed out in Sec. 5. It only remains to prove that $\mathcal{U}$ is star shaped in order to have a potential V associated with ω_F on $\mathcal{U}$, in view of Poincaré's Lemma. For, let's prove the following

Lemma. If U is a star-shaped open set in $\mathbb{R}^n$ then so is $\mathcal{U}$ as a subset of $\mathcal{M}$.

Proof. Assume without loss of generality that 0 is a center of U. Let u_0 be the linear function interpolating the boundary points in (6-1), then translate $\mathcal{U}$ by $-u_0$ in order to obtain its parallel subspace $\mathcal{S}$. Then the zero function is a center for $\mathcal{S}$. For, $u \in \mathcal{S}$ implies $tu(s) \in U$ for every $t \in [0,1]$ and every $s \in [a,b]$. Moreover $tu(a) = 0$, $tu(b) = 0$ so that $tu \in \mathcal{S}$ and $\mathcal{S}$ is star-shaped as asserted, hence the same is true of $\mathcal{U}$. Q.E.D.

Proposition. Let U be star shaped and construct $\mathcal{U}$ as detailed earlier in this section. Let F be a nonlinear operator of the second order acting on $\mathcal{U}$. Then solving equation $F(u) = 0$ amounts to extremizing a potential V on $\mathcal{U}$ if and only if $F'(u)$ is self-adjoint as an operator on $\mathcal{S}$ for each $u \in \mathcal{U}$.

Proof. It should be clear from the foregoing lemma and the remarks made so far. Q.E.D.

Note, however, that it cannot be asserted in general that the potential V associated with F can in fact be given by Lagrangian, just as in (6-2). Indeed, a potential is known to be given by

$$V(u) = \int_0^1 \langle \omega_F(u_0 + t(u-u_0)),\ u - u_0\rangle\,dt$$

$$= \int_0^1 (F(u_0 + t(u-u_0)),\ u-u_0)\,dt$$

$$= \int_a^b \left\{(u(s) - u_0(s))\cdot\int_0^1 F(u_0 + t(u-u_0))(s)\,dt\right\}ds$$

and conditions under which a C^4 function $L_F\colon U\times\mathbb{R}^n \to \mathbb{R}$ exists such that

$$L(u(s), \dot{u}(s)) = (u(s) - u_0(s))\cdot\int_0^1 F(u_0 + t(u-u_0))(s)\,dt \tag{6-8}$$

are not evident, but see below.

However, the foregoing characterization of those nonlinear second order differential operators that can be derived as conditions for the extremum of a certain functional—not necessarily one of the type (6-2)—does allow us to connect extremal problems and conservation laws. In fact, F'(u) is everywhere self-adjoint if and only if $\omega_F = dV$ (locally) for a suitable V and this happens if and only if V is conserved along the integral curves of the Pfaffian equation associated with ω_F.

The importance of F'(u) being self-adjoint in order to ensure the existence of a potential function V for F is now well established. Let's now concentrate upon characterizing this property in terms rendering it more easily verifiable. For, begin by observing that the Fréchet derivative F'(u) can be computed as usual, giving

$$[F'(u)\,v]_i = \sum_k \left\{ \frac{\partial f_i}{\partial u_k} v_k + \frac{\partial f_i}{\partial \dot{u}_k} \dot{v}_k + \frac{\partial f_i}{\partial \ddot{u}_k} \ddot{v}_k \right\} \tag{6-9}$$

for $i = 1, \ldots, n$, $u \in \mathcal{U}$, $v \in \mathcal{S}$, provided each F_i is given by

$$F_i(u)(t) = f_i(u(t), \dot{u}(t), \ddot{u}(t)) \tag{6-10}$$

Using (6-9) plus the rule for differentiating a product we obtain Santilli's version of Lagrange's identity ([26], p. 57)

$$\sum_{k,i} w_i \left[\frac{\partial f_i}{\partial u_k} v_k + \frac{df_i}{\partial \dot{u}_k} v_k + \frac{\partial f_i}{\partial \ddot{u}_k} \ddot{v}_k \right] = \frac{d}{dt} B(v,w) + \sum_{k,i} v_k \left[\frac{\partial f_k}{\partial u_i} w_i - \frac{d}{dt}\left(\frac{\partial f_k}{\partial \dot{u}_i} w_i \right) + \frac{d^2}{dt^2}\left(\frac{\partial f_k}{\partial \ddot{u}_i} w_i \right) \right]$$

where B is a bilinear form such that

$$B(v,w)\Big|_a^b = 0 \quad \text{if } v, w \in \mathcal{S}$$

The integral version of Lagrange's identity is a direct generalization of Green's ([10], p. 279; [15], p. 3) and can be written in the form

$$(F'(u)\,v,\; w) = (v,\; F'(u)^* w) \tag{6-12}$$

The star denotes here "formal adjoint" and in fact

$$[F'(u)^* v]_k = \sum \left[\frac{\partial f_k}{\partial u_i} w_i - \frac{d}{dt}\left(\frac{\partial f_k}{\partial \dot{u}_i} w_i \right) + \frac{d^2}{dt^2}\left(\frac{\partial f_k}{\partial \ddot{u}_i} w_i \right) \right] \tag{6-13}$$

Recall that $v, w \in \mathcal{S}$ are arbitrary. From (6-12), F'(u) is self-adjoint on $\mathcal{S}$ if and only if

$$[F'(u)^* v]_j = [F'(u)\,v]_j\,; \quad j = 1, \ldots, n;\; v \in \mathcal{S}$$

hence using (6-9) and (6-13) we obtain the following condition, equivalent to self-adjointness:

$$\frac{\partial f_i}{\partial \ddot{u}_k} = \frac{\partial f_k}{\partial \ddot{u}_i} \tag{6-14)'$$

$$\frac{\partial f_i}{\partial \dot{u}_k} + \frac{\partial f_k}{\partial \dot{u}_i} = \frac{d}{dt}\left(\frac{\partial f_i}{\partial \ddot{u}_k} + \frac{\partial f_k}{\partial \ddot{u}_i} \right) \tag{6-14)''$$

$$\frac{\partial f_i}{\partial u_k} - \frac{\partial f_k}{\partial u_i} = \frac{1}{2}\frac{d}{dt}\left(\frac{\partial f_i}{\partial \dot{u}_k} - \frac{\partial f_k}{\partial \dot{u}_i} \right) \tag{6-14)'''$$

These last relations are known in the literature as Helmholtz' conditions for self-adjointness [26].

The foregoing developments can be conveniently summarized in the following

Proposition. A necessary and sufficient condition for the second order nonlinear differential operator F given on $\mathcal{U}$ by (6-10) to have an everywhere self-adjoint derivative F'(u) is that (6-14) should be satisfied everywhere.

As a direct application involving some calculation, this characterization of self-adjointness of F' can be applied to the Euler operator associated with a functional like (6-2). The self-adjointness of E' follows readily, as required.

It is instructive to consider the so-called "kinematic form" [26] of the equations of motion of analytical mechanics as an illustration, namely the case

$$f_i(u, \dot{u}, \ddot{u}) = \ddot{u}_i - \phi_i(u, \dot{u}) \qquad i = 1, \ldots, n$$

It is shown by Santilli ([26], p. 67) that the corresponding nonlinear differential operator has an everywhere self-adjoint Fréchet derivative if and only if $\phi_1, \ldots, \phi_n$ are linear in $\dot{u}$, i.e.,

$$\phi_i(u, \dot{u}) = \sum_{j=1}^{n} a_{ij}(u)\,\dot{u}_j + b_i(u) \qquad i = 1, \ldots, n$$

and, moreover, conditions (6-14) specialize to

$$a_{ij} + a_{ji} = 0 \tag{6-15)'}$$

$$\frac{\partial a_{ij}}{\partial u_k} + \frac{\partial a_{jk}}{\partial u_i} + \frac{\partial a_{ki}}{\partial u_j} = 0 \tag{6-15)''}$$

$$\frac{\partial b_i}{\partial u_j} = \frac{\partial b_j}{\partial u_i} \tag{6-15)'''}$$

It is also instructive to pursue the question of the validity of (6-8) for some L in the self-adjoint kinematic case

$$\ddot{u} = A(u)\,\dot{u} + b(u) \tag{6-16}$$

the entries of A(u) and b(u) satisfying (6-15). For, assuming $\mathcal{U}$ is centered at the origin, a potential is given on those functions $u \in \mathcal{U}$ satisfying the "natural" boundary conditions

$$\dot{u}(a) = 0, \quad \dot{u}(b) = 0$$

by

$$V(u) = \int_a^b \left\{ |\dot{u}(s)|^2 + \sum_{i,j=1}^{n} \alpha_{ij}(u(s))u_i(s)\dot{u}_j(s) + \sum_{i=1}^{n} \beta_i(u(s))u_i(s) \right\} ds$$

This result follows from the general expression given in Poincaré's Lemma, upon integrating by parts and recalling (6-1). Moreover, in it

$$\alpha_{ij}(u) = \int_0^1 t a_{ij}(tu)\,dt \qquad i,j = 1, \ldots, n$$

$$\beta_{ij}(u) = \int_0^1 b_i(tu)\,dt \qquad i = 1, \ldots, n$$

Summing up, provided (6-15) holds, a variational principle can be associated with (6-16) in terms of a Lagrangian function

$$L_{kin}(u, \dot{u}) = |\dot{u}|^2 + \sum_{i,j=1}^{n} \alpha_{ij}(u)u_i\dot{u}_j + \beta(u) \cdot u$$

satisfying (6-8).

It is instructive to observe that all the self-adjointness conditions presented above are merely concrete expressions of the symmetry of V'', the second derivative of the associated potential. In turn, this symmetry amounts to the closedness of the 1-form dV, thus taking us back into the realm of reciprocity conditions and the symplectic framework presented in Sec. 4. Thus, we find that both variational principles and reciprocity are two concrete expressions of the same abstract concept.

A few words concerning the feasibility of the variational approach to differential equations (the inverse problem of the calculus of variations) appear to be in order. Let's begin by noting that self-adjointness suggests some kind of "completeness" of the formulation in question, which if absent may be guaranteed by introducing a number of modifications in the mathematical setup. At least for the case of linear differential operators, variational principles can be obtained by completing a given formulation in a variety of ways. Following Tonti [29], we can quote:

(a) Morse-Feshbach's method of the mirror equation [24]. The main idea is to consider an equation of the form

$$Lu = f \qquad \text{over } 0 \leq t \leq T$$

subject to initial conditions (at t = 0), where L is, say, the familiar heat operator. Then we supplement this equation by introducing its adjoint

$$L^*v = g \qquad \text{over } 0 \leq t \leq T$$

subject now to final conditions (at t = T) and for a suitable g. Then the new matrix differential equation

$$\begin{bmatrix} 0 & L^* \\ L & 0 \end{bmatrix} \begin{bmatrix} u \\ v \end{bmatrix} = \begin{bmatrix} g \\ f \end{bmatrix}$$

has an already self-adjoint operator, hence a variational formulation is possible.

(b) Gurtin's convolution method [18]. The main idea behind Gurtin's approach is this: many linear initial value problems (like those associated with the heat equation) are not reducible to variational form when posed on the usual L_2 inner product space, but the relevant operator may become self-adjoint if we switch to a new inner product so that a variational formulation can in fact be obtained. As an illustration, let H be the heat operator subject to homogeneous boundary conditions over a region Ω, say $H = \frac{\partial}{\partial t} - \Delta$, where Δ stands for the Laplacian in the space coordinates. The relevant inner product is now

$$(u, v) \mapsto \int_0^T \int_\Omega u(t, x)v(T - t, x)\, dx\, dt$$

and it is an easy matter to show that H is now self-adjoint. On the contrary, H is not self-adjoint with respect to the usual L_2 product over $[0, T] \times \Omega$, and in fact $H^* = -\frac{\partial}{\partial t} - \Delta$.

As we know, convolution requires a time inversion in one of the two factors, which in some sense stands for taking the adjoint of the given linear operator. This in turn explains the occurrence of final instead of initial conditions for v in method (a), thus providing a link between both approaches.

Summing up, self-adjointness is an indication of having mathematical descriptions which are complete in some sense. It is clearly inner product dependent, though. Thus any approach designed in order to ensure the existence of a variational principle for a given operator must focus either on formulation completeness (a) or inner product suitability (b).

7. CONCLUSIONS

Certain symmetry conditions occurring in many field theories provide the basis for this study. Some concrete examples were discussed in Sec. 2. The ensuing analysis is based on notions of differentiable manifolds and exterior algebra, extending over Secs. 3-5. A treatment of the inverse problem of the calculus of variations is found in Sec. 6. The results of this analysis can be summarized as follows:

a) Reciprocity is a local notion full of physical meaning. It is given a global formulation in terms of the more abstract notion of Lagrangian submanifold of the cotangent bundle of the configuration manifold associated with the physical system in question. The symplectic formulation underlying this treatment is presented very much in the spirit of Tulczyjew, the submanifold being Lagrangian as a consequence of the closedness of a differential 1-form on the configuration manifold.

b) In turn, the closedness of a differential 1-form is shown to have two very desirable consequences. On the one hand, it implies the existence of local conservation laws for the solutions of a dynamical system written in Pfaffian form. The relevant tools here are Poincaré's Lemma and Frobenius' Integrability Theory. The main area of application of this result lies in the theory of finite dimensional nonlinear systems. On the other hand, consider a vector field (an operator) defined on the infinite dimensional manifold of all paths joining two fixed points in the configuration space of a nonlinear differential system. Very often such an operator can be chosen so that it is annihilated by that path giving the system's time evolution. Closedness of a differential 1-form associated with the given vector field amounts to certain self-adjointness conditions for the Fréchet derivative of the operator (this is local), which are in turn equivalent to the existence of a variational characterization of the system's trajectories.

Thus, closedness in the exterior calculus of differential forms is bound to play a very important role in the formulation of physical theories. When it is associated with a given physical system, it implies a variety of symmetry conditions allowing us to give very convenient descriptions of such a system.

c) A third point to mention concerns the inverse problem of the calculus of variations as presented here. Self-adjointness of linear differential operators is well known to be the key property to consider in this connection. However, Tonti, Magri, and others have emphasized the role played by the inner product with regard to self-adjointness. Tonti points out that inner products can be defined which render self-adjoint some operators, even operators known to be non-self-adjoint with respect to other inner products in the same space. Now, self-adjointness is known to be equivalent to closedness of an appropriate 1-form constructed from the operator in terms of the above mentioned inner product. It can then be concluded that such 1-forms can always be guaranteed to be closed, with all the ensuing advantages for the mathematical treatment of physical theories. The far reaching implications of this generalization will be pursued elsewhere.

8. ACKNOWLEDGMENTS

One of us (DBH) would like to thank J. Morcos, R. Canales, M. España, and S. T. Guillén of the Automation Section, Engineering Institute (UNAM, MEXICO) for very stimulating discussions on differential and exterior calculus on manifolds. EAL is very much indebted to A. B. Budgor (Los Angeles, California), since the motivation and some of the seminal ideas conducting to this work were stimulated from discussions with him. Further joint papers with him are forthcoming. This research was entirely supported by the Universidad Autónoma Metropolitana—Iztapalapa through its Division of Engineering and Basic Sciences.

REFERENCES

1. Abraham, R.; Marsden, J. E.: Foundations of Mechanics. Benjamin-Cummings (1978).
2. Bird, R. B.; Stewart, W. E.; Lightfoot, E. N.: Transport Phenomena. Wiley (1960).
3. Boothby, W. M.: An Introduction to Differentiable Manifolds and Riemannian Geometry. Academic Press (1975).
4. Bourbaki, N.: Varietés Différentielles et Analytiques. Éléments de Mathématique, Fasc. XXXVI, Hermann, Paris (1971).
5. Brayton, R. K.: Nonlinear reciprocal networks, in Mathematical Aspects of Electrical Network Analysis, vol. III, SIAM-AMS Proceedings, American Mathematical Society (1971), pp. 1-15.
6. Buchdahl, H. A.: The Concepts of Classical Thermodynamics. Cambride University Press (1966).
7. Callen, H. B.: Thermodynamics. Wiley (1960).
8. Cartan, H.: Formes Différentielles. Hermann, Paris (1967).
9. Courant, R.: Differential and Integral Calculus, vol. 2. Blackie and Sons, Ltd., Glasgow (1962).
10. Courant, R.; Hilbert, D.: Methods of Mathematical Physics, vol. 1. Wiley (1966).
11. de Groot, S. R.: J. Math. Phys. 4, 147-153 (1963).
12. Dieudonné, J.: Foundations of Modern Analysis. Academic Press (1960).
13. Fermi, E.: Thermodynamics. Dover (1936).
14. Flanders, H.: Differential Forms. Academic Press (1963).
15. Friedmann, A.: Partial Differential Equations. Holt, Rinehart and Winston (1969).
16. Gelfand, I. M.; Fomin, S. V.: Calculus of Variations. Prentice-Hall (1963).
17. Guillemin, E. A.: Introductory Circuit Theory. Wiley (1953).
18. Gurtin, M. E.: Quart. Appl. Math. 22, No. 3 (1964).
19. Kijowski, J.; Tulczyjew, W. M.: A Symplectic Framework for Field Theories. Springer (1979).
20. Lanczos, C.: The Variational Principles of Mechanics. University of Toronto Press (1957).
21. Liusternik, L.; Sobolev, V.: Elements of Functional Analysis. Frederick Ungar, New York (1961).
22. Marsden, J.: Applications of Global Analysis in Mathematical Physics. Publish or Perish, Inc., Boston (1974).
23. Meixner, J.: J. Math. Phys. 4, 154-159 (1963).
24. Morse, P. M.; Feshbach, H.: Methods of Theoretical Physics, vol. 1. McGraw-Hill (1953).
25. Prigogine, I.: Thermodynamics of Irreversible Processes. Charles C. Thomas (1955).
26. Santilli, R. M.: Foundations of Theoretical Mechanics, I. Springer (1978).
27. Schultz, M. H.: Spline Analysis. Prentice-Hall (1973).
28. Thompson, W. (Lord Kelvin): Proc. Roy. Soc. Edinburgh 3, 225 (1854); Trans. Roy. Soc. Edinburgh, Part I 21, 123 (1857); Math. Phys. Papers 1, 232 (1882).
29. Tonti, E.: A systematic approach to the search for variational principles, in Proc. Intl. Conf. on Variational Methods in Engineering, C. A. Brebbia and H. Tottenham (eds.), vol. 1, Southampton University Press (1973), pp. 1-12.
30. Tulczyjew, W. M.: Symp. Math. 14, 247-258 (1974).
31. Tulczyjew, W. M.: Ann. Inst. H. Poincaré, Sec. A, 27, 101-114 (1977).
32. Vainberg, M. M.: Variational Methods for the Study of Nonlinear Operators. Holden-Day (1971).
33. von Westenholz, C.: Differential Forms in Mathematical Physics. North-Holland (1978).
34. Yosida, K.: Functional Analysis. Springer (1980).

20 VARIATIONAL PRINCIPLES IN SOLITON PHYSICS

Ernst Wolfgang Laedke and Karl H. Spatschek / Department of Physics, Universität Essen-GHS, Essen, Federal Republic of Germany

I. INTRODUCTION

In almost every branch of physics solitons became recently popular and of great importance for theoretical considerations and experiments (see, e.g., [1-29]). Although nowadays the word soliton is used in a quite general manner, the more exact distinction [1,9] from a solitary wave is quite useful: A solitary wave is a quasi-stationary solution of a nonlinear wave equation under the restriction that the physically relevant quantities are localized. Only if these solutions are stable with respect to perturbations and collisions among themselves, which might be proved mathematically or is shown in computer calculations, they should be called solitons.

The first observation of a solitary wave was reported by Scott-Russell in 1834 [30]. The theoretical explanation of its structure is due to Boussinesq, Korteweg, de Vries, and others [31,32], whereas the explanation for its great form stability was first given numerically by Kruskal and Zabusky [33] as well as analytically by Gardner, Greene, Kruskal, and Miura in 1967 [34]. Since then, many physical systems in plasma physics, astrophysics, solid state physics, field theory, biophysics, etc., have been discovered to possess solitary wave solutions, and the inverse scattering solution of Gardner et al. [34] could be generalized for some of these models [8,35,36]. Among the many important nonlinear wave equations, the Korteweg-de Vries (KdV) equation [32]

$$\partial_t u + u\partial_x u + \partial_x^3 u = 0 \tag{I.1}$$

the cubic nonlinear Schrödinger equation [18,37]

$$i\partial_t \psi + \partial_x^2 \psi + |\psi|^2 \psi = 0 \tag{I.2}$$

and the sine-Gordon equation [38]

$$\partial_t^2 \phi - \partial_x^2 \phi + \sin \phi = 0 \tag{I.3}$$

are most frequently used. The reason is that they are the simplest model equations for (1) weakly dispersive, (2) weakly nonlinear, and (3) periodic systems, respectively. Note, however, that the models presented so far are only one-dimensional in space; generalizations to two and three dimensions are known and will be discussed, in part, in the present paper.

In this work, we start from a generalization of the cubic nonlinear Schrödinger equation (I.2) by including 3d effects and allowing for a more general nonlinearity [39], i.e.,

$$i\partial_t \psi + \nabla^2 \psi + |\psi|^{\nu} \psi = 0 \tag{I.4}$$

ψ can be interpreted as a (complex) field envelope, $\nu \geq 1$, and $\nabla^2 = \partial_x^2 + \partial_y^2 + \partial_z^2$. For $\nu = 2$, this equation describes, e.g., nonlinear Langmuir waves in plasmas, whereas for $\nu = 4$ the solutions are called spikons. The reason why we include three-dimensional terms is obvious; the inclusion of arbitrary powers in the nonlinear term will be motivated later.

Let us start with Eq. (I.4) as a model equation. We have to show that solitary wave solutions exist. First, Eq. (I.4) is Galilei-invariant implying that without loss of generality we can choose a rest frame with nonmoving solutions. Secondly, since ψ is a complex envelope, we require that the quantity $|\psi|^2$ is stationary and localized in space. The independence of $|\psi|^2$ on the time variable t and the Galilei-invariance lead to quasi-periodic solutions of the form

$$\psi = \psi_s(\vec{r}) \exp(4i\eta_s^2 t/\nu^2) \tag{I.5}$$

where η_s^2 is a still arbitrary parameter.

In Sec. II, we shall investigate the existence of nontrivial solutions $\psi_s(\vec{r})$ which decrease exponentially at infinity. Strauss [40] has proved the existence of such solutions for a large class of nonlinearities which also covers the present form. The main steps of his variational procedure are briefly sketched.

The rest, but that is the main part of the paper, is devoted to a thorough discussion of the stability properties of the quasistationary solutions.

In the past, many contributions to this topic have been due to Makhankov and collaborators [15,20,39]. They use the (approximate) variation of action method (VAM). In the VAM, the test functions for the variation of action are constructed from the solitary wave solution by perturbing its shape and phase. In Sec. III, the applicability of this widely used method is discussed. We shall conclude that the VAM is appropriate for a rough estimate of stability properties, but neither instability nor stability can be predicted precisely.

In Sec. IV, we shall present an exact necessary and sufficient stability criterion. First, we shall consider the stability part: A Lyapunov functional for stability [41-48] is presented. Secondly, an energy principle for instability [49-52] is formulated. We generalize the naive linear instability concept by including nonlinear terms to show nonlinear instability. The relation to collapse [37] is also discussed.

The stability criterion, formulated in Sec. IV, will be evaluated for various one- and higher-dimensional solitons in Sec. V. The paper is concluded by a short outlook where further applications and possible generalizations are mentioned.

Although the theorems of this paper are proved for the generalized nonlinear Schrödinger equation (I.4), the reader should have in mind that most procedures used here are quite general and can be applied to other equations too.

II. STATIONARY STATES AND VARIATIONAL PRINCIPLE

We start with Eq. (I.4) as the basic equation. As has been discussed already, we are interested in the quasi-stationary solutions (I.5), where $\psi_s(\vec{r})$ is determined by

$$-4(\eta_s^2/\nu^2)\psi_s + \nabla^2\psi_s + \psi_s^{\nu+1} = 0 \tag{II.1}$$

In one dimension, the solution of Eq. (II.1) can be found by direct integration,

$$\psi_s(x) = [2(\nu+2)/\nu^2]^{1/\nu} \eta_s^{2/\nu} \operatorname{sech}^{2/\nu}(\eta_s x) \tag{II.2}$$

Radial solutions in higher dimensions are also of interest. Their existence has been shown by Strauss [40]; in the following we briefly sketch the main stages of his variational procedure. In two dimensions we are looking for cylindrically symmetric solutions, $\psi_s(\rho)$, and in three dimensions for spherically symmetric ones, $\psi_s(r)$.

For our purpose, the main conclusions can be drawn from theorem 2 of [40]: Assume the following:

$$F_1(s) \geq 0, \quad F_2(s) > 0 \quad \text{for } s > 0 \tag{II.3}$$

$$\text{As } s \to 0, \; |F_1(s)/s| \leq C < \infty \text{ and } F_2(s)/s \to 0 \tag{II.4}$$

$$\left|\frac{F_2(s)}{s^{\ell} + F_1(s)}\right| \leq C < \infty \text{ and } \frac{F_2(s)}{s^{\ell} + G_1(s)/s} \to 0 \text{ as } s \to \infty \tag{II.5}$$

where ℓ is an arbitrary positive number for $n = 2$ and $\ell = (n + 2)/(n - 2)$ for $n \geq 3$; G_1 and G_2 are indefinite integrals of F_1 and F_2. Then there exists $\lambda > 0$, and a solution u of

$$-\nabla^2 u + a_0 u + F_1(u) = \lambda F_2(u), \quad a_0 > 0 \tag{II.6}$$

which is nonnegative, belongs to the Sobolev space $H^{1,2}$, decays exponentially as $|\vec{r}| \to \infty$, and $\int d\Gamma G_1(u) < \infty$.

Let us apply this theorem to our problem. Dividing through λ, scaling the space variables and letting

$$F_1(u) = u \tag{II.7}$$

$$F_2(u) = u^{\nu+1} \tag{II.8}$$

$$(a_0 + 1)/\lambda = 4\eta_S^2/\nu^2 \tag{II.9}$$

the criteria (II.3)-(II.5) are obviously satisfied provided

$$\nu > 0 \quad \text{for } n = 2 \tag{II.10}$$

$$4 > \nu > 0 \quad \text{for } n = 3 \tag{II.11}$$

Here, we have used

$$G_1 = \frac{1}{2} u^2 \tag{II.12a}$$

$$G_2 = \frac{1}{\nu + 2} u^{\nu+2} \tag{II.12b}$$

Thus, only in the region (II.10) or (II.11) the existence of radial soliton solutions is guaranteed.

A point is worth mentioning at this stage. If, as has been just quoted, a solution of

$$-\nabla^2 u + \Lambda u - u^{\nu+1} = 0 \tag{II.13}$$

with a positive Λ exists, then because of the polynomical structure of F_1 and F_2, the eigenvalue spectrum of Λ is continuous. By simple transformations, i.e.,

$$\vec{r} = \Lambda^{-1/2} \vec{r}' \tag{II.14a}$$

$$u = (\Lambda/\Lambda')^{1/\nu} u' \tag{II.14b}$$

we obtain from (II.13)

$$-\nabla'^2 u' + \Lambda' u' - (u')^{\nu+1} = 0 \tag{II.15}$$

This symmetry will be used later.

For future purposes it is also helpful to recall the key step in the proof [40] of the existence theorem.

The existence is proved by analyzing the variational principle

$$I = \inf \int d\Gamma [(\nabla u)^2 + (a_0 + 1)u^2] \tag{II.16}$$

under the extra-condition

$$\frac{1}{\nu + 2} \int d\Gamma u^{\nu+2} = 1 \tag{II.17}$$

Note that Eq. (II.6) is the corresponding Euler equation with the Lagrange parameter λ.

Strauss [40] succeeded to show that the variational principle (II.16) is well-defined and that the infimum will be attained.

It is interesting to note that the principle (II.16, 17) cannot be used for stability calculations. For $\nu > 0$, the left hand side of (II.17) is not a constant of motion of our basic equation. Therefore the calculation of a minimum under that restriction does not imply stability under arbitrary perturbations. Later on we shall derive a variational principle for stability which has the constant of motion

$$\int d\Gamma\, u^2 = \text{const} \tag{II.18}$$

as a subsidiary condition.

Nevertheless, besides its importance for existence, the variational principle (II.16, 17) will help us to prove some of the definiteness properties of operators appearing in stability considerations.

III. VARIATION OF ACTION METHOD (VAM)

Using the method (VAM) developed by Makhankov and collaborators [15], we now study the two-dimensional stability of plane soliton solutions (II.2).

The variation of action can be written in the form

$$\delta S \equiv \delta \int dx\, dy\, dt\, (L - L_\infty) = 0 \tag{III.1}$$

where L is the Lagrangian,

$$L = -\frac{i}{2}(\psi\, \partial_t \psi^* - \psi^* \partial_t \psi) + \nabla\psi \cdot \nabla\psi^* + \frac{2}{\nu + 2} |\psi|^{\nu+2} \tag{III.2}$$

For localized solutions, one has $L_\infty = 0$.

In the VAM, the test functions are constructed [39] from the solitary wave solution (II.2) by perturbing its shape and phase. In the present case the appropriate choice, used by Katyshev et al. [39], is

$$\psi = A \operatorname{sech}^{2/\nu}(Dx) \exp(-i\phi) \tag{III.3}$$

Here, the coefficients A, D, and ϕ depend on the transverse coordinate y and the time t; the unperturbed values are $A_0 = [2(\nu + 2)/\nu^2]^{1/\nu} \eta_S^{2/\nu}$, $D_0 = \eta_S$, and $\phi_0 = 4\eta_S^2/\nu^2$.

Inserting Eq. (III.3) into the Lagrangian (III.2) and performing the x-integration in the action integral, one obtains

$$S = \frac{A^2 \partial_t \phi}{D} - \frac{4I(\nu)}{\nu^2 B\left(\frac{2}{\nu}, \frac{1}{2}\right)} \frac{A^2 (\partial_y D)^2}{D^3} + \frac{A(\partial_y A)(\partial_y D)}{D^2} - \left[\frac{(\partial_y A)^2}{A^2} + (\partial_y \phi)^2\right] \frac{A^2}{D} - \frac{4DA^2}{\nu(\nu + 4)} + \frac{8A^{\nu+2}}{(\nu + 2)(\nu + 4)D} \tag{III.4}$$

where $B\left(\frac{2}{\nu}, \frac{1}{2}\right)$ is the Euler beta function, and

$$I(\nu) = \int_{-\infty}^{+\infty} \frac{\xi^2 \tanh^2 \xi}{\cosh^{4/\nu} \xi}\, d\xi \tag{III.5}$$

Taking the variations with respect to A, D, and ϕ one obtains the Euler equations. Linearizing the latter near the soliton solution in the form

$$A = A_0 + \delta A \exp(-i\omega t + iky)$$
$$D = D_0 + \delta D \exp(-i\omega t + iky) \qquad (III.6)$$
$$\phi = \phi_0 + \delta\phi \exp(-i\omega t + iky)$$

one gets the following dispersion equation [39] describing the transverse stability ($\omega^2 > 0$) and transverse instability ($\omega^2 < 0$) of the soliton solution (II.2), i.e.,

$$\omega^2 = \left\{\left[1 - \frac{16}{\nu^2}\frac{I(\nu)}{B\left(\frac{2}{\nu}, \frac{1}{2}\right)}\right]k^6 - \frac{8A_0^{\nu}}{\nu + 4}\left[\frac{3\nu}{\nu + 2} - \frac{8I(\nu)}{\nu B\left(\frac{2}{\nu}, \frac{1}{2}\right)}\right]k^4 + \frac{32\nu^2 A_0^{2\nu}}{(\nu + 2)^2(\nu + 4)}k^2\right\}\left\{\left[1 - \frac{16}{\nu^2}\frac{I(\nu)}{B\left(\frac{2}{\nu}, \frac{1}{2}\right)}\right]k^2 + \frac{4\nu(\nu - 4)}{(\nu + 2)(\nu + 4)}A_0^{\nu}\right\}^{-1} \qquad (III.7)$$

In Sec. V, this growth rate, for $\nu = 4$, is compared with the exact values derived from a variational principle (Sec. IV). Let us first discuss when instability occurs. Taking the limit $k \to 0$, (III.7) reduces to

$$\omega^2 \approx \frac{8\nu A_0^{\nu}}{(\nu + 2)(\nu + 4)}k^2 \qquad (III.8)$$

implying that for $\nu < 4$, the solitons are unstable with respect to perturbations in the transverse direction. For $\nu = 4$, Eq. (III.8) has to be replaced by

$$\omega^2 = k^4 - 2A_0^4 k^2 - \frac{64}{45}A_0^8 \qquad (III.9)$$

meaning that solitons (spikons) with $\nu = 4$ are (exponentially) unstable in both the longitudinal and transverse direction. For $\nu > 4$, the solitons are stable in the transverse direction.

However, more exact calculations (from variational principles derived in Sec. IV) show that the growth rates (III.7) are only approximately valid and the stability results by the VAM are erroneous for $\nu \geq 4$.

Here, we only briefly comment on the variation of action method: more details can be found in [50]. In general, stability cannot be concluded by the VAM since it restricts the perturbed states to a certain subclass. On the other hand, for instability, the variational principle is for the action and not for the growth rate. Therefore, trial functions approximating the action do not necessarily imply a proper dispersion relation. In other words, it is by no means clear that those trial functions are approximate solutions of the dynamic equations, when the whole x-dependence is retained.

Another obvious reason restricts the applicability of the VAM: When constructing test functions from the solitary wave solution by perturbing its shape and phase, the the quasi-stationary solution ψ_s has to be known analytically. Obviously, in higher dimensions this is not true anymore and therefore, not only for the reason of exactness, but also for the reason of availability, another method is necessary.

IV. ENERGY PRINCIPLE

Our aim is to find exact, necessary, and sufficient stability criteria for soliton solutions in arbitrary dimensions ($d \leq 3$). First, we shall derive a sufficient criterion for stability; secondly, this will be complemented by a sufficient criterion for instability.

A. Stability

To discuss the stability properties, we construct a Lyapunov functional out of the constants of motion. For Eq. (I.4) we have the invariants

$$I_1 = \int d\Gamma\ \psi\psi^* \tag{IV.1}$$

$$I_2 = \int d\Gamma\ \psi\nabla\psi^* \tag{IV.2}$$

$$I_3 = \int d\Gamma\left[\nabla\psi\cdot\nabla\psi^* - \left(\frac{\nu}{2}+1\right)^{-1}|\psi|^{\nu+2}\right] \tag{IV.3}$$

Because of Galilei invariance, we need only I_1 and I_3 to construct

$$L = \int d\Gamma\left[\nabla\psi\cdot\nabla\psi^* - \left(\frac{\nu}{2}+1\right)^{-1}|\psi|^{\nu+2} - (\nabla\psi_s)^2 + \left(\frac{\nu}{2}+1\right)^{-1}\psi_s^{\nu+2}\right]$$

$$+ \frac{4}{\nu^2}\eta_s^2 I_1^{\Theta} I_{1s}^{-\Theta}\int d\Gamma\,[\psi\psi^* - \psi_s^2]$$

where

$$I_{1s} = \int d\Gamma\ \psi_s^2 \tag{IV.5}$$

Here, and in the following, we do not restrict the dimension of the system and write $d\Gamma$ for the infinitesimal volume element. The parameter θ is a positive number; its lower limit will be determined later.

Since L is built out of the constants of motion,

$$\partial_t L = 0 \tag{IV.6}$$

follows. Furthermore

$$L = 0, \qquad \text{for } \psi = \psi_s \exp(i4\eta_s^2 t/\nu^2) \tag{IV.7}$$

is trivially satisfied.

In order to prove stability in the sense of Lyapunov [42-44] we use his stability definition and stability theorem: An invariant set S (here: all solitons of the same form) is stable if in a neighborhood U_ε of S a functional $L(\psi)$ with the following properties exists. (i) Within U_ε, $L(\psi)$ can be estimated in terms of the distance $d(\psi, S)$ between the perturbed state ψ and the invariant set, i.e., for all $c_1 > 0$ there exists a $c_2 > 0$ such that for all $\psi \in U_\varepsilon$ with $L(\psi) < c_2$ the bound $d(\psi, S) < c_1$ follows. (ii) For all $d_1 > 0$ there exists a $d_2 > 0$ such that for all $\psi \in U_\varepsilon$ with $L(\psi) < d_2$ the bound $d(\psi, S) < d_1$ follows. (iii) $\partial_t L \leq 0$ and $L(\psi \in S) = 0$ should hold.

In order to prove that L as defined in Eq. (IV.4) is a Lyapunov functional we have to show the estimates (i) and (ii). In addition, we have to define the invariant set as consisting of all solitons of the same form which might be generated from (I.5) and (II.2) by space translations and rotations in phase. We shall come back to this requirement of form stability at the end of this subsection.

Inserting a perturbed state ψ into (IV.4), L consists of first (L_1), second (L_2), and higher order (L_3) terms in the perturbations.

Calculating the first order contribution, which is identical to the first variation of L, we find

$$L_1 = \int d\Gamma\left[-\nabla^2\psi_s - \psi_s^{\nu+1} + \frac{4}{\nu^2}\eta_s^2\psi_s\right]\delta\psi^* + \text{c.c.} \tag{IV.8}$$

Recalling (II.1), which is the corresponding Euler equation, we immediately obtain

$$L_1 = 0 \tag{IV.9}$$

For the second order term L_2, one gets after some algebra

$$L_2 = \int d\Gamma\,(aH_-a + bH_+b) + \frac{16}{\nu^2}\theta\frac{\eta_s^2}{I_{1S}}\left[\int d\Gamma\, a\psi_s\right]^2 \tag{IV.10}$$

where

$$\psi = (\psi_s + a + ib) \exp(i4\eta_s^2 t/\nu^2) \tag{IV.11}$$

$$H_+ = -\nabla^2 - \psi_s^\nu + \frac{4}{\nu^2}\eta_s^2 \tag{IV.12}$$

$$H_- = H_+ - \nu\psi_s^\nu \tag{IV.13}$$

The higher order contributions L_3 can be handled when use is made of the Sobolev inequality following from the Sobolev imbedding theorem [53]. The latter theorem allows to estimate terms of the form $\int d\Gamma\, a^\mu b^\kappa \ldots$, for $\mu + \kappa \leq \nu$, in terms of the norm N^2 of the perturbations. To be more specific, we can find appropriate estimates exactly in that region of ν-values where the existence of the quasi-stationary soliton solutions is guaranteed by the methods of Sec. II.

One then arrives at an estimate

$$L \leq L_2 + F(N^2) \tag{IV.14}$$

where $F(N^2)$ is a positive function of N^2, which, in polynomial representation, starts with a term N^μ with $\mu \geq 3$.

From (IV.14) it is also obvious to find

$$L \geq L_2 - F(N^2) \tag{IV.15}$$

Once we prospered in estimating the higher order terms, a sufficient criterion for stability is

$$L_2 > 0, \quad \text{for nonvanishing perturbations} \tag{IV.16}$$

anticipating that then L_2 can be estimated in terms of the norm.

However, L_2 can still vanish. The modes

(i) $b = \psi_s$ and $a = 0$ and (ii) $b = 0$ and $a = (\nabla\psi_s)_i$, $i = x, y, z$

will do it. These are the so-called rotation and translation modes, respectively. If we translate the original soliton in space and rotate its phase

$$\psi_s(\vec{r}) \exp\left(i\frac{4}{\nu^2}\eta_s^2 t\right) \to \psi_s(\vec{r} - \vec{\alpha}) \exp\left(i\frac{4}{\nu^2}\eta_s^2 t + i\alpha_0\right) \tag{IV.17}$$

we get a soliton of the same form. Physically, only stability with respect to form [45-47] is of interest. Since a perturbed state may show in its time evolution a translation and rotation which do not affect the form, we construct from the original stationary state an invariant set by allowing for arbitrary parameters $\vec{\alpha}$ and α_0. The distance between a perturbed state

$$\psi = [\psi_s(\vec{r} - \vec{r}_0) + a(\vec{r} - \vec{r}_0, t) + ib(\vec{r} - \vec{r}_0, t)] \exp\left[i\xi_0 + i\frac{4}{\nu^2}\eta_s^2 t\right] \tag{IV.18}$$

and the invariant set is then determined by minimizing with respect to the invariant set

$$\partial_{\alpha_i} \|\psi_s + a + ib - \psi_s(\vec{r} - \vec{\alpha}) \exp(i\alpha_0)\|^2 \Big|_{\substack{\vec{\alpha} = \vec{r}_0 \\ \alpha_0 = \xi_0}} = 0 \tag{IV.19}$$

where we use the norm

$$N^2 \equiv \|A\|^2 = \int d\Gamma \left[|\nabla A|^2 + \frac{4}{\nu^2}\eta_s^2 |A|^2\right] \tag{IV.20}$$

Evaluating (IV.19) we find

$$\langle a \,|\, \nabla \psi^{\nu+1} \rangle = 0 \tag{IV.21}$$

and

$$\langle b \,|\, \psi^{\nu+1} \rangle = 0 \tag{IV.22}$$

where here and in the following we use the scalar product notation $\langle \xi | T | \eta \rangle \equiv \int d\Gamma \, \xi T \eta$. The rotation and translation mode are now forbidden. The concept of stability with respect to form excludes $L_2 = 0$.

By allowing for translations and rotations in phase we have effectively constructed an invariant set. However, a word of caution is necessary. Since during the time-development the shift parameters $\vec{\alpha}$ and α_0 can also become time-dependent, the property (IV.6) has to be checked again. However, L is still time-independent since because of the infinite integration domain a change of integration variables eliminates the time-dependent translation parameter $\vec{\alpha}$.

A.1. Longitudinal Stability. By longitudinal stability we mean that the perturbations a and b depend in the one-dimensional case only on x, in the two-dimensional case only on ρ, and in the three-dimensional case only on r (besides the time t). (The more general case, when, for example in 3d, the perturbations can depend on r, ϕ, θ is called transverse stability and will be considered in IV.A.2.) Then, the criterion (IV.10, 16) can be further evaluated in general.

Let us first briefly discuss the definiteness properties of the operators H_+ and H_-. In one dimension, obviously H_+ is positive semidefinite since it is a Schrödinger operator with $H_+\psi_s = 0$ and ψ_s has no nodes. Because of $H_-\partial_x\psi_s = 0$, H_- has one negative eigenvalue.

In two and three dimensions, we shall consider radial (cylindrically and spherically symmetric) solutions ψ_s. Then, as long as only the longitudinal part (∇_ℓ^2) of the ∇^2-operator is considered, we have

$$\nabla_\ell^2 \equiv \begin{cases} \dfrac{1}{\rho}\partial_\rho \rho \partial_\rho , & \text{for 2d} \\[2ex] \dfrac{1}{r^2}\partial_r r^2 \partial_r , & \text{for 3d} \end{cases} \tag{IV.23}$$

The (longitudinal) operators H_+^ℓ and H_-^ℓ, e.g., in 3d,

$$\begin{aligned} H_+^\ell f &= \left(-\nabla_\ell^2 - \psi_s^\nu + \frac{4}{\nu^2}\eta_s^2\right) f \\ &= -\frac{1}{\psi_s r^2}\,\partial_r\left[r^2\psi_s^2\partial_r\left(\frac{f}{\psi_s}\right)\right] \end{aligned} \tag{IV.24}$$

and

$$H_-^\ell = H_+^\ell - \nu\psi_s^\nu \tag{IV.25}$$

have the same definiteness properties as in the 1d case. For H_+^ℓ, the statement follows from the second part of (IV.24). A discussion of H_-^ℓ is more sophisticated, but follows straightforwardly from the arguments given by Strauss [40]. His variational proof of theorem 2 (Ref. [40]) implies that $\langle a | H_-^\ell | a \rangle \geq 0$ under the one restriction (II.17),

$$\int d\Gamma \, a^{\nu+2} = 1 \tag{IV.26}$$

Thus the operator H_-^ℓ can only have one negative eigenvalue.

We now investigate L_2 as calculated in (IV.10) to derive a sufficient criterion for stability according to Lyapunov's stability theorem. For the estimate in terms of the norm of the perturbations, the sign of $\langle a | H_- | a \rangle$ is crucial; the other terms, e.g., $\langle b | H_+ | b \rangle$, are always positive.

Let us first consider perturbations with $\langle a | \psi_s \rangle = 0$. In the following we omit the superscript ℓ, for reason of simplicity. The negative eigenvalue of H_- is denoted by λ_-; the corresponding eigenfunction is e_-. In general, a function a has a component a_- parallel to e_-, and a component $a_\perp$ perpendicular to e_-. We then have

$$\langle a | H_- | a \rangle = -|\lambda_-| \langle a_- | a_- \rangle + \langle a_\perp | H_- | a_\perp \rangle \tag{IV.27}$$

Note also that, for reason of simplicity, here and in the following calculations presented in this subsection periodic boundary conditions are assumed for the integration over the perpendicular coordinates. Furthermore, abbreviating

$$F = H_-^{-1} \psi_s \tag{IV.28}$$

we get

$$0 = -|\lambda_-| \langle a_- | F_- \rangle + \langle a_\perp | H_- | F_\perp \rangle \tag{IV.29}$$

Using this and applying Schwarz inequality, one obtains

$$\langle a_\perp | H_- | a_\perp \rangle \geq |\lambda_-|^2 \frac{\langle a_- | F_- \rangle^2}{\langle F_\perp | H_- | F_\perp \rangle} \tag{IV.30}$$

For

$$\langle \psi_s | H_-^{-1} | \psi_s \rangle = \langle F | H_- | F \rangle < 0 \tag{IV.31}$$

we obtain

$$\langle F_\perp | H_- | F_\perp \rangle < |\lambda_-| \langle F_- | F_- \rangle \tag{IV.32}$$

Combining (IV. 27, 31, and 32), we get the desired result

$$\langle a | H_- | a \rangle \geq 0 \tag{IV.33}$$

Next, we allow for perturbations with $\langle a | \psi_s \rangle \neq 0$. Then

$$\langle a | H_- | a \rangle \geq \langle a | \psi_s \rangle^2 \langle \psi_s | H_-^{-1} | \psi_s \rangle^{-1} \tag{IV.34}$$

since H_- has only one discrete negative eigenvalue. The proof of relation (IV.34) uses direct variational methods and can be found in Ref. [52].

Since now

$$\langle a | H_- | a \rangle + \frac{16}{\nu^2} \theta \frac{\eta_s^2}{I_{1s}} \langle a | \psi_s \rangle^2 \geq \left[\langle \psi_s | H_-^{-1} | \psi_s \rangle^{-1} + \frac{16}{\nu^2} \theta \frac{\eta_s^2}{I_{1s}} \right] \langle a | \psi_s \rangle^2 \tag{IV.35}$$

we can always find a critical value θ_c, such that for $\theta > \theta_c$ the right hand side of (IV.35) is positive. From this rough discussion we conclude that L is nonnegative in the neighborhood of the stationary soliton state, if (IV.31) holds.

The proof that L can indeed be estimated in terms of the norm has recently been presented [54] for the one-dimensional case. The generalization to 2d and 3d can be performed in a similar way. Then, as we have seen already, the eigenfunctions related with the zero eigenvalues are forbidden and a positive estimate in terms of the norm is possible. (Note that this statement holds only for the longitudinal case.)

From [54] we therefore have

$$\varepsilon_2 N^2 \geq L_2 \geq \varepsilon_1 N^2 \tag{IV.36}$$

where ε_1 and ε_2 are positive constants and N^2 is the norm of the perturbations a and b [compare (IV.20)].

Combining (IV.14, 15, and 36) we get

$$\varepsilon_2 N^2 + 0(N^3) \geq L \geq \varepsilon_1 N^2 - 0(n^3) \tag{IV.37}$$

Within a nonlinear stability proof (IV.37) shows that for a given small initial perturbation with finite N^2, belonging to a well-defined neighborhood of the quasi-stationary point, these perturbations cannot grow infinitely. The reason is that L is a constant of motion. Prescribing an upper bound ε for the perturbations, one can always construct an upper bound for the initial perturbations $N^2(t = 0) < \delta$ in the following manner: Given ε, one can find via the left hand side of (IV.37) some $\tilde{L}$ such that perturbed states with $L \leq \tilde{L}$ have $N^2(t) < \varepsilon$. The fact that L is constant guarantees, together with the right hand side of (IV.37), the existence of some $\delta > 0$.

We want to emphasize that a nonlinear stability proof is necessary for the present case since for reversible systems the linear stable situation is always a critical case in the sense of Lyapunov.

Summarizing the findings we conclude that $L_2 > 0$ is a sufficient stability criterion, where L_2 is given by (IV.10). In this subsection IV.A.1 we have shown that

$$\langle \psi_s | H_-^{-1} | \psi_s \rangle < 0 \tag{IV.38}$$

is sufficient for $L_2 > 0$ [or the estimates (IV.36)] in the longitudinal case.

From

$$H_+ \psi_s = 0 \tag{IV.39}$$

it is straightforward to show that

$$H_- \frac{\partial \psi_s}{\partial \eta_s^2} = -\frac{4}{\nu^2} \psi_s \tag{IV.40}$$

The kernel of H_- is given by $\nabla \psi_s$ and therefore

$$H_-^{-1} \psi_s = -\frac{\nu^2}{4} \frac{\partial \psi_s}{\partial \eta_s^2} \tag{IV.41}$$

can be calculated. Combining (IV.38) and (IV.41), we get the stability condition

$$\partial \eta_s^2 I_{1s} > 0 \tag{IV.42}$$

for longitudinal stability.

It should be mentioned that (IV.42) is sometimes called Q- or N-theorem [10, 15, 23, 54]. It has been derived (in a more or less rigorous manner) by several authors, but one should carefully note that it is only a sufficient criterion for longitudinal stability. Without additional investigations (see Sec. B) nothing is known so far for the dynamical behavior outside the region (IV.42).

A.2. Transverse Stability. When considering longitudinal stability, restrictions for the allowed perturbations follow from the concept of form stability; in addition, the term $\langle a | \psi_s \rangle$ helps essentially to get L positive. When looking to the transverse stability problem, because of the periodic boundary conditions in the transverse direction and the fact that ψ_s is only a function of the longitudinal coordinate, $\langle a | \psi_s \rangle = 0$ is always trivially satisfied and does not yield a constraint on a. Similar conclusions hold for the constraints (IV.21) and (IV.22) which follow from the concept of stability with respect to form.

A new feature appears since now to the longitudinal part of the Laplace operator ∇_ℓ^2 a perpendicular part $\nabla_\perp^2$ is added. Let us look at the consequences for the 3d case with spherically symmetric zeroth order solutions $\psi_s(r)$; then a can be expanded in terms of spherical harmonics

$$a = \sum_{\ell, m} a_{\ell m}(r) Y_{\ell m}(\theta, \phi) + c.c. \tag{IV.43}$$

Using

$$\nabla^2 Y_{\ell m} = -\frac{\ell(\ell+1)}{r^2} Y_{\ell m} \tag{IV.44}$$

and

$$\int_0^{2\pi} d\phi \int_0^{\pi} d\theta \sin\theta\, Y_{\ell m} Y^*_{\ell' m'} = \frac{4\pi}{2\ell+1} \frac{(\ell+m)!}{(\ell-m)!} \delta_{\ell\ell'} \delta_{mm'} \tag{IV.45}$$

we immediately see that

$$\langle a | H_- | a \rangle = \sum_{\ell,m} \int r^2 a^*_{\ell m} \hat{H}^{\ell}_- a_{\ell m}\, dr + \text{c.c.} \tag{IV.46}$$

where

$$\begin{aligned}\hat{H}^{\ell}_- &= \left(-\frac{1}{r^2}\partial_r r^2 \partial_r - (\nu+1)\psi_s^{\nu} + \frac{\ell(\ell+1)}{r^2} + \frac{4}{\nu^2}\eta_s^2\right) \frac{4\pi}{2\ell+1}\frac{(\ell+m)!}{(\ell-m)!} \\ &\equiv \left[H^{\ell}_- + \frac{\ell(\ell+1)}{r^2} \frac{4\pi}{2\ell+1}\frac{(\ell+m)!}{(\ell-m)!}\right]\end{aligned} \tag{IV.47}$$

Two aspects are worth mentioning: (i) Compared to the longitudinal case, a sum over various modes $a_{\ell m}$ occurs; however, only a_{00} is restricted by the consistency relations. [Because of the periodic boundary conditions, no other terms appear in $\langle a | \psi_S \rangle$ and the constraints (IV.21) and (IV.22.] (ii) For finite m and ℓ, H^{ℓ}_- is shifted in the positive direction. From that follows, although now instability is not prohibited, that for large ℓ and m $\hat{H}^{\ell}_-$ becomes positive definite. Thus, any transverse instability will have a cut-off in the transverse wave-numbers. More specific conclusions can be found in Sec. V.

B. Instability and Variational Principle for the Growth Rate

A different procedure is used when a sufficient criterion for (exponential) instability is derived. Again we perturb the soliton in the form (IV.11); now we study explicitly the evolution of the perturbations a and b [51, 55–57]. As will be done here, the analysis should be nonlinear. In order to have a better insight into the rather complicated nonlinear treatment, we first give a short review of the linear background.

B.1. Linear Instability. Using the ansatz (IV.11), we obtain from Eq. (I.4) in linear approximation

$$\partial_t a = H_+ b \tag{IV.48a}$$

$$\partial_t b = -H_- a \tag{IV.48b}$$

We shall consider perturbations with $\langle b | \psi_S \rangle = 0$. A part of b being parallel to ψ_S will not grow exponentially within the linear approximation. Then we redefine $b := H_+ b$. Equations (IV.48) have the constant of motion

$$\langle \partial_t b | H_+^{-1} | \partial_t b \rangle + \langle b | H_- | b \rangle = C \tag{IV.49}$$

Let us assume [49] that we can choose some initial displacement b_0 with

$$\tilde{\gamma}^2 \equiv -\langle b_0 | H_- | b_0 \rangle \langle b_0 | H_+^{-1} | b_0 \rangle^{-1} > 0 \tag{IV.50}$$

Whether this assumption is reasonable or not will be discussed at the end. Choosing for the other initial condition $\partial_t b_0 = \tilde{\gamma} b_0$, it is obvious that for these perturbations C = 0 holds and

$$\partial_t^2 \langle b | H_+^{-1} | b \rangle = -4 \langle b | H_- | b \rangle \tag{IV.51}$$

is true. Using again (IV.49) we get

$$\partial_t^2 \langle b | H_+^{-1} | b \rangle = 4 \langle \partial_t b | H_+^{-1} | \partial_t b \rangle \tag{IV.52}$$

and by Schwarz inequality

$$\partial_t [(\partial_t \langle b | H_+^{-1} | b \rangle) / \langle b | H_+^{-1} | b \rangle] \geq 0 \tag{IV.53}$$

Inequality (IV.53) can be integrated: in the first step we get

$$\partial_t \langle b | H_+^{-1} | b \rangle \geq \tilde{\gamma} \langle b | H_+^{-1} | b \rangle \tag{IV.54}$$

and then by a second integration

$$\langle b | H_+^{-1} | b \rangle \geq \langle b_0 | H_+^{-1} | b_0 \rangle \exp (\tilde{\gamma} t) \tag{IV.55}$$

This shows already that the existence of a b_0 with $\langle b_0 | H_- | b_0 \rangle < 0$ for $\langle b_0 | \psi_s \rangle = 0$ is a sufficient criterion for (exponential) instability [compare Eq. (IV.50)]. This is a very interesting result, since it demonstrates that outside the stability region (IV.16) instability occurs.

Let us elaborate a little more on this important point. We should carefully distinguish between the longitudinal and transversal case. For transversal perturbations, always

$$\langle a | \psi_s \rangle = 0 \tag{IV.56}$$

holds because of the periodic boundary conditions. Then, of course, $\langle a | H_- | a \rangle > 0$ for all a, is a necessary and sufficient stability criterion. In the longitudinal case, however, $\langle \psi_s | H_-^{-1} | \psi_s \rangle < 0$ was sufficient for the stability criterion. Now we demonstrate that in the opposite limit, $\langle \psi_s | H_-^{-1} | \psi_s \rangle > 0$ [the equality sign will be discussed in Sec. IV.B.4], $\langle a | H_- | a \rangle < 0$ with $\langle a | \psi_s \rangle = 0$ is possible: choose

$$a = \langle \xi_- | \psi_s \rangle H_-^{-1} \psi_s - \langle \psi_s | H_-^{-1} | \psi_s \rangle \xi_- \tag{IV.57}$$

where ξ_- is a function for which $\langle \xi_- | H_- | \xi_- \rangle < 0$ holds. (Since H_- has a negative eigenvalue, the function ξ_- exists.) Then we can formulate the <u>necessary and sufficient criterion for stability</u>

$$\langle \psi_s | H_-^{-1} | \psi_s \rangle < 0, \quad \text{in the longitudinal case} \tag{IV.58a}$$

and

$$\langle a | H_- | a \rangle > 0, \quad \text{in the transverse case} \tag{IV.58b}$$

for arbitrary functions a with $\langle a | \psi_s \rangle = 0$.

Obviously, there is still one piece missing in the argumentation. This discussion of the equality signs in the above criteria will be added as a supplement at the end of this section.

In the second step we now show that γ as defined by

$$\gamma^2 = \sup_{\langle a | \psi_s \rangle = 0} \frac{-\langle a | H_- | a \rangle}{\langle a | H_+^{-1} | a \rangle} \tag{IV.59}$$

is the maximum exponential growth rate [49]. Taking arbitrary perturbations ($C \neq 0$) and defining

$$d = b \exp(-\gamma t) \tag{IV.60}$$

we find from the constant of motion (IV.49)

$$\langle \partial_t d | H_+^{-1} | \partial_t d \rangle + 2\gamma \langle \partial_t d | H_+^{-1} | d \rangle + \gamma^2 \langle d | H_+^{-1} | d \rangle + \langle d | H_- | d \rangle = C \exp(-2\gamma t) \tag{IV.61}$$

After integration, this shows $\langle d | H_+^{-1} | d \rangle$ is bounded from above for all times and γ as defined in (IV.59) is the maximum exponential growth rate.

B.2. Complementary Variational Principle. From the principal point of view as well as for numerical computations it is desirable to have a complementary variational principle. Then by choosing appropriate test functions, upper and lower bounds for the actual growth rate can be easily computed.

In the unable case, the maximum exponential growth rate can alternatively be calculated from a minimum principle [52]

$$\gamma^2 = \inf_{\phi \in M} \frac{-\langle \phi | H_- H_+ H_- | \phi \rangle}{\langle \phi | H_- | \phi \rangle} \tag{IV.62}$$

where M is the set of functions ϕ for which $\langle \phi | H_- | \phi \rangle < 0$.

The proof of (IV.62) can be briefly given in two steps. First we show that unstable perturbations with a growth rate greater than that calculated from (IV.62) exist. Let us start with another constant of motion

$$\langle b | H_- H_+ H_- | b \rangle + \langle \partial_t b | H_- | \partial_t b \rangle = C \tag{IV.63}$$

Choosing in the unstable case some (initial) displacements $b_0 \in M$ which satisfy

$$\partial_t b_0 = \tilde{\gamma} b_0 \tag{IV.64}$$

where

$$\tilde{\gamma}^2 = -\langle b_0 | H_- H_+ H_- | b_0 \rangle / \langle b_0 | H_- | b_0 \rangle > 0 \tag{IV.65}$$

we have C = 0. From (IV.48) and (IV.63) we then find

$$-\partial_t^2 \langle b | H_- | b \rangle = 4 \langle b | H_- H_+ H_- | b \rangle \geq -4\gamma^2 \langle b | H_- | b \rangle \tag{IV.66}$$

where the definition (IV.62) has been used. After integration over time one gets

$$-\langle b | H_- | b \rangle \geq -\langle b_0 | H_- | b_0 \rangle \exp(2\gamma t) \tag{IV.67}$$

i.e., growing perturbations exist whose growth rates are at least γ as calculated from Eq. (IV.62).

γ from Eq. (IV.59) is the maximum growth rate and thus γ as calculated from Eq. (IV.62) is less than or equal to γ as calculated from Eq. (IV.59). In order to show that both are identical, we have to prove the opposite statement. The latter follows from

$$\begin{aligned} \sup_{\phi \in M} \frac{-\langle \phi | H_- | \phi \rangle}{\langle \phi | H_+^{-1} | \phi \rangle} &\leq \sup \frac{-\langle \phi | H_- | \phi \rangle}{\langle \phi | H_- | f \rangle^2} \sup \frac{\langle \phi | H_- | f \rangle^2}{\langle \phi | H_+^{-1} | \phi \rangle} \\ &= -\langle f | H_- | f \rangle^{-1} \langle f | H_- H_+ H_- | f \rangle \end{aligned} \tag{IV.68}$$

for all $f \in M$ since H_- has only one negative eigenvalue. Since (IV.68) is true for any $f \in M$, it means that γ^2 as calculated from (IV.59) is equal to or less than γ^2 obtained from (IV.62).

B.3. Nonlinear Instability. Two reasons motivate us to generalize the preceding linear instability calculations into the nonlinear regime. First, it is not generally proved that a linear instability automatically implies a nonlinear one. Secondly, and physically more important, an integral norm should be used instead of the supremum norm. The latter justifies a linear treatment whereas physics only demands that integral quantities of the perturbations should be small. For the present soliton model we prove now that a nonlinear theory confirms the result of the linear calculation of Sec. IV.B.1.

Additional to the terms in (IV.48) we have now to consider the contributions due to the shift parameters $\vec{\alpha}$ and α_0 (since instability with respect to form is investigated) and the higher nonlinearities. Using

$$f - \psi_s \exp\left(i \frac{4}{\nu^2} \eta_s^2 t\right) = (a + ib) \exp\left(i \frac{4}{\nu^2} \eta_s^2 t + i\alpha_0\right)$$

one gets

$$\partial_t a = H_+ b + \dot{\vec{\alpha}} \cdot \nabla\psi_s + \dot{\vec{\alpha}} \cdot \nabla a - \dot{\alpha}_0 b + R_a \tag{IV.69a}$$

$$\partial_t b = -H_- a + \dot{\alpha}_0 \psi_s + \dot{\vec{\alpha}} \cdot \nabla b + \dot{\alpha}_0 a + R_b \tag{IV.69b}$$

where H_+ and H_- are defined by (IV.12) and (IV.13), the dots mean derivatives with respect to time, and R_a and R_b designate the higher order contributions,

$$R_a = (\psi_s^2 + 2a\psi_s + a^2 + b^2)^{\nu/2}(\psi_s + a) - \psi_s^{\nu+1} - 3\psi_s^{\nu} a \tag{IV.70}$$

$$R_b = -[(\psi_s^2 + 2a\psi_s + a^2 + b^2)^{\nu/2} b - \psi_s^{\nu} b] \tag{IV.71}$$

The operators H_+ and H_- have the kernels ψ_s and $\nabla\psi_s$ and we define the invertible parts of a and b by

$$\tilde{a} = a - \kappa\psi_s \quad \text{with } \kappa = \langle a | \psi_s \rangle / \langle \psi_s | \psi_s \rangle \tag{IV.72}$$

$$\tilde{b} = b - \vec{\sigma} \cdot \nabla\psi_s \quad \text{with } \langle \vec{\sigma} \cdot \nabla\psi_s | \nabla\psi_s \rangle = \langle b | \nabla\psi_s \rangle \tag{IV.73}$$

Knowing from the linear instability calculations that perturbed states f which have the same values for the invariants I_1, I_2, and I_3 as the solitons lead to the fastest exponential growth, we consider these perturbations here. (By an argument similar to that presented by Benjamin [45] one can here also show that other perturbations do not grow faster: If the assumption is not satisfied initially, construct a soliton whose invariants I_1 and I_2 coincide with the corresponding ones for the perturbed states. Then the following calculation applies for that situation. But by the triangle inequality it is obvious that any instability with respect to the original soliton is of the same type as one with respect to the intermediate soliton state.)

$$I_\mu(f) = I_\mu(\psi_s), \quad \text{for } \mu = 1, 2, 3 \tag{IV.74}$$

We find from (IV.74)

$$\langle a | \psi_s \rangle = -\frac{1}{2}(\langle a | a \rangle + \langle b | b \rangle) \tag{IV.75}$$

$$\langle b | \nabla\psi_s \rangle = \langle a | \nabla b \rangle \tag{IV.76}$$

$$\langle a | H_- | a \rangle + \langle b | H_+ | b \rangle \equiv R \tag{IV.77}$$

$$= \left(\frac{\nu}{2} + 1\right)^{-1} \int \left[(\psi_s^2 + 2a\psi_s + a^2 + b^2)^{\frac{\nu}{2}+1} - \psi_s^{\nu+2} - (\nu + 2)a\psi_s^{\nu+1} - (\nu + 1)\left(\frac{\nu}{2} + 1\right)a^2\psi_s^{\nu} - \left(\frac{\nu}{2} + 1\right)b^2\psi_s^{\nu} \right] d\Gamma$$

and from (IV.72), (IV.73)

$$\kappa = -\frac{1}{2}(\langle a | a \rangle + \langle b | b \rangle) / \langle \psi_s | \psi_s \rangle \tag{IV.78}$$

$$\vec{\sigma} = (|\nabla\psi_s\rangle\langle\nabla\psi_s|)^{-1}\langle a|\nabla b\rangle \tag{IV.79}$$

The dynamical equations (IV.69), the constraints (IV.72-79) and the consistency relations (IV.21), (IV.22) form the basis for the nonlinear calculations.

Let us first estimate the order of magnitude (in terms of the norm of the perturbabations) of the time derivatives of the shift parameters, i.e., $\dot{\alpha}_0$ and $\dot{\vec{\alpha}}$.

Multiplying Eq. (IV.69a) by $\nabla\psi_s^{\nu+1}$ and Eq. (IV.69b) by $\psi_s^{\nu+1}$ we obtain after integration

$$0 = \langle\nabla\psi_s^{\nu+1}|H_+b\rangle + \langle\nabla\psi_s^{\nu+1}|\dot{\vec{\alpha}}\cdot\nabla(\psi_s + a)\rangle - \dot{\alpha}_0\langle\nabla\psi_s^{\nu+1}|b\rangle + \langle\nabla\psi_s^{\nu}|R_a\rangle \tag{IV.80}$$

$$0 = -\langle\psi_s^{\nu+1}|H_-a\rangle + \dot{\alpha}_0\langle\psi_s^{\nu+1}|\psi_s + a\rangle + \langle\psi_s^{\nu+1}|\dot{\vec{\alpha}}\cdot\nabla b\rangle + \langle\psi_s^{3}|R_b\rangle \tag{IV.81}$$

Obviously, from these equations

$$\dot{\vec{\alpha}} = 0(N), \quad \dot{\alpha}_0 = 0(N) \tag{IV.82}$$

follows, where

$$N^2 = \langle\nabla a|\nabla a\rangle + \langle\nabla b|\nabla b\rangle + \langle a|a\rangle + \langle b|b\rangle \tag{IV.83}$$

In a similar way [multiplying Eqs. (IV.69) by ψ_s and $\nabla\psi_s$, respectively, and integrating] we find from (IV.72) and (IV.73) for the time derivatives of κ and $\vec{\sigma}$

$$\dot{\kappa}\langle\psi_s|\psi_s\rangle = \langle\dot{\vec{\alpha}}\cdot\nabla a|\psi_s\rangle - \dot{\alpha}_0\langle b|\psi_s\rangle + \langle R_a|\psi_s\rangle \tag{IV.84}$$

$$\langle\dot{\vec{\sigma}}\cdot\nabla\psi_s|\nabla\psi_s\rangle = \langle\dot{\vec{\alpha}}\cdot\nabla b|\psi_s\rangle + \alpha_0\langle a|\nabla\psi_s\rangle + \langle R_b|\nabla\psi_s\rangle \tag{IV.85}$$

Using (IV.82), we have

$$\dot{\kappa} = 0(N^2), \quad \dot{\vec{\sigma}} = 0(N^2) \tag{IV.86}$$

Note that because of (IV.78) and (IV.79)

$$\kappa = 0(N^2), \quad \vec{\sigma} = 0(N^2) \tag{IV.87}$$

also hold.

Combining all these findings, we get

$$\partial_t\tilde{a} = H_+\tilde{b} + \dot{\vec{\alpha}}\nabla\psi_s + \Delta_{\tilde{a}} \tag{IV.88}$$

$$\partial_t\tilde{b} = -H_-\tilde{a} + \dot{\alpha}_0\psi_s + \Delta_{\tilde{b}} \tag{IV.89}$$

where $\Delta_{\tilde{a}}$ and $\Delta_{\tilde{b}}$ are higher order terms in $\tilde{a}$, $\tilde{b}$, which are straightforward to calculate.

The energy conservation

$$\langle\tilde{a}|H_-|\tilde{a}\rangle + \langle\tilde{b}|H_+|\tilde{b}\rangle = \Delta \tag{IV.90}$$

with $\Delta = 0(N^3)$, is obtained in a similar manner.

The generalization of the linear instability calculation follows along the following lines:

$$\partial_t\langle\tilde{a}|\tilde{b}\rangle = \langle\tilde{b}|H_+|\tilde{b}\rangle + \langle\dot{\vec{\alpha}}\cdot\nabla\psi_s|\tilde{b}\rangle + \langle\Delta_{\tilde{a}}|\tilde{b}\rangle - \langle\tilde{a}|H_+|\tilde{a}\rangle + \dot{\alpha}_0\langle\tilde{a}|\psi_s\rangle + \langle\tilde{a}|\Delta_{\tilde{b}}\rangle \tag{IV.91}$$

Using the appropriate Sobolev inequalities, we obtain from (IV.91), together with (IV.90),

$$\partial_t\langle \tilde{a} | \tilde{b} \rangle = 2\langle \tilde{b} | H_+ | \tilde{b} \rangle + 0(N^3) \tag{IV.92}$$

Within a linear instability calculation we arrived at a similar equation which, because of neglect of $0(N^3)$ terms, could be integrated. Since this cannot be done anymore now, we define new quantities

$$\phi = \tilde{a} - \vec{\alpha} \cdot \nabla\psi_s \tag{IV.93}$$

$$\eta = \frac{1}{2}\langle \phi | H_+^{-1} | \phi \rangle \tag{IV.94}$$

$$\xi = \langle \tilde{a} | \tilde{b} \rangle \tag{IV.95}$$

From Eq. (IV.88) we get

$$\dot{\phi} = H_+\tilde{b} + \Delta_{\tilde{a}} \tag{IV.96}$$

Together with the definition (IV.94) this leads to

$$\dot{\eta} = \langle \phi | \tilde{b} \rangle + \langle \phi | H_+^{-1} | \Delta_{\tilde{a}} \rangle \tag{IV.97}$$

and remembering $\langle \tilde{b} | \nabla\psi \rangle = 0$ from (IV.73) we obtain by Schwarz inequality

$$\dot{\eta} \leq \xi - \eta^{\frac{1}{2}} 0(N^2) \tag{IV.98a}$$

The additional estimate

$$\dot{\eta} \geq \xi + \eta^{\frac{1}{2}} 0(N^2) \tag{IV.98b}$$

is also true.

Finally (IV.92) and (IV.96) yield

$$\dot{\xi} = 2\langle \dot{\phi} | H_+^{-1} | \dot{\phi} \rangle + 0(N^3) \tag{IV.99}$$

and thus by Schwarz inequality

$$\dot{\xi} \geq \dot{\eta}^2/\eta + 0(N^3) \tag{IV.100}$$

Equations (IV.98) and (IV.100) are the basic relations for the following calculations. We can use them explicitly after relations of the form

$$|\xi| \leq N^2 \leq \delta^2\eta \tag{IV.101}$$

have been established; $\delta^2 > 0$ is a small parameter. To show that the norm N^2 can be estimated by ξ^2 and η as stated in (IV.101) we proceed as follows:

Since H_+ is positive semidefinite and its kernel is forbidden because of the consistency relations, we can find a positive constant p_2 such that in a neighborhood of the quasi-stationary state

$$\frac{1}{p_2} N^2 \leq \langle \tilde{a} | H_+ | \tilde{a} \rangle + \langle \tilde{b} | H_+ | \tilde{b} \rangle \tag{IV.102}$$

holds. A further estimate of the right hand side, i.e., the existence of an analogue constant p_1, follows from (IV.90),

$$\langle \tilde{a} | H_+ | \tilde{a} \rangle + \langle \tilde{b} | H_+ | \tilde{b} \rangle \leq p_1\langle \tilde{a} | \tilde{a} \rangle + 0(N^3) \tag{IV.103}$$

Combining (IV.102) and (IV.103) for small $N < \varepsilon_0 < 1$ one obtains

$$N^2 \leq p_3\langle \tilde{a} | \tilde{a} \rangle, \quad \text{with } p_3 > 0 \tag{IV.104}$$

From (IV.103) with (IV.104) a relation of the form

$$\langle \tilde{a} | H_+ | \tilde{a} \rangle \leq p_4\langle \tilde{a} | \tilde{a} \rangle \tag{IV.105}$$

can be found. Then Schwarz's inequality, when applied to $\langle \tilde{a} | H_+ | \tilde{a} \rangle \geq \langle \tilde{a} | \tilde{a} \rangle^2 / \langle \tilde{a} | H_+^{-1} | \tilde{a} \rangle$, yields

$$\langle \tilde{a} | \tilde{a} \rangle \leq p_4 \langle \tilde{a} | H_+^{-1} | \tilde{a} \rangle \tag{IV.106}$$

Combining (IV.104)-(IV.106) and remembering that H_+ is bounded from below and thus H_+^{-1} is bounded from above, we have

$$N^2 \leq \tilde{p} \langle \tilde{a} | H_+^{-1} | \tilde{a} \rangle \leq \tilde{q} \langle \tilde{a} | H_+ | \tilde{a} \rangle \tag{IV.107}$$

with $\tilde{p}$, $\tilde{q} > 0$.

Calculating $\langle \phi | H_+ | \phi \rangle$ for (IV.93), we can estimate

$$\langle \phi | H_+ | \phi \rangle \geq \langle \tilde{a} | H_+ | \tilde{a} \rangle - 2 \langle \tilde{a} | H_+ | \vec{\alpha} \cdot \nabla \psi_s \rangle \tag{IV.108}$$

The last term on the right hand side of Eq. (IV.108) vanishes since

$$H_+ \nabla \psi_s = \frac{\nu}{\nu + 1} \nabla \psi_s^{\nu+1} \tag{IV.109}$$

and we have the consistency relation (IV.21). Thus

$$\langle \phi | H_+ | \phi \rangle \geq \langle \tilde{a} | H_+ | \tilde{a} \rangle \tag{IV.110}$$

Proceeding in a similar way as before, we recognize that (IV.102) and (IV.103) are also valid when $\tilde{a}$ is replaced by ψ. Therefore, we finally get

$$N^2 \leq \tilde{\tilde{p}} \langle \phi | H_+^{-1} | \phi \rangle \leq \tilde{\tilde{q}} \langle \phi | H_+ | \phi \rangle, \quad \text{with } \tilde{\tilde{p}}, \tilde{\tilde{q}} > 0 \tag{IV.111}$$

$$\langle \phi | H_+ | \phi \rangle \leq p_4 \langle \phi | \phi \rangle \leq 2p_4^2 \eta, \quad \text{with } p_4 > 0 \tag{IV.112}$$

Relations (IV.111,112) yield the right hand side of (IV.101) while the left hand side is obtained from the definition of ξ and Schwarz's inequality.

Combining (IV.98), (IV.100), and (IV.101) we get

$$\dot{\eta} \geq \xi - p\eta^{3/2} \tag{IV.113}$$

$$\dot{\eta} \leq \xi + p\eta^{3/2} \tag{IV.114}$$

$$\dot{\xi} \geq \dot{\eta}^2/\eta - q\eta^{3/2} \tag{IV.115}$$

with p, q > 0.

For $\dot{\eta} > 0$

$$\partial_t(\xi/\eta) \geq -p\dot{\eta}\eta^{-1/2} - q\eta^{1/2} \tag{IV.116}$$

Let us define

$$\gamma_0 = (\xi/\eta)_{t=0} \tag{IV.117}$$

Then, because of $\tilde{a}(t{=}0) = \phi(t{=}0)$

$$\gamma_0 = 2 \left. \frac{\langle \tilde{a} | \tilde{b} \rangle}{\langle \tilde{a} | H_+^{-1} | \tilde{a} \rangle} \right|_{t=0} \tag{IV.118}$$

and we assume that this can be a positive value in accordance with (IV.90). Integrating (IV.116) once and using (IV.113) yields

$$\dot{\eta} \geq \gamma_0 \eta - 3p\eta^{3/2} - q\eta \int_0^t dt\, \eta^{1/2}, \quad \text{for } \dot{\eta} > 0 \tag{IV.119}$$

That inequality can be integrated once more in a small neighborhood of the stationary point $\eta << \tilde{\varepsilon}_0^2$,

$$\dot{\eta} \geq \left[\gamma_0 - \left(3p + 4\frac{q}{\gamma_0}\right)\tilde{\varepsilon}_0\right]\eta \qquad \text{(IV.120)}$$

Using (IV.114),

$$\xi \geq \left[\gamma_0 - \left(4p + \frac{4q}{\gamma_0}\right)\tilde{\varepsilon}_0\right]\eta \qquad \text{(IV.121)}$$

and because of

$$\xi \leq N^2 \qquad \text{(IV.122)}$$

we find instability which is at least exponential.

Let us now discuss when this calculation applies. Obviously it should be possible to have $\gamma_0 > 0$. We take special initial values $H_+\tilde{b}_0 = \gamma\tilde{a}_0$ and insert them into the energy conservation (IV.90); the latter determines γ. Since from Eq. (IV.118) $\gamma_0 = 2\gamma$ follows, the instability proof applies once γ can be chosen positive.

Collecting the various orders one obtains from (IV.90) the possible maximum growth rate

$$\gamma^2 = \sup_{\tilde{a}_0} \frac{-\langle\tilde{a}_0 | H_- | \tilde{a}_0\rangle}{\langle\tilde{a}_0 | H_+^{-1} | \tilde{a}_0\rangle} + 0(N) \qquad \text{(IV.123)}$$

Thus, when linear instability occurs the system also grows nonlinearly, and the linear result (IV.59) is confirmed by a nonlinear calculation.

B.4. The Marginal Case. Now we consider the case $\langle\psi_S | H_-^{-1} | \psi_S\rangle = 0$. We have already seen [in Sec. IV.A.1, Eqs. (IV.41,42)] that then $\partial_{\eta_S^2}\langle\psi_S | \psi_S\rangle = 0$. Using the scaling arguments (II.14) we immediately find that this occurs for

$$\nu = 4/n, \quad n = 1, 2, 3 \qquad \text{(IV.124)}$$

where n is the dimension of the system.

Following Zakharov [37], the time development of

$$I = \frac{1}{2}\int d\Gamma\, \vec{r}^{\,2}\, |\psi|^2 \qquad \text{(IV.125)}$$

is given by

$$\frac{d^2}{dt^2} I \leq 4I_3 \qquad \text{(IV.126)}$$

which leads to the inequality

$$I \leq 2I_3 t^2 + I(0) \qquad \text{(IV.127)}$$

Thus a collapse occurs provided $I_3 < 0$.

We shall now show that in the marginal case $\nu = 4/n$, $I_{3S} = 0$. Since then we can always find perturbed states in the neighborhood of the quasi-stationary state for which $I_3 < 0$ holds, the system will be nonlinearly unstable.

For the proof of $I_{3S} = 0$ for $\nu = 4/n$, we use the relations

$$\int d\Gamma\, |\nabla\psi_S|^2 = -\int d\Gamma\left(\frac{4}{\nu^2}\eta_S^2\psi_S^2 - \psi_S^{\nu+2}\right) \qquad \text{(IV.128)}$$

and

$$\int d\Gamma\left(\frac{2}{\nu^2}\eta_S^2\psi_S^2 - \frac{1}{\nu+2}\psi_S^{\nu+2}\right) = \left(\frac{1}{2} - \frac{1}{n}\right)\int d\Gamma\left(\frac{4}{\nu^2}\eta_S^2\psi_S^2 - \psi^{\nu+2}\right) \qquad \text{(IV.129)}$$

Combining (IV.3), (IV.128), and (IV.129) the assertion follows.

The question arises how this behavior is reflected in the linear instability calculation. Three special solutions of (IV.69),

$$a_1 = H_-^{-1}\psi_S, \quad b_1 = 0 \tag{IV.130}$$

$$a_2 = H_-^{-1}\psi_S t$$

$$b_2 = H_+^{-1}H_-^{-1}\psi_S - \langle \psi_S^5 | H_+^{-1}H_-^{-1} | \psi_S \rangle \langle \psi_S^3 | \psi_S^3 \rangle^{-1} \psi_S \tag{IV.131}$$

$$a_3 = \frac{1}{2} H_-^{-1}\psi_S t^2 - H_-^{-1}H_+^{-1}H_-^{-1}\psi_S, \quad b_3 = b_2 t \tag{IV.132}$$

can be found [25].

One can also show [25] that these (and all combinations) are the only modes being linearly unstable. We do not want to go into the details of this and other questions concerning the marginal states. But the general aspect is worth emphasizing: Modes growing quadratically in time within a linear investigation can lead to a drastic nonlinear instability—the collapse.

V. EVALUATION, GENERALIZATION, AND OUTLOOK

Let us now evaluate the necessary and sufficient stability criteria presented so far. First, we consider longitudinal stability in 1d. Since ψ_S is an even function of x and monotonically decreasing for $0 \leq x < \infty$, we can evaluate

$$\int_{-\infty}^{+\infty} dx\, \psi_S^2 = -2\left(\frac{2\eta_S}{\nu}\right)^{\frac{4}{\nu}-1} \int_{\left(\frac{\nu}{2}+1\right)^{1/\nu}}^{0} \frac{z\, dz}{\left[1 - \left(\frac{\nu}{2}+1\right)^{-1} z^{\nu}\right]^{1/2}} \tag{V.1}$$

to obtain from (IV.42) stability for $0 < \nu < 4$ and instability for $\nu \geq 4$. The transverse instability of the plane soliton solutions (II.2) follows by direct evaluation of the variational principles (IV.59) and (IV.62). Using a Ritz procedure with a Wieland iteration [58, 59], for a finite number of test functions (IV.59) yields lower bounds and (IV.62) yields upper bounds for the actual growth rates. The results of our computations [28] are shown in Fig. 1 and compared with the VAM results for $\nu = 4$ in Fig. 2; the numerical uncertainties lie within drawing accuracy. For small transverse wave numbers k, we can evaluate (IV.59) and (IV.62) analytically. Inserting

$$a = H_-^{-1}\psi_S, \quad \phi = \psi_S \tag{V.2}$$

into the maximum and minimum principle, respectively, we obtain

$$\gamma^2 \approx 16\eta_S^2 k^2/\nu(4+\nu) + 0(k^4), \quad \text{for } \nu < 4 \tag{V.3}$$

For $\nu = 4$,

$$a = H_-^{-1}\psi_S - \left(\frac{\pi}{2}\right) k\psi_S/\eta_S^3 \tag{V.4}$$

$$\phi = \hat{H}_- H_-^{-1}\psi_S - \frac{2}{\pi}\eta_S^3 k H_+^{-1} H_-^{-1}\psi_S$$

with

$$\hat{H}_- = H_- + k^2 = -\frac{d^2}{dx^2} + k^2 - (\nu+1)\psi_S^{\nu} + \frac{4\eta_S^2}{\nu^2} \tag{V.5}$$

lead to

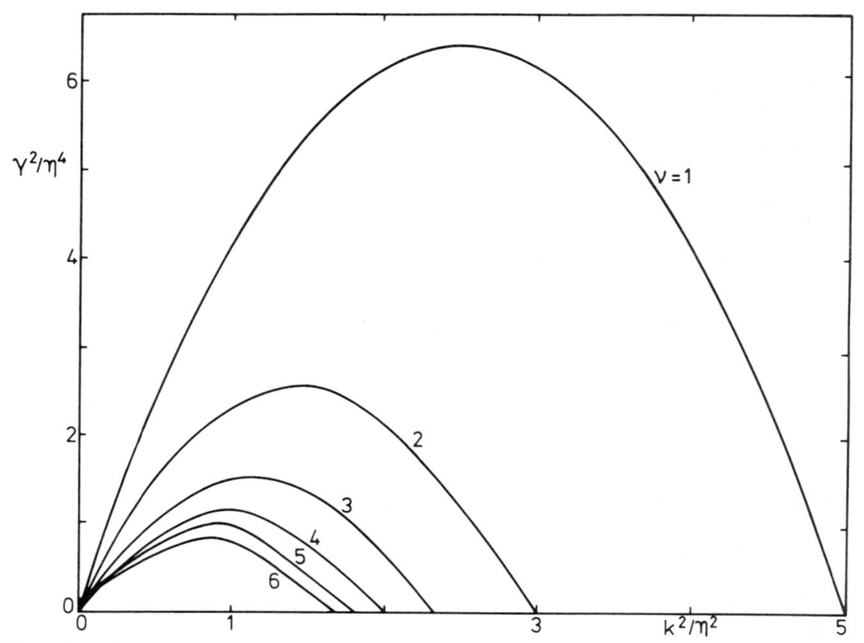

Fig. 1. γ-curves [for definition see Eqs. (IV.59) and (IV.62)] for various parameters ν vs transverse wavenumber k for plane soliton solutions.

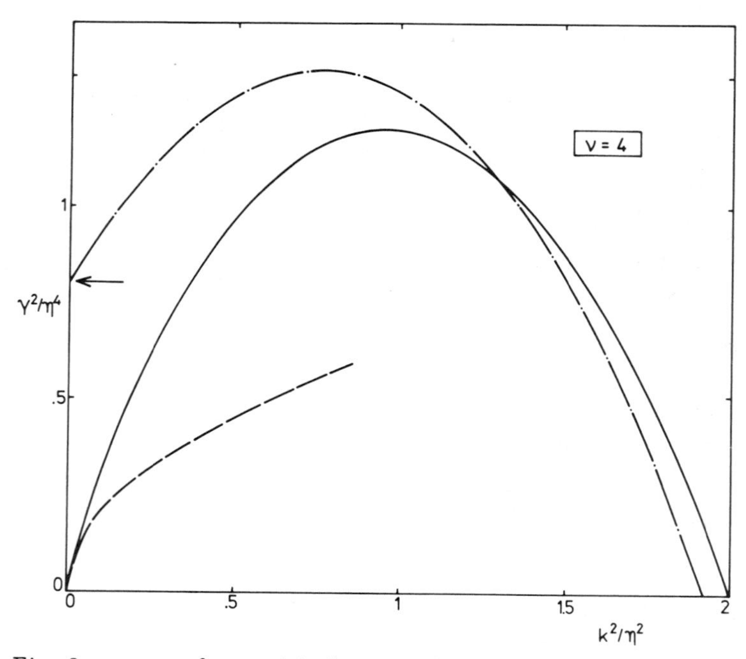

Fig. 2. γ-curve for $\nu = 4$ in the case of a plane soliton solution. The additional line --- depicts the small-k expansion [Eq. (V.6)] whereas the curve -·-·-· shows the approximate result of Ref. [39].

$$\gamma^2 \approx \frac{2}{\pi}\eta_s^3 k + 0(k^2) \tag{V.6}$$

For $\nu > 4$, spikons are both longitudinally and transversely unstable; a simple demonstration of the latter instability follows by inserting

$$a = 2x\partial_x \psi_s + \psi_s \tag{V.7}$$

into the maximum principle (IV.59). After some algebra, one gets for small k the estimate

$$\gamma^2 \geq [8(\nu - 4)\eta_s^4/\nu(\nu + 4)] \int_{-\infty}^{+\infty} dz\, \mathrm{sech}^{(4/\nu)} z \int_{-\infty}^{+\infty} dz\, z^2\, \mathrm{sech}^{(4+\nu)} z + 0(k^2) \tag{V.8}$$

Next, we investigate the 2d situation, i.e., the stability of cylindrically symmetric soliton solutions. For longitudinal instability we use again the scaling arguments to get longitudinal stability for $0 < \nu < 2$ and instability for $\nu \geq 2$.

Transverse perturbations can be characterized by two different modes: the azimuthal ϕ-dependent mode and the z-mode. Instability always occurs with respect to the latter one (by similar arguments as just presented for the plane case), whereas the ϕ-mode does not introduce any new instability. Most arguments are already given in Sec. IV.A.2; we have to add that in the cylindrically symmetric case

$$\hat{H}_-^\ell = H_-^\ell + \frac{m^2}{\rho^2} = -\frac{1}{\rho}\partial_\rho \rho \partial_\rho - (\nu + 1)\psi_s^\nu + \frac{4}{\nu^2}\eta_s^2 + \frac{m^2}{\rho^2} \tag{V.9}$$

and

$$\hat{H}_-^\ell \frac{\partial \psi_s}{\partial \rho} = 0 \quad \text{for } m = 1 \tag{V.10}$$

holds. Since $\hat{H}_-^\ell$ is a Schrödinger operator and $\partial_\rho \psi_s$ has no zeros, the transverse stability with respect to pure ϕ-dependent perturbations follows.

Finally, in 3d the system is longitudinally stable for $0 < \nu < 4/3$. A transverse instability (ϕ and θ-dependent perturbations) will not occur. The reason is again (as in the case of ϕ-dependent perturbations in the 2d case) that $\hat{H}_-^\ell$ becomes nonnegative because of the m and ℓ contributions due to angular dependence.

Summarizing these findings we clearly see that stable three-dimensional solitons for $0 < \nu < 4/3$ exist. All other soliton solutions will be unstable, either longitudinally or transversely. The first statement includes two aspects: first, the existence of three-dimensional states, and, second, the complete stability (for $0 < \nu < 4/3$); both have been discussed in detail in the preceding sections.

The Schrödinger equation (I.4) is the simplest example for which the existence of stable three-dimensional solitons could be proved and for which, in the case of instability, the growth rates could be evaluated without great numerical effort. However, the reader should have in mind that the variational principles derived here have a much larger region of applicability than just demonstrated for the Schrödinger equation (I.4).

Physics generally demands to consider more general and more realistic models than (I.4). Let us take as a simple example the Langmuir waves in a plasma. The nonlinear waves cannot be described just by (I.4), with $\nu = 2$, since then the reaction of the plasma to the wave pressure is assumed to be adiabatic. However, at least the heavy ions need some time for reacting to the wave. Another physical point is that for large amplitudes, a linear reaction of the electrons is not appropriate. All this together leads to (coupled) nonlinear equations which are much more complicated than Eq. (I.4). A similar situation occurs for other physical problems too, e.g., in field theory, solid state physics, and biophysics. There, in future, more realistic physical models should be also investigated by the basic methods presented here. Then, various topics should be attacked: in nonlinear plasma physics the existence of stable Langmuir solitons should be discussed [21,48]; in nonlinear field theory, the situation of coupled fields in three dimensions is still open [10,23,60]; in solid state physics, the anomalous resistivity in ice-like structures requires consideration of quite complicated soliton models and their stability, and so on. Several points look to us very convincing when the variational formulation is then also used: (i) The interrelation between existence and stability is transparent. (ii) The procedures are similar for quite different types of equations and some results can be easily overtaken. (iii) Numerical procedures allow to determine the relevant results in a relatively simple way since standard procedures to evaluate such variational principles are available.

If all the existence and stability properties of classical solitons are known one can proceed to base physical theories on the soliton concept. For example, after quantization, the existence of particle states might become more understandable and even in classical areas, such as turbulence theory, the soliton may turn out to be a possible candidate for a statistical theory of quasi-particles. Many areas might benefit from the soliton concept but still much has to be done in basic research.

VI. ACKNOWLEDGMENT

This work was supported in part by the Deutsche Forschungsgemeinschaft through SFB 162.

REFERENCES

1. Scott, A. C.; Chu, F. Y. F.; McLaughlin, D. W.: Proc. IEEE 61, 1443 (1973).
2. Newell, A. C. (ed.): Nonlinear Wave Motion. Am. Math. Soc. (1974).
3. Karpman, V. I.: Nonlinear Waves in Dispersive Media. Pergamon (1974).
4. Hasegawa, A.: Plasma Instabilities and Nonlinear Effects. Springer (1975).
5. Lonngren, K.; Scott, A.: Solitons in Action. Academic Press (1978).
6. Lamb, G. L., Jr.: Elements of Soliton Theory. Wiley (1980).
7. Bullough, R. K.; Caudrey, P. J. (eds.): Solitons. Springer (1980).
8. Ablowitz, M. J.; Segur, H.: Solitons and the Inverse Scattering Transforms. SIAM (1981).
9. Rajaraman, R.: Solitons and Instantons. North-Holland (1982).
10. Lee, T. D.: Particle Physics and Introduction to Field Theory. Harwood, Chur (1981).
11. Eilenberger, G.: Solitons—Mathematical Methods for Physicists. Springer (1981).
12. Rajaraman, R.: Phys. Reports 21, 227 (1975).
13. Coleman, S.: In New Phenomena in Subnuclear Physics, A. Zichichi, ed. Academic Press (1975), p. 297.
14. Jackiw, R.: Rev. Mod. Phys. 49, 681 (1977).
15. Makhankov, V. G.: Phys. Reports 35, 1 (1978).
16. Zastavenko, L. G.: J. Appl. Math. Mech. 29, 497 (1965).
17. Polyakov, A. M.: Sov. Phys. JETP 41, 988 (1975).
18. Rudakov, L. I.: Sov. Phys. Dokl. 17, 1166 (1973).
19. Anderson, D. L. T.; Derrick, G. H.: J. Math. Phys. 11, 1336 (1970).
20. Katyshev, Yu. V.; Makhankov, V. G.: Phys. Lett. 57A, 10,(1976).
21. Gibbons, J.; Thornhill, S. G.; Wardrop, M. J.; ter Haar, D.: J. Plasma Phys. 17, 153 (1977).
22. Hobart, R. H.: Proc. Phys. Soc. 82, 201 (1963).
23. Friedberg, R.; Lee, T. D.; Sirlin, A.: Phys. Rev. D13, 2739 (1976).
24. Skorić, M. M.; ter Haar, D.: Physica 98C, 211 (1980).
25. Laedke, E. W.; Spatschek, K. H.: Phys. Lett. A74, 205 (1979).
26. Laedke, E. W.; Spatschek, K. H.: Phys. Rev. Lett. 45, 993 (1980).
27. Laedke, E. W.; Spatschek, K. H.: Phys. Rev. Lett. 47, 719 (1981).
28. Laedke, E. W.; Spatschek, K. H.: Physica 5D, 227 (1982).
29. Laedke, E. W.; Spatschek, K. H.; Zocha, K.: Phys. Rev. Lett. 49, 1401 (1982).

30. Scott-Russell, J.: Report on Waves. British Association Reports (1844).
31. Boussinesq, J.: J. Math. Pur. Appl. (Ser. 2) 17, 55 (1872).
32. Korteweg, D. J.; de Vries, G.: Phil. Mag. 39, 442 (1895).
33. Zabusky, N. J.; Kruskall, M. D.: Phys. Rev. Lett. 15, 240 (1965).
34. Gardner, C. S.; Greene, J. M.; Kruskall, M. A.; Miura, R. M.: Phys. Rev. Lett. 19, 1095 (1967).
35. Zakharov, V. E.; Shabat, A. B.: Sov. Phys. JETP 34, 62 (1972).
36. Ablowitz, M. J.; Kaup, D. J.; Newell, A. C.; Segur, H.: Stud. Appl. Math. 53, 249 (1975).
37. Zakharov, V. E.: Sov. Phys. JETP 35, 908 (1972).
38. Rubinstein, J.: J. Math. Phys. 11, 258 (1970).
39. Katyshev, Yu. V.; Makhalkdiani, N. V.; Makhankov, V. G.: Phys. Lett. 66A, 456 (1978).
40. Strauss, W. A.: Commun. Math. Phys. 55, 149 (1977).
41. Velte, M.: Direkte Methoden der Varationsrechnung. Teubner (1976).
42. Malkin, J. G.: Theorie der Stabilität einer Bewegung. Oldenburg (1959).
43. Zubov, V. I.: Methods of A. M. Liapunov and Their Application. Noordhoff (1964).
44. La Salle, J.; Lefschetz, S.: Die Stabilitätstheorie von Liapunov. Bibl. Inst. Mannheim (1967).
45. Benjamin, T. B.: Proc. R. Soc. Lond. 328A, 153 (1972).
46. Benjamin, T. B.: Lectures on Nonlinear Wave Motion, in Nonlinear Wave Motion, A. C. Newell, ed. Am. Math. Soc. (1974).
47. Bona, J.: Proc. R. Soc. Lond. 344A, 363 (1975).
48. Laedke, E. W.; Spatschek, K. H.: Phys. Fluids 23, 44 (1980).
49. Laval, G.; Mercier, C.; Pellat, R.: Nucl. Fusion 5, 156 (1965).
50. Laedke, E. W.; Spatschek, K. H.: J. Math. Phys. 20, 1838 (1979).
51. Laedke, E. W.; Spatschek, K. H.: Phys. Rev. Lett. 41, 1798 (1978); 42, 1534 (1979).
52. Laedke, E. W.; Spatschek, K. H.: Phys. Lett. 82A, 335 (1981); J. Math. Phys. 23, 460 (1982).
53. Adams, R. A.: Sobolev Spaces. Academic Press (1975).
54. Laedke, E. W.; Spatschek, K. H.; Stenflo, L.: J. Math. Phys. 24, 2764 (1983).
55. Zakharov, V. E.; Rubenchik, A. M.: Sov. Phys. JETP 38, 494 (1974).
56. Infeld, E.; Rowlands, G.: Plasma Phys. 19, 343 (1977).
57. Schmidt, G.: Phys. Rev. Lett. 34, 724 (1975).
58. Collatz, L.: Funktionalanalysis und Numerische Mathematik. Springer (1968).
59. Stummel, F.; Hainer, K.: Praktische Mathematik. Teubner (1971).
60. Laedke, E. W.; Spatschek, K. H.; Wilkens, M., Jr.: Phys. Lett. A91, 378 (1982).

21 SOME FINITE CODIMENSIONAL LIE SUBGROUPS OF $\mathrm{DIFF}^{\omega}(M)$

Joshua Leslie / Department of Mathematics, Northwestern University, Evanston, Illinois

1. INTRODUCTION

In [8] we showed that the group of real analytic diffeomorphisms, $\mathrm{Diff}^{\omega}(M)$, of a compact real analytic manifold M admits the structure of a Lie group modeled on a Silva space with Lie algebra, $\mathscr{D}^{\omega}(M)$, the Lie algebra of real analytic vector fields on M. It is natural to ask which Lie subalgebras of $\mathscr{D}^{\omega}(M)$ are integrable in the sense that there exist Lie subgroups of $D^{\omega}(M)$, the connected component of the identity in $\mathrm{Diff}^{\omega}(M)$, having these algebras as their Lie algebras. In this paper we begin to address that question, and we show that there exists a large class of finite codimensional Lie subalgebras of $\mathscr{D}^{\omega}(M)$ which are integrable.

In Sec. 1 we summarize results concerning Silva spaces and Silva differentiability.

In Sec. 2 we study differential equations on Silva spaces culminating in a proof of a Frobenius Theorem.

In Sec. 3 we consider a compact manifold M real analytically embedded in C^N; that is, there exists a real analytic embedding $i: M \to R^N \subseteq C^N$. We prove that those finite codimensional Lie subalgebras $\mathscr{A} \subseteq \mathscr{D}^{\omega}(M)$, which have finite complementary spaces $F \subseteq \mathscr{D}^{\omega}(M)$ so that the closure, $\bar{\mathscr{A}}^{\infty}$, in the C^{∞} vector fields on M, have the property that $\bar{\mathscr{A}}^{\infty} \cap F = \{0\}$, are integrable.

2. SILVA SPACES AND SILVA DIFFERENTIABILITY

Definition 2.1. Let E_{ℓ}, $\ell = 1, 2, \ldots, n, \ldots$ be a sequence of Banach spaces and

$$i^{\ell}_{\ell+1} : E_{\ell} \to E_{\ell+1}$$

a sequence of continuous compact injective linear operators. $E = \varinjlim_{\ell} E_{\ell}$ will be called a Silva space where "$\varinjlim$" designates the inductive limit (see [1], pp. 61-63).

We shall suppose for convenience that $E_{\ell} \subseteq E_{\ell+1}$ and that $\| \ \|_{\ell+1} \leq \| \ \|_{\ell}$. For $\varepsilon > 0$ set $B^{\varepsilon}_{\ell} = \{x \in E_{\ell} : \|x\|_{\ell} < \varepsilon\}$.

Given an arbitrary sequence of positive numbers $\varepsilon_{\ell} > 0$, sets of the form

$$\bigcup_{k>0} \left(\sum_{\ell=1}^{k} B_{\ell}^{\varepsilon_{\ell}} \right) \qquad (*)$$

form a fundamental system of neighborhoods for the inductive limit topology on

$$E = \varinjlim_{\ell} E_{\ell}$$

In the above description of the topology on $\varinjlim_{\ell} E$ we are considering that $B_{\ell}^{\varepsilon_{\ell}} \subseteq E_{\ell}$ are canonically embedded in $\lim E_{\ell}$.

We shall need a criterion for open sets of a Silva space $E = \varinjlim_{\ell} E_\ell$ in terms of its bounded sets. Hogbe-Nlend (Chap. 7, p. 61) gives such a criterion. Let's recall that in topological vector space E a subset D centered at $a \in E$ is called bornivorous when $D - a$ absorbs every bounded $B \subseteq E$; that is, there exists $\lambda > 0$ with $\lambda B \subseteq D - a$.

Proposition 2.1. If a subset U of a Silva space E has the property that every $x \in U$ is the center of a bornivorous subset $D_x \subseteq U$, then U is an open set.

We also have

Proposition 2.2. Silva spaces are locally convex, Lindelöf (i.e., every open covering contains a countable subcovering), separable, and complete. (See [8].)

It is useful to observe

Proposition 2.3. Given a Silva space $E = \varinjlim_{n} E_n$ a subset $B \subseteq E$ is bounded if and only if it is relatively compact; further, B is bounded if and only if there exists n so that $B \subseteq E_n$ and B is bounded in E_n.

(and let's recall that a circled convex subset of a vector space is called a disk).

Proposition 4. If $E = \varinjlim_{n} E_n$ is a Silva space and $B \subseteq E$ is a compact disc, then $E_B = \bigcup_{\lambda R} \lambda B$ is a Banach space for the B-gauge norm $\|\alpha\|_B = \inf_{\lambda > 0} \{\alpha \in \lambda B\}$.

We also have

Proposition 2.5. If B is a separable Banach space and E is a Silva space, then the space of all continuous linear maps from B to E, L(B, E), with the topology of uniform convergences on bounded sets is a Silva space.

It is not difficult to prove

Proposition 2.6. Suppose E_1 and E_2 are separable Banach spaces and F is a Silva space. Let $L(E_1, E_2; F)$ be the Silva space of continuous bilinear maps $E_1 \times E_2 \to F$ with the topology of uniform convergence on bounded sets and $L(E_2; F)$ the Silva space of continuous linear maps with the topology of uniform convergence on bounded sets, then the canonical isomorphism $L(E_1, L(E_2, F)) \to L(E_1, E_2; F)$ is a homeomorphism.

Given $a \in E = \varinjlim E_n$, set $n(a) = \inf_k \{k : a \in E_k\}$.

Definition 2.2. Let $E = \varinjlim_{n} E_n$ and $F = \varinjlim_{n} F_n$ be Silva spaces and $U \subseteq E$ an open set, $f : U \to F$ is said to be Silva differentiable (resp. Silva C^n on U) at $a \in U$ if for each $n \geq n(a)$ we have $m_a(n) \geq 1$ and a neighborhood V_n of a in E_n so that $V_n \subseteq U$ and $(f | V_n) : V_n \to F_{m_{a(n)}}$ is Frechet differentiable (resp. Frechet C^n at a).

The elementary properties (e.g., chain rule, linearity, Taylor's expansion) follow from the analogous properties in the case of Frechet differentiability.

We have the following useful though obvious

Proposition 2.7. Let $E = \varinjlim_n E_n$ be a Silva space and $U_n \subseteq E_n$, $n \geq n_0$, for some n_1, a sequence of open sets so that $U_n \subseteq U_{n+1}$, then $U = \bigcup_n U_n$ is open in E.

Lemma 2.1. Let $E = \varinjlim E_n$ and $F = \varinjlim F_n$ be Silva spaces, suppose $U \subseteq E$ is open, and let $f: U \to F$ be a function. If for each $a \in U$ and $n \geq n(a)$ there exists an integral valued function $m_a(n)$ and an open set $V_n \subseteq E_n$ so that $\bar{V}_n \subseteq U$ and $(f \mid V_n): V_n \to F_{m_a(n)}$ is continuous. Then f is continuous.

Proof. Let W be an open neighborhood of f(a), there exists a bounded open neighborhood of a, $U_{n(a)} \subseteq E_{n(a)}$, so that $f(U_{n(a)}) \subseteq W \cap F_{m_a(n(a))}$, we may suppose $(f \mid U_{n(a)}): U_{n(a)} \to F_{m_a(n(a))}$ continuous. Since $\bar{U}^{n(a)+1}_{n(a)}$ is compact in $E_{n(a)+1}$ and E and the E_n are normal, there exists a finite covering of $\bar{U}^{n(a)+1}_{n(a)}$ in $E_{n(a)+1}$ by open bounded subsets $\{W_i\}$ $i = 1, \ldots, n$ so that $\bar{W}_i^{n(a)+2} \subseteq f^{-1}(W)$, where "$\bar{}^{n}$" represents the closure in E_n. Thus we construct a sequence of bounded open sets $U_i \subseteq E_i$, $i \geq n(a)$ so that $U_i \subset U_{i+1}$ and $f(U_i) \subseteq W$ for all i. Proposition 2.7 implies $U = \bigcup_i U_i$ is open, by construction $(f \mid U) \subseteq W$; since $a \in U$, we have that f is continuous at a. Q.E.D.

We are able to show

Proposition 2.8. Suppose E and F Silva spaces, $U \subseteq E$ open. If $f: U \to F$ is Silva C^n, then there exist k-multilinear symmetric continuous functions $D^k f(x): \underbrace{E \times \ldots \times E} \to F$ so that $D^k f: U \times \underbrace{E \times \ldots \times E} \to F$ are continuous and $F(v) = f(x + v) - f(x) - \frac{1}{1!} Df(x)(v) - \ldots - \frac{1}{k!} D^k f(x)(v, \ldots, v)$, $1 \leq k < n$ satisfy the property that

$$G(t, v) = \begin{cases} \dfrac{F(tv)}{t^k}, & t \neq 0 \\ 0, & t = 0 \end{cases}$$

is continuous at $(0, v)$.

The above proposition is a consequence of the definitions and the Lagrange remainder form for the Taylor's series. When E is a Banach space, F a Silva space, and $U \subseteq E$ open, the form of the theorems of the calculus are in all salient points analogous to the classical case [4].

3. FROBENIUS THEOREM

In this section we begin by proving an existence and uniqueness theorem for C^1 differential equations defined on a Silva space; we then study the dependence of solutions on parameters and initial conditions in the C^2 case, which we use following the analogy with the classical theory (see [4]) to prove the Frobenius theorem.

Theorem 3.1. Let $E = \varinjlim_n E_n$ be a Silva space, I an open interval, and $U \subseteq E$ a convex open subset. Suppose $F: I \times U \to E$ is a C^1 map so that given any $x_0 \in U$, $t_0 \in I$ there exist a neighborhood W of $0 \in E$ and a $\delta > 0$ so that for any bounded disk $B \subseteq W$ containing $0 \in E$ we have $(t_0 - \delta, t_0 + \delta) \subseteq I$ and a sequence of bounded disks B_n with $\bar{B} = B_1$ satisfying the following three hypotheses:

(a) $t \in (t_0 - \delta, t_0 + \delta)$ implies $f(t, x_0) \in B_1$ and $x \in x_0 + (\delta/1!)B_1 + \ldots + (\delta^n/n!)B_n$ implies $(\partial f/\partial x)(t, x) B_n \subseteq B_{n+1}$

(b) $x_0 + \Sigma_n (\delta^n/n!) B_n \subseteq U$

(c) There exists an integer $p \geq 1$ so that $F_n = \Sigma_{q \geq n} (\delta^q/q!) B_q$ form the base of a filter which converges to 0 in E_p.

Then there exists a unique solution $t \to \phi(t)$ defined on $(t_0 - \delta, t_0 + \delta)$ whose image is contained in $x_0 + \bar{\Gamma}(\Sigma_n (\delta^n/n!) B_n)$ with $\phi(t_0) = x_0$ and

$$\frac{d\phi(t)}{dt} = f(t, \phi(t)) \tag{*}$$

where $\bar{\Gamma}$ stands for the closure of the convex hull.

Proof. As in the classical case (see [4]) set $x_0(t) \equiv x_0$ and $x_n(t) = x_0 + \int_{t_0}^{t} f(u, x_{n-1}(u))\, du$, $n \geq 1$. From a) it follows that $x_1(t) \in x_0 + (t - t_0) B_1$. We have $x_2(t) - x_1(t) = \int_{t_0}^{t_1} (f(u, x_1(u)) - f(u, x_0))\, du$ and $f(u, x_1(u)) - f(u, x_0(u)) \in \bar{\Gamma}_{0 \leq \theta \leq 1} \{(\partial f/\partial x)(u, \theta x_0 + (1 - \theta) x_1(u)) \cdot (u - t_0) B_1 \subseteq (u - t_0) B_2\}$, so that $x_2(t) - x_1(t) \in ((t - t_0)^2/2!) B_2$. By iteration using a) we obtain

$$x_n(t) - x_{n-1}(t) \in \frac{(t - t_0)^n}{n!} B_n$$

therefore

$$x_n(t) \in x_0 + (t - t_0) B_1 + \cdots + \frac{(t - t_0)^n}{n!} B_n$$

Now, since $x_0 + (t - t_0) B_1 + \cdots + ((t - t_0)^n/n!) B_n$ is relatively compact (in view of c), there exists $N > 0$ so that $x_n(t)$ is a Cauchy sequence in E_N to a continuous function of t, $\phi(t)$. As in the classical case we may conclude

$$\phi(t) = x_0 + \int_{t_0}^{t} f(u, \phi(u))\, du$$

which implies

$$\phi'(t) = f(t, \phi(t))$$

by construction $\phi(t_0) = x_0$.

Now suppose we have a second solution to (*) $\psi(t)$ so that $\psi(t_0) = x_0$. Given any neighborhood of the origin V in E there exists a $\delta > \delta_1 > 0$ and a convex compact subset B of E containing 0 contained in V so that

$$\phi([t_0 - \delta_1, t_0 + \delta_1]) \quad \text{and} \quad \psi([t_0 - \delta_1, t_0 + \delta_1])$$

is contained in $x_0 + \delta B$. We have

$$\phi(t) - \psi(t) = \int_{t_0}^{t} [f(u, \phi(u)) - f(u, \psi(u))]\, du \tag{1}$$

Noting that

$$f(u, \phi(u)) - f(u, \psi(u)) \in \bar{\Gamma}_{x \; X_1 + B} \left\{ \frac{\partial f}{\partial x}(u, x)(\phi(u) - \psi(u)) \right\}$$

we obtain from (1) that

$$\phi(t) - \psi(t) \in 2(t - t_0) B_1$$

now we use a) and an interative argument to obtain

$$\phi(t) - \psi(t) \in 2\frac{(t - t_0)^n}{n!} B_n \tag{2}$$

Given 2) c) implies that $\phi(t) = \psi(t)$ for $t \in [t_0 - \delta_1, t_0 + \delta_1]$. Let's suppose that there exists a $t' \in (t_0 - \delta, t_0 + \delta)$ so that $\phi(t') = \psi(t')$. Now let $0 < \delta' < \delta$ be the largest number so that $\psi|[t_0 - \delta', t_0 + \delta'] = \phi([t_0 - \delta', t_0 + \delta'])$. By applying the existence portion of the present theorem we can extend both ψ and ϕ to the end points $t_0 - \delta'$ and $t_0 + \delta'$ which contradicts the definition of δ'. Q.E.D.

Now we turn to the question of dependence on parameters and initial conditions.

Theorem 3.2. Let E and P be Silva spaces, $\delta > 0$, $I = (t_0 - \delta, t_0 + \delta)$, $U \subseteq E$, $T \subseteq P$ be open subsets, $x_0 \in U$, $y_0 \in T$, and suppose that

$$F: I \times U \times T \to E$$

is C^2 mapping. Let $B_n \subseteq E_n$ and $C_n \subseteq P_n$ be the unit balls, suppose that for each n there exists a sequence $B_{n,1}, \ldots, B_{n,k}, \ldots$ of bounded disks in E and $\varepsilon_n > 0$ so that if $|t - t_0| < \delta$ and $y \in y_0 + \varepsilon_n C_n \subseteq T$, then

$$x \in x_0 + \varepsilon_n B_n + \frac{\delta}{1!} B_{n,1} + \cdots + \frac{\delta^m}{m!} B_{n,m} \subseteq U$$

implies

$$\frac{\partial f}{\partial x}(t, x, y) B_{n,k} \subseteq B_{n,k+1}$$

and

$$\sum_{q \geq n} \frac{\delta^q}{q!} B_{n,q}$$

converges to 0 in E_ℓ for some ℓ. Then there exists open neighborhoods $U_0 \subseteq U$ of x_0 and $T_0 \subseteq T$ of y_0 so that the function

$$\phi: (t_0 - \delta, t_0 + \delta) \times U_0 \times V_0 \to E$$

satisfying $\phi(t_0, x, y) = x$ and $(\partial\phi/\partial t)(t, x, y) = F(t, \phi(t, x, y), y)$ is differentiable in x and y; further, if F is C^∞ then so is $\phi(t, x, y)$ at (t_0, x_0, y_0).

The proof of the above theorem depends on the following

Lemma 3.1 (Colombeau [3]). Let E and P be Silva spaces $U \subseteq E$, $T \subseteq P$ open sets and let $F: (t_0 - \delta, t_0 + \delta) \times U \times T \to E$ be C^2, where $\delta > 0$. Given $x_0 \in U$, $y_0 \in T$ let C_0 be a bounded disk of P and B_0 a bounded disk of E so that $x_0 + B_0 \subseteq U$, $y_0 + C_0 \subseteq T$, suppose that for each pair of bounded disks B, B_1 there exists a sequence of compact disks $(B_n)_n \geq 1$ so that $x_0 + B + \frac{\delta}{1!} B_1 + \cdots + \frac{\delta^n}{n!} B_n \subseteq U$ for all n; suppose that for $|t - t_0| < \delta$, $y \in y_0 + C_0$, and suppose that $x \in x_0 + B + \frac{\delta}{1!} B_1 + \cdots + \frac{\delta}{n!} B_n$ implies

$$\frac{\partial F}{\partial x}(t, x, y) B_n \subseteq B_{n+1}$$

and

$$\sum_{q \geq n} \frac{\delta^q}{q!} B_q \to 0 \quad \text{in } F_\ell \text{ for some } \ell$$

Then there exists a mapping $\phi:(t_0 - \delta, t_0 + \delta) \times (x_0 + B_0) \times (x_0 + C_0) \to E$ satisfying the differential equation

$$\phi_t(t_1x_1y) = F(t, \phi(t,x,y), y) \tag{*}$$

and $\phi(t_0, x, y) = x$: further, $\phi(t,x,y)$ is differentiable at (t_0, x_0, y), $\phi_x \in L(E_{B_0}, E)$, $\phi_x(t_0, x_0, y_0) = I$ is the canonical injection where E_{B_0} is the Banach space generated by B_0 with the gauge norm; that is, $\|\alpha\|_{B_0} = \inf\{\lambda > 0 : \alpha \in G\}$; secondly, $\phi_x(t,x,y)$ satisfies the differential equation

$$\frac{d\phi_x}{dt}(t, x_0, y_0) = \frac{\partial F}{\partial x}(t, \phi(t,x_0,y_0), y_0)\, \phi_x(t, x_0, y_0) \tag{**}$$

and $\phi_y \in L(P_{C_0}, E)$ satisfies the differential equation

$$\frac{d}{dt}\phi_y(t, x_0, y_0) = \frac{\partial F}{\partial x}(t, (t,x_0,y_0), y_0)\, \phi_y(t, x_0, y_0) + \frac{\partial F}{\partial y}(t, (t, x_0, y_0), y) \tag{***}$$

Proof. As $[t_0 - \delta_1, t_0 + \delta_1]$, $\bar{B}_0$, and $\bar{C}_0$ are compact, it follows from the existence and uniqueness theorem that there exists $\varepsilon > 0$ and $\phi(t,x,y)$ satisfying (*). $\partial F/\partial x(t, \phi(t,x_0,y_0), y_0)$ is bounded in $L(E_{B_n}, E)$ for $|t - t_0| \leq \delta_1 \leq \delta$; it is not difficult to show that the hypotheses of the existence and uniqueness theorem are verified for (**) and thus there exists uniquely a solution A(t) of (**) for $|t - t_0| < \delta$ with $A(t_0)$ = canonical injection. Similarly (***) has a unique solution $B(t) \in L(P_{C_0}, E)$ with $B(t_0) = 0$ for $|t - t_0| < \delta$. Consider $y(t) = \phi(t, x_1, y_1) - \phi(t, x_0, y_0)$, $y(t)$ satisfies the differential equation $y'(t) = \Phi(t, y(t))$, where

$$\Phi(t, y) = F(t, y + \phi(t,x_0,y_0), y_1) - F(t, \phi(t,x_0,y_0), y_0) \tag{1}$$

One can easily verify that (1) satisfies the hypotheses of the existence and uniqueness theorem and therefore if ε_0 is sufficiently small there exists a bounded disk B'' so that

$$\phi(t,x_1,y_1) - \phi(t,x_0,y_0) \in \varepsilon_0 B'' \text{ for } |t - t_0| < \delta,\ x_1 \in x_0 + \varepsilon_0 B_0,\ y_1 \in y_0 + \varepsilon_0 C_0 \tag{2}$$

Now set

$$\begin{aligned} R(U) = {} & F(u, y_0(u,x_1,y_1), y_1) - F(u, \phi(u,x_0,y_0), y_0) \\ & - \frac{\partial F}{\partial y}(u, \phi(u,x_0,y_0), y_0)(y_1 - y_0) \\ & - \frac{\partial F}{\partial x}(u, \phi(u,x_0,y_0), y_0)(\phi(u,x_1,y_1) - \phi(u,x_0,y_1)) \end{aligned}$$

Using the Taylor series we obtain

$$\begin{aligned} R(u) \in \frac{1}{2}\bar{\Gamma}\Bigg\{ & \frac{\partial^2 F}{\partial y^2}(u, \phi(u,x_0,y_0) + t(\phi(t,x_1,y_1) - \phi(t,x_0,y_0)), y_0 + t(y_1 - y_0)(y_1 - y_0, y_1 - y_0) \\ & + 2\frac{\partial^2 F}{\partial y\, \partial x}(\ldots)((y_1 - y_0), \phi(t,x_1,y_1) - \phi(t,x_0,y_0))) \\ & + \frac{\partial^2 F}{\partial x^2}(\ldots)((\phi(t,x_1,y_1) - \phi(t,x_0,y_0)), (\phi(t,x_1,y_1) - \phi(t,x_0,y_0)) \Bigg\} \end{aligned}$$

therefore there exists a bounded disk B''' so that $R(U) \in \varepsilon_0^2 B'''$ for $y_1 - y_0 \in \varepsilon_0 C_0$ and $x_1 - x_0 \in \varepsilon_0 B_0$.

Now consider the differential equation

$$\frac{d}{dt}\delta(t) = \frac{\partial F}{\partial x}(t, \phi(t,x_0,y_0), y_0)\,\delta(t) + R(t) \tag{3}$$

(3) satisfies the hypotheses of the existence and uniqueness theorem; further, we can verify that $C(t) = \phi(t, x_1, y_1) - \phi(t, x_0, y_0) - B(t)(y_1 - y_0) - A(t)(x_1 - x_0)$ satisfies (3) with $C(t_0) = 0$. Since $R(t) \in \varepsilon_0^2 B'''$ where B''' is compact, it follows that there exists a compact K so that $C(t) \in \varepsilon^2 K$ for $|t - t_0|$ sufficiently small, this implies $A(t) = \partial\phi/\partial x\,(t, x_0, y_0)$ and $B(t) = \partial\phi/\partial y\,(t, x_0, y_0)$. Q.E.D.

Proof of Theorem 3.2. Let $B_n \subseteq E_n$ and $C_n \subseteq P_n$ be the unit balls. By the above lemma there exists ε_1 and a unique map ϕ: $(t - \delta_1, t + \delta_1) \times (x_0 + \varepsilon_1 B_1) \times (y_0 + \varepsilon_1 C_1) \to E$ satisfying (*). At each $x \in x_0 + \varepsilon_1 B_1 \subseteq E_2$ and $y \in y_0 + \varepsilon_1 C_1 \subseteq P_2$ the hypotheses of Lemma 3.1 are fulfilled relative to B_2 and C_2 with respect to the same δ. Using the compactness of $x_0 + \varepsilon_1 B_1$ and $y + \varepsilon_1 C_1$ and the uniqueness of the solution we construct open bounded subsets $U_2 \subseteq E_2$ and $V_2 \subseteq P_2$ so that $x_0 + \varepsilon_1 B_1 \subseteq U_2$ and $y_0 + \varepsilon_1 C_1 \subseteq V_2$ and there exists an extension of the flow ϕ, ϕ_2: $(t_1 - \delta_1, t_1 + \delta_1) \times U_2 \times V_2 \to E$. By iteration we may extend ϕ to ϕ_n: $(t_0 - \delta_1, t_1 + \delta_1) \times U_n \times V_n \to E$ where $\bar{U}_n$, $\bar{V}_n$ are compact and $\bar{U}_{n-1} \subseteq U_n$, $\bar{V}_{n-1} \subseteq V_n$ and $U_n \subseteq E_n$ and $V_n \subseteq P_n$ are open. Thus we obtain open sets $U = \bigcup_n U_n$ and $V = \bigcup_n V_n$ and ϕ: $(t_0 - \delta_1, t_0 + \delta_1) \times U \times V \to E$ so that (*), (**), and (***) are satisfied. Q.E.D.

Now, as in the classical case, the ground is laid in our context to state and prove the Frobenius Theorem:

Theorem 3.3. Let $E = \varinjlim_n E_n$ and $F = \varinjlim_n F_n$ be Silva spaces, $E' \subseteq E$, $F' \subseteq F$ open subsets. Suppose that $f: E' \times F' \times E \to F$ is a C^2 map linear in E. Suppose that for each $(x,y) \in E' \times F'$ and each pair $(a,b) \in E \times E$ that the map

$$(a,b) \to \frac{\partial f}{\partial x}(x,y,a;b) + \frac{\partial f}{\partial y}(x,y,a;f(x,y,b))$$

is symmetric in (a,b).

Further, suppose that given $x_0 \in E'$, $y_0 \in F'$ and bounded disks $B \subseteq E$ and $C \subseteq F$ that there exists $0 < \varepsilon < 1$ and a sequence of compact disks $C_n \subseteq F$ so that $x_0 + \varepsilon B \subseteq E'$, $y_0 + \varepsilon C \subseteq F'$, and so that

$$f(x_0 + \varepsilon B,\ y_0 + \varepsilon C,\ B) \subseteq C_1 \tag{1}$$

$$y_0 + \varepsilon C + \frac{\varepsilon}{1!} C_1 + \cdots + \frac{\varepsilon^n}{n!} C_n^n \subseteq F' \tag{2}$$

and

$$\frac{\partial f}{\partial y}(x_0 + \varepsilon B,\ y_0 + \varepsilon C + \frac{\varepsilon}{1!} C_1 + \cdots + \frac{\varepsilon^n}{n!} C_n,\ B;\ C_n) \subseteq C_{n+1}$$

there exists a positive integer p and an integer r so that

$$D_n = \Sigma_{q \geq n} (\varepsilon^q/q!)\, C_q \subseteq C_r \tag{3}$$

for $n \geq p$ converges to 0 in F_r for the F_r topology. Then there exist open neighborhoods U of x_0 in E_1 and V of y_0 in F_1 and a unique C^2-mapping

$$\alpha : U \times V \to F$$

so that $\alpha(x_0, y) = y$ and $D_1\alpha(x,y) = f(x, \alpha(x,y))$.

Proof. Without loss of generality we may suppose $x_0 = y_0 = 0$. Set

$$g(t,y,x) = f(tx, y; x) \tag{α}$$

$g(t,y,x)$ satisfies the hypotheses of Theorem 3.2. Set $g(t,y) = g(t,y,x)$. Setting $B_n = \varepsilon^n C_n$

and using the fact that $x \in \varepsilon B$ we obtain that we may set $\delta = 1$ in Theorem 3.1; thus, the solution to

$$y'(t) = g(t, y(t)) \qquad (\beta)$$

exists for $0 \leq t \leq 1$. Set $\alpha(x, y) = h(1)$ where $h(t)$ is the unique solution of (β) so that $h(0) = y$.

$$h(t) = \phi(t, y, x), \quad x \in \varepsilon B, \quad y \in \varepsilon C$$

where

$$\phi_t(t, y, x) = g(t, \phi(t, y, x), x)$$

and $\phi(0, y, x) = y$; thus, $\alpha(x, y)$ is C^2 in x and y. Now, $\alpha(0, y) = y$, since at the parameter $x = 0$ (α) is $g \equiv 0$; further, $D_1\alpha(x, y) = D_3\phi(1, y, x) = B(1)$ (see p.).

Consider $k(t) = B(t) - tf(tx, \phi(t, y, x))$, we have $k(0) = 0$, and $k(t)$ satisfies the linear differential equation

$$\frac{d}{dt} k(t) = D_2 f(tx, \phi(t, y, x), k(t); x)$$

which implies that $k(t) \equiv 0$ so that $k(1) = 0$ which gives

$$D_1 \alpha(x, y) = B(1) = F(x, \phi(1, y, x)) \equiv f(x, \alpha(x, y)) \qquad \text{Q.E.D.}$$

4. FINITE CODIMENSIONAL LIE SUBGROUPS OF $\mathrm{Diff}^{\omega}(M)$

First, let's recall the Lie group structure on $\mathrm{Diff}^{\omega}(M)$. For details and proofs see [8].

Let M^{ℓ} and N^d be compact connected real analytic manifolds. For $f \in C^{\omega}(M, N)$ = the set of all real analytic functions from M to N. Let $C^{\omega}_f(M, T(N))$ be the vector space of all real analytic liftings of f; that is, $g \in C^{\omega}_f(M, TN)$ is a real analytic function $g: M \to TN$ such that $\pi \cdot g = f$ where $\pi: TN \to N$ is the canonical projection of the tangent bundle. Cover N by a finite collection of trivializing normal relatively compact open charts V_i, $i = 1, \ldots, n$ and M by a finite collection of trivializing relatively compact open charts U_i, $i = 1, \ldots, m$ such that $\mathrm{diam}\,(f(U_i)) < \lambda/3$, where λ is the Lebesgue number of $\{V_j\}$. Define a topological structure on $C^{\omega}_f(M, TN)$ as follows: let $k_i: U_i \to U_i' \subset R^{\ell}$ and $\ell_j: V_j \to V_j' \subseteq R^d$ be homeomorphisms determining the local real analytic structure of M and N respectively. Suppose $f(U_i) \subset V_{j(i)}$, and let $\phi_{j(i)}: \pi^{-1}(V_{j(i)}) \to V'_{j(i)} \times R^d$ be a real analytic diffeomorphism.

By a uniformly bounded analytic function we intend an analytic function, $f(x)$, so that there exist ℓ and M with

$$\|f^{(k)}(x)\| \leq k!\, M\ell^k \quad \text{for all } x$$

Suppose $C^{\omega}_b(U_i', R^d)$ is the vector space of uniformly bounded real analytic maps $f: U_i' \to R^d$; as a vector space $C^{\omega}_b(U_i', R^d) = \varinjlim E_{\ell}$, where $E_{\ell}(U_i', R^d)$ is the Banach space of uniformly bounded real analytic functions $f: U_i' \to R^d$ so that $\|f(x)\|_{\ell} \leq k!\, p\ell^k$ for some p, we set $\|f(x)\|_{\ell} = \inf_p \{p : \|f(x)\|_{\ell} \leq k!\, p\ell^k \text{ for all } x \in U_i'\}$. The maps $E_{\ell} \to E_{\ell+1}$ are the canonical injections. We shall consider $C^{\omega}_b(U, R^d)$ for the rest of our paper with the above inductive limit topology. We call this topology on $C^{\omega}_b(U_i', R^d)$ the Van Hove topology.

Designate by $B^p_k(U, R^m) \subseteq E_k(U, R^m)$ the ball of radius p centered at the origin in E_k.

We have [8]

<u>Lemma 4.1.</u> $C^{\omega}_b(U, R^m)$, where $U \subseteq R^n$ is open, is a Silva space, if U is of finite diameter.

Set $A_0 = \Sigma_{i=1}^{n} C_b^{\omega}(U_i', R^d)$, with the canonical direct sum topology. A_0 is a Silva space. Define $\gamma: C_f^{\omega}(M, TN) \to A_0$ by $\gamma(g) = g_1 + \cdots + g_n$ where $g_i \in C^{\omega}(U_i', R^d)$ is the composite

$$U_i' \xrightarrow{k_i^{-1}} U_i \xrightarrow{q} \pi^{-1}(V_{j(i)}) \xrightarrow{\phi_{j(i)}} V_{j(i)}' \times R^d \xrightarrow{\text{proj}} R^d$$

Let $A = \gamma(C_f^{\omega}(M, TN)) \subseteq A_0$; A is a closed subspace of A_0 and therefore the induced topology on A is a Silva topology. By means of γ we transport the induced Silva structure to $C_f^{\omega}(M, TN)$.

When $M = N$ and $f = id_M$, A $\Gamma^{\omega}(M)$ is the vector space of real analytic vector fields on M.

From the Cauchy inequalities and classical theorems on inductive limits we obtain

Lemma 4.2. Suppose M analytically embedded in R^n, R^N canonically embedded in C^N, $U_i = \{\rho \in M^c : d(x, M) < 1/i$, where $d((x_1, \ldots, x_n), (z_1, \ldots, z_n)) = \max_i (|x_i - z_i|)\}$, where $M^c \subseteq C^N$ is the complex n dimensional Whitney-Bruhat complexification of M [10], and suppose that F_i is the vector space of all bounded complex analytic vector fields on U_i which extend real analytic vector fields on M. Then $\varinjlim_i F_i \approx C^{\omega}$ topology on the vector space of real analytic vector fields on M, $\Gamma^{\omega}(M)$, where the norm on F_i is the sup norm. The topology on $\Gamma^{\omega}(M)$ is independent of the charts (U_i, k_i).

When we consider a Riemannian manifold N as a metric space (N, p) it will always be with respect to the metric p determined by the Riemannian structure.

Description of a "Canonical" Chart at $f \in C^{\omega}(M, N)$

Let λ be the Lebesgue number of the covering V_j. For $y \in C^{\omega}(M, N)$ such that $p(f(s), y(s)) < \lambda/3$ for all $s \in M$ set $\alpha(s) = \exp_{f(s)}^{-1}(y(s))$; $\alpha \in C_f(M, TN)$; let $U_f = \{y \in C^{\omega}(M, TN) \mid p(y(s), f(s)) < \lambda/3$ for all $s \in M\}$, $\psi_f : U_f \to C_f^{\omega}(M, TN$ given by $\psi_f(y) = \alpha$ is taken as a chart of $C^{\omega}(M, N)$ at f.

Proposition 4.1. If U_x and $U_{x'}$ are overlapping coordinate charts of $x, x' \in C^{\omega}(M, N)$, then the coordinate transformation

$$\psi = \psi_{x'} \cdot \psi_x^{-1} : \psi_x(U_x \cap U_{x'}) \to \psi_{x'}(U \cap U')$$

is a Silva C^{∞} transformation.

In order to prove the above proposition one makes use of the following two lemmas.

Lemma 4.3. Let $U, U' \subseteq R^n$ be open subsets, and $B_{\ell}^{k}(U, U') = \{h : U \to U'$, h analytic with $\|D^n h(x)\| \leq n! k \ell^n$, $x \in U\}$, and $B_{\ell'}^{k'}(U', R^n) = \{h : U' \to R^n$, h analytic with $\|h^{(n)}(x)\| \leq n! k' (\ell')^n$, $x \in U'\}$. If $g \in B_{\ell'}^{k'}(U', R^n)$ and $f \in B_{\ell}^{k}(U, U')$, then $(g \cdot f) \in B_{2\ell(1+k\ell')}^{k'}(U, R^n)$.

Lemma 4.4. Let $C_b^{\omega}(U', E_3) \subseteq C^{\omega}(U', E_3)$ (resp. $C_b^{\omega}(U, U') \subseteq C^{\omega}(U, U')$) be the subspace of uniformly bounded analytic functions defined on U. Now suppose $\alpha \in C_b^{\omega}(U', E_3)$ and define $F_{\alpha} : C_b^{\omega}(U, U') \to C_b^{\omega}(U, E_3)$ by $F_{\alpha}(\phi) = \alpha \cdot \phi$. Then F_{α} is in $C^{\infty}(C_b^{\omega}(U, U'), C_b^{\omega}(U, E_3))$ and $\alpha \times \phi \to \alpha \cdot \phi$ induces a C^{∞} map from $F_b^{\omega}(U', E_3) \times C_b^{\omega}(U, U') \to C_b^{\omega}(U, E_3)$. (Note that $C_b^{\omega}(U, U')$ is open in $C_b^{\omega}(U, E_3)$ when U' is open in E_2), where E_2 and E_3 are arbitrary Euclidean spaces.

Remark 4.1. The compactness of M is essential to the above in order for the maps in question to be uniformly bounded as analytic maps, N being compact is not essential.

Remark 4.2. The differential structure on $C^{\omega}(M,N)$ does not depend on the choice of Riemannian structures on M or N.

Lemma 4.2 also implies

Proposition 4.2. Let M, N, and P be compact, connected, real analytic manifolds. Then composition $C^{\omega}(M,N) \times C^{\omega}(N,P) \to C^{\omega}(M,P)$ defines a C^{∞} function.

There is a differential structure (see [7]) on the set of C^{∞} mappings of a compact manifold M into a compact manifold N that we shall designate by $C^{\infty}(M,N)$. That manifold was defined analogously to the one we are discussing here, but was modeled on the C^{∞} vector fields with the C^{∞} topology. As the C^{∞} topology induces a topology on $C_f^{\omega}(M,TN)$ so that open sets are bornivorous when centered at each of their points, it is a coarser topology than the Van Hove topology on $(C_f^{\omega}(M,TN))$ (See Proposition 3.2) thus

Proposition 4.3. The canonical injection $C^{\omega}(M,TN) \to C^{\infty}(M,TN)$ is a C^{∞} injection of the first manifold into the second and induces the canonical injection of the tangent bundle of the first into that of the second.

Corollary. Let M be a compact connected analytic manifold. Then there exists an open neighborhood of the identity in $C^{\omega}(M,M)$ consisting entirely of diffeomorphisms.

Proposition 4.4. $x \to x^{-1}$ is C^{∞}.

In summary, we have

Theorem 4.1. Given a compact, connected, real analytic manifold M, $\mathrm{Diff}^{\omega}(M) \equiv$ the group of real analytic diffeomorphisms with the Van Hove-Silva topology is a C^{∞} group in the sense of Silva differentiation; that is, it is a C^{∞} manifold modeled on its Silva space of real analytic vector fields, and multiplication and inversion define Silva-C^{∞} maps:

$$\mathrm{Diff}^{\omega}(M) \times \mathrm{Diff}^{\omega}(M) \to \mathrm{Diff}^{\omega}(M) \quad \text{and} \quad \mathrm{Diff}^{\omega}(M) \to \mathrm{Diff}^{\omega}(M)$$

Theorem 4.2. Let $D^{\omega}(M) \subseteq \mathrm{Diff}^{\omega}(M)$ be the connected component of the identity, and let $B \subseteq TD^{\omega}(M)$ be a sub-bundle of finite codimension with fiber $E \approx \Pi_1^{-1}(e) = B_e$, suppose that B is closed under the action induced on $TD^{\omega}(M)$ by right multiplication, and suppose that the C^{∞} sections of B are closed under a bracket operation of $TD^{\omega}(M)$; further, suppose that B_e has a finite dimensional complementary space $F \subseteq \mathscr{D}^{\omega}(M)$ so that $\bar{B}_e^{\infty} \cap F = \{0\}$, where "$\overline{}^{\infty}$" is the C^{∞} closure. Then B is integrable; that is, through every point $x \in \mathscr{D}^{\omega}(M)$ there passes an integral submanifold of B.

Before proceeding, we prove

Lemma 4.5. If $E \subseteq \mathscr{D}^{\omega}(M)$ is a finite codimensional subspace with finite dimensional complementary space $F \subseteq \mathscr{D}^{\omega}(M)$ so that $\bar{E}^{\infty} \cap F = \{0\}$, then there exists $0 \leq r < \infty$ so that $\bar{E}^r \cap F = \{0\}$, where "$\overline{}^r$" is the closure in the C^r-vector fields, $\mathscr{D}^r(M)$, for the C^r-topology, and $\mathscr{D}^r(M) = \bar{E}^r \oplus F$.

Proof. It is classical (see [9]) that Theorem A implies $\overline{\mathscr{D}^{\omega}(M)}^r = \mathscr{D}^r(M)$. Let $K \subseteq F$ be a compact subset so that $\bigcup_{\lambda \in R} \lambda K = F$ and so that $0 \notin K$. We may cover K by a finite collection of open neighborhoods in $\mathscr{D}^{\infty}(M)$, U_i, $i = 1, \ldots, n$, so that $U_i \cap \bar{E}^{\infty} = \emptyset$.

By the definition of the C^{∞} topology there exist open sets $V_i \subseteq \mathscr{D}^r(M)$, $i = 1, \ldots, n$, for some r so that $V_i \cap \mathscr{D}^{\infty}(M) = U_i$. Now by the Hahn-Banach theorem there exist closed hyperplanes $E \subseteq H_i$ so that $H_i \cap U_i = \emptyset$; this shows that $\bar{E}^r \cap (\bigcup_{i=1}^n V_i) = \emptyset$, therefore, $\bar{E}^r \cap F = \{0\}$; thus $\mathscr{D}^r(M) = \bar{E}^r \oplus F$. Q.E.D.

Let $\Pi_2 : TD^{\omega}(M) \to D^{\omega}(M)$ be the canonical projection and let $W = U \times V$, $U \subseteq B_e$, $V \subseteq F$ be a trivializing chart for Π_1 and Π_2 so that $\Pi_1 | W \approx (U \times V) \times E$ by means of the trivialization determined by right multiplication and $\Pi_2 | W \approx (U \times V) \times (E \oplus F)$ by means of the trivialization determined by the manifold structure. The canonical injection of B into $TD^{\omega}(M)$ may be represented locally by an exact sequence

$$0 \to (U \times V) \times E \xrightarrow{f} (U \times V) \times (E \oplus F) \qquad (\beta)$$

where

$$f(\omega, \alpha) = (\omega, f_1(\omega, \alpha))$$

with

$$f_1(\omega, \alpha)(x) = [(\exp_x^{-1} \cdot \exp_{\omega(x)})_* (\alpha(\omega)))]$$

We shall use

Lemma 4.6. There exists an open neighborhood W_0 of $\ell = (x_0, y_0) \in W$ so that $(\Pi_E \cdot f_1)(\omega, \cdot)$ is an isomorphism for $\omega \in W_0$.

Proof. $f_1(\omega, \alpha)$ can be extended to C^{∞} map

$$f_1^r : W \times \bar{E}^r \to W x \overline{(E \oplus F)}^r = W \times \mathscr{D}^r(M)$$

linear in $\bar{E}^r$, since $(\Pi_{\bar{E}^r} \cdot f_1)(x_0, y_0), \cdot)$ is the identity there exists an open set W_0 so that $(\Pi_{\bar{E}^r} \cdot f_1)((x,y), \cdot)$ is an isomorphism for $(x,y) \in W$. This implies that $(\Pi_E \cdot f_1)(\omega, \cdot)$ is injective for $\omega \in W_0$. Lemma 4.6 would now follow from

Lemma 4.7. Let F and G be Hausdorff locally convex topological vector spaces, $\dim(G) = n$, $p_F : F \times G \to F$ the canonical projection onto F, and $E \subseteq F \times G$ a closed subspace of codimension q. If $p : E \to F$ is the restriction of p_F to E, $p = p_F | E$, then p is a Fredholm operator (i.e., with finite dimensional kernel and finite dimensional cokernel) so that its index = dim (ker(p)) - dim (coker(p)) = n - q.

Proof. Let $H = E + \{\{0\} \times G\}$ and $K = E \cap (\{0\} \times G)$. As G is finite dimensional it's complete in the sense that every Cauchy filter converges and therefore G is closed, since E is closed by hypothesis, we have that H is closed. As E has finite codimension and $E \subseteq H$, H has finite codimension. Thus, there exists a finite dimensional subspace $J \subseteq F \times \{0\}$ such that $F \times G$ can be written as the direct sum $F \times G = H \oplus J$. As G is finite dimensional and $K \subset G$, we choose closed subspaces $E_0 \subset E$ and $G_0 \subseteq \{0\} \times G$ such that E is the direct sum of E_0 and K, $E = E_0 \oplus K$ and $\{0\} \times G = K \oplus G_0$. Thus, $H = E_0 \oplus K \oplus G_0$ and

$$F \times G = E_0 \oplus K + G_0 \oplus J$$

Then p_F restricted $E_0 \oplus J$ is an isomorphism, K is the kernel of p, and $p_F(J)$ is the complement to the image of p in F. Therefore p is a Fredholm operator with index given by

$$\text{index}(p) = \dim(K) - \dim J) = \dim(K \oplus G_0) - \dim(G_0 \oplus J) = n - q \qquad \text{Q.E.D.}$$

Proof of Theorem 4.2. In Lemma 4.6 suppose $W_0 = U_0 \times V_0$, where U_0 is an open neighborhood of x_0 and V_0 an open neighborhood of y_0. By Lemma 4.6 we may write (β) as

$$0 \to (U_0 \times V_0) \times E \xrightarrow{f} (U_0 \times V_0) \times (E + F) \qquad (\gamma)$$

where $f(\omega, \alpha) = (\omega, f_1(x,y)(\alpha))$ with $f_1(x,y): E \to F$ a continuous linear map.

Let us recall that given two sections ξ and η of $T\mathscr{D}^\omega(M)$ that their bracket is given locally by

$$[\xi, \eta](x) = D\xi(x; \eta(x)) - D\eta(x; \xi(x))$$

Given C^∞ maps $\xi_1, \eta_1: U \times V \to E$ note that the C^∞ maps given by $\xi(x,y) = (\xi_1(x,y), f_1(x,y)(\xi_1(x,y)))$ and $\eta(x,y) = (\eta_1(x,y), f_1(x,y)(\eta_1(x,y)))$ determine sections of B. The closure of the sections of B under the bracket operation implies that

$$\partial f/\partial x((x,y), \eta_1(x,y); \xi_1(x,y)) + \partial f/\partial y((x,y), \eta_1(x,y); f(x,y)(\xi_1(x,y))$$

is symmetric in $\xi_1(x,y)$, $\eta_1(x,y)$.

Since F is finite dimensional and $f_1: U_0 \times V_0 \times E \to F$ is C^∞ is easy to verify that the hypotheses of Theorem 3.3 are satisfied which implies that there exist open neighborhoods $A_0 \subseteq U_0$ of x_0 and $B_0 \subseteq V_0$ of y_0, and a flow α of f_1, $\alpha: A_0 \times B_0 \to V_0$ so that $D_x\alpha(x,y) = f(x, \alpha(x,y); \cdot)$.

We shall need

Lemma 4.8. There exist open neighborhoods $O \subseteq Z_0$ of x_0 and $W \subseteq B_0$ of y_0 such that $\phi(x,y) = (x, \alpha(x,y))$ is a C^∞ diffeomorphism onto an open neighborhood of $(x_0, y_0) \in A_0 \times B_0$.

Proof. There exist open neighborhoods $\tilde{O}$ of x_0 in U_0 and $\tilde{W}$ of y_0 in V_0 so that $(\Pi_F \cdot \alpha)(x, \cdot)$ is a local diffeomorphism from $\tilde{W}$ into V_0 for $x_0 \in O$ since F being finite dimensional and $\alpha_y(x_0, y_0) =$ identity implies that for $\tilde{O}$ and $\tilde{W}$ small enough $\alpha_y(x,y)$, $x \in \tilde{O}$, $y \in \tilde{W}$, is an isomorphism near the identity.

Note that there exists p so that

$$\phi((\tilde{O} \times \tilde{W}) \cap (E_r \times F) \subseteq (A_0 \times B_0) \cap (E_r \times F)$$

for $r \geq p$ since F is finite dimensional. Now $D\phi((x,y); (\alpha, \beta)) = \alpha$, $D\alpha((x,y); (\alpha, \beta)) = (\alpha, f(x, \alpha(x,y); \alpha)) + (O, \partial\alpha/\partial y(x,y; \beta))$ which is a continuous isomorphism from $E_s \oplus F = \mathscr{D}^s(M)$ onto $E_s \oplus F$ for $s \geq p$ where $E = \varinjlim_s E_s$ and $\mathscr{D}^w(M) = \varinjlim_s \mathscr{D}^s(M)$ for $(x,y) \in (\tilde{O} \times \tilde{W}) \cap (E_s + F)$ thus there exists a bounded open neighborhood of $(x_0, y_0) \in E_s \oplus F$, O_s for some s so that $\alpha | (O_s): O_s \to E_s \oplus F$ is a local diffeomorphism. We can construct a bounded neighborhood of O_s, $O_{s+1} \subseteq E_{s+1} \oplus F$ so that $\phi | O_{s+1}: O_{s+1} \to E_{s+1} \oplus F$ is a local diffeomorphism. Now by iteration we can construct bounded open sets $O_{s'} \subseteq E_{s'} \oplus F$ for $s' > s$ so that $\phi | O_{s'}: O_{s'} \to E_{s'} \oplus F$ is a local diffeomorphism and $O_t \subseteq O_{t+1}$ for $s \leq t$. Proposition 2.1 now implies that $\bigcup_t O_t$ is an open set so that $\phi | O$ is a local diffeomorphism.

To prove our lemma it would now suffice to show that $\phi(x,y) = (x, \alpha(x,y))$ is injective. This follows from the observation that $\{x: \alpha(x, y_1) = \alpha(x, y_2)\}$ is both open and closed; open as a corollary to the Frobenius theorem, closed since both $\alpha(\ , y_1)$ and $\alpha(\ , y_2)$ are continuous. Q.E.D.

Now we are in a position to prove the principal result of this paper.

Theorem 4.3. Let $\mathscr{A} \subseteq \mathscr{D}^\omega(M)$ be a finite codimensional Lie subalgebra with a finite dimensional complementary space F so that $\bar{\mathscr{A}}^\infty \cap F = \{0\}$, where "$\overline{}\ \infty$" is the closure in the vector space of C^∞ vector fields on M for the C^∞ topology. Then there exists a Lie subgroup $G \subseteq D^\omega(M)$ having $\mathscr{A}$ as its Lie algebra.

Proof. Designate by $\mathscr{A}(x)$ the subspace of $T_x D^\omega(M)$ made up of the vectors $\xi(x)$ for $\xi \in \mathscr{A}$. We may write $T_x D^\omega(M) = \mathscr{A}(x) + R(x)$, where R(x) is a complementary subspace of $\mathscr{A}(x)$ in $T_xD^{\omega}(M)$. Put $\Sigma = \bigcup_{x \in D^{\omega}(M)} \mathscr{A}(x)$ and let $\Pi' : \Sigma \to D^{\omega}(M)$ be the natural projection (i.e., $\Pi'(\mathscr{A}(x)) = x$). We now make Π' a sub-bundle of the tangent bundle of $D^{\omega}(M)$, $\Pi : TD^{\omega}(M) \to D^{\omega}(M)$. Let (U, ϕ) be a symmetric canonical chart of $D^{\omega}(M)$ at the identity with $\phi(U) \subseteq \mathscr{D}^{\omega}(M)$ and put $U_a = aU$ and let $\sigma_U : \Pi'^{-1}(U) = \Sigma(U) \to U \times \mathscr{A}(e)$ ($\sigma_U(y) = (\Pi'(y), a\Phi^{-1}_{\Pi'(y)}(y)$, where $\Phi_{\Pi'(y)}(x) = x \cdot \Pi'(y)$) be the restriction of the bijection $[\sigma_U : \Pi^{-1}(U) = T(U) \to U \times T_e(G)]$ giving the bundle structure of $\Pi : T(D^{\omega}(M)) \to D^{\omega}(M)$ by means of right multiplication $\Phi_x(g) = gx$.

Define $\sigma_{U_a} : (\Pi')^{-1}(U_a) \to U_a \times \mathscr{A}(e)$ by $\sigma U_a = \sigma_U \cdot \Phi_a^{-1}$, σ_{U_a} is such that the following diagram is commutative

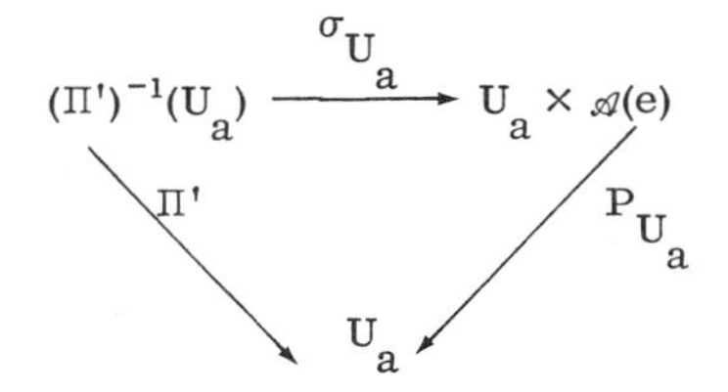

where P_{U_a} is the canonical projection.

It is useful to suppose fixed the trivializing system of charts $\mathscr{U} = \{U_a, \phi_a\}_{a \in D^{\omega}(M)}$. If one puts $\phi_{ab} = \phi_a \cdot \phi_b^{-1} : \phi_b(U_a \cap U_b) \to \phi_b(U_a \cap U_b)$ then multiplication in $D^{\omega}(M)$ being C^{∞} from the right, one obtains a C^{∞} mapping

$$I_{ba} : \phi_a(U_a \cap U_b) \times \mathscr{A}(e) \to \phi_b(U_a \cap U_b) \times \mathscr{A}(e)$$

given by

$$I_{ba}(x, v) = (\phi_{ba}(x), \phi_{ba}(x) \cdot v)$$

Under these conditions there exists a unique structure of a C^{∞} manifold on Σ such that Π' is C^{∞} and such that σ_{U_a}, $a \in D^{\omega}(M)$, are C^{∞} diffeomorphisms making $\Pi' : \Sigma \to D^{\omega}(M)$ into a vector bundle with $\{(U_a, a)\}$ as a trivializing covering.

The injection of $\mathscr{A}(x)$ into $T_xD^{\omega}(M)$ shows that Σ is a sub-bundle of $TD^{\omega}(M)$. As $\mathscr{A}$ is an algebra for the bracket operation in $\mathscr{D}^{\omega}(M)$ it follows that Σ is closed under the bracket operation in $TD^{\omega}(M)$.

Before proceeding we observe that the sub-bundle Σ may be expressed locally in a neighborhood of $e \in D^{\omega}(M)$ in the form of an exact sequence

$$0 \to U \times V \times \mathscr{A}(e) \xrightarrow{f} UxVx(\mathscr{A}(e) \times R(e))$$

when $R(e)$ is a finite dimensional complementary space to $\mathscr{A}(e)$ in $\mathscr{D}^{\omega}(M)$, where $U \subseteq \mathscr{A}(e)$ and $V \subseteq R(e)$ are open neighborhoods of x_0 and y_0, respectively.

Thus the hypotheses of Theorem 4.2 are fulfilled for the sub-bundle $\Sigma \to M$ of $TD^{\omega}(M) \to D^{\omega}M$. Let H be a maximal integral manifold of Σ in $D^{\omega}(M)$ containing the identity. As in the classical case (see [2], pp. 166-171) one has the right multiplication Φ_x permutes the maximal integral manifolds of Σ and thus H is a subgroup of $D^{\omega}(M)$. It is immediate that the Lie algebra of H is $\mathscr{A}$. Q.E.D.

REFERENCES

1. Bourbaki, N.: Espaces Vectorielles Topologiques.
2. Bourbaki, N.: Groupes et Algébres de Lie.
3. Colombeau, J. F.: Differential Calculus and Holomorphy (to appear).
4. Dieudonné, J.: Foundations of Modern Analysis. Academic Press (1960).

5. Hogbe-Nlend, H.: Théories des Bornologies et Applications. Springer (1971).
6. Jacobowitz, H.: Ann. Math. 95, 191-225 (1972).
7. Leslie, J.: Topology 6, 263-271 (1967).
8. Leslie, J.: On the group of real analytic diffeomorphisms of a compact real analytic manifold. TAMS (to appear).
9. Serre, J. P.: Applications de la théorie générale à divers problémes globaux, Seminarie H. Cartan 51/52, Exposé XX.
10. Whitney, H.; Bruhat, F.: Comm. Math. Helv. 33, Fasc. 3, 132-160 (1959).

EXTERIOR FORMS AND OPTIMAL CONTROL THEORY

L. Losco / Department of Mechanics, Ecole Nationale Superieure
de Mècanique et Microtechniques, Besançon, France

1. AN ELEMENTARY APPROACH TO PONTRYAGIN'S THEOREM

This theorem covers the whole field of optimal control theory; it is presented in the famous book [3].

We will now give it a very simplified yet intuitively satisfying treatment.

Let (S) be the differential system:

$$\frac{dx}{dt} = f(x, u, t)$$

where $x = (x_1, x_2, \ldots, x_n)$ represents n unknown functions of t, and $u = (u_1, \ldots, u_m)$ represents m functions which are called "commands." The solution x(t) or (S), with initial conditions (x_0, t_0) depends on the command u. The fundamental problem of optimal control theory is to minimize the function J, called the "cost" or the "performance index"

$$J = \varphi(x_f, t_f) + \int_{t_0}^{t_f} L(x(t), u(t), t)\, dt$$

φ is a function of the final conditions: t_f is the final instant, x_f is the value of x at t_f.

Analytical mechanics, based on the Principle of Least Action, appears as a very special case, where (S) is written $dx/dt = u$ and where $\varphi = 0$.

The command is then the velocity, and the function J is the action S:

$$S = \int_{t_0}^{t_f} L(x(t), u(t), t)\, dt$$

associated to the Lagrangian $L(x, \dot{x}, t)$.

We must therefore minimize $J = \varphi(x_f, t_f) + \int_{t_0}^{t_f} L(x(t), u(t), t)\, dt$ under the constraint $\dot{x} - f = 0$.

For this we apply the Lagrange multiplier method, by associating to the vector equation of constraint $\dot{x} - f = 0$, a vector which Pontryagin designates by ψ, other authors by λ, and which I designate by p, using the notation of analytical mechanics. This vector is such that our problem is now to minimize

$$\bar{J} = \varphi(x_f, t_f) + \int_{t_0}^{t_f} \{L + \dot{p} \cdot x - \dot{f})\}\, dt$$

with respect to x and u, considered as independent variables. Integrating by parts, we get:

$$\bar{J} = \varphi(x_f, t_f) + \{p \cdot x\}_{t_0}^{t_f} - \int_{t_0}^{t_f} \{\dot{p} \cdot x + p \cdot f - L\} dt$$

We now define Pontryagin's function

$$H(x, p, u) = p \cdot f - L$$

Then: $\bar{J} = \varphi(x_f, t_f) + [p \cdot x]_{t_0}^{t_f} - \int_{t_0}^{t_f} (\dot{p} \cdot x + H)\, dt$

Let $\mathcal{U}$ be the admissible domain of u. The optimal command is a value of u, designated by $\bar{u}$, which achieves the extremum of J. Let us assume that u achieves a relative minimum of J, which implies that $\bar{u}$ lies in the interior of $\mathcal{U}$; the minimization of $\bar{J}$ with with respect to (x, u) as well as the constraint equation are write

$$(E) \begin{cases} \dfrac{dx}{dt} = \dfrac{\partial H}{\partial p} \\ \dfrac{dp}{dt} = -\dfrac{\partial H}{\partial x} \\ \dfrac{\partial H}{\partial u} = 0 \end{cases}$$

(E) is a system of (2n + m) equations with (2n + m) unknowns. It is therefore possible for them to yield the solution of our problem, taking into account the boundary conditions at t_0 and t_f. These are obtained by minimizing $\varphi(x_f, t_f) + [p \cdot x]_{t_0}^{t_f}$

if x_f is fixed, $p(t_f)$ is arbitrary;
if x_f is free, $p(t_f) = -\partial\varphi/\partial x_f$;
if $\psi(x_f, t_f) = 0$, $\partial\varphi/\partial x_f + \nu(\partial\psi/\partial x_f) + p(t_f) = 0$;
if t_f is not fixed, we have to add $H(t_f) = \partial\varphi/\partial t_f$;
if we have a minimum time problem, we can take

$$J = t_f \;(\varphi = t_f;\ L = 0) \quad \text{or} \quad J = \int_{t_0}^{t_f} dt \;(\varphi = 0;\ L = 1)$$

The time t is usually made to play the same role as an x variable, by writing $t = x_{n+1}$ and adding to $\dot{x} - f = 0$ the equation $\dot{x}_{n+1} - 1 = 0$. One then finds that p_{n+1} (the multiplier associated with $\dot{x}_{n+1} - 1 = 0$) is equal to -H, modulo a constant.

Also, one usually considers the supplementary variable x_0 such that $dx_0/dt = L$, the function J in this case taking on the very remarkable form $J = \varphi(x_f, t_f) + (x_0)_f$. One then finds that p_0 remains constant and equal to -1. In mechanics, x_0 is the action variable [9].

2. THE LINK WITH SYMPLECTIC MECHANICS

We assume in the following that the conditions for equations (E) to apply are fulfilled. [The optimal command u lies in the interior of the admissible domain $\mathcal{U}$, and H is of class C^1 with respect to (p, x, u).] Many problems lie in this framework. We note that the relation between $\bar{J}$ and H implies that a minimum of J corresponds to a maximum of H. Finally we see that equations (E) are comprised of 2n canonical equations for the conjugate variables (x, p)

$$\frac{dx}{dt} = \frac{\partial H}{\partial p}, \quad \frac{dp}{dt} = -\frac{\partial H}{\partial x}$$

and m equations

$$\frac{\partial H}{\partial u} = 0$$

called "optimability conditions." This is not a new situation in mechanics. When one reduces a Hamiltonian system by using first integrals, one has an analogous situation. For example, if one reduces by the first integrals p_α = constant, arising from the ignorable coordinates q^α, the reduced system for $p_\alpha = c_\alpha$ is written

$$\frac{dq_{\bar{\alpha}}}{dt} = \frac{\partial \bar{H}}{\partial p_{\bar{\alpha}}}, \quad \frac{dp_{\bar{\alpha}}}{dt} = -\frac{\partial \bar{H}}{\partial q_{\bar{\alpha}}}, \quad \frac{dq_\alpha}{dt} = \frac{\partial \bar{H}}{\partial p_\alpha}$$

where $\bar{\alpha}$ is a complementary subscript to α and where $\bar{H} = H(p_\alpha, c_\alpha)$.

In the present case, we have 2(n - m) canonical equations (Routh reduction) and m differential equations. We have shown [5, 6] that this system possesses the integral invariant

$$\tilde{\omega} = \sum_{\bar{\alpha}} dp_{\bar{\alpha}} \wedge dq_{\bar{\alpha}} - d\bar{H} \wedge dt$$

The importance of Cartan's 2-form in modern analytical mechanics is well known; it allows an intrinsic characterization of the equations of motion. We therefore propose to adapt it to equations (E) of optimal control theory.

3. CHARACTERIZATION OF (E) BY A 2-FORM

We consider the 2-form:

$$\tilde{\omega} = \sum_{i=1}^{n} dp_i \wedge dx_i - dH(x, p, u) \wedge dt$$

and the characteristic vector field of optimal control theory:

$$\mathbf{X} = \sum_{i=1}^{n} \left[\dot{x}_i \frac{\partial}{\partial x_i} + \dot{p}_i \frac{\partial}{\partial p_i}\right] + \sum_{j=1}^{m} \dot{u}_j \frac{\partial}{\partial u_j} + \frac{\partial}{\partial t}$$

Let us express the fundamental relation: $i_{\mathbf{X}}\tilde{\omega} = 0$

$$i_{\mathbf{X}}\tilde{\omega} = \sum_{i=1}^{n} [\dot{p}_i dx_i - \dot{x}_i dp_i] - \frac{dH}{dt} dt + dH$$

$$= \sum_{i=1}^{n} \left\{\left[\dot{p}_i + \frac{\partial H}{\partial x_i}\right] dx_i + \left[-\dot{x}_i + \frac{\partial H}{\partial p_i}\right] dp_i\right\} + \sum_{j=1}^{m} \frac{\partial H}{\partial u_j} du_j + \left(\frac{dH}{dt} - \frac{dH}{\partial t}\right) dt$$

Equations (E) therefore express the vanishing of $i_{\mathbf{X}}\tilde{\omega}$ on dx_i, dp_i, du_j, which has as a consequence the vanishing of $i_{\mathbf{X}}\tilde{\omega}$ on dt, since:

$$\frac{dH}{dt} = \sum_i \left(\frac{\partial H}{\partial x_i} \dot{x}_i + \frac{\partial H}{\partial p_i} p_i\right) + \sum_j \frac{\partial H}{\partial u_j} + \frac{\partial H}{\partial t} = \frac{\partial H}{\partial t}$$

We have therefore established the following fundamental result: Equations (E) of optimal control theory are equivalent to the fundamental intrinsic relation $i_{\mathbf{X}}\tilde{\omega} = 0$ with

$$\tilde{\omega} = \sum_i dp_i \wedge dx_i - dH \wedge dt$$

4. REMARKS

1) H is a first degree polynomial with respect to the conjugate variables p_i, and thus has a very remarkable form.
2) In analytical mechanics, H is not the Hamiltonian, since H depends on $x = q$, $u = \dot{q}$ and p. In the case of analytical mechanics, p is actually the conjugate variable of q since:

$$H = p \cdot f - L = p \cdot u - L(q,u,t)$$

and $\partial H/\partial u = 0$ leads to $p = \partial L/\partial u = \partial L/\partial \dot{q}$. H depends therefore on the position, the velocity, and the conjugate variable. Let $\bar{u}$ be the optimal command, and K be the value of H for $\bar{u}$. Since $\partial H/\partial u = 0$ for $\bar{u}$, we immediately see that $\partial K/\partial q = \partial H/\partial q$ and $\partial K/\partial p = \partial H/\partial p$. K is therefore the classical Hamiltonian and Legendre's transformation expresses the optimality conditions.
3) Returning to the general case, where H is Pontryagin's function of an optimal control problem, we have from $dH/dt = \partial H/\partial t$ the immediate extension for the energy integral: If $\partial H/\partial t = 0$, H is a constant for the optimal solution.

5. SOME CONSEQUENCES

1) The expression $i_X\tilde{\omega}$ is intrinsic and allows the easy handling of coordinate transformations.
2) $\omega = \Sigma_i \, dp_i \wedge dx_i$ does not have maximum rank in the space, its rank being 2n in a $(2n + m)$-dimensional space. We have a degenerate symplectic structure.
3) All results concerning symmetry groups can be applied here. Indeed, if Y is a vector field associated to an infinitesimal transformation, it is characterized by $[X, Y] = \rho X$. Cartan's theorem shows that $i_Y\tilde{\omega}$ is an invariant 1-form which when it is exact corresponds to a first integral. We consider the following two fundamental examples:

a) If $\frac{\partial H}{\partial t} = 0, \quad Y = \frac{\partial}{\partial t} \quad \text{and} \quad \left[X, \frac{\partial}{\partial t}\right] = 0$

In this case $i_Y\tilde{\omega} = dH$ is an invariant 1-form associated to the generalized energy integral.

b) If $\frac{\partial H}{\partial x_\alpha} = 0, \quad Y = \frac{\partial}{\partial x_\alpha} \quad \text{and} \quad \left[X, \frac{\partial}{\partial x_\alpha}\right] = 0$

$i_Y\tilde{\omega} = -dp_\alpha$ is an invariant 1-form associated to the ignorable coordinate integral $p_\alpha = c_\alpha$, generalized to the maximum principle.

4) We remark that reduction by first integrals, as it is usually done in analytical mechanics, can also be generalized, thanks to $\tilde{\omega}$. For example, we can accomplish a generalized isoenergetic reduction [8], that is to say, we can consider the reduction on $H = h$ (when this integral exists) where H is Pontryagin's function. Solutions satisfying $H = h$ are associated with the 2-form $\bar{\omega}$, which is the restriction of $\tilde{\omega}$ and is equal to:

$$\bar{\omega} = \sum_i dp_i \wedge dx_i \quad \text{with } H(x,p,u) = h$$

In the same way, Routh's reduction for the ignorable coordinates x_α can be carried out. Reducing by $p_\alpha = c_\alpha$ $(\alpha = 1, \ldots, p)$ we have

$$\bar{\omega} = \sum_{\bar{\alpha}} dp_{\bar{\alpha}} \wedge dx_{\bar{\alpha}} - d\bar{H} \wedge dt$$

where

$$\bar{H} = H(x_{\bar{\alpha}}, p_{\bar{\alpha}}, p_\alpha = c_\alpha, u, t)$$

We obtain an equivalent problem for $2(n - p) + m$ variables instead of $2n + m$.

5) We can apply to the (x,p) variables a canonical transformation, with generating function $F(x,p,t)$, without changing the form of equations (E) or that of $\tilde{\omega}$. However, if F depends on u, the result is more complex, since $\tilde{\omega}$ changes form.

6) If one carries out a point transformation $x \to y(x)$ then $\dot{y} = J\dot{x}$ where J is the Jacobian matrix of y. There exists a simple relation between p_x and p_y since:

$$p_x \cdot (\dot{x} - f) = p_y \cdot (\dot{y} - Jf)$$

and therefore

$$p_x = {}^tJp_y$$

This relation is well known in analytical mechanics and the transformation $(x, p_x) \to (y, p_y)$ is a canonical transformation since $p_x \cdot dx = p_y \cdot dy$.

7) As J. Bryant [9] has done in analytical mechanics, we can consider the 1-form:

$$\tilde{\alpha} = dx_0 - p_i dx_i + H\,dt$$

where x_0 is the "cost" variable

$$i_X\tilde{\alpha} = L - p_i\dot{x}_i + H = 0$$

$$d\tilde{\alpha} = -\tilde{\omega}$$

Therefore, equations (E) are characterized by the fact that the 1-form $\tilde{\alpha}$ is an integral invariant. Equations (E) together with $dx_0/dt = L$ have a contact structure, to which are added m optimality equations.

6. EXTENSIONS

We have applied our theory when the optimal command achieved a relative extremum of H, with the consequence that $\partial H/\partial u = 0$. But this can be extended to very general cases, which is of great practical interest if one considers problems of the "bang-bang" type. Let us assume that $\mathcal{U}$ is defined by the inequality $F(u) \leq 0$. (There could in fact be several inequalities.)

If the maximum of H is reached in the interior of $\mathcal{U}$ then $\partial H/\partial u = 0$ and we obtain (E). If the maximum of H reached on the boundary $\partial\mathcal{U}$ we lose the condition $\partial H/\partial u = 0$. Let us then define H' by:

$$H' = H + \mu F \begin{cases} \mu = 0 & \text{in } \mathcal{U} \\ \mu < 0 & \text{on } \mathcal{U} \end{cases}$$

μ appears as a multiplier associated with the constraint $F(u) \leq 0$. We immediately see that the maximum of H, in $\mathcal{U}$ or on $\mathcal{U}$, is expressed by $\partial H'/\partial u = 0$. Therefore, replacing H by H', we get equations (E) for H'.

REFERENCES

1. Flanders, H.: Differential Forms with Applications to the Physical Sciences. Academic Press (1963).
2. Abraham, R.: Foundations of Mechanics. Benjamin (1967).
3. Pontryaguine, L.; Boltianski, V.; Gamkrélidzé, R.; Michtchenko, E.: Théorie Mathématique des Processus Optimaux. Editions Mir, Moscow (1974).
4. Losco, L.: Cours sur l'Optimisation. E.N.S.M.M., Besançon (1983).
5. Losco, L.: Solutions Particulières et Invariants Intégraux en Mécanique Céleste. Thèse de Doctorat d'Etat, Besançon (1972).
6. Losco, L.: J. Mécan. 13, No. 2 (1974).

7. Cartan, E.: Leçons sur les Invariants Intégraux. Hermann, Paris (1971).
8. Whittaker, E. T.: Analytical Dynamics of Particles and Rigid Bodies. Cambridge University Press (1937).
9. Bryant, J. Celestial Mechanics 29, 41-49 (1981).

23 THE RANGE OF RELATIVE HARMONIC DIMENSIONS

Mitsuru Nakai / Department of Mathematics, Nagoya Institute of Technology, Nagoya, Japan
Leo Sario / Department of Mathematics, University of California, Los Angeles, California

The Martin boundaries of Riemannian manifolds play an important role in the study of potential-theoretic global structures of manifolds. In order to obtain an understanding of Martin boundaries beyond formal discussion, it is necessary to examine concrete examples which show the diverse behavior of the boundaries. To this end we consider in the present paper the following example, which is also important in the theory of conformal mappings.

Let Ω be the punctured disk $\{0 < |z| < 1\}$ and $S = \{s_n\}_1^N$ $(1 \leq N \leq \infty)$ a sequence of closed line segments (slits) s_n in Ω satisfying the following three conditions:

(a) $s_n \cap s_m = \emptyset \quad (n \neq m)$,

(b) $\{n;\ s_n \cap \{|z| \geq \varepsilon\} \neq \emptyset\}$ is finite for every $\varepsilon > 0$,

(c) $\bigcup_{n=1}^{N} s_n \subset \bigcup_{\nu=1}^{k} \ell_{2\pi(\nu-1)/k}$,

where k is a positive integer and ℓ_θ stands for the half-line $\{\arg z = \theta\}$. Denote by $\mathscr{S}_k$ the totality of S satisfying the above three conditions. With each $S = \{s_n\}_1^N$ in $\mathscr{S}_k$, associate the slit punctured disk $\Omega_S = \Omega - \bigcup_1^N s_n$. Let $\Gamma = \{|z| = 1\}$ and consider the class $\mathscr{P}_S$ of positive harmonic functions u on Ω_S with boundary values zero on $\Gamma \cup (\bigcup_1^N s_n)$ and with the normalization

$$-\frac{1}{2\pi}\int_0^{2\pi}\left[\frac{\partial}{\partial r}u(re^{i\theta})\right]_{r=1} d\theta = 1$$

The class $\mathscr{P}_S$ forms a convex set. The <u>relative harmonic dimension</u> dim S of the ideal boundary point $z = 0$ with respect to the region Ω_S is the cardinal number $\#(\text{ex. } \mathscr{P}_S)$ of the set ex. $\mathscr{P}_S$ of extreme points of $\mathscr{P}_S$,

$$\dim S = \#(\text{ex. } \mathscr{P}_S) \tag{1}$$

It is known that $\dim S \geq 1$ and that dim S is the cardinal number of minimal points lying over $z = 0$ in the Martin compactification of Ω_S (cf., e.g., [3]).

We are interested in the range $W_k = \{\dim S;\ S \in \mathscr{S}_k\}$ of the mapping dim: $\mathscr{S}_k \to \{\text{cardinals}\}$. It was recently shown [6] that $W_k \supset \{1, 2, \ldots, k\}$ for each k, but it has been an open question whether the inclusion is proper or not, i.e., does $\dim S \leq k$ hold for every S in $\mathscr{S}_k$. Benedicks [2] and Ancona [1] proved that $W_2 = \{1, 2\}$, that is, $\dim S \leq 2$ for every S in $\mathscr{S}_2$. Using an idea of Benedicks [2] we will show that

$$W_k = \{1, 2, \ldots, k\} \tag{2}$$

for every positive integer $k = 1, 2, \ldots$:

Theorem. $\dim S \leq k$ for every S in $\mathcal{S}_k$ $(k = 1, 2, \ldots)$.

The proof will be given in Sec. 4, after auxiliary discussions in Secs. 1-3. A generalization will be presented in Sec. 5. We acknowledge with appreciation and pleasure the important assistance of Dr. S. Segawa in establishing the above result.

1. A Result of Domar. Take an upper semicontinuous function F defined on a region D of the complex plane $\mathbb{C}$ such that $0 \leq F(z) \leq +\infty$ on D, and consider an upper envelope U of the class $\mathcal{F}$ of subharmonic functions u on D with $0 \leq u(z) \leq F(z)$ on D, i.e.,

$$U(z) = \sup_{u \in \mathcal{F}} u(z)$$

for every z in D. Domar [4] showed that U is subharmonic on D if and only if U is bounded on each compact subset of D. He also proved the following unpublished result of Beurling: If there exists a positive number ε for each compact set K in D such that

$$\iint_K (\overset{+}{\log} F(z))^{1+\varepsilon}\, dx\, dy < +\infty \qquad (z = x + iy) \tag{3}$$

then U is bounded on each compact set in D.

It is easy to see that $F(z) = 1/|\mathrm{Im}\, z|$ and $D = \{|z| < 1\} - \bar{\ell}_\pi$ satisfy the above condition. We conclude:

Lemma 1. There exists a universal constant C such that $u(z) \leq C$ on $X = \{\mathrm{Re}\, z \geq 0, 1/3 \leq |z| \leq 2/3\}$ for any nonnegative subharmonic function u on $\{|z| < 1\} - \bar{\ell}_\pi$ satisfying

$$u(z) \leq 1/|\mathrm{Im}\, z| \tag{4}$$

2. A Theorem of Heins. Set $M(r) = M(r, v) = \sup_{|z|=r} |v(z)|$ for any function v on the complex plane $\mathbb{C}$. The order $\lambda(v)$ of v is given by

$$\lambda(v) = \limsup_{r \to +\infty} (\overset{+}{\log} M(r))/\log r$$

and the lower order $\mu(v)$ of v by

$$\mu(v) = \liminf_{r \to +\infty} (\overset{+}{\log} M(r))/\log r$$

Consider a nonnegative subharmonic function v on the entire complex plane $\mathbb{C}$. A subregion D of $\mathbb{C}$ is said to be a tract for the function v if the following two conditions are satisfied:

(α) $v > 0$ on D,

(β) $\lim_{z \to \zeta} v(z) = 0$ for every finite boundary point ζ of D.

Heins ([5], pp. 74-75) established the following inequality:

Lemma 2. If there exist $k > 0$ disjoint tracts for a nonnegative subharmonic function v on the complex plane $\mathbb{C}$, then

$$\mu(v) \geq k/2 \tag{5}$$

3. Subharmonic Functions in an Angle. Consider an angle $\Delta(\alpha, \beta) = \{\alpha < \arg z < \beta\}$ for $0 \leq \alpha < \beta \leq 2\pi$. The following auxiliary result, a revised version of an idea of Benedicks [2], plays a decisive role in the proof of our theorem:

Lemma 3. If u is a nonnegative subharmonic function on $\Delta(\alpha,\beta)$, harmonic on $\Delta(\alpha,\beta) - \overline{\ell}_{(\alpha+\beta)/2}$, then there exists a finite constant A such that

$$u(z) \leq A|z|^p \quad (p = 2\pi/(\beta - \alpha)) \tag{6}$$

when $z \to \infty$ inside the angle $\Delta(\alpha + (\beta - \alpha)/4,\ \beta - (\beta - \alpha)/4)$.

Proof. On applying a rotation about $z = 0$, if necessary, we may take $(\alpha + \beta)/2 = 0$, so that u is subharmonic on $\Delta(-\gamma,\gamma)$ and harmonic on $\Delta(-\gamma,\gamma) - \overline{\ell}_0$ with $\gamma = (\beta - \alpha)/2$. Then $p = 2\pi/(\beta - \alpha) = 2\pi(\gamma - (-\gamma))$. We have to show that (6) is valid when $z \to \infty$ inside the angle $\Delta(-\gamma/2, \gamma/2)$. Set $h(\zeta) = u(\zeta^{1/p})$. Then h is subharmonic on $\Delta(-\pi, \pi)$ and harmonic on $\Delta(-\pi, \pi) - \overline{\ell}_0$. It suffices to establish the existence of a finite constant A such that

$$h(\zeta) \leq A|\zeta| \tag{7}$$

when $\zeta \to \infty$ inside $\Delta(-\pi/2, \pi/2)$. By considering $(h(\zeta) + h(\overline{\zeta}))/2$ instead of $h(\zeta)$, if necessary, we may assume that $h(\zeta) \equiv h(\overline{\zeta})$.

Let $v_R(\zeta) = h(R\zeta)/R$ for an arbitrary $R \geq 1$. Observe that $v_R(\overline{\zeta}) = v_R(\zeta)$. Since h is a positive harmonic function on the upper half-plane $\operatorname{Im} \zeta > 0$, there exists a positive Borel measure $d\sigma$ on the real line $\operatorname{Im} \zeta = 0$ and a finite positive constant c such that the following Poisson-type representation is valid (cf., e.g., [7], p. 149):

$$h(\zeta) = c\eta + \frac{\eta}{\pi}\int_{-\infty}^{\infty} \frac{1}{|\zeta - t|^2}\, d\sigma(t) \qquad (\zeta = \xi + i\eta)$$

Replacing ζ by $R\zeta$ and then dividing both sides by R we have

$$v_R(\zeta) = c\eta + \frac{\eta}{\pi}\int_{-\infty}^{\infty} \frac{1}{|R\zeta - t|^2}\, d\sigma(t)$$

and in particular

$$v_R(i) = c + \frac{1}{\pi}\int_{-\infty}^{\infty} \frac{1}{|Ri - t|^2}\, d\sigma(t)$$

$$\leq c + \frac{1}{\pi}\int_{-\infty}^{\infty} \frac{1}{|i - t|^2}\, d\sigma(t) = h(i)$$

Therefore,

$$\sup_{R \geq 1} v_R(i) \leq h(i)$$

Let $I = \{\operatorname{Im} z = 1,\ |\operatorname{Re} z| \leq 1\}$. By the Harnack inequality there exists a universal constant C such that

$$\sup_I v_R \leq C v_R(i)$$

Take an arbitrary point $z = x + iy$ in the upper half of the unit disk, $\{|\zeta| < 1,\ \operatorname{Im} \zeta > 0\}$. From the Poisson representation of $v_R(z)$ in the disk with center $z_0 = x + i$ and radius 1 the so-called Poisson-Harnack inequality follows:

$$v_R(z) \leq [(1 + |z - z_0|)/(1 - |z - z_0|)]\, v_R(z_0)$$

Since $z - z_0 = i(y - 1)$ and $|z - z_0| = 1 - y$, we have

$$v_R(z) \leq \frac{2}{y} v_R(z_0) \leq \frac{2C}{y} v_R(i) \leq \frac{2Ch(i)}{y}$$

We again stress that $2Ch(i)/y$ is independent of $R \geq 1$. In view of $v_R(\bar{z}) = v_R(z)$ we conclude that

$$v_R(z) \leq 2Ch(i)/|\mathrm{Im}\, z|$$

for every z in $\{|z| < 1\} - \ell_\pi$. We can find by Lemma 1 a universal constant B such that $v_R(\zeta) \leq B$ or $h(R\zeta) \leq BR$ on $X = \{\mathrm{Re}\,\zeta \geq 0,\ 1/3 \leq |\zeta| \leq 2/3\}$. Let $z = R\zeta$. Then $\zeta \in X$ implies that $2|z|/3 \leq R \leq 3|z|$. Therefore, on setting $A = 3B$ we conclude that $h(z) \leq A|z|$ for every z in $\{\mathrm{Re}\, z \geq 0,\ |z| > 1/3\}$. This proves (7).

4. Proof of the Theorem. Suppose first that $S = \emptyset$. Then, by the classical Picard principle, $\dim S = 1 \leq k$. Therefore, we may assume in the sequel that $S = \{s_n\}_1^N \neq \emptyset$. We denote by $\mathscr{P}_S'$ the linear space generated by $\mathscr{P}_S$. Let $\hat{\mathbb{C}}$ be the extended complex plane and $\hat{\mathbb{C}}_0 = \hat{\mathbb{C}} - \{0\}$. In addition to the class $\mathscr{P}_S'$ of functions on $\Omega_S = \Omega - \bigcup_1^N s_n$ we consider the linear space $\tilde{\mathscr{P}}_S'$ generated by the class of positive harmonic functions on $\hat{\mathbb{C}}_0 - \bigcup_1^N s_n$ with boundary values zero on $\bigcup_1^N s_n$.

We first remark that bounded functions in $\mathscr{P}_S'$ or $\tilde{\mathscr{P}}_S'$ vanish identically. In fact, let g be the Green's function on $\hat{\mathbb{C}} - s_1$ with its pole at $z = 0$. Take a bounded function u in $\mathscr{P}_S'$ or $\tilde{\mathscr{P}}_S'$. Since the boundary values of $\varepsilon g \pm u$ are nonnegative for every $\varepsilon > 0$, the maximum principle for harmonic functions implies that $\varepsilon g \pm u > 0$ on the region of its definition. On letting $\varepsilon \to 0$ we conclude that $u \equiv 0$.

Given a continuous function φ on $\Gamma = \{|z| = 1\}$, there exists a unique bounded harmonic function $H\varphi$ on Ω_S with boundary values φ on Γ and zero on $\bigcup_1^N s_n$. The mapping $u \mapsto Tu = u - Hu$ gives a linear operator from $\tilde{\mathscr{P}}_S'$ to $\mathscr{P}_S'$. By the above remark it is readily seen that T is injective. We now show that T is surjective. Choose an arbitrary v in $\mathscr{P}_S'$. We have to find a u in $\tilde{\mathscr{P}}_S'$ such that $Tu = v$. Choose $0 < \rho < 1$ such that $\Lambda = \{|z| = \rho\}$ encircles S. Denote by $P\psi$ the harmonic function on $\{\rho < |z| \leq +\infty\}$ with given continuous boundary values ψ on Λ. With each φ in the space of $C(\Gamma)$ of continuous functions on Γ, we associate the function

$$\tau\varphi = P((H\varphi)|\Lambda)|\Gamma$$

which defines a linear operator $\varphi \mapsto \tau\varphi$ from $C(\Gamma)$ into itself. Since $\pm\varphi \leq \|\varphi\| \cdot 1$ on Γ for any $\varphi \in C(\Gamma)$ with $\|\varphi\| = \sup_\Gamma |\varphi|$, the maximum principle yields $\pm H\varphi \leq \|\varphi\|(H1)$ on Ω_S, and in particular $\sup_\Lambda |H\varphi| \leq \kappa\|\varphi\|$ with $\kappa = \sup_\Lambda |H1| \in (0,1)$. Hence, again by the maximum principle,

$$\|\tau\| = \sup\{\|\tau\varphi\|;\ \varphi \in C(\Gamma),\ \|\varphi\| = 1\} \leq \kappa < 1$$

Let 1 be the identity operator on $C(\Gamma)$. The abstract integral equation

$$(1 - \tau)\varphi = (Pv)|\Gamma$$

is solved by the Neumann series

$$\varphi = \sum_{n=0}^{\infty} \tau^n((Pv)|\Gamma)$$

since $\|\tau\| < 1$. Let $u_1 = H\varphi + v$ on Ω_S and $u_2 = P(u_1|\Lambda)$ on $\rho < |z| \leq +\infty$. Trivially $u_1 = u_2$ on Λ. Moreover, $u_2 = P(H\varphi + v) = \tau\varphi + Pv = (\varphi - Pv) + Pv = H\varphi = H\varphi + v = u_1$ on Γ. Therefore, $u_1 = u_2$ on the boundary of $\{\rho < |z| < 1\}$ and hence on $\{\rho < |z| < 1\}$. We can thus define a harmonic function u on $\hat{\mathbb{C}}_0 - \bigcup_1^N s_n$ by $u = u_1$ on Ω_S and u_2 on $\{\rho < |z| \leq +\infty\}$. It is now easily verified that $Tu = v$. We have proved that $T: \tilde{\mathscr{P}}_S' \to \mathscr{P}_S'$ is bijective.

The operators T and T^{-1} are positive, that is, $u - Hu \geq 0$ if and only if $u \geq 0$. In fact, if $u \geq 0$, then by the maximum principle, $u + \varepsilon g \geq Hu$ on Ω_S for any $\varepsilon > 0$. On letting $\varepsilon \to 0$ we obtain $u - Hu \geq 0$ on Ω_S. Conversely assume $u - Hu \geq 0$ on Ω_S. Since $\liminf_{z \to 0} u(z) \geq \liminf_{z \to 0} (Hu)(z)$, the maximum principle implies that u is bounded from below on $\hat{\mathbb{C}}_0$. For the same reason $u + \varepsilon g \geq 0$ on $\hat{\mathbb{C}}_0$ for any positive number ε. On letting $\varepsilon \to 0$, we see that $u \geq 0$ on $\hat{\mathbb{C}}_0$.

The class $\tilde{\mathscr{P}}_S = T^{-1}\mathscr{P}_S$ is thus a convex set of normalized positive functions in $\tilde{\mathscr{P}}'_S$ generating $\tilde{\mathscr{P}}'_S$. In order to show that $\dim S \leq k$, we only have to prove that the cardinal number of the set of extreme points of $\tilde{\mathscr{P}}_S$ is not greater than k.

We make one more reduction. Let $\sigma(z) = 1/z$ and consider the linear space $\mathscr{P}'$ generated by the class of positive harmonic functions v on $\mathbb{C} - \bigcup_1^N \sigma(s_n)$ which vanish on $\bigcup_1^N \sigma(s_n)$. Then $v \mapsto v \circ \sigma: \mathscr{P}' \to \tilde{\mathscr{P}}'_S$ is a bijective order-preserving linear mapping. The preimage $\mathscr{P}$ of $\tilde{\mathscr{P}}_S$ is a convex set of normalized positive harmonic functions on $\mathbb{C} - \bigcup_1^N \sigma(s_n)$ generating $\mathscr{P}'$. In order to establish $\dim S \leq k$ it suffices to show that the cardinal number of the set of extreme points in $\mathscr{P}$ is not greater than k.

Contrary to the assertion suppose that there exist k + 1 extreme points $u_1, u_2, \ldots, u_{k+1}$ of $\mathscr{P}$. Set

$$v_\nu = u_\nu - \sum_{j \neq \nu}^{1,\ldots,k+1} u_j \qquad (\nu = 1, 2, \ldots, k+1)$$

We maintain that $v_\nu \leq 0$ on $\mathbb{C}$ does not hold for any ν. If $u_\nu \leq \Sigma_{j\neq\nu}^{1,\ldots,k+1} u_j$ on $\mathbb{C}$, then there exist nonnegative c_j in the real number field $\mathbb{R}$, such that $u_\nu \equiv \Sigma_{j\neq\nu}^{1,\ldots,k+1} c_j u_j$ (cf., e.g., [3], p. 17). This contradicts the fact that u_ν is extreme. We conclude that

$$D_\nu = \{z \in \mathbb{C};\ v_\nu(z) > 0\} \neq \emptyset \qquad (\nu = 1, 2, \ldots, k+1)$$

Consider the subharmonic function v on $\mathbb{C}$ defined by

$$v(z) = \max\{v_1(z), v_2(z), \ldots, v_{k+1}(z), 0\}$$

for each $z \in \mathbb{C}$. The D_ν ($\nu = 1, 2, \ldots, k+1$) are tracts for v. In fact, $v \geq v_\nu > 0$ on D_ν, that is, requirement (α) for D_ν to be a tract for v is fulfilled. Let ζ be a finite boundary point of D_ν. Then $v_\nu(\zeta) = u_\nu(\zeta) - \Sigma_{j\neq\nu}^{1,\ldots,k+1} u_j(\zeta) = 0$ and in particular $u_j(\zeta) \leq u_\nu(\zeta)$ ($j \neq \nu$). Thus $v_\mu(\zeta) = u_\mu(\zeta) - \Sigma_{j\neq\mu}^{1,\ldots,k+1} u_j(\zeta) \leq u_\mu(\zeta) - u_\nu(\zeta) \leq 0$ ($\mu \neq \nu$), and $v(\zeta) = v_\nu(\zeta) = 0$. Therefore, $\lim_{z \in D_\nu, z \to \zeta} v(z) = v(\zeta) = 0$, so that condition ($\beta$) for D_ν to be a tract for v is satisfied. By Lemma 2 (the Heins theorem),

$$\mu(v) \geq (k + 1)/2 \tag{8}$$

Since $v_j \leq u_j$ ($j = 1, \ldots, k+1$), $v \leq \Sigma_1^{k+1} u_j$. Hence there exists a $u \in \mathscr{P}$ and a positive constant c such that $v \leq cu$. We apply Lemma 3 for u in each angle with the opening $4\pi/k$ and with the bisecting half-line $\ell_{2\pi\nu/k}$ ($\nu = 1, 2, \ldots, k$). Since $p = 2\pi/(4\pi/k) = k/2$, we conclude that there exists a constant K such that $u(z) \leq K|z|^{k/2}$ on $\mathbb{C}$ as z tends to ∞. Therefore,

$$v(z) \leq cK|z|^{k/2}$$

on $\mathbb{C}$ as z tends to ∞, i.e.,

$$\lambda(v) \le k/2 \tag{9}$$

We finally deduce a contradiction $(k+1)/2 \le k/2$ from (8), (9), and $\mu(v) \le \lambda(v)$.

5. A Generalization. Let us relax condition (c) for the class $\mathcal{S}_k$ of sets $S = \{s_n\}_1^N$ of slits s_n as follows. Fix a set of k real numbers θ_ν ($\nu = 1, \ldots, k$) satisfying

$$0 \le \theta_1 < \theta_2 < \cdots < \theta_k < 2\pi$$

In condition (c) these numbers were chosen as $\theta_\nu = 2\pi(\nu - 1)/k$ ($\nu = 1, 2, \ldots, k$) but for the present we do not require anything on the θ_ν beyond the above inequalities. Instead of (c) we consider a weaker condition

$$(c') \quad \bigcup_{n=1}^{N} s_n \subset \bigcup_{\nu=1}^{k} \ell_{\theta_\nu}$$

and the class $\mathcal{S}_k'$ of $S = \{s_n\}_1^N$ satisfying (a), (b), and (c'). By the same proof as that for $W_k \supset \{1, 2, \ldots, k\}$, we can show that

$$W_k' = \{\dim S;\ S \in \mathcal{S}_k'\} \supset \{1, 2, \ldots, k\}$$

However, we do not know whether the converse inclusion $W_k' \subset \{1, 2, \ldots, k\}$ is true or, equivalently, whether the inequality $\dim S \le k$ is valid for any $S \in \mathcal{S}_k'$.

We examine the proof for $\dim S \le k$ ($S \in \mathcal{S}_k$) to see to what extent it is also valid for $S \in \mathcal{S}_k'$. There is no change in the proof of (8): $\mu(v) \ge (k+1)/2$ since condition (c) is not used beyond (c') in the proof. Condition (c) is essentially used to show (9): $\lambda(u) \le k/2$ in the above proof. This is no longer true for a general $S \in \mathcal{S}_k'$. However, $\lambda(u) \le k/2$ is only used to deduce the contradiction $(k+1)/2 \le \mu(v) \le \lambda(v) \le \lambda(u) \le k/2$, but to deduce a contradiction only it suffices to have $\lambda(u) < (k+1)/2$.

In view of this, let us consider

$$q = \min(\theta_2 - \theta_1, \theta_3 - \theta_2, \ldots, \theta_k - \theta_{k-1}, \theta_1 + 2\pi - \theta_k)$$

By Lemma 3 for the angular region $\Delta(\theta_\nu - q, \theta_\nu + q)$ with $p = 2\pi/2q = \pi/q$, we obtain the inequality

$$\lambda(u) \le \pi/q$$

For $S \in \mathcal{S}_k$, $q = 2\pi/k$, and the above inequality is reduced to $\lambda(u) < k/2$. In the general case, q can take any value in $(0, 2\pi/k]$. Since $\lambda(u) \le \pi/q < (k+1)/2$, it suffices to assume that

$$q > 2\pi/(k+1)$$

in order to have $\lambda(u) < (k+1)/2$.

REFERENCES

1. Ancona, A.: Ann. Inst. Fourier, Grenoble 27, 71-90 (1979).
2. Benedicks, M.: Arkiv Mat. 18, 53-72 (1980).
3. Constantinescu, C.; Cornea, A.: Ideale Ränder Riemannscher Flächen. Springer (1963).
4. Domar, Y.: Arkiv Mat. 3, 429-440 (1957).
5. Heins, M.: J. d'Analyse Math. 7, 53-77 (1959).
6. Nakai, M.: Relative harmonic dimensions. Seminar Note at Research Inst. for Math. Sci. 366, 137-149 (1979) [in Japanese].
7. Tsuji, M.: Potential Theory in Modern Function Theory. Maruzen (1959).

24 THE COINCIDENCE SET FOR TWO-DIMENSIONAL AREA MINIMIZING SURFACES IN R^n WHICH AVOID A CONVEX OBSTACLE

Harold R. Parks / Department of Mathematics, Oregon State University, Corvallis, Oregon

1. Introduction. The problem of existence and regularity for surfaces which minimize area subject to the constraint of avoiding an obstacle has led to a number of interesting papers, for example, [8, 9, 10, 12]. In the nonparametric case the problem yields a variational inequality, and in recent years there has been significant progress made in the study of variational inequalities (see for example [3-6]). In this note, we present a method which allows us to apply the results of Brézis and Kinderlehrer [3] to the two dimensional parametric problem in any codimension. Our main conclusion is

Theorem. For a two dimensional integral flat chain modulo 2 in $\underline{R}^n$ which minimizes area subject to avoiding a uniformly convex obstacle with $C^{2,\alpha}$ boundary, $0 < \alpha < 1$, the boundary of the coincidence set is of Hausdorff dimension one.

Our proof relies on the regularity theory of F. J. Almgren, Jr., the existence of isothermal parameters due to Korn and Lichtenstein, the higher differentiability theory of C. B. Morrey, Jr., and, of course, the results on variational inequalities of Brézis and Kinderlehrer. The main innovation here is the use of isothermal parameters to break the problem into two parts which can be dealt with in turn rather than simultaneously. Because of the use of isothermal parameters the method is limited to two dimensional surfaces.

2. Notation and Terminology. Except when otherwise stated, we follow the notation and terminology of [7].

(1) Let $n \geq 3$ be an integer.

(2) Let Ω be a nonempty, bounded, open, convex subset of $\underline{R}^n$. Note that $\underline{R}^n \sim \Omega$ is a Lipschitz neighborhood retract. In fact, fixing $p \in \Omega$ and $\rho > 0$ such that

$$\underline{B}(p, \rho) \subset \Omega$$

we may define

$$r : \underline{R}^n \sim \underline{B}(p, \rho) \to \underline{R}^n \sim \Omega$$

by setting

$$r(x) = p + (\inf\{t : t \geq 1,\ p + t(x - p) \in \underline{R}^n \sim \Omega\})(x - p)$$

and we can then verify

$$\text{Lip}(r) \leq 2(\text{diam}\ \Omega)\rho^{-1}$$

We also note that for $1 \leq k \leq n - 1$ and $Q \in \mathscr{R}^2_{k-1}(\underline{R}^n)$ with

$$\text{spt}^2 Q \subset \underline{R}^n \sim \Omega$$

and

$$\partial S = Q$$

there exists $S \in \mathcal{R}_k^2(\underline{R}^n)$ with

$$\operatorname{spt}^2 S \subset \underline{R}^n \sim \Omega$$

and

$$\partial S = Q$$

This is seen by finding $S' \in \mathcal{R}_k^2(\underline{R}^n)$ with $\partial S' = Q$ and setting

$$S = r_{\#} S'$$

where r is defined as above with $p \in \Omega$, $\rho > 0$ chosen so that

$$\underline{B}(p,\rho) \subset \Omega \sim \operatorname{spt}^2 S'$$

(3) Fix $R \in \mathcal{R}_1^2(\underline{R}^n)$ with

$$\operatorname{spt}^2 R \subset \underline{R}^n \sim \Omega$$

$$\partial R = 0$$

(4) We will say $S \in \mathcal{R}_2^2(\underline{R}^n)$ is <u>area minimizing avoiding</u> Ω if

$$\operatorname{spt}^2 S \subset \underline{R}^n \sim \Omega$$

and

$$\underline{M}^2(S) \leq \underline{M}^2(S + W)$$

holds for each $W \in \mathcal{R}_2^2(\underline{R}^n)$ with

$$\operatorname{spt}^2 W \subset \underline{R}^n \sim \Omega$$

$$\partial W = 0$$

Note that there exists $T \in \mathcal{R}_2^2(\underline{R}^n)$ with $\partial T = R$ which is area minimizing avoiding Ω. We fix such a T.

(5) Set

$$K = \operatorname{Bdry} \Omega \cap \operatorname{spt}^2 T$$

We call K the <u>coincidence set</u>. Set

$$\Gamma = \operatorname{Bdry}_{\mathrm{rel}} K \sim \operatorname{spt}^2 \partial T$$

where $\operatorname{Bdry}_{\mathrm{rel}}(\cdot)$ is the boundary in the relative topology of Bdry Ω. We call Γ the <u>boundary of the coincidence set</u>.

3. <u>Theorem</u>. Suppose Bdry Ω is a class 2 submanifold of $\underline{R}^n$.

(1) [10] If $n = 3$, then $\operatorname{spt}^2 T \sim \operatorname{spt}^2 \partial T$ is a locally Hölder continuously differentiable submanifold of R^n.

(2) If $n = 4$, then there exists an open set U with

$$\operatorname{Bdry} \Omega \sim \operatorname{spt}^2 \partial T \subset U$$

such that $U \cap \operatorname{spt}^2 T$ is a locally Hölder continuously differentiable submanifold of $\underline{R}^n$ and

$$(\operatorname{spt}^2 T \sim \operatorname{spt}^2 \partial T) \sim U$$

consists of isolated points.

(3) [2] If $n \geq 5$, then there exists an open set U such that $U \cap \operatorname{spt}^2 T$ is a locally Hölder continuously differentiable submanifold of $\underline{R}^n$ and

$$(\mathrm{spt}^2 T \sim \mathrm{spt}^2 \partial T) \sim U$$

consists of isolated points.

Proof. As in [1] we see the tangent cone at any point, $p \in \mathrm{spt}^2 T \sim \mathrm{spt}^2 \partial T$, is a sum of discs. In case $p \in \mathrm{Bdry}\, \Omega$ those discs lie in an $(n - 1)$ dimensional hyperplane. In case $n = 4$ this leads to regularity in a neighborhood of $\mathrm{Bdry}\, \Omega \sim \mathrm{spt}^2 \partial T$.

4. Theorem. Suppose Ω is uniformly convex and $\mathrm{Bdry}\, \Omega$ is a twice Hölder continuously differentiable submanifold of $\underline{R}^n$.

(1) If $n = 3$ or $n = 4$ and

$$\mathrm{spt}^2 \partial T \cap \mathrm{Bdry}\, \Omega = \emptyset$$

then $\mathscr{H}^1(\Gamma) < \infty$.

(2) $\mathscr{H}^m(\Gamma) = 0$ whenever $1 < m \in \underline{R}$.

Proof. Let $\rho : \underline{R}^n \sim \Omega \to \mathrm{Bdry}\, \Omega$ be the nearest point retraction and define $N : \mathrm{Bdry}\, \Omega \to \underline{S}^{n-1}$ by setting $N(x)$ equal to the outward unit normal to $\mathrm{Bdry}\, \Omega$ at x. Also, define $\delta : \underline{R}^n \sim \Omega \to \underline{R}$ by setting

$$\delta(x) = |x - \rho(x)|$$

Suppose

$$p \in (\mathrm{spt}^2 T \sim \mathrm{spt}^2 \partial T) \cap \mathrm{Bdry}\, \Omega$$

is a regular point. Then there exist isothermal parameters for some neighborhood of p in $\mathrm{spt}^2 T \sim \mathrm{spt}^2 \partial T$. Accordingly, we may assume

$$f : \underline{B}(0,1) \subset \underline{R}^2 \to \underline{R}^n$$

is Hölder continuously differentiable and satisfies

$$f(0,0) = p$$

$$D_1 f(\xi, \eta) \cdot D_2 f(\xi, \eta) = 0$$

$$|D_1 f(\xi, \eta)| = |D_2 f(\xi, \eta)|$$

$$f[\underline{U}(0,1)] = U \cap (\mathrm{spt}^2 T \sim \mathrm{spt}^2 \partial T)$$

for some open $U \subset \underline{R}^n$. Because T is area minimizing avoiding Ω, we see that among all Lipschitzian functions

$$\phi : \underline{B}(0,1) \subset \underline{R}^2 \to \underline{R}^n \sim \Omega$$

with

$$\phi(\xi, \eta) = f(\xi, \eta) \quad \text{for } (\xi, \eta) \in \underline{S}^1$$

the minimum of the functional

$$\underline{D}[\phi] = \int_{\underline{U}(0,1)} (|D_1 \phi|^2 + |D_2 \phi|^2)\, d\mathscr{L}^2$$

is attained by $\underline{D}[f]$.

Set

$$u = \rho \circ f, \quad v = \delta \circ f$$

so

$$f = u + v(N \circ u)$$

We calculate

$$|D_1 f|^2 + |D_2 f|^2 = |D_1 u|^2 + |D_2 u|^2 + 2v[D_1 u \cdot \langle D_1 u, DN(u)\rangle + D_2 u \cdot \langle D_2 u, DN(u)\rangle] + v^2[|\langle D_1 u, DN(u)\rangle|^2 + |\langle D_2 u, DN(u)\rangle|^2] + (D_1 v)^2 + (D_2 v)^2$$

It follows that among all Lipschitzian functions

$$\mu : \underline{B}(0,1) \subset \underline{R}^2 \to \text{Bdry}\, \Omega$$

with

$$\mu(\xi, \eta) = u(\xi, \eta) \quad \text{for} \quad (\xi, \eta) \in \underline{S}^1$$

the minimum of the functional

$$\underline{I}[\mu] = \int_{\underline{U}(0,1)} \{|D_1\mu|^2 + |D_2\mu|^2 + 2v[D_1\mu \cdot \langle D_1\mu, DN(\mu)\rangle + D_2\mu \cdot \langle D_2\mu, DN(\mu)\rangle] + v^2[|\langle D_1\mu, DN(\mu)\rangle|^2 + |\langle D_2\mu, DN(\mu)\rangle|^2]\}\, d\mathscr{L}^2$$

is attained by $\underline{I}[u]$. Since v is Hölder continuously differentiable, we conclude by ([11], Chapter VII, Theorem 3.4) that u is twice Hölder continuously differentiable.

Now, consider u to be fixed and set

$$a = |\langle D_1 u, DN(u)\rangle|^2 + |\langle D_2 u, DN(u)\rangle|^2$$

$$b = D_1 u \cdot \langle D_1 u, DN(u)\rangle + D_2 u \cdot \langle D_2 u, DN(u)\rangle$$

Also let

$$\psi : \underline{B}(0,1) \subset \underline{R}^2 \to \underline{R}$$

be continuous, harmonic on $\underline{U}(0,1)$, and satisfy

$$\psi(\xi, \eta) = -v(\xi, \eta) \quad \text{for} \quad (\xi, \eta) \in \underline{S}^1$$

It follows that among all Lipschitzian functions

$$\nu : \underline{B}(0,1) \subset \underline{R}^2 \to \underline{R}$$

with

$$\nu \geq \psi \quad \text{on} \quad \underline{B}(0,1)$$

and

$$\nu = 0 \quad \text{on} \quad \underline{S}^1$$

the minimum of the functional

$$\underline{J}[\nu] = \int_{\underline{U}(0,1)} [(D_1\nu)^2 + (D_2\nu)^2 + a(\nu - \psi)^2 + 2b(\nu - \psi)]\, d\mathscr{L}^2$$

is attained by $\underline{J}[v + \psi]$. Consequently, $\tilde{v} = v + \psi$ satisfies the variational inequality

$$\int_{\underline{U}(0,1)} [D_1\tilde{v}D_1(\nu - \tilde{v}) + D_2\tilde{v}D_2(\nu - \tilde{v})]\, d\mathscr{L}^2 \geq \int_{\underline{U}(0,1)} h(\nu - \tilde{v})\, d\mathscr{L}^2$$

for all Lipschitzian functions

$$\nu : \underline{B}(0,1) \subset \underline{R}^2 \to \underline{R}$$

with

$$\nu \geq \psi \quad \text{on} \quad \underline{B}(0,1)$$

and

$$\nu = 0 \quad \text{on} \quad \underline{S}^1$$

where

$$h = -[a(v - \psi) + b]$$

Since h is continuously differentiable and ψ is harmonic we may apply the results of [3]. In particular, since the uniform convexity of Ω insures

$$b > 0$$

the theorem follows from 3 and ([3], Corollary 2.1).

REFERENCES

1. Almgren, F. J., Jr.: Ann. Math. 84, 277-292 (1966).
2. Almgren, F. J., Jr.: Memoirs Math. Soc. 165 (1976).
3. Brézis, H.; Kinderlehrer, D.: Indiana Univ. Math. J. 23, 831-844 (1974).
4. Caffarelli, L. A.: Acta Math. 139, 155-184 (1977).
5. Caffarelli, L. A.; Rivière, N. M.: Ann. Scuola Norm. Sup. Pisa, Sci. Fis. Math. (Ser. IV) 3, 289-310 (1976).
6. Caffarelli, L. A.; Rivière, N. M.: Ann. Math. 106, 309-317 (1977).
7. Federer, H.: Geometric Measure Theory, Springer (1969).
8. Giaquinta, M.; Pepe, L.: Ann. Scuola Norm. Sup. Pisa 25, 481-506 (1971).
9. Kinderlehrer, D.: Arch. Rat. Mech. Anal. 40, 231-250 (1970/71).
10. Miranda, M.: Ann. dell'Un. di Ferrara XVI, No. 2, 29-37 (1971).
11. Morrey, C. B., Jr.: Multiple Integral Problems in the Calculus of Variations and Related Topics. Univ. of California Publ. in Math., new ser., 1, 1-130 (1943).
12. Nitsche, J. C. C.: Arch. Rat. Mech. Anal. 35, 83-113 (1969).

25 ON THE BASINS OF ATTRACTION OF GRADIENT VECTOR FIELDS

Jean-Paul Penot / Department of Mathematics, Faculté des Sciences,
Université de Pau, Pau, France

In the following $k \geq 1$ and M denotes a complete C^{k+1} Riemannian manifold modeled on a Hilbert space E and $f: M \to \mathbb{R}$ a C^{k+1} Morse function bounded below and satisfying the well known Palais-Smale condition:

(C) Any sequence (x_n) of M such that $(\|\nabla f(x_n)\|) \to 0$ has a limit point.

If M is finite dimensional this condition is known to amount to a properness condition on f.

Let $\varphi: D \to M$ be the flow of $-\nabla f$. We wish to describe the structure of the basins of attraction of $-\nabla f$ given by

$$A(z) = \{x \in M: \lim_{t \to \omega(x)} \varphi(x,t) = z\}$$

where $D = \bigcup_{x \in M} \{x\} \times (\alpha(x), \omega(x))$. Let us observe that under our assumptions, for each $x \in M$ we have $\omega(x) = +\infty$ ([5] lemma §.10). Let K be the set of critical points of f; obviously A(z) is non void iff $z \in K$.

<u>Proposition</u>. If f is a C^{k+1} Morse function on M, bounded below and satisfying condition (C), the family $(A(z))_{z \in K}$ forms a partition of M into C^k-submanifolds of M. More precisely, if $z \in K$ has index i than A(z) is of codimension i and each A(z) is contractible.

<u>Proof</u>. Using [6] th.4.1 and the fact that the points of K are isolated we have that for any $x \in M$ $\varphi_t(x) = \varphi(x,t)$ converges in M as $t \to \infty$. Hence the family $(A(z))_{z \in K}$ form a partition of M. Let us show that for each $z \in K$ A(z) is a C^k-submanifold of M. Let $c: U \to V$ be a Morse chart of M at z [1,5]: U and V are open subsets of M and E respectively, E is the sum of two orthogonal subspaces E_1 and E_2, $V = V_1 \times V_2$ where V_i is an open neighborhood of 0 in E_i for i = 1, 2 and the expression $\tilde{f} := f \circ c^{-1}$ of f in the chart c is given by

$$\tilde{f}(x_1, x_2) = \frac{1}{2}|x_1|^2 - \frac{1}{2}|x_2|^2 + f(z)$$

Let G be the expression of the riemannian metric g_M in the chart c; $G: V \to L(E,E)$ is given by

$$g_M(u,v) = (G(x)dc(u) \mid dc(v))$$

for any $u, v \in T_m M$ with $m = c^{-1}(x) \in U$, the scalar product in E being denoted by $(\mid)$.

Let $\tilde{\varphi}$ be the expression of the flow φ of $-\nabla f$ in the chart c; it is also the flow of the expression $\tilde{X}$ of $-\nabla f$ in the chart c. This vector field is given by

$$\tilde{X}(x) = -G(x)^{-1}\nabla\tilde{f}(x)$$

as for any $x = c(m) \in V$, $y \in E$ we have

$$(G(x)\tilde{X}(x) \mid y) = g_M(-\nabla f(m),\ Tc^{-1}(x,y))$$
$$= -df(Tc^{-1}(x,y))$$
$$= -d(f o c^{-1})(x,y) = -(\nabla\tilde{f}(x) \mid y)$$

For $s \in [0,1]$ let $\tilde{X}_s$ be the vector field on V given by

$$\tilde{X}_s(x) = -((1-s)G^{-1}(0) + sG^{-1}(x))\nabla\tilde{f}(x)$$

and let $\psi(\cdot,\cdot,s): (x,t) \mapsto \psi(x,t,s)$ be its flow. As $\nabla\tilde{f}$ is the constant mapping $(x_1,x_2) \mapsto (x_1,-x_2)$ we have

$$\psi(x,t,0) = (\exp -tG(0)^{-1}x_1,\ \exp tG(0)^{-1}x_2)$$

We intend to apply the following well known result to the linear isomorphism $T = \psi(\cdot,1,0)$, observing that there exists $\lambda \in (0,1)$ such that $\| T \mid E_1 \| \leq \lambda$, $\|T^{-1} \mid E_2\| \leq \lambda$, as G(0) is a positive definite isomorphism.

<u>Lemma (Stable manifold theorem [2,4,8])</u>. Given T and $V = V_1 \times V_2$ as above there exists $\varepsilon > 0$ such that for any $r \in (0,1)$ and any C^k-mapping $h: rV \to E$ verifying $h(0) = 0$, $\sup_{v \in rV} \|h'(v) - T\| \leq \varepsilon$ there exists a C^k-mapping $g: rV_1 \to rV_2$ satisfying the following property:

(S) $x \in rV$ belongs to the graph of g iff for each $n \geq 1$ $h^n(x)$ is defined and $\lim_n h^n(x) = 0$.
Moreover, if $h'(0) = T$ then the graph of g is tangent to E_1 at 0.

Let us show how to choose h and r. We observe that for each $s \in [0,1]$, the derivative $D_1\psi(0,t,s)$ of $\psi(\cdot,t,s)$ is the solution of the differential equation

$$\frac{d}{dt}B_s(t) = D\tilde{X}_s(\psi(0,t,s))B_s(t), \quad B_s(0) = I$$

Now $\psi(0,t,s) = 0$ for each $(t,s) \in \mathbb{R} \times [0,1]$ and

$$D\tilde{X}_s(0)v = -sDG^{-1}(0)\cdot v\,\nabla\tilde{f}(0) - G^{-1}(0)D\nabla\tilde{f}(0)\cdot v$$
$$= -G^{-1}(0)D\nabla\tilde{f}(0)\cdot v$$

as $\nabla\tilde{f}(0) = 0$. Hence $D_1\psi(0,t,s) = D_1\psi(0,t,0)$ for each $t \in \mathbb{R}$. In particular, if h is the restriction to rV of $\psi(\cdot,1,1) = \tilde{\varphi}(\cdot,1)$, we have $h'(0) = D_1\psi(0,1,0) = T$. As $\tilde{\varphi}$ is of class C^k, and $k \geq 1$, we can find $r > 0$ small enough so that $\sup(\|h'(x) - T\| : x \in rV) \leq \varepsilon$.

Let $U_r = c^{-1}(rV)$. For any $x \in U_r$ we have $\lim_n h^n(c(x)) = 0$ iff $\lim_n \varphi(x,n) = z$, and, using [6] th.4.1 again, this in turn is equivalent to $\lim_{t\to+\infty} \varphi(x,t) = z$ or $x \in A(z)$. The preceding lemma can be translated as follows, yielding a local version of the announced result.

<u>Lemma</u>. $A(z) \cap U_r$ is a C^k-submanifold of M.

Let us globalize this result. Let x_0 be any point of A(z). Let $t > 0$ be large enough so that $\varphi_t(x_0) \in U_r$. As φ_t is a C^k-diffeomorphism and as A(z) is invariant under φ_t, we get that $A(z) \cap \varphi_t^{-1}(U_r) = \varphi_t^{-1}(A(z) \cap U_r)$ is a C^k-submanifold of M. The fact that A(z) is of codimension i if z is of index i follows from our local analysis at z: the index of z is the dimension of E_2, and this is the codimension of A(z).

Let us show now that A(z) is contractible. Let $\bar{\varphi}: M \times [0,+\infty] \to M$ be the extension of φ given by $\bar{\varphi}(x,\infty) = \lim_{t\to\infty} \varphi(x,t)$ for $x \in M$. Let us show that $\bar{\varphi}$ defines a continuous deformation of A(z) into itself, hence an homotopy between the identity and the constant map with value z. We only have to prove that $\bar{\varphi}$ is continuous at $(x_0, +\infty)$ for any $x_0 \in A(z)$. Let N be a neighborhood of z in M. As $\varphi(z,t) = z$ for every $t \in \mathbb{R}_+$, we can find an open neighborhood P of z such that $\varphi(p,t) \in N$ for each $(p,t) \in P \times [0,1]$ and such that for each $x \in A(z) \cap P$ and each $n \in \mathbb{N}$, $\varphi(x,n) \in P$ (we can take $P = U_r$, with $r > 0$ small enough, U_r as above). As $z = \lim_{t \to \infty} \varphi(x_0, t)$ we can find $t_0 > 0$ such that $\varphi(x_0, t_0) \in P$. As φ is continuous at (x_0, t_0) and P is open, we can find a neighborhood Q of x_0 such that $\varphi(q, t_0) \in P$ for each $q \in Q$. Then for any $q \in Q \cap A(z)$ and any $t \geq t_0$ we can write $t = t_0 + n + h$ with $n \in \mathbb{N}$, $h \in [0,1)$ and

$$\varphi(q,t) = \varphi_h(\varphi_n(\varphi_{t_0}(q))) \in N$$

as $p := \varphi_{t_0}(q) \in A(z) \cap P$, $\varphi_n(p) = \varphi_1^n(p) \in A(z) \cap P$, $\varphi_h(P) \subset N$. The result is proved.

The following example shows that the assumption that f is a Morse function is crucial.

Example. Let M be $\mathbb{R}^2$ with its canonical metric and let $f : M \to \mathbb{R}$ be given by $f(0,0) = 0$ and

$$f(x,y) = (x^2 + y^2)^3 - y^2(x^2 + y^2)^{-1}(3x^2 - y^2)^2 \qquad \text{for } (x,y) \neq (0,0)$$

Then f is of class C^3 and can be written in polar coordinates

$$f(r \cos \theta, r \sin \theta) = r^6 - r^4(\sin 3\theta)^2$$

hence is bounded below and coercive. It follows that f satisfies condition (C). However, f is not a Morse function since the origin is a degenerate critical point. It is easy to see that A(0) is the union of the three half lines emanating from 0 corresponding to $\theta = 0$, $2\pi/3$, $4\pi/3$ (for $\theta \notin 2\pi\mathbb{Z}/3$ we have $f(r \cos \theta, r \sin \theta) < 0 = f(0)$ for r small enough), hence A(0) is not a C^2-submanifold of M.

The preceding example is easily encountered: whenever three ridges meet at a peak, one gets a similar situation.

REFERENCES

1. Hirsch, M. W.: Differential Topology. Springer (1976).
2. Iooss, G.: Bifurcation of Maps and Applications. North-Holland (1979).
3. Liotard, D.; Penot, J. P.: Critical paths and passes: application to quantum chemistry, in Numerical Methods in the Study of Critical Phenomena (J. Della-Dora, J. Demongeot, B. Lacolle, eds.). Springer (1981), pp. 213-222.
4. Marsden, J. E.; McCraken, M.: The Hopf Bifurcation and Its Applications. Springer (1976).
5. Palais, R. S.: Topology 2, 299-340 (1963).
6. Palais, R. S.: Critical point theory and the minimax principle, in Global Analysis, Proc. Symp. Pure Math. vol. 15, Amer. Math. Soc. (1970), pp. 185-212.
7. Rassias, G. M.: Bull. Acad. Pol. Sci. 29, No. 5-6, 311-316 (1981).
8. Shub, M.: Stabilité globale des systèmes dynamiques. Astérisque 56 Société Mathématique de France (1978).
9. Smale, S.: Ann. Math. 74, 199-206 (1961).

26 GLOBAL ASPECTS OF THE CONTINUATION METHOD

Roy Plastock / Department of Mathematics, New Jersey Institute of Technology, Newark, New Jersey

1. INTRODUCTION

Of all mathematical problems, the most basic is that of solving a given system of equations and, of all methods for doing so, one of the most basic is the Newton or Newton-Raphson method. Being easy to learn and easy to implement on a computer (not to mention possessing a quadratic rate of convergence) the Newton scheme is almost always the method of choice. However, the widespread availability of ever more powerful computers has resulted in the increased use and development of more sophisticated and complex mathematical models, often requiring the analysis of huge systems of equations. It is at this stage where the limitations of the Newton method become pronounced. Among these are (i) the dependence for success (usually meaning convergence) on the choice of starting values, (ii) the breakdown of Newton's method near singular points (those points where the Jacobian determinant is zero), (iii) the inability to find multiple solutions in a systematic way, and (iv) the inability of the method to be adapted to problems where changing values of parameters play a role.

In short, we may say that the limitations of Newton's method are due to the fact that it is "local" in nature. The requirements of modern computing require strategies that are global in nature, methods which converge independently of the choice of starting values. In addition, these methods should enable to user to find all solutions of a given problem and be applicable to problems depending on parameters.

Recent efforts in developing global methods (in the above sense) have centered on what are called continuation or imbedding methods. (The book Continuation Methods [14] includes several useful surveys.) The reader will become aware that there are many different strategies included in the categories of continuation and imbedding methods, each with numerous implementations as numerical schemes [1,7]. It is the purpose of this paper to take a broad look at this topic, avoiding discussion of the analysis or merits of specific numerical schemes, but considering the wider aspects of the application of the continuation method. In particular, we provide a solid theoretical basis by investigating that class of mappings which allow the use of continuation methods. Interestingly, the ability to implement continuation methods has far reaching topological consequences. This allows for a deeper investigation into the nature of singular points, multiple solutions, and problems depending on parameters. Much of the work described here appears in the papers [2,8,9,10]. Additional theoretical investigations, especially into parameter dependent problems, and further references are to be found in [4,5,12].

2. THE CONTINUATION METHOD

Consider a given system of n-nonlinear equations in n unknowns. For our study we prefer to consider this system as a mapping $f: R^n \to R^n$. We shall be interested in solving the equation

$$f(x) = y_1 \tag{1}$$

Here y_1 is a given vector in R^n.

The continuation method for solving (1) consists of the following: we pick an arbitrary starting point x_0 and, with $y_0 = f(x_0)$ we joint y_0 to y_1 by the line $\ell(t) = (1-t)y_0 + ty$, $(0 \leq t \leq 1)$. The continuation method consists in finding a path p(t), with $p(0) = x_0$, so that

$$f(p(t)) = \ell(t) \tag{2}$$

Of course $x_1 = p(1)$ is then a solution of our system (1).

At first glance it appears that this formulation of the problem is considerably more difficult to solve than the original system, requiring an infinite number of solutions, p(t), to be generated. It is, however, the implementation of the continuation method which allows for a great degree of flexibility, marked by various numerical schemes. The implementation we chose to discuss (often called the Davidenko equation in [3]), is the following—we attempt to find p(t) by solving the differential equation (obtained by differentiating (2) with respect to t),

$$\begin{aligned} p'(t) &= [f'(p(t))]^{-1}(y_0 - y_1), \quad 0 \leq t \leq 1 \\ p(0) &= x_0 \end{aligned} \tag{3}$$

The tacit assumption made is the invertibility of the derivative map f'(x) at every point x in R^n. One can now solve (3), and thus (2), by applying any number of numerical schemes used in solving initial value differential equation problems. This does not resolve the question of existence of the solution of Eq. (3). Theorems for initial value problems only insure the local existence of a solution. Equation (3) is a global problem, requiring the existence of a solution over the complete interval [0, 1]. Thus the first order of business is to find criteria for the mapping f which guarantee the solution of the global problem. We prove the following:

Theorem 2.1. Suppose $f: R^n \to R^n$ is a C' map whose derivative f'(x) is an invertible linear map. If either of the conditions (i) or (ii) stated below are satisfied, then differential equation (3) possesses a solution p(t) over the entire interval $0 \leq t \leq 1$.

(i) f is a proper map (i.e., the pre-image of a compact set is compact).

(ii) $$\int_0^\infty \inf_{|x|<s} \left(\frac{1}{|[f'(x)]^{-1}|}\right) ds = \infty$$

Proof. Let ε be the largest number for which p(t) exists on the interval $0 \leq t \leq \varepsilon$ (since f'(x) is invertible, the inverse function theorem assures that ε is nonzero). We need only show that $\lim_{t \to \varepsilon} p(t)$ exists, for then we can use the inverse function theorem to extend p(t) to some interval $\varepsilon \leq t < \varepsilon + \delta$ and so conclude that $\varepsilon = 1$.

(i) If f is proper, the line $\ell(t)$, $0 \leq t \leq 1$ is a compact set and thus its pre-image under f is a compact set. Since p(t), $0 \leq t < \varepsilon$ is contained in this compact set, $\lim_{t \to \varepsilon} p(t)$ exists.

(ii) Suppose $$\int_0^\infty \inf_{|x|<s} \frac{1}{|[f'(x)]^{-1}|} ds = \infty.$$

For ease of writing let $B(x) = \frac{1}{|[f'(x)]^{-1}|}$ and $h(s) = \inf_{|x|<s} B(x)$. Choose $\gamma < \varepsilon$, and for each n, choose a partition, $0 \leq t_1 \leq \cdots \leq t_{n+1} = \gamma$, which refines the partition previously chosen at step (n - 1). Let $\bar{t}_i$ denote a number, lying in the interval $[t_i, t_{i+1}]$, that satisfies $\sup_{[t_i, t_{i+1}]} |p'(t)| = |p'(\bar{t}_i)|$.

Let $\ell(t) = (1 - t)y_0 + ty$. Then using the fact that $|p(t)|$ has bounded variation on $[0,\gamma]$, and, recalling that p(t) satisfies Eq. (3), we have:

$$\varepsilon|y_1 - y_0| \geq \int_0^\gamma \frac{1}{|[f'(p(t))]^{-1}|} |f'(p(t))(y_1 - y_0)| \, dt$$

$$= \int_0^\gamma B(p(t)) \, |p'(t)| \, dt$$

$$= \lim_{n\to\infty} \sum_0^n B(p(\bar{t}_i)) \, |p'(\bar{t}_i)| \, (t_{i+1} - t_i)$$

$$\geq \lim_{n\to\infty} \sum_0^n B(p(\bar{t}_i))(|p(t_{i+1})| - |p(t_i)|)$$

$$= \int_0^\gamma B(p(t)) d|p(t)|$$

$$\geq \int_0^\gamma \inf_{|x|<|p(t)|} B(x) d\,|p(t)|$$

$$= \int_0^\gamma h(|p(t)|) \, d|p(t)|$$

$$= \int_{|p(0)|}^{|p(\gamma)|} h(s) \, ds$$

From assumption (ii), this last inequality implies that $\{p(t)\}$ is bounded on $[0, \varepsilon]$. Also, from (ii), we have that $\sup\{s| h(s) > 0\} = \infty$. Thus B(x) is bounded from below on any bounded set (since h(s) is nonincreasing). In particular, there is a $\lambda > 0$ so that $B(p(t)) > \lambda$ for t in the interval $[0, \varepsilon]$.

We now show that $\lim_{t\to\varepsilon} p(t)$ exists. In fact, we show that for every α, there is a β so that any numbers q and r satisfying $|q - \varepsilon| < \beta$ and $|r - \varepsilon| < \beta$ also satisfy $|p(q) - p(r)| < \alpha$. Firstly, the existence of $\int_0^\varepsilon |p'(t)| \, dt$ follows from the following series of inequalities; for any $\gamma < \varepsilon$:

$$\int_0^\gamma |p'(t)| \, dt \leq \frac{1}{\lambda} \int_0^\gamma B(p(t)) \, |p'(t)| \, dt$$

$$\leq \frac{\varepsilon}{\lambda} |y_1 - y_0|$$

Finally $|p(q) - p(r)| = |\int_r^q p'(t)dt| \leq \int_r^q |p'(t)| \, dt$; that this last integral can be made less than β if q and r are sufficiently close to ε follows from the existence of the integral $\int_0^\varepsilon |p'(t)| \, dt$.

This completes the proof.

In [10, 11] other conditions are given which also guarantee the solvability of Eq. (3).

3. SINGULARITIES, MULTIPLE SOLUTIONS AND COVERING SPACES

Let us go back to the Eq. (2) of Sec. 2.

$$f(p(t)) = \ell(t)$$

From a geometrical point of view, we can think of the continuation method as the process of line lifting or line following.

Definition 3.1. f lifts lines if for any line $\ell(t) = (1 - t)y_0 + ty_1$, $(0 \leq t \leq 1)$, lying in the range of f and each point x in $f^{-1}(y_0)$ there is a path $p_x(t)$ satisfying $f(p_x(t)) = \ell(t)$ and $p\ (0) = x$.

From now on we shall look at the continuation method from the point of view of line lifting.

Definition 3.2. A map f of X onto Y is a covering space map if we can find a covering of Y by open sets $\{U_\alpha\}$ so that for each U_α, $f^{-1}(U_\alpha)$ is the disjoint union of open sets each of which is mapped homeomorphically onto U_α.

The connection between the continuation method (line lifting) and covering spaces was pointed out in ([10], Theorem 1.1):

Theorem 3.3. In Banach spaces X and Y, let $D \subseteq X$ be an open connected set, $f: D \to Y$. The following are necessary and sufficient for f to be a covering space map of D onto f(D):

(i) f is a local homeomorphism,
(ii) f lifts lines in f(D).

Thus the success of the continuation method guarantees that the system of equations f represents a covering map. The topological tools of covering space theory can now be brought to bear in developing strategies for solving systems of equations, via continuation method techniques. We shall discuss some of the implications of the coving space theorem.

Definition 3.4. The singular points B of a differentiable mapping f are the elements of the set $B = \{x \mid f'(x) \text{ is not a surjective (linear) map}\}$. The set $S = f(B)$ is the set of singular values.

Theorem 3.5. If the map $f: R^n \to R^n$ has no singular points, i.e., $f'(x)$ is an invertible linear map and either (i) f is a proper map, or (ii) $\int_0^\infty \inf_{|x|<s} \frac{1}{|[f'(x)]^{-1}|} ds = \infty$. Then, for every y in R^n, the system $f(x) = y$ has one and only one solution (which can be found by using the continuation method as implemented by differential equation (3)).

This result, which is a consequence of covering space theory and Theorem 2.1, is just a restatement of Theorems 2.2 and 3.2 of [10], where proofs and additional theorems can be found. (See also [11].)

In effect, the lack of singular points implies unique solvability. What happens if there are singular points? The next result, also a consequence of covering space theory, indicates that the structure of the singular set B plays a crucial role.

Theorem 3.6. Suppose the C' map $f: R^n \to R^n$ is a proper map whose singular set consists of isolated singularities. Then if $n \geq 3$, f is a homeomorphism of R^n onto R^n.

The proof can be found in [9].

Here we have a challenge. Theorem 3.6 tells us that (for isolated singularities) not only does the continuation path p(t) defined by Eq. (2) exist for any line $\ell(t)$ and any starting value x_0, it is also unique. The problem is to find it. We can use differential equation (3), but how should we modify it when $f'(x)$ is not invertible? This is the challenge and is an important open problem.

The following result, from [9], indicates how we can gain some insight into the structure of a mapping, and hopefully, provide some ideas how to develop strategies for finding multiple solutions.

Proposition 3.7. Let $f: R^n \to R^n$ be a C', proper mapping. If the singular values $S = f(B)$ (see Def. 3.4) satisfy (i) $f(R^n) \neq S$, and (ii) $R^n - S$ is connected, then f is a covering map of $R^n - f^{-1}(S)$ onto $R^n - S$ (and also maps R^n onto itself).

Other such structure theorems can be found in [2].

4. SYSTEMS DEPENDING UPON PARAMETERS

In this section we consider the application of the continuation method to systems of n-nonlinear equations in n variables and depending upon m parameters. We write the system as an equation $f(w, \alpha) = y$, where w and y are vectors in R^n and α lies in R^m. For a given vector y, the problem of "tuning" the system consists in finding a parameter setting α, and a solution vector w satisfying the system of equations. To facilitate our analysis, we find it convenient to view the system as a mapping $f: R^{n+m} \to R^n$.

The continuation method consists in (i) picking a starting value x_0 and letting $y_0 = f(x_0)$, and (ii) join y to y_0 by the line $\ell(t) = (1 - t)y_0 + ty$ in R^n. We now seek a path p(t) so that $f(p(t)) = \ell(t)$ and $p(0) = x_0$. If such a path can be found, then $p(1) = (w_1, \alpha_1)$ is the required solution. In order to find the path p(t), we can, as in Sec. 2, try to implement the continuation method by solving the differential equation $f'(p(t))p'(t) = y - y_0$ (with initial value $p(0) = x_0$) for $p'(t)$. We now encounter our first problem—in order to express the differential equation for $p'(t)$ in standard form, we need to invert $f'(p(t))$. However $f'(x)$ is a linear map from R^{n+m} to R^n and so cannot be one-one. In other words, there is no inverse. However if we can find a right inverse $f^{\downarrow}(x)$, then we can rephrase the differential equation as:

$$\begin{aligned} p'(t) &= f^{\downarrow}(p(t))(y - y_0), \quad 0 \le t \le 1 \\ p(0) &= x_0 \end{aligned} \tag{4}$$

The following lemma guarantees the existence of a right inverse, provided $f'(x)$ is a linear operator of maximal rank n. In other words $f'(x)$ is a surjective linear map of R^{n+m} onto R^n. (The proof can be found in [8]):

Proposition 4.1. If, for every x, the map $f'(x)$ is a surjective linear map, and is locally Lipschitz in x, then the map $f^{\downarrow}(x) = f'(x)^*[f'(x)f'(x)^*]^{-1}$ is a locally Lipschitz right inverse for $f'(x)$. (Here * denotes the adjoint map.)

With the assumptions of Prop. 4.1 we can now implement Eq. (4). The local Lipschitz assumption assures the existence of a unique solution p(t) over some interval $0 \le t < \varepsilon$. The problem now is to find criteria which guarantee the existence of p(t) over the whole interval $0 \le t \le 1$.

If we try to introduce the criterion of properness, as in Theorem 2.1, we encounter an unusual and quite unexpected situation—namely, that such a map must have singular points. In other words a proper, C' and locally Lipschitz map of R^{n+m} to R^n must have singular points (points x for which $f'(x)$ is not surjective) [2]. Thus the question of the existence of global solutions to Eq. (4) is more delicate than for Eq. (3). However we can modify condition (ii) of Theorem 2.1 to provide an appropriate criterion.

Theorem 4.2. Suppose $f: R^{n+m} \to R^n$ is a C', locally Lipschitz map whose derivative f'(x) is a surjective linear map. If furthermore $\int_0^\infty \inf_{|x|<s} \frac{1}{|f^\dagger(x)|} ds = \infty$ (where $f^\dagger(x)$ is as defined in Prop. 4.1), then differential equation (4) possesses a (unique) solution p(t) defined over the entire interval $0 \leq t \leq 1$.

The proof of this result is similar to that of Theorem 2.1.

If we take a geometric point of view and think of the continuation method from the point of line lifting (Def. 3.1), we find that there are wide-ranging implications.

Definition 4.3. A map $f: X \to Y$ is called a fiber bundle map if there is a fiber set M and a covering $\{U_\alpha\}$ of Y by open sets for that for each U_α in the covering there is a a homeomorphism, $\Phi_\alpha: f^{-1}(U_\alpha) \to U_\alpha \times M$, with the property that $f = P \cdot \Phi_\alpha$ (where P is the projection of $U_\alpha \times M$ onto U_α).

Fiber bundle maps have many important topological properties. In fact covering spaces are fiber bundles with discrete fiber. For our purposes, the connection between fiber bundle maps and the continuation method resides in the following theorem whose proof can be found in [8].

Theorem 4.4. Let $f: R^{n+m} \to R^n$ satisfy the assumptions of Theorem 4.2. Then f is a fiber bundle map of R^{n+m} onto R^n.

5. CONCLUSION

In conclusion, we have found that the numerical method called the continuation method has, in addition to computational features, far reaching topological implications. Furthermore, the global insight provided by these considerations should, in turn, help in developing strategies for dealing with the problems of calculating multiple solutions and handling singular points. Much remains to be done.

Also, many of the results in this paper extend to Hilbert and Banach space settings. This leads to the application of the continuation method to ordinary and partial differential equations.

REFERENCES

1. Abbott, J. P.; Brent, R. P.: Austral. Math. Soc. (Ser. B), 157-164 (1978).
2. Berger, M. S.; Plastock, R. A.: Proc. A.M.S. 79, No. 2, 217-221 (1980).
3. Davidenko, D. F.: Dokl. Akad. Nauk. SSSR 88, 601-602 (1953).
4. Hirsch, M. W.; Smale, S.: Comm. Pure Appl. Math. 32, 281-312 (1979).
5. Keller, H. B.: Global homotopies and Newton methods, in Recent Advances in Numerical Analysis (C. de Boor and G. H. Golub, eds.), Academic Press (1979), pp. 73-94.
6. Morse, M.: Calculus of Variations in the Large, Amer. Math. Soc. (1934).
7. Ortega, J.; Rheinboldt, W. C.: Iterative Solutions of Nonlinear Equations in Several Variables, Academic Press (1970).
8. Plastock, R.: Nonlinear mappings that are globally equivalent to a projection. Trans. A.M.S. (to appear).
9. Plastock, R.: Proc. A.M.S. 68, No. 3, 317-322 (1978).
10. Plastock, R.: Trans. A.M.S. 200, 169-183 (1974).
11. Radulescu, M.; Radulescu, S.: Nonlin. Anal., Theory, Methods and Applications 4, No. 4, 951-965 (1980).
12. Rheinboldt, W. C.: SIAM J. Numer. Anal. 17, No. 2, 221-237 (1980).
13. Spanier, E. H.: Algebraic Topology, McGraw-Hill (1966).
14. Wacker, H. (ed.): Continuation Methods. Academic Press (1978).

27 APPLICATIONS OF SMALE THEORY TO THE n-BODY PROBLEM OF MECHANICS–ASTRONOMY

George M. Rassias / International Scientific Center, Ltd., Athens, Greece

"It is a pleasant surprise to him [the pure mathematician] and an added problem if he finds that the arts can use his calculations, or that the senses can verify them, much as if a composer found that the sailors could heave better when singing his songs."

—LÉONHARD EULER

INTRODUCTION

In 1687, Isaac Newton [112] proposed a fundamental theory for studying the orbits of the planets (of our solar system) that today is expressed by the term Celestial Mechanics. He succeeded this by applying the concept of force of Galileo [58-60] to the motions of the Stars. Among Newton's predecessors are Archimedes, Copernicus, Galileo, Kepler, Descartes, etc.

In particular, Newton proposed the law of Universal Gravitation according to which the Heavenly Bodies act on one another by using the laws of the motion of the Planets. These laws were expressed by Kepler [82-83], known as Kepler's three laws of the planetary motions.

In 1543, Copernicus [42] enunciated the great idea that all the planets, including the Earth, rotated around the Sun, and not the other way around. Of course, the basic idea was known to the Ancient Greek astronomers, in particular to Aristarchus of Samos. Celestial mechanics, broadly interpreted, is involved in the entire field of astronomy of the present time. The meaning of the term usually adopted refers only to those problems in which Newton's universal gravitation law plays the chief or only part, and more particularly to those which deal with the motion of heavenly bodies about one another and with their rotations (Birkhoff [28]). Newton's law of universal gravitation underlies many natural phenomena, for example the orbits of the planets around the Sun, the motion of the Moon and artificial satellites about the Earth, the paths drawn by charged particles in atomic physics, etc. In fact Kepler's laws of the motion of the Planets are deduced from Newton's law.

The idealized model for the solar system has explained almost completely all known observational results of astronomy and enables one to make a correct prediction for the future motion of heavenly bodies.

Using Newton's theory one is able to calculate the motions of the planets for hundreds, thousands, or even millions of years ahead or back.

Newton's theory has been shown to be right in countless cases of observations. However, this theory does not hold for the case of the planet nearest the Sun, Mercury, for which the rotation of the perihelion shows a small deviation from that found by this theory, namely 43 seconds of arc every hundred years.

The calculation of this deviation was done by the astronomer Leverrier (1845), who predicted also the existence of the planet Neptune.

This result is of great importance because it shows that the Kepler's laws hold for the planets far removed from the Sun.

The general theory of relativity of Einstein [48-50] explained the motion of Mercury's perihelion, and in fact this constitutes the only confirmation of the general theory of relativity in the domain of mechanics. The motion of a planet about the Sun regarded as a geodesic line in four-dimensional space-time according to Einstein and Schwarzchild is in a very high degree approximation to that predicted by Newton.

In 1609, Kepler [82] published his first two laws of the motions of the planets in his book Astronomia Nova. These are the following.

(1) Each planet moves in an ellipse, with the Sun at one of its foci.
(2) The radius vector drawn from the Sun to a Planet covers equal areas in equal times.

In 1619, Kepler [83] published his third law in his book The Harmonies of the World. This is the following.

(3) The squares of the periods of the different Planets are proportional to the cubes of their respective major semiaxes.

In 1686, Newton [112] discovered that the gravity potential attracting a planet to the Sun is of the form $U = -k/r$, where r is the distance of a Planet to the Sun, and k is a constant for the planet.

The central potential satisfies all three Kepler's laws simultaneously [106].

The two-body problem of a planet and the Sun can be reduced to the one-body problem.

The famous n-body problem of celestial mechanics is an extension of the Kepler's problem.

It was Euler [51] who invented the conservation laws of angular momentum, linear momentum, and energy for the n-body problem.

The foundations of mechanics had been invented by Newton, but Euler was the principal architect. Euler has been characterized as "the Shakespeare of mathematics."

Many great mathematicians such as Isaac Newton [112], Léonhard Euler [51], Joseph-Louis Lagrange [87-89], Pierre-Simon de Laplace, Simeon-Denis Poisson, Dirichlet, Karl F. Gauss, William Rowan Hamilton, and of course Johannes Kepler and Galileo, have been involved in the development of mechanics and especially for solving the n-body problem. Many of them claimed to have proved that the solar system is stable using series expansions.

Henri Poincaré [116-118] showed that these series expansions diverged, and succeeded to prove that the solar system is stable through geometrical methods. Poincaré showed the great importance of topology for proving the existence of periodic solutions in the three-body problem of celestial mechanics that had kept many scientists working for its solution from the time of Newton without any significant success.

Also, his work on the n-body problem led him to invent the very important and interesting field of qualitative dynamics.

Important work has been done recently by Vladimir Arnold [18-21], G. D. Birkhoff [27-31], Andrei N. Kolmogorov [85], Jürgen Moser [109-110], and Stephen Smale [129-134].

An important example of the n-body problem is the solar system with the Sun and planets being the n-bodies. In the case that the system consists of the Sun, Earth, and the Moon, we have the 3-body problem, i.e., the problem of determining the motion of a small third body moving under the influence of the first two and lying in the same plane.

> The interest of the pure mathematician in the problem of three or more bodies has been stimulated by its importance for an understanding of the past and future of the stellar universe. The entrance upon the field of the theory of relativity of Einstein has altered this situation considerably. If the relativistic point of view prevails there can be little doubt that new factors of the utmost importance will be introduced in astronomical speculation concerning great lapses of time, although for limited intervals of time the problem of three or more bodies will maintain its importance. Only the very simplest features of the modifications required by the theory of relativity have as yet been determined, mainly those for a very small body in the presence of a central body.

(See [31].)

Acknowledgments. It is my pleasure to express my gratitude to Professor Stephen Smale for introducing me to the fields of topology and mechanics and for his constant encouragement and inspiration. I am also grateful to Professors R. Abraham and M. Mizushima for their kind permission to use certain of the contents of their books [106, 2]. Also, I would like to thank Professors S. S. Chern, J. Marsden, E. Spanier, and J. R. Stallings for helpful discussions while I was at the University of California at Berkeley.

1. THE LAGRANGIAN AND HAMILTONIAN THEORIES

Using the generalized coordinates (spherical coordinates r, θ, and ϕ or cylindrical coordinates ρ, z, and ϕ), given by

Spherical coordinates		Cylindrical coordinates
$x = r \cdot \sin\theta \cdot \cos\phi$		$x = \rho \cdot \cos\phi$
$y = r \cdot \sin\theta \cdot \sin\phi$	or	$y = \rho \cdot \sin\phi$
$z = r \cdot \cos\theta$		$z = z$

respectively, such that the motion of each generalized coordinate is independent from the other coordinates, one can use these coordinates toward the solution of a many-body problem.

Isaac Newton in his Principia [112] proposed the two of his (three) laws of motion. These two laws are expressed by Newton's equation of motion.

$$F = m \cdot a \equiv m \cdot \dot{V} = m \cdot \frac{dV}{dt} \tag{1}$$

where a is the acceleration, and the proportionality constant m is the mass of the particle.

Joseph Louis Lagrange [75-77] found the so called Lagrange equations, which are equivalent to the (above) Newton equations (1).

Lagrange's equations have the nice property that they do not change their form regardless of the coordinate system (Cartesian or generalized). That property, however, is not possessed by Newton's equations.

For an n-body system one can use a set of 3n generalized coordinates r_1, r_2, ..., r_{3n} for determining the position of all bodies.

Consider first a 1-body system. The kinetic energy is

$$T = \frac{1}{2}m(\dot{x}^2 + \dot{y}^2 + \dot{z}^2) \tag{2}$$

Thus,

$$\frac{\partial T}{\partial \dot{r}_i} = m\left(\dot{x}\frac{\partial \dot{x}}{\partial \dot{r}_i} + \dot{y}\frac{\partial \dot{y}}{\partial \dot{r}_i} + \dot{z}\frac{\partial \dot{z}}{\partial \dot{r}_i}\right) \tag{3}$$

However,

$$\dot{x} = \sum_i \frac{\partial x}{\partial r_i}\dot{r}_i \rightarrow \frac{\partial \dot{x}}{\partial \dot{r}_i} = \frac{\partial x}{\partial r_i} \tag{4}$$

From (4) and (3) one obtains

$$\begin{aligned}\frac{d}{dt}\cdot\frac{\partial T}{\partial \dot{r}_i} &= m\frac{d}{dt}\left(\dot{x}\frac{\partial x}{\partial r_i} + \dot{y}\frac{\partial y}{\partial r_i} + \dot{z}\frac{\partial z}{\partial r_i}\right) \\ &= m\left(\ddot{x}\frac{\partial x}{\partial r_i} + \ddot{y}\frac{\partial y}{\partial r_i} + \ddot{z}\frac{\partial z}{\partial r_i}\right) + m\left(\dot{x}\frac{\partial \dot{x}}{\partial r_i} + \dot{y}\frac{\partial \dot{y}}{\partial r_i} + \dot{z}\frac{\partial \dot{z}}{\partial r_i}\right)\end{aligned} \tag{5}$$

If the potential U exists, Newton's equation (1) can be expressed as

$$F = -\nabla U = m\dot{V} \tag{6}$$

Using the new form of Newton's equation (6), in the first terms of (5), one obtains

$$\frac{d}{dt}\cdot\frac{\partial T}{\partial \dot{r}} = -\frac{\partial U}{\partial r_i} + \frac{\partial T}{\partial r_i} \tag{7}$$

Assuming that U is independent of $\dot{r}_i$, (7) can be expressed as

$$\frac{d}{dt}\cdot\frac{\partial L}{\partial \dot{r}_i} - \frac{\partial L}{\partial r_i} = 0 \tag{8}$$

where $L = T - U$ is called the Lagrangian function, and Eq. (8) is called Lagrange's equation for a 1-body system.

Generalizing the above process for an n-body system, one finally obtains

$$\frac{d}{dt}\cdot\frac{\partial L}{\partial \dot{r}_i} - \frac{\partial L}{\partial r_i} = 0 \tag{10}$$

$$L = \sum_{j=1}^{n} T_j - U \tag{11}$$

where (11) is called the Lagrange function, and the equation (10) is called the Lagrange equation for an n-body system.

Define the canonical momentum, P_i, corresponding to a generalized coordinate r_i as

$$p_i \equiv \frac{\partial L}{\partial \dot{r}_i} \tag{12}$$

Note that if U is independent of $\dot{x}$, one obtains

$$P_x \equiv \frac{\partial L}{\partial \dot{x}} = \frac{\partial L_1}{\partial \dot{x}} = \frac{\partial T_1}{\partial \dot{x}} = m\dot{x} \tag{13}$$

that is the ordinary momentum in the x direction. Then (12) and (10) give

$$\dot{p}_i = \frac{\partial L}{\partial \dot{r}_i}$$

which shows that if r_i does not appear explicitly in L, then the corresponding p_i is a constant of motion.

The Hamiltonian function is defined as

$$H = \sum_{i}^{\ell} p_i \cdot \dot{r}_i - L(r_1, \ldots, r_\ell, \dot{r}_1, \ldots, \dot{r}_\ell, t) \tag{14}$$

From (14), one obtains

$$dH = \sum_i \left(p_i d\dot{r}_i + \dot{r}_i dp_i - \frac{\partial L}{\partial r_i} dr_i - \frac{\partial L}{\partial \dot{r}_i} d\dot{r}_i\right) - \frac{\partial L}{\partial t} dt$$

$$= \sum_i (\dot{r}_i dp_i - \dot{p}_i dr_i) - \frac{\partial L}{\partial t} dt \tag{15}$$

Now, if r_i are expressed with respect to p_i, the Hamiltonian is

$$H = H(r_1, \ldots, r_k, p_1, \ldots, p_k, t) \tag{16}$$

However,

$$dH = \sum_i \left(\frac{\partial H}{\partial r_i} dr_i + \frac{\partial H}{\partial p_i} dP_i \right) + \frac{\partial H}{\partial t} dt \tag{17}$$

From (15) and (17), one obtains

$$\frac{\partial H}{\partial r_i} = -\dot{p}_i \tag{18}$$

$$\frac{\partial H}{\partial p_i} = \dot{r}_i \tag{19}$$

$$\frac{\partial H}{\partial t} = -\frac{\partial L}{\partial t} \tag{20}$$

Equations (18), (19), (20) are called <u>Hamilton's equations</u>.

If T is of the form

$$T = \sum_{i,j} a_{ij} \dot{r}_i \dot{r}_j \tag{21}$$

then

$$p_i = \frac{\partial L}{\partial \dot{r}_i} = \frac{\partial T}{\partial \dot{r}_i} = 2 \sum_j a_{ij} \dot{r}_j \tag{22}$$

where $a_{ij} = a_{ji}$.

From (14) and (9) one obtains

$$H = 2 \cdot \sum_{i,j} a_{ij} \dot{r}_j \dot{r}_i - T + U = T + U \tag{23}$$

The Hamiltonian function written in cylindrical coordinates is

$$H = p_\rho \cdot \dot{\rho} + p_\phi \cdot \dot{\phi} - \frac{1}{2} m(\dot{\rho}^2 + \rho^2 \dot{\phi}^2) + U \tag{24}$$

however,

$$\dot{\rho} = \frac{P_\rho}{m}, \qquad \dot{\phi} = \frac{P_\phi}{m_\rho^2} \tag{25}$$

Therefore,

$$H = \frac{1}{2m} P_\rho^2 + \frac{1}{2m_\rho^2} P_\phi^2 + U \tag{26}$$

taking the form of (16). Now, since

$$T = \frac{1}{2} m(\dot{x}^2 + \dot{y}^2) = \frac{1}{2} m(\dot{\rho}^2 + \rho^2 \dot{\phi}^2) \tag{27}$$

is of the form of (21), one obtains

$$H = T + U \tag{28}$$

Hence, Hamilton's equations are the following:

$$\begin{gathered} -\frac{P_\phi^2}{m_\rho^3} + \frac{\partial U}{\partial \rho} = -\dot{p}_\rho , \qquad \frac{\partial U}{\partial \phi} = -\dot{p}_\phi \\ \frac{P_\rho}{m} = \dot{\rho} , \qquad \frac{P_\phi}{m\rho^2} = \dot{\phi} , \qquad \frac{\partial H}{\partial t} = 0 \end{gathered} \tag{29}$$

Given a function

$$f = f(r_1, \ldots, r_k, p_1, \ldots, p_k, t) \tag{30}$$

then

$$\frac{df}{dt} = \sum_i \left(\frac{\partial f}{\partial r_i} \dot{r}_i + \frac{\partial f}{\partial p_i} \dot{p}_i \right) + \frac{\partial f}{\partial t} \tag{31}$$

Hamilton's equations combined with (31) give

$$\frac{df}{dt} = \sum_i \left(\frac{\partial f}{\partial r_i} \cdot \frac{\partial H}{\partial p_i} - \frac{\partial f}{\partial p_i} \cdot \frac{\partial H}{\partial r_i} \right) + \frac{\partial f}{\partial t} \tag{32}$$

Poisson's bracket, introduced by Simeon Denis Poisson, is

$$|f, g|_p \equiv \sum_i \left(\frac{\partial f}{\partial r_i} \cdot \frac{\partial g}{\partial p_i} - \frac{\partial f}{\partial p_i} \cdot \frac{\partial g}{\partial r_i} \right) \tag{33}$$

Now, (32) can be written as

$$\frac{df}{dt} = |f, H|_p + \frac{\partial f}{\partial t} \tag{34}$$

which expresses that if f is independent of time explicitly and its Poisson bracket with H is zero, then f is a constant of motion. In this way, constants of motion can be calculated.

The Hamiltonian function

$$H = \frac{1}{2m} P_\rho^2 + \frac{1}{2m\rho^2} P_\phi^2 + U \tag{35}$$

So,

$$|P_\phi, H|_p = \frac{1}{2m} |P_\phi, P_\rho^2|_p + \frac{1}{2m} \left| P_\phi, \frac{P_\phi^2}{\rho^2} \right| + |P_\phi, U|_p = -\frac{\partial U}{\partial \phi} \tag{36}$$

Therefore, if U is independent of ϕ, then P_ϕ is a constant of motion. Similarly,

$$|P_\rho, H|_p = \frac{P_\phi^2}{2m} |P_\rho, \rho^{-2}|_p + |P_\rho, U|_p = \frac{P_\phi^2}{m\rho^3} - \frac{\partial U}{\partial \rho} \tag{37}$$

Consider an n-body system interacting one another.

Then there exists a Hamiltonian function describing the n-body system even if their mass and way of interaction is arbitrary.

By displacement Δx_i of the ith body along the x axis, the change of the Hamiltonian function H is

$$\Delta_i H = \frac{\partial H}{\partial x_i} \Delta x_i \equiv -\dot{p}_{ix} \cdot \Delta x_i \tag{38}$$

So, by the same displacement Δx of all bodies along the x axis, the total change of the Hamiltonian function H is

$$\begin{aligned} \Delta H &= \sum_{i=1}^{n} \frac{\partial H}{\partial x_i} \Delta x = \Delta x \sum_{i=1}^{n} \frac{\partial H}{\partial x_i} \\ &= -\Delta x \sum_{i=1}^{n} \dot{p}_{ix} = -\Delta x \cdot \frac{d}{dt} \left(\sum_{i=1}^{n} p_{ix} \right) \end{aligned} \tag{39}$$

If there is no external field, one should expect $\Delta H = 0$. But the displacement Δx is arbitrary, so

$$\frac{d}{dt}\Big(\sum_{i=1}^{n} p_{ix}\Big) = 0 \tag{40}$$

Therefore, one obtains the conservation of total linear momentum $\Sigma_{i=1}^{n} p_{ix}$.

Applying the same process for each other component, one finally obtains the conservation of the total linear momentum of the n-body system, assuming that the space is homogeneous.

If the space is homogeneous with respect to a rotation around some axis, then the corresponding total angular momentum $\Sigma_{i=1}^{n} p_{i\theta}$ is conserved.

Suppose Lagrange's function L is a function of $r_1, \ldots, r_i, \ldots, r_k, \dot{r}_1, \ldots, \dot{r}_k$, and t.

By considering the integral $\int_{t_1}^{t_2} L dt$, where t_1, t_2 are the initial and final times, Lagrange's equation is obtained from

$$\Delta \int_{t_1}^{t_2} L dt = 0 \tag{41}$$

under the following conditions

$$\Delta r_i = 0 \quad \text{for all i at } t_1 \text{ and } t_2 \tag{42}$$

and

$$\Delta t = 0 \tag{43}$$

where Δ denotes variation.

Then, (41) becomes after consideration of (42), (43),

$$\Delta \int_{t_1}^{t_2} L dt = \int_{t_1}^{t_2} (\Delta L) dt = \int_{t_1}^{t_2} \sum_i \Big(\frac{\partial L}{\partial r_i} \Delta r_i + \frac{\partial L}{\partial \dot{r}_i} \Delta \dot{r}_i\Big) dt \tag{44}$$

From (43) and (42) one obtains

$$\int_{t_1}^{t_2} \frac{\partial L}{\partial \dot{r}_i} \Delta \dot{r}_i dt = - \int_{t_1}^{t_2} \Big(\frac{d}{dt} \cdot \frac{\partial L}{\partial \dot{r}_i}\Big) \Delta r_i dt \tag{45}$$

Combining (45) and (44), one can see that the variational equation (41) is expressed as

$$\int_{t_1}^{t_2} \sum_i \Big(\frac{\partial L}{\partial r_i} - \frac{d}{dt} \frac{\partial L}{\partial \dot{r}_i}\Big) \Delta r_i dt = 0 \tag{46}$$

For the above variational equation (46) to be true for an arbitrary variation Δr_i, the coefficient of Δr_i must be equal to zero for each (time) t.

Therefore, one can see that the variational equation (41) together with (42) and (43) is equivalent to Lagrange's equations (10) and (11).

With auxiliary conditions (42) and (43) one obtains

$$\Delta \int_{t_1}^{t_2} \Big(\sum_i p_i \dot{r}_i - H\Big) dt = \int_{t_1}^{t_2} \sum_i \Big\{p_i \Delta \dot{r}_i + \dot{r}_i \Delta p_i - \Big(\frac{\partial H}{\partial p_i} \Delta p_i + \frac{\partial H}{\partial r_i} \Delta r_i\Big)\Big\} dt \tag{47}$$

taking instead of (41), its equivalent form

$$\Delta \int_{t_1}^{t_2} \Big(\sum_i p_i \dot{r}_i - H\Big)\, dt = 0 \tag{48}$$

and finding the variations of Δp_i and of Δr_i.

From (43), one can see that

$$\Delta \dot{r}_i = \Delta \cdot \frac{dr_i}{dt} = \frac{d}{dt}(\Delta r_i) \tag{49}$$

Thus, because of (42) and (49) it is true that

$$\int_{t_1}^{t_2} p_i \Delta \dot{r}_i\, dt = -\int_{t_1}^{t_2} \dot{p}_i \Delta r_i\, dt \tag{50}$$

and because of (47) and (48),

$$\int_{t_1}^{t_2} \sum_i \left\{ \Big(\dot{r}_i - \frac{\partial H}{\partial p_i}\Big) \Delta p_i - \Big(\dot{p}_i + \frac{\partial H}{\partial r_i}\Big) \Delta r_i \right\} dt = 0 \tag{51}$$

So, the variational equation (48) is equivalent to Hamilton's equations (18) and (19).

Hence, nature chooses among all possible paths between t_2 and t_1, that path for which the integral

$$\int_{t_1}^{t_2} L dt = \int_{t_1}^{t_2} \Big(\sum_i p_i \dot{r}_i - H\Big)\, dt \tag{52}$$

assumes its extremum value. This is the so called Hamilton's principle. In this way, William Rowan Hamilton gave the above interpretation of Lagrange's equation, or Newton's equation.

It can be shown that the variational equation (41) under the conditions (42), (43), and

$$\Delta \dot{r}_i = 0 \quad \text{at } t_1 \text{ and } t_2$$

implies

$$\frac{d^2}{dt^2} \cdot \frac{\partial L}{\partial \ddot{r}_i} - \frac{d}{dt} \frac{\partial L}{\partial \dot{r}_i} + \frac{\partial L}{\partial r_i} = 0$$

if the Lagrangian L depends on the acceleration $\ddot{r}_i$.

The integral $\int_{t_1}^{t_2} L dt$ is sometimes called action. However, the integral

$$\int_{t_1}^{t_2} \sum_i p_i \dot{r}_i\, dt \tag{53}$$

is more often called action.

In 1744, Maupertuis claimed that the above integral (53), i.e., the action, takes an extremum value when Newton's equation holds.

It remained, however, for Euler and Lagrange to give a proof of the above Maupertuis variational principle, that sometimes is called the Maupertuis principle or least action principle.

The least action principle is the variational principle

$$\Delta \int_{t_1}^{t_2} \sum_i p_i \dot{r}_i\, dt = 0 \tag{54}$$

under the following two conditions

$$\Delta r_i = 0 \quad \text{for all i at } t_1 \text{ and } t_2 \tag{55}$$

$$\Delta H = 0 \quad \text{or} \quad H = \text{constant } (= C) \tag{56}$$

Note that the first condition is exactly the same as that of Hamilton's principle, but the second condition (56), which expresses that the total energy remains constant, replaces (43). Thus,

$$\Delta \int_{t_1}^{t_2} \sum_i p_i \dot{r}_i \, dt = \Delta \int_{t_1}^{t_2} (L + H)\, dt$$

$$= \int_{t_1}^{t_2} (\Delta L)\, dt + \int_{t_1}^{t_2} (L + H)\, d(\Delta t) \tag{57}$$

under the new conditions,

$$\Delta \dot{r} = \frac{d(r + \Delta r)}{d(t + \Delta t)} - \frac{dr}{dt} = \frac{d}{dt}(\Delta r) - \dot{r}\frac{d}{dt}(\Delta t) \tag{58}$$

So

$$\int_{t_1}^{t_2} (\Delta L)\, dt = \int_{t_1}^{t_2} \sum_i \left(\frac{\partial L}{\partial r_i} \Delta r_i + \frac{\partial L}{\partial \dot{r}_i} \Delta \dot{r}_i \right) dt$$

$$= \int_{t_1}^{t_2} \sum_i \left(\frac{\partial L}{\partial r_i} \Delta r_i + \frac{\partial L}{\partial \dot{r}_i} \frac{d}{dt} (\Delta r_i) \right) dt - \int_{t_1}^{t_2} \sum \frac{\partial L}{\partial \dot{r}_i} \dot{r}_i \frac{d(\Delta t)}{dt}\, dt$$

$$= \int_{t_1}^{t_2} \sum_i \left(\frac{\partial L}{\partial r_i} - \frac{d}{dt} \cdot \frac{\partial L}{\partial \dot{r}_i} \right) \Delta r_i \, dt - \int_{t_1}^{t_2} \sum_i p_i \dot{r}_i \, d(\Delta t)$$

$$= \int_{t_1}^{t_2} \sum_i \left(\frac{\partial L}{\partial r_i} - \frac{d}{dt} \cdot \frac{\partial L}{\partial \dot{r}_i} \right) \Delta r_i \, dt - \int_{t_1}^{t_2} (L + H)\, d(\Delta t) \tag{59}$$

Combining (59) and (57) one can see that (54) implies Lagrange's equation. When (21) and (22) hold then (54) can be expressed as

$$\Delta \int_{t_1}^{t_2} T\, dt = 0 \tag{60}$$

In case, there is no external force, then (56) reduces to

$$\Delta T = 0 \tag{61}$$

So, (60) means

$$\Delta(t_2 - t_1) = 0 \tag{62}$$

expressing that a system chooses among all possible paths between given end points compatible with a fixed energy, that path of least transit time. This is the so-called Maupertuis principle.

It can be shown that if 1 is the length of a path that a body may follow, then the least action principle can also be expressed as

$$\Delta \int_a^b \sqrt{C - U}\, dl = 0 \tag{63}$$

where C is the total energy mentioned in (56). This is the so-called Jacobi's form of the least action principle.

Max Planck, the founder of quantum theory, expressed that the highest and most coveted aim of physical science is to condense all natural phenomena which have been observed and are still to be observed into one simple principle. Amid the more or less general laws which mark the achievements of physical science during the course of the last centuries, the principle of least action is perhaps that which, as regards form and content, may claim to come nearest to this ideal final claim of theoretical research.

Lagrange's equations have the advantage of being invariant under transformations among all possible generalized coordinates. Hamilton's equations are also invariant under the same transformations.

Hamilton's equations, however, are invariant not only under transformations among generalized coordinates, but also between generalized coordinates and generalized momenta.

Let $r_1, \ldots, r_i, \ldots, p_1, \ldots, p_i, \ldots$, be the original set of generalized coordinates and their canonically conjugate momenta, for which Hamilton's equations hold. If under a transformation

$$R_i = R_i(r,p,t), \qquad P_i = P_i(r,p,t) \tag{64}$$

where r and p stand for all r_i's and p_i's, respectively, there exists a function K(R, P, t) such that one still has Hamilton's equations

$$\dot{R}_i = \frac{\partial K}{\partial P_i}, \qquad \dot{P}_i = -\frac{\partial K}{\partial R_i} \tag{65}$$

then the transformation is said to be canonical.

Note that K is a new Hamiltonian function, which may not be equal to the original Hamiltonian function H.

Now, since Hamilton's equations are obtained from Hamilton's principle or equation (48),

$$\Delta \int \left(\sum_i P_i \mathring{R}_i - K \right) dt = 0 \tag{66}$$

holds for the new set of canonical variables.

From (48) and (66) one obtains

$$\sum_i p_i \dot{r}_i - H = \lambda \left(\sum_i P_i \dot{R}_i - K \right) + \dot{G} \tag{67}$$

where λ is a constant and G is a function called the generating function of the canonical transformation, defined by $\Delta G = 0$ at t_1 and t_2. The constant λ is mostly taken to be equal to 1.

The generating function G may be a function of r, R, and t:

$$G = G(r,R,t) \tag{68}$$

So,

$$\dot{G} = \sum_i \frac{\partial G}{\partial r_i} \dot{r}_i + \sum_i \frac{\partial G}{\partial R_i} \dot{R}_i + \frac{\partial G}{\partial t} \tag{69}$$

when

$$P_i = \frac{\partial G}{\partial r_i} \tag{70}$$

$$\lambda P_i = -\frac{\partial G}{\partial R_i} \tag{71}$$

$$\lambda K = H + \frac{\partial G}{\partial t} \tag{72}$$

then (67) holds. For example

$$G = \sum_i r_i R_i \tag{73}$$

can be taken as such a generating function. Formulas (70) through (72) for this case reduce to

$$R_i = P_i, \quad \lambda P_i = -r_i \quad \text{and} \quad \lambda K = H \tag{74}$$

These relations combined with (65) give the original Hamilton equations (18) and (19).

The above example shows the nondistinguishability of coordinates and conjugate momenta. So p and r will be simply called canonically conjugate quantities.

Note that the Poisson bracket defined by (64) is invariant under a canonical transformation.

In other words, if r, p and R, P are two canonically conjugate sets, then

$$\sum_i \left(\frac{\partial f}{\partial r_i}\frac{\partial g}{\partial p_i} - \frac{\partial f}{\partial p_i}\frac{\partial g}{\partial r_i}\right) = \sum_i \left(\frac{\partial f}{\partial R_i}\frac{\partial g}{\partial P_i} - \frac{\partial f}{\partial P_i}\frac{\partial g}{\partial R_i}\right) \tag{75}$$

for any pair of functions f and g, provided that their variables are transformed from r, p to R, P, as one is going from the left- to the right-hand side of this equation.

Using the notation $| \;\; |$ for the Poisson bracket by r, p on the left-hand side of relation (75), one obtains

$$|f,g|_p = \sum_{i,j} \left\{\frac{\partial f}{\partial r_i}\left(\frac{\partial g}{\partial R_j}\cdot\frac{\partial R_j}{\partial P_i} + \frac{\partial g}{\partial P_j}\cdot\frac{\partial P_j}{\partial p_i}\right) - \frac{\partial f}{\partial p_i}\left(\frac{\partial g}{\partial R_j}\cdot\frac{\partial R_j}{\partial P_i} + \frac{\partial g}{\partial P_j}\cdot\frac{\partial P_j}{\partial r_i}\right)\right\}$$

$$= \sum_j \left(\frac{\partial g}{\partial R_j}|f,R_j|_P + \frac{\partial g}{\partial R_j}|f,P_j|_P\right) \tag{76}$$

For g = H, this relation (76) becomes

$$|f,H|_P = \sum_j \left(\frac{\partial H}{\partial R_j}|f,R_j|_P + \frac{\partial H}{\partial P_j}|f,P_j|_P\right)$$

$$= \sum_j \left(\frac{\partial K}{\partial R_j}|f,R_j|_P + \frac{\partial K}{\partial P_j}|f,P_j|_P\right)$$

$$= \sum_j (-\dot{P}_j|f,R_j|_P + \dot{R}_j|f,P_j|_P) \tag{77}$$

from Hamilton's equations (65) and the relations

$$\frac{\partial H}{\partial R_j} = \frac{\partial K}{\partial R_j} - \frac{\partial^2 G}{\partial R_j \partial t} = \frac{\partial K}{\partial R_j} + \frac{\partial P_j}{\partial t} = \frac{\partial K}{\partial R_j} \tag{78}$$

$$\frac{\partial H}{\partial P_j} = \frac{\partial K}{\partial P_j} - \frac{\partial}{\partial t}\left(\frac{\partial G}{\partial P_j}\right) = \frac{\partial K}{\partial P_j} \tag{79}$$

where G is the generating function defined in (68), (71), and (72).

The left-hand side of (77) is

$$|f,H|_P = \dot{f} - \frac{\partial f}{\partial t}$$

because of relation (34). So (77) requires that

$$|f, R_j|_P = -\frac{\partial f}{\partial P_j} \quad \text{and} \quad |f, P_j|_P = \frac{\partial f}{\partial R_j} \tag{80}$$

Now relation (80) and (76) give the proof of (75). It can be shown that under the canonical transformation generated by the generating function defined in (68), the equation of motion (34) is transformed into

$$\frac{df}{dt} = |f, K|_P - \left| f, \frac{\partial G}{\partial t} \right|_P + \frac{\partial f}{\partial t}$$

If one takes the generating function so that

$$\frac{\partial G}{\partial t} = -H\left(r_i, \frac{\partial G}{\partial r_i}, t\right) \tag{81}$$

then, from (72), the new Hamiltonian is identically zero

$$K = 0 \tag{82}$$

and so

$$\dot{R}_i = \dot{P}_i = 0 \tag{83}$$

because of (64), i.e., the new canonical variables are all constants of motion. Therefore, this canonical transformation gives us the complete set of constants of motion for a given system.

2. THE TWO-BODY PROBLEM

We work with the following model for the planar two-body problem. This is a system (M, H, m) where

(i) $M = T^*(R^2 \setminus \{0\})$ with canonical symplectic structure;

(ii) $m \in M$ (initial conditions);

(iii) $H: M \to R$ is C^∞ defined by

$$H(r, \theta . p_r, p_\theta) = \frac{1}{2}\left(p_r^2 + \frac{p_\theta^2}{r^2}\right) - \frac{1}{r}$$

where $(r, \theta) \in (0, \infty) \times S^1$, $(P_r, P_\theta) \in R^2$.

Definition. The mechanical system with symmetry for the two-body problem is a quadruple (M, K, V, G), where

(i) $M \approx S^1 \times (0, \infty)$ is a Riemannian manifold with the metric

$$\langle (r_1, \theta_1, \dot{r}_1, \dot{\theta}_1), (r_2, \theta_2, \dot{r}_2, \dot{\theta}_2) \rangle = \dot{r}_1 \dot{r}_2 + r_1 r_2 \theta_1 \dot{\theta}_2$$

M is called the configuration space, and the cotangent bundle T^*M with its canonical symplectic structure, the phase space of the system;

(ii) $K: T^*M \to R$, being C^∞, is the kinetic energy of the system, defined by

$$K(r, \theta, p_r, p_\theta) = \frac{1}{2}(P_r^2 + P_\theta^2/r^2)$$

(iii) $V: M \to R$, being C^∞, is the potential energy given by

$$V(r, \theta) = -\frac{1}{r}$$

$G = SO(2) \cong S^1$ is the Lie group acting on M by rotations.

(iv) The Hamiltonian of the system is

$$H(r, \theta, p_r, p_\theta) = \frac{1}{2}(P_r^2 + P_\theta^2/r^2) - 1/r$$

The momentum mapping $J : T^*M \to R$ is given by

$$J(r, \theta, p_r, p_\theta) = p_\theta$$

and J has no critical points on T^*M. The effective potential is given by

$$V_\mu(r, \theta) = (H \circ \alpha_\mu)(r, \theta) = \frac{\mu^2}{2r^2} + V(r) = \frac{\mu^2}{2r^2} - \frac{1}{r}$$

where α_μ is the one-form defined by the conditions

$$\alpha_\mu(x) \in J^{-1}(\mu) \cap T_x^*M = J_x^{-1}(\mu)$$

$$K(\alpha_\mu(x)) = \inf_{\alpha \in J_x^{-1}(\mu)} K(\alpha)$$

So $\alpha_\mu(x)$ is the minimum of $\frac{1}{2}(p_r^2 + \mu^2/r^2)$ with respect to p_r, that is

$$\alpha_\mu(x) = (r, \theta, 0, \mu)$$

Smale [129] proved the following important theorem and corollary.

Theorem. (a) If $\mu \neq 0$, the invariant manifolds are

(i) if $h \geq 0$, then $I_{h,\mu} \approx S^1 \times R$

(ii) if $-1/2\mu^2 < h < 0$, then $I_{h,\mu} \approx S^1 \times S^1$

(iii) if $h = -1/2\mu^2$, then $I_{h,\mu} \approx S^1$

(iv) if $h < -1/2\mu^2$, then $I_{h,\mu} = \emptyset$

(b) If $\mu = 0$, the invariant manifolds are:

(i) if $h < 0$, then $I_{h,0} \approx S^1 \times R$

(ii) if $h \geq 0$, then $I_{h,0} \approx S^0 \times S^1 \times R$

Corollary. The bifurcation set $\Sigma_{H\times J}$ is the following:

$$\Sigma_{H\times J} = \Sigma'_{H\times J} \cup \{(h, 0) \mid h \in R\} \cup \{(0, \mu) \mid \mu \in R\}$$

The above characterize the invariant manifolds $I_{h,\mu}$ (for all h, μ), and the bifurcation set $\Sigma_{H\times J}$.

3. THE THREE-BODY PROBLEM

Consider, for simplicity, two bodies (say, the Earth (E) and the Moon (M)) moving in circles about their center of mass, and a third small body (say, a spaceship) moving under the influence of the first two and lying in the same plane.

This problem is the so-called restricted problem of three bodies [23, 24;2].

Definitions. The Lagrangian model for the (restricted) three-body problem is (TN, L, x, μ), where

(i) $N = R^2 \setminus \{(-\mu, 0), (1 - \mu, 0)\}$

(ii) $L: TN \to R$ is the Lagrangian defined by

$$L(q, \dot{q}) = \frac{\|\dot{q}\|^2}{2} + \dot{q}^1 q^2 - \dot{q}^2 q^2 + \frac{\mu}{\rho(q)} + \frac{1-\mu}{\sigma(q)} + \frac{\|q\|^2}{2}$$

(iii) $x \in TN$

(iv) $0 < \mu < 1$

The symplectic structure on TN is defined by the symplectic form

$$\omega_L = L_{\dot{q}^i q^j}\, dq^i \wedge dq^j + L_{\dot{q}^i \dot{q}^j}\, dq^i \wedge d\dot{q}^j$$

The prediction of the Lagrangian model is determined by the Lagrangian equations and by the energy integral

$$E = \frac{\|\dot{q}\|^2}{2} - \frac{\|q\|^2}{2} - \frac{\mu}{\rho(q)} - \frac{1-\mu}{\sigma(q)}$$

and is the integral curve of the Lagrangian vector field X_S at x.

The critical points in the restricted three-body problem correspond to periodic orbits of period 2π in the following (the first) Hamiltonian model for the (restricted) three-body problem. This is a system (M', H, m, μ) where

(i) $M' \subset R \times T^*R^2$ (phase space) defined by

$$M' = Rx \mid R^2 \times (R^2)^* \backslash (E_* U M_*) \times (R^2)^* \mid$$

endowed with the standard contact structure

(ii) $H: M' \to R$ (being C^∞) is the Hamiltonian defined by

$$H(t, q, p) = \frac{\|p\|^2}{2} - \frac{\mu}{\rho(t, q)} - \frac{1-\mu}{\sigma(t, q)}$$

where $\rho(t, q) = \|q - M_t\|$, and $\sigma(t, q) = \|q - E_t\|$ where $\|\cdot\|$ is the Euclidean norm, and

$$E_t = (-\mu \cdot \cos t, \ -\mu \cdot \sin t) \in R^2$$

$$M_t = ((1-\mu) \cos t, \ (1-\mu) \sin t) \in R^2$$

for $t \in R$ and $0 < \mu < 1$ and

$$E_* = U\{(t, E_t) \mid t \in R\}$$

$$M_* = U\{(t, M_t) \mid t \in R\}$$

(iii) $m \in M'$ (initial conditions)

(iv) $\mu \in R$, $0 < \mu < 1$ is the reduced mass, where

mass of $E = 1 - \mu$, and mass of $M = \mu$

The prediction of the first Hamiltonian model consists of the orbit of m in the time dependent Hamiltonian system.

The second Hamiltonian model for the restricted three-body problem is a system (M', H, m, μ), where

(i) $M' \in T^*R^2$ (phase space) defined by $M' = T^*N$, where $N = R^2 \backslash \{(-\mu, 0), (1-\mu, 0)\}$, endowed with the natural symplectic structure.

(ii) $H: M \to R$ (being C^∞), the Hamiltonian is defined by

$$H(q, p) = \frac{\|p\|^2}{2} + q^1 p_2 - q^2 p_1 - \frac{\mu}{\rho(q)} - \frac{1-\mu}{\sigma(q)}$$

The prediction of the model is the integral curve of X_H at m, where H may be taken to be the rotational energy of a coordinate system.

(iii) $m \in M'$ (initial conditions)

(iv) $0 < \mu < 1$ is the reduced mass.

Observation. The point $(q^1, q^2, \dot{q}_1, \dot{q}_2) \in TN$ is a critical point of X_S if and only if $\dot{q} = 0$ and q is a critical point of the function $V: N \to R$ (being C^∞) given by

$$V(q^1, q^2) = -\frac{1}{2}[(q^1)^2 + (q^2)^2] - \frac{\mu}{\rho} - \frac{1-\mu}{\sigma}$$

Note that $m \in TN$ is a critical point if and only if $X_S(m) = 0$.

The three critical points corresponding to $q^2 = 0$, are the collinear solutions (m_1, m_2, m_3) of Euler, and there are two more critical points, namely, the two equilateral solutions (m_4, m_5) of Lagrange.

In fact,

$$\left.\begin{aligned} m_4 &= \left(\frac{1}{2} - \mu, \frac{\sqrt{3}}{2}\right) \\ m_5 &= \left(\frac{1}{2} - \mu, -\frac{\sqrt{3}}{2}\right) \end{aligned}\right\} \text{ and } \sigma = \rho = 1$$

These are critical points, and it can be shown that the collinear critical points exist and that these are the only such critical points.

The following theorem shows that the three collinear critical points m_1, m_2, m_3 are unstable [119, 2].

Theorem 1. The three collinear points m_1, m_2, m_3 of Euler are unstable.

Proof. Consider the characteristic polynomial

$$Q(x) = x^4 + (4 + V_{11} + V_{22})x^2 + (V_{11}V_{22} - V_{12}^2)$$

where

$$-V_1(q) = -V_{q^1}(q) = q^1 - \frac{\mu(q^1 - 1 + \mu)}{|\rho(q)|^3} - \frac{(1-\mu)(q^1 + \mu)}{|\sigma(q)|^3},$$

$$-V_2(q) = -V_{q^2}(q) = q^2 - \frac{(1-\mu)q^2}{|\sigma(q)|^3} - \frac{\mu q^2}{|\rho(q)|^3}, \text{ and}$$

$$V_{ij} = V_{q^i, q^j}.$$

The roots of Q(x) are of the form r_1, r_2, $-r_1$, $-r_2$.

Consider the polynomial

$$R(y) = y^2 + (4 + V_{11} + V_{22})y + (V_{11}V_{22} - V_{12}^2)$$

It can be shown that the roots of R are real if

$$\frac{\mu}{\rho^3} + \frac{1-\mu}{\sigma^3} > \frac{8}{9}$$

It will be shown that $-V_{22} < 0$ at the three collinear critical points of Euler.

Consider the potential U in the variables ρ, σ

$$U(\rho, \sigma) = -\frac{1}{2}\left|\mu\rho^2 + (1-\mu)\sigma^2 - \mu(1-\mu)\right| - \frac{\mu}{\rho} - \frac{1-\mu}{\sigma}$$

Then,

$$-\frac{\partial U}{\partial \rho} = \mu\left(\rho - \frac{1}{\rho^2}\right), \quad -\frac{\partial U}{\partial \sigma} = (1 - \mu)\left(\sigma - \frac{1}{\sigma^2}\right)$$

However,

$$V_2 = \frac{\partial U}{\partial \rho}\frac{\partial \rho}{\partial q^2} + \frac{\partial U}{\partial \sigma}\frac{\partial \sigma}{\partial q^2} = q^2\left(\frac{1}{\rho}\cdot\frac{\partial U}{\partial \rho} + \frac{1}{\sigma}\cdot\frac{\partial U}{\partial \sigma}\right)$$

So,

$$V_{22} = \frac{1}{\rho}\cdot\frac{\partial U}{\partial \rho} + \frac{1}{\sigma}\cdot\frac{\partial U}{\partial \sigma} \quad \text{at } q^2 = 0$$

Also,

$$\frac{\partial \rho}{\partial q^1} + \frac{\partial \sigma}{\partial q^1} = 0$$

along the axis between the masses m_2 and m_1, m_3.

Then,

$$V_1 = \frac{\partial U}{\partial \rho}\cdot\frac{\partial \rho}{\partial q^1} + \frac{\partial U}{\partial \sigma}\cdot\frac{\partial \sigma}{\partial q^1}$$

So, if $V_1 = 0$,

$$\mu\left(\rho - \frac{1}{\rho^2}\right) = (1 - \mu)\left(\sigma - \frac{1}{\sigma^2}\right)$$

and thus

$$-V_{22} = \mu\cdot\left(\frac{1}{\rho} + \frac{1}{\sigma}\right)\cdot\left(\rho - \frac{1}{\rho^2}\right) < 0$$

because $0 < \rho < 1$ and $\sigma > \rho$. If $\rho = \sigma + 1$,

$$\frac{\partial \rho}{\partial q^1} = \frac{\partial \sigma}{\partial q^1}$$

and

$$\frac{\partial U}{\partial \rho} = -\frac{\partial U}{\partial \sigma} \quad \text{if } V_1 = 0$$

Thus

$$-V_{22} = \mu\left(\frac{1}{\rho} - \frac{1}{\sigma}\right)\cdot\left(\rho - \frac{1}{\rho^2}\right) < 0$$

because $\sigma < 1 < \rho$.

Therefore $-V_{22} < 0$ for all three collinear critical points of Euler. However, $-V_{11} > 0$ and so $V_{11} \cdot V_{22} < 0$ which implies that the roots of R are real and of opposite sign, and so Q(x) does not have all its roots imaginary at m_1, m_2, m_3 implying that some eigenvalue must have positive real part and so the fixed point has an unstable manifold.

Corollary. The three collinear critical points m_1, m_2, m_3 of Euler have one-dimensional stable and unstable manifolds and a two-dimensional center manifold.

The following is a corresponding stability theorem for the equilateral solutions m_4 and m_5 of Lagrange ([93, 2]).

Theorem 2. The two equilateral solutions m_4 and m_5 of Lagrange are unstable for $\mu' < \mu \le 1/2$, where $\mu' = 1/2 - \sqrt{69}/18 = .03852\ldots$, and are stable for all μ, such that $0 < \mu < \mu'$, except possibly those in a set of measure zero.

In fact, Deprit and Deprit-Bartolome [45] showed that the two equilateral solutions m_4 and m_5 are stable for all except three values of μ in the interval $0 < \mu < \mu'$.

All five critical points $m_1, \ldots, m_5$ of Euler and Lagrange are mapped into periodic orbits $\gamma_1, \ldots, \gamma_5$ of period 2π in the first Hamiltonian model for the restricted three-body problem by the canonical transformation which relates the first and second Hamiltonian models.

Note that γ_1, γ_2, γ_3 are unstable periodic orbits and also γ_4, γ_5 for $0 < \mu < \mu'$.

Lyapunov [95] has shown the following.

Theorem 3. In M'_A there are five closed orbits of period 2π corresponding to the critical points in M'_B, and for almost all μ, $0 < \mu < \mu'$, every neighborhood of any of these contains infinitely many closed orbits of arbitrary high period.

The closed orbits obtained by "analytic continuation" from $\mu = 0$ are first those discovered by Poincaré [117] that are close to the circular Keplerian orbits in the second Hamiltonian model for the restricted three-body problem, and those discovered by Arenstorf [15-17] that are close to Keplerian orbits of arbitrary (positive) eccentricity.

These results of Poincaré and Arenstorf are expressed respectively by the two parts of the following theorem.

Theorem 4. a. In the cotangent Poincaré model for the restricted three-body problem, the closed orbit γ_0 of the vector field $X_{\bar{\Omega}^0}$ containing the initial conditions $P_0 = (0, 0, 0, \bar{\Lambda}_0)$ with $\bar{\Lambda}_0^{-3} = m/\kappa$ and period $2\kappa\pi$, is preserved under perturbation of the mass ratio μ away from zero.

That is, there is an $\varepsilon > 0$ and a C^∞ function $f: (-\varepsilon, \varepsilon) \to R$ such that if $\mu \in (-\varepsilon, \varepsilon)$, then p^μ is in a closed orbit γ_μ of the vector field $X_{\bar{\Omega}^\mu}$ of period $2\kappa\pi$, where $p^m = (0, 0, 0, f(\mu))$, and $f(0) = \bar{\Lambda}_0$.

b. In the cotangent Poincaré model for the restricted three-body problem, the closed orbit γ_0 containing $p_0 = (0, 0, \bar{\xi}_0, \bar{\Lambda}_0)$, where $\bar{\Lambda}_0^{-3} = m/\kappa$, $(m, \kappa) = 1$, $\bar{\xi}_0 \neq 0$, and the period is $2\kappa\pi$) is preserved under perturbation of the mass ratio μ away from zero.

The existence of the closed orbits of the second kind of Moser and the 0^+-stability of the closed orbits of the first kind of Poincaré is shown by the following theorem of Moser.

Theorem 5. In the cotangent Poincaré model for the restricted three-body problem, let γ_0 be a closed orbit of $X_{\bar{\Omega}^0}$ containing $p^0 = (0, 0, 0, \bar{\Lambda}_0)$ with $\bar{\Lambda}_0^{-3} = m/\kappa$, $(m, \kappa) = 1$. Let γ_μ be the closed orbit of the first kind of $X_{\bar{\Omega}^\mu}$ containing $p^m = (0, 0, 0, \bar{\Lambda}_\mu)$ given by part (a) of the previous theorem (due to Poincaré), $\bar{\Lambda}_\mu = f(\mu)$. Then, there is an $\varepsilon > 0$ such that

(i) if $|\mu| < \varepsilon$, γ_μ is $0^\pm$-stable, and

(ii) if $|\mu| < \varepsilon$, V is a neighborhood of $\gamma_\mu \subset P_1^{*\mu}$ the configuration space of the cotangent Poincaré model for the restricted three-body problem, and N is a positive integer, there exists a closed orbit of $X_{\bar{\Omega}^\mu}$ in V with period greater than N, if $\kappa/(m + \kappa) \neq p/q$ for $q = 1, 2, 3, 4$ and any integer p.

4. THE n-BODY PROBLEM

Consider n bodies moving in the plane R^2, under Newton's law of gravitation, without collisions.

The configuration space is the 2n-dimensional manifold

$$M = \{x \in (R^2)^n \mid x \notin \Delta\}$$

where

$$\Delta = \cup\{\Delta_{ij} \mid 1 \leq i < j \leq n\}$$

and

$$\Delta_{ij} = \{x \in (R^2)^n \mid x_i = x_j\}$$

Define a Riemannian metric on M by

$$B_x(u, v) = \langle u, v \rangle_x = \sum_{i=1}^{n} m_i u_i \cdot v_i$$

where $x \in M$, $u, v \in TM = Mx(R^2)^n$, $u = (u_1, \ldots, u_n)$, $v = (v_1, \ldots, v_n)$ and $u_i v_i$ is the dot product in R^2.

The kinetic energy is a function $K: MxM \rightarrow R$ given by

$$K(x, v) = \frac{1}{2} \sum_{i=1}^{n} m_i \|v_i\|^2$$

where $\|\cdot\|$ denotes the norm in R^2.

The potential energy is a function V on M defined by

$$V(x) = -\sum_{1 \leq i < j \leq n} \frac{m_i m_j}{\|x_i - x_j\|}$$

Note that this function is not defined at any "collision" (where $x_i = x_j$).

The total energy $E: TM \rightarrow R$ is given by

$$E = K + Vop$$

where $p: TM \rightarrow M$ is the projection mapping.

The total energy E can also be written as follows:

$$E(x, v) = K(v) + V(x)$$

The law of conservation of energy holds true in the mechanics of heavenly bodies. The kinetic energy in cotangent formulation is

$$K(x, r) = \frac{1}{2} \sum_{i=1}^{n} \frac{1}{m_i} \|r_i\|^2$$

where $(x, r) \in Mx(R^2)^n \cong T^*M$, and $\|\cdot\|$ is the Euclidean norm on R^2. The Hamiltonian H is

$$H = K + V o\, \tau_M^*$$

$$H(x, r) = \frac{1}{2} \sum_{i=1}^{n} \frac{1}{m_i} \|r_i\|^2 - \sum_{1 \leq i < j \leq n} \frac{m_i m_j}{\|x_i - x_j\|}$$

An $x = (x_1, \ldots, x_n) \in N = \{x \in M \mid \Sigma_{i=1}^n m_i x_i = 0\}$ is called a central configuration if the force which acts on x_i computed at x is proportional to $m_i x_i$ for $1 \leq i \leq n$, in other words, if there exists $\nu(x) \in R$ such that

$$\text{grad}_i V(x) = \nu m_i x_i, \quad 1 \leq i \leq n$$

where

$$\mathrm{grad}_i\, V(x) = \left(\frac{\partial V(x)}{\partial x_i^1}, \frac{\partial V(x)}{\partial x_i^2}\right), \quad x_i = (x_i^1, x_i^2)$$

Note that $\nu(x)$ is uniquely determined by x.

The number of equivalence classes of collinear central configurations is given by the following theorem.

Theorem 1 (Moulton, Smale). For any choice of the masses in the planar n-body problem there are exactly $n!/2$ equivalence classes of collinear central configurations.

Note that for noncollinear central configurations the above theorem does not hold. S. Smale [129-134], however, has shown that the set

$$\hat{C}_n = C_n/S^1 \times (R\backslash\{0\})$$

of equivalence classes of central configurations is finite. Especially, Smale proved the following fundamental theorem.

Theorem 2 (Smale [129]). If for a given choice of the masses $m_1, \ldots, m_n$ in the planar n-body problem,

$$\hat{V}_s : CP^{n-2}\backslash\hat{\Delta} \to R$$

has all its critical points nondegenerate, then $\hat{C}_n$ is finite, where

$$\hat{\Delta} = \pi_n(E^{2n-3} \cap \Delta)$$

and $\pi_n : S_Q^{2n-3} \to S_Q^{2n-3}/S^1$ is the canonical projection,

$$S_Q^{2n-3} = \{x \in M \mid Q(x) = 1\}$$

with

$$Q(x) = \sum_{i=1}^{n} m_i \|x_i\|^2$$

Note that Δ is invariant under the action of S^1 on $(R^2)^n$ by rotations. Therefore S_Q^{2n-3} is diffeomorphic to the $(2n-3)$-dimensional sphere S^{2n-3} in the $(2n-2)$-dimensional subspace

$$\left\{x \in (R^2)^n \mid \sum_{i=1}^{n} m_i x_i = 0\right\}$$

of $(R^2)^n$ with all the points of Δ removed.

Given $(m_1, \ldots, m_n)$, a position $x = (x_1, \ldots, x_n) \in M$ of the n-bodies is said to be a relative equilibrium if there is a 1-parameter group of rotations of R^2 inducing a motion of the x_i,

$$\Phi_t(x) = (\Phi_t(x_1), \ldots, \Phi_t(x_n))$$

satisfying Newton's equations of motion. Smale's theorem estimates the number of relative equilibria, counting critical points.

The system $(m_1, \ldots, m_n)$ of masses is called critical if $\hat{V}_S$ has at least one degenerate critical point. Denote by $\Sigma_n \subset (0, \infty) \times \cdots \times (0, \infty)$ (n times) the set of all such critical masses in the planar n-body problem.

Smale [130] asked to determine the structure of Σ_n. It has been shown that $\Sigma_3 = \emptyset$ and that $\Sigma_n \neq \emptyset$ for all $n \geq 4$ and the Lebesgue measure of Σ_n is zero.

The following important theorem of Smale and Palmore answers the above problem.

Theorem 3. The set of $(m_1, \ldots, m_n) \in (0, \infty)^n$ for which the set $\hat{C}_n$ of relative equilibria classes in the planar n-body problem is not finite has Lebesgue measure zero.

The following theorem of Lagrange is in the direction of determining $\hat{C}_3$.

Theorem 4. For any choice of the masses m_1, m_2, m_3 in the planar three-body problem, $x \in C_3$ is a noncollinear central configuration if and only if

$$\|x_1 - x_2\| = \|x_1 - x_3\| = \|x_2 - x_3\| = \left(-m \frac{Q(x)}{V(x)}\right)^{1/3}$$

where $m = m_1 + m_2 + m_3$. Thus, the three bodies move in circles forming a fixed equilateral triangle.

The momentum mapping $J : T^*M \to R$ is the usual angular momentum given by

$$J(x, \alpha) \cdot \lambda = \alpha(\lambda_M(x)) = (\alpha_1 \cdot Ax_1 + \cdots + \alpha_n \cdot Ax_n)\lambda$$

where

$$x = (x_1, \ldots, x_n) \in M \subset (R^2)^n$$

$$\alpha = (\alpha_1, \ldots, \alpha_n) \in (R^2)^n$$

$$A = \begin{pmatrix} 0 & -1 \\ 1 & 0 \end{pmatrix}, \quad \lambda \in R$$

and "$\cdot$" is the dot product in R^2.

Let $V_S : S_Q^{2n-3} \to R$ be the restriction of the potential V to $S_Q^{2n-3} \subset M$. Then, define

$$S'_{H\times J} = \bigcup_{z \in \sigma(V_S)} \{(h, \mu) \in R^2 \mid 2h\mu^2 = -V_S^2(z)\}$$

where $\sigma(V_S)$ determines the relative equilibria, central configurations, and the critical part $S'_{H\times J}$ of the bifurcation set $S_{H\times J}$.

The following important theorems and corollaries of Smale [129] are in the direction of determining the topology of the invariant and reduced invariant manifolds $I_{h,m}$ and $\hat{I}_{h,m}$ of the energy-momentum mapping,

$$H \times J : T^*M \to R \times R$$

Note that $I_{h,m} = (H \times J)^{-1}(h, \mu)$ will be called invariant manifolds, despite the fact that for some values $(h, \mu) \in S_{H\times J}$ (the bifurcation set), they do not possess a manifold structure.

The isotropy group

$$G_\mu = \{g \in G \mid Ad^*_{g^{-1}}\mu = \mu\}$$

of the Ad*-equivariant momentum mapping action, acting on $J^{-1}(\mu)$, where G is a connected Lie group acting on M. Then,

$$\hat{I}_{h,\mu} = I_{h,\mu}/G_\mu$$

are submanifolds of the symplectic manifolds $J^{-1}(\mu)/G\mu$ and will be called reduced manifolds. There are Hamiltonian vector fields

$$(P_\mu)_*(X_H/J^{-1}(\mu)) = X_{H_\mu}$$

defined on $J^{-1}(\mu)/G\mu$, and the vector fields

$$(P_{h,\mu})_*(X_H/I_{h,\mu})$$

defined on $\hat{I}_{h,\mu}$, where

$$P_{h,\mu}: I_{h,\mu} \to \hat{I}_{h,\mu}$$

is the canonical projection.

Theorem 5. In the planar n-body problem for zero total momentum, the invariant and reduced manifolds are given by

$$h \geq 0 \tag{i}$$

$$I_{h,0} = (S^{2n-3}\backslash\Delta) \times R \times S^{2n-4}$$

$$\hat{I}_{h,0} = (CP^{n-2}\backslash\hat{\Delta}) \times R \times S^{2n-4}$$

$$h < 0 \tag{ii}$$

$$I_{h,0} = (S^{2n-3}\backslash\Delta) \times R^{2n-3}$$

$$\hat{I}_{h,0} = (CP^{n-2}\backslash\hat{\Delta}) \times R^{2n-3}$$

Corollary. In the planar three-body problem, the invariant and reduced invariant manifolds for zero momentum are given by:

$$h \geq 0 \tag{i}$$

$$I_{h,0} = S^1 \times (S^2\backslash\{p_1, p_2, p_3\}) \times R \times S^2$$

$$\hat{I}_{h,0} = (S^2\backslash\{p_1, p_2, p_3\}) \times R \times S^2$$

$$h < 0 \tag{ii}$$

$$I_{h,0} = S^1 \times (S^2\backslash\{p_1, p_2, p_3\}) \times R^3$$

$$\hat{I}_{h,0} = (S^2\backslash\{p_1, p_2, p_3\}) \times R^3$$

where $p_1 = p^{-1}(\zeta_1)$, $p_2 = p^{-1}(\zeta_2)$, $p_3 = (0,0,1)$.

Theorem 6. In the planar n-body problem for $\mu \neq 0$, $h \geq 0$

$$I_{h,\mu} = (S^{2n-3}\backslash\Delta) \times R^{2n-3}$$

$$\hat{I}_{h,\mu} = (CP^{n-2}\backslash\hat{\Delta}) \times R^{2n-3}$$

Corollary. In the planar three-body problem, for $\mu \neq 0$, $h \geq 0$

$$I_{h,\mu} = S^1 \times (S^2 \backslash \{p_1, p_2, p_3\}) \times R^3$$

$$\hat{I}_{h,\mu} = (S^2 \backslash \{p_1, p_2, p_3\}) \times R^3$$

It remains the case for which $\mu \neq 0$ and $h < 0$. Define

$$M_{h,\mu} = \{x \in M \mid V_\mu(x) \leq h\}$$

where

$$V_\mu(x) = V(x) + \frac{\mu^2}{2 \cdot \sum_{i=1}^{n} m_i \|x_i\|^2}$$

Then,

$$M_{h,m} = \left\{(z,t) \in S_Q^{2n-3} \times (0,\infty) \;\middle|\; \frac{1}{t} V_s(z) + \frac{\mu^2}{2t^2} \leq h\right\}$$

Define

$$g_\mu(z) = \inf_{t \in R} \left(\frac{V_s(z)}{t} + \frac{\mu^2}{2t^2}\right) = -\frac{V_s^2(z)}{2\mu^2}$$

where $\mu = \pm\sqrt{-V(x_0) \cdot Q(x_0)}$. Then,

$$M_{h,m} = \left\{(z,t) \in g_\mu^{-1}((-\infty, h)] \times (0,\infty) \;\middle|\; \frac{V_s(z)}{t} + \frac{\mu^2}{2t^2} \leq h\right\}$$

The following theorems of Smale [129] give an answer for the case $\mu \neq 0$, $h \geq 0$.

Theorem 7. In the planar n-body problem, if $h < 0$, $\mu \neq 0$, then

$$I_{h,\mu} = f_{2n-3}(\Phi'(J_{h,\mu}))$$

$$\hat{I}_{h,\mu} = f_{2n-3}(\Phi'(\hat{J}_{h,\mu}))$$

where $J_{h,\mu} = g_\mu^{-1}((-\infty, h)]$, and $\hat{J}_{h,\mu} = J_{h,\mu}/S^1$. Note that $\Phi'(J_{h,\mu}) = \Phi(J_{h,\mu} \times R)$, the reduced unit disk bundle of the vector bundle $J_{h,\mu} \times R$, and $\hat{J}_{h,\mu} = J_{h,\mu}/S^1$.

Theorem 8. In the planar three-body problem for $(h,\mu) \in R^2 \backslash S'_{H \times J}$, $h < 0$, $\mu \neq 0$, and the additional hypothesis $d_1 < d_2 < d_3 < d_4$ (< 0), where $d_i = \bar{V}_S(\varepsilon_i)$, $i = 1, 2, 3$, $d_4 = \bar{V}_S(\lambda_1) = \bar{V}_S(\lambda_2)$, ε_i, $i = 1, 2, 3$ being the Euler critical points and λ_j, $j = 1, 2$ the Lagrange critical points the following hold (note that all unions are disjoint unions and the points subtracted are all interior to the manifolds).

$a < 2h\mu^2 < b$		$\hat{J}_{h,\mu}$
$a = -\infty$	$b = -d_1^2$	$(D^2 \backslash \{0\}) \cup (D^2 \backslash \{0\}) \cup (D^2 \backslash \{0\})$
$a = -d_1^2$	$b = -d_2^2$	$(D^2 \backslash \{\text{two points}\}) \cup (D^2 \backslash \{0\})$
$a = -d_2^2$	$b = -d_3^2$	$D^2 \backslash \{\text{three points}\}$
$a = -d_3^2$	$b = -d_4^2$	$(S^1 \times I) \backslash (\text{three points})$
$a = -d_4^2$	$b = 0$	$S^2 \backslash \hat{\Delta} = S^2 \backslash \{\text{three points}\}$

$a < 2h\mu^2 < b$		$I_{h,\mu}$; $\hat{I}_{h,\mu}$
$a = -\infty$	$b = -d_1^2$	$S^1 \times ((S^5\backslash S^3) \cup (S^5\backslash S^3) \cup (S^5\backslash S^3))$ $(S^5\backslash S^3) \cup (S^5\backslash S^3) \cup (S^5\backslash S^3)$
$a = -d_1^2$	$b = -d_2^2$	$S^1 \times (((S^5\backslash (S^3 \cup S^3)) \cup (S^5\backslash S^3)))$ $(S^5\backslash (S^3 \cup S^3)) \cup (S^5\backslash S^3)$
$a = -d_2^2$	$b = -d_3^2$	$S^1 \times (S^5\backslash (S^3 \cup S^3 \cup S^3))$ $S^5\backslash (S^3 \cup S^3 \cup S^3)$
$a = -d_3^2$	$b = -d_4^2$	$S^1 \times ((S^1 \times S^4)\backslash (S^3 \cup S^3 \cup S^3))$ $(S^1 \times S^4)\backslash (S^3 \cup S^3 \cup S^3)$
$a = -d_4^2$	$b = 0$	$S^1 \times (S^2\backslash\{\text{three points}\}) \times S^3$ $(S^2\backslash\{\text{three points}\}) \times S^3$

Corollary. The bifurcation set $\Sigma_{H\times J}$ contains the coordinate axes of R^2.

Theorem 9. (i) $H \times J \mid (H \times J)^{-1}(\{(h,\mu) \in R^2 \mid h > 0,\ \mu \neq 0\})$ is a trivial fiber bundle. $(H \times J)^{-1}(\{(h,\mu) \in R^2 \mid h > 0,\ \mu \neq 0\}) \cong \{(h,\mu) \in R^2 \mid h > 0,\ \mu \neq 0\} \times \mid (S^{2n-3}\backslash\Delta) \times R^{2n-3} \mid$

(ii) If $(h_0, \mu_0) \in R^2\backslash S'_{H\times J}$, $h_0 < 0$, $\mu_0 \neq 0$, then there exists a neighborhood U of (h_0, μ_0) in R^2 and a diffeomorphism $\tau_{h,\mu} : I_{h,\mu} \to I_{h_0,\mu_0}$ that is smooth.

Corollary. The bifurcation set in the planar n-body problem is

$$S_{H\times J} = S'_{H\times J} \cup (R \times \{0\}) \cup (\{0\} \times R)$$

The following conjecture of Smale would complete the study of the planar three-body problem.

Conjecture (Smale [114]). For almost all choices of m_1, m_2, m_3 in the planar three-body problem, the numbers d_i, $i = 1, 2, 3, 4$ are distinct.

The above conjecture of Smale has been verified by Palmore [115, 114].

McGehee [103] gives an analysis of the behavior of the flow near triple collisions. In this way, McGehee extends the work of Sundman [138] and Siegel [128'].

Supplementary results may be found in the works of Saari [126'] and Mather and McGehee [101'].

For further work the reader is referred to [1-3, 18-21, 40, 101, 103, 114, 115, 120-125, 129-134].

REFERENCES

1. Abraham, R.: Lectures of Smale on Differential Topology, mimeographed notes, Columbia (1962).
2. Abraham, R.; Marsden, J.: Foundations of Mechanics. Benjamin (1978).
3. Abraham, R.; Robbin, J.: Transversality of Mappings and Flows. Benjamin (1967).
4. Alonso, M.; Finn, E. J.: Mechanics (Fundamental University Physics). Addison-Wesley (1967).

5. Arons, A.: The Development of the Concepts of Physics. Addison-Wesley (1965).
6. Airault, H.; McKean, H. P.; Moser, J.: Comm. Pure Appl. Math. 30(1), 95-148 (1977).
7. Alekseev, V.: Actes Congr. Intern. Math. 2, 893-907 (1970).
8. Andronov, A. A.; Pontriagin, L.: Dokl. Akad. Nauk SSSR 14, 247-251 (1937).
9. Anosov, D.: Proc. Steklov Math. Inst. 90 (1967).
10. Appell, P.: Traité de mechanique rationelles, vols. I and II. Gauthier-Villars.
11. Arens, R.: Nuovo Cim. 21, 395-409 (1974).
12. Arens, R.: J. Math. Phys. 16, 1191-1198 (1975).
13. Arens, R.: J. Math. Phys. 7, 1341-1348 (1966).
14. Arens, R.; Babbit, D. G.: Pac. J. Math. 28, 243-274 (1969).
15. Arenstorf, R. F.: Am. J. Math. 85, 27-35 (1963).
16. Arenstorf, R. F.: Astron. J. 68, 548-555 (1963).
17. Arenstorf, R. F.: Periodic trajectories passing near both masses of the restricted three-body problem, in Proc. Fourteenth Intern. Astronautical Congr., Paris, vol. 4 (1963), pp. 85-97.
18. Arnold, V. I.: Russian Math. Surveys 18, 9-36 (1963).
19. Arnold, V. I.: Mathematical Methods of Mechanics, Springer Graduate Texts in Math. No. 60, Springer.
20. Arnold, V. I.: Russian Math. Surveys 18, 85-192 (1963).
21. Arnold, V. I.; Avez, A.: Théorie ergodique des systemes dynamiques. Gauthier-Villars (English edition: Benjamin-Cummings, 1968).
22. Ball, K. I.; Osborne, G. F.: Space Vehicle Dynamics, Claremont Press (1967).
23. Barrar, R.: Math. Ann. 160, 363-369 (1965).
24. Barrar, R.: Astron. J. 70, 3-5 (1965).
25. Bass, R. W.: Adv. Eng. Sci. 1, 323-330 (1969).
26. Berger, M.: J. Math. Anal. Appl. 29(3), 512-522 (1970); Am. J. Math. 93, 1-10 (1971).
27. Birkhoff, G. D.: Rend. Circ. Mat. Palermo 39, 1-70 (1915).
28. Birkhoff, G. D.: Collected Mathematical Papers, 3 vols. Amer. Math. Soc. (1950); Dover reprint (1968).
29. Birkhoff, G. D.: Relativity and Modern Physics. Harvard Univ. Press (1927).
30. Birkhoff, G. D.: Trans. Am. Math. Soc. 14, 14-22 (1913).
31. Birkhoff, G. D.; Brown, E. H.; Leuschner, A. O.; Russell, H. N.: Bull. Nat. Res. Council 19, 1-22 (1922).
32. Bott, R.: Lectures on Morse Theory. Bonn (1960).
33. Blumenthal, O., ed.: Das Relativitatsprinzip—Lorentz, Einstein, Minkowski. Teubner (1920).
34. Browder, F., ed.: Mathematical Developments Arising from Hilbert's Problems. Amer. Math. Soc. (1976).
35. Brown, E. W.: Mon. Not. R. Astron. Soc. 71, 438-492 (1911).
36. Carathéodory, C.: Variationsrechnung und Partielle Differentialgleichungen erster Ordnung. Teubner (1935).
37. Carathéodory, C.: Calculus of Variations and Partial Differential Equations. Holden-Day (1965).
38. Cerf, J.: Travaux de Smale sur la structure des variétés, Seminaire Burbaki No. 230, Paris (1961-1962).
39. Chalmers, B.: Energy. Academic Press (1963).
40. Chern, S. S.; Smale, S., eds.: Proceedings of the Symposium in Pure Mathematics XIV, XV, Global Analysis (1970).
41. Christie, D. E.: Vector Mechanics. McGraw-Hill (1964).
42. Copernicus, N. (trans. C. G. Wallis): On the Revolutions of the Heavenly Spheres (1543). Encyclopaedia Britannica, Great Books, vol. 16 (1952).
43. Courant, R.; Hilbert, D.: Methods of Mathematical Physics. Interscience (1965).
44. Darboux, G.: Leçons sur la theorie générale des surfaces et les applications géometriques du calcul infinitesimal, 1re Partie, 2me éd. Paris (1914).

45. Deprit, A.; Deprit-Bartolomé, A.: Astron. J. 72, 173-179 (1967).
46. Douglas, J.: Ann. R. Scuola Norm. Sup. Pisa, Ser. II, 8 (1939).
47. Duncan, J. C.: Astronomy. Harper and Row (1955).
48. Einstein, A.: The Meaning of Relativity. Princeton University Press (1956).
49. Einstein, A.: Franklin Inst. J. 221, 313-357 (1936).
50. Einstein, A.; Podolsky, B.; Rosen, N.: Phys. Rev. 47, 777-780 (1935).
51. Euler, L.: Novi Comm. Acad. Sci. Imp. Petrop. 11, 144-151.
52. Feynman, R. P.; Leighton, R. B.; Sands, M.: The Feynman Lectures in Physics. Addison-Wesley (1963).
53. Fermi, L.; Bernardini, G.: Galileo and the Scientific Revolution. Basic Books (1961).
54. Fischer, A.; Marsden, J.: Gen. Rel. Grav. 5, 89-93 (1974).
55. Fischer, A.; Marsden, J.: Proc. Symp. Pure Math. 23, 309-328 (1973).
56. Fischer, A.; Marsden, J.: J. Grav. Gen. Rel. 7, 915-920 (1976).
57. Flanders, H.: Differential Forms with Applications to the Physical Sciences. Academic Press (1963).
58. Galilei, G. (trans. S. Drake): The starry messenger (1610), in Discoveries and Opinions of Galileo. Doubleday Anchor (1957).
59. Galilei, G. (trans. S. Drake): Dialogue Concerning the Two Chief World Systems (1632). University of California (1953).
60. Galilei, G. (trans. H. Grew and A. Di Salvio): Two New Sciences (1638). Dover (1952) (1952).
61. Gallissot, F.: Ibid. 4, 145-297 (1952).
62. Gelfand, I. M.; Fomin, S.: Calculus of Variations. Prentice-Hall (1963).
63. Germain, P.; Nayroles, B., eds.: Methods of Functional Analysis to Problems in Mechanics (1976).
64. Godbillon, C.: Géometrie différentielle et mécanique analytique. Hermann, Paris (1969).
65. Goldstein, H. Mechanics. Addison-Wesley (1950).
66. Gromoll, D.; Meyer, W.: J. Diff. Geom. 3, 493-510 (1969).
67. Guillemin, V.; Pollack, A.: Differential Topology. Prentice-Hall (1974).
68. Hagihara, Y. Celestial Mechanics (5 vols.). M.I.T. (1970).
69. Halmos, P. R.; Von Neumann, J.: Ann. Math. 43, 332-350 (1942).
70. Hermann, R. Lectures on Mathematical Physics (I, II). Benjamin-Cummings (1971).
71. Hermann, R.: Geometry, Physics and Systems. Marcel Dekker (1973).
72. Herivel, J. W.: The Background to Newton's Principia. Clarendon Press (1965).
73. Hilbert, D.: Mathematische Probleme. Nachr. Acad. Wiss., Göttingen (1900).
74. Hill, G.: Am. J. Math. 1, 5-26, 129-147, 245-260 (1878).
75. Hilton, P. (ed.). Structural Stability, the Theory of Catastrophes, and Applications in the Sciences. Springer (1976).
76. Hirsch, M. W.: Differential Topology. Springer (1976).
77. Hirsch, M. W.; Smale, S.: Differential Equations, Dynamical Systems, and Linear Algebra. Academic Press (1974).
78. Hörmander, L.: Acta Math. 127, 79-183 (1971), 128, 83 (1972).
79. Hurt, N. E.: J. Math. Phys. 11, 539-551 (1970).
80. Iacob, A.: Topological Methods in Mechanics [in Romanian]. Editura Acad. Repub. Social Romania (1971).
81. Jacobi, C. G. J.: Vorlesungen über Dynamik. G. Reimer (1884).
82. Kepler, J. Astronomia Nova, in Johannes Kepler, Gesammelte Werke, vol. 3. C. H. Beck (1937).
83. Kepler, J.:(trans. C. G. Wallis): The Harmonies of the World. Encyclopaedia Britannica, Great Books, vol. 16, pp. 1009-1085 (1952).
84. Kervaire, M. A.: La methode de Smale pour le denombrement des équilibres relativs dans le problème des n Corps, in The Greek Mathematical Society, C. Caratheodory Symposium, September 3-7, 1973, pp. 296-305.

85. Kolmogorov, A. N.: General theory of dynamical systems and mechanics, in Proc. of the 1954 Intern. Congress Math. North-Holland, pp. 315-333.
86. Kuiper, N. H. (ed.): Manifolds-Amsterdam 1970. Springer (1971).
87. Lagrange, J. L.: Mécanique Analytique. Mallet-Bachelier, Paris (1853).
88. Lagrange, J. L.: Mem. Cl. Sci. Math. Phys. Ins. France, 1-72 (1808).
89. Lagrange, J. L.: Mem. Cl. Sci. Math. Phys. Inst. France, 343-352 (1809).
90. Lanczos, C.: The Variational Principles of Mechanics, 2nd ed. Univ. of Toronto Press (1962).
91. Landau, L. D.; Lifshitz, E. M. (trans. J. B. Sykes and J. S. Bell): Mechanics. Pergamon (1960).
92. Lefschetz, S.: Differential Equations: Geometric Theory, 2nd ed. Wiley (1963).
93. Leontovich, A. M.: Dokl. Akad. Nauk SSSR 143, 525-528 (1962).
94. Levi-Civita, T.: Acta-Math. 42, 99-144 (1920).
95. Lyapunov, A. M.: Problème general de la stabilite du movement. Ann. Math. Studies No. 17. Princeton Univ. Press (1947).
96. Littlewood, J. E.: Meddel. Lunds Univ. Mat. Sem. Suppl. M. Riesz, 143-151 (1952).
97. Lusternik and Schnirelman. Méthodes topologiques dans les problèmes variationelles. Paris (1934).
98. Mach, E.: The Science of Mechanics, 6th American edition. Open Court Publishing Co. (1960).
99. Maclane, S.: Geometrical Mechanics (2 parts). Dept. of Math., Univ. of Chicago (1968).
100. Maclane, S.: Am. Math. Monthly 77, 570-586 (1970).
101. Marsden, J.: Applications of Global Analysis in Mathematical Physics. Lect. Notes #3. Boston (1974).
101'. Mather, J. N.; McGehee, R.: Solutions of the Collinear Four Body Problem Which Becomes Unbounded in a Finite Time. Springer (1975).
102. Maxwell, J. C.: Matter and Motion (1879). Dover (1954).
103. McGehee, R.: Inv. Math. 27, 191-227 (1974).
103'. McGehee, R.: Triple Collision in Newtonian Gravitational Systems. Lect. Notes in Math., vol. 38. Springer (1975).
104. Milnor, J.: Morse Theory. Princeton Univ. Press (1963).
105. Misner, C. W.; Thorne, K.; Wheeler, J. A.: Gravitation. W. H. Freeman (1973).
106. Mizushima, M.: Theoretical Physics. Wiley (1972).
107. Morse, M.: The Calculus of Variations in the Large. Amer. Math. Soc. (1934).
108. Morse, M.; Cairns, S.: Critical Point Theory in Global Analysis and Differential Topology. Academic Press (1969).
109. Moser, J.: Math. Ann. 126, 325-335 (1953).
110. Moser, J.: Stability and nonlinear character of differential equations in nonlinear problems, in Nonlinear Problems (R. Langer, ed.). Univ. Wisconsin Press, pp. 139-149.
111. Moulton, F.: An Introduction to Celestial Mechanics. Macmillan (1902).
112. Newton, I.: Principia (Mathematical Principles of Natural Philosophy) (1687). F. Cajori's revision of A. Motte's translation (1729). Univ. of California Press (1960).
113. Palais, R.: Seminar on the Atiyah-Singer Index Theorem. Princeton Univ. Press (1965).
114. Palmore, I.: Lett. Math. Phys. 1, 119-123 (1976).
115. Palmore, J.: Ann. Math. 104, 421-429 (1976).
116. Poincaré, H.: Oeuvres. Gauthier-Villars (1928).
117. Poincaré, H.: Les méthodes nouvelles de la mécanique céleste 1, 2, 3. Gauthier-Villars (1892); Dover (1957).
118. Poincaré, H.: Analysis situs, in Collected works. Gauthier-Villars (1953).

119. Plummer, H. C.: Mon. Not. Roy. Astron. Soc. 62, 6-12 (1901).
120. Rassias, G. M.: Archiv der Mathematik 38, 366 (1982).
121. Rassias, G. M.: Counterexamples to a Conjecture of René Thom. Springer (1980), p. 415; see also Bull. de l'Academie Polon. des Sciences, Ser. des Scien. Math. XXIX, No. 5-6, 311 (1981).
122. Rassias, G. M.: Bull. Austr. Math. Soc. 20, 281 (1979).
123. Rassias, G. M.: Bull. de la Société Royale des Sciences de Liége 49, no. 3-4, 125 (1980).
124. Rassias, G. M.: Proceedings of the National Academy of Athens 53, 191 (1978).
125. Rassias, G. M.: Homotopy properties of CW-complexes, in Selected Studies: Physics-Astrophysics, Mathematics, History of Science, A Volume Dedicated to the Memory of Albert Einstein. North-Holland (1982), p. 257 (Rassias, ed.).
125'. Rassias, G. M.: Algebraic and Differential Topology—Global Differential Geometry. Teubner (1984).
126. Rutherford, D. E.: Mechanics. Oliver and Boyd (1957).
126'. Saari, D.: Arch. Rat. Mech. Anal. 49, 311-320.
127. Schiefele, G.; Stiefel, E.: Linear and Regular Celestial Mechanics. Springer (1971).
128. Siegel, C. L.; Moser, J. K.: Lectures on Celestial Mechanics. Springer (1971).
128'. Siegel, C. L.: Ann. Math. 42, 127-168 (1941).
129. Smale, S.: Inv. Math. 10, 305-331; 11, 45-64 (1970).
130. Smale, S.: Problems on the nature of relative equilibria in celestial mechanics, in Manifolds (N. H. Kuiper, ed.). Springer (1971).
131. Smale, S.: Personal perspectives on mathematics and mechanics, in Intern. Union of Pure and Applied Physics Conference on Statistical Mechanics, Chicago (1971).
132. Smale, S.: The Mathematics of Time. Springer (1980).
133. Smale, S.: Topology and Mechanics. Springer (1971).
134. Smale, S.: Ann. Math. 74(2) (1961).
134'. Smale, S.: Bull. Am. Math. Soc. 73 (1967).
135. Sommerfeld, A.: Mechanics. Academic Press (1964).
136. Spanier, E.: Algebraic Topology. McGraw-Hill (1966).
137. Sternberg, S.: Celestial Mechanics, vols. I, II. Benjamin-Cummings (1969).
138. Sundmann, K.: Acta Math. 36, 105-179 (1913).
139. Van De Kamp, P.: Elements of Astromechanics. Freeman (1964).
140. Vertregt, M.: Principles of Astronautics. Elsevier (1965).
141. Von Neumann, J.: Ann. Math. 33, 587-648 (1932).
142. Whittaker, E. T.: A Treatise on the Analytical Dynamics of Particles and Rigid Bodies. Cambridge Univ. Press (1959).
143. Wintner, A.: The Analytical Foundations of Celestial Mechanics. Princeton University Press (1941).
144. Wintner, A.: Proc. Nat. Acad. Sci. U.S.A. 22, 435-439 (1936).

28 ON THE MORSE-SMALE INDEX THEOREM FOR MINIMAL SURFACES

Themistocles M. Rassias / Department of Mathematics, University of La Verne, Kifissia, Athens, Greece

INTRODUCTION

In this paper we present:

I. A brief (historical) account of some of the fundamental problems of the calculus of variations with their meaning in Mechanics.

II. The Morse-Smale Index theorem for a global analysis of variational problems in several variables with applications to the computation of the Morse-Smale Index of the catenoid and Enneper's minimal surface.

III. A list of a number of research problems in the fields of differential geometry and calculus of variations in the large. These problems are still open and most of these are well known to active researchers in the respective fields.

1. AN APPLICATION OF THE VARIATIONAL METHODS IN MECHANICS

It is Euler who started a new field in mathematics, now known as variational analysis. Euler in 1744 deduced the first general rule, now known as Euler's differential equation, for the characterization of the maximizing or minimizing arcs.

Much of the terminology of variational analysis was introduced shortly thereafter by Lagrange. The years 1929-1936 proved to be very essential for the development of the calculus of variations in the large of integrals depending on a curve.

One of the most important questions concerning the problem of Plateau, and especially the existence and non-uniqueness theory of the problem is the one studied by Morse and Tompkins [11] and Shiffman [19]. Their main theorem is stated as follows: If Γ is a Jordan curve in R^3 (the Euclidean 3-space) which bounds two minimal surfaces which are proper relative minima, it bounds at least one minimal surface which is not a proper relative minimum. This result was proved for a special class of curves Γ, a class which includes curves having a continuously turning tangent line and of bounded variation (i.e., the boundary curve Γ has a finite length).

In [18] we have proved a generalized version of Shiffman's main result and we have investigated a new approach for Plateau's problem in the spirit of Smale's work (see for example [21, 22]).

This stability problem is used very frequently in the applications of the calculus of variations in mechanics. It deals with the determination of the equilibrium shape of a soap film suspended between two parallel coaxial circular rings. The problem can be solved by minimizing the surface energy of the film, or equivalently, by finding the figure of revolution with minimum surface area which has its boundaries on the rings. The solution to the problem relates the radius of the film r to the displacement z along the axis of symmetry by the equation of a catenary,

$$r = a \cosh \frac{z - b}{a} \tag{1}$$

where a and b are constants that are to be determined such that r be equal to the fixed radii of the rings for $z = 0$ and h, where h is the separation of the rings. If the rings have equal radius r_0, the surface is symmetrical about $z = h/2$, b is equal to $h/2$, and a, the minimum radius of the film satisfies the equation

$$r_0 = a \cosh \frac{h}{2a} \tag{2}$$

It can be proved that there are two solutions for $h/2r_0 < 0.66274 \ldots$, only one of which is stable (see for example [3,4,8]).

One is provided a complete analysis of the problem of the stability and instability of the film by using eigenvalue methods.

In fact one can see that the dynamical stability of the film is determined by the sign of the lowest (the first) eigenvalue λ_1 of an associated Sturm-Liouville problem. The film then is stable when $\lambda_1 > 0$ and unstable when $\lambda_1 < 0$ (see [8]).

Let us assume that the film is attached to two plane parallel coaxial rings with radii r_1 and r_2, separated by a distance h, and have no other boundaries. The equilibrium surfaces are axially symmetric, with a surface energy given by

$$E_S = 2\sigma \int_S dS = 4\pi\sigma \int_S r\sqrt{dr^2 + dz^2} \tag{3}$$

where S is the surface area of an ideal static soap film and σ is the surface tension, neglecting gravity. It can be proved that E_S vanishes if r(z) satisfies the Euler equation

$$\frac{1}{r_z} \frac{d}{dz}\left(\frac{r}{\sqrt{1 + r_z^2}}\right) = 0 \tag{4}$$

where $r_z \equiv \frac{d}{dz} r(z)$. However, this can happen if either

$$\frac{r}{\sqrt{1 + z_r^2}} = a \tag{5}$$

where a is a positive constant, or

$$r_z \text{ is infinite} \tag{6}$$

Integrating (5) we obtain

$$r(z) = a \cosh \frac{z - b}{a} \tag{7}$$

such that b is a constant of integration.

From (6) it follows that $z_r = 1/r_z$ vanishes, and the surface S consists, for our boundary conditions, of two disconnected plane disks which fill the rings.

2. COMPUTATION OF THE MORSE-SMALE INDEX FOR MINIMAL SURFACES IN R^3

In the following we apply the Morse-Smale Index Theorem to the problem of a global analysis of variational problems in several variables. Particularly, the following examples of complete minimal surfaces are studied: the plane $R^2 C R^3$, the catenoid, and Enneper's minimal surface.

We answer various questions on the Morse-Smale Index and nullity of those surfaces. We conclude with a global description of the Index as a function of the radius of the disk under consideration. We verify that Enneper's surface is unstable. In this way we study the geometry of surfaces by applying the methods of elliptic partial differential equations and more precisely the eigenvalue formulation of the Morse-Smale Index. The above examples cover some of the most famous surfaces studied in the last century.

Let Γ be a Jordan curve in the ordinary Euclidean space R^3 represented by a continuous, one-to-one mapping

$$a : \partial D_r^2 \to R^3$$

where

$$D_r^2 = \{(u,v) \in R^2 : u^2 + v^2 < r^2\}$$

and ∂D_r^2 is the boundary of D_r^2. A minimal surface $q : D_r^2 \to R^3$ spanned by Γ is a map of class

$$C^0(\overline{D_r^2}, R^3) \cap C^2(D_r^2, R^3)$$

satisfying the differential equations

$$\Delta q = 0, \quad \left|\frac{\partial q}{\partial u}\right|^2 = \left|\frac{\partial q}{\partial v}\right|^2, \quad \frac{\partial q}{\partial u} \cdot \frac{\partial q}{\partial v} = 0$$

in D_r^2, where q maps ∂D_r^2 onto Γ in a topological manner. For a surface in isometric representation, the Dirichlet integral equals twice the surface area. The Dirichlet integral is defined as

$$D[q] = \iint_{D_r^2} |\nabla q|^2 \, dudv, \qquad \nabla q = \left(\frac{\partial q}{\partial u}, \frac{\partial q}{\partial v}\right)$$

The second variation (derivative) of the area function for a variation whose deformation vector field is given by $g = fN$ is then

$$\Pi(g,g) = \iint_{D_r^2} f(-\Delta f + 2KWf) \, dudv \tag{1}$$

where f is a smooth function (e.g., C^2 function), $f : \overline{D_r^2} \to R$(reals) with the additional condition that $f \equiv 0$ on ∂D_r^2, and N is the unit normal field along $q(D_r^2)$. K is the Gaussian curvature of the minimal immersion and W is the discriminant of the first fundamental form. The operator $L = -\Delta + 2KW$, used to define the second variation of the area function, is a symmetric differential operator. It is also strongly elliptic. The operator L is known as the Jacobi operator for the minimal immersion $q : \overline{D_r^2} \to R^3$ (given in a conformal parametric representation). The general theory of such operators shows that L can be diagonalized with eigenvalues

$$\lambda_1 < \lambda_2 < \lambda_3 < \cdots \to \infty$$

where each eigenspace V_{λ_i} is finite dimensional. The quadratic form (1) is the Hessian form of second derivatives of the area function at this point. In analogy with standard critical point theory we give the following definition:

Definition. Let $q : D_r^2 \to R^3$ be a minimal immersion, given in parametric representation,

$$q(u,v) = (x(u,v), y(u,v), z(u,v))$$

where $(u,v) \in D_r^2$ for $r \in R$.

a) The Morse-Smale Index of q is $I(q; D_r^2)$ = the sum of the dimensions of those V_λ where V_λ is an eigenspace corresponding to the negative eigenvalue λ of the Jacobi operator for q. It is frequently written as

$$I(q; D_r^2) = \dim\left(\bigoplus_{\lambda < 0} V_\lambda\right)$$

b) The nullity of q is

$$N(q\,;D_r^2) = \dim V_0$$

where V_0 is the eigenspace corresponding to the zero eigenvalue of the Jacobi operator L for q (see Th. M. Rassias [14-17], Simons [20], Smale [23]).

Definitions. A minimal immersion is defined to be stable if $\Pi(g,g) > 0$ for all g $(\neq 0)$.
A minimal immersion is defined to be unstable if $\Pi(g,g) < 0$ for some g $(\neq 0)$.
A Jacobi field on $\overline{D_r^2}$ is a normal field fN, where

$$f: \overline{D_r^2} \to R$$

has the property of being a smooth function, satisfying the Jacobi equation $-\Delta f + 2KWf = 0$.

Remark. Clearly if $N(q\,;D_r^2) = 0$, then

a) a minimal immersion q is stable $\Leftrightarrow$ $I(q\,;D_r^2) = 0$

b) a minimal immersion q is unstable $\Leftrightarrow$ $I(q\,;D_r^2) > 0$

Consider the eigenvalue problem:

(i) $-\Delta f + 2KWf = \lambda f$ where $f = f(u,v)$ on the disk $\bar{D}^2_{r_0} = \{(u,v) \in R^2 : u^2 + v^2 \leq r_0^2,\ r_0 \in R\}$

(ii) $f \equiv 0$ on $\partial D^2_{r_0} = \{(u,v) \in R^2 : u^2 + v^2 = r_0^2,\ r_0 \in R\}$ (boundary condition) (2)

Definitions. A conjugate boundary of a minimal immersion is a boundary $r = r_0$ such that zero is an eigenvalue of the Jacobi operator, for the disk $\overline{D^2_{r_0}}$, in the eigenvalue problem (2).

A first conjugate boundary of a minimal immersion is a boundary $r = \bar{r}$ such that zero is the first eigenvalue of the Jacobi operator for the disk $\overline{D^2_{\bar{r}}}$, in the eigenvalue problem (2), i.e., $\lambda_1(\bar{r}) = 0$.

The multiplicity of a conjugate boundary $r = r_0$ is the number of linearly independent Jacobi fields defined on $\bar{D}^2_{r_0}$ and vanishing on $\partial D^2_{r_0}$. That says that the dimension of the eigenspace corresponding to the zero eigenvalue of the Jacobi operator for the disk $\overline{D^2_{r_0}}$ in the eigenvalue problem (2).

Theorem 1. The multiplicity of a first conjugate boundary of any minimal immersion

$$q: \overline{D^2_{r_0}} \to R^3$$

is always one.

Proof. See for example the paper by Th. M. Rassias [16].

Theorem 2. The multiplicity of a first conjugate boundary of any minimal immersion

$$q: \overline{D^n_r} \to R^{n+1}, \quad \text{for all } n \geq 2$$

is always one, where

$$\overline{D^n_r} = \{(x_1, x_2, \ldots, x_n) \in R^n : x_1^2 + x_2^2 + \cdots + x_n^2 \leq r^2\}$$

Proof. The argument used to prove Theorem 1 goes through without any modification to prove Theorem 2.

Q.E.D.

An Application of the Morse-Smale Index to the Study of Minimal Surfaces in R^3

We are going to investigate the main topological and stability properties of some of the most important examples of complete minimal surfaces in R^3, by making use of the Morse-Smale Index Theorem (see [23]). It is the main intention to obtain global results for certain submanifolds in R^3 and so be able to point out a global variational analysis of the theory of minimal submanifolds in the spirit of Smale's approach. We consider the eigenvalue formulation of the Index because this is the best way to get global results for complete minimal surfaces and in addition to being in a better situation to examine the eigenvalues of certain elliptic operators on manifolds. This leads to the study of the eigenvalues of the Laplacian which play a very significant role in mathematics (see [1, 2, 5-7, 10, 23, 26]). It is our purpose to analyze the following examples of complete minimal surface in R^3:

1. the plane, $R^2 \subset R^3$;
2. the catenoid;
3. Enneper's minimal surface.

We are going to study the following eigenvalue problem:

$$\text{(E)} \quad \begin{cases} -\Delta f + 2KWf = \lambda f, & f = f(u,v) \text{ defined on } \overline{D^2_{r_0}} \\ f \equiv 0 \text{ on } \partial D^2_{r_0} & \text{(boundary condition)} \end{cases}$$

where f is a variation in the direction of the surface normal, Δ is the Laplacian of the surface, K is the Gaussian curvature of the minimal surface, and W is the discriminant of the first fundamental form, i.e., $W = (EG - F^2)^{\frac{1}{2}}$, where E, F, G are the coefficients of the first fundamental form of the minimal surface.

Remark. The eigenvalue problem (E) will be used to find an upper bound for the Morse-Smale Index of the complete minimal surfaces, mentioned above.

Example 1—The Plane as a Complete Minimal Surface in R^3. The plane in R^3 is a surface $S: ax + by + cz = d$ where the numbers a, b, c are necessarily not all zero. It is straightforward to see that S is a complete minimal surface whose Gaussian curvature $K = 0$.

Theorem. The Morse-Smale Index of the plane S as a minimal surface in R^3 is identically zero.

Proof. After some straightforward computation the eigenvalue problem (E) for the plane S becomes

$$\text{(I)} \quad \begin{cases} \nabla^2 f + \lambda f = 0, & f = f(u,v) \text{ on } \overline{D^2_{r_0}} \\ f \equiv 0 \text{ on } \partial D^2_{r_0} & \text{(boundary condition)} \end{cases}$$

where f is a variation in the direction of the surface normal. However, $\nabla^2 f + \lambda f = 0 \Leftrightarrow \nabla^2 f = -\lambda f$ and $-\nabla^2$ is a positive definite operator, so it is impossible to find negative eigenvalues for (I).

This implies the Morse-Smale Index of S is zero. Q.E.D.

Example 2—The Catenoid as a Complete Minimal Surface in R^3. We define as catenoid the surface obtained by revolving the curve $y = \cosh(z)$ about the z-axis. This way we get a global expression of the surface. It can be easily shown that its mean curvature H is zero, for this reason it is a minimal surface. It can be proved that catenoids are the only complete nonplanar surfaces of revolution which are minimal (see [9, 13]). From the topological viewpoint the catenoid is doubly connected. It is easy to see that it is properly embedded in R^3. It can also be remarked that the catenoid is not an algebraic minimal surface. The reason why we study this surface is because of its interesting properties. In fact the following holds: The catenoid and Enneper's minimal surface are the only surfaces whose normal map is one-to-one. The catenoid has total curvature -4π.

A conformal parametric representation of a portion of the catenoid in R^3 is a mapping

$$q : D^2_{r_0} \to R^3, \quad \text{for each } r_0 > 0$$

given by

$$q(u,v) = (x(u,v),\ y(u,v),\ z(u,v))$$

where $(u,v) \in D^2_{r_0}$ and

$$x(u,v) = \cosh u \cos v$$

$$y(u,v) = \cosh u \sin v$$

$$z(u,v) = u$$

Theorem. The Morse-Smale Index $I(q; D^2_{r_0})$ of the catenoid $q : D^2_{r_0} \to R^3$ is bounded above by the sum of the dimensions of the eigenspaces corresponding to those eigenvalues λ of the Bessel equation which are $< 2r_0^2$, for all $r_0 \in R$. (Note: The Bessel equation is an ordinary differential equation given by

$$\frac{d^2y}{dp^2} + \frac{1}{p}\cdot\frac{dy}{dp} + \left(1 - \frac{n^2}{p^2}\right)y = 0 \quad \text{for all } n = 0, 1, 2, \ldots$$

where

$$y = y(p), \quad p \in [0,\infty) \text{ and}$$

$$p = rr_0\sqrt{\lambda + 2} \quad \text{for } 0 \le r \le 1 \text{ and } r_0 \text{ given real number with } r_0 \in [0,\infty).)$$

Proof. We are going at first to prove the following lemma.

Lemma. The eigenvalue problem (E) of the catenoid $q : D^2_{r_0} \to R^3$ is given by

$$\begin{cases} \dfrac{\partial^2 f}{\partial u^2} + \dfrac{\partial^2 f}{\partial v^2} + 2f + \lambda(\cosh^2 u) f = 0 & \text{where } f = f(u,v) \text{ on } \overline{D^2_{r_0}} \\ f \equiv 0 \quad \text{on } \partial D^2_{r_0} & \text{(boundary condition)} \end{cases}$$

Proof. We are going to find the Gaussian curvature K, as well as the discriminant of the first fundamental form W, of the catenoid. We will derive the Laplacian of the catenoid and therefore the corresponding eigenvalue problem (E). At first we want to find E, F, G, L, M, N, K, W, however,

$$q_u = (\sinh u \cos v,\ \sinh u \sin v,\ 1)$$

$$q_{uu} = (\cosh u \cos v,\ \cosh u \sin v,\ 0)$$

$$q_v = (-\cosh u \sin v,\ \cosh u \cos v,\ 0)$$

$$q_{vv} = (-\cosh u \cos v,\ -\cosh u \sin v,\ 0)$$

We get

$$q_{uu} + q_{vv} = 0$$

Therefore q is a harmonic function.

$$E = q_u \cdot q_u = \cosh^2 u$$

therefore $E = \cosh^2 u$.

$$F = q_u \cdot q_v = 0$$

therefore $F = 0$.

$$G = q_v \cdot q_v = \cosh^2 u$$

therefore $G = \cosh^2 u$. Hence

$$E = G, \ F = 0$$

It is clear that $q = q(u,v)$ as defined is a complete minimal surface in R^3.

We know that $L = \xi \cdot q_{uu}$, $M = \xi \cdot q_{uv}$, $N = \xi \cdot q_{vv}$ where $\xi = (X, Y, Z)$ is the unit normal vector to the surface under consideration

$$X = \frac{1}{W} \cdot \frac{\partial(y,z)}{\partial(u,v)} = \frac{1}{W} \cdot \begin{vmatrix} \frac{\partial y}{\partial u} & \frac{\partial y}{\partial v} \\ \frac{\partial z}{\partial u} & \frac{\partial z}{\partial v} \end{vmatrix} = -\frac{\cos v}{\cosh u}$$

Therefore $X = -(\cos v/\cosh u)$. Similarly, we obtain

$$Y = \frac{1}{W} \cdot \frac{\partial(z,x)}{\partial(u,v)} = \frac{1}{W} \cdot \begin{vmatrix} \frac{\partial z}{\partial u} & \frac{\partial z}{\partial v} \\ \frac{\partial x}{\partial u} & \frac{\partial x}{\partial v} \end{vmatrix} = -\frac{\sin v}{\cosh u}$$

Therefore $Y = -(\sin v/\cosh u)$. Similarly, we obtain

$$Z = \frac{1}{W} \cdot \begin{vmatrix} \frac{\partial x}{\partial u} & \frac{\partial x}{\partial v} \\ \frac{\partial y}{\partial u} & \frac{\partial y}{\partial v} \end{vmatrix} = \frac{\sinh u}{\cosh u}$$

Hence $Z = (\sinh u/\cosh u)$. Therefore

$$\xi = (X, Y, Z) = \left(-\frac{\cos v}{\cosh u}, -\frac{\sin v}{\cosh u}, \frac{\sinh u}{\cosh u}\right)$$

Therefore

$$L = \xi \cdot q_{uu} = -1 \qquad \text{therefore } L = -1$$

$$M = \xi \cdot q_{uv} = 0 \qquad \text{therefore } M = 0$$

$$N = \xi \cdot q_{vv} = 1 \qquad \text{therefore } N = 1$$

Hence

$$K = \frac{LN - M^2}{EG - F^2} = -\frac{1}{\cosh^4 u}$$

Therefore

$$K = -\frac{1}{\cosh^4 u}$$

Similarly we get $W = \sqrt{EG - F^2} = \cosh^2 u$. Hence $W = \cosh^2 u$. We are going to derive the Laplacian for the catenoid.

We know that the Laplacian of a Riemannian manifold is given by

$$\Delta f = \sum_{i,j=1}^{n} g^{ij} \frac{\partial^2 f}{\partial x^i \partial x^j} + \sum_{i,j=1}^{n} \left(\frac{\partial g^{ij}}{\partial x^i} + \frac{1}{2} g^{ij} \frac{\partial \log \tilde{G}}{\partial x^i} \right) \frac{\partial f}{\partial x^j}$$

where (g^{ij}) denotes the inverse matrix of (g_{ij}), i.e.,

$$\sum_{j=1} g^{ij} \cdot g_{j\ell} = \delta^i_\ell \quad \text{for all } i \text{ and } \ell$$

and

$$\tilde{G} = \det(g_{ij})$$

For the catenoid in R^3, we set $n = 2$. However, $g_{11} = E$, $g_{12} = g_{21} = F$, $g_{22} = G$, and $\tilde{G} = EG - F^2 = \det(g_{ij})$.

$$\begin{pmatrix} E & F \\ F & G \end{pmatrix}^{-1} = \begin{pmatrix} g^{11} & g^{12} \\ g^{21} & g^{22} \end{pmatrix}$$

but $E = G$, $F = 0$ for the catenoid, therefore $g^{11} = 1/E$, $g^{12} = g^{21} = 0$, $g^{22} = 1/E$, and $\tilde{G} = EG - F^2 = E^2$. Therefore we obtain:

$$\begin{aligned} \Delta f &= g^{11} \frac{\partial^2 f}{\partial x^1 \partial x^1} + 2g^{12} \frac{\partial^2 f}{\partial x^1 \partial x^2} + g^{22} \frac{\partial^2 f}{\partial x^2 \partial x^2} + \left(\frac{\partial g^{11}}{\partial x^1} + \frac{1}{2} g^{11} \frac{\partial \log \tilde{G}}{\partial x^1} \right) \frac{\partial f}{\partial x^1} \\ &\quad + \left(\frac{\partial g^{12}}{\partial x^1} + \frac{1}{2} g^{12} \frac{\partial \log \tilde{G}}{\partial x^1} \right) \frac{\partial f}{\partial x^2} + \left(\frac{\partial g^{21}}{\partial x^2} + \frac{1}{2} g^{21} \frac{\partial \log \tilde{G}}{\partial x^2} \right) \frac{\partial f}{\partial x^1} \\ &\quad + \left(\frac{\partial g^{22}}{\partial x^2} + \frac{1}{2} g^{22} \frac{\partial \log \tilde{G}}{\partial x^2} \right) \frac{\partial f}{\partial x^2} \\ &= \frac{1}{E} \frac{\partial^2 f}{\partial u^2} + \frac{1}{E} \frac{\partial^2 f}{\partial v^2} + \left\{ \frac{\partial}{\partial u}\left(\frac{1}{E}\right) + \frac{1}{2} \cdot \frac{1}{E} \cdot \frac{\partial \log E^2}{\partial u} \right\} \frac{\partial f}{\partial u} + \left\{ \frac{\partial}{\partial v}\left(\frac{1}{E}\right) + \frac{1}{2} \cdot \frac{1}{E} \frac{\partial \log E^2}{\partial v} \right\} \frac{\partial f}{\partial v} \end{aligned}$$

where we have set $u = x^1$, $v = x^2$. Therefore by setting $E = \cosh^2 u$ we obtain the Laplacian of the catenoid to be

$$\Delta f = \frac{1}{\cosh^2 u} \left(\frac{\partial^2 f}{\partial u^2} + \frac{\partial^2 f}{\partial v^2} \right)$$

which is conformally invariant. Hence the eigenvalue problem (E) for the catenoid becomes

$$\begin{cases} \dfrac{\partial^2 f}{\partial u^2} + \dfrac{\partial^2 f}{\partial v^2} + 2f + \lambda(\cosh^2 u) f = 0, & \text{where } f = f(u,v) \text{ on } \overline{D^2_{r_0}} \\ f \equiv 0 \text{ on } \partial D^2_{r_0} & \text{(boundary condition)} \end{cases}$$

Q.E.D.

<u>Proof of the Theorem</u>. Consider now the following eigenvalue problems:

$$\begin{cases} -\Delta f - 2f = \lambda(\cosh^2 u) f & \text{on } \overline{D^2_{r_0}} \\ f \equiv 0 & \text{on } \partial D^2_{r_0} \end{cases} \tag{1}$$

and

$$\begin{cases} -\Delta f - 2f = \mu f & \text{on } \overline{D^2_{r_0}} \\ f \equiv 0 & \text{on } \partial D^2_{r_0} \end{cases} \tag{2}$$

where $\Delta = \dfrac{\partial^2}{\partial u^2} + \dfrac{\partial^2}{\partial v^2}$.

Then the number of eigenvalues λ of (1) less than or equal to zero is equal to the number of nonpositive eigenvalues μ of (2). The proof of this fact goes in the following way. Let

$$D(g) = \iint_{D^2_{r_0}} (|\nabla g|^2(u,v) - 2g^2(u,v))\, dudv$$

Then (1) is associated with the Rayleigh quotient

$$R_1(g) = \frac{D(g)}{\iint_{D^2_{r_0}} (\cosh^2 u)\, g^2(u,v)\, dudv}, \quad g \in C_0^\infty(D^2_{r_0})$$

and (2) to the Rayleigh quotient

$$R_2(g) = \frac{D(g)}{\iint_{D^2_{r_0}} g^2(u,v)\, dudv}$$

Since $D^2_{r_0}$ is bounded, we have $1 \le \cosh^2 u \le A^{-1}$ for $(u,v) \in D^2_{r_0}$ hence, for any $g \in C_0^\infty(D^2_{r_0})$, $AR_2(g) \le R_1(g) \le R_2(g)$. From the variational characterization of eigenvalues it follows that

$$A\lambda_i^{(2)} \le \lambda_i^{(1)} \le \lambda_i^{(2)} \quad \text{(with obvious notations)}$$

Hence

$$\#\{i : \lambda_i^{(2)} \le \lambda\} \le \#\{i : \lambda_i^{(1)} \le \lambda\} \le \#\{i : \lambda_i^{(2)} \le \lambda A^{-1}\}$$

Now the assertion follows if we take $\lambda = 0$.

Applying the above lemma to our situation we denote by S_0 the negative eigenspace of the problem

$$\begin{cases} -\nabla^2 f - 2f = \lambda K f, \quad f = f(u,v) \text{ on } \overline{D^2_{r_0}} \\ f \equiv 0 \text{ on } \partial D^2_{r_0} \end{cases} \tag{3}$$

where K is a positive constant.

Applying standard theory of eigenvalue problems we consider the problem of finding the eigenvalues Λ of the problem (4).

$$\begin{cases} -\nabla^2 f = \Lambda f, \quad f = (u,v) \text{ on } \overline{D^2_{r_0}} \\ f \equiv 0 \text{ on } \partial D^2_{r_0} \text{ (boundary condition)} \end{cases} \tag{4}$$

We obtain

dim S_0 = the number of those eigenvalues of the problem (4) which are < 2

Hence,

dim S_0 < the number of those eigenvalues λ of the Bessel equation which are less than $2r_0^2$, for all $r_0 \in R$. Q.E.D.

Corollary. The Morse-Smale Index $I(q; D^2_{r_0})$ of the catenoid $q: D^2_{r_0} \to R^3$ varies as follows:

1. $I(q; D^2_{r_0}) = 0$ for $r_0 = 0$
2. $I(q; D^2_1) = 0$
3. $I(q; D^2_{\sqrt{3}}) \leq 1$
4. $I(q; D^2_2) \leq 1$
5. $I(q; D^2_3) \leq 2$
6. $I(q; D^2_4) \leq 4$
7. $I(q; D^2_5) \leq 6$
8. $I(q; D^2_{r_0}) \to \infty$ as $r_0 \to \infty$

Proof. 1. If $r_0 = 0$, then $2r_0^2 = 0$, therefore

$I(q; D^2_0) = 0$

2. If $r_0 = 1$, then $2r_0^2 = 2$, therefore

$I(q; D^2_1) = 0$

3. If $r_0 = \sqrt{3}$, then $2r_0^2 = 6$.

By making use of the tables of the zeros of Bessel functions we have: From $j_{0,n}$ we get only one eigenvalue. Therefore $I(q; D^2_{\sqrt{3}}) \leq 1$.

4. If $r_0 = 2$, then $2r_0^2 = 8$. From $j_{0,n}$ we get only one eigenvalue. Therefore $I(q; D^2_2) \leq 1$.

5. If $r_0 = 3$, then $2r_0^2 = 18$. From $j_{0,n}$ we get only one eigenvalue. From $j_{1,n}$ we get only one eigenvalue. From $j_{k,n}$ we do not obtain eigenvalues, for $k \geq 2$. Therefore $I(q; D^2_3) \leq 2$.

6. If $r_0 = 4$, then $2r_0^2 = 32$. From $j_{0,n}$ we get only two eigenvalues. From $j_{1,n}$ we get only one eigenvalue. From $j_{2,n}$ we get only one eigenvalue. From $j_{k,n}$ we do not obtain eigenvalues for $k \geq 3$. Therefore $I(q; D^2_4) \leq 4$.

7. If $r_0 = 5$, then $2r_0^2 = 50$. From $j_{0,n}$ we get only two eigenvalues. From $j_{1,n}$ we get only two eigenvalues. From $j_{2,n}$ we get only one eigenvalue. From $j_{3,n}$ we get only one eigenvalue. From $j_{4,n}$ we get zero eigenvalues. From $j_{k,n}$ we do not obtain eigenvalues for $k \geq 4$. Therefore $I(q; D^2_5) \leq 6$.

8. $I(q; D^2_{r_0}) \to \infty$ as $r_0 \to \infty$, as proved in [20].

Table of Zeros of Bessel Functions

$j_{0,1}$	= 2.4048256	$j_{1,1}$	= 3.8317060
$j_{0,2}$	= 5.5200781	$j_{1,2}$	= 7.0155867
$j_{0,3}$	= 8.6537279	$j_{1,3}$	= 10.1734681
$j_{2,1}$	= 5.1356223	$j_{3,1}$	= 6.3801619
$j_{2,2}$	= 8.4172441	$j_{3,2}$	= 9.7610231
$j_{2,3}$	= 11.6198412	$j_{3,3}$	= 13.0152007

$$j_{4,1} = 7.5883427 \qquad j_{5,1} = 8.7714838$$
$$j_{4,2} = 11.0647095 \qquad j_{5,2} = 12.3386042$$
$$j_{4,3} = 14.3725367 \qquad j_{5,3} = 15.7001741$$

(See for example [25].)

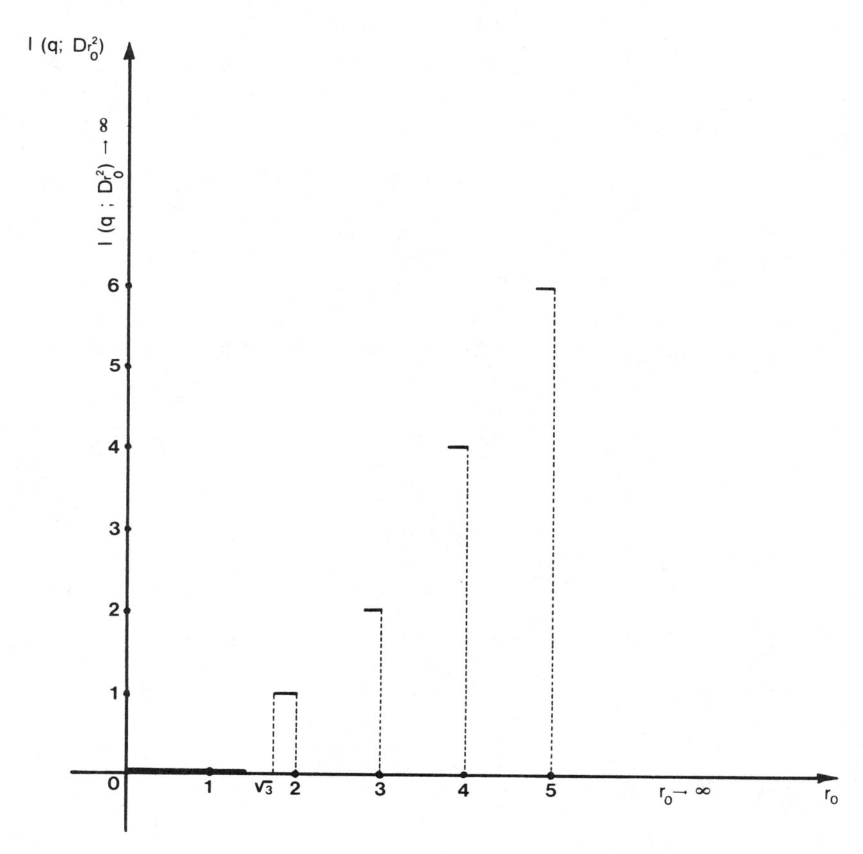

Fig. 1. Behavior of the Morse-Smale Index $I(q; D^2_{r_0})$ of the catenoid $q : D^2_{r_0} \to R^3$ as a function of r_0 for all $r_0 \in R$.

Theorem. The first conjugate boundary of the catenoid $q : D^2_{r_0} \to R^3$ (in the sense of the eigenvalue problem (E)) occurs for

$$r_0 = \frac{\mu_1^{(0)}}{\sqrt{2}} \simeq 1,7$$

where $\mu_1^{(0)}$ is the first zero of the Bessel equation for the unit disk.

Note. $\mu_1^{(0)} \simeq 2.4048256$, $\sqrt{2} \simeq 1.4142135$.

Proof. Consider the eigenvalue problem

$$\begin{cases} \Delta f + \lambda f = 0 & \text{where } f = f(u,v) \text{ on } \overline{D^2_{r_0}} \\ f \equiv 0 & \text{on } \partial D^2_{r_0} \end{cases}$$

This problem can be formulated as follows:

$$\frac{1}{r}\frac{\partial}{\partial r}\left(r\frac{\partial f}{\partial r}\right) + \frac{1}{r^2}\frac{\partial^2 f}{\partial \Theta^2} + \lambda f = 0 \tag{1}$$

with

$$f(r,\Theta) = 0 \text{ at } r = r_0 \tag{2}$$

If we set $f(r,\Theta) = R(r)\,\Phi(\Theta)$ and carry out the separation of variables, we obtain

$$\Phi'' + \nu\Phi = 0 \tag{3}$$

$$\frac{1}{r}\frac{d}{dr}\left(r\frac{dR}{dr}\right) + \left(\lambda - \frac{\nu}{r^2}\right)R = 0 \tag{4}$$

with boundary condition $R(r_0) = 0$.

If we let $\Phi(\Theta)$ be periodic, then we get $\nu = n^2$, where n is an integer. The function R(r), therefore, satisfies the Bessel equation

$$L[R] + \lambda r R = 0 \quad \text{where} \quad L[R] = \frac{d}{dr}\left(r\frac{dR}{dr}\right) - \frac{n^2}{r}R \tag{5}$$

with the boundary condition

$$R(r_0) = 0 \tag{6}$$

and the boundedness condition

$$|R(0)| < \infty \tag{7}$$

at the point $r = 0$. If we substitute

$$x = \sqrt{\lambda}r, \quad y(x) = R(r) = R\left(\frac{x}{\sqrt{\lambda}}\right) \tag{8}$$

we get the differential equation

$$\frac{1}{x}\frac{d}{dx}\left(x\frac{dy}{dr}\right) + \left(1 - \frac{n^2}{x^2}\right)y = 0 \tag{9}$$

with the conditions

$$y(\sqrt{\lambda}x_0) = 0 \tag{10}$$

and

$$|y(0)| < \infty \tag{11}$$

Therefore we obtain

$$y(x) = AJ_n(x) \tag{12}$$

Because of the boundary conditions $y(r_0\sqrt{\lambda}) = 0$, $r_0\sqrt{\lambda} = \mu$, the following

$$J_n(\mu) = 0 \tag{13}$$

is valid

The above transcendental equation possesses infinitely many real roots

$$\mu_1^{(n)}, \mu_2^{(n)}, \mu_3^{(n)}, \ldots, \mu_m^{(n)}, \ldots$$

in other words Eq. (1) has infinitely many eigenvalues given by

$$\lambda_m^{(n)} = \left(\frac{\mu_m^{(n)}}{r_0}\right)^2 \qquad (m = 1, 2, 3, \ldots) \tag{14}$$

which correspond to the eigenfunctions

$$R(r) = AJ_n\left(\frac{\mu_m^{(n)}}{r_0} r\right) \tag{15}$$

of the problem (1), (2).

It is directly seen from the type and manner in which eigenfunctions have been constructed that all nontrivial solutions of the boundary value problem considered are represented by formula (15). We want to find the first conjugate boundary of the catenoid (in the sense of the eigenvalue problem E). This means find the first boundary r_0 such that the first eigenvalue of the eigenvalue problem (E) for the catenoid in the disk $r \leq r_0$ is zero, i.e., $\lambda_1(r_0) = 0$. However, this is equivalent to finding the smallest r_0 in such a way that the first eigenvalue of the eigenvalue problem (16) is zero.

$$\begin{cases} -\nabla^2 f - 2f = \lambda f, \quad f = f(u,v) \text{ on } \overline{D^2_{r_0}} \\ f \equiv 0 \text{ on } \partial D^2_{r_0} \end{cases} \tag{16}$$

This is equivalent to finding the smallest r_0 in such a way that a solution of the problem

$$\begin{cases} -\nabla^2 f - 2f = 0, \quad f = f(u,v) \text{ on } \overline{D^2_{r_0}} \\ f \equiv 0 \text{ on } \partial D^2_{r_0} \end{cases} \tag{17}$$

is never vanishing inside the disk $r < r_0$. Therefore we take $y(x) = AJ_0(x)$ and by setting $m = 1$, $n = 0$

$$\lambda_1^{(0)} = \left(\frac{\mu_1^{(0)}}{r_0}\right)^2 = 2$$

Therefore

$$\mu_1^{(0)} = r_0\sqrt{2}$$

therefore

$$r_0 = \frac{\mu_1^{(0)}}{\sqrt{2}} \simeq \frac{2.4048256}{1.4142135} \simeq 1.7 \qquad \text{Q.E.D.}$$

Example 3—Enneper's Minimal Surface as a Complete Minimal Surface in R^3. A conformal parametric representation of a portion of Enneper's minimal surface in R^3 is a mapping

$$q : D^2_{r_0} \to R^3, \quad \text{for each } r_0 > 0$$

given by

$$q(u,v) = (x(u,v),\ y(u,v),\ z(u,v))$$

where

$$(u,v) \in D^2_{r_0} = \{(u,v) \in R^2 : u^2 + v^2 < r_0^2\}$$

and

$$\begin{aligned} x(u,v) &= u + uv^2 - \frac{1}{3}u^3 \\ y(u,v) &= -v - u^2 v + \frac{1}{3}v^3 \\ z(u,v) &= u^2 - v^2 \end{aligned} \tag{1}$$

This way we get a global expression of Enneper's minimal surface in R^3.

The boundary of the portion of Enneper's minimal surface described above is a mapping

$$\alpha : \partial D^2_{r_0} \to R^3$$

given in a parametric representation as follows:

$$\Gamma_{r_0} = \alpha(r_0, \Theta) = (x(r_0,\Theta),\ y(r_0,\Theta),\ z(r_0,\Theta))$$

where

$$\begin{aligned} x(r_0,\Theta) &= r_0 \cos \Theta - \frac{1}{3} r_0^3 \cos 3\Theta \\ y(r_0,\Theta) &= -r_0 \sin \Theta - \frac{1}{3} r_0^3 \sin 3\Theta \\ z(r_0,\Theta) &= r_0^2 \cos 2\Theta, \quad \Theta \in [0, 2\pi] \end{aligned}$$

It is easy to show that the portion of Enneper's minimal surface given by (1) in the disk $D^2_{r_0}$ can also be written (in an equivalent way) as

$$\begin{aligned} S_{r_0} = \{(x,y,z) : x &= r_0 u + r_0^3 uv^2 - \frac{1}{3} r_0^3 u^3 \\ y &= -r_0 v - r_0^3 u^2 v + \frac{1}{3} r_0^3 v^3 \\ z &= r_0^2 (u^2 - v^2) ; u^2 + v^2 \leq 1\} \end{aligned}$$

It can be shown that its mean curvature H is zero; for this reason it is a minimal surface. From the topological point of view Enneper's surface is simply connected. It is properly immersed in R^3, but this is not an embedding because as we will indicate later the surface has self-intersections. It can be easily remarked that Enneper's surface is given in isothermal coordinates. The above surface is algebraic, in fact a polynomial minimal surface of degree 3. Moreover, as a surface defined in the $w = u + iv$-plane has a pole of order 3 at $w = \infty$ and also exactly one zero (of order 1) at the point $w = 0$. It was mentioned at the time we worked on the catenoid that these two surfaces, the catenoid and Enneper's minimal surface, are the only surfaces whose normal map is one-to-one.

For $r_0 = 3/2$, a picture of Γ_{r_0} as well as its projections onto two coordinate planes, is given in the first of the following figures.

In Figs. 2-6 we give the geometry of the surface as $r_0 \uparrow \infty$.

It has been proved by Nitsche [12] that if $\bar{r} < r_0 < \sqrt{3}$, $\bar{r} = 1,681475$ the curve Γ_{r_0} bounds two distinct solutions of Plateau's problem, both having least area among all surfaces of disk bounded by Γ_{r_0}.

However, these two solutions of Plateau's problem are proper, i.e., isolated. This follows by an article of Tomi [24], who has proved that if the boundary curve Γ is a regular curve of class $C^{4+\alpha}$, $0 < \alpha < 1$, and assume that every solution of Plateau's

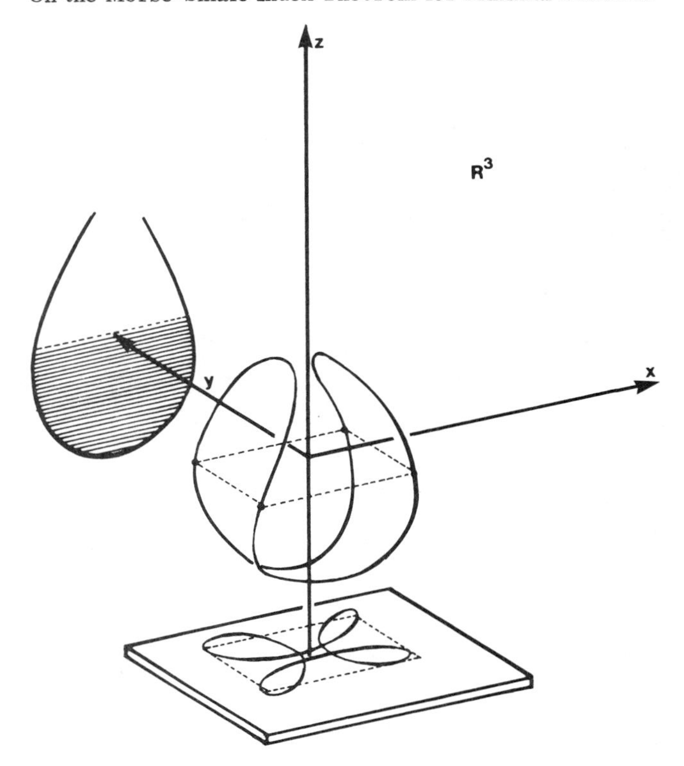

Fig. 2. Diagram of Γ_{r_0} as well as its projections onto two coordinate planes for $r_0 = 3/2$.

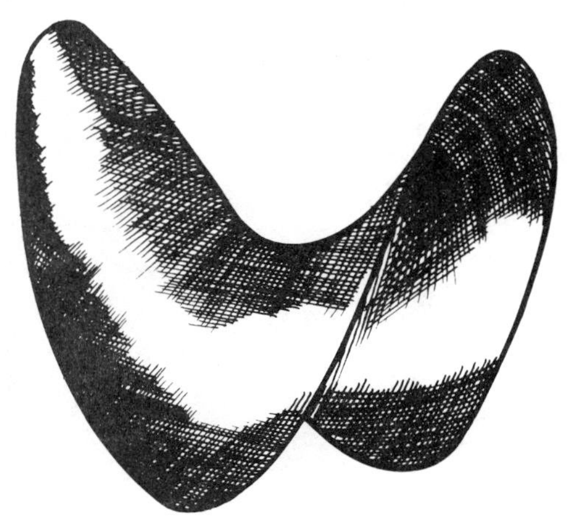

Fig. 3

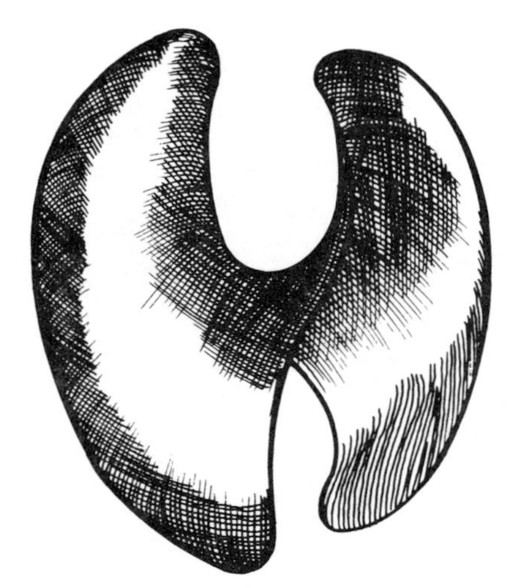

Fig. 4

Fig. 5

Fig. 6

problem is regular up to the boundary, i.e., regular at any point of the closed unit disk,

$$\overline{D_1^2} = \{(u,v) \in R^2 : u^2 + v^2 \leq 1\}$$

then the set of the solutions of Plateau's problem is finite.

Definition. A surface $f \in C^1(D_1^2; R^3)$ is called "regular at x" if the rank of $\nabla f(x)$ is two.

Definition. A Shiffman curve is a Jordan curve Γ given in a proper parametric representation $q(\theta) = (\alpha(\theta), \beta(\theta), \gamma(\theta))$, $q: \partial D_1^2 \to R^3$ such that

I) $q(\theta)$ is of bounded variation.
II) There is a δ such that $dq(\theta) \cdot dq(\phi) \geq 0$ (a product of vectors) for all θ, ϕ for which $|\theta - \phi| < \delta$.

In (II), the relation $dq(\theta) \cdot dq(\phi) \geq 0$ shall be shorthand for the statement that there is an ε such that

$$\{q(\theta + \Delta\theta) - q(\theta)\} \cdot \{q(\phi + \Delta\phi) - q(\phi)\} \geq 0$$

for all positive (or all negative) $\Delta\theta$, $\Delta\phi$ such that $|\Delta\theta| < \varepsilon$, $|\Delta\phi| < \varepsilon$.

Note. $|\theta - \phi| < \delta$ means the length of the shorter arc joining θ, ϕ.

Theorem 1. (1) The boundary curve Γ_{r_0} for the portion of Enneper's minimal surface S_{r_0} is a Jordan curve for $0 < r_0 < \sqrt{3}$. The multiple points of Γ_{r_0} appear for $r_0 > 0$ and $0 \leq \theta_1 < \theta_2 < 2\pi$ at the following places:

For $r_0 = \sqrt{3}$:

$$\theta_1 = 0, \quad \theta_2 = \pi; \quad \text{or} \quad \theta_1 = \frac{\pi}{2}, \quad \theta_2 = \frac{3\pi}{2}$$

For $r_0 > \sqrt{3}$:

$$r_0 = \frac{1}{\sqrt{\frac{1}{3}\cos^2\theta_1 - \sin^2\theta_1}}$$

$$0 < \theta_1 < \frac{\pi}{6} \quad \text{and} \quad \theta_2 = \pi - \theta_1$$

or

$$r_0 = \frac{1}{\sqrt{\frac{1}{3}\sin^2\theta_1 - \cos^2\theta_1}}$$

$$\frac{\pi}{3} < \theta_1 < \frac{\pi}{2} \quad \text{and} \quad \theta_2 = 2\pi - \theta_1$$

or

$$r_0 = \frac{1}{\sqrt{\frac{1}{3}\sin^2\theta_1 - \cos^2\theta_1}}$$

$$\frac{\pi}{2} < \theta_1 < \frac{2\pi}{3} \quad \text{and} \quad \theta_2 = 2\pi - \theta_1$$

or

$$r_0 = \frac{1}{\sqrt{\frac{1}{3}\cos^2\theta_1 - \sin^2\theta_1}}$$

$$\pi < \theta_1 < \frac{7\pi}{6} \quad \text{and} \quad \theta_2 = 3\pi - \theta_1$$

(2) In particular Γ_{r_0} is a Shiffman's curve for $0 < r_0 < \sqrt{3}$.

Proof. The proof of part (1) is a matter of straightforward computation. We are interested to indicate the proof of part (2). Condition (I) follows by the fact that Γ_{r_0} is of bounded variation and condition (II) is a consequence of the fact that Γ_{r_0} is C^∞ with continuous and nonzero tangent. Moreover, from part (1) we know that Γ_{r_0} is a Jordan curve for $0 < r_0 < \sqrt{3}$, therefore Γ_{r_0} is a Shiffman's curve for $0 < r_0 < \sqrt{3}$. Q.E.D.

Theorem 2. The Morse-Smale Index $I(q\,;D^2_{r_0})$ of Enneper's minimal surface $q: D^2_{r_0} \to R^3$, is bounded above by the sum of the dimensions of the eigenspaces corresponding to those eigenvalues λ of the Bessel equation which are $< 8r_0^2$, for all $r_0 \in R$.

Note. The Bessel equation is an ordinary linear differential equation given by

$$\frac{d^2y}{dp^2} + \frac{1}{p}\cdot\frac{dy}{dp} + \left(1 - \frac{n^2}{p^2}\right)y = 0 \qquad \text{for all } n = 0, 1, 2, \ldots$$

where

$$y = y(p), \qquad p \in [0, \infty)$$

and

$$p = rr_0\sqrt{\lambda + 8} \text{ for } 0 \leq r \leq 1 \text{ and } r_0 \text{ a given real number with } r_0 \in [0, \infty)$$

Proof. Following the same method as we did for the catenoid and applying comparison techniques which we have also indicated at the time we studied the catenoid we reduce the eigenvalue problem (E) for Enneper's minimal surface to

$$\begin{cases} \dfrac{\partial^2 f}{\partial u^2} + \dfrac{\partial^2 f}{\partial v^2} + 8f + \lambda f = 0 & \text{where } f = f(u,v) \text{ defined on } \overline{D^2_{r_0}} \\ f \equiv 0 & \text{on } \partial D^2_{r_0} \text{ (boundary condition)} \end{cases} \tag{1}$$

Set S_0 = the negative eigenspace of the problem

$$\begin{cases} -\dfrac{\partial^2 f}{\partial u^2} - \dfrac{\partial^2 f}{\partial v^2} - 8f = \lambda f, \quad f = f(u,v) \text{ on } \overline{D^2_{r_0}} \\ f \equiv 0 \text{ on } \partial D^2_{r_0} \end{cases}$$

Consider the problem of finding the eigenvalues Λ of (2),

$$\begin{cases} -\dfrac{\partial^2 f}{\partial u^2} - \dfrac{\partial^2 f}{\partial v^2} = \Lambda f, \quad f = f(u,v) \text{ on } \overline{D^2_{r_0}} \\ f \equiv 0 \text{ on } \partial D^2_{r_0} \end{cases} \tag{2}$$

We obtain

$\dim S_0$ = the number of those eigenvalues of the problem (2) which are < 8

Hence,

$\dim S_0 <$ the number of those eigenvalues for the corresponding Bessel equation which are less than $8r_0^2$, for all $r_0 \in R$. Q.E.D.

Corollary. The Morse-Smale Index $I(q; D^2_{r_0})$ of Enneper's minimal surface $q: D^2_{r_0} \to R^3$ varies as follows:

(1) $I(q; D^2_{r_0}) = 0$ for $r_0 = 0$

(2) $I(q; D^2_1) \leq 1$

(3) $I(q; D^2_{\sqrt{3}}) \leq 2$

(4) $I(q; D^2_2) \leq 4$

(5) $I(q; D^2_{r_0}) \to \infty$ as $r_0 \to \infty$

Proof. (1) If $r_0 = 0$, then $8r_0^2 = 0$, therefore $I(q, D^2_{r_0}) = 0$.

(2) If $r_0 = 1$, then $8r_0^2 = 8$.

However, to find an upper bound for $I(q; D^2_1)$ we ask to determine the eigenvalues of the Bessel equation on the unit disk which are less than or equal to 8. It is well known that the eigenvalues of the Bessel equation are the squares of its zeros.

By making use of the tables of the zeros of Bessel functions we obtain that $I(q; D^2_1) \leq 1$.

(3) If $r_0 = \sqrt{3}$, then $8r_0^2 = 24$.

From $j_{0,n}$ we obtain only one eigenvalue. From $j_{1,n}$ we obtain only one eigenvalue. From $j_{k,n}$ we do not obtain eigenvalues for $k \geq 2$.

Therefore $I(q; D^2_{\sqrt{3}}) \leq 2$.

(4) If $r_0 = 2$, then $8r_0^2 = 32$.

From $j_{0,n}$ we obtain only two eigenvalues. From $j_{1,n}$ we obtain only one eigenvalue. From $j_{2,n}$ we obtain only one eigenvalue. From $j_{3,n}$ we obtain zero eigenvalues. From $j_{k,n}$ we do not obtain eigenvalues for $k \geq 3$. Therefore $I(q; D^2_2) \leq 4$.

(5) $I(q; D^2_{r_0}) \to \infty$ for $r_0 \to \infty$.

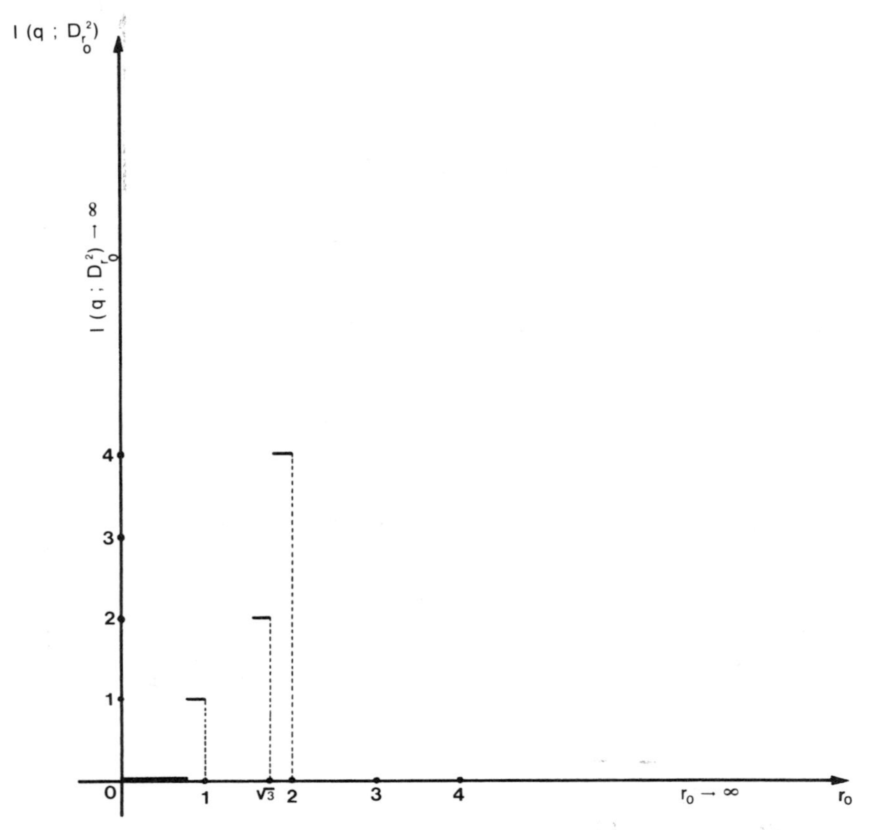

Fig. 7. Behavior of the Morse-Smale Index $I(q;D^2_{r_0})$ of Enneper's minimal surface $q : D^2_{r_0} \to R^3$ as a function of r_0 for all $r_0 \in R$.

Theorem 3. The portion of Enneper's minimal surface, $q : D^2_{r_0} \to R^3$ for each $r_0 > 0$, given by

$$q(u,v) = (x(u,v),\ y(u,v),\ z(u,v))$$

where $(u,v) \in D^2_{r_0}$ and

$$\begin{cases} x(u,v) = u + uv^2 - \dfrac{1}{3}u^3 \\ y(u,v) = -v - u^2v + \dfrac{1}{3}v^3 \\ z(u,v) = u^2 - v^2 \end{cases}$$

is an unstable extremal (i.e., a nonrelative minimum of the area functional) for all r_0 with $\bar{r} < r_0 < \sqrt{3}$ and $\bar{r} = 1.681475$.

Proof. Consider P to be the minimizing surface whose general representation is given by $r = r(u,v)$ where $(u,v) \in D$, D is the region under consideration. Assume $\lambda(u,v)$ to be a C^1 function in D which vanishes on the boundary of the region D and ε to be a parameter.

The area of the surface $r(u,v) + \varepsilon\lambda(u,v)\xi(u,v)$ where $\xi(u,v)$ is the unit normal vector of P, is a function of ε denoted by $A(\varepsilon)$. This function attains its minimum for $\varepsilon = 0$.

Therefore $A'(0) = 0$ and $A''(0) \geq 0$ where by A' and A'' we denote the first and second variation, respectively.

The condition $A'(0) = 0$ can be written as follows

$$A'(0) = \iint_D HW\lambda \, dudv = 0 \tag{1}$$

where H stands for the mean curvature of the surface P and W for the discriminant of the first fundamental form of the surface P.

Assume that in the $u + iv = w$-plane, the Jordan region D is bounded by an analytic Jordan curve. We designate by $\tau(w)$ a function which is analytic and different from zero in D. This function $\tau(w)$ characterizes the surface.

It is well known that the equations in (2)

$$\begin{aligned} x &= \mathrm{Re} \int_D (1 - w^2)\, \tau(w)\, dw \\ y &= \mathrm{Re} \int_D i(1 + w^2)\, \tau(w)\, dw \\ z &= \mathrm{Re} \int_D 2w\tau(w)\, dw \end{aligned} \tag{2}$$

determine a regular minimal surface. The condition $A''(0) \geq 0$ becomes

$$A''(0) = \iint_D \left\{ \lambda_u^2 + \lambda_v^2 - \frac{8\lambda^2}{(1 + u^2 + v^2)^2} \right\} dudv \geq 0 \tag{3}$$

By making use of (3) it is easy to verify that some surfaces have area which is not a minimum. This criterion depends only on the region D and not on the function $\tau(w)$.

H. Schwarz's theorem states that if the region D contains the unit disk $\overline{D_1^2}$ in its interior then $A''(0)$ can become negative, i.e., $A''(0) < 0$.

Now we come to our example of Enneper's surface $q: D^2_{r_0} \to R^3$ for the contour Γ_{r_0}.

By setting $\tau(w) = 1$, formula (2) is reduced to the following form:

$$\begin{aligned} x &= \mathrm{Re} \int_D (1 - w^2)\, dw \\ y &= \mathrm{Re} \int_D i(1 + w^2)\, dw \\ z &= \mathrm{Re} \int_D 2w\, dw \end{aligned} \tag{4}$$

where $w = re^{i\theta}$, $w^3 = r^3 e^{3i\theta}$. Therefore we obtain

$$\begin{aligned} x &= \mathrm{Re} \left(w - \frac{w^3}{3} \right) + c_1 \\ y &= \mathrm{Re} \left(i \left(w + \frac{w^3}{3} \right) \right) + c_2 \\ z &= \mathrm{Re} \left(2 \cdot \frac{w^2}{2} \right) + c_3 \end{aligned} \tag{5}$$

Thus

$$\begin{aligned} x &= r \cos \theta - \frac{1}{3} r^3 \cos 3\theta = u + uv^2 - \frac{1}{3} u^3 \\ y &= -r \sin \theta - \frac{1}{3} r^3 \sin 3\theta = -v - u^2 v + \frac{1}{3} v^3 \\ z &= r^2 \cos 2\theta = u^2 - v^2 \end{aligned} \tag{6}$$

where $u = r \cos \theta$, $v = r \sin \theta$, $0 \leq \theta \leq 2\pi$. Then the equations

$$\begin{aligned} x &= u + uv^2 - \frac{1}{3}u^3 \\ y &= -v - u^2 v + \frac{1}{3}v^3 \qquad (7) \\ z &= u^2 - v^2, \quad u^2 + v^2 \leq r^2, \quad 1 < r < \sqrt{3} \end{aligned}$$

give an example of a minimal surface in parametric representation which is bounded by a Jordan curve and having the property that its area is not a minimum. It is understood, however, that the area of the surface given by (7) is finite and according to J. Douglass's solution we can find a minimal surface which furnishes a minimum and which is different to the one given by (7). The reason for that is because the surface given by (7) does not minimize the area functional. We have proved that Γ_{r_0}, $\bar{r} < r_0 < \sqrt{3}$ and $\bar{r} = 1.681475$, being a Jordan curve in the sense of Shiffman, bounds two distinct proper solutions of Plateau's problem both having least area among all surfaces of disk-bounded by Γ_{r_0} and therefore Γ_{r_0} bounds at least one minimal surface which is not a proper relative minimum (as well as not a relative minimum) and Enneper's minimal surface for $\bar{r} < r_0 < \sqrt{3}$ and $\bar{r} = 1.681475$ is one such minimal surface. This minimal surface is given by (8)

$$\begin{aligned} x &= u + uv^2 - \frac{1}{3}u^3 \\ y &= -v - u^2 v + \frac{1}{3}v^3 \qquad (8) \\ z &= u^2 - v^2; \; u^2 + v^2 \leq r_0^2 \text{ and } \bar{r} < r_0 < \sqrt{3}, \; \bar{r} = 1.681475 \end{aligned}$$

Q.E.D.

3. RESEARCH PROBLEMS

In this section we list a number of unsolved problems in the fields of Differential Geometry and Calculus of Variations. The name associated with a problem is almost always the person who suggested it. Most of the problems are well known and some of these have attracted the attention of several research mathematicians.

Problem 1. Is it true that for any C^1 Jordan curve Γ in the Euclidean space R^3 the number of minimal surfaces bounded by Γ is odd?

Remarks. I have given several examples where this number is odd, and I believe that an affirmative answer should be given for the general case.

This will justify a problem by T. Rado on the characterization of the number of minimal surfaces which span a given contour in R^3.

Problem 2 (E. Calabi). Does there exist a complete minimal surface of R^3 which is a subset of the unit ball?

Problem 3 (M. O. Reade). If S is a minimal surface of the type of the disk, and if the boundary of S is a simple closed analytic curve that is symmetric with respect to the xy-plane, does it follow that S is symmetric with respect to the xy-plane?

Problem 4. Does there exist a harmonic mapping $q: D \to R^3$, such that $\partial q/\partial u \cdot \partial q/\partial u = \partial q/\partial v \cdot \partial q/\partial v$, $\partial q/\partial u \cdot \partial q/\partial v = 0$ and $q|_{\partial D}: \partial D \to \Gamma$ is not a homeomorphism but it is homotopic to a homeomorphism?

Here Γ stands for a C^1 Jordan curve in R^3 and D for the open unit disk in R^2, i.e., $D = \{(u, v) \in R^2 : u^2 + v^2 < 1\}$.

Problem 5. Does every smooth, regular Jordan curve in R^n, $n \geq 3$, bound only a finite number of stable minimal surfaces?

Problem 6 (M. Shiffman). Characterize the Morse type of a minimal surface by properties similar to the number of conjugate points (or the number of negative characteristic roots in the associated problem) in the case of single integral problem in the calculus of variations.

Problem 7 (M. Shiffman). Prove that the k-th type number of a minimal surface bounded by a Jordan curve Γ is either zero for all k, or zero for all $k \neq j$ and 1 for $k = j$.

Problem 8 (Th. M. Rassias). Give an example of a Jordan curve Γ in R^3 spanning five different minimal surfaces of the type of the disk, all of which are explicitly known. Is there any characterization of this phenomenon in the sense of Morse theory on Hilbert (or Banach) manifolds as developed by R. Palais and S. Smale?

Problem 9 (S. T. Yau). Does there exist in R^n a complete hypersurface with Ricci curvature less than -1?

Problem 10 (Th. M. Rassias). Consider the eigenvalue problem $r^2R'' + rR' + \lambda((r^4 + 1)^2/r^2)R = (n^2 - 32)R$ where $R = R(r)$, $r \in [0, r_0]$, r_0 a fixed real number, and $n = 1, 2, 3, \ldots$, with boundary condition $R(r_0) = 0$ and $R \in C^2_p[0, r_0]$ where C^2_p stands for piecewise C^2 functions. Is anything known about the sum (or an upper bound of the sum) of the dimensions of the eigenspaces corresponding to the negative eigenvalues of this eigenvalue problem, as a function of r_0?

Problem 11 (J. Simons). Let $f: R^{n-1} \to R^n$ be an immersion as a complete minimal variety. Then either the image is a hyperplane, or, for sufficiently large r, the sphere of radius r in R^{n-1} is mapped in as a conjugate boundary. (Conjecture)

Problem 12. Does there exist a closed minimal surface in S^n with negative curvature?

Problem 13. Examine if the first eigenvalue for the Laplace-Beltrami operator on an embedded minimal surface of S^3 is 2?

Problem 14. Let $f: R^n \to R$ be a convex function, $n \geq 6$, satisfying the condition

$$\mathrm{Det}\left(\frac{\partial^2 f}{\partial x_i \partial x_j}\right) \equiv 1$$

where $i, j = 1, 2, \ldots, n$ and $(x_1, x_2, \ldots, x_n) \in R^n$. Is, then, f a quadratic polynomial?

Remark. This is true for $n = 2$ (E. Heinz and J. C. C. Nitsche) and for $n \leq 5$ (E. Calabi).

Problem 15 (Spherical Bernstein problem). Let the (n - 1)-sphere be imbedded as a minimal hypersurface in $S^n(1)$. Is it necessarily an equator?

Remark. Wu-Yi has constructed infinitely many distinct new examples of minimal imbeddings of S^{n-1} into $S^n(1)$ for the case $n = 4, 5, 6$.

Problem 16. Let M and N be two manifolds each with negative curvature. If $\pi_1(M) = \pi_1(N)$, is then M diffeomorphic to N?

Problem 17 (The Hopf Conjecture). Does $S^2 \times S^2$ admit a metric with positive sectional curvature?

Problem 18. Let M be an n-dimensional manifold having nonnegative sectional curvature. Is the k-th Betti number of M at most equal to the k-th Betti number of the n-torus T^n?

Problem 19 (J. Cheeger-D. Gromoll). If M is a complete, noncompact manifold with nonnegative sectional curvature K; and if $K(x) > 0$ for some point $x \in M$, then M is diffeomorphic to R^n. (Conjecture)

Problem 20 (H. Yamabe). Prove that any metric on a compact manifold can be conformally deformed to a metric of constant scalar curvature.

Problem 21. Consider $H: M^4 \to R$ to be a C^2 Hamiltonian function and p a strict local minimum of H. Are there two families of periodic orbits emanating from p?

Remark. If p is a nondegenerate minimum the answer is yes (A. Weinstein, 1973).

Problem 22 (Th. M. Rassias). Do there exist necessary and sufficient conditions on a complete Riemannian C^∞-manifold M modelled on a separable Hilbert space H so that two points $p,\ q \in M$ can be joined by a geodesic whose length is exactly the Riemannian distance $\delta(p, q)$?

Problem 23 (Carathédory Conjecture). Every closed convex surface in R^3 has at least two umbilical points.

Problem 24. Does every continuous area-preserving map of S^2 have at least two fixed points?

Problem 25. Can one approximate any C^r, $r \geq 1$, volume-preserving diffeomorphism in the C^r topology by a C^{r+1} volume-preserving diffeomorphism?

Problem 26 (Th. M. Rassias). Does there exist a harmonic homeomorphism of the open unit ball B in R^3 onto R^3? I.e., do there exist harmonic functions f_1, f_2, f_3 defined in $B = \{z = (z_1, z_2, z_3) : |z| < 1\}$, such that the mapping $z \to (f\ , f\ , f_3)$ is a homeomorphism of B onto all of R^3?

Problem 27 (L. Bers). Consider f to be a harmonic function (solution of the ordinary Laplace equation) defined in the unit ball of Euclidean n-space $(n > 2)$. Assume that f is infinitely differentiable in the closed unit ball. Assume also that f and its first partial derivatives vanish on a set on the surface of the ball of positive two-dimensional Lebesgue measure. Examine if f must vanish identically.

Remark. For $n = 2$ the theorem is due to F. and M. Riesz.

Problem 28 (Th. M. Rassias). Consider $p(x_1, x_2, \dots, x_n)$ to be a polynomial of degree k, where $x_1, x_2, \dots, x_n \in R$. Let $Z_p = \{(\bar{x}_1, \bar{x}_2, \dots, \bar{x}_n) \in R^n : p(\bar{x}_1, \bar{x}_2, \dots, \bar{x}_n) = 0\}$. Find the number of connected components in R^n of the set $R^n - Z_p$.

Remark. I have solved the problem for $n = 2, 3$.

Problem 29 (P. Révész). Let $f(x, y)$ be a continuous function defined on the unit square $[0, 1] \times [0, 1] \subset R^2$ so that

$$f(x,y) \geq 0, \quad \int_0^1 f(x,y)\,dx = \int_0^1 f(x,y)\,dy = 1$$

Does there exist a measurable and measure preserving transformation T defined on [0, 1], mapping this interval into itself and such that $f(x, Tx) > 0$ for almost every x?

Acknowledgment. It is my pleasure to express my thanks to the organizing committee of the Colloquium on Global Differential Geometry and Global Analysis, Technische Universität Berlin, Berlin, June 10-14, 1984, for inviting me to deliver this lecture.

REFERENCES

1. Atiyah, M. F.: Eigenvalues and Riemannian Geometry. Manifolds, Tokyo, 5-9 (1973).
2. Atiyah, M. F.; Singer, I. M.: Ann. Math. (1968).
3. Bolza, O.: Vorlesungen über Variationsrechnung. Teubner, Leipzig and Berlin (1909).
4. Carathéodory, C.: Variationsrechnung und Partielle Differentialgleichungen Erster Ordnung. Teubner, Leipzig and Berlin (1935).
5. Choquet-Bruhat, Y.: Problems and Solutions in Mathematical Physics. Holden-Day (1967).
6. Courant, R.: Dirichlet's Principle, Conformal Mapping and Minimal Surfaces. Interscience (1950).
7. Courant, R.; Hilbert, D.: Methods of Mathematical Physics. Interscience (1965).
8. Durand, L.: Am. J. Phys. 49, 334-343 (1981).
9. Kreyszig, E.: Introduction to Differential Geometry and Riemannian Geometry. University of Toronto Press (1967).
10. Morrey, C. B.: Multiple Integrals in the Calculus of Variations. Springer (1966).
11. Morse, M.; Tompkins, C.: Ann. Math. 40, 443-472 (1939).
12. Nitsche, J. C. C.: Arch. Rat. Mech. Anal. 30, 1-11 (1968).
13. Osserman, R.: A Survey of Minimal Surfaces. Van Nostrand (1969).
14. Rassias, Th. M.: C. R. Acad. Sci. Paris 284, 497-499 (1977).
15. Rassias, Th. M.: Tamkang J. Math., Taiwan, 9(2), 251-257 (1978).
16. Rassias, Th. M.: On certain properties of eigenvalues and the Poincaré inequality, in Th. M. Rassias, ed., Global Analysis—Analysis on Manifolds. Teubner (1983), pp. 282-300.
17. Rassias, Th. M.: Eleftheria 2, 233-300 (1980).
18. Rassias, Th. M.: Morse theory and Plateau's problem, in Th. M. Rassias and G. M. Rassias, eds., Selected Studies: Physics-Astrophysics, Mathematics, History of Science, a Volume Dedicated to the Memory of Albert Einstein. North-Holland (1982), pp. 261-292.
19. Shiffman, M.: Ann. Math. 40, 834-854 (1939).
20. Simons, J.: Ann. Math. 88, 62-105 (1968).
21. Smale, S.: Ann. Math. 2(80), 382-396 (1964).
22. Smale, S.: Bull. Am. Math. Soc. 83, 683-693 (1977).
23. Smale, S.: J. Math. Mech. 14, 1049-1056 (1965).
24. Tomi, F.: Arch. Rat. Mech. Anal. 52, 312-318 (1973).
25. Watson, G. N.: A Treatise on the Theory of Bessel Functions, 2nd ed. Cambridge University Press (1944).
26. Weyl, H.: J. R. Angew. Math. 141, 1-11 (1912).

29 A CARTAN FORM FOR THE FIELD THEORY OF CARATHÉODORY IN THE CALCULUS OF VARIATIONS OF MULTIPLE INTEGRALS

Hanno Rund / Program in Applied Mathematics, University of Arizona, Tucson, Arizona

1. INTRODUCTION

In the calculus of variations of single integrals a role of fundamental significance is played by a certain 1-form which in recent years has often been referred to as the Cartan form. The importance of this form is due inter alia to the fact that the analysis of the conditions which ensure that it be exact leads, on the one hand, to the Euler-Lagrange equations in canonical variables, and, on the other, to the notion of Lagrange and Poisson brackets. Moreover, the integrand of the well known first relative integral invariant may be identified with the Cartan form.

When one seeks to extend this concept to the calculus of variations of multiple integrals, one is confronted with several problems. First, in view of the existence of a hierarchy of distinct field theories, which includes the theories of Carathéodory [1] and Weyl [9], it is conceivable that several different generalizations of the Cartan form may be appropriate, namely one for each field theory. Second, the definition of the Cartan form for single integral variational problems is best formulated in terms of canonical variables $(q^1, \ldots, q^N; p_1, \ldots, p_N)$ since this gives rise to the homogeneous representation $p_1 dq^1 + \cdots + p_N dq^N + p_t dt$, in which the component p_t of the canonical momentum is to be identified with the negative of the Hamiltonian function: however, this particular aspect of the classical canonical formalism has no immediate counterpart for the case of multiple integrals.

It is the object of this note to touch upon these problems to some limited extent, the main emphasis being on the field theory of Carathéodory. Accordingly this introduction is devoted to a brief sketch of this theory and of an associated canonical formalism. The concept of an independent Hilbert integral is used in the second section to motivate the introduction of an n-form Π as a Cartan form, where n denotes the number of independent variables. In the third section it is shown that Π is nothing other than the exterior product of n canonical momentum 1-forms. This in turn allows one to infer that the necessary and sufficient conditions that ensure the existence of a complete figure (in the sense of Carathéodory) are characterized by the hypothesis that Π be closed. In the concluding section the relationship between the n-form Π and (tentative) Cartan forms for field theories other than that of Carathéodory are briefly discussed.

The independent and dependent variables are denoted by x^j and ψ^A respectively; they are regarded as local coordinates of n- and M-dimensional manifolds X_n and X_M. (Here, and in the sequel, lower case Latin indices j, k, ℓ, ... range from 1 to n, while capital indices A, B, C, ... range from 1 to M; the summation convention applies to both types.) A set of M equations of the form $\psi^A = \psi^A(x^j)$ defines an n-dimensional subspace C_n of the product manifold $X_{n+M} = X_n \times X_M$, the tangent planes of C_n being determined by the derivatives $\psi^A_j = \partial\psi^A/\partial x^j$. An n-fold integral variational problem is specified when one is given a Lagrangian L, that is, a differentiable function of n + M + nM variables (x^j, ψ^A, ψ^A_j), and a region R of X_n, it being assumed that R is bounded by a smooth, closed hypersurface ∂R. The substitution in L of the values of ψ^A_j and ψ^A on C_n gives rise to a functional, whose value will in general depend on the choice of C_n. This functional, then, is the action integral

$$I[C_n] = \int_R L(x^i, \psi^A, \psi^A_j)\, d(x) \tag{1.1}$$

where we have used the notation

$$d(x) = dx^1 \wedge \cdots \wedge dx^n \tag{1.2}$$

By the means of derivatives

$$\Lambda^j_A = \frac{\partial L}{\partial \psi^A_j} \tag{1.3}$$

the following entities are constructed:

$$\Lambda^j_h = L\delta^j_h - \Lambda^j_A \psi^A_h, \qquad \Lambda^A_B = L\delta^A_B - \Lambda^h_B \psi^A_h \tag{1.4}$$

These give rise to two sets of vector fields on X_{n+M}:

$$\Lambda^j = (\Lambda^j_h, \Lambda^j_A), \qquad \Lambda_A = (-\Lambda^h_A, \Lambda^B_A) \tag{1.5}$$

By virtue of their construction these fields are linearly independent [6].

Instead of restricting ourselves to a single subspace such as C_n, we shall consider an M-parameter family of n-dimensional subspaces $C_n(w^B)$, whose parametric representation is of the form

$$\psi^A = \psi^A(x^h, w^B) \tag{1.6}$$

in which w^B denotes the M parameters. It will be supposed that

$$\det\left(\frac{\partial \psi^A}{\partial w^B}\right) \neq 0 \tag{1.7}$$

so that the family $C_n(w^B)$ covers a region G of X_{n+M} simply; it will be tacitly assumed that our considerations are restricted to G. In the tangent space $X_P(X_{n+M})$ of X_{n+M} at any point $P \in G$ the tangent plane $t_n(P)$ of the subspace $C_n(w^B)$ that passes through P is spanned by the n vectors

$$B_j = (\delta^h_j, \psi^A_j(x^\ell, w^B)) \tag{1.8}$$

where ψ^A_j denotes the derivatives of the functions (1.6) with respect to x^j.

A vector $X \in T_P(X_{n+M})$ is said to be transversal to $t_n(P)$ if its components X^h, X^B satisfy the conditions

$$\Lambda^j_h X^h + \Lambda^j_B X^B = 0 \tag{1.9}$$

(This is the most natural generalization of the notion of transversality as it occurs in the calculus of variations of single integrals. See [3], p. 13.) Since the construction (1.4) implies that

$$\Lambda^j_h \Lambda^h_A - \Lambda^j_B \Lambda^B_A = 0 \tag{1.10}$$

identically, it is evident that the M fields Λ_A as defined in (1.5) are transversal to $t_n(P)$: these fields span an M-dimensional subspace $T_M(P)$ in each tangent space $T_P(X_{n+M})$. The totality of these subspaces constitutes an M-dimensional distribution D_M on X_{n+M}, the so-called <u>transversal distribution</u>. It may be shown that D_M is integrable if and only if the conditions

$$\Pi^{j}_{A\,B} = 0 \tag{1.11}$$

are satisfied, where

$$\Pi^{j}_{A\,B} = \Lambda^{j}_{h}(\Lambda_{A}\Lambda^{h}_{B} - \Lambda_{B}\Lambda^{h}_{A}) - \Lambda^{j}_{C}(\Lambda_{A}\Lambda^{C}_{B} - \Lambda_{B}\Lambda^{C}_{A}) \tag{1.12}$$

in which Λ_A, Λ_B are to be interpreted in accordance with (1.5) as

$$\Lambda_{A} = -\Lambda^{h}_{A}\frac{\partial}{\partial x^{h}} + \Lambda^{B}_{A}\frac{\partial}{\partial \psi^{B}} \tag{1.13}$$

The conditions (1.11) are also necessary and sufficient for the existence of n independent functions S^j on X_{n+M} that satisfy the partial differential equations

$$\Lambda^{h}_{A}\frac{\partial S^{j}}{\partial x^{h}} - \Lambda^{B}_{A}\frac{\partial S^{j}}{\partial \psi^{B}} = 0 \tag{1.14}$$

and, under these circumstances, the M-dimensional integral surfaces of the distribution D_M admit the following representation in terms of n parameters Σ^j:

$$S^{j}(x^{h}, \psi^{B}) = \Sigma^{j} \tag{1.15}$$

These functions give rise to the entities

$$c^{j}_{h} = S^{j}_{h} + S^{j}_{A}\psi^{A}_{h} \tag{1.16}$$

in which

$$S^{j}_{h} = \frac{\partial S^{j}}{\partial x^{h}}, \qquad S^{j}_{A} = \frac{\partial S^{j}}{\partial \psi^{A}} \tag{1.17}$$

and ψ^A_h refers to the family (1.6). A central role is played by the determinant

$$\Delta = \det(c^{j}_{h}) \tag{1.18}$$

It will be assumed that $\Delta \neq 0$ on the region G; also, the cofactor of the entry c^j_h in Δ will be denoted by C^j_h. In the terminology of Carathéodory [1] the family (1.15) is a <u>geodesic field</u> if the following conditions are satisfied:

$$\Lambda^{j}_{A}(x^{h}, \psi^{B}, \psi^{B}_{h}) = C^{j}_{\ell}S^{\ell}_{A} \tag{1.19}$$

and

$$L(x^{h}, \psi^{B}, \psi^{B}_{h}) = \Delta \tag{1.20}$$

These relations imply that the subspaces (1.6) are extremals (that is, they represent solutions of the Euler-Lagrange equations); moreover, these subspaces afford extreme values to the action integral (1.1) if, in addition, an appropriate Weierstrass condition is satisfied. The configuration consisting of a geodesic field and the associated family of extremal subspaces is said to constitute a <u>complete figure</u> of the variational problem at hand. In the course of this construction it is generally assumed that the Lagrangian L is such that

$$\det(L^{jh}_{AB}) \neq 0 \tag{1.21}$$

where L^{jh}_{AB} denotes the <u>Legendre form</u>

$$L^{jh}_{AB} = \frac{\partial^2 L}{\partial \psi^A_j \partial \psi^B_h} - L^{-1}\left(\frac{\partial L}{\partial \psi^A_j}\frac{\partial L}{\partial \psi^B_h} - \frac{\partial L}{\partial \psi^A_h}\frac{\partial L}{\partial \psi^B_j}\right) \quad (1.22)$$

The analysis on which the entire theory is based may be simplified considerably by means of a canonical formalism ([4,5]). The introduction of this formalism is quite independent of the validity or otherwise of the integrability condition (1.11); however, it must be assumed that (1.21) is satisfied. The n canonical momentum 1-forms

$$\pi^j = \pi^j_h dx^h + \pi^j_A d\psi^A \quad (1.23)$$

are, by definition, required to span the orthogonal complements $T_n(P)$ of the planes $T_M(P)$ of the transversal distribution, that is, $\pi^j(X) = 0$ for any vector field X that is tangential to D_M. This is tantamount to the relations

$$\pi^j_h = L^{-1} p^j_\ell \Lambda^\ell_h, \qquad \pi^j_A = L^{-1} p^j_\ell \Lambda^\ell_A \quad (1.24)$$

where

$$p^j_h = \pi^j_h + \pi^j_A \psi^A_h \quad (1.25)$$

it being assumed also that $D \neq 0$, where

$$D = \det(p^j_h) \quad (1.26)$$

In addition to (1.24) it is required that the components of the 1-forms (1.23) be normalized such that

$$D = L \quad (1.27)$$

By virtue of (1.21) the second member of (1.24) can be solved for the nm variables ψ^A_j as functions of the canonical momenta:

$$\psi^A_j = \chi^A_j(x^h, \psi^B, \pi^h_k, \pi^h_B) \quad (1.28)$$

This is used to construct the Hamiltonian function

$$H(x^h, \psi^B, \pi^h_k, \pi^h_B) = -L(x^h, \psi^B, \chi^B_h) + \det(\pi^j_h + \pi^j_A \chi^A_h) \quad (1.29)$$

For our present purposes it is important to observe that the conditions (1.24) and (1.27) do not specify the canonical momenta uniquely (unless $n = 1$); the latter may be subjected to a unimodular transformation

$$\pi^j \to \pi'^j = a^j_h \pi^h, \qquad \det(a^j_h) = 1 \quad (1.30)$$

without affecting the above formalism. In particular, whenever the integrability conditions (1.11) are satisfied, so that n independent solutions of (1.14) exist, the canonical momenta can thus be adjusted in order to admit the identifications

$$\pi^j_h = S^j_h, \qquad \pi^j_A = S^j_A \quad (1.31)$$

These relations serve to characterize the complete figure whenever the condition (1.20) is also satisfied. In the present context the latter requirement assumes the form of a Hamilton-Jacobi equation, namely

$$H\left(x^h, \psi^B, \frac{\partial S^h}{\partial x^k}, \frac{\partial S^h}{\partial \psi^B}\right) = 0 \quad (1.32)$$

as is immediately evident from (1.20), (1.18), (1.16), (1.31), and (1.29).

2. THE INDEPENDENT HILBERT INTEGRAL AND THE CARTAN FORM

For the purposes of the present section it is assumed that we have at our disposal a complete figure. Under these circumstances it is a simple matter to construct an integral that is independent in a sense that is directly analogous to the independence of the Hilbert integral of the calculus of variations of single integrals.

Accordingly we shall consider a differentiable subspace $\tilde{C}_n$ of X_{n+M}, described by the equations $\psi^A = \tilde{\psi}^A(x^j)$, for which we shall write $\tilde{\psi}^A_j = \partial\tilde{\psi}^A/\partial x^j$; as in (1.16) and (1.18) we shall put

$$\tilde{c}^j_h = S^j_h + S^j_A \tilde{\psi}^A_h \tag{2.1}$$

and

$$\tilde{\Delta} = \det(\tilde{c}^j_h) \tag{2.2}$$

it being supposed that $\tilde{\Delta} \neq 0$. The substitution of $\tilde{\psi}^A$ as arguments of the function S^j that specify the geodesic field gives rise to a map from X_n to the configuration space Σ_n of the parameters Σ^j:

$$S^j(x^h, \tilde{\psi}^A(x^h)) = \Sigma^j \tag{2.3}$$

the Jacobian of this map being

$$\det\left(\frac{\partial \Sigma^j}{\partial x^h}\right) = \det\left(\frac{\partial S^i}{\partial x^h} + \frac{\partial S^j}{\partial \psi^A}\tilde{\psi}^A_h\right) = \tilde{\Delta} \tag{2.4}$$

Thus, if we denote by $\tilde{R}$ the image in Σ_n of the region R of X_n under the map (2.3), we have

$$\int_R \tilde{\Delta} d(x) = \int_R \det\left(\frac{\partial \Sigma^j}{\partial x^h}\right) d(x) = \int_{\tilde{R}} d\Sigma^1 \wedge \cdots \wedge d\Sigma^n \tag{2.5}$$

which can be interpreted as the n-dimensional volume of $\tilde{R}$, whose value is determined uniquely by the boundary $\partial\tilde{R}$ of $\tilde{R}$. But $\partial\tilde{R}$ depends on the values of the functions $\tilde{\psi}^A$ solely on boundary ∂R of R: the integral (2.5) is therefore independent in the sense that it assumes the same value for all subspaces $\tilde{C}_n$ that coincide on ∂R. It is this property that is characteristic of the concept of an independent Hilbert integral; it should be remarked, however, that the common value of these integrals depends on the choice of the geodesic field at hand.

The integrand of (2.5) may be expressed in a manner that does not involve the functions S^j explicitly. Because of (1.20) the relation (1.19) is equivalent to

$$S^j_A = L^{-1} c^j_\ell \Lambda^\ell_A \tag{2.6}$$

and thus we have, by (1.16) and (2.1),

$$\tilde{c}^j_h = c^j_h + S^j_A(\tilde{\psi}^A_h - \psi^A_h) = c^j_h + L^{-1} c^j_\ell \Lambda^\ell_A(\tilde{\psi}^A_h - \psi^A_h) = L^{-1} c^j_\ell [L\delta^\ell_h + \Lambda^\ell_A(\tilde{\psi}^A_h - \psi^A_h)]$$

With the aid of (1.18) and (1.20) it therefore follows that

$$\tilde{\Delta} = L^{1-n} \det[L\delta^j_h + \Lambda^j_A(\tilde{\psi}^A_h - \psi^A_h)] \tag{2.7}$$

in which there is no reference to the functions S^j.

This result suggests the introduction of the n-form

$$\tilde{\Pi} = L^{1-n} \det [L\delta^j_h + \Lambda^j_A \phi^A_h]\, d(x) \tag{2.8}$$

in which we have put

$$\phi^A_h = \tilde{\psi}^A_h - \psi^A_h \tag{2.9}$$

By means of an expansion formula for a characteristic determinant ([2], p. 116), one can express (2.8) as

$$\tilde{\Pi} = L^{1-n}[L^n + \sum_{p=1}^{n} \frac{1}{p!} L^{n-p} \delta^{h_1 \cdots h_p}_{j_1 \cdots j_p} \Lambda^{j_1}_{A_1} \cdots \Lambda^{j_p}_{A_p} \phi^{A_1}_{h_1} \cdots \phi^{A_p}_{h_p}]\, d(x) \tag{2.10}$$

A more elegant representation of this n-form can be given in terms of the so-called dual bases of (n - p)-forms on X_n, these bases being defined by

$$(n - p)!\, \theta_{j_1 \cdots j_p} = \varepsilon_{j_1 \cdots j_p j_{p+1} \cdots j_n} dx^{j_{p+1}} \wedge \cdots \wedge dx^{j_n} \tag{2.11}$$

For, if one puts p = 0 in the formula (A_1) of the Appendix, one obtains the identities

$$\delta^{h_1 \cdots h_p}_{j_1 \cdots j_p} d(x) = dx^{h_1} \wedge \cdots \wedge dx^{h_p} \wedge \theta_{j_1 \cdots j_p} \tag{2.12}$$

and consequently (2.10) is equivalent to

$$\tilde{\Pi} = L^{1-n}\left[L^n d(x) + \sum_{p=1}^{n} \frac{1}{p!} L^{n-p} \Lambda^{j_1}_{A_1} \cdots \Lambda^{j_p}_{A_p} \phi^{A_1}_{h_1} \cdots \phi^{A_p}_{h_p} dx^{h_1} \wedge \cdots \wedge dx^{h_p} \wedge \theta_{j_1 \cdots j_p}\right] \tag{2.13}$$

This suggests the introduction of the contact forms associated with the subspaces $C_n(w^B)$:

$$C^A = d\psi^A - \psi^A_j dx^j \tag{2.14}$$

whose restriction to the subspace $\tilde{C}_n$ we shall denote by $\tilde{C}^A$, so that, by (2.9)

$$\phi^A_h dx^h = d\tilde{\psi}^A - \psi^A_h dx^h = \tilde{C}^A \tag{2.15}$$

Accordingly (2.13) can be written as

$$\tilde{\Pi} = L\, d(x) + \sum_{p=1}^{n} \frac{1}{p!} L^{1-p} \Lambda^{j_1}_{A_1} \cdots \Lambda^{j_p}_{A_p} \tilde{C}^{A_1} \wedge \cdots \wedge \tilde{C}^{A_p} \wedge \theta_{j_1 \cdots j_p} \tag{2.16}$$

By virtue of (2.7) and (2.8) <u>it is this n-form that appears in the representation</u>

$$\int_R \tilde{\Delta}\, d(x) = \int_R \tilde{\Pi} \tag{2.17}$$

<u>of the independent Hilbert integral</u> (2.5).

This conclusion provides the motivation for the following construction: irrespective of whether the conditions that ensure the existence of a complete figure are satisfied, we shall associate with any Lagrangian L of an n-fold integral variational problem an n-form defined by

$$\Pi = L\, d(x) + \sum_{p=1}^{n} \frac{1}{p!} L^{1-p} \Lambda^{j_1}_{A_1} \cdots \Lambda^{j_p}_{A_p} C^{A_1} \wedge \cdots \wedge C^{A_p} \wedge \theta_{j_1 \cdots j_p} \tag{2.18}$$

where Λ^j_A, C^A, $\theta_{j_1 \cdots j_p}$ are defined by (1.3), (2.14), and (2.11), respectively. We shall refer to this as the Cartan form associated with the field theory of Carathéodory. We observe that (2.18) is not necessarily identical with the n-forms that have also been called Cartan forms (see, for instance, Shadwick [8], and also Sec. 4 below).

3. THE CARTAN FORM IN CANONICAL VARIABLES

It will now be shown that the n-form (2.18) admits an extremely simple representation in terms of the canonical momenta introduced at the end of Sec. 1. The derivation of this result is based on first principles; and the conditions which ensure the existence of a complete figure need not be assumed for this purpose. However, it is supposed that the determinant condition (1.21) is satisfied, since this requirement is indispensable for the construction of our canonical formalism.

We begin by expressing the canonical momentum 1-forms (1.23) in a slightly different manner with the aid of (1.24), (1.4), and (2.14):

$$\pi^j = L^{-1} p^j_\ell [\Lambda^\ell_h dx^h + \Lambda^\ell_A d\psi^A]$$

$$= L^{-1} p^j_\ell [L dx^\ell + \Lambda^\ell_A (d\psi^A - \psi^A_h dx^h)]$$

$$= L^{-1} p^j_\ell [L dx^\ell + \Lambda^\ell_A C^A] \tag{3.1}$$

This suggests that we introduce the 1-forms

$$\omega^j = L dx^j + \Lambda^j_A C^A \tag{3.2}$$

in terms of which (3.1) can be written as

$$\pi^j = L^{-1} p^j_\ell \omega^\ell$$

The 1-forms defined in (3.2) are related directly to the n-forms (2.18). In order to see this, we consider the product

$$\omega^1 \wedge \cdots \wedge \omega^n = \frac{1}{n!} \varepsilon_{j_1 \cdots j_n} \omega^{j_1} \wedge \cdots \wedge \omega^{j_n}$$

$$= \frac{1}{n!} \sum_{p=0}^{n} \binom{n}{p} L^{n-p} \varepsilon_{j_1 \cdots j_p j_{p+1} \cdots j_n} \Lambda^{j_1}_{A_1} \cdots \Lambda^{j_p}_{A_p} C^{A_1} \wedge \cdots \wedge C^{A_p} \wedge dx^{j_{p+1}} \wedge \cdots \wedge dx^{j_n}$$

In terms of the (n - p)-forms (2.11), this is

$$\omega^1 \wedge \cdots \wedge \omega^n = \sum_{p=0}^{n} \frac{1}{p!} L^{n-p} \Lambda^{j_1}_{A_1} \cdots \Lambda^{j_p}_{A_p} C^{A_1} \wedge \cdots \wedge C^{A_p} \wedge \theta_{j_1 \cdots j_p}$$

$$= L^n d(x) + L^{n-1} \sum_{p=1}^{n} \frac{1}{p!} L^{1-p} \Lambda^{j_1}_{A_1} \cdots \Lambda^{j_p}_{A_p} C^{A_1} \wedge \cdots \wedge C^{A_p} \wedge \theta_{j_1 \cdots j_p}$$

When this is compared with (2.18) it is seen that

$$\omega^1 \wedge \cdots \wedge \omega^n = L^{n-1} \Pi \tag{3.4}$$

Moreover, from (3.3) it follows with the aid of (1.26) and (1.27) that

$$\pi^1 \wedge \cdots \wedge \pi^n = L^{-n} \det (p^j_\ell) \omega^1 \wedge \cdots \wedge \omega^n = L^{1-n} \omega^1 \wedge \cdots \wedge \omega^n \tag{3.5}$$

and thus (3.4) yields

$$\Pi = \pi^1 \wedge \cdots \wedge \pi^n \tag{3.6}$$

Thus the Cartan form is the exterior product of the n momentum 1-forms.

We shall now investigate some consequences of this representation. Let us observe first that when all the conditions for the existence of a complete figure are satisfied, the equations (1.31) imply that $\pi^j = dS^j$, so that under these circumstances

$$\Pi = dS^1 \wedge \cdots \wedge dS^n \tag{3.7}$$

and consequently

$$d\Pi = 0 \tag{3.8}$$

Conversely, let us regard this relation as an hypothesis. Because of (3.6) one then has

$$d\pi^1 \wedge \pi^2 \wedge \cdots \wedge \pi^n - \pi^1 \wedge d\pi^2 \wedge \cdots \wedge \pi^n + \cdots + \cdots + (-1)^{n-1}\pi^1 \wedge \cdots \wedge \pi^{n-1} \wedge d\pi^n = 0$$

and, when the exterior products of this expression with $\pi^1, \pi^2, \ldots, \pi^n$ are successively evaluated, it is found that

$$d\pi^j \wedge \pi^1 \wedge \cdots \wedge \pi^n = 0 \qquad (j = 1, \ldots, n) \tag{3.9}$$

However, these n relations constitute a set of necessary and sufficient conditions for the complete integrability of the system of n total differential equations

$$\pi^j = 0 \tag{3.10}$$

Thus there exist n independent functions S^j, and a nonsingular matrix with entries λ^j_h, such that

$$\pi^j = \lambda^j_h \, dS^h \tag{3.11}$$

In view of (1.23) this is tantamount to

$$\pi^j_h = \lambda^j_\ell \frac{\partial S^\ell}{\partial x^h}, \qquad \pi^j_A = \lambda^j_\ell \frac{\partial S^\ell}{\partial \psi^A} \tag{3.12}$$

This is substituted in (1.24); in terms of the inverse $(\hat{\lambda}^j_h)$ of (λ^h_j) it is thus found that

$$\frac{\partial S^j}{\partial x^h} = L^{-1}\hat{\lambda}^j_m p^m_\ell \Lambda^\ell_h, \qquad \frac{\partial S^j}{\partial \psi^B} = L^{-1}\hat{\lambda}^j_m p^m_\ell \Lambda^\ell_B \tag{3.13}$$

and consequently

$$\Lambda^h_A \frac{\partial S^j}{\partial x^h} - \Lambda^B_A \frac{\partial S^j}{\partial \psi^B} = L^{-1}\hat{\lambda}^j_m p^m_\ell (\Lambda^\ell_h \Lambda^h_A - \Lambda^\ell_B \Lambda^B_A) = 0$$

where, in the second step, we have used the identity (1.10). Thus the functions S^j are solutions of the system (1.14), which in turn requires that the integrability conditions (1.11) be satisfied (as may also be verified by direct calculation). It is therefore inferred that the hypothesis (3.8) implies that the transversal distribution be integrable. (In passing, we observe that this conclusion could have been deduced from an hypothesis that is somewhat weaker than (3.8), namely

$$d\Pi = \mu \wedge \Pi \tag{3.14}$$

for some 1-form μ, since this is already sufficient to guarantee the complete integrability of the system (3.10). Moreover, exterior differentiation of (3.11) yields

$$d\pi^j + \omega^j_h \wedge \pi^h = 0 \tag{3.15}$$

in which

$$\omega^j_h = \lambda^j_\ell \, d\hat{\lambda}^\ell_h$$

The relation (3.15) can be regarded as an equation of parallelism relative to the "connection" 1-forms ω^j_h; these have vanishing "curvature" 2-forms, since their definition entails that

$$d\omega^j_h + \omega^j_\ell \wedge \omega^\ell_h = 0$$

identically.)

Now in terms of (3.11) we can express the Cartan form (3.6) as

$$\Pi = \lambda dS^1 \wedge \cdots \wedge dS^n \tag{3.16}$$

where

$$\lambda = \det(\lambda^j_h) \tag{3.17}$$

and the condition (3.8) requires that

$$d\lambda \wedge dS^1 \wedge \cdots \wedge dS^n = 0 \tag{3.18}$$

In order to exploit this result we introduce new coordinates $(\bar{x}^j, \bar{\psi}^A)$ on X_{n+M} by putting

$$\left.\begin{aligned} \bar{x}^j &= \bar{x}^j(x^h, \psi^B) = S^j(x^h, \psi^B) \\ \bar{\psi}^A &= \bar{\psi}^A(x^h, \psi^B) = \psi^A \end{aligned}\right\} \tag{3.19}$$

The Jacobian of this transformation is

$$\frac{\partial(\bar{x}^j, \bar{\psi}^A)}{\partial(x^h, \psi^B)} = \det \begin{bmatrix} \dfrac{\partial S^j}{\partial x^h}, & \dfrac{\partial S^j}{\partial \psi^B} \\ 0 & \delta^A_B \end{bmatrix}$$

and in order to ensure that (3.19) be locally invertible, we shall assume that the functions S^j have been chosen such that this determinant does not vanish. Thus, with $x^h = x^h(\bar{x}^j, \psi^A)$, we define a function f of the new coordinates by putting

$$\lambda(x^h, \psi^B) = \lambda(x^h(\bar{x}^j, \psi^A), \psi^B) = f(\bar{x}^j, \psi^B) \tag{3.20}$$

so that, by (3.19),

$$d\lambda = \frac{\partial f}{\partial \bar{x}^j} dS^j + \frac{\partial f}{\partial \psi^A} d\psi^A$$

When this is substituted in (3.18) it is found that

$$\frac{\partial f}{\partial \psi^A} d\psi^A \wedge dS^1 \wedge \cdots \wedge dS^n = \frac{\partial f}{\partial \psi^A} d\psi^A \wedge d\bar{x}^1 \wedge \cdots \wedge d\bar{x}^n = 0$$

from which it follows that $\partial f/\partial\psi^A = 0$, and (3.20) gives

$$\lambda = f(\bar{x}^j) = f(S^j) \tag{3.21}$$

Let F^h denote a set of n class C^2 functions of n independent variables t^j, these functions having been chosen such that $\det(F^h_j) \neq 0$, where $F^h_j = \partial F^h/\partial t^j$. The n functions S'^j defined by

$$S'^j = F^j(S^1, \ldots, S^n) \tag{3.22}$$

are solutions of the partial differential equations (1.14) whenever this is the case for S^j: they therefore also describe the integral surfaces of the transversal distribution D_M. Moreover, if $(\hat{F}^h_j)$ denotes the inverse of (F^j_h), one has

$$\frac{\partial S^j}{\partial x^h} = \hat{F}^j_\ell \frac{\partial S'^\ell}{\partial x^h}, \qquad \frac{\partial S^j}{\partial \psi^A} = \hat{F}^j_\ell \frac{\partial S'^\ell}{\partial \psi^A} \tag{3.23}$$

and (3.12) can be expressed as

$$\frac{\partial S'^j}{\partial x^h} = a^j_\ell \pi^\ell_h, \qquad \frac{\partial S'^j}{\partial \psi^A} = a^j_\ell \pi^\ell_A \tag{3.24}$$

where we have put

$$a^j_\ell = F^j_m \hat{\lambda}^m_\ell \tag{3.25}$$

Thus, by (3.17),

$$\det(a^j_\ell) = \lambda^{-1} \det(F^j_m)$$

and we now choose the functions $F^j(S^1, \ldots, S^n)$ such that $\det(F^j_h) = \lambda$, which is possible by virtue of (3.21). We then have

$$\det(a^j_\ell) = 1 \tag{3.26}$$

which allows us to subject the canonical momenta to the transformation (1.30); according to (3.24) the new canonical momenta are such that

$$\pi'^j_h = \frac{\partial S'^j}{\partial x^h}, \qquad \pi'^j_A = \frac{\partial S'^j}{\partial \psi^A} \tag{3.27}$$

From (1.27) and (3.26) it then follows that

$$L = D = D' = \det(\pi'^j_h + \pi'^j_A \psi^A_h)$$

$$= \det\left(\frac{\partial S'^j}{\partial x^h} + \frac{\partial S'^j}{\partial \psi^A} \psi^A_h\right) = \Delta' \tag{3.28}$$

which, according to the concluding remarks of the introduction, indicates that the construction of a complete figure has been accomplished. One may therefore conclude that <u>the complete figure of the field theory of Carathéodory is characterized by the requirement that the Cartan form (3.6) be closed</u>.

4. CARTAN FORMS ASSOCIATED WITH THE METHOD OF EQUIVALENT INTEGRALS

The method of equivalent integrals gives rise directly to a hierarchy of field theories, the differences between these theories resulting directly from the choice of distinct integrands by means of which the various equivalent integrals are constructed. This state of affairs, which can be quite complicated analytically, will now be briefly discussed in terms of the results developed above.

To this end we shall adopt a slightly more general point of view. Instead of considering an M-parameter family of integrable subspaces $C_n(w^B)$ such as (1.6), we shall now suppose that an n-dimensional distribution D_n is defined by a set of planes, one such plane being given in each tangent space of X_{n+M}. These planes are supposed to be spanned by n prescribed vector fields

$$B_j = (\delta^h_j, B^A_j) \tag{4.1}$$

as in (1.8); clearly the proposed extension consists merely of the replacement of the derivatives ψ^A_j by a set of nM entities B^A_j that are essentially arbitrary for our present purposes. The concepts introduced in Sec. 1, such as transversality, are readily reformulated in accordance with this prescription. In particular, for a given set of n independent functions S^j on X_{n+M} we shall write

$$c^j_h = S^j_h + S^j_A B^A_h \tag{4.2}$$

instead of (1.16), so that

$$dS^j = c^j_h dx^h + S^j_A C^A \tag{4.3}$$

the contact forms C^A now being defined by

$$C^A = d\psi^A - B^A_h dx^h \tag{4.4}$$

The representation (3.7) of the Cartan form in the presence of a complete figure suggests the examination of the following set of n distinct n-forms:

$$r!\overset{(r)}{\omega}(S) = dS^{j_1} \wedge \cdots \wedge dS^{j_r} \wedge \theta_{j_1 \cdots j_r}, \qquad r = 1, \ldots, n \tag{4.5}$$

where $\theta_{j_1 \cdots j_r}$ denotes as before the dual basis (2.11) of (n - r)-forms on Z_{n+M}; when r = n in (4.5), the resulting n-form clearly coincides with (3.7). In order to obtain an explicit expression for these forms we substitute from (4.3) to obtain

$$r!\overset{(r)}{\omega}(S) = \sum_{p=0}^{r} \binom{r}{p} S^{j_1}_{A_1} \cdots S^{j_p}_{A_p} c^{j_{p+1}}_{h_{p+1}} \cdots c^{j_r}_{h_r} C^{A_1} \wedge \cdots \wedge C^{A_p} \wedge dx^{h_{p+1}} \wedge \cdots \wedge dx^{h_r} \wedge \theta_{j_1 \cdots j_r} \tag{4.6}$$

Because of the formula (A_1) of the Appendix this is equivalent to

$$r!\overset{(r)}{\omega}(S) = \sum_{p=0}^{r} \frac{1}{p!} \binom{r}{p} S^{j_1}_{A_1} \cdots S^{j_p}_{A_p} \delta^{h_1 \cdots h_p h_{p+1} \cdots h_r}_{j_1 \cdots j_p j_{p+1} \cdots j_r} c^{j_{p+1}}_{h_{p+1}} \cdots c^{j_r}_{h_r} C^{A_1} \wedge \cdots \wedge C^{A_p} \wedge \theta_{h_1 \cdots h_p} \tag{4.7}$$

It is now recalled that the n integrands that generate the aforementioned n distinct field theories via the method of equivalent integrals are defined by ([4])

$$r!\,\overset{(r)}{\Phi} = \delta^{h_1 \cdots h_r}_{j_1 \cdots j_r} c^{j_1}_{h_1} \cdots c^{j_r}_{h_r} \tag{4.8}$$

Clearly this expression occurs as the term corresponding to $p = 0$ in the sum on the right-hand side of (4.7), and accordingly we write the latter as

$$r!\,\overset{(r)}{\omega}(S) = \overset{(r)}{\Phi}\,d(x) + \sum_{p=1}^{r} \frac{1}{(p!)^2} \overset{(r)}{C}{}^{h_1 \cdots h_p}_{j_1 \cdots j_p} S^{j_1}_{A_1} \cdots S^{j_p}_{A_p} C^{A_1} \wedge \cdots \wedge C^{A_p} \wedge \theta_{h_1 \cdots h_p} \tag{4.9}$$

where we have put

$$(r-p)!\,\overset{(r)}{C}{}^{h_1 \cdots h_p}_{j_1 \cdots j_p} = \delta^{h_1 \cdots h_p h_{p+1} \cdots h_r}_{j_1 \cdots j_p j_{p+1} \cdots j_r} c^{j_{p+1}}_{h_{p+1}} \cdots c^{j_r}_{h_r} \tag{4.10}$$

An application of the formula (A_2) of the Appendix now shows that the <u>n-forms (4.9) can be expressed entirely in terms of the functions (4.8) and their derivatives</u>, namely as

$$\overset{(r)}{\omega}(S) = \overset{(r)}{\Omega}(\overset{(r)}{\Phi}) \tag{4.11}$$

where

$$\overset{(r)}{\Omega}(\overset{(r)}{\Phi}) = \overset{(r)}{\Phi}\,d(x) + \sum_{p=1}^{r} \frac{1}{(p!)^2} \frac{\partial^p \overset{(r)}{\Phi}}{\partial B^{A_1}_{h_1} \cdots \partial B^{A_p}_{h_p}} C^{A_1} \wedge \cdots \wedge C^{A_p} \wedge \theta_{h_1 \cdots h_p} \tag{4.12}$$

This construction clearly suggests that one should associate with any class C^n Lagrangian L and any integer r, $1 \leq r \leq n$, the n-form

$$\overset{(r)}{\Omega}(L) = L\,d(x) + \sum_{p=1}^{r} \frac{1}{(p!)^2} \frac{\partial^p L}{\partial B^{A_1}_{h_1} \cdots \partial B^{A_p}_{h_p}} C^{A_1} \wedge \cdots \wedge C^{A_p} \wedge \theta_{h_1 \cdots h_p} \tag{4.13}$$

This can be regarded as an alternative generalization of the Cartan forms of the single integral theory; we shall therefore call (4.13) a <u>Cartan form of the second kind</u>. [When one puts $r = n$ in (4.13), one obtains a form first suggested and examined by I. M. Anderson (private communication).]

The relationship between (4.13) and the n-form Π as given in (2.18) is best illustrated for the case when $r = n$, which characterizes the field theory of Carathéodory. From (4.2) and (4.8) it is evident that

$$\overset{(n)}{\Phi} = \det(c^j_h) = \Delta \tag{4.14}$$

Also, according to the formulae (A_2) and (A_3) of the Appendix we have

$$\frac{\partial^p \Delta}{\partial B^{A_1}_{j_1} \cdots \partial B^{A_p}_{j_p}} = \Delta^{1-p} \delta^{j_1 \cdots j_p}_{\ell_1 \cdots \ell_p} C^{\ell_1}_{h_1} \cdots C^{\ell_p}_{h_p} S^{h_1}_{A_1} \cdots S^{h_p}_{A_p} \tag{4.15}$$

and, when this is substituted for the expression for $\Omega(\Delta)$ as defined in accordance with (4.13), it is found that

$$\overset{(n)}{\Omega}(\Delta) = \Delta\, d(x) + \sum_{p=1}^{n} \frac{1}{(p!)^2} \Delta^{1-p} \delta^{j_1 \cdots j_p}_{\ell_1 \cdots \ell_p} C^{\ell_1}_{h_1} \cdots C^{\ell_p}_{h_p} S^{h_1}_{A_1} \cdots S^{h_p}_{A_p} C^{A_1} \wedge \cdots \wedge C^{A_p} \wedge \theta_{j_1 \cdots j_p}$$

$$= \Delta\, d(x) + \sum_{p=1}^{n} \frac{1}{p!} \Delta^{1-p} C^{j_1}_{h_1} \cdots C^{j_p}_{h_p} S^{h_1}_{A_1} \cdots S^{h_p}_{A_p} C^{A_1} \wedge \cdots \wedge C^{A_p} \wedge \theta_{j_1 \cdots j_p} \tag{4.16}$$

However, when the conditions (1.19) and (1.20) that characterize a geodesic field are satisfied, this expression assumes the form

$$\overset{(n)}{\Omega}(\Delta) = L\, d(x) + \sum_{p=1}^{n} \frac{1}{p!} L^{1-p} \Lambda^{j_1}_{A_1} \cdots \Lambda^{j_p}_{A_p} C^{A_1} \wedge \cdots \wedge C^{A_p} \wedge \theta_{j_1 \cdots j_p} \tag{4.17}$$

which is nothing other than the Cartan form (2.18). (This result could, of course, have been anticipated by the construction of Sec. 3.)

We shall conclude with two observations. First, a Cartan form for the field theory of Weyl is obtained directly from (4.13) when one puts $r = 1$:

$$\overset{(1)}{\Omega}(L) = L\, d(x) + \Lambda^{j}_{A} C^{A} \wedge \theta_{j} \tag{4.18}$$

This is the n-form that occurs most frequently in the literature on Cartan forms ([8]). It may be used to construct an integral invariant ([7]). Second, for $r \geq 1$, the first two terms on the right-hand sides of (4.13) and (2.18) coincide, these terms being displayed in (4.18). It is remarkable that the difference between the third terms ($p = 2$) is represented precisely by the Legendre form (1.22). This would suggest that one should introduce generalized Legendre forms

$$L^{j_1 \cdots j_p}_{A_1 \cdots A_p} = \frac{\partial^p L}{\partial B^{A_1}_{j_1} \cdots \partial B^{A_p}_{j_p}} - L^{1-p} \delta^{j_1 \cdots j_p}_{\ell_p \cdots \ell_p} \Lambda^{\ell_1}_{A_1} \cdots \Lambda^{\ell_p}_{A_p} \tag{4.19}$$

These vanish for $p = 1$, and reduce to (1.22) for $p = 2$, while for $r = n$ one thus has an explicit formula for the difference between two types of Cartan forms:

$$\overset{(n)}{\Omega}(L) = \Pi(L) + \sum_{p=2}^{n} \frac{1}{(p!)^2} L^{j_1 \cdots j_p}_{A_1 \cdots A_p} C^{A_1} \wedge \cdots \wedge C^{A_p} \wedge \theta_{j_1 \cdots j_p} \tag{4.20}$$

5. APPENDIX

In this Appendix the proofs of several identities that have been used in the text are briefly sketched.

Let r, p be integers with $1 \leq r \leq n$, and $0 \leq p \leq r$. Then, according to the definition (2.11)

$$\frac{1}{p!} \delta^{h_1 \cdots h_p h_{p+1} \cdots h_r}_{j_1 \cdots j_p j_{p+1} \cdots j_r} \theta_{h_1 \cdots h_p}$$

$$= \frac{1}{p!(n-p)!} \delta^{h_1 \cdots h_p h_{p+1} \cdots h_r}_{j_1 \cdots j_p j_{p+1} \cdots j_r} \varepsilon_{h_1 \cdots h_p \ell_{p+1} \cdots \ell_n} dx^{\ell_{p+1}} \wedge \cdots \wedge dx^{\ell_n} \tag{5.1}$$

We now use the identities ([2], pp. 111-113)

$$\delta^{h_1\cdots h_p h_{p+1}\cdots h_r h_{r+1}\cdots h_n}_{j_1\cdots j_p j_{p+1}\cdots j_r h_{r+1}\cdots h_n} = (n-r)!\,\delta^{h_1\cdots h_p h_{p+1}\cdots h_r}_{j_1\cdots j_p j_{p+1}\cdots j_r} \tag{5.2}$$

and

$$\varepsilon^{h_1\cdots h_n}\varepsilon_{j_1\cdots j_n} = \delta^{h_1\cdots h_n}_{j_1\cdots j_n} \tag{5.3}$$

to express the right-hand side of (5.1) as

$$\frac{1}{(n-p)!\,(n-r)!}\varepsilon_{j_1\cdots j_p j_{p+1}\cdots j_r h_{r+1}\cdots h_n}\delta^{h_{p+1}\cdots h_r h_{r+1}\cdots h_n}_{\ell_{p+1}\cdots \ell_r \ell_{r+1}\cdots \ell_n} dx^{\ell_{p+1}}\wedge\cdots\wedge dx^{\ell_n}$$

$$= \frac{1}{(n-r)!}\varepsilon_{j_1\cdots j_p j_{p+1}\cdots j_r h_{r+1}\cdots h_n} dx^{h_{p+1}}\wedge\cdots\wedge dx^{h_r}\wedge dx^{h_{r+1}}\wedge\cdots\wedge dx^{h_n}$$

To this expression one may apply the definition (2.11). This, combined with (5.1), establishes the required identity

$$dx^{h_{p+1}}\wedge\cdots\wedge dx^{h_r}\wedge\theta_{j_1\cdots j_r} = \frac{1}{p!}\delta^{h_1\cdots h_p h_{p+1}\cdots h_r}_{j_1\cdots j_p j_{p+1}\cdots j_r}\theta_{h_1\cdots h_p} \tag{A$_1$}$$

We now turn to the definition (4.2), from which it follows that

$$\frac{\partial c^m_\ell}{\partial B^A_j} = \delta^j_\ell S^m_A \tag{5.4}$$

Thus differentiation of (4.8) gives

$$r!\frac{\partial \overset{(r)}{\Phi}}{\partial B^A_j} = r\delta^{\ell j_2\cdots j_r}_{mh_2\cdots h_r}\frac{\partial c^m_\ell}{\partial B^A_j}c^{h_2}_{j_2}\cdots c^{h_r}_{j_r} = r\delta^{jj_2\cdots j_r}_{mh_2\cdots h_r}c^{h_2}_{j_2}\cdots c^{h_r}_{j_r}S^m_A$$

and similarly

$$(r-1)!\frac{\partial \overset{(r)}{\Phi}}{\partial B^A_j\,\partial B^B_h} = (r-1)\delta^{jhj_3\cdots j_r}_{mph_3\cdots h_r}c^{h_3}_{j_3}\cdots c^{h_r}_{j_r}S^m_A S^p_B$$

By repeating this process it is found that the pth derivative of the functions (4.8) can be expressed in terms of the entities (4.10) as

$$\frac{\partial^p \overset{(r)}{\Phi}}{\partial B^{A_1}_{j_1}\cdots\partial B^{A_p}_{j_p}} = \overset{(r)}{C}{}^{j_1\cdots j_p}_{h_1\cdots h_p} S^{h_1}_{A_1}\cdots S^{h_p}_{A_p}, \qquad 1\le p\le n \tag{A$_2$}$$

Similarly it is readily verified that

$$\frac{\partial}{\partial B^A_j}\overset{(r)}{C}{}^{j_1\cdots j_p}_{h_1\cdots h_p} = \overset{(r)}{C}{}^{j_1\cdots j_p j}_{h_1\cdots h_p m} S^m_A \tag{5.5}$$

Finally, it is asserted that

$$\Delta^{p-1}\overset{(n)}{C}{}^{j_1\cdots j_p}_{h_1\cdots h_p} = \delta^{j_1\cdots j_p}_{\ell_1\cdots \ell_p} C^{\ell_1}_{h_1}\cdots C^{\ell_p}_{h_p}, \qquad (1 \leq p \leq n) \tag{A$_3$}$$

This is established as follows. When $p = 1$, the identity (A_3) is merely the general formula for the cofactor C^j_h of the element c^h_j in the determinant Δ. The case $p = 2$ is deduced by differentiation with respect to B^A_k by means of the usual expressions for the derivatives of cofactors. This gives rise to an inductive proof in which repeated use is made of (5.5).

ACKNOWLEDGMENT

It is a pleasure to thank Dr. Ian M. Anderson of Utah State University for valuable discussions.

REFERENCES

1. Carathéodory, C.: Acta Sci. Math. Szeged 4, 193-216 (1929). [Collected Works, Vol. 1 (Beck, Munich, 1954), 401-426.]
2. Lovelock, D.; Rund, H.: Tensors, Differential Forms, and Variational Principles. Wiley (1975).
3. Morse, M.: The Calculus of Variations in the Large. American Mathematical Society Colloquium Publications 18 (1934).
4. Rund, H.: Aeq. Math. 3, 44-63 (1969).
5. Rund, H.: The Hamilton-Jacobi Theory in the Calculus of Variations. Van Nostrand (1966); augmented and revised edition, Krieger Publications (1973).
6. Rund, H.: Aeq. Math. 10, 236-261 (1973).
7. Rund, H.: Aeq. Math. 13, 121-149 (1975).
8. Shadwick, W. F.: Lett. Math. Phys. 5, 137-141 (1981).
9. Weyl, H.: Ann. Math. (2) 36, 607-629 (1935).

30 THE KY FAN MINIMAX PRINCIPLE, SETS WITH CONVEX SECTIONS, AND VARIATIONAL INEQUALITIES

Mau-Hsiang Shih / Department of Mathematics, Chung Yuan University, Chung-Li, Taiwan

Kok-Keong Tan / Department of Mathematics, Statistics and Computing Science, Dalhousie University, Halifax, Nova Scotia, Canada

1. THE KY FAN MINIMAX PRINCIPLE

Among the various applications of the KKM-mapping principle [8,9,12-16], a both simple and nontrivial application is the celebrated 1972 minimax inequality of Ky Fan [15]. The Ky Fan minimax inequality is of a topological nature and has important repercussions both in convex analysis and in nonlinear functional analysis [1,2,15].

Among the various generalizations of the Ky Fan minimax inequality, the simplest and most convenient form was established by Ben-El-Mechaiekh, Deguire and Granas [3]. Their result is

Theorem A. Let X be a nonempty compact convex subset of a Hausdorff topological vector space and let f and g be real-valued functions defined on $X \times X$ such that:

(i) $f \leq g$ on $X \times X$;

(ii) $y \to f(x,y)$ is lower semicontinuous on X for each fixed $x \in X$;

(iii) $x \to g(x,y)$ is quasiconcave on X for each fixed $y \in X$.

Then for each $\lambda \in \mathbb{R}$ one of the following properties holds:

(a) There exists a point $x_0 \in X$ such that $g(x_0, x_0) > \lambda$;
(b) There exists a point $y_0 \in X$ such that $f(x, y_0) \leq \lambda$ for all $x \in X$.

A real-valued function ϕ defined on a convex set X is said to be quasiconcave if for every real number t the set $\{x \in X: \phi(x) > t\}$ is convex. Equivalently, ϕ is said to be quasiconcave if and only if $\phi(tx + (1-t)y) \geq \min\{\phi(x), \phi(y)\}$ for all x, $y \in X$ and for all $t \in (0,1)$.

A proof of Theorem A given in [3] is based on the KKM-mapping principle. We shall give an alternative proof based on the partition of unity argument initiated by Browder [6]. The partition of unity argument occurs very often when we apply the Ky Fan minimax principle.

Proof of Theorem A. Suppose that for each $y \in X$ there exists $x \in X$ such that $f(x,y) > \lambda$. For each $x \in X$, let

$$Q(x) := \{y \in X: f(x,y) > \lambda\}$$

then Q(x) is open in X by lower semicontinuity of $y \to f(x,y)$, and $X = \bigcup_{x \in X} Q(x)$. By compactness of X, there exists $\{x_1, \ldots, x_n\} \subset X$ so that $X = \bigcup_{i=1}^{n} Q(x_i)$. Let $\alpha_1, \ldots, \alpha_n$ be a continuous partition of unity subordinated to the covering $\{Q(x_i): 1 \leq i \leq n\}$ of X, that is, $\alpha_1, \ldots, \alpha_n$ are nonnegative real-valued functions on X such that for each

*This work was partially supported by NSERC of Canada under grant A8096.

$1 \leq i \leq n$, α_i vanishes on $X \backslash Q(x_i)$ and $\Sigma_{i=1}^{n} \alpha_i(\zeta) = 1$ for all $\zeta \in X$. We define a map $q: X \to X$ by setting

$$q(\zeta) := \sum_{i=1}^{n} \alpha_i(\zeta)\, x_i$$

Then q is continuous on X and q maps the convex hull $\operatorname{conv}\{x_1, \ldots, x_n\}$ of $\{x_1, \ldots, x_n\}$ into itself. By Brouwer's fixed point theorem, there exists a point $x_0 \in \operatorname{conv}\{x_1, \ldots, x_n\} \subset X$ such that

$$q(x_0) = \sum_{i=1}^{n} \alpha_i(x_0) x_i = x_0$$

For each $\alpha_i(x_0) > 0$ $(i = 1, \ldots, n)$, we have $x_0 \in Q(x_i)$ and hence

$$g(x_i, x_0) \geq f(x_i, x_0) > \lambda \quad (i = 1, \ldots, n)$$

By quasiconcavity of $x \to g(x,y)$ we have

$$\begin{aligned} g(x_0, x_0) &= g\Big(\sum_{i=1}^{n} \alpha_i(x_0)x_i, x_0\Big) \\ &\geq \min\{g(x_i, x_0) : \alpha_i(x_0) > 0,\ i = 1, \ldots, n\} \\ &> \lambda \end{aligned}$$

This concludes the proof of Theorem A. Q.E.D.

As an immediate consequence, we obtain

Theorem 1. Let X be a nonempty compact convex subset of a Hausdorff topological vector space and let ϕ and ψ be real-valued functions defined on $X \times X$ such that

(i) $\phi \leq \psi$ on the diagonal $\Delta = \{(x,x): x \in X\}$ and $\phi \geq \psi$ on $(X \times X)\backslash\Delta$;

(ii) $y \to \phi(y,y) - \phi(x,y)$ is lower semicontinuous on X for each fixed $x \in X$;

(iii) $x \to \psi(x,y)$ is quasiconvex on X for each fixed $y \in X$. Then there exists a point $y_0 \in X$ such that

$$\phi(y_0, y_0) \leq \phi(x, y_0) \text{ for all } x \in X$$

Proof. Define f and g on $X \times X$ by setting

$$f(x,y) := \phi(y,y) - \phi(x,y)$$

$$g(x,y) := \psi(y,y) - \psi(x,y)$$

Then f and g satisfy the hypothesis of Theorem A. Since $g(x,x) = 0$ for all $x \in X$, Theorem A implies that there exists a point $y_0 \in X$ such that $f(x, y_0) \leq 0$ for all $x \in X$. This completes the proof. Q.E.D.

When $\phi \equiv \psi$, Theorem 1 was given by Ky Fan ([11], p. 118) and ([15], Corollary 1).

2. EQUIVALENT FORMULATIONS

We shall extend Ky Fan's idea [15] to give various formulations of Theorem A.

Theorem 2 (Generalized Ky Fan Minimax Inequality). Let X be a nonempty compact convex subset of a Hausdorff topological vector space and let f and g be real-valued functions defined on $X \times X$ such that:

(i) $f \leq g$ on $X \times X$;

(ii) $y \to f(x,y)$ is lower semicontinuous on X for each fixed $x \in X$;

(iii) $x \to g(x,y)$ is quasiconcave on X for each fixed $y \in X$.

Then the minimax inequality

$$\min_{y \in X} \sup_{x \in X} f(x,y) \leq \sup_{x \in X} g(x,x)$$

holds.

Theorem A $\Rightarrow$ Theorem 2. Let $\lambda := \sup_{x \in X} g(x,x)$. We may assume that $\lambda < +\infty$. Since there is no $x_0 \in X$ such that $g(x_0, x_0) > \lambda$, Theorem A implies that there exists a point $y_0 \in X$ such that $f(x, y_0) \leq \lambda$ for all $x \in X$. The conclusion follows.

Theorem 2 $\Rightarrow$ Theorem A. Obvious.

Theorem 2 is due to Yen [21]. When $f \equiv g$, Theorem 2 reduces to the Ky Fan minimax inequality [14].

Theorem 3 (First Geometric Form). Let X be a non-empty compact convex subset of a Hausdorff topological vector space. Let $A, B \subset X \times X$ be such that

(i) $A \subset B$;

(ii) For each fixed $x \in X$, the set $\{y \in X: (x,y) \in A\}$ is open in X;

(iii) For each fixed $y \in X$, the set $\{x \in X: (x,y) \in B\}$ is convex.

Then one of the following properties holds:

(a) There exists a point $x_0 \in X$ such that $(x_0, x_0) \in B$;

(b) There exists a point $y_0 \in X$ such that $\{x \in X: (x, y_0) \in A\} = \phi$.

Theorem A $\Rightarrow$ Theorem 3. Let f and g be two characteristic functions defined on A and B, respectively. Since $A \subset B$, we have $f \leq g$ on $X \times X$. For each fixed $x \in X$ and any $t \in \mathbb{R}$, we have

$$\{y \in X: f(x,y) > t\} = \begin{cases} X & \text{if } t < 0 \\ \phi & \text{if } t \geq 1 \\ \{y \in X: (x,y) \in A\} & \text{if } 0 \leq t < 1 \end{cases}$$

Thus $y \to f(x,y)$ is lower semicontinuous on X for each fixed $x \in X$. For each fixed $y \in X$ and any $t \in \mathbb{R}$, we have

$$\{x \in X: g(x,y) > t\} = \begin{cases} X & \text{if } t < 0 \\ \phi & \text{if } t \geq 1 \\ \{x \in X: (x,y) \in B\} & \text{if } 0 \leq t < 1 \end{cases}$$

Thus $x \to g(x,y)$ is quasiconcave on X for each fixed $y \in X$. Let $0 \leq \lambda < 1$. By Theorem A, there exists a point $x_0 \in X$ such that $g(x_0, x_0) > \lambda$ or there exists a point $y_0 \in X$ such that $f(x, y_0) \leq \lambda$ for all $x \in X$. This means that there exists a point $x_0 \in X$ such that $(x_0, x_0) \in B$ or there exists a point $y_0 \in X$ such that $\{x \in X: (x, y_0) \in A\} = \phi$.

Q.E.D.

Theorem 3 ⇒ Theorem A. Assume the hypothesis of Theorem A. Let $\lambda \in \mathbb{R}$. Define

$$A := \{(x,y) \in X \times X : f(x,y) > \lambda\}$$

$$B := \{(x,y) \in X \times X : g(x,y) > \lambda\}$$

Then we find all the requirements in Theorem 3 are satisfied, and the assertion of Theorem A follows by applying Theorem 3. Q.E.D.

When $A \equiv B$, Theorem 3 is due to Ky Fan [15]. Theorem 3 can be formulated in terms of the complement M of A and the complement N of B as follows:

Theorem 4 (Second Geometric Form). Let X be a nonempty compact convex subset of a Hausdorff topological vector space. Let $M, N \subset X \times X$ be such that

(i) $N \subset M$;

(ii) The set $\{y \in X: (x,y) \in M\}$ is closed in X for each fixed $x \in X$;

(iii) The set $\{x \in X: (x,y) \notin N\}$ is convex (possibly empty) for each fixed $y \in X$.

Then one of the following properties holds:

(a) There exists a point $x_0 \in X$ such that $(x_0, x_0) \notin N$;

(b) There exists a point $y_0 \in X$ such that $X \times \{y_0\} \subset M$.

When $M \equiv N$, Theorem 4 is due to Ky Fan ([12], Lemma 4). We can easily see the following fixed point theorem is equivalent to Theorem 3 or Theorem 4.

Theorem 5 (Fixed Point Version). Let X be a nonempty compact convex subset of a Hausdorff topological vector space. Let $F, G : X \to 2^X$ be set-valued maps such that

(i) $F(x) \subset G(x)$ for each $x \in X$;

(ii) $G^{-1}(y)$ is convex for each $y \in X$;

(iii) $F(x)$ is open in X for each $x \in X$.

Then one of the following properties holds:

(a) There exists a point $w \in X$ such that $w \in G(w)$;

(b) There exists a point $y_0 \in X$ such that $F^{-1}(y_0) = \phi$.

When $F \equiv G$, Theorem 5 was formulated by Browder [6].

3. SETS WITH CONVEX SECTIONS

Given a Cartesian product $X := \Pi_{i=1}^n X_i$ of topological spaces, let

$$\hat{X}_i := \prod_{j \neq i} X_j \quad \text{and} \quad \hat{x}_i := (x_1, \ldots, x_{i-1}, x_{i+1}, \ldots, x_n) \in \hat{X}_i$$

For $x := (x_1, \ldots, x_n)$, $y := (y_1, \ldots, y_n) \in X$, we let

$$(y_i, \hat{x}_i) := (x_1, \ldots, x_{i-1}, y_i, x_{i+1}, \ldots, x_n)$$

Theorems of Ky Fan [13-15] concerning sets with convex sections can be extended in a natural way.

Theorem 6. Let $n \geq 2$ and let $X_1, X_2, \ldots, X_n$ be nonempty compact convex sets, each in a Hausdorff topological vector space. Let $f_1, \ldots, f_n, g_1, \ldots, g_n$ be real-valued functions on the product space $X := \Pi_{i=1}^n X_i$ such that

(i) $f_i \leq g_i$ on X for $i = 1, \ldots, n$;

(ii) For each $i = 1, \ldots, n$ and for each fixed x_i of X_i, the map $\hat{x}_i \to f_i(x_i, \hat{x}_i)$ is lower semicontinuous on $\hat{X}_i$;

(iii) For each $i = 1, \ldots, n$ and for each fixed $\hat{x}_i \in \hat{X}_i$, the map $x_i \to g_i(x_i, \hat{x}_i)$ is quasiconcave on X_i.

Let $t_1, \ldots, t_n$ be n real numbers and suppose that for each $i = 1, \ldots, n$ and for each point $\hat{x}_i \in \hat{X}_i$ there exists a point $x_i \in X_i$ such that $f_i(x_i, \hat{x}_i) > t_i$. Then there exists a point $u \in X$ such that $g_i(u) > t_i$ for $i = 1, \ldots, n$.

Proof. For any two points $x := (x_1, \ldots, x_n)$ and $y := (y_1, \ldots, y_n)$ of X we define

$$\phi(x,y) := \min\{ f_i(y_1,\ldots,y_{i-1}, x_i, y_{i+1},\ldots,y_n) - t_i : i = 1,\ldots,n\}$$

$$\psi(x,y) := \min\{g_i(y_1,\ldots,y_{i-1}, x_i, y_{i+1},\ldots,y_n) - t_i : i = 1,\ldots,n\}$$

Then:

1. $\phi \leq \psi$ on $X \times X$ by (i);
2. $y \to \phi(x,y)$ is lower semicontinuous on X for each fixed $x \in X$ by (ii);
3. $x \to \psi(x,y)$ is quasiconcave on X by (iii).

Since the last part of the hypothesis asserts that for each $y \in X$ there exists a point $x \in X$ such that $\phi(x,y) > 0$, according to Theorem A, there exists a point $u \in X$ such that $\psi(u,u) > 0$. Thus,

$$g_i(u) > t_i \quad \text{for } i = 1, \ldots, n$$

Q.E.D.

When $f_i \equiv g_i$ for $i = 1, \ldots, n$, Theorem 6 is due to Ky Fan [13-15]. The following theorem is a geometric formulation of Theorem 6.

Theorem 7. Let $n \geq 2$ and let $X_1, \ldots, X_n$ be non-empty compact convex sets, each in a Hausdorff topological vector space, and let $A_1, \ldots, A_n, B_1, \ldots, B_n$ be subsets of the product space $X = \Pi_{i=1}^n X_i$. Suppose that

(i) $A_i \subset B_i$ for $i = 1, \ldots, n$;

(ii) For each $i = 1, \ldots, n$ and for each point $x_i \in X_i$, the set

$$A_i(x_i) := \{\hat{x}_i \in \hat{X}_i : (x_i, \hat{x}_i) \in A_i\}$$

is open in $\hat{X}_i$;

(iii) For each $i = 1, \ldots, n$ and for each point $\hat{x}_i \in \hat{X}_i$, the set

$$B_i(\hat{x}_i) := \{x_i \in X_i : (x_i, \hat{x}_i) \in B_i\}$$

is convex and the set

$$A_i(\hat{x}_i) := \{x_i \in X_i : (x_i, \hat{x}_i) \in A_i\} \neq \phi$$

Then $\bigcap_{i=1}^n B_i \neq \phi$.

Theorem 6 ⇒ Theorem 7. For each $i = 1, \ldots, n$ let f_i and g_i be characteristic functions defined on A_i and B_i, respectively, that is

$$f_i(x_1, \ldots, x_n) := \begin{cases} 1 \text{ if } (x_1, \ldots, x_n) \in A_i \\ 0 \text{ if } (x_1, \ldots, x_n) \in X \backslash A_i \end{cases}$$

$$g_i(x_1, \ldots, x_n) := \begin{cases} 1 \text{ if } (x_1, \ldots, x_n) \in B_i \\ 0 \text{ if } (x_1, \ldots, x_n) \in X \backslash B_i \end{cases}$$

Since $A_i \subset B_i$ $(i = 1, \ldots, n)$, $f_i \leq g_i$ on X $(i = 1, \ldots, n)$. For each fixed $x_i \in X_i$ $(i = 1, \ldots, n)$ and any $t \in \mathbb{R}$, we have

$$\{\hat{x}_i \in \hat{X}_i : f_i(x_i, \hat{x}_i) > t\} = \begin{cases} X & \text{if } t < 0 \\ A_i(x_i) & \text{if } 0 \leq t < 1 \\ \phi & \text{if } t \geq 0 \end{cases}$$

This implies the condition (ii) of Theorem 6 is satisfied. Similarly,

$$\{x_i \in X_i : g_i(x_i, \hat{x}_i) > t\} = \begin{cases} X & \text{if } t < 0 \\ B_i(\hat{x}_i) & \text{if } 0 \leq t < 1 \\ \phi & \text{if } t \geq 1 \end{cases}$$

This implies the condition (iii) of Theorem 6 is satisfied. Let $0 \leq t < 1$. Since $A_i(\hat{x}_i) \neq \phi$ $(i = 1, \ldots, n)$, it follows that the final hypothesis of Theorem 6 is satisfied. Therefore, according to Theorem 6, there exists $u \in X$ such that $g_i(u) > t$ for $i = 1, \ldots, n$. So $g_i(u) = 1$ for $i = 1, \ldots, n$ and hence $u \in \bigcap_{i=1}^n B_i$. Q.E.D.

Theorem 7 ⇒ Theorem 6. For $i = 1, \ldots, n$ we define the subsets A_i and B_i of X to be

$$A_i := \{u \in X: f_i(u) > t_i\}$$

$$B_i := \{u \in X: g_i(u) > t_i\}$$

Since $f_i \leq g_i$ on X $(i = 1, \ldots, n)$, $A_i \subset B_i$ $(i = 1, \ldots, n)$. The condition (ii) of the hypothesis of Theorem 6 is equivalent to the condition that for each $i = 1, \ldots, n$ and for each fixed $x_i \in X_i$, the set

$$A_i(x_i) := \{\hat{x}_i \in \hat{X}_i : (x_i, \hat{x}_i) \in A_i\}$$

is open in $\hat{X}_i$.

The condition (iii) of the hypothesis of Theorem 6 is equivalent to that for each $i = 1, \ldots, n$ and for each fixed $\hat{x}_i \in \hat{X}_i$, the set

$$B_i(\hat{x}_i) := \{x_i \in X_i : (x_i, \hat{x}_i) \in X_i\}$$

is convex. The condition of the final statement of Theorem 6 tells us that for each $i = 1, \ldots, n$ and for each fixed $\hat{x}_i \in \hat{X}_i$ the set $A_i(\hat{x}_i) \neq \phi$.

Hence the conditions of Theorem 7 are all satisfied, and we conclude that $\bigcap_{i=1}^n B_i \neq \phi$ by applying Theorem 7. Let $u \in \bigcap_{i=1}^n B_i$; then $g_i(u) > t_i$ $(i = 1, \ldots, n)$. Q.E.D.

When $A_i \equiv B_i$ $(i = 1, \ldots, n)$, Theorem 7 is due to Ky Fan [13-15]. Since the case $n = 2$ of Theorem 7 is most helpful, we state the case $n = 2$ of Theorem 7 explicitly below as Theorem $\hat{7}$.

Theorem $\hat{7}$. Let X, Y be nonempty compact convex sets, each in a Hausdorff topological vector space, and let A_1, A_2, B_1, B_2 be subsets of $X \times Y$. Suppose that

(i) $A_i \subset B_i$ $(i = 1, 2)$;

(ii) For each fixed $x \in X$, $A_1(x) := \{y \in Y: (x,y) \in A_1\}$ is open in Y, and for each fixed $y \in X$, $A_2(y) := \{x \in X: (x,y) \in A_2\}$ is open in X;

(iii) For each fixed $y \in Y$, $B_1(y) := \{x \in X: (x,y) \in B_1\}$ is convex and $A_1(y) := \{x \in X: (x,y) \in A\} \neq \phi$, and for each fixed $x \in X$, $B_2(x) := \{y \in Y: (x,y) \in B_2\}$ is convex and $A_2(x) := \{y \in Y: (x,y) \in A_2\} \neq \phi$.

Then $B_1 \cap B_2 \neq \phi$.

4. MINIMAX INEQUALITY OF THE VON NEUMANN TYPE

Minimax inequalities treated in this section evolve from the von Neumann minimax principle. We shall show that such kind of minimax type inequalities are simple consequences of the Ky Fan minimax principle. (Hence the Ky Fan minimax principle implies the von Neumann minimax principle.)

We begin with the following symmetric and more stronger form due to Ben-El-Mechaiekh, Degurie, and Granas [4]. The aim is to offer a direct intrinsic proof by making use of the Ky Fan minimax principle.

Theorem 8 (Ben-El-Mechaiekh-Degurie-Granas). Let X and Y be non-empty compact convex sets, each in a Hausdorff topological vector space. Let f, s, t, g be four real-valued functions on $X \times Y$ satisfying

(i) $f \leq s \leq t \leq g$ on $X \times Y$;

(ii) $y \to f(x,y)$ is lower semicontinuous on Y for each fixed $x \in X$;

(iii) $x \to s(x,y)$ is quasiconcave on X for each fixed $y \in Y$;

(iv) $y \to t(x,y)$ is quasiconvex on Y for each fixed $x \in X$;

(v) $x \to g(x,y)$ is upper semicontinuous on X for each fixed $y \in Y$.

Then for each $\lambda \in \mathbb{R}$ one of the following properties holds:

(a) There exists a point $y_0 \in Y$ such that $f(x,y_0) \leq \lambda$ for all $x \in X$;

(b) There exists a point $x_0 \in X$ such that $g(x_0,y) \geq \lambda$ for all $y \in Y$.

Proof. Let $\lambda \in \mathbb{R}$ and

$$A_1 := \{(x,y) \in X \times Y: f(x,y) > \lambda\}$$

$$A_2 := \{(x,y) \in X \times Y: g(x,y) < \lambda\}$$

$$B_1 := \{(x,y) \in X \times Y: s(x,y) > \lambda\}$$

$$B_2 := \{(x,y) \in X \times Y: t(x,y) < \lambda\}$$

Suppose that the assertion of the theorem is false. Then for each fixed $y \in Y$, $A_1(y) := \{x \in X: f(x,y) > \lambda\} \neq \phi$ and for each fixed $x \in X$, $A_2(x) := \{y \in Y: g(x,y) < \lambda\} \neq \phi$. Other conditions of Theorem $\hat{7}$ are easily seen from the hypotheses of Theorem 8. Thus, according to Theorem $\hat{7}$, $B_1 \cap B_2 \neq \phi$. Let $(x_0,y_0) \in B_1 \cap B_2$; then

$$\lambda < s(x_0, y_0) \leq t(x_0, y_0) < \lambda$$

which is a contradiction. This completes the proof. Q.E.D.

As Ben-El-Mechaiekh, Degurie, and Granas have seen in [4], Theorem 8 implies the following:

Theorem 9. Let X and Y be non-empty compact convex sets, each in a Hausdorff topological vector space. Let f, s, t, g be four real-valued functions on $X \times Y$ satisfying

(i) $f \leq s \leq t \leq g$ on $X \times Y$;

(ii) $y \to f(x,y)$ is lower semicontinuous on Y for each fixed $x \in X$;

(iii) $x \to s(x,y)$ is quasiconcave on X for each fixed $y \in Y$;

(iv) $y \to t(x,y)$ is quasiconvex on Y for each fixed $x \in X$;

(v) $x \to g(x,y)$ is upper semicontinuous on X for each fixed $y \in Y$.

Then the minimax inequality

$$\inf_{y \in Y} \sup_{x \in X} f(x,y) \leq \sup_{x \in X} \inf_{y \in Y} g(x,y)$$

holds.

Theorem 9 is due to Ben-El-Mechaiekh, Deguire, and Granas [4]. When $f \equiv s$, $t \equiv g$, Theorem 9 reduces to a result of Liu [18]. When $f \equiv s \equiv t \equiv g$, we obtain a general version of the von Neumann minimax principle due to Sion [21].

Theorem 10 (Minimax Principle of von Neumann). Let X and Y be two non-empty compact convex sets, each in a Hausdorff topological vector space. Let $f: X \times Y \to \mathbb{R}$ be such that

(i) $y \to f(x,y)$ is quasiconvex and lower semicontinuous for each fixed $x \in X$;

(ii) $x \to f(x,y)$ is quasiconcave and upper semicontinuous for each fixed $y \in Y$.

Then the minimax equality

$$\max_{x \in X} \min_{y \in Y} f(x,y) = \min_{y \in Y} \max_{x \in X} f(x,y)$$

holds.

5. VARIATIONAL INEQUALITIES

The theory of variational inequality appears in a natural way in the calculus of variations when a function is over a convex set of constraints. In this case the classical Euler equation must be replaced by a set of inequalities. A very basic variational inequality was established by Hartman and Stampacchia in 1966 [17]. Our aim in this section is to give a multivalued version of the Hartman-Stampacchia variational inequality. The main feature is the use of the Ky Fan minimax principle. (Multivalued transformations occur very often in partial differential equations, dynamical systems, mathematical programming, game theory.)

Let E be a Hausdorff topological vector space. We shall denote by E' the dual space of E (i.e., the vector space of all continuous linear functionals on E). We denote the pairing between E' and E by $\langle w,x \rangle$ for $w \in E'$ and $x \in E$. Let X be any non-empty subset of E. A set-valued map $T: X \to 2^{E'}$ is said to be monotone on X ([7], p. 79) if for all x and y in X, each u in T(x), and each w in T(y), $\mathrm{Re}\,\langle w-u, y-x \rangle \geq 0$. Let X and Y be two topological spaces. A set-valued map $T: X \to 2^Y$ is said to be lower semicontinuous on X [5, p. 109] if for every $x_0 \in X$ and any open set G in Y such that $T(x_0) \cap G \neq \phi$, there is a neighborhood U of x_0 in X such that $T(x) \cap G \neq \phi$ for every $x \in U$. In other words, $T: X \to 2^Y$ is lower semicontinuous on X if for every open set G in Y, the set $\{x \in X: T(x) \cap G \neq \phi\}$ is open in X.

We shall establish the following:

Theorem 11. Let X be a non-empty weakly compact convex set in a Hausdorff topological vector space E and $T: X \to 2^{E'}$ be a set-valued map such that for each $x \in X$, T(x) is a non-empty subset of E', and that T is monotone. Assume that for each one-dimensional flat $L \subset E$, $T|L \cap X$ is lower semicontinuous from the topology of E to the weak*-topology $\sigma(E', E)$ of E'. Then there exists a point $\hat{y} \in X$ such that

$$\sup_{w \in T(\hat{y})} \text{Re}\,\langle w, \hat{y}-x\rangle \leq 0 \quad \text{for all } x \in X$$

Proof. By monotonicity of T, for each $x, y \in X$, $u \in T(x)$ and $w \in T(y)$, we have

$$\text{Re}\,\langle u, y-x\rangle \leq \text{Re}\,\langle w, y-x\rangle$$

Thus,

$$\sup_{u \in T(x)} \text{Re}\,\langle u, y-x\rangle \leq \inf_{w \in T(y)} \text{Re}\,\langle w, y-x\rangle \quad \text{for all } x, y \in X$$

For each $x, y \in X$, define

$$f(x,y) := \sup_{u \in T(x)} \text{Re}\,\langle u, y-x\rangle$$

$$g(x,y) := \inf_{w \in T(y)} \text{Re}\,\langle w, y-x\rangle$$

Then:

(i) $f \leq g$ on $X \times X$;

(ii) $y \to f(x,y)$ is weakly lower semicontinuous on X for each fixed $x \in X$;

(iii) $x \to g(x,y)$ is quasiconcave on X for each fixed $y \in X$.

We now equip E with the weak topology and we find that all the conditions in Theorem A are satisfied. Since $g(x,x) = 0$ for all $x \in X$, according to Theorem A, there exists a point $\hat{y} \in X$ such that $f(x,\hat{y}) \leq 0$ for all $x \in X$; in other words,

$$\sup_{u \in T(z_t)} \text{Re}\,\langle u, \hat{y}-x\rangle \leq 0 \quad \text{for all } x \in X \tag{*}$$

Let $x \in X$ be arbitrarily fixed, let $z_t := tx + (1-t)\hat{y} \equiv \hat{y} - t(\hat{y}-x)$ for $t \in [0,1]$. As X is convex, we have $z_t \in X$ for $t \in [0,1]$. Therefore, by (*) we have

$$\sup_{u \in T(x)} \text{Re}\,\langle u, \hat{y}-z_t\rangle \leq 0 \quad \text{for all } t \in [0,1]$$

so that $t \cdot \sup_{u \in T(z_t)} \text{Re}\,\langle u, \hat{y}-x\rangle \leq 0$ for all $t \in [0,1]$ and it follows that

$$\sup_{u \in T(z_t)} \text{Re}\,\langle u, \hat{y}-x\rangle \leq 0 \quad \text{for all } t \in (0,1] \tag{**}$$

Let $w_0 \in T(\hat{y})$ be arbitrarily fixed. For each $\varepsilon > 0$, let

$$U_{w_0} := \{w \in E' : |\langle w_0 - w, \hat{y}-x\rangle| < \varepsilon\}$$

then U_{w_0} is a $\sigma(E', E)$-neighborhood of w_0. Since T is lower semicontinuous, and $U_{w_0} \cap T(\hat{y}) \neq \phi$, there exists a neighborhood $N(\hat{y})$ of $\hat{y}$ such that $z \in N(\hat{y})$ implies $T(z) \cap U_{w_0} \neq \phi$. Note that $z_t \to \hat{y}$ as $t \to 0^+$, thus there exists $\delta \in (0,1)$ such that for all $t \in (0,\delta)$, we have $z_t \in N(\hat{y})$. But then $T(z_t) \cap U_{w_0} \neq \phi$ for $t \in (0,\delta)$; take any $u \in T(z_t) \cap U_{w_0}$, we have

$$|\langle w_0 - u, \hat{y}-x\rangle| < \varepsilon$$

This implies

$$\text{Re}\,\langle w_0, \hat{y}-x\rangle < \text{Re}\,\langle u, \hat{y}-x\rangle + \varepsilon$$

By (**), Re $\langle w_0, \hat{y}-x\rangle \leq \varepsilon$. Since $\varepsilon > 0$ is arbitrary, Re $\langle w_0, \hat{y}-x\rangle \leq 0$. As $w_0 \in T(\hat{y})$ is arbitrary,

$$\sup_{w \in T(\hat{y})} \text{Re}\,\langle w, \hat{y}-x\rangle \leq 0 \quad \text{for all } x \in X$$

This concludes the proof of our theorem. Q.E.D.

In a reflexive Banach space E we may consider a convex lower semicontinuous function $j: X \to \mathbb{R}$, where X is a non-empty closed bounded convex subset of E, and ask the question: Find $\hat{y} \in X$ such that

$$\sup_{w \in T(\hat{y})} \text{Re}\,\langle w, \hat{y}-x\rangle \leq j(x) - j(\hat{y}) \quad \text{for all } x \in X$$

Such a problem occurs for example in the theory of non-Newtonian fluids [10]. The following result gives a solution of the above problem. However, the proof is a slight modification of Theorem 11.

Theorem 12. Let X be a nonempty closed bounded convex subset in a reflexive Banach space E and $T: X \to 2^{E'}$ be a set-valued map such that for each $x \in X$, $T(x)$ is a nonempty subset of E', and that T is monotone. Let $j: X \to \mathbb{R}$ be a convex lower semicontinuous function. Assume further that for each one-dimensional flat $L \subset E$, $T|L \cap X$ is lower semicontinuous. Then there exists a point $\hat{y} \in X$ such that

$$\sup_{w \in T(\hat{y})} \text{Re}\,\langle w, \hat{y}-x\rangle \leq j(x) - j(\hat{y}) \quad \text{for all } x \in X$$

Proof. For each $x, y \in X$, define

$$f(x,y) := \sup_{u \in T(x)} \text{Re}\,\langle u, y-x\rangle + j(y) - j(x)$$

$$g(x,y) := \inf_{w \in T(y)} \text{Re}\,\langle w, y-x\rangle + j(y) - j(x)$$

By monotonicity of T, we have $f \leq g$ on $X \times X$. Since j is convex lower semicontinuous on X in the reflexive Banach space E, j is weakly lower semicontinuous on X. Thus f and g satisfy the conditions (ii) and (iii) in Theorem A. Since $g(x,x) = 0$ for all $x \in X$, according to Theorem A, there exists a point $\hat{y} \in X$ such that $f(x,\hat{y}) \leq 0$ for all $x \in X$; in other words,

$$\sup_{u \in T(x)} \text{Re}\,\langle u, \hat{y}-x\rangle \leq j(x) - j(\hat{y}) \quad \text{for all } x \in X$$

The remaining part of the proof may be obtained along lines parallel to the proof of Theorem 11. Q.E.D.

Notice that a more complicated form of Theorem 11 was given in Shih and Tan [19]. For related results of Theorem 11 or 12, see also Browder [6], Dugundji and Granas [8], and Yen [22]. Further results on variational inequalities will be given in [20].

REFERENCES

1. Aubin, J.-P.: Applied Functional Analysis. Wiley (1979).
2. Aubin, J.-P.: Mathematical Methods of Game and Economic Theory. North-Holland (1979).
3. Ben-El-Mechaiekh, H.; Deguire, P.; Granas, A.: C.R. Acad. Sci. Paris. Sér. I 295 (1982), 257-259.

4. Ben-El-Mechaiekh, H.; Deguire, P.; Granas, A.: C.R. Acad. Sci. Paris (Sér. I) 292, 381-384 (1982).
5. Berge, C.: Topological Spaces. Oliver and Boyd (1963).
6. Browder, F. E.: Math. Ann. 17, 283-301 (1968).
7. Browder, F. E.: Nonlinear operators and nonlinear equations of evolution in Banach spaces, Proc. Sympos. in Pure Math., Vol. 18, Part 2. Amer. Math. Soc. (1976).
8. Dugundji, J.; Granas, A.: Ann. Scuolo, Norm Sup., Pisa (Ser. 4) 5, 679-682 (1978).
9. Dugundji, J.; Granas, A.: Fixed Point Theory, Vol. 1, Monografie Matematyczne Tom 61. PWN-Polish Scientific Publ. (1982).
10. Duvaut, G.; Lions, J. L.: Les inequations en mecanique et en physique. Dunod (1972).
11. Fan, K.: Convex Sets and Their Applications, Summer Lectures. Appl. Math. Div., Argonne Nat. Lab. (1959).
12. Fan, K.: Math. Ann. 142, 305-310 (1961).
13. Fan, K.: C.R. Acad. Sci. Paris, I 259, 3925-3928 (1964).
14. Fan, K.: Math. Ann. 163, 189-203 (1966).
15. Fan, K.: A minimax inequality and applications, in O. Shisha, ed., Inequalities III, Proc. of the Third Symposium on Inequalities. Academic Press (1972).
16. Granas, A.: KKM-maps and their applications to nonlinear problems, in R. D. Mauldin, ed., The Scotish Book. Birkhauser (1982).
17. Hartman, P.; Stampacchia, G.: Acta Math. 115, 271-310 (1966).
18. Liu, F.-C.: Bull. Inst. Math. Acad. Sinica 6, 512-524 (1978).
19. Shih, M.-H.; Tan, K.-K.: A further generalization of Ky Fan's minimax inequality and its applications (to appear in Studia Math.).
20. Shih, M.-H.; Tan, K.-K.: The Ky Fan minimax principle: Theory and applications (to appear).
21. Sion, M.: Pacific J. Math. 8, 171-176 (1958).
22. Yen, C.-L.: Pacific J. Math. 97, 477-481 (1981).

31 SOME REMARKS ABOUT VARIATIONAL PROBLEMS WITH CONSTRAINTS

Gerhard Ströhmer / Institut für Mathematik der Rheinisch-Westfälischen Technische Hochschule, Aachen, Federal Republic of Germany

1. INTRODUCTION

By a variational problem we mean a pair (Z, I) consisting of a set Z of admissible objects and a function $I: Z \to \mathbb{R}$. An element $u_0 \in Z$ with $I(u_0) \leq I(u)$ for all $u \in Z$ is called a solution of this variational problem.

Now let $\Omega \subset \mathbb{R}^n$ be a bounded domain and $F^0 : \bar{\Omega} \times \mathbb{R} \times \mathbb{R}^n \to \mathbb{R}$ a function ($F^0 = F^0(x, z, p)$). We assume $\partial\Omega$ to belong to $C^{2+\alpha}$ and F^0 to $C^{2+\alpha}_{loc}(\bar{\Omega} \times \mathbb{R} \times \mathbb{R}^n)$ for some $\alpha \in (0,1)$. For $u \in C^1(\bar{\Omega})$ we define

$$I_0(u) = \int_\Omega F^0(x, u, \nabla u)\, dx$$

With $\varphi \in C^{2+\alpha}(\bar{\Omega})$ and $Z_\varphi = \{u \in C^1(\bar{\Omega}) : u|\partial\Omega = \varphi|\partial\Omega\}$ we consider the variational problem (Z_φ, I_0). If F^0 is sufficiently nice (see, e.g., [5], Theorem 1.10.4, and [3], chapter 5, pp. 318-337), the set of solutions of (Z_φ, I_0) is nonempty, its elements u satisfy the Euler equation

$$\sum_{i=1}^n \frac{\partial}{\partial x_i}(F^0_{p_i}(x, u, \nabla u)) = F^0_z(x, u, \nabla u) \tag{E}$$

in the classical sense, and the $C^{2+\alpha}(\bar{\Omega})$-norm of all bounded classical solutions of E can be estimated from above. If F^1 is of the same type as F^0 we define

$$I_1(u) = \int_\Omega F^1(x, u, \nabla u)\, dx$$

and $Z_\varphi(C) = \{u \in Z_\varphi : I_1(u) \leq C\}$. The solutions of the problem $(Z_\varphi(C), I_0)$ satisfy its Euler-Lagrange equation

$$\sum_{i=1}^n \frac{\partial}{\partial x_i}(a_i^\lambda(x, u, \nabla u)) = a^\lambda(x, u, \nabla u) \tag{EL}$$

with

$$a_i^\lambda(x, z, p) = \lambda F^1_{p_i}(x, z, p) + F^0_{p_i}(x, z, p) \quad \text{for } i = 1, \ldots, n$$

$$a^\lambda(x, z, p) = \lambda F^1_z(x, z, p) + F^0_z(x, z, p) \quad \text{and } \lambda \geq 0$$

It now seems natural to ask whether estimates independent of λ similar to those known for (E) can be derived for (EL). This question is answered in the affirmative in Sec. 2 if $\lambda \geq 0$. (The case $\lambda < 0$ can obviously lead to equations of varying type.) Apart from being interesting in themselves, such estimates can be used to study variational problems without a constraint. Taking a regularity theorem for weak solutions of (EL) proved in Sec. 3 into account, they imply that for $u \in Z_\varphi$ with $I_1(u_0) = C$ there always is a $v \in C_0^\infty(\Omega)$ such that

$$\frac{d}{d\varepsilon}\bigg|_{\varepsilon=0} I_i(u + \varepsilon v) < 0 \qquad (i = 0,\ 1)$$

if C is sufficiently large. This means we can confine many of the investigations carried out in the calculus of variations to the sets $Z_\varphi(C)$. If we now choose

$$F^1(x, z, p) = (1 + |p|^2)^{q/2}$$

with $q > n$ these are compact with regard to weak convergence in $H_q^1(\Omega) \subset C^0(\bar{\Omega})$. The situation sketched here is used to prove the solvability of the problem (Z_φ, I_0) in Sec. 4. (The hypotheses differ from those of the references already given.) A somewhat more sophisticated type of reasoning leads to a mountain-pass lemma in [6].

Roughly the same idea was used by Lewy [4] and later by other authors [1, 2, 8, 9]. The difference is that they proceed by constraining the H_∞^1-norm, which is not given by an integral. So instead of solutions of (EL) they have to cope with so-called quasisolutions of (E). As H_∞^1 is not reflexive, this constraint also turns out to be inappropriate for the purpose of [6].

1.1 Notation

Let $B_r^n(x_0) = \{x \in \mathbb{R}^n : |x - x_0| < r\}$ and $e_k = (\delta_k^\nu)_{\nu=1}^n$. $\|\cdot\|_B$ denotes the norm of the Banach space B, constants are usually denoted by K or C. The meaning of K with arbitrary indices stays the same throughout the paper, while the meaning of C may differ in the proofs of different theorems. $C^\infty(\mathbb{R}^n)$ is the set of all functions defined on $\mathbb{R}^n$ that are differentiable infinitely often. Then for arbitrary open sets $O \subset \mathbb{R}^n$ we define

$$C^\infty(\bar{O}) = \{u : \bar{O} \to \mathbb{R} : \text{There is a } \bar{u} \in C^\infty(\mathbb{R}^n) \text{ with } u = \bar{u}|\bar{O}\}$$

For $u : O \to \mathbb{R}$ let $\operatorname{supp} u = \overline{\{x \in O : u(x) \neq 0\}}$. Then we define

$$C_0^\infty(O) = \{u \in C^\infty(\bar{O}) : \operatorname{supp} u \subset\subset O\}$$

The following Banach spaces are the completions of $C^\infty(\bar{O})$ with regard to the norms given:

$$C^0(\bar{O}) \ : \ \|u\|_{C^0(\bar{O})} = \max_{x\in\bar{O}} |u(x)|$$

$$C^k(\bar{O}) \ : \ \|u\|_{C^k(\bar{O})} = \|u\|_{C^{k-1}(\bar{O})} + \max_{x\in\bar{O}} |\nabla^k u| \quad (k \geq 1)$$

$$C^\beta(\bar{O}) \ : \ \|u\|_{C^\beta(\bar{O})} = \|u\|_{C^0(\bar{O})} + \sup_{x,x'\in\bar{O}} \frac{|u(x) - u(x')|}{|x - x'|^\beta}$$

$$C^{k+\beta}(\bar{O}) \ : \ \|u\|_{C^{k+\beta}(\bar{O})} = \|u\|_{C^k(\bar{O})} + \|\nabla^k u\|_{C^\beta(\bar{O})} \quad (k \geq 1,\ 0 < \beta < 1)$$

$$H_p^1(O) \ : \ \|u\|_{H_p^1(O)} = \|u\|_{L_p(O)} + \|\nabla u\|_{L_p(O)}$$

$$H_p^2(O) \ : \ \|u\|_{H_p^2(O)} = \|u\|_{L_p(O)} + \|\nabla u\|_{H_p^1(O)} \quad (1 \leq p < \infty)$$

$H^1_{p0}(O)$ is the completion of $C^\infty_0(O)$ with regard to $\|\cdot\|_{H^1_p(O)}$. If O is the set Ω mentioned in the introduction it is usually dropped, e.g., $C^0 = C^0(\bar{\Omega})$. The lower index loc indicates the set of all functions that belong to the corresponding spaces for all $O' \subset\subset O$. If $A \subset \mathbb{R}^n$ is measurable, mes (A) denotes its Lebesgue measure. Finally $f^+ = \max(f, 0)$ for arbitrary numeric functions f and $\mathbb{R}^+ = \mathbb{R}^+ \cup \{0\}$.

2. A PRIORI ESTIMATES

We assume that F^0 and F^1 generate regular variational problems with exponents m_0 and m_1, a property defined as follows:

2.1 Definition. A function $F: \bar{\Omega} \times \mathbb{R} \times \mathbb{R}^n \to \mathbb{R}$ generates a regular variational problem with an exponent $m > 1$ if $F \in C^{2+\alpha}_{loc}(\bar{\Omega} \times \mathbb{R} \times \mathbb{R}^n)$ and there are two monotonous functions $\nu_1, \nu_2 : \mathbb{R}^+ \to \mathbb{R}^+ \backslash \{0\}$ such that for all $\xi \in \mathbb{R}^n$

$$(1) \quad \nu_1(|z|)(1 + |p|)^{m-2}|\xi|^2 \le \sum_{i,j=1}^{n} F_{p_i p_j}(x, z, p)\xi_i \xi_j \le \nu_2(|z|)(1 + |p|)^{m-2}|\xi|^2$$

$$(2) \quad \sum_{i=1}^{n} (|F_{p_i z}| + |F_{p_i}|)(1 + |p|) + \sum_{i,j=1}^{n} |F_{p_i x_j}| + |F_z| \le \nu_2(|z|)(1 + |p|)^m$$

(ν_1 decreases, ν_2 increases).

Without loss of generality we assume ν_1 and ν_2 to be the same for both functions. For $\lambda \in \mathbb{R}$ and $t \ge 0$ we also define $f_1(t) = t^{m_0} + \lambda t^{m_1}$ and $f_2(t) = (1 + t)^{m_0 - 2} + \lambda(1 + t)^{m_1 - 2}$.

In this section we give a priori estimates for solutions of (EL) independent of $\lambda \in [0, \infty)$. For each individual λ, e.g., the theory developed in chapter 4 (pp. 244-291) of [3] can be applied to this equation, but the estimates thus obtained go to infinity as λ goes to zero. In spite of being stated for zero boundary values, the following theorems can be carried over to sufficiently regular nonzero ones by subtracting a suitably chosen function. This also is how we are going to use them later.

2.2 Theorem. There are two numbers $K_1 < \infty$ and $\alpha_1 > 0$ depending only on M, ν_1, ν_2, m_0, and m_1, such that for all $u \in C^2$ with $\|u\|_{C^0} \le M$ and $u|\partial\Omega = 0$ that fulfill equation (EL) in Ω and estimate $\|u\|_{C^{\alpha_1}} \le K_1$ applies.

Proof. The proof is similar to that given on pp. 249-251 in [3]. We also use roughly the same notation. Without loss of generality we suppose $0 \le \lambda \le 1$. If this is not true we divide the equation (EL) by λ and replace λ by $1/\lambda$, F^0 by F^1 and vice versa. We prove our assertion by showing u to belong to a class $\mathscr{B}_{m-}(\bar{\Omega}, M, c^*, \delta, 0)$ (see [3], p. 90) where $m- = \min(m_0, m_1, n)$. c^* and δ only depend on ν_1, ν_2, M.

Once ν_1, ν_2, m_0, and m_1 are given, there are two monotonous functions $\bar{\nu}_1, \bar{\nu}_2 : \mathbb{R}^+ \to \mathbb{R}^+ \backslash \{0\}$ such that

$$\sum_{i=1}^{n} a_i^\lambda(x, z, p)p_i \ge \bar{\nu}_1(|z|)f_1(|p|) - \bar{\nu}_2(|z|)$$

and

$$\sum_{i=1}^{n} |a_i^\lambda(x, z, p)|(1 + |p|) + |a^\lambda(x, z, p)| \le \nu_2(|z|)f_1(1 + |p|)$$

($\bar{\nu}_1$ decreases, $\bar{\nu}_2$ increases). For the following calculations note

$f_2(t)(1+t)^2 \leq 2^{m+}(2+f_1(t))$, $f_1(1+t) \leq 2^{m+}(2+f_1(t))$ and

$f_1(1+t) \leq f_2(t)(1+t)^2$ $(m+ = \max(m_0, m_1))$

We start by choosing a fixed $x_0 \in \bar{\Omega}$ and defining

$$A_{k,\rho} = \{x \in \bar{\Omega} : u(x) \geq k \text{ and } |x - x_0| < \rho\}$$

Let ξ belong to $C_0^\infty(B_\rho^n(x_0))$ and satisfy $0 \leq \xi \leq 1$. We also assume $k \geq u(x)$ for all $x \in \partial\Omega \cap B_\rho^n(x_0)$. With the test function $\eta(x) = \xi^{m+}(u-k)^+$ we obtain from the weak form of (EL)

$$\int_{A_{k,\rho}} \sum_{i=1}^n a_i^\lambda(x,u,\nabla u)u_{x_i}\xi^{m+}\,dx$$

$$= -\int_{A_{k,\rho}} \sum_{i=1}^n a_i^\lambda(x,u,\nabla u)(m+)\xi^{m+-1}\xi_{x_i}(u-k)^+ + a^\lambda(x,u,\nabla u)(u-k)^+\xi^{m+}\,dx$$

The left side of this equation can be estimated by

$$\bar{\nu}_1(M)\int_{A_{k,\rho}} f_1(|\nabla u|)\xi^{m+}\,dx - \bar{\nu}_2(M)\int_{A_{k,\rho}} \xi^{m+}\,dx$$

from below, the right one by

$$\nu_2(M)\int_{A_{k,\rho}} f_2(|\nabla u|)(1+|\nabla u|)(m+)\xi^{m+-1}|\nabla\xi||u-k| + f_1(1+|\nabla u|)\xi^{m+}|u-k|\,dx$$

from above. By the definition of $\mathscr{B}_{m-}(\bar{\Omega}, M, c^*, \delta, 0)$ we only have to consider the case $u(x) \leq k + \delta$ for all $x \in A_{k,\rho}$. We choose $\delta > 0$ small enough to enable us to subtract the second term of the expression estimating the right side from the left side in order to obtain

$$\int_{A_{k,\rho}} f_1(|\nabla u|)\xi^{m+}\,dx \leq C_1\int_{A_{k,\rho}} \xi^{m+} + |u-k|\xi^{m+-1}|\nabla\xi|(|\nabla u|^{m_0-1} + \lambda|\nabla u|^{m_1-1})\,dx$$

C_1 only depends on $\bar{\nu}_1$, $\bar{\nu}_2$, and $m+$. As a consequence of Young's inequality we have for $\varepsilon \in (0,1)$ and $i = 0, 1$

$$|u-k|\,|\nabla\xi|\xi^{m+-1}|\nabla u|^{m_i-1}$$

$$\leq \varepsilon^{-m_i}m_i^{-1}|u-k|^{m_i}|\nabla\xi|^{m_i} + \varepsilon(m_i-1)^{-1}m_i|\nabla u|^{m_i}\xi^{(m+-1)m_i(m_i-1)^{-1}}$$

$$\leq \varepsilon^{-m+}(m-)^{-1}|u-k|^{m_i}|\nabla\xi|^{m_i} + \varepsilon(m+-1)^{-1}(m+)\xi^{m+}|\nabla u|^{m_i}$$

The second inequality is due to the fact that $(m+-1)m_i(m_i-1)^{-1} \geq m+$ and $0 \leq \xi \leq 1$. Applying this to the last integral inequality we get

$$\int_{A_{k,\rho}} f_1(|\nabla u|)\xi^{m+}\,dx \leq C_2\int_{A_{k,\rho}} \xi^{m+} + \varepsilon^{-m+}f_1(|u-k||\nabla\xi|) + \varepsilon f_1(|\nabla u|)\xi^{m+}\,dx$$

With $\varepsilon > 0$ sufficiently small we obtain

$$\int_{A_{k,\rho}} (|\nabla u|^{m_0} + \lambda|\nabla u|^{m_1})\xi^{m+}\,dx \le C_3(\mathscr{I}_0 + \lambda\mathscr{I}_1)$$

where

$$\mathscr{I}_0 = \int_{A_{k,\rho}} \xi^{m+} + (u-k)^{m_0}|\nabla\xi|^{m_0}\,dx \quad \text{and} \quad \mathscr{I}_1 = \int_{A_{k,\rho}} (u-k)^{m_1}|\nabla\xi|^{m_1}\,dx$$

Let us first consider the case $\mathscr{I}_0 \ge \lambda\mathscr{I}_1$ and choose ξ as in ([3], p. 250, immediately after 1.12). We obtain thus $(0 < \sigma \le 1)$

$$\int_{A_{k,\rho-\sigma\rho}} |\nabla u|^{m_0}\,dx \le 2C_3\mathscr{I}_0 \le 2C_3((\sigma\rho)^{-m_0} \max_{x\in A_{k,\rho}} (u-k)^{m_0} + 1)\,\mathrm{mes}\,(A_{k,\rho})$$

So we have

$$\int_{A_{k,\rho-\sigma\rho}} |\nabla u|^{m-}\,dx \le \Big(\int_{A_{k,\rho-\sigma\rho}} |\nabla u|^{m_0}\,dx\Big)^{\frac{m-}{m_0}}\Big(\int_{A_{k,\rho}} 1\,dx\Big)^{1-\frac{m-}{m_0}}$$

$$\le (2C_3((\sigma\rho)^{-m_0} \max_{x\in A_{k,\rho}} (u-k)^{m_0} + 1)\,\mathrm{mes}\,(A_{k,\rho}))^{\frac{m-}{m_0}}(\mathrm{mes}\,(A_{k,\rho}))^{1-\frac{m-}{m_0}}$$

$$\le (2C_3)^{\frac{m-}{m_0}}((\sigma\rho)^{-m-} \max_{x\in A_{k,\rho}} (u-k)^{m-} + 1)\,\mathrm{mes}\,(A_{k,\rho})$$

In the other case we get

$$\lambda\int_{A_{k,\rho}} |\nabla u|^{m_1}\xi^{m+}\,dx \le \lambda 2C_3\mathscr{I}_1$$

After dividing this inequality by λ and proceeding as before, we put the results of both cases together and arrive at the following one:

$$\int_{A_{k,\rho-\sigma\rho}} |\nabla u|^{m-}\,dx \le C_4((\sigma\rho)^{-m-} \max_{x\in A_{k,\rho}} (u-k)^{m-} + 1)\,\mathrm{mes}\,(A_{k,\rho})$$

The argument just described also being valid for $-u$, u belongs to the class $\mathscr{B}_m(\bar{\Omega}, M, C_4, \delta, 0)$.

<u>2.3 Theorem</u>. There is a constant K_2 depending only on M, ν_1, and ν_2 such that for all $u \in C^2$ with $u|\partial\Omega = 0$ and $\|u\|_{C^0} \le M$ satisfying the equation (EL) in Ω, the estimate $|\nabla u|\,\big|_{\partial\Omega} \le K_2$ is valid.

<u>Proof</u>. Calculating the divergence occurring in (EL) and dividing the result by $f_2(|p|)$ we find that u satisfies the equation

$$\sum_{i,j=1}^{n} a_{ij}^{\lambda}(x,u,\nabla u)u_{x_i x_j} = \hat{a}^{\lambda}(x,u,\nabla u)$$

with

$$a_{ij}^{\lambda}(x,z,p) = f_2^{-1}(|p|)a_{ip_j}^{\lambda}(x,z,p)$$

and

$$\hat{a}^{\lambda}(x,z,p) = f_2^{-1}(|p|)\left[\sum_{i=1}^{n}(-a_{iz}^{\lambda}(x,z,p)p_i - a_{ix_i}^{\lambda}(x,z,p)) + a^{\lambda}(x,z,p)\right]$$

One easily sees that

$$\nu_1(|z|) \leq \sum_{i,j=1}^{n} a_{ij}^{\lambda}(x,z,p)\,\xi_i\xi_j \leq \nu_2(|z|) \quad \text{for } |\xi| = 1 \quad \text{and}$$

$$|\hat{a}^{\lambda}| \leq \nu_2(|z|)(1+|p|)^2$$

The assertion can now be proved by using Lemma 2.1 (p. 351) in [3].

<u>2.4 Theorem</u>. Depending on M, ν_1, ν_2, m_1, m_2, and $\bar{q}$ we can choose a number K_3 such such that for $\bar{q} \in (1,+\infty)$ and all solutions $u \in C^2$ of (EL) with $u|\partial\Omega = 0$ and $\|u\|_{C^0} \leq M$ the inequality $\|u\|_{H^1_{\bar{q}}} \leq K_3$ is true.

<u>Proof</u>. Without loss of generality we suppose $0 \leq \lambda \leq 1$. Replacing $|Du|^{\tau}$ and $(1+|Du|)^{\tau}$ by $|\nabla u|^{m_0-2} + \lambda|\nabla u|^{m_1-2}$ and $(1+|\nabla u|)^{m_0-2} + \lambda(1+|\nabla u|)^{m_1-2}$ in the proof of 14.4(ii) in [1] (pp. 316-319) up to the last inequality on page 318, we obtain

$$\int_{\Omega} \eta^2(1+|\nabla u|)^{2\beta+4}((1+|\nabla u|)^{m_0-2} + \lambda(1+|\nabla u|)^{m_1-2}))\,dx \leq C_1(\mathcal{I}_0 + \lambda\mathcal{I}_1)$$

where

$$\mathcal{I}_0 = \int_{\Omega} \eta^2(1+|\nabla u|)^{2(\beta+1)} + |\nabla\eta|^2(1+|\nabla u|)^{2\beta+m_0}\,dx \quad \text{and}$$

$$\mathcal{I}_1 = \int_{\Omega} |\nabla\eta|^2(1+|\nabla u|)^{2\beta+m_1}\,dx$$

For $\mathcal{I}_0 \geq \lambda\mathcal{I}_1$ we get

$$\int_{\Omega} \eta^2(1+|\nabla u|)^{2\beta+m_0+2}\,dx \leq 2C_1\int_{\Omega} \eta^2(1+|\nabla u|)^{2(\beta+1)} + |\nabla\eta|^2(1+|\nabla u|)^{2\beta+m_0}\,dx$$

From this point we can continue the argument as given in [1] on page 319 and get

$$\int_{\Omega} [\eta(1+|\nabla u|)]^{2\beta+m_0+2}\,dx \leq C_2$$

In the other case we have

$$\lambda\int_{\Omega} \eta^2(1+|\nabla u|)^{2\beta+m_1+2}\,dx \leq 2\lambda C_1\int_{\Omega} |\nabla\eta|^2(1+|\nabla u|)^{2\beta+m_1}\,dx$$

After dividing by λ we proceed as before to obtain

$$\int_{\Omega} [\eta(1+|\nabla u|)]^{2\beta+m_1+2}\,dx \leq C_3$$

In both cases the estimate

$$\int_\Omega [\eta(1 + |\nabla u|)]^{2\beta+\bar{m}+2}\, dx \leq C_4$$

is true with $\bar{m} = \min(m_0, m_1)$. Although this only leads to an interior one, an estimate up to the boundary can be derived replacing the function v appearing in [1] by $\tilde{v} = (|\nabla u| - 2 \max_{\partial\Omega} |\nabla u|)^+$.

2.5 Theorem. There is a constant K_4 that only depends on M, ν_1, ν_2, m_0, and m_1 such that for all $u \in C^2$ with $u|\partial\Omega = 0$ and $\|u\|_{C^0} \leq M$ fulfilling (EL) in Ω the inequality $\|u\|_{C^1} \leq K_4$ is satisfied.

Proof. Without loss of generality suppose $0 \leq \lambda \leq 1$. We proceed as in the proof of theorem 14.4(i) (p. 316) in [1]. Instead of the function used there we choose

$$\bar{v}(x) = \int_0^{v(x)} (1 + \sqrt{t})^{m_0 - 2} + \lambda(1 + \sqrt{t})^{m_1 - 2}\, dt$$

The assertion then follows from 8.15 (p. 179) in [1], 2.3 and 2.4.

2.6 Theorem. For $r > 0$ let $G^*(r) = \Sigma_{i=0}^{1} \|F^i|\bar{B}_r^{2n+1}\|_{C^{2+\alpha}(\bar{B}_r^{2n+1})}$. Once G^*, M, ν_1, ν_2, m_0, and m_1 are given, there is a constant K_5 such that for all solutions $u \in C^{2+\alpha}$ of (EL) with $u|\partial\Omega = 0$ and $\|u\|_{C^0} \leq M$ we have $\|u\|_{C^{2+\alpha}} \leq K_5$.

Proof. Once again let $0 \leq \lambda \leq 1$ By virtue of 2.5 and theorem 12.2 (p. 268) of [1] an estimate for $\|u\|_{C^{1+\alpha_2}}$ with some $\alpha_2 \in (0,1)$ can be concluded. The linear theory then gives the remainder of the assertion. (See 6.7 (p. 94) in [1].)

3. A REGULARITY THEOREM

We now derive a regularity theorem for weak solutions of (EL), using the notation of Sec. 2 and its hypotheses concerning F^0 and F^1. In addition we suppose $F^1_z = 0$,

$$\sum_{i,j=1}^{n} |F^1_{x_i p_j}| \leq K_6(1 + |p|)^{m_1 - 1}$$

and $m_1 > \max(n, 4m_0)$. The proof combines the ideas of [3] (chapter 4.5, pp. 270-277) and [1] (chapter 8.3, pp. 173-177), making use of the fact $a^\lambda(x, u, \nabla u) \in L_4$. So a result is obtained that is more suitable for our present purpose than what could be achieved by means of theorems 5.1 and 5.2 (pp. 276, 277) in [3].

3.1 Theorem. If $u \in H^1_{m_1 0}$ and

$$\int_\Omega \sum_{i=1}^{n} a_i^\lambda(x, u, \nabla u)\, \psi_{x_i}\, dx = - \int_\Omega a^\lambda(x, u, \nabla u)\, \psi\, dx$$

for all $\psi \in C_0^\infty(\Omega)$ then $u \in C^{2+\alpha}$.

Remark. 3.1 is also valid if u has sufficiently regular nonzero boundary values.

To prove 3.1 we assume 3.2 which we shall prove later.

3.2 Lemma. With the same hypotheses as in 3.1, u belongs to $H^2_{\bar{m}}$ ($\bar{m} = \min(m_0, 2)$) and

$$\int_\Omega f_2(|\nabla u|) \sum_{k,j=1}^{n} u^2_{x_k x_j}\, dx < +\infty$$

Proof of 3.1. As a consequence of Lemma 4.5 (p. 63) in [3] we have

$$\int_\Omega f_2(|\nabla u|)(1 + |\nabla u|)^4\, dx < +\infty$$

Making use of theorem 4.1 (p. 270) in [3] we can then conclude $u \in H^2_2 \cap H^1_\infty$. Hence we get $u \in H^2_2 \cap C^{1+\alpha^*}$ with an $\alpha^* \in (0,1)$ by 6.1 (p. 281) in [3] and—by 6.3 (p. 283) in [3]—the statement of the theorem.

Proof of 3.2. The constants appearing below may depend on λ, but not on h. Note that $H^1_{m_1} \subset C^0$ because $m_1 > n$. We fix $k \in \{1,\ldots,n\}$ and define with $h_k = he_k$ wherever possible

$$(\Delta^h_k u)(x) = \frac{u(x + h_k) - u(x)}{h}$$

$$\bar{a}^h_{ij}(x) = \int_0^1 a^\lambda_{ip_j}(x + h_k, u(x + h_k), (1 - \tau)\nabla u(x) + \tau\nabla u(x + h_k))\, d\tau$$

$$a^h_i(x) = \int_0^1 a^\lambda_{iz}(x + h_k, (1 - \tau)u(x) + \tau u(x + h_k), \nabla u(x))\, d\tau$$

$$\tilde{a}^h_i(x) = \int_0^1 a^\lambda_{ix_k}(x + \tau h_k, u(x), \nabla u(x))\, d\tau \quad \text{and}$$

$$g^h(x) = \int_0^1 f_2(|(1 - \tau)\nabla u(x) + \tau\nabla u(x + h_k)|)\, d\tau$$

As is easily seen there are positive constants C_1 and C_2 satisfying

$$C_2|\xi|^2 g^h(x) \le \sum_{i,j=1}^{n} \bar{a}^h_{ij}\xi_i\xi_j \le C_1 g^h(x)|\xi|^2$$

$$|\bar{a}^h_i(x)| \le C_1(1 + |\nabla u|)^{m_0 - 1}, \quad |\tilde{a}^h_i(x)| \le C_1 f_2(|\nabla u|)(1 + |\nabla u|) \quad \text{and}$$

$$f_2(|\nabla u| + |\nabla u(x + h_k)|) \le C_1 g^h(x) \quad \text{for } \xi \in \mathbb{R}^n$$

(Note $a^\lambda_{iz} = F^0_{zp_j} + \lambda F^1_{zp_j} = F^0_{zp_j}$.) Let $\eta \in C^\infty(\mathbb{R}^n)$ $(0 \le \eta \le 1)$ and $h^0 > 0$ be such that for $|h| < h^0$ and $x \in A = \operatorname{supp}\eta \cap \bar{\Omega}$ the point $x + h_k$ also belongs to $\bar{\Omega}$. With $v = \Delta^h_k u\eta^2$ we obtain, using "integration by parts,"

$$\int_\Omega a^\lambda(x, u, \nabla u)\Delta^{-h}_k v\, dx = \int_\Omega \sum_{i=1}^{n} \Delta^h_k(a^\lambda_i(x, u, \nabla u))v_{x_i}\, dx$$

$$= \int_\Omega \sum_{i,j=1}^{n} \bar{a}^h_{ij}\Delta^h_k u_{x_j} v_{x_i} + \sum_{i=1}^{n} (\bar{a}^h_i(x)\Delta^h_k u\, v_{x_i} + \tilde{a}^h_i v_{x_i})\, dx$$

Now we estimate all but one of the appearing expressions. For all $\varepsilon > 0$ we have

$$\int_\Omega |\tilde{a}_i^h v_{x_i}| \, dx \le C_1 \int_\Omega f_2(|\nabla u|)(1 + |\nabla u|)|\nabla v| \, dx$$

$$\le C_1 \int_\Omega f_2(|\nabla u(x)| + |\nabla u(x + h_k)|)(1 + |\nabla u(x)| + |\nabla u(x + h_k)|)|\nabla v| \, dx$$

$$\le C_1 \varepsilon^{-1} \int_A f_2(|\nabla u(x)| + |\nabla u(x + h_k)|)(1 + |\nabla u(x)| + |\nabla u(x + h_k)|)^2 \, dx$$

$$+ C_1 \varepsilon \int_\Omega f_2(|\nabla u(x)| + |\nabla u(x + h_k)|)|\nabla v|^2 \, dx$$

$$\le C_3' \varepsilon^{-1} + \varepsilon C_1^2 \int_\Omega g^h(x)|\nabla v|^2 \, dx$$

The last inequality is due to the fact that the integrand in the first integral is bounded in L_1 independent of h. We thus obtain

$$\int_\Omega |\tilde{a}_i^h v_{x_i}| \, dx \le C_3 \varepsilon^{-1} + \varepsilon C_3 \int_\Omega g^h(x)|\nabla v|^2 \, dx$$

In addition we have

$$\int_\Omega |\bar{a}_i^h(x)\Delta_k^h u v_{x_i}| \, dx \le \frac{1}{\varepsilon} \int_A \frac{(\bar{a}_i^h(x)\Delta_k^h u)^2}{g^h(x)} \, dx + \varepsilon \int_\Omega g^h(x)|\nabla v|^2 \, dx$$

$$\le C_4 \varepsilon^{-1} + \varepsilon \int_\Omega g^h(x)|\nabla v|^2 \, dx$$

because for $\lambda \neq 0$ or $m_0 \ge 2$ ($m^* = m_1(m\;\; - 1)^{-1}$, $\check{m} = m_2(2 - m_1)^{-1}$)

$$\|(g^h(x))^{-1}\|_{L_\infty(A)} + \|\bar{a}_i^h(x)\|_{L_{m^*}(A)} + \|\Delta_k^h u\|_{L_{m_2}(A)} \le C_4'$$

and for $\lambda = 0$ and $m_0 < 2$ at least $\|(g^h(x))^{-1}\|_{L_{\check{m}}(A)} \le C_4''$.

Last but not least

$$\int_\Omega |a^\lambda(x, u, \nabla u)\Delta_k^{-h} v| \, dx \le \|a^\lambda\|_{L_4} \|\Delta_k^{-h} v\|_{L_{4/3}} \le C_5' \|\nabla v\|_{L_{4/3}}$$

$$\le C_5' \|(g^h)^{-\frac{1}{2}}\|_{L_4(A)} \|(g^h)^{\frac{1}{2}}|\nabla v|\|_{L_2} \le C_5'' \|(g^h)^{\frac{1}{2}}|\nabla v|\|_{L_2}$$

$$\le C_5 \varepsilon^{-1} + C_5 \varepsilon \int_\Omega g^h |\nabla v|^2 \, dx$$

Putting these inequalities together we get

$$\int_\Omega \sum_{i,j=1}^n \bar{a}_{ij}^h(x)\Delta_k^h(u_{x_i}) v_{x_j} \, dx \le C_6 \varepsilon^{-1} + \varepsilon C_6 \int_\Omega g^h |\nabla v|^2 \, dx$$

for all $\varepsilon > 0$. Restricting ourselves to $\varepsilon \in (0, 1]$ and replacing v by $\Delta_k^h(u)\eta^2$ on the right side of the inequality we find

$$\int_\Omega \sum_{i,j=1}^n \bar{a}_{ij}^h \Delta_k^h(u_{x_j}) v_{x_i} \, dx \le C_7 \varepsilon^{-1} + \varepsilon C_7 \int_\Omega g^h(x)|\Delta_k^h \nabla u|^2 \eta^2 \, dx$$

We now do the same on the left side.

$$\int_\Omega \sum_{i,j=1}^n \bar{a}_{ij}^h(x)\Delta_k^h(u_{x_j})\Delta_k^h(u_{x_i})\eta^2\, dx$$

$$\leq \varepsilon^{-1}C_7 + \varepsilon C_7 \int_\Omega g^h(x)|\Delta_k^h\nabla u|^2\eta^2\, dx + \int_\Omega \left|\sum_{i,j=1}^n \eta\bar{a}_{ij}^h\Delta_k^h(u_{x_j})\eta_{x_i}\Delta_k^h(u)\right| dx$$

$$\leq \varepsilon^{-1}C_7 + \varepsilon C_7 \int_\Omega g^h(x)|\Delta_k^h\nabla u|^2\eta^2\, dx + C_1\int_\Omega g^h(x)|\nabla\Delta_k^h(u)||\nabla\eta||\eta||\Delta_k^h u|\, dx$$

$$\leq \varepsilon^{-1}C_7 + \varepsilon C_7 \int_\Omega g^h(x)|\Delta_k^h\nabla u|^2\eta^2\, dx + \varepsilon^{-1}C_1\int_\Omega g^h(x)|\nabla\eta|^2|\Delta_k^h(u)|^2\, dx$$

$$+ C_1\varepsilon\int_\Omega g^h(x)|\nabla\Delta_k^h(u)|^2\eta^2\, dx$$

Choosing ε sufficiently small we get

$$\int_\Omega \eta^2|\Delta_k^h\nabla u|g^h(x)\, dx \leq C_8\left(1 + \int_\Omega g^h(x)|\nabla\eta|^2|\Delta_k^h(u)|^2\, dx\right) \leq C_9$$

for all h with $|h| < h^0$. If $\eta \in C_0^\infty(\Omega)$ we can always find an $h^0 > 0$ fulfilling the conditions stated at the beginning of the proof. For every domain $\Omega' \subset\subset \Omega$ therefore

$$\|(g^h)^{\frac{1}{2}}|\Delta_k^h\nabla u|\|_{L_2(\Omega')} \leq C_{10}(\Omega')$$

independent of h. In addition

$$\|\Delta_k^h\nabla u\|_{L_{\bar{m}}(\Omega')} \leq \|(g^h)^{-\frac{1}{2}}\|_{L_{\tilde{m}^*}(\Omega')}\|(g^h)^{\frac{1}{2}}\Delta_k^h\nabla u\|_{L_2(\Omega')} \leq C_{11}(\Omega')$$

as $(g^h)^{-\frac{1}{2}}$ is bounded on $L_{\tilde{m}^*}(\Omega')$. ($\tilde{m}^* = \infty$ for $\bar{m} = 2$ and $\tilde{m}^* = 2\bar{m}(2 - \bar{m})^{-1}$ for $\bar{m} < 2$). The function u therefore belongs to $H^2_{\bar{m}\ \mathrm{loc}}(\Omega)$ and $\Delta_k^{h_\nu}\nabla u$ converges to $(\nabla u)_{x_k}$ almost everywhere for a suitably chosen sequence h_ν. So we have

$$\int_\Omega \eta^2 f_2(|\nabla u|)|(\nabla u)_{x_k}|^2\, dx \leq C_{12}(\eta) \quad \text{and}$$

$$f_2(|\nabla u|)\sum_{i,k=1}^n u_{x_ix_k}^2 \in L_{1\ \mathrm{loc}}$$

As the classes of functions that F^0 and F^1 belong to are invariant under changes of coordinates in the x-space, we only have to consider boundaries of the type $x_n = 0$. In this case we can estimate as before the second derivatives of u with the exception of $u_{x_nx_n}$. To bridge this gap we have to take into account that u satisfies (EL) almost everywhere.

4. ABSOLUTE MINIMA OF I_0

This section aims at solving the problem (Z_φ, I_0). In order to do this we have to assume that F^0 generates a regular variational problem with an exponent $m_0 > 1$. Then we choose

$$F^1(x,z,p) = (1 + |p|^2)^{q/2}$$

with $q > \max(n, 4m_0)$. It is easy to verify that F^1 generates a regular variational problem with the exponent q and satisfies the additional requirements of Sec. 3. In order to prove the following theorem we have to suppose in addition that the set

$$\{\|u\|_{C^0} : u \in C^2 \text{ and solves } (Z_\varphi(C), I_0) \text{ for some } C \in \mathbb{R}^+\}$$

is bounded. Then we have

4.1 Theorem. There is a solution u_0 of (Z_φ, I_0) which belongs to $C^{2+\alpha}$ and therefore solves (E).

Theorem 4.1 is not obviously implied by the results given in [3] (chapter 5, pp. 318-337), [5] (theorem 1.10.4) and [8] as these make additional assumptions about the growth of certain derivatives of F^0 or impose other conditions. Neither can it be deduced easily using [7] because boundary regularity is not treated there. We finally should not fail to mention [1] and [8]. On the other hand the results mentioned above are not contained in 4.1 on account of our boundedness assumption. This restriction, however, does not seem too substantial, as, e.g., the arguments given in [3] (chapter 5.3, pp. 327-330, chapter 4.7, pp. 285-290) concerning L_∞-estimates can be adapted to the situation given here.

The following convergence lemma is going to be used to prove 4.1. It goes beyond what is necessary for 4.1 in order to provide for the needs of [6]. The usual lower semi-continuity theorems cannot be applied here as we do not assume $F^0 \geq \phi$ with $\phi \in L_1$.

4.2 Lemma. Let $F \in C^2_{loc}(\bar{\Omega} \times \mathbb{R} \times \mathbb{R}^n)$ $(F = F(x, z, p))$ and suppose there is an increasing function $C_1 : \mathbb{R}^+ \to \mathbb{R}^+$ such that

$$\sum_{i=1}^n |F_{p_i}|(1 + |p|) + |F_z| \leq C_1(|z|)(1 + |p|)^q$$

In addition we assume that $(F_{p_ip_j}(x, z, p))$ is a positive definite matrix for all $x \in \bar{\Omega}$, $z \in \mathbb{R}$, and $p \in \mathbb{R}^n$. For $u \in H^1_q$ we then define

$$I(u) = \int_\Omega F(x, u, \nabla u)\, dx$$

Let $\{u_k\}$ be a bounded sequence in H^1_q converging in C^0 to $u \in C^0$. Then $u \in H^1_q$ and $I(u) \leq \liminf_{k\to\infty} I(u_k)$. If $I(u_k) \to I(u)$ then $u_k \to u$ in $H^1_{q'}$ for all $q' \in [1, q)$. If $u_k \to u$ in H^1_q then $I(u_k) \to I(u)$.

Proof. Obviously $u_k \rightharpoonup u$ in H^1_q and $\|u\|_{H^1_q} < \infty$. There is a positive decreasing function $C_2 : \mathbb{R}^+ \to \mathbb{R}^+$ such that

$$\sum_{i,j=1}^n F_{p_ip_j}(x, z, p)\xi_i\xi_j \geq C_2(|z| + |p|)|\xi|^2$$

for all $x \in \bar{\Omega}$, $z \in \mathbb{R}$, $\xi, p \in \mathbb{R}^n$. Now for $p, p' \in \mathbb{R}^n$ and $\tau \in [0, 1]$ we consider the function $f(\tau) = F(x, z, \tau p + (1 - \tau)p')$. Then we have

$$F(x, z, p) - F(x, z, p') = f(1) - f(0) = \int_0^1 f'(\tau)\, d\tau = f'(0) + \int_0^1 (1 - \tau) f''(\tau)\, d\tau$$

$$= \sum_{i=1}^n F_{p_i}(x, z, p')(p_i - p_i') + \int_0^1 (1 - \tau) \sum_{i,j=1}^n [F_{p_ip_j}(x, z, \tau p + (1 - \tau)p')(p_i - p_i')(p_j - p_j')]\, d\tau$$

$$\geq \sum_{i=1}^{n} F_{p_i}(x,z,p')(p_i - p'_i) + \frac{1}{2}C_2(|z| + |p| + |p'|)|p - p'|^2$$

With $f_k(x) = |u(x)| + |\nabla u(x)| + |\nabla u_k(x)|$ this leads to the estimate

$$I(u_k) - I(u) \geq \int_\Omega \sum_{i=1}^{n} F_{p_i}(x,u,\nabla u)((u_k)_{x_i} - u_{x_i})\, dx + \frac{1}{2}\int_\Omega C_2(f_k)|\nabla u_k(x) - \nabla u(x)|^2\, dx$$

$$- C_1(\|u\|_{C^0} + \|u_k\|_{C^0})\|u - u_k\|_{C^0}\|(1 + |\nabla u_k|)\|_{L_q}$$

Thus we can conclude $\liminf_{k\to\infty}(I(u_k) - I(u)) \geq 0$. If $I(u_k) \to I(u)$ then

$$\lim_{k\to\infty} \int_\Omega C_2(f_k)|\nabla u_k(x) - \nabla u(x)|^2\, dx = 0 \quad \text{and}$$

$$\lim_{k\to\infty} \int_{\{x\in\Omega : f_k \leq M\}} |\nabla u_k(x) - \nabla u(x)|^2\, dx = \lim_{k\to\infty} \int_{\{f_k\leq M\}} |\nabla u_k(x) - \nabla u(x)|^{q'}\, dx = 0$$

Also

$$\int_{\{f_k>M\}} |\nabla u_k - \nabla u|^{q'}\, dx \leq \int_{\{f_k>M\}} (f_k + 1)^{q'-q}(f_k + 1)^q\, dx$$

$$\leq (M + 1)^{q'-q} \int_{\{f_k>M\}} (f_k + 1)^q\, dx \leq C_3(M + 1)^{q'-q}$$

as f_k is a bounded sequence in L_q. So finally

$$\limsup_{k\to\infty} \int_\Omega |\nabla u_k - \nabla u|^{q'}\, dx \leq C_3(M + 1)^{q-q'}$$

for all $M \in \mathbb{R}^+$, i.e., $\lim_{k\to\infty} \|u_k - u\|_{H^1_{q'}} = 0$.

To finish the proof we only have to show that I is continuous on H^1_q. For $u, u' \in H^1_q$ we have

$$I(u) - I(u') \leq \int_\Omega |F(x,u,\nabla u) - F(x,u',\nabla u)| + |F(x,u',\nabla u) - F(x,u',\nabla u')|\, dx$$

$$\leq C_1(\|u\|_{C^0} + \|u'\|_{C^0})\|(1 + |\nabla u|)\|^q_{L_q} \|u - u'\|_{C^0}$$

$$+ C_1(\|u'\|_{C^0}) \int_\Omega (1 + |\nabla u| + |\nabla u'|)^{q-1}|\nabla u - \nabla u'|\, dx$$

$$\leq C_4(\|u\|_{H^1_q} + \|u'\|_{H^1_q})\|u - u'\|_{H^1_q}$$

To prove 4.1 let us extend I_0 and I_1 to H^1_q and consider $\tilde{Z}_\varphi = \{u \in H^1_q : u|\partial\Omega = \varphi|\partial\Omega\}$ and $\tilde{Z}_\varphi(C) = \{u \in \tilde{Z}_\varphi : I_1(u) \leq C\}$. On account of 4.2 there is an $a \in \mathbb{R}^+$ such that $\tilde{Z}_\varphi(C) \neq \emptyset$ for $C \geq a$ and $\tilde{Z}_\varphi(C) = \emptyset$ for $C < a$. If $\tilde{Z}_\varphi(C) \neq \emptyset$ there also is a solution u^C of $(I_0, \tilde{Z}_\varphi(C))$. We shall now look for solutions of (I_0, Z_φ) among the functions u^C. This enables us to stay inside the space H^1_q which would not be possible if we were to use the direct methods of the calculus of variations: Then we should have to deal with functions which only belong to $H^1_{m_0}$.

Now for $i = 0, 1$ and $u, v \in H^1_q$ let

$$V_i(u,v) = \int_\Omega \sum_{\ell=1}^n F^i_{p_\ell}(x,u,\nabla u)v_{x_\ell} + F^i_z(x,u,\nabla u)\, v\, dx$$

If $\psi \in C_0^\infty(\Omega)$ and $u \in H^1_q$ we have

$$I_i(u + \varepsilon\psi) = I_i(u) + \int_0^\varepsilon V_i(u + \tau\psi, \psi)\, d\tau$$

If, in addition, $V_1(u, \psi) \neq 0$ on account of Lebesgue's theorem there is an $\varepsilon_0 > 0$ such that

$$|I_i(u + \varepsilon\psi) - I_i(u) - \varepsilon V_i(u,\psi)| \le \frac{|\varepsilon|}{2}|V_i(u,\psi)| \qquad (|\varepsilon| < \varepsilon_0)$$

because

$$|F^i_{p_1}(x, u + \tau\psi, \nabla u + \tau\nabla\psi)| + |F^i_z(x, u + \tau\psi, \nabla u + \tau\nabla\psi)| \le C_1(1 + |\nabla u|)^q$$

for $|\tau| \le 1$. In consequence of this $V_1(u^a, \psi) = 0$ for all $\psi \in C_0^\infty(\Omega)$ and therefore all $\psi \in H^1_{q0}$.

We now show that u^C belongs to $C^{2+\alpha}$ and fulfills equation (E) for C sufficiently large. We consider two cases.

Case 1. If $V_1(u^C, \cdot)|H^1_{q0} = 0$ and $C > a$ then

$$0 = V_1(u^C, u^C - u^a) - V_1(u^a, u^C - u^a) = \int_\Omega \sum_{i=1}^n (F^1_{p_i}(\nabla u^C) - F^1_{p_i}(\nabla u^C))(u^C_{x_i} - u^a_{x_i})\, dx$$

This integral can only be zero if $u^C = u^a$. So $I_1(u^C) \le a < C$ and therefore $V_0(u^C, \cdot)|H^1_{q0} = 0$. By 3.1, u^C also has to belong to $C^{2+\alpha}$ and to satisfy (E).

Case 2. Let us now assume $V_1(u^C, \cdot)|H^1_{q0} \neq 0$. Then for every $\psi \in C_0^\infty$ with $v_1(u^C, \psi) = 0$ the number $V_0(u^C, \psi)$ must also equal zero. Otherwise we could find a $\psi_1 \in C_0^\infty(\Omega)$ with $V_1(u^C, \psi_1) < 0$ and $V_0(u^C, \psi_1) < 0$. Now we choose $\psi_2 \in C_0^\infty(\Omega)$ with $V_1(u^C, \psi_2) = 1$. Then $\lambda = -V_0(u^C, \psi_2)$ must be nonnegative, otherwise $V_1(u^C, -\psi_2) < 0$ and $V_0(u^C, -\psi_2) < 0$. So for all $\psi \in C_0^\infty$ we have

$$V_0(u^C, \psi) = V_0(u^C, V_1(u^C, \psi)\psi_2 + (\psi - V_1(u^C, \psi)\psi_2))$$

$$= V_1(u^C, \psi)\, V_0(u^C, \psi_2) = -\lambda V_1(u^C, \psi)$$

as

$$V_1(u^C, \psi - V_1(u^C, \psi)\psi_2) = V_1(u^C, \psi) - V_1(u^C, \psi) = 0$$

Thus we have found $V_0(u^C, \psi) + \lambda V_1(u^C, \psi) = 0$ for all $\psi \in C_0^\infty$. As $u^C \in C^{2+\alpha}$ by 3.1, u^C also solves (EL). Obviously u^C solves $(I_0, Z_\varphi(C))$ as well, so by means of the assumption made at the beginning of Sec. 4 $\|u^C\|_{C^0} \le C_2$ independent of C. On account of 2.6 we also have $\|u^C\|_{C^{2+\alpha}} \le C_3$ and $I_1(u^C) \le C_4$. So $V_0(u^C, \cdot)|H^1_{q0}$ must be zero for $C > C_4$.

In both cases $\|u^C\|_{C^0} \leq C_5$, so we have $\|u^C\|_{C^{2+\alpha}} \leq C_6$ for $C > \max(a, C_4)$. As $\{u^n\}_{n=1}^{\infty}$ also is a minimizing sequence for (I_0, Z_φ) the solvability of this problem is now obvious.

REFERENCES

1. Gilbarg, D.; Trudinger, N. S.: Elliptic Partial Differential Equations of Second Order. Springer (1977).
2. Hartmann, P.; Stampacchia, G.: Acta Math. 115, 271-310 (1966).
3. Ladyzhenskaya, O. A.; Ural'tseva, N. N.: Linear and Quasilinear Elliptic Equations. Academic Press (1968).
4. Lewy, H.: Math. Ann. 98, 107-124 (1928).
5. Morrey, C. B., Jr.: Multiple Integrals in the Calculus of Variations. Springer (1966).
6. Ströhmer, G.: Math. Z. 186, 179-199 (1984).
7. Tolksdorff, F.: J. Differ. Equations 51, 126-150 (1984).
8. Williams, G. H.: Math. Z. 154, 51-65 (1977).
9. Williams, G. H.: J. Math. Pures Appl. IX Sér. 60, 213-226 (1981).

32 INVERSE PROBLEM: ITS GENERAL SOLUTION

Enzo Tonti / Istituto di Scienza della Costruzioni, Università di Trieste, Trieste, Italy

1. INTRODUCTION

Every book on the calculus of variations starts with the typical phrase: "Let us consider a functional . . . ." Today much interest is given to the inverse problem: "Given an equation, does a functional exist that admits the given equation as its Euler-Lagrange equation?" This is known as the inverse problem of the calculus of variations.

In the literature the statement of such inverse problem varies greatly from one author to another. Let us examine these different statements.

The first distinction lies in the fact that some people limit the variational formulation to differential equations ignoring initial or boundary conditions while others take account of these.

In the first case the study is made on the mathematical form of the differential equation and the main interest is finding the lagrangian: we shall call this the formal inverse problem.

In the second case the kinds of additional conditions (initial and/or boundary conditions) are an essential part of the problem: we shall call this simply the inverse problem.

In both cases one must say whether the equation may be transformed in another by an integrating factor or not. We shall speak of extended and restricted variational formulation, respectively.

We shall deal with the inverse problem in its full meaning (not in the formal sense): a precise definition of it will be given in the next section.

We want to deal with the problem from a very general point of view, without distinguishing among ordinary or partial differential equations; among equations containing first, second, ... order derivatives; single equations or systems of them; among differential, integral, integrodifferential equations or equations with retarded arguments, etc. To make this possible we must use the operatorial notation.

A Brief Historical Survey

The history of the inverse problem has a curious feature: it is formed by two branches that developed separately for about eighty years (Fig. 1).

Both branches started in the same year (1887): one with Helmholtz, which deals with the formal inverse problem; the other with Volterra, which uses functional analysis. The singular fact is that no papers of one branch quote a paper of the other branch: the developments of the two branches are entirely separate.

Let us consider the first branch.

In 1887 Helmholtz [15] gave the necessary conditions in order that a single ordinary equation of second order may be considered as an Euler-Lagrange equation.

In 1897 Hirsch [16] gave the analogous condition for an equation of order n.

In 1928 Davis [3] gave the conditions for a partial differential equation to be of variational kind: he introduced the integrating factor.

In 1941 Douglas [6] developed Davis theory.

In 1957 an expository article of Havas [14] appeared dealing with the search for integrating factors.

HISTORY OF THE (LOCAL) INVERSE PROBLEM OF THE CALCULUS OF VIBRATIONS

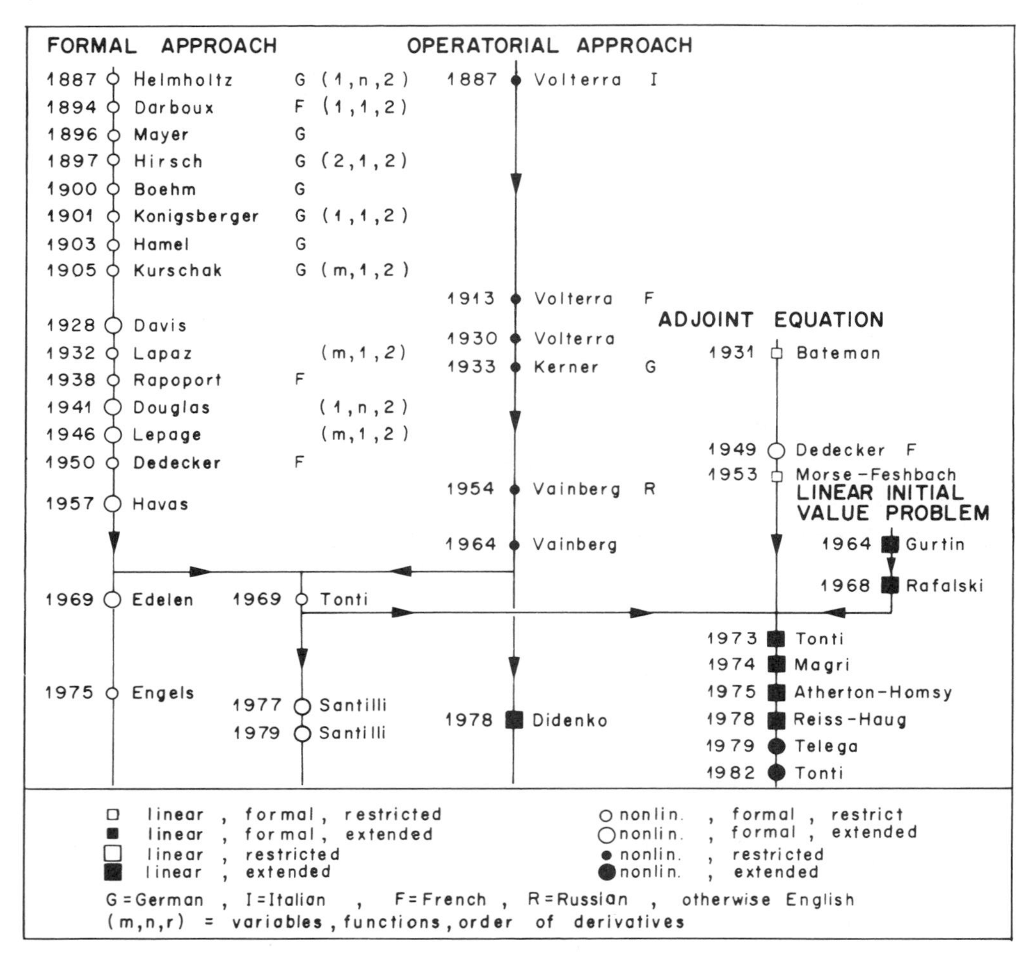

Fig. 1

The second branch starts in 1887 with three papers on the theory of functionals [42]. In paper I (p. 104) the symmetry of the second derivative of a functional appears.

In 1913 Volterra published a book in French [41] in which he gave the condition for the variational formulation and the formula to find the functional (p. 43).

In 1930 Volterra published another book in English [42] that contains the same results (Ch. V, Sect. II, Sec. 2).

In 1933 Kerner [19] published a paper giving the result of Volterra (its reference 4).

In 1954 a book of Vainberg [40] appeared (in Russian) that contained the theorem on the variational formulation. He quoted Kerner (its reference 42b).

In 1954 a book of Volterra and Pérès in French appeared [44] in which the theorem on the variational formulation was given (p. 98).

In 1959 Volterra's book [43] was reprinted by Dover.

In 1964 Vainberg's book was translated into English.

In 1969 the paper [39] appeared connecting the two branches. With the intention of giving an elementary exposition of the operatorial approach (the second branch), the author obtained the integrability conditions of Helmholtz and of other authors of the first branch. (See [28], p. 14 and p. 204.)

A third branch originated in more recent times: it is centered on the problem of giving a variational formulation to equations (including the initial/boundary conditions) that do not admit one in a classical context, as, for example, the Fourier equation of heat transfer. The inverse problem is here involved in a nonformal sense.

One of the first methods devised was that of adding the adjoint equation. This method, inaugurated by Morse and Feshbach in their book ([23], p. 298), consists of the arbitrary addition to a given linear equation of the adjoint homogeneous equation. The system of the two equations is symmetric. The adjoint function has no physical meaning. This fact and the fact that the adjoint problem introduces adjoint boundary conditions make the method artificial.

In the above-mentioned book (p. 299) the authors say: "By this arbitrary trick we are able to handle dissipative systems as though they were conservative. This is not very satisfactory if an alternate method of solution is known . . . ." Today an alternate method is known (see below) and the Morse and Feshbach technique may be abandoned.

Since 1953 many pseudovariational formulations have been devised: they have been named "quasi" or "almost" or "restricted" variational formulations. The reader may see [10] for a critical review.

The method that opened a new era is the one introduced by Gurtin in 1964 [11, 12]. He showed how to give variational formulation to linear initial value problems. This means that the initial conditions were taken into consideration from the beginning.

Gurtin's idea was the preliminary transformation of an equation into an integro-differential equation and the introduction of the convolution product of two functions.

This method opened the way to giving variational formulation to many linear initial value problems and a large number of papers appeared, mainly in engineering reviews.

The method of Gurtin was simplified in 1973 by the present author [38] who showed that the preliminary transformation of the differential equation into an integrodifferential equation is not necessary. The essential point is the introduction of a convolutive bilinear functional to give a variational formulation to a linear initial value problem whose equation has constant coefficients. In this paper Gurtin's method was included in the operatorial approach (see Fig. 1).

The idea of adapting the bilinear functional to the given operator was brought up to its apex by Magri, in 1974 [21]: he showed that every linear equation (not only those with constant coefficients) admits a variational formulation giving the explicit way to obtain the functional.

This result overthrows the common belief that equations admitting a variational formulation constitute a privileged class. At this date to every linear problem one may associate many functionals whose stationary value is attained at the solution of the problem. In general these functionals are not extremum at that point.

In 1978 Reiss and Haug [26] using Magri's result explored the possibility of finding among the many functionals those that give an extremum principle for linear initial value problems.

What about the larger class of nonlinear problems? Some attempts have been made to extend the method of adjoint equation to nonlinear problems [9]: the method suffers the same drawbacks of the linear case.

In 1979 Telega [31] first tried to extend Magri's result to nonlinear problems: the class of operators that was included was severely limited.

In 1982 the present author [32] showed that every nonlinear problem admits a variational formulation giving the explicit form of the functional. This result was further developed toward practical applications in the paper [33]: in particular nonlinear initial value problems have been solved using the Ritz method and the gradient method.

In the present paper I summarize the results of papers [32, 33] stressing the conceptual framework that makes possible the variational formulation to (practically) every nonlinear problem.

2. OPERATORIAL NOTATION

For better adherence to physical applications we shall consider operators with domain and range in two different spaces U and V.

Let us remember that an operator A is said to be equal to an operator B if $D(A) = D(B)$ and if $A(u) = B(u)$ for every $u \in D(A)$. An operator A is called a restriction of an operator B if $D(A) \subset D(B)$ and if $A(u) = B(u)$ for every $u \in D(A)$. The operator B is called extension of the operator A. Every element u_0 of D(A) such that $A(u) = O_V$ is called a null element of A ([2], p. 91), the set of null elements is called the null manifold of A.

A differential equation is formed by a differential expression equated to zero. The differential expression is called formal differential operator ([41], p. 146; [38]). To form a full operator we must select a set of functions, for example those of an assigned functional class. The operator so defined has a very large domain: any supplementary boundary or initial condition gives rise to a new operator that is a restriction of the initial operator.

To give a variational formulation to a problem, say $N(u) = O_V$, we need a real-valued, bilinear functional $V \times U \to R$, denoted $\langle v,u \rangle$, that is nondegenerate: we shall call it the scalar product of $v \in V$ and $u \in U$. The two spaces U and V are said to be put in duality by the bilinear functional $\langle v,u \rangle$ and V is called the dual of U and is denoted by U^*.

It is at this stage that we introduce norms on U and V such that the bilinear functional be continuous in both arguments.

We call the adjoint of a linear operator L with respect to a given bilinear functional $\langle v,u \rangle$ the linear operator L^* that satisfies the relation

$$\langle Lp,q \rangle = \langle L^*q,p \rangle \tag{2.1}$$

for every $p \in D(L)$ and every possible q which will form the domain $D(L^*)$.

An operator L is said to be symmetric if

$$\langle Lp,q \rangle = \langle Lq,p \rangle \tag{2.2}$$

for every $p,q \in D(L)$. Comparing (2.2) with (2.1) we see that in general it will be $D(L^*) \supset D(L)$, i.e., the operator L^* is an extension of the operator L. If the two domains coincide we have $L = L^*$ and the operator is called self-adjoint.

Given a formal differential operator $\mathscr{L}$, the formal differential operator $\mathscr{L}^*$ that satisfies the relation

$$\int u\,\mathscr{L}\bar{u}\,dx = \int \bar{u}\,\mathscr{L}^*u\,dx + \{\text{boundary terms}\} \tag{2.3}$$

is called formal adjoint. If $\mathscr{L} = \mathscr{L}^*$ the formal operator is called formally symmetric or formally self-adjoint, the two names are equivalent ([18], p. 274).

If F is a real functional, i.e., $F : D(F) \subset U \to R$ and if N is an operator $N : D(N) \subset U \to V = U^*$ and if

$$\delta F\{u\} = F'_u\{u;\delta u\} = \langle N(u), \delta u \rangle \tag{2.4}$$

is true with the condition $D(F) = D(N)$, then the operator N is called the gradient of the functional F and F is called the potential of N; N is said to be a potential operator. The symbol δ is the usual one of the calculus of variations: δF coincides with the Gâteaux differential of the functional.

To say that the functional F is stationary at u_0 means that $\delta F\{u\} = 0$ at $u_0 \in D(F)$. The elements u_0 for which the functional is stationary are called critical points and the set formed by them is called the critical manifold.

Let $N'_u(u,\delta u)$ denote the Gâteaux differential of N and $N'_u(u;\cdot)$ the (linear) Gâteaux derivative calculated in u.

At this point we are able to enunciate two forms of the inverse problem.

Inverse Problem in the Restricted Sense. Given a problem $N(u) = O_V$ with $D(N) \subset U$ and $R(N) \subset V = U^*$ find a functional F, if any, whose gradient is the operator N.

Inverse Problem in the Extended Sense. Given a problem $N(u) = O_V$ with $D(N) \subset U$ and $R(N) \subset V = U^*$ find a functional F, if any, whose critical points are the solutions of the problem and vice versa.

The inverse problem in the "extended" sense requires only the coincidence of the critical manifold of the functional F with the null manifold of the operator N. The inverse problem in the "restricted" sense requires a stricter link between N and F: N must be the gradient of F with respect to a given bilinear functional. In this case it follows that the domain of N coincides with the domain of F and moreover the null manifold of N coincides with the critical manifold of F.

Stated in another way: in the extended sense the gradient of the functional F will be an operator $\bar{N}$ linked in some way with N and with the same null manifold while in the restricted sense the gradient of F must coincide with N.

The inverse problem in the restricted sense was solved for the first time in 1913 by Volterra [41,43] with the following theorem.

Theorem. The necessary and sufficient condition in order that an operator $N: D(N) \subset U \to R(N) \subset V = U^*$, whose domain is simply connected, be the gradient of a functional is that

$$\langle N'_u(u;p), q\rangle = \langle N'_u(u;q), p\rangle \tag{2.5}$$

Putting $w(s) = su + (1 - s)u_0$ the functional is

$$F\{u\} = F\{u_0\} + \int_0^1 \langle N(w(s)), \partial w(s)/\partial s\rangle\, ds \tag{2.6}$$

The condition (2.5) expresses the symmetry of the Gâteaux derivative of N and we shall call it the Volterra symmetry condition.

The condition (2.5) is necessary: the hypothesis that D(N) be simply connected (an implicit hypothesis in the original Volterra formulation) is almost always satisfied in practice. In fact, usually the domain is either a linear or a convex set. In the particular case in which the operator is linear, condition (2.5) becomes

$$\langle Lp, q\rangle = \langle Lq, p\rangle \tag{2.7}$$

i.e., the linear operator must be symmetric (not necessarily self-adjoint).

In the linear case the functional (2.6) may be cast in the closed form

$$F\{u\} = F\{u_0\} + 1/2\,\langle Lu, u\rangle - \langle f, u\rangle \tag{2.8}$$

Historical Remark

Volterra's theorem is usually called Vainberg's theorem or Kerner's theorem. But Vainberg [40] quoted Kerner (its reference 42.b) and Kerner [19] quoted Volterra (its reference 4). The theorem was contained in the book published by Volterra in 1913 [41] (in French) and in the book (1930) [43] (in English); the latter was reprinted by Dover in 1959.

3. RELATIVITY OF SYMMETRY

The keystone for giving variational formulation of problems lies in the observation that the symmetry of an operator is relative to the bilinear functional considered. Then, contrary to a common belief, the adjoint of an operator is not necessarily unique, and there can be many different possible bilinear functionals. Moreover, if a given operator does not satisfy the Volterra symmetry condition with respect to a given bilinear functional we may look for other bilinear functionals with respect to which this condition is satisfied. The following example will clarify this statement. The linear operator

$$D = \{d/dt,\ u(0) = 0, u \in C^1(0,T)\} \tag{3.1}$$

is not symmetric with respect to the cartesian bilinear functional

$$\langle v,u\rangle_0 = \int_0^T v(t)\, u(t)\, dt \tag{3.2}$$

because its adjoint is

$$D^* = \{-d/dt,\ v(T) = 0,\ v \in AC(0,T)\} \tag{3.3}$$

But if we consider the convolutive bilinear functional

$$\langle v,u\rangle_C = \int_0^T v(T-t)\, u(t)\, dt \tag{3.4}$$

the adjoint becomes

$$D^* = \{d/dt,\ v(0) = 0,\ v \in AC(0,T)\} \tag{3.5}$$

Since $D \subset D^*$ the operator is symmetric with respect to the convolutive bilinear functional.

This observation opens the way to the variational formulation for problems that do not admit one in the "classical" sense.

At this point a question arises: given an operator, does a bilinear functional exist such that it lets the given operator satisfy the Volterra symmetry condition?

The answer was given for linear operators in 1975 by Magri [21] and for nonlinear operators in 1982 by the present author [32].

Magri has shown how to find a bilinear functional that makes symmetric a given linear operator: the operative rule is the following.

Let L be a linear invertible operator with domain in a vector space U and range in a vector space V. Let

$$Lu = f \tag{3.6}$$

be the given problem.

Let us consider the cartesian bilinear functional on $V \times U$

$$\langle v,u\rangle_0 = \int_0^T v(t)\, u(t)\, dt \tag{3.7}$$

and let us suppose that the spaces U and V be such that the functional (3.7) is nondegenerate.

Let us introduce a real bilinear functional (v, v) on $V \times V$

$$(v,v) = \langle v, Kv\rangle = \int_0^T v(t) \int_0^T k(t,s)\, v(s)\, ds\, dt \tag{3.8}$$

that is symmetric and nondegenerate. These two conditions are satisfied if $k(t,s) = k(s,t)$ and if the integral transform

$$w(t) = \int_0^T k(t,s)\, v(s)\, ds \tag{3.9}$$

is invertible.

Let us define a new bilinear functional $\langle v,u \rangle$ on $V \times U$ by

$$\langle v,u \rangle = (v, Lu) = \langle v, KLu \rangle_0 = \int_0^T v(t) \int_0^T k(t,s)Lu(s)\, ds\, dt \tag{3.10}$$

The bilinear functional $\langle v,u \rangle$ is nondegenerate because L is invertible. The operator L is symmetric with respect to this bilinear functional and the solution of problem (3.6) is the critical point of the functional

$$F\{u\} - F\{u_0\} = 1/2 \langle Lu, u \rangle - \langle f,u \rangle = 1/2 (Lu, Lu) - (f, Lu)$$
$$= 1/2 \langle Lu, KLu \rangle_0 - \langle f, KLu \rangle_0 \tag{3.11}$$

To extend this result to nonlinear operators we need to introduce the notion of integrating operator.

4. INTEGRATING OPERATOR

An observation of capital importance in our problem is the following: the change of the bilinear functional is equivalent to the application (on the left) to the given operator L of a suitable linear operator R.

Let us show this equivalence in a particular case referring to the operator D defined by Eq. (3.1). Let us define the convolution operator

$$Cv(t) = v(T - t) \tag{4.1}$$

and the convolutive bilinear functional

$$\langle v,u \rangle_C = \langle Cv, u \rangle_0 \tag{4.2}$$

We have the symmetry of D with respect to $\langle v,u \rangle_C$, and thus

$$\langle CDu, \bar{u} \rangle_0 = \langle Du, \bar{u} \rangle_C = \langle D\bar{u}, u \rangle_C = \langle CD\bar{u}, u \rangle_0 \tag{4.3}$$

that proves the symmetry of CD with respect to the bilinear functional $\langle v,u \rangle_0$.

More in general if $L: U \to V$ is a linear operator symmetric with respect to the bilinear functional

$$\langle v,u \rangle_R = \langle Rv, u \rangle_0 \tag{4.4}$$

where $R: V \to V$ is a linear, invertible operator, then $RL: U \to V$ is symmetric with respect to the cartesian bilinear functional $\langle v,u \rangle_0$.

The operator R has the same role of the integrating factor used with differential equations: we shall call it an integrating operator. The requirement that R be invertible (i.e., kernel-free) assures that the bilinear functional be nondegenerate.

5. HOW TO FIND THE INTEGRATING OPERATOR

Let us start with a linear system of algebraic equations

$$Ax = b \tag{5.1}$$

If we apply to both members the adjoint matrix A* we obtain the system

$$A^*Ax = A^*b \tag{5.2}$$

whose matrix is now symmetric with respect to the cartesian bilinear form

$$\langle x,x \rangle = \sum_k x_k x^k \tag{5.3}$$

and then the solutions of problem (5.1) make stationary the function

$$f(x) = 1/2 \langle A^*Ax, x\rangle - \langle A^*f, x\rangle \tag{5.4}$$

Then the matrix A^* is an integrating operator, provided that it is invertible; if it were not there would be critical values of f(x) that are not solutions of problem (5.1).

Is it possible to extend this procedure to general linear operators, say to differential operators? Let us consider the operator

$$Du = f \quad D = \{d/dt,\ u(0) = 0,\ u \in C^1(0,T),\ f \in C(0,T)\} \tag{5.5}$$

If we consider the adjoint with respect to the cartesian bilinear functional (3.2) given by (3.3), we see that it is applicable to both members of problem (5.5) only if the domain of D^* contains the given function f. This implies that $f \in AC(0,T)$ and that $f(T) = 0$.

While the derivability requirement is satisfied, because $C(0,T) \in AC(0,T)$, the second condition is not, a priori, satisfied.

One may be tempted to add to f(t) the additional condition $f(T) = 0$. But <u>we take as a fundamental principle that of imposing no supplementary conditions on the functions entering a problem different from the ones that are assigned to the problem</u>. In fact, if the given problem expresses a physical law or a geometric condition or a technical process, every additional condition imposed would exclude possible source distributions or possible configurations. Even the simple condition that the unknown functions be of class $C_0^\infty(0,T)$ would result in an inadmissible restriction of the domain and consequently a restriction of the range.

Observing that the main hindrance to the application of the adjoint operator D^* is the final condition $f(T) = 0$, the idea arises of performing a preliminary transformation, like the following:

$$\bar{f}(t) = \int_0^T k(t,s)\, f(s)\, ds \tag{5.6}$$

in which the kernel k(t, s) must satisfy the following conditions:

$$k(T,s) = 0 \qquad k(t,s) = k(s,t)$$

and moreover be such that the integral operator K defined by (5.6) be invertible. In such a way the final condition $\bar{f}(T) = 0$ is satisfied.

One then obtains the integrodifferential equation

$$\int_0^T k(t,s)\, d/ds\, u(s)\, ds = \int_0^T k(t,s)\, f(s)\, ds \tag{5.7}$$

that has the same solution of the given problem. Now we may apply the operator D^* (adjoint of D with respect to the ordinary cartesian bilinear functional):

$$-d/dt \int_0^T k(t,s)\, d/ds\, u(s)\, ds = -d/dt \int_0^T k(t,s)\, f(s)\, ds \tag{5.8}$$

Let K denote the integral operator (5.6). We may write problem (5.8) as follows:

$$D^*KDu = D^*Kf \tag{5.9}$$

We see that the operator D^*K is the integrating operator we were searching for. In fact,

$$\langle (D^*KD)u, \bar{u}\rangle_0 = \langle (D\bar{u}), K(Du)\rangle_0 = \langle (Du), K(D\bar{u})\rangle_0 = \langle (D^*KD)\bar{u}, u\rangle_0 \tag{5.10}$$

The corresponding functional is

$$\bar{F}\{u\} = \bar{F}\{u_0\} + 1/2 \langle Du, KDu\rangle_0 - \langle f, KDu\rangle_0 \tag{5.11}$$

If the operator K is also positive definite then the critical points of the functional are points of minimum.

The integrating operator is

$$Rv(t) = D^*Kv(t) = -d/dt \int_0^T K(t,s)\, v(s)\, ds \qquad (5.12)$$

that is, an integrodifferential operator.

It is not difficult to find integral operators K meeting these requirements: all Green functions of linear positive definite operators may be used. For example, the inverse of the operator

$$S = \{-d^2/dt^2,\ u(0) = 0,\ u(T) = 0;\ u \in C^2(0,1)\} \qquad (5.13)$$

is

$$w(t) = Kv(t) = \int_0^T \{-(t - s)\, H(t - s) - (T - s)\}\, v(s)\, ds \qquad (5.14)$$

being $H(t)$, the Heaviside function. It is $w(T) = 0$.

Then, at least for linear operators, we have succeeded in finding a variational formulation under the hypothesis that L is invertible.

In particular, if L is symmetric with respect to the cartesian bilinear functional putting $K = L^{-1}$ the functional

$$\bar{F}\{u\} = \bar{F}\{u_0\} + 1/2 \langle Lu, KLu \rangle - \langle f, KLu \rangle \qquad (5.15)$$

reduces itself to

$$F\{u\} = F\{u_0\} + 1/2 \langle Lu, u \rangle - \langle f, u \rangle \qquad (5.16)$$

that is, the classical one.

Then, in the linear case, the extended variational formulation contains the restricted one when this exists.

6. NONLINEAR PROBLEMS

It is possible to extend this procedure to nonlinear problems. We have the following [32]

Theorem. Let us consider two linear spaces U and V such that a nondegenerate, real valued, bilinear functional $\langle v, u \rangle$ may be defined; let the two spaces be endowed with a norm that makes $\langle v, u \rangle$ continuous in both arguments. Let

$$N(u) = O_V \qquad (6.1)$$

be a problem, whose operator $N: D(N) \in U \rightarrow R(N) \in V$ is such that its domain is simply connected and it admits (linear) Gâteaux derivative $N'_u(u;\cdot)$ for every $u \in D(N)$. Let $N'_u{}^*(u;\cdot)$ be its adjoint with respect to the bilinear functional $\langle v, u \rangle$: if $N'_u{}^*(u;\cdot)$ is invertible for every $u \in D(N)$ and $D(N'_u)$ is dense in U, then for every operator $K: D(K) \subset V \rightarrow U$ such that

(1) $D(K) \supset R(N)$,

(2) $R(K) \subset D(N'_u{}^*)$,

(3) is linear,

(4) is invertible, i.e., kernel-free,

(5) is symmetric, i.e., $\langle v, Kv \rangle = \langle v, Kv \rangle$,

(6) is positive definite, i.e., $\langle v, Kv \rangle > 0 \quad (v \neq O_V)$,

the operator $\bar{N}$ defined by

$$\bar{N}(u) = N_u'^*(u;KN(u)) \tag{6.2}$$

has the following properties:

(a) it has the same domain of N;
(b) it has the same null manifold of N;
(c) it is potential (i.e., it satisfies the Volterra condition);
(d) it is a gradient of the functional

$$\bar{F}\{u\} = \bar{F}\{u_0\} + 1/2 \langle N(u),\ KN(u) \rangle \tag{6.3}$$

(e) the functional is minimum at the critical points.

Proof. If $u \in D(N)$, for the properties (1) and (2) also $u \in D(\bar{N})$. Contrarily, if $u \in D(\bar{N})$ it follows from (6.2) that $u \in D(N)$: this proves property (a).

If u_0 is a solution of $N(u) = O_V$, on account of the linearity of $N_u'^*(u;\cdot)$ and of K we have

$$\bar{N}(u) = N_u'^*(u_0;KN(u_0)) = N_u'^*(u_0;KO_V) = N_u'^*(u_0;O_u) = O_V \tag{6.4}$$

i.e., it is a solution of $\bar{N}(u) = O_V$, too. Contrarily, if u_0 is a solution of $\bar{N}(u) = O_V$, since $N_u'^*$ and K are invertible, it is

$$K^{-1}(N_u'^*)^{-1}(u_0;\bar{N}(u_0)) = O_V \tag{6.5}$$

i.e., u_0 is also a solution on $N(u) = O_V$. This proves property (b).

Since $N_u'^*(u;w)$ is linear on w from Eq. (6.2) we have

$$\bar{N}_u'(u;\delta u) = (N_u'^*)_u'(u;KN(u),\ \delta u) + N_u'^*(u;KN_u'(u;\delta u)) \tag{6.6}$$

Now

$$\langle \bar{N}_u'(u;\delta u),\ w\rangle = \langle (N_u'^*)_u'(u;KN(u),\ \delta u),\ w\rangle + \langle N_u'^*(u;KN_u'(u;\delta u)),\ w\rangle \tag{6.7}$$

From the relation that defines the adjoint

$$\langle N_u'(u;p),\ q\rangle = \langle N_u'^*(u;q),\ p\rangle \tag{6.8}$$

by differentiation with respect to u we obtain

$$\langle N_{uu}''(u;p,\ \delta u),\ q\rangle = \langle (N_u'^*)_u'(u;q,\ \delta u),\ p\rangle \tag{6.9}$$

Relation (6.7) becomes

$$\langle N_u'(u;\delta u),\ w\rangle = \langle N_{uu}''(u;w,\ \delta u),\ KN(u)\rangle + \langle N_u'(u;w),\ KN_u'(u;\delta u)\rangle \tag{6.10}$$

The second Gâteaux derivative is symmetric:

$$N_{uu}''(u;p,q) = d^2/dadb\,[N(u + ap + bq)] = d^2/dbda\,[N(u + ap + bq)] = N_{uu}''(u;q,p) \tag{6.11}$$

From this property and from the symmetry of K it follows that

$$\langle \bar{N}_u'(u;\delta u),\ w\rangle = \langle \bar{N}_u'(u;w),\ \delta u\rangle \tag{6.12}$$

which proves the symmetry of $\bar{N}_u'$. Then property (c) is proved.

Putting $w(s) = su + (1 - s)u_0$ the functional is given by the general formula

$$F\{u\} = \int_0^T \langle \bar{N}(w(s)),\ \partial w(s)/\partial s\rangle\, ds = \int_0^T \langle N_u'^*(w;KN(w)),\ \partial w/\partial s\rangle\, ds$$

$$= \int_0^T \langle N'_u(w\,;\,\partial w/\partial s),\ KN(w)\rangle\, ds = \int_0^T \langle \delta N(w),\ KN(w)\rangle$$

$$= \int_0^T \delta\,[(1/2)\,\langle N(w),\ KN(w)\rangle] = (1/2)\,[\langle N(w),\ KN(w)\rangle]_0^1$$

$$= F\{u_0\} + (1/2)\langle N(u),\ KN(u)\rangle \tag{6.13}$$

Taking account of the symmetry of K we obtain

$$\delta F\{u\} = \langle \delta N(u),\ KN(u)\rangle = \langle N'_u(u\,;\delta u),\ KN(u)\rangle = \langle N'^{*}_u(u\,;KN(u)),\ \delta u\rangle \tag{6.14}$$

and then if

$$\delta F\{u\} = \langle N'^{*}_u(u\,;\,KN(u)),\ \delta u\rangle = 0 \tag{6.15}$$

since $\delta u \in D(N'_u)$ and this domain is dense in U, and $\langle v,u\rangle$ is nondegenerate and continuous, it follows that

$$N'^{*}_u(u\,;KN(u)) = O_V \tag{6.16}$$

which proves property (d).

Since

$$F\{u\} - F\{u_0\} = (1/2)\,\langle N(u),\ KN(u)\rangle > 0 \tag{6.17}$$

for every $N(u) \neq O_V$ it follows that $F\{u\}$ is minimum at the solution: this proves property (e).

We remark that the continuity of the operator N is not required and that the Gâteaux derivative (not the Fréchet one) is involved. Q.E.D.

The theorem shows how, under mild conditions on the operator, one may give variational formulation in the extended sense to nonlinear problems without changing the initial or boundary conditions or even the functional class of the functions entering the problem. Moreover, it is always possible to find a functional that is minimum at the critical points.

In the particular case that N is a potential operator the functional F does not reduce to the potential of N. To have this inclusion a further generalization of the theorem is necessary; this has been done in [33].

7. CRITICAL REMARKS

The preceding theorem enables us to give many variational formulations to a problem whose operator satisfies few requirements. This means that one may actually characterize the solutions of the problem $N(u) = O_V$ as those elements of the domain of N that make minimum some functional.

From this it follows that the common belief that the existence of a variational principle for a given problem may be used as a criterion to accept or to refuse possible physical laws is meaningless.

But the usual belief, often expressed, sometimes written and never proved, that dissipative phenomena cannot be described by a variational principle is invalid even in a classical context.

For example, if we throw a book on a table its motion is uniformly retarded according to the law

$$m\,\ddot{q}(t) = -k\,m\,g \tag{7.1}$$

where k is the dynamic frictional coefficient. Nothing is more dissipative than this motion! Yet the Lagrangian exists: it is

$$L(q, \dot{q}) = 1/2\ m\ \dot{q}^2(t) - k\ m\ g\ q(t) \tag{7.2}$$

Another well-known example is that of the equation of the harmonic oscillator with damping term, i.e.,

$$m\ \ddot{q}(t) + h\ \dot{q}(t) + k\ q(t) = 0 \tag{7.3}$$

It admits the integrating factor exp (h/mt). The corresponding lagrangian is

$$L(q, \dot{q}) = 1/2\ \exp\ (h/mt)\{m\dot{q}^2(t) + kq^2(t)\} \tag{7.4}$$

Yet the motion is dissipative.

These two examples show that even on the classical ground a selection criterion does not exist.

The fact is that the variational formulation is based on the form of the equation. We know that every linear second order ordinary differential equation can be cast in self-adjoint form. The transformation to a self-adjoint form changes the form of the equation, not the solution set, i.e., the substance. Then a mathematical requirement (variational formulation) that is based on the form (self-adjointness) when we let it be altered by an integrating factor cannot be reasonably used as a discriminating criterion.

8. CRITIQUE OF THE HAMILTON PRINCIPLE

Let us consider the Hamilton principle in mechanics. This principle ignores the initial condition on the velocity and adds a fictitious final condition. So the initial value problem

$$m\ddot{q} + kq = 0 \qquad q(0) = a;\ \dot{q}(0) = b \tag{8.1}$$

has an operator that is formally symmetric (with respect to the cartesian bilinear form). The term arising from the integration by parts is

$$m\{q(T)\ \delta q(T) - q(0)\ \delta q(0)\} \tag{8.2}$$

The given initial conditions imply $\delta q(0) = 0$ and $\delta\dot{q}(0) = 0$ and then only the second term vanishes. Of course we do not know $q(T)$. What do we do? We add a fictitious final condition $q(T) = c$ so that $\delta q(T) = 0$. The whole boundary term vanishes and the operator is now symmetric. The functional is

$$A\{q\} = 1/2 \int_0^T \{m\dot{q}^2 - kq^2\}\ dt \tag{8.3}$$

We arrive in this way at the Hamilton principle. In its statement the natural motion is compared with those motions that respect the same initial and final conditions.

It is evident that in order to obtain a variational formulation we have had to alter the problem: we have forgotten the initial condition on the velocities and added a fictitious final condition. In operatorial language this means that the domain of the functional is not the domain of the operator.

This expedient has invaded all physics: it is used in all evolution phenomena, in field theories of classical, relativistic, and quantum physics.

There is no longer a reason to use the mathematical trick necessary for the Hamilton principle when we know that the solution of the problem of motion makes minimum a functional like

$$F\{q\} = 1/2 \int_0^T \{\ddot{q}(t) - f(t;q,\dot{q})\} \int_0^T k(t,s)\{\ddot{q}(s) - f(s;q,\dot{q})\}\ ds\ dt \tag{8.4}$$

The lagrangian of this functional is no longer a function of q, $\dot{q}$ as in the classical case, but is an operator. Many formalisms used dealing with the calculus of variations, like the theory of exterior forms, cohomology theory, the notions of jets, spray, etc., cannot be applied.

Faced with such a loss one may be tempted to reject the extended variational formulation considering in some way unacceptable an integrating operator.

But where is the line of demarcation between integrating factor and integrating operator?

May we accept the integrating factor $\exp(h/mt)$ to make formally symmetric the operator of the damped harmonic oscillator and refuse an integral operator that makes the operator symmetric only because it destroys the differential nature of the equation?

What is more important: to change the form of the problem keeping intact the solution manifold or to change the solution manifold to preserve the form of the problem?

A mathematician is free to change the additional conditions or the functional class if the equation is for him a pretext to utilize a given algorithm. But a mathematician, a physicist, an engineer cannot do the same if the equation describes a phenomenon or a process that he must study or solve.

When we form an equation and the additional conditions to describe a process, what we have in mind is to characterize the solutions of the problem among all functions of a certain set. The form of the equation is immaterial. Two mathematical formulations that lead to the same set of solutions are equally acceptable. So when we know the Green function of an operator we may transform a differential equation into an integral one without changing the solution. For example, the problem of the harmonic oscillator (8.1) is equivalent to the integral equation

$$q(t) = -k/m \int_0^T (t - s)\, H(t - s)\, q(s)\, ds + a + bt \tag{8.5}$$

The form is changed but the content is the same. The passage from (7.1) to (8.5) is equivalent to the application of the integral operator

$$R(\cdot) = \int_0^T (t - s)\, H(t - s) \cdots ds + a + bt \tag{8.6}$$

in which the kernel is the propagator for the problem

$$\ddot{q}(t) = f(t) \qquad q(0) = 0 \qquad \dot{q}(0) = 0 \tag{8.7}$$

This is an example of application of an operator to a given problem that changes the form but not the content of the problem.

There is one last thing to be said on the Hamilton principle: we cannot apply to it the direct methods of the calculus of variations, say the Ritz method, because we don't know the final value $q(T)$. On the contrary, a direct method has been applied with success to the functional (8.4): see [33].

ACKNOWLEDGMENT

The author is grateful to the student F. Beltram for his invaluable help during the redaction of the paper.

REFERENCES

1. Boehm, K.: J. Reine Angew. Math. 121, 124.
2. Collatz, L.: Functional Analysis and Numerical Mathematics. Academic Press (1966).
3. Davis, D. R.: Trans. Am. Math. Soc. 30, 716, 736 (1928).
4. Dedecker, P.: Bull. Acad. Roy. Belg. Sci. 36, 63-70 (1950).
5. Didenko, V. P.: Dokl. Akad. Nauk SSSR 240, 736-740 (1978).
6. Douglas, J.: Trans. Am. Math. Soc. 50, 71-128 (1941).
7. Dunford, N.; Schwartz, J. T.: Linear Operators in Hilbert Spaces, Vol. II. Interscience (1964).

8. Edelen, D. G. B.: Nonlocal Variations and Local Invariance of Fields. Elsevier (1969).
9. Finlayson, B. A.: Methods of Weighted Residuals and Variational Principles. Academic Press (1972).
10. Finlayson, B. A.; Scriven, L. E.: J. Heat Mass Transfer 10, 799-821 (1957).
11. Gurtin, M. E.: Quart. Appl. Math. 22, 252-256 (1964).
12. Gurtin, M. E.: Arch. Rat. Mech. Anal. 13, 179-197 (1963).
13. Hamel, G.: Math. Ann. 57, 231.
14. Havas, P.: Suppl. Nuovo Cim. V, Ser. X, 363-388 (1957).
15. Helmholtz, H. von: J. Reine Angew. Math., 137-166, 213-222 (1886).
16. Hirsch, A.: Math. Ann. 49, 49-72 (1897).
17. Horndneski, G. W.: Tensor, New Ser. 28, 203.
18. Kato, T.: Perturbation Theory for Linear Operators. Springer (1966).
19. Kerner, M.: Ann. Math. 34, 546-572 (1933).
20. Konigsberger, L.: Die Prinzipien der Mechanik, Teubner, Leipzig.
21. Magri, F.: Int. J. Eng. Sci. 12, 537-549 (1974).
22. Mayer, A.: Ber. Ges. Wiss. Leipzig, Phys. Cl., p. 519.
23. Morse, M.; Feshbach, H.: Methods of Theoretical Physics. McGraw-Hill (1953).
24. Rafalski, P.: Int. J. Eng. Sci. 6, 465 (1968).
25. Rapoport, I.M.: C. R. Acad. Sci. USSR 18, 131.
26. Reiss, R.; Haug, E. J.: Int. J. Eng. Sci. 16, 231-251 (1978).
27. Santilli, R. M.: Foundations of Theoretical Mechanics I: The Inverse Problem of Newtonian Mechanics. Springer (1978).
28. Santilli, R. M.: Foundations of Theoretical Mechanics II: Generalization of the Inverse Problem in Newtonian Mechanics. Springer (1979).
29. Stone, M. H.: Linear Transformation in Hilbert Spaces. A.M.S. Coll. Publ. XV (1932).
30. Takens, F.: J. Diff. Geom. 14, 543-562 (1979).
31. Telega, J. J.: J. Inst. Maths. Appl. 24, 175-195 (1979).
32. Tonti, E.: Hadronic J. 5, 1404-1450 (1982).
33. Tonti, E.: Variational formulation for every nonlinear problem, Int. Journ. Engn. Sci. 22, No. 11/12, 1343-1371 (1984).
34. Tonti, E.: Rend. Acc. Lincei LII, 175-181, 350-356 (1972).
35. Tonti, E.: Rend. Acc. Lincei LII, 39-56 (1972).
36. Tonti, E.: Rend. Seminario Matematico Fisico Milano XLVI, 163-257 (1976) (preprint: On the formal structure of physical theories, Consiglio Nazionale delle Ricerche, 1975).
37. Tonti, E.: Appl. Math. Modelling I, 37-60 (1976).
38. Tonti, E.: Ann. Mat. Pura Appl. XCV, 331-360 (1972).
39. Tonti, E.: Bull. Acad. Roy. Belg. LV, Scr. 5, 137-165, 262-278 (1969).
40. Vainberg, M. M.: Variational Methods for the Study of Nonlinear Operators. Holden-Day (1964).
41. Volterra, V.: Lecons sur les fonctions de ligne. Gauthier-Villars (1913).
42. Volterra, V.: Rend. Acc. Lincei III, 97-105, 141, 153-158 (1887).
43. Volterra, V.: Theory of Functionals and Integrodifferential Equations, London (1929) (reprinted in 1959 by Dover).
44. Volterra, V.; Pérès, J.: Theorie generale des fonctionnelles. Gauthier-Villars (1936).

33 ON THE STABILITY OF A FUNCTIONAL WHICH IS APPROXIMATELY ADDITIVE OR APPROXIMATELY QUADRATIC ON A-ORTHOGONAL VECTORS

Hamid Drljević / Ekonomski Fakultet, Mostar, Yugoslavia

Let X be a complex Hilbert space, and A a continuous self-adjoint operator from X into X with the property that dim $A(X) \neq 1, 2$.

We say that a continuous functional f defined on X is additive on A-orthogonal vectors if the following holds:

$$(x, Ay) = 0 \quad \text{implies} \quad f(x + y) = f(x) + f(y) \qquad (x, y \text{ in } X)$$

We say that a continuous functional h defined on X is quadratic on A-orthogonal vectors if the following holds:

$$(x, Ay) = 0 \quad \text{implies} \quad h(x + y) + h(x - y) = 2h(x) + 2h(y) \qquad (x, y \text{ in } X)$$

By analogy with the theorem in [1] the following theorem can be proved.

Theorem 1. Let X be a Banach space, h a functional on X with h(tx) continuous in t for each fixed x. Let there exist $\theta \geq 0$ and $p \in [0, 2)$ such that

$$\frac{|h(x + y) + h(x - y) - 2h(x) - 2h(y)|}{\|x\|^p + \|y\|^p} \leq \theta \quad \text{for each } x, y \text{ in } X$$

Then there exists a unique quadratic functional h_1 on X such that

$$\frac{|h(x) - h_1(x)|}{\|x\|^p} \leq \frac{2^2 \theta}{2^2 - 2^p} \quad \text{for each } x \text{ in } X$$

Theorem 2 (Drljević [2]). Let φ be a functional defined on a complex Hilbert space X and let $\varphi(tx)$ be continuous in t for each fixed x in X. Let there exist $\theta \geq 0$ and $p \in [0, 1)$ such that

$$|\varphi(x + y) - \varphi(x) - \varphi(y)| \leq \theta[|Ax, x)|^{p/2} + |(Ay, y)|^{p/2}]$$

for x, y in X for which $(Ax, y) = 0$, where A is a continuous self-adjoint operator from X into X with the property that dim $A(x) \neq 1, 2$. Then there exists a unique continuous functional φ_1 which is additive on A-orthogonal pairs such that

$$|\varphi(x) - \varphi_1(x)| \leq \theta_1 \cdot |(Ax, x)|^{p/2}$$

for each x in X, where θ_1 is a constant.

From the proof of theorem 2 it can be seen that the following theorem holds also.

Theorem 3. Let X and Y be complex Hilbert spaces, and φ an operator from X into Y with the property that $\varphi(tx)$ is a continuous function for each fixed x in X. Let there exist $\theta \geq 0$ and $p \in [0, 1)$ such that

$$\|\varphi(x + y) - \varphi(x) - \varphi(y)\| \leq \theta[|(Ax, x)|^{p/2} + |(Ay, y)|^{p/2}]$$

for x, y in X for which $(Ax, y) = 0$, where A is given continuous self-adjoint operator from X into Y with the property that $\dim A(X) \neq 1, 2$. Then there exists only one continuous operator φ_1 which is additive on A-orthogonal pairs and such that

$$\| \varphi(x) - \varphi_1(x) \| \leq \theta_1 |(Ax, x)|^{p/2}$$

for each x in X.

By theorem 2 in [4], operator φ_1 is of the form

$$\varphi_1(x) = Bx + Cx + a \cdot (Ax, x) \quad \text{for each x in X}$$

where B is a continuous linear operator, C is a continuous nonlinear operator (from X into Y), and a in Y is a fixed vector.

We will present a sketch of the proof of the following theorem. One can find the whole proof in [3].

Theorem 4 (Drljević [3]). Let X be a complex Hilbert space, $\dim X \geq 3$, $A: X \to X$ a bounded self-adjoint operator with $\dim A(X) \neq 1, 2$, $\theta \geq 0$ and $p \in [0, 2)$ numbers, and h a continuous functional defined on X. If

$$|h(x + y) + h(x - y) - 2h(x) - 2h(y)| \leq \theta[|(Ax, x)|^{p/2} + |(Ay, y)|^{p/2}]$$

whenever $(Ax, y) = 0$, then

$$h_1(x) = \lim_{n \to \infty} 2^{-2n} h(2^n x)$$

defines a continuous function on X such that

$$h_1(x + y) + h_1(x - y) = 2h_1(x) + 2h_1(y)$$

whenever $(Ax, y) = 0$. Furthermore there exists a real number $\varepsilon > 0$ such that

$$|h(x) - h_1(x)| \leq |(Ax, x)|^{p/2} \varepsilon$$

Proof (sketch). First we prove inequality

$$|[h(2^n x)/2^{2n} - h(x)] + [h(2^n y)/2^{2n} - h(x)]| \leq \theta |(Ax, x)|^{p/2} (1 + 2^{p/2-1}) \cdot 2^2/(2^2 - 2^p) \quad (*)$$

for x, y in X such that $(Ax, y) = 0$ and $(Ay, y) = (Ax, x)$.

Then we prove that the sequence $\{h(2^n x)/2^{2n} + h(2^n y)/2^{2n}\}_{n=0}^{\infty}$ converges. For this purpose there is found z in X such that $(Ay, y) = (Az, z) = \pm (Ax, x)$, and it is proved that sequences $\{h(2^n x)/2^{2n} + h(2^n z)/2^{2n}\}_{n=0}^{\infty}$ and $\{h(2^n y)/2^{2n} + h(2^n z)/2^{2n}\}_{n=0}^{\infty}$ converge, and from that, convergence of the sequence $h_1(x) = \lim_{n \to \infty} 2^{-2n} h(2^n x)$ follows for all x in X. From (*) and that result,

$$|h(x) - h\ (x)| \leq \varepsilon \cdot |(Ax, x)|^{p/2}$$

for all x in X follows, where $\varepsilon = (3/2)\theta(1 + 2^{p/2-1})2^2/(2^2 - 2^p)$.

For the proof of continuity of the functional h_1 the functionals

$$\hat{h}_n^{\pm}\{x, y\} = h(2^n x)/2^{2n} \pm h(2^n y)/2^{2n}$$

are considered on the set

$$\varnothing^{\pm} \equiv \{\{x,y\} \mid (Ax,y) = 0,\ (Ay,y) = \pm(Ax,x),\ x,\ y \text{ in } X\}$$

which uniformly converge on $\varnothing^{\pm} \cap S$ (S is any sphere in $X \times X$). So, the functionals $h^{\pm}\{x,y\} = \lim_{n\to\infty} [h(2^n x)/2^{2n} \pm h(2^n y)/2^{2n}]$ are continuous on $\varnothing^{\pm}$.

Let $x_n \to x_0$ $(n \to \infty)$. Two cases are considered: (1) $(Ax_0, x_0) \neq 0$, (2) $(Ax_0, x_0) = 0$.

(1) In the first case the sequences

$$y_n = \sqrt{\frac{(Ax_n, x_n)}{(A\tilde{y}_n, \tilde{y}_n)}}\,\tilde{y}_n \quad \text{and} \quad z_n = \sqrt{\frac{(Ay_n, y_n)}{(A\tilde{z}_n, \tilde{z}_n)}}\,\tilde{z}_n \quad (n = 1, 2, \ldots)$$

are formed, where $\tilde{y}_n = y_0 - x_n/[(Ax_n, x_n)(Ay_0, x_n)]$, and $\tilde{z}_n = z_0 - y_n/[(Ay_n, y_n)(Az_0, y_n)] - x_n/[(Ax_n, x_n)(Az_0, x_n)]$ $(n = 1, 2, \ldots)$. It is obvious that $\{y_n, z_n\}$, $\{x_n, z_n\}$ are in $\varnothing^+$ and that

$$\{x_n, z_n\} \to \{x_0, z_0\} \text{ in } \varnothing^+ \quad \text{and} \quad \{y_n, z_n\} \to \{y_0, z_0\} \text{ in } \varnothing^+ \quad (n \to \infty)$$

From the above as well as from the continuity of the functional h^+, it follows that $h_1(x_n) \to h_1(x_0)$ $(n \to \infty)$.

(2) In the case $(Ax_0, x_0) = 0$ and $Ax_n = 0$ $(n = 1, 2, \ldots)$, it is easy to prove that $h_1(x_n) \to h_1(x_0)$ $(n \to \infty)$.

In the case $Ax_n \neq 0$ $(n = 1, 2, \ldots)$, from the hypothesis that $h_1(x_n) \not\to h_1(x_0)$ $(n \to \infty)$ we get the sequence $\bar{x}_k$ $(k = 1, 2, \ldots)$ so that

$$|h_1(\bar{x}_k) - h_1(x_0)| > \varepsilon_0 > 0 \tag{**}$$

from which $A\bar{x}_k \neq 0$. From the sequence $\{z_k\}_{k=1}^{\infty} = \{A\bar{x}_k/\|A\bar{x}_k\|\}_{k=1}^{\infty}$ $(\|z_k\| = 1,$ $k = 1, 2, \ldots)$ the subsequence $\{z_{k_p}\}_{p=1}^{\infty}$ is separated, which weakly converges to some z_0. We form the sequence $y_p = y_0 - (A\bar{x}_k, y_0)/\|A\bar{x}_{k_p}\|^2 A\bar{x}_{k_p}$ $(p = 1, 2, \ldots)$ where $y_0 \neq 0$, $y_0 \perp z_0$, and $(Ay_0, y_0) \neq 0$. It is shown that $y_p \to y_0$ $(p \to \infty)$, $(Ax_0, y_0) = 0$, $(A(x_0 \pm y_0),$ $(A(x_0 \pm y_0), x_0 \pm y_0) \neq 0$, and that

$$(A(\bar{x}_{k_p} \pm y_p), \bar{x}_{k_p} \pm y_p) \to (A(x_0 \pm y_0), x_0 \pm y_0) \quad (p \to \infty)$$

From that it is obtained that h_1 is continuous at the points $x_0 \pm y_0$ and y_0. From this using the properties of the functional h_1 we obtain that $h_1(\bar{x}_{k_p}) \to h_1(x_0)$ $(p \to \infty)$, which contradicts (**). From that the continuity of the functional h_1 follows.

It is easy to prove the uniqueness of the functional h_1.

REFERENCES

1. T. M. Rassias, On the stability of the linear mapping in Banach spaces, Proc. Amer. Math. Soc. 72 (1978), 297-300.
2. H. Drljević and Z. Mavar, About the stability of a functional approximately additive on A-orthogonal vectors, Akad. Nauka Umjet. Bosne i Herceg. Rad. Odjelj. Prirod. Mat. Nauka 69 (1982), 155-172.
3. H. Drljević, On the stability of the functional quadratic on A-orthogonal vectors, Publications de l'Institut Mathématique, 36(50), 1984, 111-118.
4. F. Vajzović, On a functional which is additive on A-orthogonal pairs, Glasnik matematički 1(21), 1966, 75-81.

INDEX

A

B

C

T

U

V